Wheat Quality For Improving Processing And Human Health

Gilberto Igrejas • Tatsuya M. Ikeda
Carlos Guzmán
Editors

Wheat Quality For Improving Processing And Human Health

Springer

Editors
Gilberto Igrejas
Department of Genetics and Biotechnology
University of Trás-os-Montes and Alto Douro
Vila Real, Portugal

Tatsuya M. Ikeda
Western Region Agricultural Research Cente
NARO
Fukuyama, Japan

Carlos Guzmán
Departamento de Genética
Escuela Técnica Superior de Ingeniería Agronómica y de Montes
Edificio Gregor Mendel
Campus de Rabanales
Universidad de Córdoba
Córdoba, Spain

ISBN 978-3-030-34162-6 ISBN 978-3-030-34163-3 (eBook)
https://doi.org/10.1007/978-3-030-34163-3

This Springer imprint is published by the registered company Springer Nature Switzerland AG
The registered company address is: Gewerbestrasse 11, 6330 Cham, Switzerland

Preamble

Wheat is one of the most consumed, produced and stored food crops worldwide. The paramount importance of wheat to human populations can be attributed to the remarkable work done by wheat breeders, who have improved wheat varieties keeping them in the spotlight of global agriculture. The recently completed annotation of the entire genome of bread wheat ended 13 years of collective effort to crack the wheat genetic code. The genome sequence can be used to study gene expression at any point in the life cycle of the plant, and to define which genes to target to improve yield and stress resistance. This massive advance not only allowed better understanding of relevant genes for agricultural applications but also for end-use quality traits.

During the last four years, wheat quality scientists from different countries have worked to develop the Expert Working Group (EWG) on Improving Wheat Quality for Processing and Health under the Wheat Initiative umbrella. This joint effort provides a framework to establish strategic research and organisation priorities for wheat improvement at the international level in both developed and developing countries. This EWG aims to maintain and improve wheat quality for processing and health under varying environmental conditions. The EWG has been focused on wheat quality in the broad sense, including seed proteins, carbohydrates, nutritional quality, grain processing and food safety. Bioactive compounds are also being considered, both those with negative effects, such as allergens and mycotoxins, that cause serious problems that need to be resolved, and those with positive effects, such as antioxidants or fibers, that can potentially be exploited. The EWG also works in the development of germplasm sets and other tools that can be deployed in wheat quality research.

The preparation of this book covering the whole range of grain quality topics is one of the important activities that the EWG is doing nowadays. The book should serve to identify possible gaps in important areas of wheat quality research and to position the EWG as an initial point of reference for the global wheat community regarding the different topics covered in depth here. Forty EWG members worked on 21 chapters of the book. This book adheres to the same policies that the EWG promotes such as using unified nomenclature to name the different alleles and providing correct information about materials (accession name, Germplasm Bank of origin, etc.) so other researchers know exactly what is being described and how to obtain the same materials or information.

The present book brings together a group of leading researchers from all over the world who describe different aspects of wheat quality for processing and health. During the meetings of the EWG different topics have been identified in recent years that need close attention or updating so more oriented and ordered research can be carried out in the years to come. The chapters on this topic seek to address this question while capitalizing on outputs of other international initiatives, wheat organizations and other EWGs, namely:

1. The importance of wheat
2. Wheat gluten protein structure and function: is there anything new under the sun?
3. Starch and starch-associated proteins: impacts on wheat grain quality
4. Contribution of genetic resources to grain storage protein composition and wheat quality
5. Durum wheat storage protein composition and the role of LMW-GS in quality
6. Gluten analysis
7. Proteomics as a tool in gluten protein research
8. Genotypic and environmental effects on wheat technological and nutritional quality
9. Improving wheat nutritional quality through biofortification
10. Phenolic compounds in wheat kernels: genetic and genomic studies of biosynthesis and regulations
11. Wheat cell wall polysaccharides (Dietary Fibre)
12. Grain quality in breeding
13. High throughput testing of key wheat quality traits in hard red spring wheat breeding programs
14. Molecular marker development and application for improving qualities in bread wheat
15. Durum wheat products, couscous
16. Understanding the mechanics of wheat grain fractionation and the impact of puroindolines on milling and product quality
17. The impact of processing on potentially beneficial wheat grain components for human health
18. *Fusarium* species infection in wheat: impact on quality and mycotoxin accumulation
19. Effects of environmental changes on the allergen content of wheat grain
20. Health hazards associated with wheat and gluten consumption in susceptible individuals and status of research on dietary therapies
21. FODMAPs in wheat
22. Epilogue: The main activities of the International collaboration on wheat quality and safety

In conceiving and compiling this book, we intend to make all these data and recent findings related to the advances on research of wheat quality genomics, proteomics, and other topics accessible to the general scientific community. Considering the importance of this crop in the human diet and its potential to promote health, all the wheat quality research and breeding community will be interested in the topics addressed by the book. Professionals working on the wheat value chain (millers, food manufacturers) or in nutrition and healthcare may also find this book a useful resource to increase and update their knowledge about wheat quality, nutrition and health issues.

Vila Real, Portugal — Gilberto Igrejas
Fukuyama, Japan — Tatsuya M. Ikeda
Córdoba, Spain — Carlos Guzmán

Acknowledgement

This work was supported by the project "Optimizing natural low toxicity of wheat for celiac patients through a nano/microparticles detoxifying approach". GLUTEN2TARGET (PTDC/BAA-AGR/29068/2017). FCT, Portugal 2020, 02/SAICT/2017. We greatly appreciate the Wheat Initiative to provide a platform for us and support our Expert Working Group.

Contents

The Importance of Wheat

Gilberto Igrejas and Gérard Branlard

Abstract The history of wheat domestication and use is closely linked to the efforts of humans to protect themselves from hunger and gain control over their food supply. Now grown worldwide wheat has become the most important source of food. For centuries bread wheat (*Triticum aestivum*) and durum wheat (*Triticum durum*) have been cultivated in the West to provide humans with energy and essential nutrients. Today China and India are the top two wheat-producing countries, largely because wheat has the advantage of requiring less water for cultivation than other comparable crops while being the main ingredient of a variety of processed foods valued in modern, mainly urban life. For more than a century, breeders have continuously improved wheat focusing on factors affecting grain yield and, more recently, technological quality. The properties of wheat that are ideal for processing into different food products have been greatly improved since the 1960s thanks to detailed research on storage proteins, which constitute the gluten. Most of these genetic successes are referred to in this book but many important goals remain to be achieved. Today further progress is crucial in the use of shared genetic resources, common analytical protocols for allele identification and technological processing, and dedicated tools for analysing polymer formation and characterisation particularly in response to climatic and other environmental factors. Technological properties are not the only wheat quality attributes, as consumers are increasingly aware and concerned about the nutritional value (the content in fiber, minerals, macro- and micro-nutrients, vitamins) and the health impact, whether positive or negative. For example, research on several pathologies associated with the consumption of gluten-based products will require collaboration between allergy specialists and wheat protein geneticists.

G. Igrejas (✉)
Department of Genetics and Biotechnology, University of Trás-os-Montes and Alto Douro, Vila Real, Portugal

Functional Genomics and Proteomics Unit, University of Trás-os-Montes and Alto Douro, Vila Real, Portugal

LAQV-REQUIMTE, Faculty of Science and Technology, Nova University of Lisbon, Lisbon, Portugal
e-mail: gigrejas@utad.pt

G. Branlard
INRAE, UCA UMR1095 GDEC, Clermont-Ferrand, France

G. Igrejas et al. (eds.), *Wheat Quality For Improving Processing And Human Health*, https://doi.org/10.1007/978-3-030-34163-3_1

1 Wheat and Humans, a Shared History

Wheat is one of the most important food crops to human populations as it is consumed worldwide. The history of wheat is closely linked to the history of the changing relationship of humans to their environment, and especially the efforts to protect families, tribes and populations from hunger and to master food supply and use. The reader can discover this common past in "The Saga of Wheat" (Bonjean 2016). Long before becoming settled, nomadic peoples ate cereals. Hunter-gatherers would have begun to cultivate the wild emmer (*Triticum dicoccoides*) 10,000 years BP. Einkorn (*Triticum monococcum*), the first domesticated wheat (Weiss and Zohary 2011), was cultivated at that time in the Fertile Crescent in a limited area between the Tigris and the Euphrates. Einkorn cultivation appeared in Greece and Balkans around 6000 BC. Later spelt wheat (*Triticum spelta*), and progressively free-threshing wheat (*Triticum turgidum* subsp. *durum*) and hexaploid bread wheat (*Triticum aestivum*) spread out from the Mediterranean Basin towards eastern, then western Europe, around 4000 and 3500 BC respectively (Bonjean 2016). The spread of wheat to Asia took several routes, including "the steppe route" and "the silk road", around 2500 BC. The Sumerian, Egyptian, Greek and Ancient Roman civilizations gradually came to favour cereal based foods derived from wheat meal and flour (Bruyérin-Champier 1560). The Egyptians taught breadmaking to Greeks, who invented dry yeast and perfected ovens, and after the Persian war Greek prisoners who were bakers would have developed breadmaking in Rome. The Roman Empire took wheat from conquered countries and spread its own culinary habits among the provinces (Flandrin and Montanari 1996). Wheat played such a dominant role in the Roman Empire that it is frequently described as a wheat empire (Shellengerger in Pomeranz 1971). Much earlier in China, the Shang Dynasty (16–11th centuries BC) considered wheat as one of the five sacred plants alongside millet, rice, barley and soybean. Changes in wheat use over the last two millennia clearly shows how people in western countries progressively perfected flour milling, particularly during the industrial revolution in the nineteenth century, as well as flour sieving and mill flow diagrams to get high yielding refined white flour for breadmaking (Branlard and Chiron 2016). These efforts contributed to the rise in the importance of bread wheat for human nutrition compared to other cereals like barley and rye. But the eventual dominance of bread wheat over rye bread was also the result of consumer demand for white bread, the relative increase in wheat production in numerous western countries, the evolution of transport using sea routes and railroad (Braudel 1985), and the improvement in germplasm achieved by agronomists and wheat geneticists. The first hybridizations were developed by de Vilmorin in France from the mid-nineteenth century and by Strampelli in Italy at the beginning of the twentieth century (Bonjean et al. 2011), thanks to the rediscovery of Mendel's laws of genetics.

2 Wheat Yield Improvement

Wheat breeders firstly focused their efforts on characteristics associated to grain yield like lodging resistance, frost resistance, disease resistance of roots, stems and leaves, and grain yield components. World wheat production progressively increased in the twentieth century and particularly after World War II to meet the demands of population growth. Agronomic and genetic advances, particularly through the Green Revolution, made wheat an essential crop to humankind. The total land area given over to bread and durum wheat in the world increased by only 6.8% from 204 Mha to 218 Mha between 1961 and 2013, while world production increased by 321% from 222 MT (a worldwide yield of 1 T/ha) to 713 MT (3.2 T/ha) (FAOstat 2014). Wheat production in 2017 was 757 MT with a harvested area of over 220 M ha, and wheat ranked third in terms of total cereal production behind maize and rice. About 95% of world wheat production is from the hexaploid (*T. aestivum*) bread wheat. According to FAO 2018 figures, China, India and Russia were the top 3 producers in 2017 with respectively 134, 98.5 and 85.9 MT of wheat produced.

3 Wheat Gluten Quality

These remarkable grain yield performances, mainly achieved since the 1960s, were not so successful for quality improvement. At the beginning of the twentieth century, the only measure breeders could use to assess grain quality was the Kjeldahl assay for nitrogen content. Several empirical tools were designed, like the Extensograph, Alveograph, Mixograph, and Farinograph, to indirectly test grain quality by measuring flour and dough properties and these were adopted for use in breeding programs to follow and select quality traits (Branlard and Chiron 2016). Breadmaking quality, which combines several characteristics like rheological dough properties, dough fermentation, gas retention, crumb texture, loaf volume, crust color, is highly polygenic and of low heritability. Moderate progress was made in genetic improvement of cultivars after the 1950s using indirect tests. Since the 1980s, progress accelerated thanks to the breakthroughs resulting from genetic analysis of wheat storage proteins, the components of gluten (Biesiekierski 2017).

The processing properties of wheat are largely determined by gluten proteins. Beccari was the first to successfully isolate gluten proteins around the mid-eighteenth century (Bailey 1941). Thomas Osborne (1907) later classified grain storage proteins based on their solubility. Albumins are water soluble, globulins are salt soluble, prolamins are soluble in aqueous ethanol and glutenins remain in the flour residue. Storage proteins became the focus of many studies and biochemical, genetic and molecular approaches greatly helped to decipher the major roles played by glutenins and gliadins in determining gluten properties. These proteins form a complex network during dough processing, giving the unique property of viscoelasticity to the dough. Different types of food can be made depending on the particular balance of functional properties of the dough, because the relative composition and variations in glutenins and gliadins have important effects on gluten behaviour (Wang et al. 2017). Since 1980 thirteen International Gluten Workshops have been held around the world acting as milestones that show the great progress achieved in all aspects of gluten research for bread and durum wheat uses. For instance, the gluten proteins were among the first genetic markers employed in breeding for bread wheat and durum wheat quality.

4 Wheat for Industrial Uses

Industry has developed specific processes for starch extraction from corn and wheat. Modern manufacturing plants can extract 40–50 T of flour per hour in a low water input centrifugation process. More than 30 countries today have private wheat starch industries but publicly available statistics are rather scarce. Several hundred products are currently prepared from wheat starch like:

- Food additives like sweetener in beverages, binding agent in soups and sauces, moistening agent in bakery, texture agent in many dairy products, etc.
- Green chemistry (fermentation), adhesives, bioplastics, paper industry, ethanol etc.
- Baby food, energy drinks, emulsifier, etc.
- Animal feed (milk powder), piglet starter feed, aquaculture feed pellets, etc.

The wheat starch industry generates a “first-class byproduct”, gluten. The increase in wheat starch production worldwide has made gluten the cheapest “green protein” now available for any food or feed producer.

In developed countries millers have turned to adding gluten powder to flour to improve the rheological and technological properties to the levels required by the baking industry. Between 0.2% and 10% of gluten can be added to flour according to the characteristics sought for the numerous food products that can be made like steamed buns, toast breads, crusty breads, sweet breads, leavened and laminated sweet goods, laminated puff pastries, rolls and buns, crackers, cookies, sponge cakes, wafers, and snacks (Branlard and Chiron 2016). These products are much more compatible with the “western lifestyle” as they are easily produced and consumed, making them preferable to the traditional ones (Shewry and Hey 2015). Wheat gluten is also largely used as a protein binder in a variety of food preparations like for meat mixes in sausages.

For the feed industry, the gluten now available can be used in a variety of ways. Gluten’s insolubility in water and its binding properties are an advantage in aquaculture in reducing pellet breakdown and providing fish with ‘green’ proteins. Gluten is also used in the preparation of biopolymers, in which the genetic diversity of glutenin can be exploited to tailor specific polymers (Johansson et al. 2013).

Wheat is also used for animal feedstuff, mainly for poultry diets (Bushuk and Rasper 1994). The European Common Agricultural Policy greatly helped this development by granting premium for cereals incorporated into feedstuffs. As an example, each year around 12% of wheat produced in France is incorporated into feedstuffs. This usage as a feed grain is directly dependent on the price relationship between wheat and other crops meaning that in years where harvests are negatively affected by climatic conditions and there is excess grain unsuitable for human consumption, this will be used to feed livestock. Other uses of low-grade grain are in the production of alcohol, adhesives, paper additives, soaps, rubbers, cosmetics and varnishes, the wide range of uses contributing to its increasing demand and production (Peña-Bautista et al. 2017). More transparency may be necessary to inform customers and consumers of the myriad products which contain ingredients derived from wheat.

5 Important Questions to Be Addressed

The diversity of storage proteins remains a central aspect to study for the decades ahead. To further our understanding of gluten properties, international cooperation on the following topics related to storage proteins will be of prime importance for geneticists, breeders, scientists and nutritionists interested in wheat quality. (1) Wheat genetic resources must be managed such that any known cultivars with specific allelic compositions are made available to the community of scientists. (2) The analytical protocols for allele identification and nomenclature (using electrophoresis, chromatography, DNA sequencing or proteomics tools) and for technological properties assessment have to be shared among the scientific community to ensure comparability of results. (3) Molecular mechanisms involved in the polymerization of storage proteins in protein bodies need to be elucidated. Specific tools for analysing polymer characteristics (mass, size, dispersity index), particularly to measure responses to climatic and environmental factors, will need to be used.

Wheat storage proteins are also responsible for celiac disease as they trigger an immune response when eaten by susceptible individuals leading to inflammation and small intestine damage (prevalence 1–3%), and are associated to several pathologies like wheat dependent exercise-induced anaphylaxis, an immediate hypersensitivity (prevalence <0.1%) (Laurière et al. 2007). A recent health trouble (Alessio et al., 2015) has been attributed to gluten, non-celiac gluten sensitivity (NCGS), the prevalence of which could be up to 6% in the US population (Igbinedion et al. 2017). Candidate proteins responsible for NCGS are actively sought. Amylase-trypsin inhibitors (ATIs) known to be nutritional activators of innate immunity and resistant to proteases could be possible candidates as they have been shown to increase intestinal inflammation (Zevallos et al. 2017). It is worth pointing out here that the characteristics of polymers are largely influenced by high temperature during storage protein accumulation in the grain, but this aspect has yet to be investigated in relation to the human pathologies (Branlard et al. 2019a, b).

Wheat grain is today the most important source of food on earth. It contains 75–80% carbohydrates, 9–18% protein, fiber, many vitamins (especially B vitamins), calcium, iron and many macro- and micro-nutrients. Heritability estimates of the content of several vitamins and nutritional constituents, including arabinoxylans, have shown that there is potentially useful genetic variability that can be exploited in breeding new varieties (Saulnier et al. 2007, Shewry et al. 2012). The importance of wheat for human nutrition and health is not sufficiently prioritized today by breeders and nutritionists. The recently completed annotation of the entire complex bread wheat genome (15,961 megabases) ended 13 years of collective effort from multiple researchers to crack the genetic code of this cereal. This massive step forward will not only improve our understanding of the role of relevant genes throughout plant development, but also, combined with proteomics and metabolomics, help to target end-use quality traits and valuable wheat grain components for better nutrition and health.

References

Alessio F, Sapone A, Zevallos V, Schuppan D (2015). Nonceliac gluten sensitivity, Gastroenterology 148: 1195–1204.

Bailey C 1941 A translation of Beccari's lecture 'Concerning Grain' (1728). Cereal Chem 18:555–561.

Biesiekierski JR (2017) What is gluten? J Gastroenterol Hepatol 32:78–81.

Bonjean AP (2016). The saga of wheat the successful story of wheat and human interaction. The world wheat book: A history of wheat breeding, volume 3," Alain P. Bonjean, William J. Angus, Maarten Van Ginkel eds. Edition Lavoisier Tec&Doc, ISBN 978-2-7430-2091-0, pp. 1672, Chapter 1, 1–90.

Bonjean AP, Angus WJ, van Ginkel M (2011). The world wheat book volume 2. A history of wheat breeding (Bonjean AP, Angus WJ, van Ginkel M eds) Lavoisier Publishing. ISBN: 978-2-7430-1102-4, pp. 1201.

Branlard G, Chiron H (2016). History and milestones of wheat processing and end-uses around the world (East Asia excepted): past, present and prospectives. In: "The world wheat book: a history of wheat breeding, volume 3," Alain P. Bonjean, William J. Angus, Maarten Van Ginkel eds. Edition Lavoisier Tec&Doc, ISBN 978-2-7430-2091-0, pp. 1672, Chapter 42, 1297–1343.

Branlard G, Faye A, Méléard B, Le Gouis J, Oury F-X, Ravel C (2019a). Quelle hypothèse approfondir pour appréhender la sensibilité non céliaque au gluten: 1. L'abondance du gluten et la diversité des prolamines Industries des Céréales 211: 34–37.

Branlard G, Rhazi L, Méléard B, Le Gouis J, Ravel C (2019b). Quelle hypothèse approfondir pour appréhender la sensibilité non céliaque au gluten: 2. Les polymères des gluténines. Industries des Céréales 212: 26–29.

Braudel F (1985). Civilisation matérielle, économie et capitalisme XV, XVIII siècles,1. Les structures du quotidien. Poche Editions. Bruneton – Governatori, 1984. Le pain de bois, ethnohistoire de la châtaigne et du châtaignier; Eché Editions. pp. 547.

Bruyérin C (1560). De re Cibaria, L'alimentation de tous les peuples et de tous les temps jusqu'au XVI siècle. Traduit du latin par Sigurd Amundsen, 1998. Editions Intermédiaire des Chercheurs et curieux. pp. 666.

Bushuk W, Rasper VF (Eds.) (1994). Wheat. Production, properties and quality. Springer: Blackie/Chapman and Hall, London, 239 pp.

FAO (2018). World food and agriculture – Statistical pocketbook 2018. Rome. 254 pp.

FAOSTAT (2014). Food and agriculture organization of the United Nations. Statistics Division, Crops data. http://faostat3.fao.org/download/Q/QC/E

Flandrin JL, Montanari M (1996). Histoire de l'alimentation. Fayard Editions, Paris. 392–426.

Igbinedion SO, Ansari J, Vasikaran A, Gavins FN, Jordan P, Boktor M, Alexander JS (2017) Non-celiac gluten sensitivity: all wheat attack is not celiac. World J Gastroenterol. 23(40): 7201–7210.

Johansson E, Malik AH, Hussain A, Rasheed F, Newson WR, Plivelic T,. Hedenqvist MS, Gällstedt M, Kuktaite R (2013). Wheat gluten polymer structures: the impact of genotype, environment, and processing on their functionality in various applications. Cereal Chem 90: 367–376.

Laurière M, Pecquet C, Boulenc E, Bouchez-Mahiout I, Snégaroff J, Choudat D, Raison-Peyron N, Vigan M, Branlard G (2007). Genetic differences in omega-gliadins involved in two different immediate food hypersensitivities to wheat. Allergy, 62. 890–896.

Osborne TB (1907) The proteins of the wheat kernel, by Thomas B. Osborne, Washington, Carnegie Inst.

Peña-Bautista RJ, Hernandez-Espinosa N, Jones JM, Guzmán C, Braun, HJ (2017). CIMMYT Series on carbohydrates, wheat, grains, and health: wheat-based foods: their global and regional importance in the food supply, nutrition, and health. Cereal Foods World 62(5): 231–249.

Pomeranz Y (1971). Wheat chemistry and technology. Published by AACC. 1–2.

Saulnier L, Sado P-E, Branlard G, Charmet G, Guillon F (2007). Wheat arabinoxylans: Exploiting variation in amount and composition to develop enhanced varieties. J Cereal Sci. The Contribution of Cereals to a Healthy Diet, 46: 261–281.

Shewry PR, Hey SJ (2015). The contribution of wheat to human diet and health. Food Energy Secur 4(3): 178–202.

Shewry PR, Charmet G, Branlard G, Lafiandra D, Gergely S, SalgóA, Saulnier L, Bedő Z, Mills C, Ward JL (2012). Developing new types of wheat with enhanced health benefits. Trends Food Sci Technol 25: 70–77.

Wang Z, Li Y, Yang Y, Liu X, Qin H, Dong Z, Wang D (2017). New insight into the function of wheat glutenin proteins as investigated with two series of genetic mutants. Sci Rep 7(1): 3428.

Weiss E, Zohary D, 2011 The neolithic Southwest Asian founder crops: their biology and archaeobotany. Curr Anthropol, 52(S4): S239-S240.

Zevallos VF, Raker V, Tenzer S, Jimenez-Calvente C, Ashfaq-Khan M, Rüssel N, Pickert G, Schild H, Steinbrink K, Schuppan D (2017). Nutritional wheat amylase-trypsin inhibitors promote intestinal inflammation via activation of Myeloid cells. Gastroenterology 152(5): 1100–1113.

Wheat Gluten Protein Structure and Function: Is There Anything New under the Sun?

Ramune Kuktaite and Catherine Ravel

Abstract This chapter focuses on wheat gluten protein and how its protein components, gliadin and glutenin, interact at the molecular level to produce structures, which contribute to particular functional properties. The aspects of gluten protein are highlighted in wheat gluten, in both, food and non-food products. Factors impacting wheat gluten protein chemistry and structure under various processing conditions and in different end-use products are discussed. The influence of the genetic make-up of wheat grain on the molecular structure and functional performance of gluten protein in the end-use products is discussed. The main factors steering wheat gluten protein structure-function relationships are thus summarised in the context of traditional and innovative applications.

1 Introduction

Wheat (*Triticum aestivum* L.), cultivated at latitudes spanning from Scandinavia to Argentina, is one of the most widely grown crops in the world. Around 90% of wheat is used for human consumption in various its forms such as, bread, cookies and pasta etc. Therefore, the most important aspects of wheat grain quality are the nutritional value due to the grain bioactive components, dietary fibers, minerals and vitamins, notably B vitamins (Hussain et al. 2012a, 2012b, 2012d, Shewry and Hey, 2015) and its breadmaking quality characteristics, such as milling, processing and baking performance (Hernandez-Espinosa et al. 2018; Guzman et al. 2015). The remaining 10% is used as seed or flour for industrial production of gluten, starch and other products. The major factors determining the quality of processed wheat food products are grain storage protein (about 80% of the grain total protein) content

R. Kuktaite (✉)
Department of Plant Breeding, Swedish University of Agricultural Sciences, Skåne, Sweden
e-mail: ramune.kuktaite@slu.se

C. Ravel
INRAE, UCA UMR1095 GDEC, Clermont-Ferrand, France

G. Igrejas et al. (eds.), *Wheat Quality For Improving Processing And Human Health*, https://doi.org/10.1007/978-3-030-34163-3_2

and composition (Shewry et al. 2002). These proteins are able to interact with each other and form the proteinaceous network when the wheat flour is mixed with water. The gluten protein fraction impacts the end-use quality of wheat-derived foods in specific ways. These proteins have also recently been evaluated for their applicability in innovative non-food materials, such as bio-based plastics (Rasheed et al. 2015, 2016; Johansson et al. 2013; Kuktaite et al. 2011). The physicochemical, structural and functional properties of processed wheat gluten products are to a large extent determined by the presence and proportions of gliadins and glutenins and the molecular interactions they foster through disulfide bonds and non-covalent Van de Waals forces, as well as through hydrogen and isopeptide bonds (Rombouts et al. 2013; Rasheed et al. 2018; Kuktaite et al. 2004). Many factors are known to contribute to the structural and functional properties of gluten protein during processing, as for example in dough. When wheat dough is mixed for an optimum time, more structurally ordered, homogeneous and elastic gluten is formed (Kuktaite et al. 2004). Different structure-function relationships are observed in dough produced from different wheat varieties (Shewry et al. 2001). In the processing of bio-based plastics from gluten, hierarchical structures are formed in the presence of specific additives or when specific conditions are applied (Kuktaite et al. 2011, Johansson et al. 2013; Muneer et al. 2015; Türe et al. 2011). The genetic composition of wheat gluten protein also has an impact (Rasheed et al. 2016). The characteristic properties of gluten, particularly viscoelasticity and extensibility, in relation to the structures formed in the processing of diverse wheat products are very important and have been the focus of basic and applied research for more than 260 years (Wieser and Kieffer 2001). Then how can wheat gluten structures and properties be fine-tuned to modify and improve end-use quality of wheat products in all their diversity? Understanding more about the basis of wheat gluten functionality is directly relevant to optimize gluten processing in various industrial applications, and could be used by breeders to improve wheat varieties suited to desired end-use requirements. This chapter summarizes some of the latest studies of food and non-food gluten systems to review the factors influencing the structure-function relationships of wheat gluten and includes genetics, chemistry, structure and processing technology.

2 Wheat Grain Proteins Are Key Factors for Functionality of Wheat Flour Processing to Food and Non-food Products

Wheat grain proteins are divided into functional proteins (albumins, globulins) and grain storage proteins that provide nutrients such as amino acids to the growing plant (Day et al. 2006). Grain storage proteins also called prolamins consist of monomeric gliadins and polymeric glutenins. Gluten protein are known for their impressive level of diversity. Gliadin proteins are divided into the four main types α-, β-, γ- and ω-gliadins according to their electrophoretic mobility in acid conditions, the α- and β-gliadins being closely related in terms of structure. Glutenins are either high molecular weight (HMW) or low molecular weight (LMW) types. The β-spiral structure of HMW glutenins explains their intrinsic elasticity (Shewry et al. 2001). As sulfur plays a special role, they are also divided into groups that differ by their S-amino acid composition (Shewry et al. 2001). S-rich gluten proteins consist of α/β- and γ-gliadins and LMW glutenins. S-poor gluten proteins correspond to ω-gliadins and HMW glutenins.

The gluten confers special characteristics to bread made from the common wheat (*Triticum aestivum* L.) flour (Shewry et al. 2002), to pasta made from durum wheat flour (*Triticum turgidum* L. var. *durum*) and to bio-plastics made from industrially produced gluten after starch has been removed (Rasheed et al. 2015). The gluten proteins have unique viscoelastic properties, which are distinct from their structural and functional aspects. Roughly, glutenins confer the gluten elasticity and tenacity while gliadins confer viscosity (MacRitchie 1999; Shewry et al. 2002). Gluten proteins can easily polymerize and form large complex polymers with low solubility, which hampers research efforts to determine their structures. Despite this, several studies have proposed some interpretations of the structure-function relationships of gluten, where HMW glutenins and LMW glutenins interact *via* disulfide and other types of bonding to form its backbone. The formation of disulfide bonds and their interchange reactions, as well as non-covalent Van der Waals interactions have been well studied in gluten food systems (Belton 1999; Wieser 2007).

When gluten is used for non-food applications such as bioplastics, films, or foams, and it has clearly been shown that the molecular organization of the protein includes disulfide crosslinks, hydrogen bonding, as well as non-reducible isopeptide bonding, and lanthionine and lysinoalanine interactions (Rombouts et al. 2010, 2013; Kuktaite et al. 2016; Blomfeldt et al. 2011; Türe et al. 2011) (Fig. 1). The characteristics and end-uses of both food and non-food wheat gluten products mainly depend on gluten strength. Therefore improving gluten strength and protein

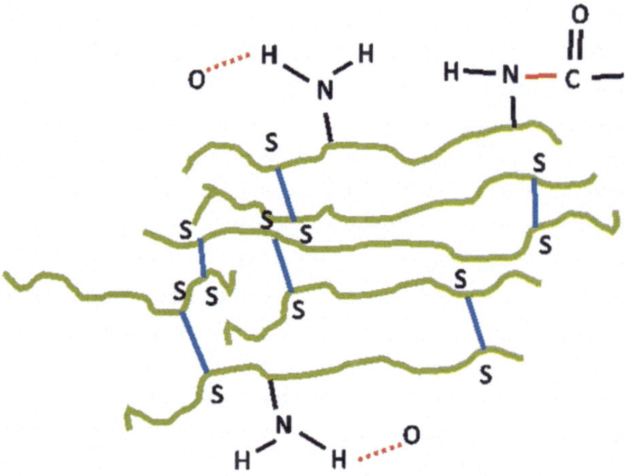

Fig. 1 Schematic representation of possible crosslinking in wheat gluten protein (green lines) in processed food or non-food systems through disulfide (blue lines), peptide (solid red line) and hydrogen (red dotted lines) bonds

content in the grain have been emphasized as among the most important targets in breeding of different types of wheat (Shewry et al. 2002; Patil et al. 2009).

There are significant challenges in the wheat sector arising from a growing demand for food driven by the world's growing population, rapidly changing food preferences, and challenging climate variation and the need for healthier food to improve diet and combat diseases. In addition, wheat protein provides about 20% of the total protein in the human diet. Therefore, wheat protein could play an important role towards a more sustainable alimentation with a smaller part of animal protein. The production of this higher quality wheat must also take into account concerns over the impact on agricultural systems and on the environment, as well as the effect of climate change on the stability of wheat quality. Most of these concerns require wheat with a specific quality profile in terms of protein concentration and composition, which mainly define gluten strength that is most closely suited to its intended end-use.

In conclusion, wheat quality is a very complex characteristic to deal with and depends on many parameters as grain yield because it is negatively correlated to protein concentration. Similarly genetically determined protein composition (Malik et al. 2013) also influence this trait, and for end-use value, several studies have ranked gliadin or glutenin alleles in order of their influence on flour quality as decribed in the pionner studies of Branlard and Dardevet (1985) or Payne in 1987. Wheat grain quality directly impacts mixed dough (Kuktaite et al. 2004, 2005) and end-product such as, bread quality (Hussain et al. 2012b, 2012c, 2013).

3 Wheat Gluten Protein Structure During Processing

3.1 Impact in Food Processing

Wheat gluten proteins in the presence of water form a viscoleastic protein network and with the starch in dough can produce structures suitable for various end-uses. The importance of the viscoleastic properties of wheat dough in breadmaking and other wheat product processes have been well described (Shewry et al. 2002). Thermal polymerization of gliadins and glutenins occurs, for example, during bread baking or pasta making processes, and is related to the formation of covalent bonds between polypeptides or different parts of a polypeptide (Lagrain et al. 2007; Cubadda et al. 2007). Gliadins and glutenins under alkaline conditions are known to form unreducible covalent crosslinks, such as those formed through lanthionine or lysinoalanine (Rombouts et al. 2010). Isopeptide bond formation involving glutamine residues and heat treatment has been suggested to occur in bread, pasta and gluten films (Sakamoto et al. 1995; Petitot et al. 2009; Rombouts et al. 2013).

Baking quality is usually determined through a number of tests, including dough rheology and mixing tests that assess the viscosity and elasticity during dough preparation, together with complex tests of extensibility and baking (Guzman et al. 2015, Wang and Kovacs 2002; Li et al. 2015). Mixing behaviour, including parameters such as the optimal mixing time to form a gluten network, has been found to impact the gluten structure and rheological behaviour of the resultant dough (Kuktaite et al. 2005). However, performing all these tests is time consuming and expensive. The genetic determination of baking quality of flour and dough have been studied (Bordes et al. 2011) and some molecular markers could be used to improve quality. On another hand, genomic selection could be used to predict quality traits by taking into account the effect of a large number of markers (Meuwissen et al. 2001) as was done in a recent research attempts by Guzman et al. (2016a). For example, CIMMYT has used genomic selection to predict all of the processing and end-use quality traits in the spring wheat breeding program (Guzman et al. 2016b) while Michel et al. (2018) has used this method to improve baking quality.

3.2 *Impact in Non-Food Processing*

The unique properties of wheat gluten in forming polymeric protein matrices made it an interesting subject of applied studies in the area of bioplastics (Kuktaite et al. 2011; Türe et al. 2011; Blomfeldt et al. 2011; Rasheed et al. 2015). Another reason for growing interest in wheat gluten protein is the worldwide move to replace synthetic plastics with polymers from renewable agro-resources as a solution to pollution caused by non-biodegradable synthetic polymers (Payne and Corfield 1979). In addition, gluten is a cheap product from starch industries, which search outlets for this sub-product. Previous studies have revealed the importance of key parameters such as genetic make-up and cultivation inputs to the plant (Rasheed et al. 2015; 2016), as well as chemical and physical treatments of gluten protein during processing. The latter treatments applied to gluten were used in making films, foams and composite forming materials, respectively (Kuktaite et al. 2012, 2014; Muneer et al. 2015, 2016; Rasheed et al. 2014; Johansson et al. 2013). The reactive chemistry of gluten originates from the available modifiable protein side groups, which make it possible to obtain three-dimensional polymeric protein networks with appropriate strengths and functional properties (Yu et al. 2016; Kuktaite et al. 2011, 2014, 2016; Andrade et al. 2018; Rasheed et al. 2016). Processing methods such as solution casting and foaming, including those related to temperature and pressure conditions like extrusion and compression molding, have been widely used to process gluten (Gontard et al. 1992; Gennadios et al. 1993; Kuktaite et al. 2011, 2012, 2014, 2016). In various studies, aggregation or pre-aggregation of gluten proteins was observed to take place at earlier stages of processing. Crosslinking between gluten protein molecules through disulfide/sulfydryl interchange reactions, hydrophobic interactions and iso-peptide bonding occurs according to the chosen temperature, additives used or processing method. The protein secondary structures, and the micro- and nano-structures in gluten, gliadins and glutenins, processed alone or with starch are summarised in Table 1.

Wheat gluten protein structure formation at various molecular levels (from macro through micro to nano) has been correlated with functional properties in a

Table 1 Wheat gluten structures observed in different non-food systems.

Protein type and processing conditions	Secondary structure	Microstructure	Molecular distances d, Å
Wheat gluten *Gluten blend with modified potato starch, temperature (plasticizer)*			*SAXS data from* Muneer et al. 2015
Protein:starch ratio 30:70, 110 ° C (glycerol)			$d_1 = 89.4$, $d_2 = 55$, $d_3 = 16.2$
Protein:starch ratio 50:50, 110 ° C (glycerol)			$d_1 = 85.5$, $d_2 = 55$, $d_3 = 16.1$
Protein:starch ratio 70:30, 110 ° C (glycerol)			$d_1 = 83.6$, $d_2 = 55$, $d_3 = 15.9$
Protein:starch ratio 30:70, 130 ° C (glycerol)		non-homogeneous	$d_1 = 88.0$, $d_2 = 63.0$, $d_3 = 16.3$
Protein:starch ratio 50:50, 130 ° C (glycerol)		homogeneous	$d_1 = 81.9$, $d_2 = 63.6$, $d_3 = 16.3$
Protein:starch ratio 70:30, 130 ° C (glycerol)			$d_1 = 73.2$, $d_2 = 54.4$, $d_3 = 16.2$
Protein:starch ratio 50:50, 110 ° C (water + glycerol)	β-sheets		$d_1 = 74.6$, $d_2 = 53.0$, $d_3 = 16.1$
Protein:starch ratio 50:50, 130 ° C (water + glycerol)	β-sheets		$d_1 = 71.0$, $d_2 = 58.9$ $d_3 = 16.2$
Gliadin			*SAXS data from* Muneer et al. 2016
Gliadin only, 110 ° C (glycerol)			$d_{broad} = 106.7$, $d_1 = 55.9$, $d_2 = 32.3$, $d_3 = 28.1$
Gliadin:starch ratio 30:70, 110 ° C (glycerol)	β-turns, β-sheets and unordered		$d_{broad} = 94.6$, $d_1 = 57.9$, $d_2 = 33.6$, $d_3 = 29.2$
Gliadin:starch ratio 50:50, 110 ° C (glycerol)	β-sheets, weak β-sheet interactions, β-turns		$d_{broad} = 92.1$, $d_1 = 57.3$, $d_2 = 33.2$, $d_3 = 28.8$
Gliadin:starch ratio 70:30, 110 ° C (glycerol)	β-sheets, weak β-sheet interactions, β-turns	non-homogeneous	$d_{broad} = 96.3$, $d_1 = 56.6$, $d_2 = 32.8$, $d_3 = 28.4$
Gliadin only, 130 ° C (glycerol)			$d_{broad} = 97.1$, $d_1 = 56.1$, $d_2 = 32.5$, $d_3 = 28.2$
Gliadin:starch ratio 30:70, 130 ° C (glycerol)	β-turns, β-sheets and unordered		$d_{broad} = 93.8$, $d_1 = 58.2$, $d_2 = 33.8$, $d_3 = 29.3$
Gliadin:starch ratio 50:50, 130 ° C (glycerol)	β-turns, β-sheets and weak β-sheet interactions		$d_{broad} = 99.2$, $d_1 = 59.3$, $d_2 = 34.4$, $d_3 = 29.8$
Gliadin:starch ratio 70:30, 130 ° C (glycerol)	weak β-sheet interactions, α-helix & random coil, β-turns		$d_{broad} = 96.6$, $d_1 = 57.9$, $d_2 = 33.7$, $d_3 = 29.2$
Glutenin			*WAXS data from* Rasheed et al. 2018
Genotype 2 + 12, A∗	strong β-sheet interactions, α-helices & random coils, β-turns		$d_1 = 6.8$, $d_2 = 4.5$, $d_3 = 3.9$, $d_4 = 2.7$, $d_A = 9.6$
Genotype 5 + 10, B∗∗	strong β-sheet interactions, α-helices & random coils, β-turns		$d_1 = 6.8$, $d_2 = 4.5$, $d_3 = 3.9$, $d_4 = 2.7$, $d_A = 9.7$

SAXS, small angle X-ray scattering. WAXS, wide-angle X-ray scattering.

A∗ and B∗∗ refer to growing environments from Rasheed et al. 2018

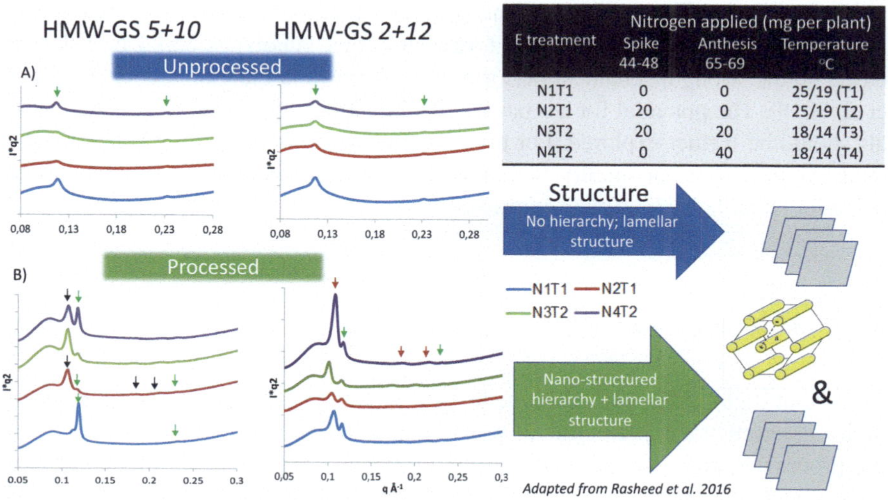

Fig. 2 Wheat gluten protein nanostructure studied by small-angle X-ray scattering. Gluten was extracted from wheat lines with genetically different HMW glutenin compositions (HMW-GS 5 + 10 and HMW-GS 2 + 12) grown in 4 different environments as specified in the table (insert). Variation in nanostructural profiles of (A) unprocessed and (B) processed gluten showing the detection of lamellar (small green arrows) and hexagonal structural arrangements (red arrows) shown schematically in diagrams (bottom right). Modified from Rasheed et al. 2016

number of studies (Kuktaite et al. 2011, 2014, 2016; Muneer et al. 2015, 2016; Rasheed et al. 2014, 2016, 2018; Johansson et al. 2013; Andrade et al. 2018). In particular, the formation of hierarchically arranged nano-structures of gluten proteins (Kuktaite et al. 2011) or complex hierarchical hexagonal and lamellar structures can be highlighted (Fig. 2, modified from Rasheed et al. 2016). The formation of hexagonal structures was also observed in gliadins that had been temperature processed in a blend with glycerol (Kuktaite et al. 2016; Muneer et al. 2016).

4 Structure-Function Relationships in End-Products from Processed Wheat

Being able to combine genetic, structural and functional information into a model that can predict end-use characteristics of processed wheat gluten protein is still a valid goal. Some attempts to correlate the molecular structure of gluten protein with the mechanical behavior of bio-based materials have been made. Crosslinking, structure and baking performance in baked wheat products, such as bread, have been broadly studied, but more research is needed to account for other factors such as genetics, growing environment, and structure-function relationships. By extracting gluten in mild conditions, it is foreseeable that genotypes could be selected based on the strong mechanical performance of non-food gluten material (Rasheed et al. 2018). The potential for a broad range of qualities for non-food gluten materials should be further explored. For prediction of wheat bread baking performance, methods to screen for quality including structural characteristics and functional behavior should also be further improved.

5 Conclusions and Summary

The use of multiscale and multistage factors, including genomic selection, to assist in the assessment and prediction of wheat quality, as well as the prediction of all of the processing and end-use quality traits in wheat is a priority. In this context, a good understanding of the relationship between gluten structure and function is needed to tailor wheat gluten proteins to specific food and non-food end-uses.

References

Andrade FD, Newson WR, Bernardinelli OD, Rasheed F, Cobo MF, Plivelic TS, de Azevedo ER, Kuktaite R (2018) An insight into molecular motions and phase composition of gliadin/glutenin glycerol blends studied by ^{13}C Solid-state and ^{1}H Time-domain NMR. Journal of Polymer Science Part B: Polymer Physics 56: 739–750.

Belton PS (1999) On the elasticity of wheat gluten. Journal of Cereal Science 29: 103–107.

Blomfeldt T, Kuktaite R, Azhdar B, E Johansson, Hedenqvist MS (2011) Mechanical properties and network structure of wheat gluten foams. Biomacromolecules 12: 1707–1715.

Bordes J, Ravel C, Le Gouis J, Lapierre A, Charmet G, Balfourier F (2011) Use of a global core collection for association analysis of flour and dought quality traits. Journal of Ccereal Science 54: 137–147.

Branlard G, Dardevet M (1985) Diversity of grain protein and bread wheat quality .2. Correlation between high molecular-weight subunits of glutenin and flour quality characteristics. Journal of Cereal Science 3: 345–354.

Cubadda RE, Carcea M, Marconi E, Trivisonno MC (2007) Influence of gluten proteins and drying temperatures on the cooking quality of durum wheat pasta. Cereal Chemistry 84: 48–55.

Day L, Augustin MA, Batey IL, Wrigley CW (2006) Wheat-gluten uses and industry needs. Trends in Food Science and Technology 17: 82–90.

Gennadios A, Weller CL, Testin RF (1993) Modification of physical and barrier properties of edible wheat gluten-based films. Cereal Chemistry 70: 426–429.

Gontard N, Guilbert S, Cuq J (1992) Edible wheat gluten films: Influence of the main process variables on the film properties using response surface methodology. Journal of Food Science 57: 190–199.

Guzman C, Mondal S, Govindan V, Autrique JE, Posadas-Romano G, Carvantes F, Crossa J, Vargas M, Singh RP, Peña R J (2016a) Use of rapid tests to predict quality traits of CIMMYT bread wheat genotypes grown under different environments. LWT-Food Science and Technology 69: 327–333.

Guzman C, Peña R J, Singh R, Autrique E, Dreisigacker S, Crossa J, Rutkoski J, Poland J Battenfield S (2016b) Wheat quality improvement at CIMMYT and the use of genomic selection. Applied and Translational Genomics 11: 3–8.

Guzman C, Posadas-Romano G, Hernandez-Espinoza N, Morales-Dorantes A, Peña R J (2015) A new standard water absorption criteria based on solvent retention capacity (SRC) to determine dough mixing properties, viscoelasticity, and bread-making quality. Journal of Cereal Science 66: 59–65.

Hernandez-Espinosa N, Mondal S, Autrique E (2018) Milling, processing and end-use quality traits of CIMMYT spring bread wheat germplasm under drought and heat stress. Field Crops Research 215: 104–112.

Hussain A, Larsson H, Kuktaite R, Johansson E (2012a) Concentration of some heavy metals in organically grown primitive, old and modern wheat genotypes: Implications for human health. Journal of Environmental Science and Health, Part B 47(7): 751–758.

Hussain A, Larsson H, Kuktaite R, Johansson E (2012b) Healthy food from organic wheat: choice of genotypes for production and breeding. Journal of the Science of Food and Agriculture 92: 2826–2832.

Hussain A, Larsson H, Kuktaite R, Johansson E (2012c) Towards the understanding of bread-making quality in organically grown wheat: Dough mixing behaviour, protein polymerisation and structural properties. Journal of Cereal Science 56: 659–666.

Hussain A, Larsson H, Kuktaite R, Prieto-Linde ML, Johansson E (2013) Amount and size distribution of monomeric and polymeric proteins in the grain of organically produced wheat. Cereal Chemistry 90: 80–86.

Hussain A, Larsson H, Olsson ME, Kuktaite R, Grausgruber H, Johansson E (2012d) Is organically produced wheat a source of tocopherols and tocotrienols for health food? Food Chemistry 132: 1789–1795.

Johansson E, Malik AH, Hussain A, Rasheed F, Newson WR, Plivelic T, Hedenqvist MS, Gälstedt M, Kuktaite R (2013) Wheat gluten polymer structures: The impact of genotype, environment and processing on their functionality in various applications. Cereal Chemistry 90: 367.

Kuktaite R, Larsson H, Johansson E (2004) Variation in protein composition and its relationship to dough mixing behavior in wheat flour. Journal of Cereal Science 40: 31–39.

Kuktaite R, Larsson H, Marttilla S, Johansson E (2005) Effect of mixing time on gluten recovered by ultracentrifugation studied by microscopy and rheological measurements. Cereal Chemistry 82: 375–384.

Kuktaite R, Plivelic T, Cerenius Y, Hedenqvist MS, Gällstedt M, Marttila S, Ignell R, Popineau I, Tranquet O, Shewry P, Johansson E (2011) Structure and morphology of wheat gluten films: from polymeric protein aggregates toward superstructure arrangements. Biomacromolecules 12: 1438–1448.

Kuktaite R, Plivelic TS, Türe H, Hedenqvist MS, Gällstedt M, Marttila S, E Johansson (2012) Changes in the hierarchical protein polymer structure: urea and temperature effects on wheat gluten films. RSC Advances 2: 11908–11914.

Kuktaite R, Türe H, Hedenqvist MS, Gällstedt M, Plivelic TS (2014) Gluten biopolymer and nanoclay-derived structures in wheat gluten-urea-clay composites: relation to barrier and mechanical properties. ACS Sustainable Chem. Eng 2(6): 1439–1445.

Kuktaite R, Newson WR, Rasheed F, Plivelic TS, Hedenqvist MS, Gällstedt M, Johansson E (2016) Monitoring nano-structure dynamics and polymerization in glycerol plasticized wheat gliadin and glutenin films: relation to mechanical properties. ACS Sustainable Chemistry and Engineering 4 (6): 2998–3007.

Lagrain B, Thewissen BG, Brijs K, Delcour JA (2007) Impact of redox agents on the extractability of gluten protein during bread bread making. Journal of Agricultural and Food Chemistry 55: 5320–5325.

Li FL, Wu Y, Hernandez-Espinoza N, Peña R J (2015) Comparing small scale testing methods for predicting wheat gluten strength across environments. Cereal Chemistry 92: 231.

MacRitchie F (1999) Wheat proteins: Characterization and role in flour functionality. Cereal Food World 44(4):183–193.

Malik AH, Kuktaite R, Johansson E (2013) Combined effect of genetic and environmental factors on the accumulation of proteins in the wheat grain and their relationship to bread-making quality. Journal of Cereal Science 57: 170–174.

Meuwissen THE, Hayes BJ, Goddard ME (2001) Prediction of total genetic value using genome-wide dense marker maps. Genetics 157: 1819–1829.

Michel S, Kummer S, Gallee M, Hellinger J, Ametz C, Akgöl B, Epure D, Güngör H, Löschenberger F, Buerstmayr H (2018) Improving the baking quality of bread wheat by genomic selection in early generations. Theoretical Applied Genetics. 131: 477 493.

Muneer F, Andersson M, Koch K, Menzel C, Hedenqvist MS, Gällstedt M, Plivelic TS, R Kuktaite (2015) Nanostructural morphology of plasticized wheat gluten and modified potato starch composites: relationship to mechanical and barrier properties. Biomacromolecules 16 (3): 695–705.

Muneer F, Andersson M, Koch K, Hedenqvist M, Gällstedt M, Plivelic TS, Menzel C, Rhazi L, Kuktaite R (2016) Innovative gliadin/glutenin and modified potato starch green composites: Chemistry, structure and functionality induced by processing. ACS Sustainable Chemistry and Engineering 4(12): 6332–6343.

Patil RM, Oak MD, Tamhankar SA, Rao VS (2009) Molecular mapping of QTLs for gluten strength as measured by sedimentation volume and mixograph in durum wheat (*Triticum turgidum* L. spp *durum*). J Cereal Sci. 49(3): 378–386.

Payne PI, Corfield KG (1979) Subunits composition of wheat glutenin proteins isolated by gel filtration in a dissociating medium. Planta 14: 83–88.

Petitot M, Brossad C, Barron C, Larre C, Morel MH, Micard V (2009) Modification of pasta structure induced by high drying temperaturas. Effects on the *in vitro* digestability of protein and starch fractions and the potential allergenicity of protein hydrolysates. Food Chemistry, 116: 401–412.

Rasheed F, Newson WR, Plivelic TS, Kuktaite R, Hedenqvist MS, Gällstedt M, Johansson E (2014) Structural architecture and solubility of native and modified gliadin and glutenin proteins: non-crystalline molecular and atomic organization. RSC Advances 4: 2051.

Rasheed F, Newson WR, Plivelic TS, Kuktaite R, Hedenqvist MS, Gällstedt M, Johansson E (2015) Wheat protein nano-scale structure, disulphide crosslinks and functionality upon chemical modification. International Journal of Biological Macromolecules 79: 151–159.

Rasheed F, Kuktaite R, Hedenqvist MS, Gällstedt M, Plivelic TS, E Johansson (2016) The use of plants as a "green factory" to produce high strength gluten-based materials. Green Chemistry 18: 2782–92.

Rasheed F, Plivelic TS, Kuktaite R, Hedenqvist MS, Johansson E (2018) Unravelling the structural puzzle of the giant glutenin polymer - An interplay between protein polymerization, nanomorphology and functional properties. ACS Omega 3(5): 5584–5592.

Rombouts I, Lagrain B, Brijs K, Delcour JA (2010) ß-Elimination reactions and formation of covalent cross-links in gliadin during heating at alkaline pH. Journal of Cereal Science 52: 362–367.

Rombouts I, Lagrain B, Delcour JA, Türe H, Hedenqvist MS, Johansson E, Kuktaite R (2013) Crosslinks in wheat gluten films with hexagonal close-packed protein structures. Industrial Crops and Products 51: 229–235.

Sakamoto H, Kumazawa Y, Kawajiri H, Motoki M (1995) E-(y-Glutamyl)lysine cross-link distribution in foods determined by improved method. Journal of Food Science 60: 416–419.

Shewry PR, Popineau Y, Lafiandra D, Belton P, Lellis SCD (2001) Wheat glutenin subunits and dough elasticity: findings of the EUROWHEAT project. Trends in Food Science & Technology 11: 433–441.

Shewry PR, Halford NG, Belton PS, Tatham, AS (2002) The structure and properties of gluten: an elastic protein from wheat grain. Phil Trans R Soc Lond B 357: 133–142.

Shewry PR, Hey S (2015) The contribution of wheat to human diet and health. Food Energy Security 4: 178–202.

Türe H, Gällstedt M, Johansson E, Kuktaite R and MS Hedenqvist (2011) Protein network structure and properties of wheat gluten extrudates using a novel solvent-free approach with urea as a combined denaturant and plasticizer. Soft Matter 7: 9416–9423.

Wang C, Kovacs MIP (2002) Swelling index of glutenin Test. I. Method and comparison with sedimentation, gel-protein, and insoluble glutenin tests. Cereal Chemistry. 79: 183.

Wieser H (2007) Chemistry of gluten proteins. Food Microbiology 24: 115–119.

Wieser H, Kieffer R (2001) Correlations of the amount of gluten protein types to the technological properties. Journal of Cereal Science 43: 19–27.

Yu S, Chen F, Wu Q, Roth S, Brüning K, Schneider K, Kuktaite R, MS Hedenqvist (2016) The structural changes of gluten/glycerol plastics at dry and moist conditions and during tensile tests. ACS Sustainable Chemistry and Engineering 4(6): 3388–3397.

Starch and Starch-Associated Proteins: Impacts on Wheat Grain Quality

Ahmed Regina and Carlos Guzmán

Abstract Wheat storage proteins have been historically examined and periodically established to be the major determinant of wheat quality. Gluten proteins largely contribute to the formation of viscoelastic network in a dough, enabling processing of wheat to food products including bread. More recently starch, the major component constituting 60–70% of wheat grain, is understood to play key roles in flour quality, dough functionality and end product and nutritional quality. Starch is composed of two neutral macromolecules of glucose, amylose and amylopectin. The structural differences between amylose and amylopectin are predominantly dependent on the extent and distribution of α-1,4 and α-1,6 linkages that connect the glucose units to form these two polymers. The functional properties of starch as governed by its structure, molecular organisation, granule morphology and size distribution influence dough behaviour during processing, differentially impacting the end product qualities. Also, varyingly important are the roles of starch granule associated proteins, comprised of both surface proteins and granule-integral proteins with enzyme functions, in driving starch responses in a complex dough matrix system. This chapter aims to provide an extensive review on how starch, associated proteins and starch-protein interactions influence functional properties of food systems.

A. Regina (✉)
International Rice Research Institute-South Asia Regional Centre (IRRI-SARC)NSRTC Campus, Varanasi, India
e-mail: a.regina@irri.org

C. Guzmán
Departamento de Genética, E.T.S.I.A.M., Universidad de Cordoba, Campus de Rabanales, Edificio Gregor Mendel C5, Córdoba, Spain

G. Igrejas et al. (eds.), *Wheat Quality For Improving Processing And Human Health*, https://doi.org/10.1007/978-3-030-34163-3_3

1 Introduction

Our planet occupies a myriad of edible plants. Amongst these, more than 50% of energy intake by humans is from the three so called mega crops, rice, wheat and maize. Ancient domestication, wider geographic and climatic adaptability, and versatility in grain utilization have made wheat the most widely grown crop and the most traded cereal in the world. For more than a third of the world's population wheat is a staple food. Wheat primarily laid the foundation for most of the foods enjoyed by humans over centuries across the globe. Wheat flour is an essential ingredient for an umpteen number of food products developed to suit the diverse taste buds of human race. It inherently possesses quality attributes (that are also constantly being improved through breeding) that enable processing to produce a range of products including, but not limited to breads, cakes, cookies, biscuits, pastries breakfast cereals and noodles. Wheat quality attributes required for different products are different and generally, wheat is categorised into different classes to reflect the different end use quality. The classification standards followed by different countries vary and these are generally dependent on various parameters such as protein content, morphological features, milling and end use quality and the region and season of growing.

With globalisation and diversification of diets, wheat based product consumption has spread widely across the world to a large extent. Moreover, innovative products are being developed to suit the fast changing lifestyle and convenience of the growing population. While the primary breeding target for wheat improvement continues to be yield enhancement, increasing demand is there to incorporate targets for grain quality trait enhancement in breeding program to cater for unique and distinct quality feature requirements for various types of wheat based products. Understanding the influence of each of the grain components on quality is essential to drive the efforts on improving wheats for superior quality. In this chapter, we discuss how starch, the major component of wheat grain, is synthesised, the impact of altered starch biosynthetic pathway on starch structure and functionality, the influence of starch granule associated proteins on quality, and the impact of abiotic stresses on starch properties.

2 Starch Biosynthesis in Wheat

The quantity and structure of starch synthesised and deposited in plant tissues are dictated by the participating enzyme machinery. The process of starch synthesis fundamentally involves; (1) provision of carbon flux to produce ADP glucose, the precursor molecule of starch, (2) building linear glucan chains with α-1,4 linkages (3) creating α-1,6 branches by cleaving specific portions of the linear chains and re-attaching at specific intervals along the chains and (4) removal of excess branches to generate the semi-crystalline insoluble homopolymer of starch. While there is a significant level of redundancy in the nature of enzymes involved in the synthesis of both transitory starch contained in the vegetative tissues and reserved starch deposited in storage tissues such as the wheat grain endosperm, fine differences exist in the biosynthesis of the two. As the focus of this book is on wheat grain quality, this chapter will cover only starch synthesis in the endosperm of wheat grain.

Supply of a hexose phosphate (glucose 1-phosphate (G1P) being the preferred) is important for starch synthesis to take place (Emes et al. 2003; Keeling et al. 1988; Zi et al. 2018). This supply comes from the initial conversion of sucrose, the widely transported photoassimilate in wheat, into uridine diphosphate glucose (UDPG) and fructose by sucrose synthase (SUS: EC 2.4.1.13). G1P is subsequently synthesised from these carbon precursors, catalysed by enzymes such as UDP-glucose pyrophosphorylase, hexokinase and phosphoglucomutase. A recent study on waxy wheat has highlighted the importance of sucrose conversion for starch synthesis, through the demonstration that the low kernel weight and total starch accumulation of the waxy wheat is caused by the reduced conversion of sucrose to starch in the late grain filling stage (Zi et al. 2018).

Starch is committed to be synthesised when G1P is converted to adenosine diphosphate glucose (ADPG) by ADPG pyrophosphorylase (AGPase). AGPase is a heterotetrameric protein complex with two small subunits and two large subunits with mostly catalytic and regulatory functions respectively. Wheat grain has AGPase present both in the cytoplasm and in the amyloplast, however the cytosolic isoform accounts for >90% of activity (Burton et al. 2002; Emes et al. 2003; Tetlow and Emes 2017). Evidence suggests that different genes encode the large and small subunits present in the cytosol and the amyloplast in wheat endosperm (Burton et al. 2002; Zhang et al. 2017b). AGPase is commonly found to be allosterically regulated, with activation by 3 phosphoglyceric acid (3PGA) and inhibition by inorganic phosphate (Pi) (Preiss et al. 1991). However the level of sensitivity to such regulation varies depending on the plant species and the enzyme localisation within the plant tissues. AGPase encoding genes are associated with grain attributes such as thousand kernel weight, total starch content and yield (Batra et al. 2017; Hou et al. 2017). Allosterically insensitive AGPase potentially could hasten starch accumulation and grain filling in wheat and help mitigate yield reduction under high temperature stress (Kaur et al. 2017).

As most of the glucosyl donor for starch synthesis comes from extraplastidially synthesised ADPG, there remains the need for transporters to transfer the ADPG from the cytosol into the amyloplast where most other starch biosynthetic enzymes are located, ready to activate the remaining stages of starch synthesis. ADPG transporters in maize (ZmBT1) and barley (HvNST1) are characterised and the respective defective proteins led to reduced uptake of ADPG into the amyloplast (Cao and Shannon 1997; Patron et al. 2004). Kinetic properties of a wheat ADPG transporter have also been characterised from isolated amyloplast (Bowsher et al. 2007).

The ADPG thus transported into the amyloplast now becomes the source of glucose moiety to be attached to the non-reducing end of a glucan primer through α-1,4 linkage resulting in the extension of glucan chains. This reaction is catalysed by a set of single polypeptide enzymes known as starch synthases (SSs) belonging to the glycosyl transferase family. SSs fall into two groups, one localised more or less exclusively to the starch granule matrix known as granule bound starch synthase (GBSS) or waxy protein and the other set present in the endosperm in a soluble form in the stroma as well as in starch granule, loosely termed as soluble SSs. Multiple isoforms are detected for most starch synthases however not all of them are involved in endosperm starch synthesis. GBSS is critical for synthesising amylose, although involvement of other enzymes such as starch branching enzyme (SBE) are indicated in some plant species (Regina et al. 2012). On the other hand, soluble SSs are involved in the synthesis of amylopectin. In wheat endosperm there are four types of soluble SSs, SSI, SSII, SSIII and SSIV, out of which the first three are known to be catalysing amylopectin chain elongation (Konik-Rose et al. 2007; Kosar-Hashemi et al. 2007; Li et al. 2011; Li et al. 1999; Li et al. 2000). Each of the SSs involved in chain elongation have specificity in the length of the chain it synthesises. While SSI generates shorter chains of ~8 to 12 degree of polymerisation (DP), optimal preferences of SSIIa and SSIIIa (the SSII and SSIII isoforms in wheat endosperm) are to produce much longer chains of ~11 to 30 DP and > 50 DP respectively (McMaugh et al. 2014). SSIV is functionally assigned to priming of starch granule formation influencing the number and size of the starch granules, rather than any involvement in the chain elongation, according to studies in other plant species and in wheat chloroplast (Guo et al. 2017a; Roldan et al. 2007). An extensive bioinformatics on SSs provided insight into sequence specificities that underscore the unique catalytic features and functional differences of various SSs in wheat (Leterrier et al. 2008).

Both amylose and amylopectin components of starch are branched molecules, although the frequency of branches are approximately six times more in amylopectin compared to amylose. Hence amylose is more often referred to as 'more or less a linear glucan'. Starch branching enzymes (SBEs) catalyse the building of α-1,6 linkages through cleavage of an internal α-1,4 chain and attaching the released reducing end to C6 hydroxyl group of a chain, thus converting linear chains into a branched structure. SSs then act on the non reducing ends created by SBEs to further elongate the chains, resulting in the growth of the molecule (Tetlow and Emes 2017). Two broad types of SBEs are present in cereals including wheat, SBEI and SBEII. In wheat endosperm, only one isoform of SBEI is detected while there are two isoforms of SBEII, SBEIIa and SBEIIb. The two classes of SBE enzymes functionally differ in cereals with regard to the length of the chains they transfer, the minimum chain length requirement to create a branch and the substrate affinity (Morell et al. 1997). SBEI prefers to act on longer chains than SBEII and has more affinity to amylose, while amylopectin is the preferred substrate for SBEII. Also, the two classes temporally differ in their time of expression during grain development and also spatially differ in their localisation within the amyloplast. In wheat SBEI is more expressed towards the later stage of endosperm development (>20 days post anthesis (DPA)), while SBEII is highly expressed earlier in the developmental stage (<15DPA) (Morell et al. 1997; Regina et al. 2005). Amongst the two isoforms of SBEII, there is 2–3 times more SBEIIa in the amyloplast stroma of wheat than SBEIIb, unlike in maize and rice where there is a predominance of SBEIIb over SBEIIa in the endosperm. SBEII in cereals, in general, plays more substantial roles than SBEI as revealed by studies in mutants of these enzymes (see following section of this chapter). A third class of SBE, SBEIII is also detected in wheat (as in some other plant species), which is constitutively expressed during the whole grain filling period. The function of SBEIII is not very clear, although an association with the synthesis of A and B starch granules is speculated (Kang et al. 2013a).

Role of debranching enzyme in starch biosynthesis is mainly to trim excess branches formed in the growing amylopectin molecules to establish an organised semi crystalline structure to starch. Out of the two (isoamylase and pullulanase) types of debranching enzymes detected in plants, the isoamylase types act on amylopectin to remove unfitting α-1, 6 linkages. Functionally, ISA1 and ISA2 are the isoforms of isoamylase that are involved in amylopectin trimming either in homomeric or heteromeric complex forms, while a third isoform ISA3 is mostly involved in starch degradation (Kubo et al. 2010; Lin et al. 2013; Nielsen et al. 2002; Utsumi et al. 2011; Yun et al. 2011). Involvement of more enzymes in reserve starch biosynthesis (that are not covered here) such as starch phophorylase, disproportionating enzyme, glucan water dikinase and phospho glucan water dikinase are suggested, although their precise roles are yet to be defined.

Concerted action of the multiple isoforms of starch biosynthetic enzymes are essential in the temporal and spatial management of starch biosynthesis. Evidences of protein-protein interactions and protein complexes existing in starch synthesising organelles supports the need for enzymes to act collaboratively to regulate starch biosynthesis (Crofts et al. 2017; Fushan et al. 2012; Tetlow et al. 2008). Also more insights are emerging on the regulation of starch synthesis and accumulation influenced by transcription factors and enzymes outside the core pathway. For example, TaRSR1 (reduced sugar response 1) in wheat appears to negatively regulate most of the key starch biosynthetic enzymes including AGPase, SSs and SBEs (Kang et al. 2013b). Stimulation of starch synthesis and its accumulation by Trehalose 6-phosphate, the precursor of the disaccharide trehalose, has been more recently demonstrated in plants (Kolbe et al. 2005; Paul et al. 2018; Zhang et al. 2017a). The effect is mainly through the post translational redox dependent activation of AGPase, the rate limiting enzyme of starch synthesis. In wheat external application of precursors of T6P during grain filling stage has shown to increase the grain size and total starch content (Griffiths et al. 2016). TaGW2-6A, a weight related gene allele that encodes a functional E3 RING-type ubiquitin ligase, is recently speculated to affect starch-related genes promoting accumulation of starch (Geng et al. 2017).

Readers are also referred to recent review articles on starch biosynthesis for further insights and understanding (Crofts et al. 2017; Jeon et al. 2010; MacNeill et al. 2017; Regina et al. 2016; Tetlow and Emes 2017).

3 Impact of Starch Enzyme Mutations on Wheat Grain and Quality

Spontaneous occurrence of starch biosynthetic enzyme mutations in nature leading to a specific phenotype in wheat is hindered by the complex genetic make up of wheat. Being an allohexaploid by nature, most of the known genes encoding starch biosynthetic enzymes are triplicated with one copy present in each of the three genomes in wheat. The effect of mutation in any one of the three allelic forms will most often be masked by the functional redundancy of homoeologs, leading to lack of manifestation of a phenotypic effect in the grain. Multiple studies have been carried out to identify mutations in the wheat gene pool of starch synthesis enzymes. Probably, the waxy proteins have been the most intensively examined, with scores of studies describing lot of alleles in several wheat species (see Guzman and Alvarez 2016 for a review). Variability for SSII was also detected in wheat cultivars (Yamamori and Endo 1996) using SDS-PAGE gels. Simultaneous mutations of the three homoeo alleles of any specific gene to occur naturally is highly unlikely. The situation is more complicated with the presence of multiple isoforms for several of the starch biosynthetic enzymes (as detailed earlier). Lessons from diploid cereal species like maize, rice and barley led to speculative reflections on starch enzyme functionality in wheat; however traditional and advanced breeding and genetic techniques allowed further wheat specific insights on starch biosynthetic enzyme functionality and mutant phenotypes. These include, but not limited, induced mutagenesis using chemical mutagens such as ethyl methane sulphonate (EMS), sodium azide, followed by mutant detection and hybridisation, and transgenic technologies such as RNAi. High throughput allele detection techniques such as targeting induced lesions in genomes (TILLING) is increasingly being used in identifying allele variants in starch biosynthetic genes created through induced mutations (Regina et al. 2015a; Slade et al. 2012). The most modern techniques of targeted mutagenesis through gene editing (eg CRISPR/CAS9 mediated gene editing) is a powerful strategy for polyploid plants like wheat to create allelic variations (Uauy et al. 2017), and their use in modifying starch biosynthesis in wheat is optimistically a matter of time.

A short review on the impact of functionally impaired key enzymes in wheat starch biosynthetic pathway is presented in Table 1.

4 Starch and Nutritional Quality of Wheat

Glycemic carbohydrate (or available carbohydrate) is important for humans to support the normal functioning of several organs including brain, red blood cells and reproductive tissues (Hardy et al. 2015). Starch is the most abundant form of dietary carbohydrate and as such the main source of dietary energy. The most important functionality of starch that permits manipulation to serve a nutritional purpose is its digestibility. Approximately >99% of naturally occurring wheat starch is digestible and only the remaining very small proportion is resistant to digestion in the small intestine by human digestive enzymes (referred to as resistant starch (RS)). Within the digestible portion of starch, the rapidly digestible starch (RDS), the fraction that is digested within 20 min results in a rapid increase in postprandial blood glucose level, while the slowly digestible starch (SDS) that completely digests in the small intestine at a slower rate than RDS results in a sustained postprandial glucose level (Lehmann and Robin 2007). The relative proportion of RDS, SDS and RS in starch varies depending on the genotype, the form of the material containing starch (grain/flour/product) and the type of processing followed to transform the grain into a food product.

Table 1 Impact of impaired starch biosynthetic enzyme activity on wheat grain and quality

Target starch enzyme/s	Mode of impairment (transgenic/ non transgenic)	Phenotype	Reference
AGPase	Non transgenic-mutation	Reduced total starch content	Guo et al. (2017b)
Isoamylase 1	Transgenic-down regulation	Reduced starch content, elevated phytoglycogen and beta-glucan, altered amylopectin chain length distribution, disrupted semi crystalline structure	Sestili et al. (2016)
GBSS1 (waxy wheat)	Non transgenic-mutation	Low to zero amylose, A-type X-ray diffraction pattern, higher crystallinity, lower gelatinisation temperature, higher peak viscosity and swelling power (lower peak viscosity and setback in some studies), higher resistance to retrogradation, higher flour water absorption, lower dough stability, desired for high quality white-salted noodles. Null4A GBSS desired for Udon noodles	Ahuja et al. (2013); Hayakawa et al. (1997); Hoshino et al. (1996); Miura et al. (1994); Nakamura et al. (1993a); Nakamura et al. (1993b); Yamamori et al. (1995)
GBSS1 & SSIIa (sweet wheat)	Non transgenic-mutation	Elevated maltose, sucrose and fructan, small and misshapen starch granules, altered crystallinity, lowered gelatinisation temperature, lower molecular weight for amylopectin, increased malt oligosaccharides and very short chains of DP 2 and 3	Nakamura et al. (2006); Shimbata et al. (2011); Vrinten et al. (2012)
SSIIa (SGP-1)	Non transgenic	Moderately elevated amylose content, higher protein content, lower starch content, increase in the amount of short amylopectin chains (6–10 DP) and decrease in intermediate chains of 11–25 DP, higher total dietary fibre, lower gelatinisation temperature, altered starch crystallinity, lower peak viscosity, lower kernel weight and flour swelling power, higher flour water absorption, lower bread loaf volume	Berky et al. (2016); Hogg et al. (2013); Hogg et al. (2017); Hung et al. (2006); Konik-Rose et al. (2007); Shimbata et al. (2005)
SBEII (predominantly SBEIIa, in combination with SBEIIb)	Non transgenic-mutation	Higher amylose, higher resistant starch, reduced total starch content, increased grain hardness	Hazard et al. (2012); Hazard et al. (2015); Regina et al. (2015a); Schonhofen et al. (2017); Sestili et al. (2015); Slade et al. (2012)
SBE 1	Non transgenic	No substantial alteration in any of the starch structural or functional properties	Regina et al. (2004)

Rapid digestion of starch is important whenever a rapid energy conversion is required, as in the cases of infant diets and addressing severe undernutrition. In the current era of increasing incidence of diet related non communicable diseases such as Type II diabetes, gut diseases and cardiovascular disorders, SDS and RS are of much nutritional value, due to their physiological ability to address some of these health conditions. SDS is the fraction of starch that is converted into glucose only after 120 min of enzymatic digestion (Englyst et al. 1992). SDS enriched foods help in addressing diseases like metabolic syndrome and diabetes, due to the prolonged digestion and slow release of glucose (El Hindawy et al. 2018; Herrmann et al. 1995; Seal et al. 2003). SDS form of starch could be produced *in vitro* through several modification means such as cross-linking and enzymatic modification (Raigond et al. 2015; Shin et al. 2004). Amylopectin structure is significantly associated with SDS levels in starch. Studies on maize mutants showed a parabolic relationship between SDS content and weight ratio of amylopectin short chains ($DP < 13$, named SF) to long chains ($DP \geq 13$, named LF), suggesting starches with higher and lower SF/LF ratio producing high SDS levels compared to that with a medium SF/LF ratio (Zhang et al. 2008).

The potential of RS in preventing and retarding the most common diet-related diseases is increasingly being demonstrated (Bird and Regina 2017). Being resistant to enzymatic digestion, RS escapes digestion in the small intestine, thus not contributing directly to the spike of glucose in the blood stream assisting in Type II diabetes prevention and management (Behall et al. 2006). Once reaching the large bowel where it gets fermented by colonic microbiota resulting in the production of short chain fatty acid (SCFA). There is a growing body of evidence on the diverse benefits of SCFA for human health (Bird and Regina 2017; Bird et al. 2010; Conlon et al. 2012; Guilloteau et al. 2010; Keenan et al. 2015; Kim et al. 2016; Topping and Clifton 2001; Vetrani et al. 2018). The benefits include maintenance of bowel health through proliferation of beneficial gut microbial population at the same time suppressing the pathogenic bacterial species. Luminal SCFA levels, particularly the level of butyrate, are associated with protection of DNA damage caused by unhealthy diets, promotion of gut barrier function and suppression of epithelial inflammation, all of which are important in protection against bowel diseases such as colorectal cancer. RS also influences metabolism of skeletal muscle, adipose tissue depots and liver. The positive effect of RS on weight management is also suggested through promoting satiety and reduced food intake. Involvement of SCFA in regulating immune system function and responses to infection is also recently highlighted.

RS is significantly correlated with the contents of amylose and long chained amylopectin that is functionally similar to amylose (hence also called as amylose like molecule) (Regina et al. 2012). Thus enabling the necessary structural changes in starch to elevate one or the other or both these fractions in the grain is a major strategy adopted in generating RS enriched wheats. Mainly two mechanisms are demonstrated in wheat to elevate the amylose and amylose like molecules in wheat, once is suppressing SSIIa activity and the other is suppressing SBEIIa activity in combination with some of SBEIIb activity (Regina et al. 2015b; Regina et al. 2006; Yamamori et al. 2000). While SSIIa mediated elevation of amylose is only moderate

(<50% amylose), a higher level of amylose increase (>75%) is achieved through the SBEIIa mediated approach. The high amylose wheat developed through selective inhibition of SBEII is also enriched with RS, with a > ten-fold increase compared to a standard wheat. Flour from this wheat is suitable to produce bakery products such as bread that are of comparable quality to those from standard wheat flours, with the added advantage of having significantly elevated RS and TDF in the products (Berbezy et al. 2015). A high amylose durum wheat developed through SBEII mediated strategy was shown to produce pasta of acceptable quality with positive effects on pasta firmness (Hazard et al. 2015).

5 Starch Granule Associated Proteins

Major macronutrient components in a cereal based food system are starch and proteins. While each of these individually influence the functional properties of the system, the interaction between these two components, also mediated by other compounds present such as lipids, plays an important role in determining the final quality of the grain and, in turn, the products. In wheat, starch granules are embedded in a protein matrix within the endosperm tissue (Marshall and Chrastil 1992).

There are mainly two types of proteins associated with starch in cereals (Baldwin 2001); (a) storage proteins such as the gluten proteins that remain adsorbed to the starch granule surface (storage proteins are outside the scope of this chapter, but reviewed comprehensively elsewhere in this book); and (b) starch granule associated proteins (SGAPs) that are either bound to the surface of starch granules or present as integral component of starch granules. SGAPs are mostly distinct from the storage proteins both structurally and functionally. Higher levels of basic and hydrophobic amino acids are characteristic features of SGAPs, which are attributable to their binding ability to starch granules. In a very broad sense, surface proteins are of low molecular weight (5 to 30 kDa) that are easily extractable using salt solutions or aqueous buffers, while integral proteins are of high molecular weights (60 to 149 kDa) that require stronger detergents and heat swelling of starch granules for their extraction. However, exceptions to this broad classification have also been demonstrated with certain low molecular weight proteins found integral to starch granules and vice versa. Major starch granule associated integral proteins and some surface proteins have enzymatic functions and are involved in starch biosynthesis (Rahman et al. 1995), taking either biosynthetic or degradative roles. Starch biosynthetic enzymes, their roles and their impacts on starch structure and quality are already covered in previous sections of this chapter.

Surface SGAPs, their quantities, orientation and the nature are known to influence several starch properties such as gelatinisation, viscosity, damage and enzyme resistance (Baldwin et al. 1997; Hamaker and Griffin 1993). Kernel texture of wheat grain is an important characteristic that determines its quality, milling performance, market classification and end use. The presence of ~13 kDa SGAPs known as friabilins or puroindolines (Pins) on the surface of starch granules modulates the adhesiveness of the protein-starch complex, and provides wheat kernel with a soft texture. Flour from such wheats with soft kernel texture is used for biscuits and cakes, while wheats with a harder kernel texture that are lacking or are lower in the levels of puroindolines on starch granule surface are better suited for breads, noodles and pasta. Two genes encoding puroindolines, *Pina-D1* and *Pinb-D1* when both present in the wild type forms in the endosperm imparts a soft kernel texture, while deletions or diverse mutations in either or both of the *Pin-D1* genes result in a harder kernel texture (Morris and Bhave 2008). A suggested reference for a detailed review on SGAPs is Baldwin (2001).

6 Environmental Effects and Starch Properties

Wheat grain starch properties, as most of the other grain characteristics, are related to the genotype and also are influenced by environmental factors during crop growth. In this regard, a significant number of research studies have been carried out to understanding the influence of specific environmental factors on wheat starch characteristics and properties (Vignola et al. 2016). In the current context of the Climate Change, understanding how different environmental conditions and, particularly, how abiotic stresses such as drought or heat modify starch composition and properties and accumulation in cereal endosperm is key. This knowledge will lead to improve the predictions of grain and flour quality and will be useful to breeding programs when deploying new germplasm that can tolerate extremes of environment (Thitisaksakul et al. 2012) and to the production of high quality wheat.

Different studies have analysed what starch characteristics are modified with different environmental conditions in general. Geera et al. (2006) targeted to identify the starch characteristics most influenced by the environment (under field conditions) that modulate the flour/starch properties, particularly the flour pasting properties. They found that total starch and A/B-type granule contents were affected by the environment while amylose content was minimally affected. These changes appeared to explain part of the environment-induced fluctuations in the flour pasting properties observed. Labuschagne et al. (2007), in an experiment conducted with grain from three different irrigated fields, also found that the total starch content was strongly affected by the environment but not the amylose content, which was more dependent on the genotype or genotype x environment interaction. Jing et al. (2003) did not report any significant changes in amylose content due to the environment too. On the other hand, Nhan and Copeland (2014) found the environment and genotype x environment as significant sources of variability in amylose content and amylopectin chain lengths. Starch pasting properties were also found to be affected by the environment (Konik et al. 1993; Morris et al. 1997). But, as described by Graybosch et al. (2003) in an experiment conducted with waxy wheat lines, stable starch properties (recorded in this case with the Rapid Visco Analyzer) responses over diverse environments are expected. This is contrary to what Vignola et al. 2016 showed in their study in which environment had a much greater impact on the starch properties than genotype for starch pasting parameters except for the pasting temperature. Ansari et al. (2010) showed that around 34% of the variation found in starch swelling power of wheat genotypes grown in different locations was determined by the environmental factors. In this study the differences in starch swelling power at different locations did not affect the ranking of the genotypes, which was quite consistent, which also agrees with the results of Nhan and Copeland (2014). In summary, it seems that in general, the environment affects the starch and its properties, but the magnitude of the impact will depend a lot on how contrasting are the different environments tested. Therefore, a breeding strategy considering multiples sites for evaluation of starch properties may not be always required.

One of the abiotic stresses most studied regarding its effect on starch and its properties is heat or elevated temperature during grain filling. It is well established that high temperatures during grain filling decrease starch accumulation in the grain leading to a reduction in the grain weight and yield (Gibson & Paulsen 1999; Hurkman et al. 2003; Matsuki et al. 2003; Wang et al. 2017). This is probably due to the reduction in the activity of enzymes involved in the starch synthesis pathway such as the soluble starch synthase activity (Keeling et al. 1993), the starch branching enzyme (Keeling et al. 1994) or the ADP-glucose pyrophosphorylase (Smidansky et al. 2002; Altenbach et al. 2003), and to reduced grain filling period. Other studies have showed that elevated temperature during grain filling may result in a qualitative change in starch: Shi et al. (1994) showed that amylose content was slightly increased and starch gelatinization temperature increased when wheat is grown at 40 °C compared with 15 °C while A-type granules concentration was reduced. In agreement with this, Panozzo and Eagles (1998) reported that environmental variation with accumulated temperatures above 30 °C led to an increase in amylose percentage and to an increase in A-type starch granules. Vignola et al. (2016) found higher amylose contents in the environments with higher temperatures during the grain filling period and also an increase in A-type starch granules. Opposite to these results, Stone and Nicolas (1995) reported a reduction in amylose content by heat treatment but limited and confined to small proportion of the genotypes of the experiment. On the other hand, Matsuki et al. (2003) reported that amylose contents were not significantly affected by elevated maturation temperature in several cultivars when wheat was grown at 15 °C and at 30 °C. Similar results were obtained by Wang et al. (2017), although in this last case the heat stress only represented 2.2 °C higher temperature than the normal environment. Therefore, the intensity of the heat stress and the susceptibility of the genotypes to heat determine the effect on amylose content (Thitisaksakul et al. 2012). In terms of end-use quality governed by starch properties, no significant changes in noodles swelling power in response to high temperature were found (Stone and Nicolas 1995).

The effect of drought stress on starch has been also studied. Wheat plants under drought stress have reduced yield, and this is due, in part, to alterations in starch biosynthetic enzyme activity (Jenner et al. 1991), mainly because of a loss of activity in starch synthases, which are the most sensitive enzymes to drought in the starch biosynthesis pathway (Ahmadi and Baker 2001). Amylose content is reduced in the case of drought stress and the proportion of A-type starch granules is increased (Fabian et al. 2011; Singh et al. 2008)

7 Conclusions

Starch is the major component of the wheat grain and it is an important determinant of both industrial and nutritional quality. Significant progress has been made in the last thirty years in understanding how starch is synthetized in the wheat grain and how starch composition can be manipulated through different approaches including classical breeding or more modern tools such as TILLING. This has allowed the development of commercial cultivars with modified starch properties, which are already available for cultivation and could emerge as protagonists in the coming years in farmers' fields due to their unique functional properties and nutritional profile.

Acknowledgments Carlos Guzman would like to greatly acknowledge to the Spanish Ministerio de Ciencia, Innovación y Universidades for a Ramon y Cajal grant (RYC-2017-21891).

References

Ahuja G, Jaiswal S, Hucl P, Chibbar RN (2013) Genome-Specific Granule-Bound Starch Synthase I (GBSSI) Influences Starch Biochemical and Functional Characteristics in Near-Isogenic Wheat (Triticum aestivum L.) Lines. Journal of Agriculture and Food Chemistry 61:12129–12138.

Ahmadi A, Baker DA (2001) The effect of water stress on the activities of key regulatory enzymes of the sucrose to starch pathway in wheat. Plant Growth Regulation 35: 81–91.

Altenbach SB, DuPont FM, Kothari KM, Chan R, Johnson EL, Lieu D (2003) Temperature, water and fertilizer influence the timing of key events during during grain development in a US spring heat. Journal of Cereal Science 37: 9–20.

Ansari O, Båga M, Chibbar RN, Sultana N, Howes NK (2010) Analysis of starch swelling power in Australian breeding lines of hexaploid wheat (Triticum aestivum L.). Field Crops Research 115: 171–178.

Baldwin PM, Melia CD, Davies MC (1997) The surface chemistry of starch granules studied by time-of-flight secondary ion mass spectrometry. Journal of Cereal Science 26: 329–346.

Baldwin PM (2001) Starch granule-associated proteins and polypeptides: A review. Starch-Starke 53: 475–503.

Batra R, Kumar P, Jangra MR, Passricha N, Sikka VK (2017) High Precision Temperature Controlling AGPase in Wheat Affecting Yield and Quality Traits. Cereal Research Communications 45: 610–620.

Behall KM, Scholfield DJ, Hallfrisch JG, Liljeberg-Elmstahl HGM (2006) Consumption of both resistant starch and beta-glucan improves postprandial plasma glucose and insulin in women. Diabetes Care 29: 976–981.

Berbezy P, Regina A, Chapron S, Bird T, Duperrier B, Chanliaud E (2015) Making high amylose wheat flour to get high resistant starch content bread. In: 6th International Dietary Fibre Conference.

Berky R, Sipko E, Balazs G, Harasztos AH, Kemeny S, Fekete J (2016) Coupled-Column RP-HPLC in Combination with Chemometrics for the Characterization and Classification of Wheat Varieties. Chromatographia 79: 811–821.

Bird A, Regina A (2017) High amylose wheat: A platform for delivering human health benefits. Journal of Cereal Science 82: 99–105.

Bird AR, Conlon MA, Christophersen CT, Topping DL (2010) Resistant starch, large bowel fermentation and a broader perspective of prebiotics and probiotics. Beneficial Microbes 1: 423–431.

Bowsher CG, Scrase-Field EF, Esposito S, Emes MJ, Tetlow IJ (2007) Characterization of ADP-glucose transport across the cereal endosperm amyloplast envelope. Journal of Experimental Botany 58: 1321–1332.

Burton RA, Johnson PE, Beckles DM, Fincher GB, Jenner HL, Naldrett MJ, Denyer K (2002) Characterization of the genes encoding the cytosolic and plastidial forms of ADP-glucose pyrophosphorylase in wheat endosperm. Plant Physiology 130: 1464–1475.

Cao HP, Shannon JC (1997) BT1, a possible adenylate translocator, is developmentally expressed in maize endosperm but not detected in starchy tissues from several other species. Physiologia Plantarum 100: 400–406.

Conlon MA, Kerr CA, McSweeney CS, Dunne RA, Shaw JM, Kang S, et al. (2012) Resistant Starches Protect against Colonic DNA Damage and Alter Microbiota and Gene Expression in Rats Fed a Western Diet. Journal of Nutrition 142: 832–840.

Crofts N, Nakamura Y, Fujita N (2017) Critical and speculative review of the roles of multi-protein complexes in starch biosynthesis in cereals. Plant Science 262: 1–8.

El Hindawy M, Kim CY, Hamaker BR (2018) Intestinal Simulation of Gut-brain Axis by Dietary Slowly Digestible Starch Regulates Satiety and Glucose Homeostasis. In Vitro Cellular & Developmental Biology – Animal 54: S25-S25.

Emes MJ, Bowsher CG, Hedley C, Burrell MM, Scrase-Field ESF, Tetlow IJ (2003) Starch synthesis and carbon partitioning in developing endosperm. Journal of Experimental Botany 54: 569–575.

Englyst HN, Kingman SM, Cummings JH (1992) Classification and measurement of nutritionally important starch fractions. European Journal of Clinical Nutrition 46: S33-S50.

Fabian A, Jager K, Rakszegi M, Barnabas B (2011) Embryo and endosperm development in wheat (Triticum aestivum L.) kernels subjected to drought stress. Plant Cell Reports 30: 551–563.

Fushan L, Romanova N, Lee EA, Ahmed R, Evans M, Gilbert EP, Morell MK, Emes MJ, Tetlow IJ (2012) Glucan affinity of starch synthase IIa determines binding of starch synthase I and starch-branching enzyme IIb to starch granules. Biochemical Journal 448: 373–387.

Geera BP, Nelson JE, Souza E, Huber KC (2006) Flour Pasting Properties of Wild-Type and Partial Waxy Soft Wheats in Relation to Growing Environment-Induced Fluctuations in Starch Characteristics. Cereal Chemistry 83: 558–564.

Geng J, Li LQ, Lv Q, Zhao Y, Liu Y, Zhang L, Li XJ (2017) TaGW2-6A allelic variation contributes to grain size possibly by regulating the expression of cytokinins and starch-related genes in wheat. Planta 246: 1153–1163.

Gibson LR, Paulsen GM (1999) Yield components of wheat grown under high temperature stress during reproductive growth. Crop Science 39: 1841–1846.

Griffiths CA, Sagar R, Geng Y, Primavesi LF, Patel MK, Passarelli MK, et al. (2016) Chemical intervention in plant sugar signalling increases yield and resilience. Nature 540, 574–578.

Graybosch RA, Souza E, Berzonsky W, Baenziger PS, Chung O (2003) Functional properties of waxy wheat flours: genotypic and environmental effects. Journal of Cereal Science 38: 69–76.

Guilloteau P, Martin L, Eeckhaut V, Ducatelle R, Zabielski R, Van Immerseel F (2010) From the gut to the peripheral tissues: the multiple effects of butyrate. Nutrition Research Reviews 23: 366–384.

Guo HJ, Liu YC, Li X, Yan ZH, Xie YD, Xiong HC, et al. (2017a) Novel mutant alleles of the starch synthesis gene TaSSIVb-D result in the reduction of starch granule number per chloroplast in wheat. BMC Genomics 18: 358.

Guo HJ, Yan ZH, Li X, Xie YD, Xiong HC, Liu YC, et al. (2017b) Development of a High-Efficient Mutation Resource with Phenotypic Variation in Hexaploid Winter Wheat and Identification of Novel Alleles in the TaAGP.L-B1 Gene. Frontiers in Plant Science 8: 9.

Guzman C, Alvarez JB (2016) Wheat waxy proteins: polymorphism, molecular characterization and effects on starch properties. Theoretical and Applied Genetics 129: 1–16.

Hamaker BR, Griffin VK (1993) Effect of disulfide bond-containing protein on rice starch gelatinization and pasting. Cereal Chemistry 70: 377–380.

Hardy K, Brand-Miller J, Brown KD, Thomas MG, Copeland L (2015) The importance of dietary carbohydrate in human evolution. The Quarterly Review of Biology 90: 251–268.

Hayakawa K, Tanaka K, Nakamura T, Endo S, Hoshino T (1997) Quality characteristics of waxy hexaploid wheat (Triticum aestivum L.): Properties of starch gelatinization and retrogradation. Cereal Chemistry 74: 576–580.

Hazard B, Zhang X, Colasuonno P, Uauy C, Beckles DM, Dubcovsky J (2012) Induced Mutations in the Starch Branching Enzyme II (SBEII) Genes Increase Amylose and Resistant Starch Content in Durum Wheat. Crop Science 52: 1754–1766.

Hazard B, Zhang XQ, Naemeh M, Hamilton MK, Rust B, Raybould HE, et al. (2015) Mutations in Durum Wheat SBEII Genes affect Grain Yield Components, Quality, and Fermentation Responses in Rats. Crop Science 55: 2813–2825.

Herrmann C, Göke R, Richter G, Fehmann HC, Arnold R, Göke B (1995) Glucagon-Like Peptide-1 and Glucose-Dependent Insulin-Releasing Polypeptide Plasma Levels in Response to Nutrients. Digestion 56: 117–126.

Hogg AC, Gause K, Hofer P, Martin JM, Graybosch RA, Hansen LE, Giroux MJ (2013) Creation of a high amylose durum wheat through mutagenesis of starch synthase II (SSIIa). Journal of Cereal Science 57: 377–383.

Hogg AC, Martin JM, Giroux MJ (2017) Novel ssIIa Alleles Produce Specific Seed Amylose Levels in Hexaploid Wheat. Cereal Chemistry 94: 1008–1015.

Hoshino T, Ito S, Hatta K, Nakamura T, Yamamori M (1996) Development of waxy common wheat by haploid breeding. Breeding Science 46: 185–188.

Hou J, Li T, Wang YM, Hao CY, Liu HX, Zhang XY (2017) ADP-glucose pyrophosphorylase genes, associated with kernel weight, underwent selection during wheat domestication and breeding. Plant Biotechnology Journal 15: 1533–1543.

Hung PV, Maeda T, Morita N (2006) Waxy and high-amylose wheat starches and flours - characteristics, functionality and application. Trends in Food Science & Technology 17: 448–456.

Hurkman WJ, McCue KF, Altenbach SB, Korn A, Tanaka CK, Kothari KM, et al. (2003) Effect of temperature on expression of genes encoding enzymes for starch biosynthesis in developing wheat endosperm. Plant Science 164: 873–881.

Jenner CF, Ugalde TD, Aspinall D (1991) The physiology of starch and protein deposition in the endosperm of wheat. Australian Journal of Plant Physiology 18: 211–226.

Jeon JS, Ryoo N, Hahn TR, Walia H, Nakamura Y (2010) Starch biosynthesis in cereal endosperm. Plant Physiology and Biochemistry 48: 383–392.

Jing Q, Jiang D, Dai T, Cao W (2003) Effects of genotype and environment on wheat grain quality and protein components. Ying Yong Sheng Tai Xue Bao 14: 1649–1653.

Kang GZ, Li SY, Zhang MQ, Peng HF, Wang CY, Zhu YJ, Guo TC (2013a) Molecular Cloning and Expression Analysis of the Starch-branching Enzyme III Gene from Common Wheat (Triticum aestivum). Biochemical Genetics 51: 377–386.

Kang GZ, Xu W, Liu GQ, Peng XQ, Guo TC (2013b) Comprehensive analysis of the transcription of starch synthesis genes and the transcription factor RSR1 in wheat (Triticum aestivum) endosperm. Genome 56: 115–122.

Kaur V, Madaan S, Behl RK (2017) ADP-glucose Pyrophosphorylase Activity in Relation to Yield Potential of Wheat: Response to Independent and Combined High Temperature and Drought Stress. Cereal Research Communications 45: 181–191.

Keeling PL, Bacon PJ, Holt DC (1993) Elevated temperature reduces starch deposition in wheat endosperm by reducing the activity of soluble starch synthase. Planta 191: 342–348.

Keeling PL, Banisadr R, Barone L, Wasserman BP, Singletary GW (1994) Effect of temperature on enzymes in the pathway of starch biosynthesis in developing wheat and maize grain. Australian Journal of Plant Physiology 21: 807–827.

Keeling PL, Wood JR, Tyson RH, Bridges IG (1988) Starch biosynthesis in developing wheat-grain - evidence against the direct involvement of triose phosphates in the metabolic pathway. Plant Physiology 87: 311–319.

Keenan MJ, Zhou J, Hegsted M, Pelkman C, Durham HA, Coulon DB, Martin RJ (2015) Role of Resistant Starch in Improving Gut Health, Adiposity, and Insulin Resistance. Advances in Nutrition 6: 198–205.

Kim M, Qie YQ, Park J, Kim CH (2016) Gut Microbial Metabolites Fuel Host Antibody Responses. Cell Host Microbe 20: 202–214.

Kolbe A, Tiessen A, Schluepmann H, Paul M, Ulrich S, Geigenberger P (2005) Trehalose 6-phosphate regulates starch synthesis via posttranslational redox activation of ADP-glucose pyrophosphorylase. Proceedings of the National Academy of Sciences 102: 11118–11123.

KoniK CM, Miskelly DM, Gras PW (1993) Starch swelling power, grain hardness and protein: relationship to sensory properties of Japanese noodless. Starch/Staerke 45: 139–144.

Konik-Rose C, Thistleton J, Chanvrier H, Tan I, Halley P, Gidley M, et al. (2007) Effects of starch synthase IIa gene dosage on grain, protein and starch in endosperm of wheat. Theoretical and Applied Genetics 115: 1053–1065.

Kosar-Hashemi B, Li ZY, Larroque O, Regina A, Yamamori M, Morell MK, Rahman S (2007) Multiple effects of the starch synthase II mutation in developing wheat endosperm. Functional Plant Biology 34: 431–438.

Kubo A, Colleoni C, Dinges JR, Lin Q, Lappe RR, Rivenbark JG, et al. (2010) Functions of heteromeric and homomeric isoamylase-type starch-debranching enzymes in developing maize endosperm. Plant Physiology 153: 956–969.

Labuschagne MT, Geleta N, Osthoff G (2007) The influence of environment on starch content and amylose to amylopectin ratio in wheat. Starch/Staerke 59: 234–238.

Lehmann U, Robin F (2007) Slowly digestible starch – its structure and health implications: a review. Trends in Food Science & Technology 18: 346–355.

Leterrier M, Holappa LD, Broglie KE, Beckles DM (2008) Cloning, characterisation and comparative analysis of a starch synthase IV gene in wheat: functional and evolutionary implications. BMC Plant Biology 8: 98.

Li Z, Rahman S, Kosar-Hashemi B, Mouille G, Appels R, Morell MK (1999) Cloning and characterization of a gene encoding wheat starch synthase I. Theoretical and Applied Genetics 98: 1208–1216.

Li Z, Li D, Du X, Wang H, Larroque O, Jenkins CLD, et al. (2011) The barley amo1 locus is tightly linked to the starch synthase IIIa gene and negatively regulates expression of granule-bound starch synthetic genes. Journal of Experimental Botany 62: 5217–5231.

Li ZY, Mouille G, Kosar-Hashemi B, Rahman S, Clarke B, Gale KR, et al. (2000) The structure and expression of the wheat starch synthase III gene. Motifs in the expressed gene define the lineage of the starch synthase III gene family. Plant Physiology 123: 613–624.

Lin Q, Facon M, Putaux JL, Dinges JR, Wattebled F, D'Hulst C, et al. (2013) Function of isoamylase-type starch debranching enzymes ISA1 and ISA2 in the Zea mays leaf. New Phytologist 200: 1009–1021.

Mac Neill GJ, Mehrpouyan S, Minow MAA, Patterson JA, Tetlow IJ, Emes MJ (2017) Starch as a source, starch as a sink: the bifunctional role of starch in carbon allocation. Journal of Experimental Botany 68: 4433–4453.

Marshall WE, Chrastil J (1992) Interaction of Food Proteins with Starch. In: Biochemistry of Food Proteins (Hudson, B.J.F. ed) pp. 75–97. Boston, MA: Springer US.

Matsuki J, Yasui T, Kohyama K, Sasaki T (2003) Effects of environmental temperature on structure and gelatinization properties of wheat starch. Cereal Chemistry 80: 476–480.

McMaugh SJ, Thistleton JL, Anschaw E, Luo JX, Konik-Rose C, Wang H, et al. (2014) Suppression of starch synthase I expression affects the granule morphology and granule size and fine structure of starch in wheat endosperm. Journal of Experimental Botany 65: 2189–2201.

Miura H, Tanii S, Nakamura T, Watanabe N (1994) Genetic-control of amylose content in wheat endosperm starch and differential-effects of 3 wx genes. Theoretical and Applied Genetics 89: 276–280.

Morell MK, Blennow A, Kosar Hashemi B, Samuel MS (1997) Differential expression and properties of starch branching enzyme isoforms in developing wheat endosperm. Plant Physiology 113: 201–208.

Morris CF, Shackley BJ, King GE, Kidwell KK (1997) Genotypic and enviromental variation for flour swelling volume in wheat. Cereal Chemistry 74: 16–21.

Morris CF, Bhave M (2008) Reconciliation of D-genome puroindoline allele designations with current DNA sequence data. Journal of Cereal Science 48: 277–287.

Nakamura T, Yamamori M, Hirano H, Hidaka S (1993a) Decrease of waxy (wx) protein in 2 common wheat cultivars with low amylose content. Plant Breeding 111: 99–105.

Nakamura T, Yamamori M, Hirano H, Hidaka S (1993b) Identification of three Wx protein in wheat (Triticum aestivum L). Biochemical Genetics 31: 75–86.

Nakamura T, Shimbata T, Vrinten P, Saito M, Yonemaru J, Seto Y, et al. (2006) Sweet wheat. Genes & Genetic Systems 81: 361–365.

Nhan MT, Copeland L (2014) Effects of growing environment on properties of starch from five Australian wheat varieties. Cereal Chemistry 91: 587–594.

Nielsen TH, Baunsgaard L, Blennow A (2002) Intermediary glucan structures formed during starch granule biosynthesis are enriched in short side chains, a dynamic pulse labeling approach. Journal of Biological Chemistry 277: 20249–20255.

Panozzo JF, Eagles HA (1998) Cultivar and environmental effects on quality characters in wheat. I. Starch. Australian Journal of Agricultural Research 49: 757–766.

Patron NJ, Greber B, Fahy BE, Laurie DA, Parker ML, Denyer K (2004) The lys5 mutations of barley reveal the nature and importance of plastidial ADP-Glc transporters for starch synthesis in cereal endosperm. Plant Physiology 135: 2088–2097.

Paul MJ, Gonzalez-Uriarte A, Griffiths CA, Hassani-Pak K (2018) The Role of Trehalose 6-Phosphate in Crop Yield and Resilience. Plant Physiology 177: 12–23.

Preiss J, Ball K, Smithwhite B, Iglesias A, Kakefuda G, Li L (1991) Starch biosynthesis and its regulation. Biochemical Society Transactions 19: 539–547.

Raigond P, Ezekiel R, Raigond B (2015) Resistant starch in food: a review. Journal of the Science of Food and Agriculture 95: 1968–1978.

Rahman S, Kosarhashemi B, Samuel MS, Hill A, Abbott DC, Skerritt JH, et al. (1995) The major proteins of wheat endosperm starch granules. Australian Journal of Plant Physiology 22: 793–803.

Regina A, Kosar-Hashemi B, Li ZY, Rampling L, Cmiel M, Gianibelli MC, et al. (2004) Multiple isoforms of starch branching enzyme-I in wheat: lack of the major SBE-I isoform does not alter starch phenotype. Functional Plant Biology 31: 591–601.

Regina A, Kosar-Hashemi B, Li ZY, Pedler A, Mukai Y, Yamamoto M, et al. (2005) Starch branching enzyme IIb in wheat is expressed at low levels in the endosperm compared to other cereals and encoded at a non-syntenic locus. Planta 222: 899–909.

Regina A, Bird A, Topping D, Bowden S, Freeman J, Barsby T, et al. (2006) High-amylose wheat generated by RNA interference improves indices of large-bowel health in rats. Proceedings of the National Academy of Sciences 103: 3546–3551.

Regina A, Blazek J, Gilbert E, Flanagan BM, Gidley MJ, Cavanagh C, et al. (2012) Differential effects of genetically distinct mechanisms of elevating amylose on barley starch characteristics. Carbohydrate Polymers 89: 979–991.

Regina A, Berbezy P, Kosar-Hashemi B, Li S, Cmiel M, Larroque O, et al. (2015a) A genetic strategy generating wheat with very high amylose content. Plant Biotechnology Journal 13: 1276–1286.

Regina A, Berbezy P, Kosar-Hashemi B, Li S, Cmiel M, Larroque O, Bird AR, et al. (2015b) A genetic strategy generating wheat with very high amylose content. Plant Biotechnology Journal 13: 1276–1286.

Regina A, Rahman S, Li Z, Morell MK (2016) Starch, Synthesis. In: Reference Module in Food Science. Elsevier.

Roldan I, Wattebled F, Mercedes Lucas M, Delvalle D, Planchot V, Jimenez S, et al. (2007) The phenotype of soluble starch synthase IV defective mutants of Arabidopsis thaliana suggests a novel function of elongation enzymes in the control of starch granule formation. Plant Journal 49: 492–504.

Schonhofen A, Zhang XQ, Dubcovsky J (2017) Combined mutations in five wheat STARCH BRANCHING ENZYME II genes improve resistant starch but affect grain yield and bread-making quality. Journal of Cereal Science 75: 165–174.

Seal CJ, Daly ME, Thomas LC, Bal W, Birkett AM, Jeffcoat R, Mathers JC (2003) Postprandial carbohydrate metabolism in healthy subjects and those with type 2 diabetes fed starches with slow and rapid hydrolysis rates determined in vitro. British Journal of Nutrition 90: 853–864.

Sestili F, Palombieri S, Botticella E, Mantovani P, Bovina R, Lafiandra (2015) TILLING mutants of durum wheat result in a high amylose phenotype and provide information on alternative splicing mechanisms. Plant Science 233: 127–133.

Sestili F, Sparla F, Botticella E, Janni M, D'Ovidio R, Falini G, et al. (2016) The down-regulation of the genes encoding Isoamylase 1 alters the starch composition of the durum wheat grain. Plant Science 252: 230–238.

Shi Yong Cheng, Seib PA, Bernardin JE (1994) Effects of temperature during grain-filling on starches from six wheat cultivars. Cereal Chemistry 71: 369–383.

Shimbata T, Nakamura T, Vrinten P, Saito M, Yonemaru J, Seto Y, Yasuda H (2005) Mutations in wheat starch synthase II genes and PCR-based selection of a SGP-1 null line. Theoretical and Applied Genetics 111: 1072–1079.

Shimbata T, Inokuma T, Sunohara A, Vrinten P, Saito M, Takiya T, Nakamura T (2011) High Levels of Sugars and Fructan in Mature Seed of Sweet Wheat Lacking GBSSI and SSIIa Enzymes. Journal of Agricultural and Food Chemistry 59: 4794–4800.

Shin SI, Hea Jin C, Koo MC, Hamaker BR, Park KH, Moon TW (2004) Slowly Digestible Starch from Debranched Waxy Sorghum Starch: Preparation and Properties. Cereal Chemistry 81: 404–408.

Singh S, Singh G, Singh P, Singh N (2008) Effect of water stress at different stages of grain development on the characteristics of starch and protein of different wheat varieties. Food Chemistry 108: 130–139.

Slade AJ, McGuire C, Loeffler D, Mullenberg J, Skinner W, Fazio G, et al. (2012) Development of high amylose wheat through TILLING. BMC Plant Biology 12: 69.

Smidansky ED, Clancy M, Meyer FD, Lanning SP, Blake NK, Talbert LE, Giroux MJ (2002) Enhanced ADP-glucose pyrophosphorylase activity in wheat endosperm increases seed yield. Proceedings of the National Academy of Sciences 99: 1724–1729.

Stone PJ, Nicolas ME (1995) A survey of the effects of high temperature during grain filling on yield and quality of 75 wheat cultivars. Australian Journal of Agricultural Research 46: 475–492.

Tetlow IJ, Beisel KG, Cameron S, Makhmoudova A, Liu F, Bresolin NS, Wait R, Morell MK, Emes MJ (2008) Analysis of protein complexes in wheat amyloplasts reveals functional interactions among starch biosynthetic enzymes. Plant Physiology 146: 1878–1891.

Tetlow IJ, Emes MJ (2017) Starch Biosynthesis in the Developing Endosperms of Grasses and Cereals. Agronomy 7: 81.

Thitisaksakul M, Jiménez RC, Arias MC, Beckles DM (2012) Effects of environmental factors on cereal starch biosynthesis and composition. Journal of Cereal Science 56: 67–80.

Topping DL, Clifton PM (2001) Short-chain fatty acids and human colonic function: Roles of resistant starch and nonstarch polysaccharides. Physiological Reviews 81: 1031–1064.

Uauy C, Wulff BBH, Dubcovsky J (2017) Combining Traditional Mutagenesis with New High-Throughput Sequencing and Genome Editing to Reveal Hidden Variation in Polyploid Wheat. Annual Review of Genetics 51: 435–454.

Utsumi Y, Utsumi C, Sawada T, Fujita N, Nakamura Y (2011) Functional diversity of isoamylase oligomers: the ISA1 homo-oligomer is essential for amylopectin biosynthesis in rice endosperm. Plant Physiology 156: 61–77.

Vetrani C, Sestili F, Vitale M, Botticella E, Giacco R, Griffo E, et al. (2018) Metabolic response to amylose-rich wheat-based rusks in overweight individuals. European Journal of Clinical Nutrition 72: 904–912.

Vignola MB, Baroni V, Pérez GT (2016) Genotypic and environmental effects on starch properties of Argentinean wheat flours. Starch/Staerke 68: 1065–1072.

Vrinten PL, Shimbata T, Yanase M, Sunohara A, Saito M, Inokuma T, et al. (2012) Properties of a novel type of starch found in the double mutant "sweet wheat". Carbohydrates Polymer 89: 1250–1260.

Wang S, Li T, Miao Y, Zhang Y, He Z, Wang S (2017) Effects of heat stress and cultivar on the functional properties of starch in Chinese wheat. Cereal Chemistry 94: 443–450.

Yamamori M, Nakamura T, Nagamine T (1995) Inheritance of waxy endosperm character in a common wheat lacking 3 Wx proteins. Breeding Science 45: 377–379.

Yamamori M, Endo TR (1996) Variation of starch granule proteins and chromosome mapping of their coding genes in common wheat. Theoretical and Applied Genetics 93: 275–281.

Yamamori M, Fujita S, Hayakawa K, Matsuki J, Yasui T (2000) Genetic elimination of a starch granule protein, SGP-1, of wheat generates an altered starch with apparent high amylose. Theoretical and Applied Genetics 101: 21–29.

Yun MS, Umemoto T, Kawagoe Y (2011) Rice debranching enzyme isoamylase3 facilitates starch metabolism and affects plastid morphogenesis. Plant Cell Physiology 52: 1068–1082.

Zhang G, Ao Z, Hamaker BR (2008) Nutritional Property of Endosperm Starches from Maize Mutants: A Parabolic Relationship between Slowly Digestible Starch and Amylopectin Fine Structure. Journal of Agricultural and Food Chemistry 56: 4686–4694.

Zhang PF, He ZH, Tian XL, Gao FM, Xu DG, Liu JD, et al. (2017a) Cloning of TaTPP-6AL1 associated with grain weight in bread wheat and development of functional marker. Molecular Breeding 37: 78.

Zhang XW, Wang Q, Zhang LL, Zhong XJ, Jiang QT, Ma J, et al. (2017b) Cloning and characterization of Agp1, the gene encoding the small subunit of ADP-glucose pyrophosphorylase from wheat and its relatives. Biologia 72: 1446–1453.

Zi Y, Ding JF, Song JM, Humphreys G, Peng YX, Li CY, et al. (2018) Grain Yield, Starch Content and Activities of Key Enzymes of Waxy and Non-waxy Wheat (Triticum aestivum L.). Scientific Reports 8: 12.

Contribution of Genetic Resources to Grain Storage Protein Composition and Wheat Quality

Gérard Branlard, Patricia Giraldo, Zhonghu He, Gilberto Igrejas, Tatsuya M. Ikeda, Michela Janni, Maryke T. Labuschagne, Daowen Wang, Barend Wentzel, and Kunpu Zhang

Abstract The technological quality of wheat flour is defined by a range of dough characteristics relevant to the breadmaking processes and practices of individual countries and for particular products. The influence of storage protein diversity on wheat quality has been widely documented in the last three decades. The present chapter focuses on several aspects of wheat quality that merit more attention. The huge genetic diversity of wheat storage proteins means that all the possible allelic combinations and their interactions are too numerous to be tested in terms of their

G. Branlard (✉)
INRAE, UCA UMR1095 GDEC, Clermont-Ferrand, France
e-mail: gerard.branlard@clermont.inra.fr

P. Giraldo
Department of Biotechnology-Plant Biology, School of Agricultural, Food and Biosystems Engineering, Universidad Politécnica de Madrid, Madrid, Spain

Z. He
CIMMYT/CAAS, Beijing, China

G. Igrejas
Department of Genetics and Biotechnology, University of Trás-os-Montes and Alto Douro, Vila Real, Portugal

T. M. Ikeda
Western Region Agricultural Research Center, NARO, Fukuyama, Japan

M. Janni
CNR-IBBR, Bari, Italy

M. T. Labuschagne
Department of Plant Sciences, University of the Free State, Bloemfontein, South Africa

D. Wang · K. Zhang
State Key Laboratory of Wheat and Maize Crop Science, College of Agronomy, Henan Agricultural University, Zhengzhou, China

B. Wentzel
Agricultural Research Council – Small Grain, Bethlehem, Republic of South Africa

G. Igrejas et al. (eds.), *Wheat Quality For Improving Processing And Human Health*, https://doi.org/10.1007/978-3-030-34163-3_4

influence on the major quality parameters. However it is still relevant to describe the variation in rheological and viscoelastic properties of gluten in relation to its component proteins, glutenin and gliadin. Although gluten plays a major role in determining the properties of dough, the abundance of the two major storage protein fractions does not solely explain the observed variation in those properties. We therefore examine the influence of some genetic factors, including those affecting the protein composition, on the variation in the glutenin polymer sizes. Some examples will be given to illustrate how end-use quality can be improved by taking advantage of the available genetic resources in parallel with molecular genome analyses with the dual aim of widening the scope of characteristics that can be harnessed in breeding and ensuring consistent wheat quality in changing agro-climatic situations. The known alleles of the major genes are highlighted in the context of the challenges that the research community is facing regarding wheat allele nomenclature, exchange of gene bank material and the numerous quality attributes of interest. Finally, important research objectives are proposed for breeding future wheats with grain protein quality and technological properties tailored for different food products.

1 Introduction

The storage proteins (SPs) in cereal grain have no other known role than to serve as a reservoir of amino acids for the developing seedling. These proteins, namely gliadins and glutenins, are highly hydrophobic so are not readily soluble or hydrolysable during grain formation. As humans evolved, they learned to exploit the unique properties of SPs as they turned to cereal grain as a natural source of energy. For more than two millennia humans have influenced which cereal species are grown and have progressively adapted techniques to mill and bake flour. For several centuries it has been recognized that the flour fraction obtained by water-washing makes the flour suitable for loaf bread due to the formation of gluten. Glutenins were shown to confer elasticity whereas gliadins mainly influence the extensibility and viscosity of the dough (Finney 1943; Wall 1979). Since the 1970s the impact of these main gluten components on the breadmaking process has received a great deal of attention. Russian scientists extensively analysed gliadin diversity using starch gel electrophoresis to identify quality traits for wheat breeding (Sozinov et al. 1974). Their efforts were pursued by Dr. Konarev's group in St Petersburg and Dr. Metakovsky in Moscow. The genetic control of gliadins was extensively analysed in many countries by E. Metakovsky, who identified and catalogued numerous alleles using acid-PAGE (Metakovsky 2015; Metakovsky et al. 2018). Since the pioneering work of Payne et al. (1979) and Burnouf and Bouriquet (1980) showing the influence of the high molecular weight glutenin subunits (HMW-GS) on technological quality, many aspects of gluten have been studied all around the world. As well as low molecular weight subunits (LMW-GS), among the subjects that have been investigated are genetic determinism, molecular structures of the genes and loci, diversity in many wheat related species, allelic variants, molecular structure and properties of the subunits, interaction between subunits, genetic engineering of the glutenin sequences and influences on gluten or on dough properties or on end-use wheat quality tests (Ribeiro et al. 2013; Shewry et al. 2003b). The International Gluten Workshop, inaugurated at INRA Nantes in 1980, has been a fruitful opportunity to gather scientists from many countries every three years for informal discussions on gluten and all aspects of gluten-related wheat quality. Since then, the thirteen International Gluten Workshops have become milestones showing the great progress achieved in all aspects of gluten research and particularly for bread and durum wheat uses. The gluten proteins were among the first genetic markers employed in wheat breeding and are today actively studied in old wheat cultivars and related cereal species as sources of variability to broaden the genetic basis of wheat quality. This aim will be well illustrated here by examples from three articles in the first topic of focus. For our second topic the impacts of mutagenesis on wheat SP diversity will be discussed in relation to some major uses in different cultures like those in China, South Africa and Europe, particularly considering the polymer characteristics associated with current climatic conditions. In the third topic the numerous community challenges of dealing with allele nomenclature, exchange of gene bank material, and defining attributes for health and technological quality will be highlighted. Some important future research objectives will be proposed to conclude.

2 Genetic Resources and Potential from Old Cultivars and Related Species

2.1 *Molecular and Functional Diversities of HMW-GS Alleles from Wild* Triticeae *Species*

Molecular and biochemical studies have shown that the x- and y-type HMW-GSs share a conserved primary structure composed of a signal peptide that is removed upon protein maturation, an N-terminal domain (ND), a repetitive domain (RD) and a C-terminal domain (CD). Four and seven conserved cysteine residues are found in typical x- and y-type HMW-GSs, respectively, e.g. 1Dx2 and 1Dy12 (Fig. 1). It is these cysteine residues located in the ND, RD or CD that form the inter- and intra-molecular disulfide bonds required for gluten assembly in dough (Shewry et al. 2003b; Wrigley et al. 2009). To date, HMW-GSs have been mined and characterized in a large number of wild *Triticeae* species (Wang et al. 2018). In general, three categories of HMW-GS variations have been revealed in wild *Triticeae* species.

Differences in the Size of the RD Orthologous and allelic HMW-GSs often vary in the size of the RD (Fig. 1). In wheat, the 1Dx2.2 and 1Dx2.2* subunits have substantially longer RDs than their allelic counterpart 1Dx2 due to the presence of

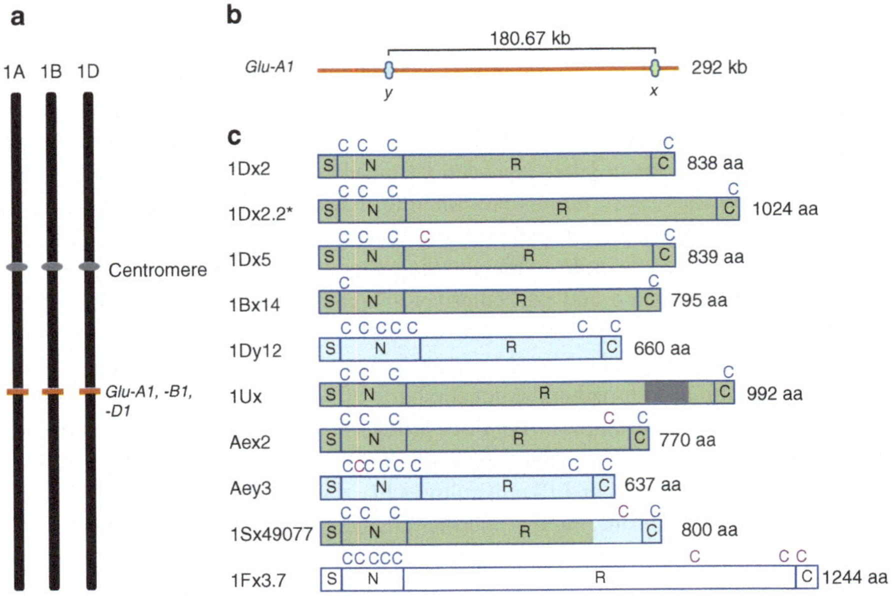

Fig. 1 Schematic representation of homoeologous *Glu-1* loci on wheat group 1 chromosomes (**a**), distribution of x- and y-type HMW-GS genes in a typical *Glu-1* locus (exemplified by *Glu-A1*) (**b**), and primary structure of representative HMW-GSs identified in wheat and wild *Triticeae* species (**c**). (**a**) The *Glu-A1*, *-B1* and *-D1* loci are located on the long arms of 1A, 1B and 1D chromosomes, respectively, (**b**) The size of the genomic DNA segment sequenced for *Glu-A1* is 292 kb, while 180.67 kb denotes the physical distance between x- and y-type HMW-GS genes, (**c**) The black capital letters S, N, R and C indicate the signal peptide, N-terminal domain, repetitive domain and C-terminal domain, respectively. Conserved and non-conserved cysteine residues are represented by blue and purple C letters, respectively. The grey shading represents the amino acid insertion in 1Ux. The C-terminal part of 1Sx49077 (light blue shading) resembles the C-terminal portion of typical y-type HMW-GSs. 1Fx3.7 is not shaded because it differs from typical HMW-GSs. GenBank accession numbers are X03346 (1Dx2), AY893508 (1Dx2.2*), X12928 (1Dx5), AY367771 (1Bx14), X03041 (1Dy12) for the wheat sequences and AF476961 (1Ux), DQ478576 (Aex2), EF190196 (Aey3), AY611723 (1Sx49077) or KC796696 (1Fx3.7) for the sequences from wild *Triticeae* species. The diagram is adapted from Wang et al. (2018)

internally duplicated segments (Wan et al. 2005). This kind of subunit has also been identified in wild *Triticeae* species. For example, relative to 1Ax1, 1Bx7 and 1Dx5 subunits, the 1Ux subunit in the *Aegilops umbellulata* accession IG46953 has an extended RD because of the insertion of 111 amino acids (Liu et al. 2003). The RDs of several x- and y-type HMW-GSs from *Aegilops longissima* are also longer due to the insertion of extra amino acids (Jiang et al. 2012; Wang et al. 2013). These subunits have been found to increase dough functionality and breadmaking quality when transferred to the common wheat landrace Chinese Spring (Wang et al. 2013), demonstrating the potential of such subunits in improving wheat end-use quality.

Differences in the Number of Conserved Cysteine Residues Differences in the number of cysteine residues in wheat HMW-GSs are known. For example, an extra cysteine residue is located in the RD of the 1Dx5 subunit, which is associated with increased gluten and dough functionalities (Lafiandra et al. 1993; Wrigley et al. 2009). Two conserved cysteine residues in the ND of 1Bx14 and 1Bx20 are replaced by tyrosines, which may be linked to the reduced function of these subunits in dough strength control (Li et al. 2004; Shewry et al. 2003a). HMW-GSs with different numbers of cysteine residues are also present in wild *Triticeae* species (Wang et al. 2018) (Fig. 1).

This is best illustrated by the five HMW-GSs characterized from decaploid *Agropyron elongatum*, Aex2, Aex4, Aey1, Aey3 and Aey7 (Liu et al. 2008). Aex2 and Aex4 each have one extra cysteine residue in their respective RDs, Aey1 and Aey3 carry one extra cysteine residue in the RD and ND, respectively, and Aey7 lacks one of the five cysteine residues conserved in the ND of typical y-type subunits.

More Complex Differences Many HMW-GS like prolamin proteins identified from wild *Triticeae* species show more complex or drastic differences in their primary structure compared to typical HMW-GSs (Fig. 1). For example, hybrid HMW-GS like proteins, with x-type ND and y-type CD or vice versa, have been isolated from *Aegilops searsii* (1S^{s}x49077, Sun et al. 2006), *Thinopyrum intermedium* (1Aix1, Cao et al. 2014), and *Pseudoroegneria stipifolia* (*Glu-1St1*, Li et al. 2008). 1Aix1 also harbors two extra cysteine residues in its RD, and the *Glu-1St1* RD is unusually short. One unusual HMW-GS like protein from *Erymopyrum bonaepartis*, 1Fx3.7, resembles y-type HMW-GSs in its ND, but its RD is the longest of all reported HMW-GSs and is unique in having multiple copies of the PGQQ tetrapeptide and two cysteine residues at unconserved locations (Jiang et al. 2014). Finally, several y-type HMW-GS like proteins found in certain *Elymus* and *Leymus* species lack five or six amino acid residues in their NDs, while their CDs have the LAAQLPAMCRL peptide conserved in the CD of typical x-type HMW-GSs (Sun et al. 2014). Overall, these prolamin proteins are more distantly related to wheat HMW-GSs in structure, and their value in wheat end-use quality improvement awaits further study.

2.2 Storage Proteins in Durum Wheat Landraces and Old Varieties: Characterization and Allele Mining

Durum wheat is the tenth most valuable crop on a global scale (Giraldo et al. 2016). Since the mid-1990s only part of the genetic diversity available for this species has been captured in modern varieties through breeding progress (Kabbaj et al. 2017; Pignone et al. 2015). Most of the allelic variation of genes found in original wild relatives, which has gradually been lost through domestication and breeding, can be recovered only by going back to landraces (Lopes et al. 2015; Tester and Langridge 2010). Durum wheat quality as related to the value of its end products has been demonstrated to be strongly dependent on its allelic composition of gliadins and glutenins. Variation at the LMW-GS loci is associated with significant differences in dough strength while variation at the HMW-GS loci is associated with increased gluten polymer size and thus gluten strength (Southan and MacRitchie 1999). Unlocking the favorable genetic diversity in germplasm resources is a pivotal strategy for enhancing durum wheat yield potential and quality (Longin and Reif 2014). A number of studies have recently investigated the genetic diversity of durum wheat landraces, old varieties and modern varieties in terms of their glutenin and gliadin profiles related to quality (Table 1). The genetic material investigated is from the Mediterranean area (Bellil et al. 2012, 2014; Cherdouh et al. 2005; Hamdi et al. 2010; Ribeiro et al. 2011) or specific countries, like Greece (Xynias et al. 2011), Bulgaria (Melnikova et al. 2010), Morocco (Henkrar et al. 2017), Spain (Aguiriano et al. 2008; Carrillo. 1995, Ruiz et al. 2012), Portugal (Brites and Carrillo 2001; Igrejas et al. 1999; Ribeiro et al. 2011), Syria (Mir Ali et al. 1999) or Tunisia (Sourour et al. 2016). A few studies extend the analyses to cover wider geographic origins and distribution (Janni et al. 2018; Katyal et al. 2018; Moragues et al. 2006; Nieto-Taladriz et al. 1997; Raciti et al. 2003).

HMW-GS Characterization The genetic polymorphism of SP alleles *Glu-A1 and Glu-B1* for HMW-GS and *Glu-A3, Glu-B3* and *Glu-B2* for LMW-GS has been used as a measure of genetic diversity in durum wheat germplasms. A few papers focused specifically on HMW-GS (Branlard et al. 1989; Janni et al. 2018) or LMW-GS alone (Melnikova et al. 2010; Nieto-Taladriz et al. 1997) and only a few focused on gliadin diversity (Gregová et al. 2012; Melnikova et al. 2010; Ribeiro et al. 2011; Xynias et al. 2011; Zilić et al. 2011).

Several common, rare and new alleles encoding HMW-GS were identified. At the *Glu-A1* locus, four different alleles were identified with *Glu-A1c,* the null allele, being the most frequent in almost all the materials analyzed (Henkrar et al. 2017; Igrejas et al. 1999; Janni et al. 2018; Mir Ali et al. 1999; Moragues et al. 2006; Naghavi et al. 2009; Raciti et al. 2003; Xynias et al. 2011) and the *Glu-A1a* (subunit 1) found at low frequency (Hamdi et al. 2010; Janni et al. 2018). The *Glu-A1b and Glu-A1VI* alleles, encoding the subunits 2* and 2*** respectively, were also considered rare (Hamdi et al. 2010). Conversely, *Glu-A1a* and *Glu-A1b* alleles were predominant in Moroccan genotypes (Henkrar et al. 2017).

More extensive variation was found at the locus *Glu–B1*, with three very frequent alleles, *Glu-B1b*, *Glu-B1d* and *Glu-B1e*, encoding the 7 + 8, 6 + 8 and 20 subunits respectively, with *Glu-B1e* and *GluB1b* being prevalent (Bellil et al. 2012, 2014; Carrillo 1995; Moragues et al. 2006). By comparison, in Algerian germplasm the *Glu-B1e* and *Glu-B1d* alleles were prevalent (Cherdouh et al. 2005; Hamdi et al. 2010; Henkrar et al. 2017; Janni et al. 2018), together with *Glu-B1f* (13 + 16) as reported by Hamdi and co-workers (2010). The *Glu-B1d* allele was predominant in germplasm from the Iberian peninsula and Spain (Carrillo 1995; Moragues et al. 2006). The *Glu-B1e* allele was absent from Bulgarian genotypes (Moragues et al. 2006). In Syrian genotypes, *Glu-B1b* and *Glu-B1d* were the most frequent, and the 6 + 15 subunit was found at high frequencies (Mir Ali et al. 1999). Alleles *Glu-B1d and Glu-B1b* were not present in Iranian landraces, where the *Glu-B1a* (7), the *Glu-B1e* (20) and the *Glu-B1i* (17 + 18) alleles predominated (Naghavi et al. 2009).

Table 1 Summary of glutenin and gliadin allelic composition of durum wheat genetic resources analyzed to date

Countries or regions of origin	Number of accessions	Germplasm source	Glutenin/gliadin loci analysed	Method of analysis	Reference
Spain	52 92	CRF-INIA, Spain Spanish core collection	*Glu-A1, Glu-B1,Glu-A3, Glu-B3,Glu-B2*	SDS-PAGE A-PAGE	Aguiriano et al. (2008) Ruiz et al. (2018)
Sahara	n.a.	INRAA-Adrar, ITDAS-Adrar, Djamaa-El Oued and Biskra, CRSTRA-Biskra, Algeria	*Glu-A1, Glu-B1,Glu-A3, Glu-B3, Glu-B2*	SDS-PAGE	Bellil et al. (2012)
Algeria	120	ITGC, Constantine, Algeria	*Glu-A1, Glu-B1,Glu-A3, Glu-B3, Glu-B2*	SDS-PAGE	Bellil et al. (2014)
Several countries	502	n.a.	*GluA1, Glu-B1*		Branlard et al. (1989)
Portugal	F2 population originating from 6 cultivars	n.a.	*Glu-A1, Glu-B1, Glu-B3, Glu-B2*	SDS-PAGE	Brites and Carrillo (2001)
Spain	201		*Glu-A1, Glu-B1, Glu-A3, Glu-B3*		Carrillo (1995)
Algeria	45	National plant Germplasm system, Beltsville, MP, USA; Institut des sciences de la nature, Constantine, Algeria	*Glu-A1, Glu-B1, Glu-A3, Glu-B3, Glu-B2*	SDS-PAGE	Cherdouh et al. (2005)
Italy	15	n.a.	*Glu-A1, Glu-B1, Glu-B3*	SDS-PAGE	De Santis et al. (2017)
Several countries	108	Gene Bank of the Slovak Republic	*Glu-A1, Glu-B1*; LMW-1, LMW-2	SDS-PAGE	Gregová et al. (2012)
Algeria	856	ITGC, Constantine, Algeria; international Center for Agricultural research in dry areas, Aleppo, Syria	*Glu-A1, Glu-B1, Glu-A3, Glu-B3, Glu-B2*	SDS-PAGE	Hamdi et al. 2010
Morocco	26	n.a.	*Glu-A1, Glu-B1, Glu-A3, Glu-B3*	SDS-PAGE	Henkrar et al. (2017)
Portugal	21	n.a.	*Glu-A1, Glu-B1, Glu-A3, Glu-B3, Glu-B2; Gli-A1, Gli-B1*	SDS-PAGE	Igrejas et al. (1999)

(continued)

Table 1 (continued)

Countries or regions of origin	Number of accessions	Germplasm source	Glutenin/gliadin loci analysed	Method of analysis	Reference
Several countries	152	Mediterranean Germplasm Bank of IBBR-CNR, Bari, Italy	*Glu-A1, Glu-B1*	SDS-PAGE, LoAC, PCR	Janni et al. (2018)
Several countries	42	NBPGR, New Delhi, India	HMW; LMW	SDS-PAGE	Katyal et al. (2018)
Bulgaria	98	Institute for plant genetic resources "K. Malkov", Sadovo, Bulgaria	*Gli-A1, Gli-A2, Gli-B1, Gli-B2*	A-PAGE	Melnikova et al. (2010)
	21	Cham1 EMS mutant Population	*Glu-A1, Glu-B1, Glu-A3, Glu-B3, Glu-B2*	SDS-PAGE	Elyadini et al. (2014)
Syria	140	n.a.	*Glu-A1, Glu-B1*; LMW-1, LMW-2	SDS-PAGE	Mir Ali et al. (1999)
Iberian peninsula, Mediterranean countries	63	n.a.	*Glu-A1, Glu-B1, Glu-A3, Glu-B3, Glu-B2*	SDS-PAGE	Moragues et al. (2006)
Iran	96	Gene bank of the agricultural College at the University of Tehran, Iran	*Glu-A1, Glu-B1*	SDS-PAGE	Naghavi et al. (2009)
Several countries	88	Instituto Nacional de Semillas y Plantas de Vivero, Spain	*Gli-B1; Glu-A3, Glu-B3, Glu-B2*	SDS-PAGE	Nieto-Taladriz et al. (1997)
Several countries	119	ICARDA	*Glu-A1, Glu-B1*; LMW-1, LMW-2	SDS-PAGE	Raciti et al. (2003)
Portugal	128	Portuguese national seed registration Centre	*Glu-A1, Glu-B1, Glu-A3, Glu-B3, Glu-B2*; *Gli-A1, Gli-B1*	SDS-PAGE	Ribeiro et al. (2011)
Tunisia	12	National agronomic institute of Tunisia	*Glu-A1, Glu-B1*; LMW-1, LMW-2	SDS-PAGE	Sourour et al. (2016)
Greece	29	Hellenic Gene Bank	*Glu-A1, Glu-B1; Gli-A1, Gli-A2, Gli-B1, Gli-B2*	SDS-PAGE	Xynias et al. (2011)
Serbia/Italy	5	MRIZP, Serbia	*Gli-A1, Gli-A2, Gli-B1, Gli-B2*	SDS-PAGE	Zilić et al. (2011)

.a., not available; LoAC, lab-on-a-chip; EMS, ethyl methanesulfonate; PAGE, polyacrylamide electrophoresis; SDS, sodium dodecyl sulfate; A, acid

Some cases of heterogeneity were reported in one or both loci of HMW-GS (Janni et al. 2018; Ribeiro et al. 2011) or of gliadins (Melnikova et al. 2010).

LMW-GS Characterization The LMW glutenin loci *Glu-A3, Glu-B3 and Glu-B2* and variation in their respective frequencies in durum wheat germplasm have been the subject of many studies (Table 1). At the *Glu-A3* locus several alleles have been identified. *Glu-A3a* is the most common in germplasm from Mediterranean countries (Igrejas et al. 1999; Ribeiro et al. 2011) and in the Algerian landraces (Bellil et al. 2014; Cherdouh et al. 2005; Hamdi et al. 2010; Nieto-Taladriz et al. 1997). The *Glu-A3c* allele was predominant in Saharan durum wheats (Bellil et al. 2012) which coincides with the high frequency of this allele in Moroccan genotypes (Henkrar et al. 2017), while the *Glu-A3h* allele was the most frequently observed in samples from the Iberian peninsula (Moragues et al. 2006). An update on LMW glutenins in Spanish durum wheat relevant to breeding for quality was recently reported (Ruiz et al. 2018).

The allele *Glu-B3a* was predominant in Algerian germplasm (Bellil et al. 2014; Cherdouh et al. 2005; Hamdi et al. 2010) and in Spanish, Portuguese and Mediterranean collections (Aguiriano et al. 2008; Igrejas et al. 1999; Moragues et al. 2006). Rare LMW alleles (*Glu-A3e, f, g, i, Glu-B3d, e, f, g, h, i, j, k*) were detected in old Spanish and Portuguese cultivars, and also in Moroccan genotypes (Aguiriano et al. 2008; Bellil et al. 2012; Brites and Carrillo 2001; Henkrar et al. 2017; Nieto-Taladriz et al. 1997) but not in Algerian germplasm (Cherdouh et al. 2005). Some of these rare alleles were also observed in Saharan and Mediterranean genotypes (Bellil et al. 2012; Moragues et al. 2006). A number of newly identified alleles were also reported, highlighting the high level of polymorphism encountered at this locus. LMW-1 and LMW-2 patterns have been analyzed, with the LMW-2 pattern, which endows semolina with better properties, emerging as the most frequent (Aguiriano et al. 2008; Carrillo 1995; De Santis et al. 2017; Gregová et al. 2012; Mir Ali et al. 1999; Raciti et al. 2003; Sourour et al. 2016). At the *Glu-B2* locus, the *Glu-B2a* allele has been found to be the most frequent in all reports (Table 1). Conversely, the *Glu-B2b* allele has been detected at high frequencies in some Spanish accessions (Aguiriano et al. 2008) and in the Saharan germplasm (Bellil et al. 2012).

Gliadin Characterization A few reports describe the gliadin composition of durum wheat germplasm. The *Gli-A1e* (N) allele was described as the most common in Portuguese samples (Igrejas et al. 1999) as well as in a more comprehensive collection (Ribeiro et al. 2011), while the *Gli-A1r* allele was predominant in material from Greece (Xynias et al. 2011).

At the *Gli-B1* locus, *Gli-B1e* (d5) and the pattern d5 + d5'11′ were predominant in Portuguese germplasm and in world genotypes (Igrejas et al. 1999; Ribeiro et al. 2011). *Gli-B1h* was the most frequent in the Greek populations (Xynias et al. 2011), while the Bulgarian genotypes were highly variable (Melnikova et al. 2010).

Other Considerations Most, if not all, the glutenin alleles found in modern elite varieties are also found in germplasm collections of landraces and ancient or traditional varieties. On the contrary, much of the diversity observed in the latter genotypes is absent from modern varieties. This is further confirmation that genetic diversity has been depleted in recent years as modern breeding techniques and procedures have developed. With respect to quality, most of the germplasm accessions have been shown to possess HMW glutenin alleles related to high grain quality, such as *Glu-A1a* and *Glu-B1d* associated with good gluten strength (Brites and Carrillo 2001). LMW glutenin alleles related to high gluten strength and extensibility, such as *Glu-A3a, Glu-A3c, Glu-A3h, Glu-B3a,* and *Glu-B3c*, were observed in many landraces and old varieties (Carrillo et al. 2000). A targeting induced local lesions in genomes (TILLING) approach was used to generate new allelic variation at the *Glu-A1*, *Glu-B1*, *Glu-A3*, *Glu-B3*, and *Glu-B2* loci (Elyadini et al. 2014).

The vast assortment of diversity encountered, in most cases known to be correlated with good quality, reinforces the idea that the use of landraces and of old or traditional varieties in breeding programs is a viable strategy to improve the quality of modern varieties, for a better yield and yield stability, especially under stress and future climate change conditions (De Vita et al. 2007; Jaradat 2013).

Up to now only a few papers demonstrate a direct correlation between gluten allele variability and quality properties (Brites and Carrillo 2001; De Santis et al. 2017; Katyal et al. 2018), so further evaluation of the effect on quality of the alleles and allele combinations for which no data are available is a prerequisite to exploiting landraces in breeding programs.

2.3 Storage Protein Alleles of Ancient Hexaploid Varieties and Landraces

In the last fifty years, there have been several investigations into the genetic diversity and genetic structure of ancient varieties or landraces of hexaploid wheat, especially in view of developing new cultivars to meet specific end product requirements. The long-term efforts of breeders and the fall in the cultivation of landraces and old cultivars are associated with decreasing genetic variability in wheats and loss of several HMW-GS alleles.

HMW-GS variation has been analysed among wheat landraces and obsolete cultivars of winter wheat originating from 14 European countries (Gregová et al. 1997, 1999, 2006). High HMW-GS variability was detected, the most frequent combination being [1, 7 + 9, 2 + 12], while one novel HMW-GS allele was identified. Two major HMW-GS genotypes, [1, 13 + 16, 2 + 12] and [1, 6.1 + 22.1, 2 + 12], occur in Central European spelt wheat cultivars and landraces at higher frequencies (35% and 28%, respectively). Iranian spelt varieties differ from European spelts and have similar HMW-GS alleles to those of common wheat (An et al. 2005). A new mutant A1x-type HMW glutenin allele was identified in an old genotype from Hungary called Bánkúti 1201 (a heterogeneous population with special quality traits), which encodes an extra cysteine residue and has a moderate positive effect on gluten properties (Juhász et al. 2003).

Wheat genetic resources from the Iberian peninsula have been characterized. The genetic variability of the Portuguese 'Barbela' wheat SPs was described by Igrejas et al. (1997). Eleven patterns for HMW-GS, twelve for LMW-GS and seven for ω gliadins were identified (Igrejas et al. 1997). In another study of Portuguese landraces, none were found to have the $2^{\circ\circ}$, 1.1, or 13 subunits (Rodriguez-Quijano et al. 1998), which can be contrasted with a study of Spanish landraces, where the frequency of the $2^{\circ\circ}$ glutenin subunit was close to 1 and Null subunits were also detected (Giraldo et al. 2010). In Spanish landraces of *Triticum aestivum* ssp. *spelta,* a subunit designated 13 was observed (Rodriguez-Quijano et al. 1990) that is similar to subunit 13 described in Portuguese 'Barbela' wheat lines (Igrejas et al. 1997). The high frequency of the subunit pairs 20x + 20y and 13 + 16 is a strong difference between Iberian landraces and landraces from other world regions. In previous works these two combinations were found at a very low frequency between 1% and 3% (Morgunov et al. 1993; Payne and Lawrence 1983; Tohver 2007), or not at all (Fang et al. 2009; Li et al. 2009).

Extrusion energy, length, density and surface appearance in the biscuit test of flour milled from landrace 'Barbela' were significantly influenced by variation at the *Glu-1* locus. The lines encoding subunit 1.1 produced flour with considerably higher dough extensibility than the others (Igrejas et al. 2002). A very high level of identity was found between the sequence encoding 1Ax1.1 and other alleles, and the molecular difference was found to be an insertion of 36 amino acids in the central RD (Ribeiro et al. 2013a).

In a recent study, 46 landraces from Southern Spain were characterized for their HMW-GS and LMW-GS composition and two new alleles were observed, one for the *Glu-A1* locus and the other for the *Glu-B3* locus (Ayala et al. 2016). In a study of 100 lines of Creole Mexican wheats, which are derived from old Spanish wheat varieties and have been cultivated in Mexico for centuries by small-scale traditional farmers, the allelic variation at the *Glu-1* loci was wide, but the low frequency of some alleles revealed a clear risk of genetic erosion (Caballero et al. 2010).

Other studies have addressed the HMW-GS allelic variation of wheats from regions of Asia. The composition of each HMW loci in 1068 wheat landraces from East and West Asian revealed that the most common alleles were *Glu-A1c* (73.6%), *Glu-B1b* (60.2%), and *Glu-D1a* (68.5%) (Lee et al. 2018). Protein heterogeneity for HMW-GS has been analysed in common wheat cultivars from specific countries like Pakistan (Tahir et al. 1996). The allelic combinations including 2∗, 5 + 10, and 17 + 18 that have high quality scores were frequent among Pakistan landraces, indicating their potential usefulness in future crop improvement and breeding programs (Yasmeen et al. 2015). In another study of more than 500 Indian landraces (Goel et al. 2018), five novel subunits from the *Glu-D1* locus were reported, and a total of 33 *Glu-1* alleles (3 at *Glu-A1*, 15 at *Glu-B1*, and 15 at *Glu-D1*) were detected. Among 174 hexaploid Japanese wheat landraces examined, three alleles were identified at the *Glu-A1* locus, six at the *Glu-B1* locus and four at the *Glu-D1* locus. HMW-GS 2.2 controlled by the *Glu-D1f* allele is frequently found in Japanese cultivars and landraces (Nakamura 2001). The *Glu-B1al* ($Bx7^{OE}$ + By8) allele is important for breadmaking quality and was found in a Korean wheat landrace using specific DNA markers (Cho et al. 2017, 2018).

Chinese wheat germplasm has also been exhaustively studied. Guo et al. (1993) found that in Chinese landraces the null allele is the most frequent allele at the *Glu-A1* locus in 97% of cases. In another collection of landraces from the same origin, 89% of the lines had this allele (Redaelli et al. 1997). Subunits 7 and 7 + 8 are the most common at the *Glu-B1* locus in both collections while the 2 + 12 subunits are the most common at the *Glu-D1* locus. Several novel HMW-GS alleles have been described, including 1Ax5*, 1Bx6* (Dai et al. 2004), 1Bx7*, 1By8*, 1By8**, the rare allele 1Dx2 + null (Liu et al. 2007), 1Dx2.6 (Cong et al. 2007), 1Bx7** (Fang et al. 2009), 1Dx1.5*, 1Dy12.2* (Guo et al. 2010), 1Bx14* and 1By15* (Shao et al. 2015). Among these, the relationship of the 1Dx1.5* and 1Dy12.2* subunits with dough quality has been studied (Guo et al. 2010). In another study of HMW allelic variation in 485 Chinese wheat landraces, 16 additional novel HMW glutenin subunits were identified by matrix-assisted laser desorption/ionization time-of-flight mass spectrometry (MALDI-TOF) technology (Zheng et al. 2011). Two novel y-type HMW-GS genes were characterized in Chinese wheat landraces from the Yangtze River region, 1Dy12.6 and 1Dy12.7 (Peng et al. 2015). The same procedure was used to characterize samples of 27 landraces of Tibetan wheat collected from farms in the hilly areas of the Himalayas (Lan et al. 2013). The landraces were more diverse for HMW-GS coded by *Glu-B1*, with distinct subunit combinations 6 + 8, 7 + 9, 13 + 16, and five novel LMW-GS genes were isolated. A newly established MALDI-TOF procedure was also used to analyse the LMW-GS in 478 landraces of bread wheat collected from the Yangtze River region in China (Peng et al. 2016).

Gliadin alleles have proved to be a very valuable criterion to analyse the genetic polymorphism of wheat collections from different countries. In a study of 59 Spanish landraces of common wheat belonging to different agro-climatic types, the loci *Gli-B1*, *Gli-A2*, *Gli-D2* and *Gli-A1* were shown to be the most polymorphic and useful in discriminating between the accessions studied. As a result, 22 new gliadin alleles were found and 12 of them were catalogued, evidence that Spanish wheat germplasm is highly polymorphic and rather unique (Ruiz et al. 2002a, 2002b). When 117 cultivars or landraces of common wheat from Japan were studied, 27 different patterns were identified, 13 corresponding to ω-gliadin, and 8 to β, γ-gliadin and 6 to α-gliadin (Tanaka et al. 2003). A considerable polymorphism in gliadin genes was detected in wheat landraces from West Siberia.

Recently, a new and improved version of the catalogue of alleles at the six Gli loci of common wheat (*T. aestivum* L.) has been achieved by analysing 1060 cultivars and lines bred in the twentieth century (Metakovsky et al. 2018).

To sum up, landraces and cultivated genotypes resulting from natural selection are considerable sources of unique alleles that may be well adapted to adverse biotic and abiotic factors (Gepts 1993; Zhang 1995; Zeven 1998). It is noted that developing or less developed countries are socio-economic stakeholders in the important reservoirs of biodiversity they can access.

3 Storage Proteins with Major Roles in Wheat Quality

3.1 HMW and LMW Composition of Chinese Wheats and their Association with Qualities of Pan Bread, Noodles, and Steamed Bread

3.1.1 Allelic Variation of HMW-GS and LMW-GS in Chinese Wheat

Allele-specific functional markers have been widely used to identify allelic variations in germplasm for quality related genes. Liang et al. (2010) surveyed functional markers for grain quality in 273 advanced CIMMYT lines and identified several genes that were positively selected. Jin et al. (2011) identified HMW-GS and LMW-GS in cultivars from 20 different countries, and described some geographical patterns in how quality alleles are deployed. Some alleles like $Bx7^{OE}$ were extremely rare and were only identified in 12.1% of cultivars from Argentina, 4.1% from Australia, 30.0% from Japan, 25.0% from Canada, and 20.0% from Iran (Jin et al. 2011). The frequencies of Ax2∗ were 39.7%, 83.3%, 20.4% and 72.2% in Australian, Canadian, Chinese and USA lines, respectively. The highest frequency of By8 (35.6%) was in cultivars from Australia. For By9, Canadian cultivars had the highest frequency (50.0%). By16 is present in only 2.0% of Chinese cultivars, whereas it was found in 19.0% of USA cultivars, 4.1% of Australian cultivars, but not at all in Canadian cultivars (Jin et al. 2011). Liu et al. (2005) screened Chinese wheat germplasm and concluded that *Glu-A3a*, *Glu-A3d*, *Glu-B3j* (1BL.1RS) and Glu-B3d were dominant in Chinese wheat germplasm at frequencies of 37.1%, 31.7%, 44.6% and 20.3%, respectively. We have recently developed high-throughput Kompetitive allele-specific (KASP) markers for the $Bx7^{OE}$ and *wheat breadmaking* (*wbm*) genes and screened more than 300 global wheat accessions. We reached the same conclusion that these genes are extremely rare in global wheat germplasm and almost absent in Chinese winter wheat germplasm despite their known positive effect on wheat breadmaking quality (Rasheed et al. 2019).

3.1.2 Association of Glutenin Proteins with Pan Bread and Noodle Quality

Knowing the composition of HMW-GS and LMW-GS and their respective associations with Chinese pan bread and noodle quality can guide the genetic improvement of the processing quality of Chinese bread wheats. He et al. (2005) analysed 158 winter and facultative cultivars and advanced lines to understand glutenin allele effects on dough properties, and the quality of pan bread and dry white noodles. Gluten strength was significantly explained by HMW-GS alleles 1, 7 + 8, 5 + 10, and *Glu-A3d*. For dry white noodle quality, no significant differences were observed between HMW-GS alleles at *Glu-1*, while *Glu-A3d* and *Glu-B3d* performed slightly better than other alleles. Similarly, pan bread and white salted noodle quality were significantly explained by the quantity of gliadins and the ratio of HMW-GS to LMW-GS (Zhang et al. 2007). Therefore, not only the allelic variation but also the quantity of each SP fraction is responsible for pan bread and Chinese noodle quality. In another study, near-isogenic wheat lines derived from Aroona were evaluated for raw white Chinese noodle and northern style Chinese steamed bread quality (Jin et al. 2013). The strong gluten-encoding HMW-GS alleles 17 + 18 and 5 + 10 and LMW-GS alleles *Glu-A3b*, *Glu-A3d*, *Glu-B3g* and *Glu-D3f* significantly affected Mixograph properties, but there was no significant effect on most parameters relevant to the production of raw white noodles or northern style steamed bread. However, near-isogenic lines with HMW-GS alleles 1, 7 + 9, 2 + 12, and LMW-GS alleles *Glu-A3c*, *Glu-B3d*, and *Glu-D3c* conferred superior viscoelasticity to raw white Chinese noodles. Near-isogenic lines with HMW-GS alleles 1, 7 + 9, and 2 + 12, and LMW-GS alleles *Glu-A3e*, *Glu-B3b*, and *Glu-D3c* had the highest total quality score for northern style steamed bread. These results provide useful information for breeding strategies to improve the qualities of traditional Chinese wheat products.

3.2 Contribution of Wheat Protein Composition to the Quality of South African Dryland Winter Wheat

Wheat production areas in South Africa can be divided into the winter rainfall area, the summer rainfall area (Free State), and irrigated areas. The breadmaking quality of winter wheat produced in the Free State is inconsistent and therefore diminishes its market value. Variation in protein content is a primary cause of this inconsistency. For example, climatic conditions during grain filling influence protein content and mixing behavior of dough (Van Lill and Smith 1997). This should be taken into account when testing and releasing new cultivars.

Ten South African winter dryland wheat cultivars, BettaDN, Caledon, Elands, Gariep, Komati, Limpopo, Matlabas, PAN3118, PAN3349 and PAN3377, were studied to observe environmental influences on protein composition, and the fluctuating breadmaking quality of genotypes. Trials were planted over two seasons at localities in the North Western Free State (NW-FS) and the Eastern Free State (E-FS). Approved AACC methods were used to determine flour quality (AACC 2000). Size-exclusion high-performance liquid-chromatography (SE-HPLC) was used to determine the molecular weight distribution of SDS-soluble and SDS-insoluble protein fractions (Labuschagne et al. 2014).

Comparing the two trial areas, there were significant differences for all the protein fraction and quality parameters. The overall analysis of variance model explained between 77% and 97% of the variation in protein fractions ($0.77 \leq R^2 \leq 0.97$), which suggests that the chosen model adequately described the observed variability.

Environment (in this case the locality) made the largest contribution to variation in flour protein content (FPC). Protein composition changed with changes in FPC but

Table 2 Combined protein content and quality traits of ten wheat cultivars grown in two regions of South Africa

	FPC	MPT	AlvP	AlvL	LFV
	12% mb	min	mm H_2O	mm	cm^3
Mean[a]	12.2	3.3	80.5	115.4	944
Range[a]	8.7-16.4	1.8-4.4	51.0–145.0	39.0–186.0	710–1030
STD Error[a]	0.13	0.05	1.19	2.59	6.08
Mean[b]	11.3	3.5	92.5	104.6	881
Range[b]	7.9–14.4	2.0–5.9	50.0–167.0	35.0–178.0	685–1030
STD Error[b]	0.09	0.04	1.37	1.94	5.11

[a]North Western Free State
[b]Eastern Free State. FPC, flour protein content; MPT, Mixograph peak time; AlvP, Alveograph tenacity; AlvL, Alveograph extensibility; LFV, loaf volume; mb, moisture basis; STD, standard deviation

there was no direct relationship between the two. Increased FPC resulted in higher amounts of gliadin, and lesser increases in polymeric protein then in albumin and globulin. Genotype made the largest contribution to variation in Mixograph peak time (43%) and Alveograph tenacity (31%) for both regions. For all the quality parameters, genotype contributed more to the variation observed for wheat grown in NW-FS than for wheat grown in E-FS. Stepwise multiple linear regression analysis indicated the contribution of FPC and protein fractions made to four selected variables for individual cultivars. The average values for all 10 cultivars are presented in Table 2. FPC explained the variation in loaf volume for PAN3118 and the variation in Alveograph extensibility for Matlabas in E-FS but it did not contribute significantly to variables in NW-FS. The regression model explained a larger percentage of the variation in Alveograph extensibility for BettaDN, Caledon, Elands, Gariep, Komati and Limpopo grown in E-FS compared to the same genotypes grown in NW-FS. By contrast, the regression model explained a larger percentage of the variation in Alveograph extensibility for Matlabas, PAN3118, PAN3349 and PAN3377 grown in NW-FS than for the same genotypes grown in E-FS. The opposite was true for Mixograph peak time, where the regression model explained a larger percentage of the variation of BettaDN, Caledon, Elands, Gariep, Komati and Limpopo grown in NW-FS than for their counterparts in E-FS, but a larger percentage of the variation for Matlabas, PAN3118, PAN3349 and PAN3377 grown in E-FS than for their counterparts in NW-FS.

Albumin and globulin fractions contributed more to the variation in Alveograph tenacity, Alveograph extensibility and loaf volume for wheat grown in NW-FS than in E-FS. Furthermore, these contributions were larger for BettaDN, Caledon, Elands, Gariep, Komati and Limpopo than for Matlabas, PAN3118, PAN3349 and PAN3377. The positive contribution of specific albumin and globulin fractions to strength related parameters and loaf volume needs to be further investigated.

General conclusions about the effect of protein fractions and FPC on quality parameters cannot be deduced when average values for genotypes are used, especially when entries showed significant differences for quality traits (Li et al. 2013).

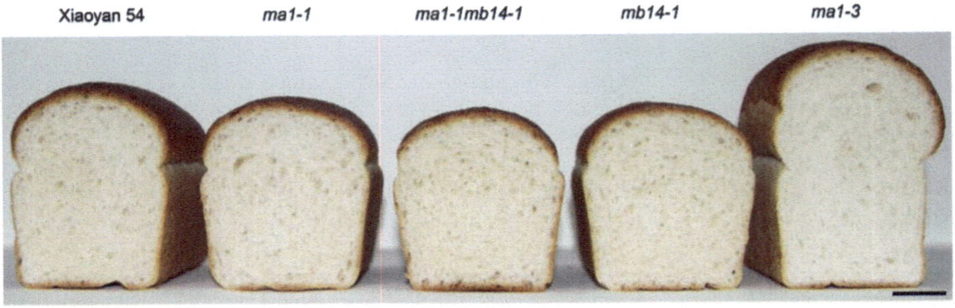

Fig. 2 Comparison of bread loaves among Xiaoyan 54 and four isogenic HMW-GS mutant lines. A transverse sectional view of the loaves prepared for the five genotypes. Xiaoyan 54 is the wild type control expressing five HMW-GSs (1Ax1, 1Bx14, 1By15, 1Dx2 and 1Dy12). The single knockout mutants *ma1–1* and *mb14–1* lack 1Ax1 and 1Bx14, respectively, the double knockout mutant *ma1–1mb14–1* is devoid of both 1Ax1 and 1Bx14, and the missense mutant *ma1–3* expresses a mutated 1Ax1 with the substitution G330E. Scale bar, 2 cm. The photograph shown is representative of 18 independent experiments (Li et al. 2015)

3.3 Functional Analysis of Gluten Genes and Chromosomal Loci Using Chemical and Radiation Mutants

It is now generally believed that HMW-GSs and LMW-GSs are the key determinants of dough elasticity and extensibility through forming glutenin macropolymers (GMPs) via inter- and intra-molecular disulfide bonding, whereas gliadins may exist mainly as monomers in the dough, and can modulate dough elasticity and extensibility through interacting with glutenins by hydrogen bonding (MacRitchie 2014). The functionality and end-use property of a dough are largely shaped by collective interactions among HMW-GSs, LMW-GSs and gliadins, and variations in the composition and amount of gluten proteins are mainly responsible for different end-use properties. However, despite extensive studies, direct genetic evidence for the functions of different gluten genes and chromosomal loci in wheat end-use quality control based on the analysis of defined mutants is still rare (Wang et al. 2018).

Recently, several studies have started to address the above problem by developing and characterizing chemical mutants for HMW-GSs and radiation induced deletion mutants for glutenin and gliadin loci. An analysis involving the knockout mutants of the HMW-GSs 1Ax1 and 1Bx14 in Xiaoyan 54, a winter type wheat cultivar, demonstrates that the two subunits differ clearly in how they contribute to important gluten, dough and breadmaking quality parameters (Li et al. 2015). 1Ax1 contributes more to gluten and dough strength and bread volume than does 1Bx14, but the latter exerts a larger effect on dough extensibility (Fig. 2). Interestingly, a missense mutant of 1Ax1, caused by a single amino acid substitution (G330E), significantly improves dough and breadmaking qualities compared to the wild-type control (Li et al. 2015) (Fig. 2).

Using Xiaoyan 81 as a progenitor, ion beam induced locus deletion mutants have been developed for the homoeologous *Glu-1* loci (*Glu-A1*, *-B1* and *-D1*) specifying

HMW-GSs, the three composite *Gli-1/Glu-3* loci (*Gli-A1/Glu-A3*, *Gli-B1/Glu-B3* and *Gli-D1/Glu-D3*) specifying γ-, δ- and ω-gliadins and LMW-GSs, and the three *Gli-2* loci (*Gli-A2*, *-B2* and *-D2*) controlling α-gliadins (Wang et al. 2017a; 2017b; Yang et al. 2014). Comparison of a complete series of deletion mutants lacking one, two or all three *Glu-1* loci (Wang et al. 2017b; Yang et al. 2014) confirmed the different contributions of these loci (i.e., *Glu-D1* > *Glu-B1* > *Glu-A1*) to breadmaking quality originally proposed by Lawrence et al. (1988) based on a correlative analysis of wheat recombinant lines. Furthermore, a combined examination of a *Glu-D1a* locus deletion mutant and the knockout mutants of 1Dx2 and 1Dy12 indicates that 1Dx2 contributes more potently to breadmaking quality than 1Dy12, because the former enables more efficient incorporation of HMW-GSs and LMW-GSs into functional GMPs (Wang et al. 2017b). Lastly, functional genomic analysis of *Gli-1/Glu-3* and *Gli-2* deletion mutants has shed new light on the range of gliadin transcripts and proteins that accumulate in mature wheat grains (Wang et al. 2017a), which facilitated the subsequent identification of a *Gli-D2* deletion mutant as a valuable germplasm for simultaneously enhancing wheat end use and health related quality traits (Li et al. 2018).

The results outlined above suggest that mutagenic studies can generate new information and novel resources for understanding and improving the functions of gluten genes and chromosomal loci, but further work with mutants lacking appropriate combinations of gluten proteins is still needed to unravel the complex interactions among HMW-GSs, LMW-GSs and gliadins. As the composition of these proteins often differ among wheat cultivars, additional mutants for more gluten protein alleles will need to be generated and analysed in order to promote a thorough dissection of the roles of different gluten genes and chromosomal loci in wheat end-use quality control.

3.4 Respective Part of Storage Protein Alleles, Grain Hardness and Environmental Factors on Rheological Properties and Bread Loaf Volume

It is well known that high temperatures shorten the grain filling period, often dramatically reducing starch accumulation relative to SP accumulation resulting in increased protein concentration at harvest (Altenbach et al. 2003; Blumenthal et al. 1991, Triboï et al. 2003). The higher protein concentration resulting from nitrogen fertilisation is generally associated with better dough properties, but this is not often the case when it results from high temperature (Johansson et al. 2008; Malik et al. 2013). The response of wheat grain to high temperatures during grain filling has been studied using proteomics and transcriptomics but few studies have addressed how current high temperatures associated with global warming affect SP alleles chosen for the basic rheological dough characteristics and breadmaking quality they confer.

To better understand the genetic and molecular bases of several dough characteristics, a large multi-local study was carried out in France in two consecutive years, 2009 and 2010. Briefly a total of 68 diverse wheat cultivars provided by INRA and by 11 private wheat breeding companies were tested in six locations. The cultivars were grown in conventional conditions with full mineral supply and fungicide protection. Climatic parameters in the experimental locations were also recorded. For quality phenotyping, 14 technological tests were performed on the 240 samples (40 cultivars x 6 environments) (Branlard et al. 2013). Some of these results were recently published in a report highlighting the usefulness of a proteomics approach for investigating the phenomena occurring in grain during SP accumulation (Branlard et al. 2015). The performance of genetically diverse wheat in different environmental conditions are useful sources of variation helping us to understand the phenomena occurring in the current cultivars grown in France and Europe. A total of 28 SP alleles were identified with 3, 8, 4, 3, 8 and 2 alleles respectively encoded at *Glu-A1*, *Glu-B1*, *Glu-D1*, *Glu-A3*, *Glu-B3* and *Glu-D3*. Here the influence of SPs (glutenin alleles and polymers) and of some grain characteristics (test weight, thousand kernel weight (TKW), grain hardness, and the relative viscosity of water-extractable arabinoxylan (AX)) on dough rheological characteristics like strength, tenacity and extensibility for standardized evaluation of dough and bread loaf volume are examined.

Large variations in all the phenotypic values were observed. For example, values ranged from 36.4 g to 64.9 g for TKW, from 9.0 to 14.9%dm for GPC and from 1 to 113 for grain hardness. The quality characteristics also exhibited very large variations of from 64 to 400 10^{-4} J for dough strength, from 33 to 156 mmH2O for dough tenacity, from 21 to 195 mm for dough extensibility and from 1024 to 1941 cm^3 for loaf volume. All these phenotypic values including the size of the glutenin polymers (which varied from 5.4×10^3 to 4.9×10^4 kDa) were highly responsive to both environmental and genetic effects. The broad heritability coefficients were rather high (0.67 to 0.89) for TKW, grain hardness, the relative viscosity of water-extractable AX and tenacity as expected, were moderate (0.35 to 0.78) for test weight, dough strength and extensibility, and low (0.17 to 0.44) for GPC, polymer size and loaf volume.

Several results presented in Table 3 were reported in a previous proteomics study (Branlard et al. 2015). Here additional comments are given to highlight the genetic and environmental factors that are today influencing the quality of wheat.

1. GPC and grain hardness have a major influence on dough strength and loaf volume. This positive effect is confirmed with grain hardness for tenacity and with GPC for extensibility. As the Alveograph is used at constant hydration, dough viscosity increases as well as tenacity because starch damage increases with grain hardness. For hard cultivars the stronger link between starch granules and the protein matrix also increases dough tenacity.
2. Surprisingly GPC has a negative effect on dough tenacity. This results from the higher temperature in June ($\sum$temp.June) when SPs have yet to fully accumulate, which limits the accumulation of starch and probably also glutenin but not gliadin which is favorable to loaf volume.

Table 3 Main factors and SP alleles associated with effects on phenotypic variations in strength, tenacity, extensibility and loaf volume as revealed through partial least squares (PLS) regression

	Dough strength W		Tenacity P	
	R^2 PLS 78.1%		R^2 PLS 73.7%	
	Negative effect	**Positive effect**	**Negative effect**	**Positive effect**
Major factors		GPC > Ghard	∑temp.June	Ghard > ∑temp. July
Other factors	∑temp.July	TW ≥ ∑temp.June > R.Visco	GPC ≥ R.Visco	Mw2/Mn2 ≥ TW ≥ TKW
Glu-A1	n	1, 2*	2*	1
Glu-B1	6–8 > 6.1–22	7–8 > 7–9	17–18 ≥ 6–8 ≥ 6.1–22	7–8 > 13–16 ≥ 7–9
Glu-D1	2–12 = 4–12 > 3–12	5–10	2–12 > 3–12 ≥ 4–12	5–10
Glu-A3	ef	d	ef	a
Glu-B3	d ≥ c' ≥ c	b > g	b' > b > j	c ≥ c' ≥ d = f
Glu-D3	b	c	c	b
	Extensibility L		Loaf volume	
	R^2 PLS 77.9%		R^2 PLS 43.6%	
	Negative effect	**Positive effect**	**Negative effect**	**Positive effect**
Major factors	∑temp.July	GPC > R.Visco	∑temp.July ≥ Mw2	GPC > Ghard
Other factors	TW > Mw2		TKW	∑temp.June > R. Visco
Glu-A1		2*	ns	ns
Glu-B1	7–8 > 13–16	17–18 > 14–15	6.1–22 ≥ 14–15 = 13–16	6–8 > 7 = 7–9
Glu-D1	5–10	2–12 > 3–12 = 4–12	3–12 > 4–12	2–12 > 5–10
Glu-A3	ns	ns	a	ef > d
Glu-B3	f ≥ c = j	b ≥ b'	b = b' = c	j >> d
Glu-D3	ns	ns	b	c

GPC, grain protein content, Temp, monthly temperature; TW, test weight; TKW, thousand kernel weight, Ghard, grain hardness, R.Visco, relative viscosity of water-extractable arabinoxylan, ns, not significant.

3. Water soluble AX are highly hydrophilic compounds. When AX are more abundant, less water is available to hydrate the gluten, so tenacity P is negatively influenced when Alveograph performed at constant hydration. The AX-associated water makes dough less viscous and more extensible (extensibility increases). During bread baking some of the water associated with AX is vaporized in alveoli, which increases the loaf volume.
4. During starch and protein accumulation in June, the high temperature in some locations impacted starch accumulation more than SP accumulation, and consequently GPC and dough strength increased. The higher temperature in June did not impact polymer size significantly. This was not the case for the high temperature occurring in July when endosperm usually progressively dehydrates. July heat stress (temperature > 35 °C) in some locations accelerated endosperm

dehydration, causing oxidative stress and the unfolded protein responses in the endoplasmic reticulum (Lesage et al. 2012, 2013) resulting in abnormal sizes of MW2 polymers (here up to 49,000 kDa instead of the usual 3000–5000 kDa). These higher polymer masses were favorable for dough tenacity, but unfavorable for extensibility and loaf volume.

5. While the HMW-GS and LMW-GS alleles associated with the rheological properties W, P and L remain globally identical to what was known twenty years ago, their contribution to phenotypic variation is today drastically reduced. In addition, for loaf volume some alleles like *Glu-B1* 14–15 or 13–16, *Glu-A3a* or *Glu-B3b* that were positively associated with breadmaking quality and loaf volume are today negatively influencing loaf volume. More surprisingly the glutenin alleles *Glu-B1* 6–8, *Glu-D1* 2–12, *Glu-A3e* or *f* and *Glu-B3j*, which would have been rejected in breeding for good loaf volume twenty-five years ago, are today having the most positive influence on loaf volume. The excessively high MW2 polymer masses obviously hamper dough swelling and increased loaf volume, so any alleles less involved in polymer formation like the *Glu-B3j* allele (marker of the 1BL.1RS translocation) have a very favorable influence. We also note that the higher amount of AX associated with the 1BL.1RS translocation positively influences loaf volume, as indicated above.

The multi-local trials performed ten years ago and discussed above, those reported by Li et al. (2013) and by B. Wentzel and M. Labuschagne in the present book chapter, provide evidence that the genetic basis of wheat quality is different today than in the past. These results also show that genetic factors as well as regulatory mechanisms involved in glutenin polymer formation, in response to environmental stresses (like drought and temperature) need to be investigated.

4 Reference Cultivars and Lines with Useful Storage Protein Alleles

A Case Study of Checking and Sharing Common Wheat *Glu-3* Alleles The very high degree of allelic variation in HMW-GS and LMW-GS alleles in common wheat and durum wheat has been comprehensively listed by curators in the Wheat Gene Catalog, alongside reference papers and representative germplasm. Unfortunately not all germplasm in the list is available to researchers worldwide. Another problem is that some cultivars do not seem to have the same genotype in different laboratories. Experimental conditions also differ between laboratories. This explains why researchers in different laboratories could not identify the same alleles. Therefore, it is very critical to share the same reference germplasms among researchers internationally. As an attempt to solve these problems, we set up an international collaboration unifying *Glu-3* nomenclature systems in common wheat among researchers from Argentina, China, France, Japan and Mexico (CIMMYT) (Ikeda et al. 2008; Liu et al. 2010).

Table 4 Standard cultivars for *Glu-3* alleles of common wheat

Locus	Allele	Cultivar	Locus	Allele	Cultivar
Glu-A3	***a***	Chinese Spring	*Glu-B3*	*j*	Clement, Seri
	b	Gabo, Pavon		*m*	Soissons
	c	Thesee, Seri, Cheyene		*ab*	Klein Proteo
	d	Cappelle-Desprez, Wilgoyne		*ac*	Thesee, Acta 801
	e	Marquis, Neepawa		*ad*	Ruso, Opata, Heilo
	f	Clement, Insignia, Heilo	*Glu-D3*	*a*	Chinese Spring
Glu-B3	*a*	Chinese Spring		*b*	Seri
	b	Gabo, Marquis		*c*	Insignia
	c	Insignia, Halberd		*d*	Brasil
	d	Pepital, Eshimashinriki		*f*	Cheyenne
	e	Cheyenne		*h*	India 115
	f	Magali-Blondeau		*k*	Ernest
	g	Brimstone, Cappelle-Desprez		*l*	Amadina, Heilo
	h	Petrel, Pavon		*m*	Darius
	i	Demai3, Norin61		*n*	Fengmai 27

Materials and Methods A total of 103 cultivars from each participating group were shared. The routine methods including sodium dodecyl sulfate polyacrylamide gel electrophoresis (SDS-PAGE), two-dimensional gel electrophoresis (isoelectric focusing × SDS-PAGE), and polymerase chain reaction (PCR). It became clear that each group used different SDS-PAGE conditions for the concentration and pH of separation gels, acrylamide/bis-acrylamide ratio, and running current, for example. It was found to be important to alkylate SH groups with 4-vinylpyridine, and use a longer separation gel using less cross linker.

Allele Assignments We also found that allele assignment was different among groups using identical materials. To identify some alleles, we need to use 2D or PCR methods. There is no single method to identify all alleles. We assigned *Glu-3* alleles based on the Wheat Gene Catalog and added several new alleles (e.g. *Glu-B3ab*, *Glu-B3ac*, *Glu-B3ad*, *Glu-D3l* and *Glu-D3m*, and *Glu-D3n*).

Selecting a Standard Set After confirming their homogeneity, we were able to select cultivars as a standard cultivar set representing *Glu-3* alleles in common wheat as shown in Table 4.

Sharing the Standard Cultivar Set Through a Public Gene Bank Through international collaboration under the Wheat Initiative, the expert working group for improving wheat quality for processing and health (http://www.wheatinitiative.org/activities/expert-working-groups/improving-wheat-quality-processing-and-health), this master set has become available in the CIMMYT gene bank (http://wgb.cimmyt.org/gringlobal/search.aspx, search for "%GluStd"). It provides 98 cultivars including 10 durum wheat cultivars as candidates for a durum *Glu-3* master set.

Learning from the process to select the master set for SPs, it became clear that the following conditions should be applied:

- Explore alleles worldwide.
- Select each cultivar from a single source and check homogeneity.
- Select methods to identify alleles and compare the results among laboratories.
- Multiple methods might be needed to characterize all alleles.
- Allele assignments should be consistent with the Wheat Gene Catalog.
- New alleles should be reported to curators of the Wheat Gene Catalog.
- Seeds should be made available to other researchers (with due consideration of breeder's rights, where necessary).
- Multiply seeds and distribute them through public gene banks.

Unifying *Glu-3* Nomenclature With Durum Wheat Different aspects of durum wheat *Glu-3* alleles were studied (Babay et al. 2015; Brites and Carrillo 2000; Igrejas et al. 2009; Muccilli et al. 2010; Ruiz and Carrillo 1993). Durum alleles could be useful resources to increase genetic diversity of common wheat *Glu-3* alleles and improve gluten quality. Although a few alleles seem to be in common, *Glu-3* nomenclature still needs to be unified between common wheat and durum wheat.

Connection With Genomic Data It is important to characterize *Glu-3* alleles further using emerging wheat genome data. Ibba et al. (2017) reported on *Glu-3* gene family members. Each *Glu-3* allele was represented by a specific haplotype, but some alleles were indistinguishable or differentiated. This approach will add to the data on *Glu-3* gene variation.

Further Development of Other Quality Related Seed Protein Alleles Gliadin proteins consist of α/β/γ/ω-gliadins. It has been difficult to clarify the effects of gliadin alleles on wheat quality partly due to the tight linkage between *Glu-3* and *Gli-1* loci. However, gliadin composition also affects dough quality (Branlard and Metakovsky 2006 for review). For example, there are reports showing positive associations of null and Cheyenne type *Gli-D1* alleles to quality (Branlard and Dardevet 1994; Johansson 1996; Brönneke et al. 2000). Therefore, it is important to establish the master set for gliadin alleles for further study. Gliadin alleles (*Gli-1* and *Gli-2*) have been studied and classified by Acid-PAGE (Metakovsky 1991a; Metakovsky and Novoselskaya 1991b). With this one-dimensional gel electrophoresis, they demonstrated the huge genetic diversity of gliadins. We need to adopt the acid-PAGE method and other complementary techniques to confirm these alleles using the same standard cultivars. We list in Table 5 the lowest number of wheat cultivars having the 167 different gliadin alleles described by Metakovsky and Graybosch (2006). Metakovsky et al. (2018) recently reported 182 alleles at the six *Gli* loci. Some but not all of the cultivars are available in a gene bank in INRA. We aim to collect these cultivars as important genetic resources for improving wheat quality, and share them for further analyses.

Table 5 List of cultivars having the 167 gliadin alleles as described by Metakovsky and Graybosch (2006)

Variety	Country	*Gli-A1*	*Gli-B1*	*Gli-D1*	*Gli-A2*	*Gli-B2*	*Gli-D2*	Variety	Country	*Gli-A1*	*Gli-B1*	*Gli-D1*	*Gli-A2*	*Gli-B2*	*Gli-D2*
Magnif 27	Argentina	b	k	b	b	ap	a	Kirgizskaya Yubileinaya	Kyrgyzstan	m	e	g	u	u	f
Javelin	Australia	f	b	o	a	at	w	Norrona	Norway	c	b	a	ak	g	a
Mokoan	Australia	f	c	b	a	ao	q	Dankowska	Poland	d	f	f	g	al	a
Insignia	Australia	f	i	i	a	i	i	F 168/62	Romania	b	d	j	b	p	b
Gabo	Australia	g	b	f	c	c	t	Cluj	Romania	o	j	a	j	j	a
Bungulla	Australia	m	b	h	a	an	w	Saratovskaya 27	Russia	c	i	a	q	o	New
Spear	Australia	m	d	i	m	aq	w	Uralochka	Russia	f	b	a	m	am	q
Giurgana	Azerbaijan	o	b	a	v	h	b	Omskaya	Russia	f	b	a	u	o	o
251/83–89	Bulgaria	b	b	b	b	x	b	Donskaya Poluintens.	Russia	f	b	b	n	m*	k
Leader	Canada	m	d	j	f	ak	h	Kamchadalka	Russia	f	m	c	u	k	m
Chinese Spring	China	a	a	a	a	a	a	Caezium	Russia	f	m	i	j	t	ae
Open	France	e	b	g	n	m	e	Erytrospermum	Russia	i	x	i	s	q	s
Orepi	France	f	f	b	r	ab	a	Saratovskaya 33	Russia	j	e	a	q	o	m
Balthasar	France	f	f	j	l	ad	a	Tarskaya	Russia	o	k	a	af*	c	q
Genial	France	k	f	b	r	r	n	Akmolinka	Russia	q	m	g	k	d	s
Heurtebise	France	m	b	m	l	g	g	Strela	Russia	t	q	a	m	as	a
Roazon	France	o	b	b	g	ar	v	Navarro	Spain	a	e	a	ab	h	w
Darius	France	o	f	Null	l	g	g	Escualo	Spain	a	e*	b	c	c	c
Albatros	France	o	g	b	g	l	ac	Navarro	Spain	a	k	a	aa	o	j
Japhet	France	v	h	d	j	n	n	Ablaca	Spain	f	b	g	a	h*	j
Floreal	Germany	a	h	l	j	au	a	Montjuich	Spain	f	v	a	h	af	x
Solo	Germany	f	e	d	x	n	a	Blanquillo	Spain	f	v	i	a	ai	x
Basalt	Germany	f	f	j	h	f	g	Ardica	Spain	f	w	b	e	ag	b

(continued)

Table 5 (continued)

Variety	Country	*Gli-A1*	*Gli-B1*	*Gli-D1*	*Gli-A2*	*Gli-B2*	*Gli-D2*	Variety	Country	*Gli-A1*	*Gli-B1*	*Gli-D1*	*Gli-A2*	*Gli-B2*	*Gli-D2*
New Pusa	India	a*	p	a	t	New	p	Aragon	Spain	m	o	f	u	o	z
Fiorello	Italy	a	e	f	g	v	u	San Rafael	Spain	m	o	i	a	ac	a
Centauro	Italy	a	e	k	g	y	j	Maestro	Spain	o	f	b	p	z	h
Granarolo	Italy	a	g	k	g	j	r	Campeador	Spain	o	f	j	g	ae	h
Ardito (sample 3)	Italy	f	k	d	e	e	j	Candeal de Arévalo	Spain	o	o	i	u*	h	a
Spada	Italy	k	n	b	c	o	q	Sevillano	Spain	o	r	b	e	p*	q
Salmone	Italy	l	s	b	e	aa	aa	Barbilla	Spain	r	g	b	f	h	ak
Giuliana	Italy	m	q	a	z	w	m	Jeja del	Spain	t	t	a	u	ag	a
Cologna	Italy	o	e	a	y	j	c	Candeal Alkalá	Spain	u	f	i	q	o	y
Grano del	Italy	t	o	a	o	ah	a	Admiral	UK	o	l	l	r	g	g
Lutescens	Kazakhstan	p	e	i	q	s	s	Krimka Mestnaya	Ukraine	e	b	g	d	o	q
								Lesostepka	Ukraine	l	q	g	i	h	e

5 Conclusion

Important progress has been made in the three last decades on the major aspects of wheat quality. Details of genetic determinism, gene mapping, allelism, and the expression of SPs and numerous enzymes involved in the synthesis of other grain components like starch, AX, cell walls, some lipids and vitamins have been described by the scientific community. This progress has been continuous thanks to the way experimental tools and approaches have developed. However many questions remain to be addressed for the major aspects of wheat quality.

1. Both genotypes and methodological tools used for SP identification and nomenclature of alleles must be made available to all scientists interested in wheat quality. It is imperative that wheat standards possessing the reference glutenin and gliadin alleles are preciously maintained. As clearly shown above for *T. aestivum* and *T. durum*, the effort to identify new SP alleles in *Triticum* and related species must be pursued. With efficient gene sequencing, knowing individual protein sequences is today of prime importance when analyzing their possible effects on gluten and dough properties. Apparatus like gene sequencers and mass spectrometers are thus needed for complete description of alleles and possible quality attributes. Ways to curate those new alleles and deposit the germplasm in public gene banks must be reconsidered.
2. The expression of SPs always results from the balance between energy requirement for plant metabolism and grain component accumulation and this aspect needs to be fully investigated. The known environmental effects of nitrogen, sulfur and other mineral supplies must be extended to all soil and climatic parameters governing plant development. For example, the proteomic approaches already undertaken to study heat stress should be combined with transcriptome analysis for wheat grown in a fully controlled environment where parameters including temperature, availability of water, mineral fertilizers and fungicide are known. To know the influence of the numerous regulatory genes on quantitative expression of SPs relies on such a global effort, especially by benefiting from the use of mutants and isogenic lines with different SP composition.
3. The technological tools used to describe wheat quality parameters have to be improved in light of today's biological findings. In the 1980s wheat breeders were among the first to select SP alleles in genotypes suited to the technology requirements for wheat uses. These demands were addressed using rheological tools that were mostly designed in the first half of the twentieth century. Several issues have been raised over the two last decades. For example, dough strength does not vary linearly with gluten strength. In addition, an optimum gluten strength can be found for an optimum bread loaf volume. So what is the use of continuing to combine favorable SP alleles in a genetic background if it is not required for grain uses? The question of gluten properties in terms of dough properties is today more pertinent than ever due to global warming. The accurate measurement of wheat grain polymer size in a set of diverse European cultivars revealed that the temperature during grain accumulation was far more important

than the SP alleles. The polymer characteristics are not yet taken into account either by breeders or by food technologists. The glutenin polymers impact the rheological properties and end use value in such a way that glutenin alleles that were unfavorable to dough strength and breadmaking quality thirty years ago are today, due to an increasingly warm environment, associated with better loaf volume.

4. The structure of the glutenin polymers which also involves some bound α-, β- and γ- gliadins obviously needs to be deciphered. The complex structure influenced by available nutritional elements like nitrogen and sulfur was also reported to result from the unfolded protein response to oxidative stress that occurs in the endoplasmic reticulum. Glutenin polymer sizes ten to twenty times bigger than usual have been observed in wheat grown under heat stress. The study of the redox mechanisms together with oxygen and water availability in the grain should be considered when seeking the genetic factors involved in environmental stability of glutenin polymer characteristics and dough quality.
5. Re-assessing the main rheological tests cannot be performed without the cooperation of scientists and professionals involved in wheat grain milling and flour use. A clear understanding of what occurs in wheat grain and what is required for flour and dough properties will undoubtedly be a foundation for mutual progress. Both wheat geneticists and technologists have to be more driven by societal demands. Today people pay less attention to bread volume but require tasty and nutritional foods that are good for health. The pyramiding of alleles favoring gluten strength into a wheat genotype obviously does not correspond to any customer demand. Wheat geneticists are becoming eager to create genotypes with fewer epitopes that trigger immune responses that can occur in genetically predisposed people, as in celiac disease. The catalogue of more than 180 different gliadin alleles, made by Dr. E. Metakovsky et al. (2018), offers today a great opportunity to identify those harboring the fewest epitopes. With the known glutenin alleles wheat scientists can now cumulate SP alleles with sequences having the lowest number of gluten-allergy epitopes into useful genotypes.
6. Non celiac gluten sensitivity is also a serious concern expressed by a growing number of wheat consumers in many countries. That question has to be connected to the abnormal size of glutenin polymers resulting from high temperature and probably also from drought stress occurring during protein accumulation. Gluten which is generously added to many flour products including several types of bread is today composed of large polymers that are not fully hydrolyzed during the fermentation time commonly practiced. Joint research by wheat geneticists, nutritionists and technologists promises to solve this important issue that too many wheat consumers are facing.

References

AACC (2000). Approved methods of the American Association of Cereal Chemists, 10th Edition. American Association of Cereal Chemists, Incorporated, St. Paul, Minnesota.

Aguiriano E, Ruiz M, Fité R, Carrillo, JM (2008) Genetic variation for glutenin and gliadins associated with quality in durum wheat *Triticum turgidum* L. ssp. *turgidum* landraces from Spain. Spanish Journal of Agricultural Research 6: 599.

Altenbach SB, DuPont FM, Kothari KM et al (2003). Temperature, water and fertilizer influence the timing of key events during grain development in a US spring wheat. Journal of Cereal Science 37: 9–20.

An X, Li Q, Yan Y, Xiao Y, Hsam, SLK, Zeller FJ (2005) Genetic diversity of European spelt wheat *Triticum aestivum* ssp *spelta* L. em. Thell. revealed by glutenin subunit variations at the *Glu-1* and *Glu-3* loci. Euphytica 146: 193–201.

Ayala M, Guzmán C, Peña RJ, Alvarez JB (2016) Diversity of phenotypic plant and grain morphological and genotypic glutenin alleles in *Glu-1* and *Glu-3* loci traits of wheat landraces *Triticum aestivum* from Andalusia Southern Spain. Genetic Resources and Crop Evolution 63: 465–475.

Babay E, Hanana M, Mzid R, Slim-Amara H, Carrillo JM, Rodríguez-Quijano M (2015) Influence of allelic prolamin variation and localities on durum wheat quality. Journal of Cereal Science 63: 27–34.

Bellil I, Chekara Bouziani, M, Khelifi, D (2012) Genetic diversity of high and low molecular weight glutenin subunits in Saharan bread and durum wheats from Algerian oases. Czech Journal of Genetics and Plant Breeding 48: 23–32.

Bellil I, Hamdi O, Khelifi D (2014) Diversity of five glutenin loci within durum wheat *Triticum turgidum* L. ssp. *durum* Desf. Husn. germplasm grown in Algeria. Plant Breeding 133: 179–183.

Blumenthal CS, Bekes F, Batey IL, Wrigley CW (1991) Interpretation of grain quality results from wheat variety trials with reference to high temperature stress. Australian Journal of Agricultural Research 42: 325–334.

Branlard G, Autran JC, Monneveux P (1989) High molecular weight glutenin subunit in durum wheat *T. durum*. Theoretical and Applied Genetics 78: 353–358.

Branlard G, Dardevet M. (1994) A null *Gli-D1* allele with a positive effect on bread wheat quality. Journal of Cereal Science 20: 235–244.

Branlard G, Lesage VS, Bancel E, Martre P, Méleard B, Rhazi L (2015) Coping with wheat quality in a changing environment – proteomics evidence for stress caused by environmental changes. In "Advances in Wheat Genetics: From Genome to Field. Proceedings of the 12th International Wheat Genetics Symposium" Y. Ogihara, S. Takumi, H. Handa Eds, Yokohama, Japan ISBN: 978-4-431-55674-9, 255–264.

Branlard G, Méléard B, Oury FX, Rhazi L, Boinot N (2013) Compréhension du rapport Ténacité/Extensibilité et du volume du pain. In : « Synthèse du programme de recherche FSOV, actes de la rencontre scientifique 15 mars 2013, Paris 18–26.

Branlard G, Metakovsky EV (2006) Chapter 4. Some *Gli* alleles related to common-wheat dough quality in 'Gliadin and Glutenin: The Unique Balance of Wheat Quality', AACC St Paul MN USA, 115–139.

Brites C, Carrillo JM (2000) Inheritance of gliadin and glutenin proteins in four durum wheat crosses. Cereal Research Communication, 28: 239–246.

Brites C, Carrillo JM (2001) Influence of High Molecular Weight HMW and Low Molecular Weight LMW glutenin subunits controlled by *Glu-1* and *Glu-3* loci on durum wheat quality. Cereal Chemistry 78: 59–63.

Brönneke V, Zimmermann G, Killermann B, 2000. Effect of high molecular weight glutenins and D-zone gliadins on bread making quality in German wheat varieties. Cereal Res Commun. 28:187–194.

Burnouf T, Bouriquet R (1980) Glutenin subunits of genetically related European hexaploid wheat cultivars: their relation to bread-making quality. Theoretical and Applied Genetics 58: 107–111.

Caballero L, Peña RJ, Martín LM, Alvarez JB (2010) Characterization of Mexican Creole wheat landraces in relation to morphological characteristics and HMW glutenin subunit composition. Genetic Resources and Crop Evolution 57: 657–665.

Cao S, Li Z, Gong C, Xu H, Yang R, Hao S, et al. (2014) Identification and characterization of high-molecular-weight glutenin subunits from *Agropyron intermedium*. PLoS One 9: e87477.

Carrillo JM (1995) Variability for glutenin proteins in spanish durum wheat landraces. in durum wheat quality in the Mediterranean region, CIHEAM, 143–147.

Carrillo JM, Martinez MC, Moita Brites C, Nieto-Taladriz MT, Vázquez JF (2000) Relationship between endosperm proteins and quality in durum wheat *Triticum turgidum* L. var. *durum*. Options Méditerranéennes 40: 463–467.

Cherdouh A, Khelifi D, Carrillo JM, Nieto-Taladriz MT (2005) The high and low molecular weight glutenin subunit polymorphism of Algerian durum wheat landraces and old cultivars. Plant Breeding 124: 338–342.

Cho SW, Cho K, Bang G, Park CS (2018) Molecular profiling of a y-type high molecular weight glutenin subunit at *Glu-D1* locus from a North Korean landrace wheat *Triticum aestivum* L., Plant Biotechnology Reports 12:139–148.

Cho SW, Roy SK, Chun J-B, Cho K, Park CS (2017) Overexpression of the Bx7 high molecular weight glutenin subunit on the *Glu-B1* locus in a Korean wheat landrace. Plant Biotechnology Reports 11:97–105.

Cong H, Takata K, Ikeda T, Yanaka M, Fujimaki H, Nagamine T (2007) Characterization of a novel high-molecular-weight glutenin subunit pair 2.6+12 in common wheat landraces in the Xinjiang Uygur autonomous district of China. Breeding Science 57: 253–255.

Dai S, Yan ZH, Wei YM, Zheng YL (2004) Allelic variations of high molecular weight glutenin subunits HMW-GS in Tibetan wheat. Acta Agriculturae Boreali-Occidentalis Sinica 17: 5–11.

De Santis MA, Giuliani MM, Giuzio L, De Vita P, Lovegrove A, Shewry PR, et al. (2017) Differences in gluten protein composition between old and modern durum wheat genotypes in relation to 20th century breeding in Italy. European Journal of Agronomy 87: 19–29.

De Vita P, Li Destri Nicosia O, Nigro F, Platani C, Riefolo C, Di Fonzo N, et al. (2007) Breeding progress in morpho-physiological, agronomical and qualitative traits of durum wheat cultivars released in Italy during the 20th century. European Journal of Agronomy 26: 39–53.

Elyadini M, Labhilili M, Bentata F, Azeqour M, Taghouti M, Kahama I, et al. (2014) Characterization of new allelic variation for glutenin in EMS—mutant durum wheat population (*Triticum turgidum* L. subsp. *durum* (Desf.)) Journal of Life Sciences 8: 880–888.

Fang J, Liu Y, Luo J, Wang Y, Shewry PR, He G (2009) Allelic variation and genetic diversity of high molecular weight glutenin subunit in Chinese endemic wheats *Triticum aestivum* L. Euphytica 166: 177–182.

Finney KF (1943) Fractionating and reconstituting techniques as tools in wheat flour research. Cereal Chemistry 20: 381–396.

Gepts P (1993) The use of molecular and biochemical markers in crop evolution studies. Evolutionary Biology 27: 51–94.

Giraldo P, Rodriguez-Quijano M, Simon C, Vazquez JF, Carrillo JM (2010) Allelic variation in HMW glutenins in Spanish wheat landraces and their relationship with bread quality. Spanish Journal of Agricultural Research 8: 1012–1023.

Giraldo P, Royo C, González M, Carrillo JM, Ruiz M (2016) Genetic diversity and association mapping for agromorphological and grain quality traits of a structured collection of durum wheat landraces including subsp. *durum*, *turgidum* and *diccocon*. PLoS One 11.

Goel S, Yadav M, Singh K, Jaat RS, Singh NK (2018). Exploring diverse wheat germplasm for novel alleles in HMW-GS for bread quality improvement. Journal of Food Science and Technology -Mysore 55: 3257–3262.

Gregová E, Tisová V, Kraic J (1997) Genetic variability at the *Glu-1* loci in old and modern wheats *Triticum aestivum* L. cultivated in Slovakia. Genetic Resources and Crop Evolution 44: 301–306.

Gregová E, Hermuth J, Kraic J, Dotlačil L (1999) Protein heterogeneity in European wheat landraces and obsolete cultivars. Genetic Resources and Crop Evolution 46: 521–528.

Gregová E, Hermuth J, Kraic J, Dotlačil L. (2006) Protein heterogeneity in European wheat landraces and obsolete cultivars: Additional information II. Genetic Resources and Crop Evolution 53: 867–871.

Gregová E, Medvecká E, Jómová K, Sliková S (2012) Characterization of durum wheat *Triticum durum* desf. quality from gliadin and glutenin protein composition. Journal of Microbiology, Biotechnology and Food Sciences 1: 610.

Guo BH, Wang ZN, Fang R, Li HJ, Pei CJ (1993) HMW glutenin variation in landraces of wheat in Northern China. In 8th International Wheat Genetics Symposium, 20–25 July, Beijing, China, Vol. II, 725–729.

Guo X, Guo J, Li X, Yang X, Li L (2010) Molecular characterization of two novel *Glu-D1*-encoded subunits from Chinese wheat *Triticum aestivum* L. landrace and functional properties of flours possessing the two novel subunits. Genetic Resources and Crop Evolution 57: 1217–1225.

Hamdi O, Bellil I, Branlard G, Khelifi D (2010). Genetic Variation and Geographical Diversity for Seed Storage Proteins of Seventeen Durum Wheat Populations Collected in Algeria. Notulae Botanicae Horti Agrobotanici Cluj-Napoca 38(2) special Issue, 22–32. doi: https://doi.org/10.15835/nbha3824738.

He ZH, Liu L, Xia XC, Liu JJ, Peña RJ (2005) Composition of HMW and LMW Glutenin Subunits and Their Effects on Dough Properties, Pan Bread, and Noodle Quality of Chinese Bread Wheats. Cereal Chemistry 82: 345–350.

Henkrar F, El-Haddoury J, Iraqi D, Bendaou N, Udupa SM (2017) Allelic variation at high-molecular weight and low-molecular weight glutenin subunit genes in Moroccan bread wheat and durum wheat cultivars.3 Biotech 7.doi: https://doi.org/10.1007/s13205-017-0908

Ibba MI, Kiszonas, AM, Guzmán C, Morris CF (2017) Definition of the low molecular weight glutenin subunit gene family members in a set of standard bread wheat (*Triticum aestivum* L.) varieties. Journal of Cereal Science 74: 263–271.

Igrejas G, Branlard G, Gateau I, Carnide V, Guedes-Pinto H. (1997) Storage protein diversity within the Old Portuguese bread wheat 'Barbela' population. Journal of Genetics and Plant Breeding 51: 167–173.

Igrejas G, Guedes-Pinto H, Carnide V, Branlard G (1999) The high and low molecular weight glutenin subunits and ω-gliadin composition of bread and durum wheats commonly grown in Portugal. Plant Breeding 118: 297–302.

Igrejas G, Guedes-Pinto H, Carnide V, Clement J, Branlard G (2002) Genetical, biochemical and technological parameters associated with biscuit quality. II. Prediction using storage proteins and indirect tests in a soft wheat population. Journal of Cereal Science 36: 187–197.

Igrejas G, Juhász A, Gianibelli MC, Gale KR, Rahman S. (2009) Low-molecular-weight glutenins in durum wheat: analysis of Glu-A3 alleles using PCR markers. Plant Breeding 129: 574–577.

Ikeda TM, Branlard G, Peña RJ, Takata K, Liu L, He Z, et al. (2008) International collaboration for unifying *Glu-3* nomenclature system in common wheat in: "The 11th International Wheat Genetics Symposium 2008" Brisbane Aust. Sydney University Press.

Janni M, Cadonici S, Bonas U, Grasso A, Dahab AAD, Visioli G, et al. (2018) Gene-ecology of durum wheat HMW glutenin reflects their diffusion from the center of origin. Scientific Reports 8, 1–9.

Jaradat AA (2013) Wheat landraces: a mini review. Emirates Journal of Food and Agriculture 1: 20–29.

Jiang QT, Ma J, Wei YM, Liu YX, Lan XJ, Dai SF, et al. (2012) Novel variants of HMW glutenin subunits from *Aegilops* section *Sitopsis* species in relation to evolution and wheat breeding. BMC Plant Biology 12: 73.

Jiang QT, Zhang XW, Ma J, Wei L, Zhao S, Zhao QZ, et al. (2014) Characterization of high-molecular-weight glutenin subunits from *Eremopyrum bonaepartis* and identification of a novel variant with unusual high molecular weight and altered cysteine residues. Planta 239: 865–875.

Jin H, Yan J, Pena RJ, Xia XC, Morgounov A, Han LM, et al. (2011) Molecular detection of high- and low-molecular-weight glutenin subunit genes in common wheat cultivars from 20 countries using allele-specific markers. Crop and Pasture Science 62: 746–754.

Jin H, Zhang Y, Li GY, Mu PY, Fan ZR, Xia XC, et al. (2013) Effects of allelic variation of HMW-GS and LMW-GS on Mixograph properties and Chinese noodle and steamed bread qualities in a set of Aroona near-isogenic wheat lines. Journal of Cereal Science 57: 146–152.

Johansson E 1996. Quality evaluation of D-zone omega gliadins in wheat. Plant Breed 115:57–62.

Johansson E, Prieto-Linde ML, Gissén C (2008) Influences of weather, cultivar and fertilizer rate on grain protein polymer accumulation in field-grown winter wheat, and relations to grain water content and falling number. Journal of the Science of Food and Agriculture 8811: 2011–2018.

Juhász A, Larroque OR, Tamás L, Hsam SLK, Zeller FJ, Békés F, et al. (2003) Bánkúti 1201 - an old Hungarian wheat variety with special storage protein composition. Theoretical and Applied Genetics 107: 697–704.

Kabbaj H, Sall AT, Al-Abdallat A, Geleta M, Amri A, Filali-Maltouf A, et al. (2017) Genetic diversity within a global panel of durum wheat *Triticum durum* landraces and modern germplasm reveals the history of alleles exchange. Frontiers in Plant Science 8. 1277.

Katyal M, Virdi AS, Singh N, Kaur A, Rana JC, Kumari J (2018) Diversity in protein profiling, pasting, empirical and dynamic dough rheological properties of meal from different durum wheat accessions. Journal of Food Science and Technology 55: 1256–1269.

Labuschagne MT, Mkhatywa N, Wentzel B, Johansson E, van Biljon A (2014) Tocochromanol concentration, protein composition and baking quality of white flour of South African wheat cultivars. Journal of Food Composition and Analysis 33: 127–131.

Lafiandra D, D'Ovidio R, Porceddu E, Margiotta B, Colaprico G (1993) New data supporting high Mr glutenin subunit 5 as the determinant of quality differences among the pairs 5+10 vs 2+12. Journal of Cereal Science 18: 197–205.

Lan QX, Lan Q, Feng B, Xu Z, Zhao G, Wang T (2013) Molecular cloning and characterization of five novel low molecular weight glutenin subunit genes from Tibetan wheat landraces *Triticum aestivum* L. Genetic Resources and Crop Evolution 60: 799–806.

Lawrence GJ, MacRitchie F, Wrigley CW (1988) Dough and baking quality of wheat lines deficient in glutenin subunits controlled by the *Glu-A1*, *Glu-B1* and *Glu-D1* loci. Journal of Cereal Science 7: 109–112.

Lee S, Choi Y-M, Lee M-C, Hyun DY, Oh S, Jung Y 2018. Geographical comparison of genetic diversity in Asian landrace wheat (Triticum aestivum L.) germplasm based on high-molecular-weight glutenin subunits.). Genet Resour Crop Evol 65:1591–1602.

Lesage VS, Merlino M, Chambon C, Bouchet, B, Marion, D, Branlard, G. (2012) Proteomes of hard and soft near-isogenic wheat lines reveal that kernel hardness is related to the amplification of a stress response during endosperm development. Journal of Experimental Botany 63: 1001–1011.

Lesage V, Rhazi L, Aussenac A, Meleard B, Branlard G (2013) Effects of HMW- & LMW-glutenins and grain hardness on size of gluten polymers. In: He Z, Wang D eds Wheat Gluten 2012, Proceedings of the 11th international wheat gluten workshop, Beijing, 200–205.

Li D, Jin H, Zhang K, Wang Z, Wang F, Zhao Y, et al. (2018) Analysis of the *Gli-D2* locus identifies a genetic target for simultaneously improving the breadmaking and health-related traits of common wheat. The Plant Journal 95: 414–426.

Li W, Wan Y, Liu Z, Liu K, Liu X, Li B, et al. (2004) Molecular characterization of HMW glutenin subunit allele 1Bx14: further insights into the evolution of *Glu-B1-1* alleles in wheat and related species. Theoretical and Applied Genetics 109: 1093–1104.

Li Y, Huang C, Sui X, Fan Q, Li G, Chu X (2009) Genetic variation of wheat glutenin subunits between landraces and varieties and their contributions to wheat quality improvement in China. Euphytica 169: 159–168.

Li Y, An X, Yang R, Guo X, Yue G, Fan R, et al. (2015) Dissecting and enhancing the contributions of high-molecular-weight glutenin subunits to dough functionality and bread quality. Molecular Plant 8: 332–334.

Li Y-F, Wu Y, Hernandez-Espinosa N, Peña RJ (2013) The influence of drought and heat stress on the expression of end-use quality parameters of common wheat. Journal of Cereal Science 57: 73–78.

Li ZX, Zhang XQ, Zhang HG, Cao SH, Wang D, Hao S, et al. (2008) Isolation and characterization of a novel variant of HMW glutenin subunit gene from the St genome of *Pseudoroegneria stipifolia*. Journal of Cereal Science 47: 429–437.

Liang D, Tang J, Pena RJ, Singh RP, He X, Shen X, et al. (2010) Characterization of CIMMYT bread wheats for high and low-molecular weight glutenin subunits and other quality-related genes with SDS-PAGE, RP-HPLC and molecular markers. Euphytica 172: 235–250.

Liu L, He Z, Yan J, Zhang Y, Xia X, Peña RJ (2005) Allelic variation at the *Glu-1* and *Glu-3* loci, presence of the 1B.1R translocation, and their effects on Mixographic properties in Chinese bread wheats. Euphytica 142: 197–204.

Liu L, Ikeda TM, Branlard G, Pena RJ, Rogers WJ, Lerner SE, et al. (2010) Comparison of low molecular weight glutenin subunits identified by SDS-PAGE, 2-DE, MALDI-TOF-MS and PCR in common wheat. BMC Plant Biology 10: 1–24.

Liu S, Gao X, Xia G (2008) Characterizing HMW-GS alleles of decaploid *Agropyron elongatum* in relation to evolution and wheat breeding. Theoretical and Applied Genetics 116: 325–334.

Liu Y, Xiong Z-Y, He Y-G, Shewry PR, He G-Y (2007) Genetic diversity of HMW glutenin subunit in Chinese common wheat *Triticum aestivum* L. landraces from Hubei province. Genetic Resources and Crop Evolution 54: 865–874.

Liu Z, Yan Z, Wan Y, Liu K, Zheng Y, Wang D (2003) Analysis of HMW glutenin subunits and their coding sequences in two diploid *Aegilops* species. Theoretical and Applied Genetics 106: 1368–1378.

Longin CFH, Reif JC (2014) Redesigning the exploitation of wheat genetic resources. Trends in Plant Science 19: 631–636.

Lopes MS, El-Basyoni I, Baenziger PS, Singh S, Royo C, Ozbek K, et al. (2015) Exploiting genetic diversity from landraces in wheat breeding for adaptation to climate change. Journal Of Experimental Botany 66: 3477–3486.

MacRitchie F (2014) Theories of glutenin/dough systems. Journal of Cereal Science 60: 4–6.

Malik AH, Kuktaite R Johansson E (2013) Combined effect of genetic and environmental factors on the accumulation of proteins in the wheat grain and their relationship to bread-making quality. Journal of Cereal Science 572: 170–174.

Melnikova NV, Ganeva GD, Popova ZG, Landjeva SP, Kudryavtsev AM (2010) Gliadins of Bulgarian durum wheat *Triticum durum* Desf. landraces: genetic diversity and geographical distribution. Genetic Resources and Crop Evolution 57: 587–595.

Metakovsky, E.V. 1991a. Gliadin allele identification in common wheat. 2.Catalogue of gliadin alleles in common wheat. J. Genet. & Breed. 45:325–344.

Metakovsky EV, Graybosch RA, 2006. Chapter 3. Gliadin Alleles in Wheat: Identification and Applications in 'Gliadin and Glutenin: The Unique Balance of Wheat Quality', AACC St Paul MN USA, Page 85–114.

Metakovsky, E.V. Novoselskata AY, 1991b Gliadin allele identification in common wheat. 1 Methodological aspects of the analysis of gliadin pattern by one-dimensional polyacrylamide gel electrophoresis. J. Genet. & Breed. 45:317–324.

Metakovsky EV (2015) Wheat storage proteins: genes, inheritance, variability, mutations, phylogeny, seed production, flour, quality, LAP Lambert Acad. Publishing, Saarbrücken, Deutschland, Germany, pp. 320 (in Russian with English abstract).

Metakovsky EV, Melnik V, Rodriguez-Quijano M, Upelniek V, Carrillo JM (2018) A catalog of gliadin alleles: Polymorphism of 20th-century common wheat germplasm. Crop Journal 6: 628–641.

Mir Ali N, Arabi MIE, Al- Safadi B (1999) Frequencies of high and low molecular weight glutenin subunits in durum wheat grown in Syria. Cereal Research Communications 27: 301–305.

Moragues M, Zarco-Hernández J, Moralejo MA, Royo C (2006) Genetic diversity of glutenin protein subunits composition in durum wheat landraces [*Triticum turgidum* ssp. *turgidum* Convar. *durum* Desf. MacKey] from the Mediterranean basin. Genetic Resources and Crop Evolution 53: 993–1002.

Morgunov AI, Peña RJ, Crossa J, Rajaram S. (1993) Worldwide distribution of *Glu-1* alleles in bread wheat. Journal of Genetics and Plant Breeding 47: 53–60.

Muccilli V, Cunsolo V, Saletti R, Foti S, Margiotta B, Scossa F, et al. (2010) Characterisation of a specific class of typical low molecular weight glutenin subunits of durum wheat by a proteomic approach. Journal of Cereal Science 51:134–139.

Naghavi MR, Monfared SR, Ahkami AH, Ombidbakhsh MA (2009) Genetic variation of durum wheat landraces and cultivars using morphological and protein markers. World Academy of Science, Engineering and Technology 3: 33–35.

Nakamura H (2001) Genetic diversity of high-molecular-weight glutenin subunit compositions in landraces of hexaploid wheat from Japan. Euphytica 120: 227–234.

Nieto-Taladriz MT, Ruiz M, Martínez MC, Vázquez JF, Carrillo JM (1997) Variation and classification of B low-molecular-weight glutenin subunit alleles in durum wheat. Theoretical and Applied Genetics 95: 1155–1160.

Payne PI, Corfield KG, Blackman JA, (1979) Identification of a high-molecular-weight subunit of glutenin whose presence correlates with bread-making quality in wheats of related pedigree. Theoretical and Applied Genetics 55: 153–159.

Payne PI, Lawrence GJ (1983) Catalogue of alleles for the complex gene loci, *Glu-A1*, *Glu-B1* and *Glu-D1* which code for high-molecular-weight subunits of glutenin in hexaploid wheat. Cereal Research Communications 11: 29–35.

Peng YC, Yu K, Zhang Y, Islam S, Sun D, Ma W (2015) Two novel y-type high molecular weight glutenin genes in Chinese wheat landraces of the Yangtze-River region. Plos One, 1011: e0142348.

Peng YC, Yu Z, Islam S, Zhang Y, Wang X, Lei Z, et al. (2016) Allelic variation of LMW-GS composition in Chinese wheat landraces of the Yangtze-River region detected by MALDI-TOF-MS. Breeding Science 66: 646–652.

Pignone D, De Paola D, Rapanà N, Janni M (2015) Single seed descent: a tool to exploit durum wheat *Triticum durum* Desf. genetic resources. Genetic Resources and Crop Evolution 62: 1029–1035.

Raciti CN, Doust MA, Lombardo GM, Boggini G, Pecetti L (2003) Characterization of durum wheat mediterranean germplasm for high and low molecular weight glutenin subunits in relation with quality. European Journal of Agronomy 19: 373–382.

Rasheed A, Jin H, Xiao Y, Zhang Y, Hao Y, Zhang Y, et al. (2019) Allelic effects and variations for key bread-making quality genes in bread wheat using high-throughput markers. Journal of Cereal Science 85: 305–309.

Redaelli R, Ng PKW, Ward RW (1997) Electrophoretic characterization of storage proteins of 37 Chinese landraces of wheat. Journal of Genetics and Breeding 51: 239–246.

Ribeiro JM, Bancel E, Faye A, Dardevet M, Ravel C, Branlard G, et al. (2013a) Proteogenomic characterization of novel x-type high molecular weight glutenin subunit 1Ax1.1. International Journal of Molecular Sciences 14: 5650–5667.

Ribeiro M, Carvalho C, Carnide V, Guedes-Pinto H, Igrejas G (2011) Towards allelic diversity in the storage proteins of old and currently growing tetraploid and hexaploid wheats in Portugal. Genetic Resources and Crop Evolution 58: 1051–1073.

Ribeiro M, Nunes-Miranda JD, Branlard G, Carrillo JM, Rodriguez-Quijano M, Igrejas G (2013) One hundred years of grain Omics: Identifying the glutens that feed the world. Journal of Proteome Research 12: 4702–4716.

Rodriguez-Quijano M, Vásquez JF, Carrillo JM (1990) Variation of high molecular weight glutenin subunits in Spanish landraces of *Triticum aestivum* ssp. *vulgare* and ssp. *spelta*. Journal of Genetics and Breeding 44: 121–126.

Rodriguez-Quijano M, Vásquez JF, Moita-Brites C, Carrillo JM (1998) Allelic variation of HMW glutenin subunits in Portuguese landraces of *Triticum aestivum* ssp *vulgare*. Journal of Genetics and Breeding 52: 95–98.

Ruiz M, Bernal G, Giraldo P (2018) An update of low molecular weight glutenin subunits in durum wheat relevant to breeding for quality. Journal of Cereal Science, 83: 236–244.

Ruiz M, Carrillo JM (1993) Linkage relationships between prolamin genes on chromosome 1A and chromosome 1B of durum wheat. Theoretical and Applied Genetics 87: 353–360.

Ruiz M, Metakovsky EV, Rodriguez-Quijano M, Vazquez JF, Carrillo JM (2002a) Assessment of storage protein variation in relation to some morphological characters in a sample of Spanish landraces of common wheat *Triticum aestivum* L. ssp *aestivum*. Genetic Resources and Crop Evolution 49: 371–382.

Ruiz M, Metakovsky EV, Rodriguez-Quijano M, Vazquez JF, Carrillo JM (2002b) Polymorphism, variation and genetic identity of Spanish common wheat germplasm based on gliadin alleles. Field Crops Research 79: 185–196.

Ruiz, M., Giraldo P, Royo C, Villegas D, Aranzana M J, Carrillo J M (2012) Diversity and genetic structure of a collection of Spanish durum wheat landraces. Crop Science, 52: 2262–2275.

Shao H, Liu T, Ran C-F, Li L-Q, Yu J, Gao X, et al. (2015) Isolation and molecular characterization of two novel HMW-GS genes from Chinese wheat *Triticum aestivum* L. landrace Banjiemang. Genes & Genomics 37: 45–53.

Shewry PR, Gilbert SMA, Savage WJ, Tatham AS, Wan YF, Belton PS, et al. (2003a) Sequence and properties of HMW subunit 1Bx20 from pasta wheat *Triticum durum* which is associated with poor end use properties. Theoretical and Applied Genetics 106: 744–750.

Shewry PR, Halford NG, Lafiandra D (2003b) Genetics of wheat gluten proteins. Advanced Genetics 49: 111–184.

Sourour A, Salah B, Afef O, Zoubeir C, Younes B (2016) Variability of HMW and LMW glutenin subunits in durum wheat *Triticum Durum* Desf. Journal of the Institute of Agriculture and Animal Science 4: 10–13.

Southan M, MacRitchie F (1999) Molecular weight distribution of wheat proteins. Cereal Chemistry 76: 827–836.

Sozinov IA, Poperelya FA, Stacanova AI (1974) Use of electrophoresis of gliadin for selection of wheat by quality. In Russian, Vestnik Navki, 7: 99–108.

Sun X, Hu S, Liu X, Qian W, Hao S, Zhang A, et al. (2006) Characterization of the HMW glutenin subunits from *Aegilops searsii* and identification of a novel variant HMW glutenin subunit. Theoretical and Applied Genetics 113: 631–641.

Sun Y, Pu Z, Dai S, Pu X, Liu D, Wu B, et al. (2014) Characterization of y-type high-molecular-weight glutenins in tetraploid species of *Leymus*. Development Genes And Evolution 224: 57–64.

Tahir M, Turchetta T, Anwar R, Lafiandra D (1996) Assessment of genetic variability in hexaploid wheat landraces of Pakistan based on polymorphism for HMW glutenin subunits. Genetic Resources and Crop Evolution 43: 211–220.

Tanaka H, Tomita M, Tsujimoto H, Yasumuro Y (2003) Limited but specific variations of seed storage proteins in Japanese common wheat *Triticum aestivum* L. Euphytica 132: 167–174.

Tester M, Langridge P (2010) Breeding technologies to increase crop production in a changing world. Science 327: 818–822.

Tohver M (2007) High Molecular Weight HMW glutenin subunit composition of some Nordic and Middle European wheats. Genetic Resources and Crop Evolution 54: 67–81.

Triboï E, Martre P, Triboï-Blondel AM (2003) Environmentally-induced changes of protein composition for developing grains of wheat are related to changes in total protein content. Journal Of Experimental Botany 54: 1731–1742.

Van Lill D, Smith MF (1997) A quality assurance strategy for wheat *Triticum aestivum* L. where growth environment predominates. South African J Plant Soil 14: 183–191.

Wall JS (1979) The role of wheat protein in determining the baking quality. In: Recent advances in biochemistry of cereals. Laidman D.L. and Wyn-Jones R.G. eds Academic, London, 275–311.

Wan YF, Yan Z, Liu K, Zheng YL, D'Ovidio R, Shewry PR, et al. (2005) Comparative analysis of the D genome-encoded high molecular weight subunits of glutenin. Theoretical and Applied Genetics 111: 1183–1190.

Wang D, Zhang K, Dong L, Dong Z, Li Y, Hussain A, et al. (2018) Molecular and genomic analysis of wheat milling and end-use traits in China: Progress and perspectives. Crop Journal 6: 68–81.

Wang DW, Li D, Wang J, Zhao Y, Wang Z, Yue G, et al. (2017a) Genome-wide analysis of complex wheat gliadins, the dominant carriers of celiac disease epitopes. Scientific Reports 7: 44609.

Wang Z, Li Y, Yang Y, Liu X, Qin H, Dong Z, et al. (2017b) New insight into the function of wheat glutenin proteins as investigated with two series of genetic mutants. Scientific Reports 7: 3428.

Wang S, Yu Z, Cao M, Shen X, Li N, Li X, et al. (2013) Molecular mechanisms of HMW glutenin subunits from 1Sl genome of *Aegilops longissima* positively affecting wheat breadmaking quality. PLoS One 8: e58947.

Wrigley C, Asenstorfer R, Batey I, Cornish G, Day L, Mares D, et al. (2009) The biochemical and molecular basis of wheat quality. In: Carver BF ed Wheat science and trade. Wiley, Iowa, USA.

Xynias IN, Kozub NA, Sozinov IA (2011) Analysis of hellenic durum wheat *Triticum turgidum* L. var. *durum* germplasm using gliadin and high-molecular-weight glutenin subunit loci. Cereal Research Communications 39: 415–425.

Yang Y, Li S, Zhang K, Dong Z, Li Y, An X, Chen J, et al. (2014). Efficient isolation of ion beam-induced mutants for homoeologous loci in common wheat and comparison of the contributions of *Glu-1* loci to gluten functionality. Theoretical and Applied Genetics 127: 359–372.

Yasmeen F, Khurshid H, Ghafoo A (2015) Genetic divergence for high-molecular weight glutenin subunits HMW-GS in indigenous landraces and commercial cultivars of bread wheat of Pakistan. Genetics and Molecular Research 14: 4829–4839.

Zeven AC (1998) Landraces: A review of definitions and classifications. Euphytica 104: 127–139.

Zhang PP, He ZH, Zhang Y, Xia XC, Liu JJ, Yan J, et al. (2007) Pan bread and Chinese white salted noodle qualities of Chinese winter wheat cultivars and their relationship with gluten protein fractions. Cereal Chemistry 84:370–378.

Zhang ZJ (1995) Evidence of durable resistance in nine Chinese land races and one Italian cultivar of *Triticum aestivum* to *Puccinia striiformis*. European Journal of Plant Pathology 101: 405–409.

Zheng W, Peng Y, Ma J, Appels R, Sun D, Ma W (2011) High frequency of abnormal high molecular weight glutenin alleles in Chinese wheat landraces of the Yangtze-River region. Journal of Cereal Science 54: 401–408.

Zilić S, Barać M, Pešić M, Dodig D, Ignjatović-Micić D (2011) Characterization of proteins from grain of different bread and durum wheat genotypes. International Journal of Molecular Sciences 12: 5878–5894.

Durum Wheat Storage Protein Composition and the Role of LMW-GS in Quality

Patricia Giraldo, Magdalena Ruiz, M. Itria Ibba, Craig F. Morris, Maryke T. Labuschagne, and Gilberto Igrejas

1 History of the Study of Durum Wheat Storage Proteins

Durum wheat (*Triticum turgidum* L. var. *durum* (Desf.) Husn., a tetraploid species, is an important staple food mainly used for pasta and bread making in Europe and North Africa. It represents only approximately 8% of wheat production worldwide, 90% of which is produced in the Mediterranean region (Ammar et al. 2000). It is widely accepted that durum wheat was introduced into North Africa and the Iberian Peninsula from the south of Italy (MacKey 2005). However, recent findings based on the genetic similarities between landraces from the Maghreb countries and those from Spain and Portugal have suggested North Africa was an additional route for the introduction of wheat into the Iberian Peninsula (Moragues et al. 2006, 2007).

P. Giraldo (✉)
Department of Biotechnology-Plant Biology, School of Agricultural, Food and Biosystems Engineering, Universidad Politécnica de Madrid, Madrid, Spain
e-mail: patricia.giraldo@upm.es

M. Ruiz
Centro de Recursos Fitogenéticos (CRF-INIA), Madrid, Spain

M. I. Ibba
Global Wheat Program, International Maize and Wheat Improvement Center (CIMMYT), CP, Mexico

C. F. Morris
USDA ARS Western Wheat Quality Laboratory, Washington State University, Pullman, WA, USA

M. T. Labuschagne
Department of Plant Sciences, University of the Free State, Bloemfontein, South Africa

G. Igrejas
Department of Genetics and Biotechnology, University of Trás-os-Montes and Alto Douro, Vila Real, Portugal

G. Igrejas et al. (eds.), *Wheat Quality For Improving Processing And Human Health*, https://doi.org/10.1007/978-3-030-34163-3_5

In wheat grain development, the initial synthesis of biologically active proteins (enzymes) takes place in the aleurone layer, while the storage proteins, which are considered biologically inactive, are synthesized in the endoplasmic reticulum of endosperm cells. Storage proteins then pass into the Golgi complex where they are included in vacuoles, which go on to form protein bodies (Cheftel et al. 1985; Shewry et al. 1995; Shewry 1999). The first systematic study of plant proteins was carried out by Osborne (1907). In 1924, Osborne classified plant proteins into four groups according to how they were extracted, and hence their solubility. Albumins are soluble in water, globulins are soluble in saline solutions, prolamins are soluble in 70% ethanol, and glutenins are soluble only in acidic or reducing solutions. The proteins of the endosperm, a source of amino acids required by the embryo, are essentially comprised of the prolamins, which represent about 75–95% of the total proteins of the cereals (Bushuk 1981; Cheftel et al. 1985; Shulka 1975). Prolamins are well known in the agro-food sector because of their role in the formation of gluten, but in cereals there is no known function other than storage (Shewry and Tatham 1990). Payne et al. (1980) classified the storage proteins according to their aggregation characteristics. Gliadins are a complex mixture of polypeptides, while glutenins form disulfide bridges between subunits. In the absence of enzymatic activity, gliadins and glutenins make up a reserve of nitrogen, carbon, and sulfur that can be mobilized during seed germination (Ciaffi et al. 1999; Kreis et al. 1985).

Gliadins represent about 40–45% of wheat endosperm protein (Branlard et al. 1990). These proteins were separated in an electrophoretic system with an aluminum lactate buffer by Jones et al. (1959) and according to their mobility in a starch gel system at acidic pH (3.1) were classified by Woychik et al. (1961) as α, β, γ, and ω gliadins. The α, β, and γ gliadins have molecular weights ranging from 36,000 Da to 44,000 Da, whereas ω gliadins range from 65,000 Da to 79,000 Da (Bietz and Wall 1972). Bushuk and Zillman (1978) proposed a nomenclature based on relative mobility and presumed molecular mass of the four gliadin groups, namely: ω (<40.4 Da), γ (40.4–53.2 Da), β (53.2–68.6 Da), and α (>68.6 Da).

Glutenins, high molecular weight aggregates of several million daltons in the native state, account for between 40% and 50% of the protein in wheat flour (Melas et al. 1993, Payne et al. 1984a). These proteins are insoluble in aqueous solutions and ethanol, and can be separated by the action of reducing agents on disulfide bonds into high molecular weight glutenin subunits (HMW-GS) and low molecular weight glutenin subunits (LMW-GS) (Payne et al. 1979). According to their estimated molecular weight in sodium dodecyl sulfate polyacrylamide gel electrophoresis (SDS-PAGE), glutenins are classified into three groups: A (95,000–136,000 Da), B (42,000–51,000 Da), and C (31,500–35,500 Da) (Payne and Corfield 1979). Using two-dimensional electrophoresis, Jackson et al. (1983) identified a fourth group, D, with mobility similar to that of the ω gliadins (Branlard et al. 1992; Khelifi and Branlard 1991). The B, C, and D groups are considered to be LMW-GS, representing approximately 40% of the total gluten protein, whereas group A are the HMW-GS represent about 10% of the total gluten protein (Melas et al. 1994). The use of the SDS-PAGE system at discontinuous pH, developed by Laemmli (1970), contributed to a better separation of the different groups and improved knowledge of the genetic and biochemical variability of glutenins.

Table 1 Composition and classification of proteins in wheat grain according to Osborne (1907) and Shewry et al. (1986)

Osborne (1907)	% of proteins	Shewry et al. (1986)		
Albumins	10	Functional and structure proteins		
Globulins	10			
Gliadins	45	ω gliadins	Sulfur-poor prolamins	Monomeric
		α, β, γ gliadins	Sulfur-rich prolamins	
Glutenins	35	LMW glutenins		Aggregative
		HMW glutenins	HMW prolamins	

The N-terminal sequences of LMW-GS were used to divide the protein subunits into two main groups (Ikeda et al. 2002). The first group corresponded to typical LMW-GS, LMW-i (the first amino acid is isoleucine) and LMW-m (the first amino acid is methionine) types, and the second group, named gliadin-like sequences (Tao and Kasarda 1989) based on the similarity of N-terminal sequences to α-, γ-, and ω- gliadins. Comparison of the amino acid sequences of gliadins and glutenins allowed Shewry et al. (1986) to propose a new classification. They classified the storage proteins as high molecular weight prolamins, sulfur-rich prolamins (α, β, and γ gliadins) and low-sulfur prolamins (ω gliadins). Table 1 shows the correspondence between the different classification systems.

The amino acid composition of glutenins and gliadins is quite similar. Both groups are rich in proline and glutamine, but have a low lysine content (unlike the albumins and globulins). Glutamine and proline play important metabolic or physiological roles during the germination process, glutamine in the synthesis of other amino acids, and proline for tolerance to drought and cold (Branlard and Rousset 1987). Higher levels of non-polar amino acids, such as phenylalanine, valine, leucine, isoleucine and methionine, makes glutenins particularly hydrophobic (Melas et al. 1993; Shewry et al. 1983). HMW-GS contain more glycine than gliadins do, although the latter have higher levels of glutamine, proline and cysteine (MacRitchie et al. 1990; Shewry et al. 1986). Compared to HMW-GS, LMW-GS contain more cysteine. The cysteine residues are involved in the formation of disulfide bonds among glutenins. The number and nature of these types of connections influence the viscoelasticity of the dough and, consequently, their industrial use. The amino acid composition of gliadins, mainly its different levels of cysteine, methionine, glutamine, proline, and phenylalanine, subdivides them into two distinct groups: α, β and γ gliadins, and ω gliadins (MacRitchie et al. 1990).

Discovering the structure of the storage proteins has been made possible through the isolation and sequencing of the genes that encode them (MacRitchie 1992). Cassidy and Dvorak (1991) reported that the LMW-GS were comprised of two domains. The hydrophobic domain 1 includes repetitive sequences and is rich in proline and glutamine but poor in cysteine, and the hydrophilic domain 2 is subdivided into three domains (A, B and C) interspersed by four regions (I1, I2, I3 and I4), has no repetitive sequences, and is rich in cysteine but low in proline. According to Benmoussa et al. (2000) deletions and insertions are the basis for the evolution of prolamin genes and so responsible for allelic variations in LMW-GS. The main step in the evolution of HMW-GS and sulfur-rich prolamins is the insertion of repetitive sequences among the three subdomains that constitute them. On the other hand, the sulfur-poor prolamins evolved by amplifying the region consisting of repetitive sequences, with deletion of all non-repetitive sequences, except for a remaining trace of the C domain. It seems likely that the three groups of prolamins previously defined evolved from an ancestral protein of about 90 residues with three distinct domains (Kreis et al. 1985; Shewry et al. 1995; Shewry and Tatham 1990; Shewry and Tatham 1997; Tatham and Shewry 1995). Some of the HMW-GS were studied by D'Ovidio et al. (1994, 1995, 1996) using restriction fragment length polymorphism (RFLP) and polymerase chain reaction (PCR), with the conclusion that the variation in subunit length is due to differences in the length of the central repetitive domain.

2 Genetic Control and Nomenclature

The nucleotide sequences of α, β, γ and ω gliadins and LMW-GS genes show close homologies between the gene families, but the HMW-GS gene family is not so similar.

In durum wheat, the *Glu-1* loci located on the long arms of chromosome 1A (*Glu-A1* locus) and 1B (*Glu-B1 l*ocus) 30 cM from the centromere encode HMW-GS. Each locus contains two genes tightly linked, *Glu-1-1* and *Glu-1-2*, the products of which were described as 'x-type'and 'y-type' based on different molecular weights and isoelectric points (Payne et al. 1981). At the Glu-A1 locus, the 'y-type' protein is not expressed in hexaploid wheat although it has been found in diploid wheats (Levy et al. 1988; Waines and Payne 1987) and the 'x-type' subunit is rarely expressed (Payne et al. 1981). The subunit nomenclature commonly used is that proposed by Payne et al. (1981) where each subunit has a numerical reference. In addition to subunit nomenclature, alleles for the *Glu-1* loci are formally defined and compiled in the Catalogue of Gene Symbols for Wheat (McIntosh et al. 2013).

The main genes controlling gliadin synthesis are located at *Gli-1* loci on the short arm of chromosomes 1A and 1B for most γ and ω gliadins, and at *Gli-2* loci on chromosomes 6A and 6B for α and β gliadins (du Cros et al. 1983; Joppa et al. 1983). There are other minor loci, which have limited allelic variability, located on the homoeologous group 1 chromosomes (Mecham et al. 1978). The *Gli-A3* locus, located between *Gli-A1* and *Glu-A1*, encodes ω-gliadins and some minor γ-gliadins, and the *Gli-B3*, which maps at 19.5 cM from *Gli-B1* and encodes some ω-gliadins (Ruiz and Carrillo 1993). Another locus, *Gli-B5*, has been mapped in durum wheat at 4.7 cM from *Gli-B1* (Mazza et al. 1996). Each gliadin locus consists of several closely linked genes with multiple alleles, which makes allelic classification quite difficult. The nomenclature for gliadin alleles was established by Metakovsky in 1991 and is based on the results of fractionation of gliadins by acid polyacrylamide gel electrophoresis (A-PAGE) and the identification of jointly inherited gliadin bands called 'blocks'.

Several molecular and proteomic studies have shown that LMW-GS subunit loci are clearly multigenic. The number of LMW-GS genes is variable depending on the locus and the variety and up to 40 complete genes, partial genes, or pseudogenes of LMW-GS have been sequenced from *Triticum* species (e.g. Beom et al. 2018;

Cassidy et al. 1998; Dupont et al. 2011; Qi et al. 2009). In durum wheat, the main loci that control the synthesis of the B-LMW glutenin subunits are *Glu-A3*, *Glu-B3* and *Glu-B2* (Gupta and Shepherd 1988; Liu 1995; Ruiz and Carrillo 1993), which are tightly linked to the loci *Gli-A1*, *Gli-B1* and *Gli-B3*, respectively (Liu and Shepherd 1995; Pogna et al. 1990; Ruiz and Carrillo 1993). Both *Glu-3* loci are located at the distal end of the short arm of the group 1 chromosomes, whereas the *Glu-B2* is in the middle of the short arm of chromosome 1B. Liu and Shepherd (1995) found another locus, *Glu-B4*, at a distance of 3 cM from *Glu-B3*, which encodes a B-LMW glutenin subunit. In *T. durum*, the genetic distance between *Glu-A3* and *Gli-A1* loci has been estimated as 1.3 +/− 0.4 cM, and that between *Glu-B3* and *Gli-B1* as 2.0 +/− 0.4 cM (Ruiz and Carrillo 1993). The *Glu-B2* locus, which controls a B-LMW glutenin subunit, is tightly linked to the *Gli-B3* locus and it was mapped at a distance of 20 cM from *Gli-B1* and 14 cM from *Glu-B3* (Liu and Shepherd 1995; Martínez et al. 2004; Ruiz and Carrillo 1993). Although the majority of the LMW-GS are controlled by genes on the chromosomes of homoeologous group 1, some of the LMW-C subunits must be controlled by loci elsewhere in the genome (Gupta and Shepherd 1993).

The most important characteristic of the genes encoding LMW-GS is the conservation of codons for eight cysteine residues, which may be involved in the secondary or tertiary structure and the formation of disulfide bonds in the gene product (Ciaffi et al. 1999). Another feature is that the genes at each *Glu-3* locus do not contain any introns and are separated by large intergenic regions of up to 748 Kb between two neighboring genes at the *Glu-B3* locus (Ibba et al. 2017). By using molecular markers, intralocus recombination has been detected at the *Glu-A3*, *Glu-B3*, and *Glu-D3* loci (Dong et al. 2010; Espí et al. 2014; Ibba et al. 2017).

The group of LMW-GS encoded at the same locus are considered to be allelic variants. This designation replaces the inadequate LMW pattern system of classification (LMW-1 and LMW-2) used in earlier studies (Payne et al. 1984b). Nieto-Taladriz et al. (1997) analysed the genetic control of the B-LMW subunits and identified 5 subunits controlled at the *Glu-A3* locus, 14 at the *Glu-B3* locus, and 1 at the *Glu-B2* locus. The LMW-GS were named with numbers that more or less correlate with their mobility in SDS-PAGE. Later, several studies identified new subunits using the same nomenclature system (Aguiriano et al. 2008; Bellil et al. 2014; Brites and Carrillo 2000; Martínez et al. 2004; Lerner et al. 2004; Rodríguez-Quijano et al. 2010). In the latest edition of the Catalogue of Gene Symbols for Wheat the nomenclature has been changed to new allele names (McIntosh et al. 2013) with the earlier durum designation given as synonyms. According to the catalogue, in durum wheat there are 11, 11 and 3 different allelic variants at the *Glu-A3, Glu-B3* and *Glu-B2* loci, respectively, although the designation of some subunits as alleles from the *Glu-3* locus has been deduced only from their electrophoretic mobility and awaits confirmation through genetic studies (McIntosh et al. 2013; Ruiz et al. 2018).

To assist in the identification of LMW-GS alleles it is important to have accessible standard cultivars to be included in SDS-PAGE analyses. In 2005, a cooperative international program was developed to establish a set of standard cultivars for identifying LMW-GS alleles in bread wheat. A set of 103 cultivars from 12 countries was used for the identification of LMW-GS and allelic assignment (Liu et al.

2010). In durum wheat, Nieto-Taladriz et al. (1997) proposed a standard set of 11 varieties for LMW-GS, however, new subunits have been described since so the definition of new standard varieties for these subunits is required in order to expand the master set to cover as much variability as possible (Ruiz et al. 2018).

3 Biochemical Studies and Proteomics

Proteomics is the study of the full complement of polypeptides expressed by the genes of an organism in a specific tissue, at a particular stage of development and under specified growth conditions (Dunn 2000; Humphery-Smith et al. 1997; Pennington and Dunn 2001; Williams 1999). The term 'proteome' was originally coined by Wasinger et al. (1995) to emphasize the functional aspects of genomic studies involving the elucidation of downstream effects of the genome and taking into account the interplay of environment with the genome to determine the characteristics of the organism. While there is only one genome in a specific organism, proteomes are different in various tissues in an organism. Therefore, proteomes do not correspond exactly to the genome in an organism because of different posttranscriptional and posttranslational modifications. In the post-genomic era, proteomics promises to analyze dynamic molecular changes occurring during the life of an organism that are not predictable at genomic level (Park 2004).

3.1 Methods for Prolamin Proteomic Research

Gluten proteins are among the most complex protein networks in nature due to their numerous different components and sizes, and due to variability caused by genotype, growing conditions and technological processes (Wieser 2007). Over the years a number of different electrophoretic methods and techniques have been adopted for the characterization and study of this interesting group of proteins. Four different methods can be used to characterize the wheat prolamins: SDS-PAGE, two-dimensional gel electrophoresis (2DE) which is isoelectric focusing in one dimension followed by SDS-PAGE in the second dimension, reversed-phase high-performance liquid chromatography (RP-HPLC), and matrix-assisted laser desorption/ionization time-of-flight mass spectrometry (MALDI-TOF-MS).

3.1.1 Sodium Dodecyl Sulphate Polyacrylamide Gel Electrophoresis

The introduction of SDS-PAGE to the study of prolamins was a major milestone for the overall improvement of wheat quality. This method not only allowed a deeper understanding of prolamin composition, but also provided consistent and reliable markers useful for the prediction of gluten strength. Depending on the SDS-PAGE glutenin profile, most of the HMW-GS and LMW-GS alleles present in both durum and common wheat varieties could be identified and characterized in relation to different quality parameters (Payne 1987; Payne et al. 1984b). Consequently, SDS-PAGE analysis became routine in wheat breeding programs allowing a more accurate selection for end-use quality and an overall improvement in the pasta-making quality of the durum wheat varieties released during the twentieth century (De Santis et al. 2017; De Vita et al. 2007; Subira et al. 2014). To date, SDS-PAGE is still the most commonly used method for the analysis of the glutenin alleles in durum wheat (Ammar et al. 2000; Dreisigacker et al. 2016; Hailegiorgis et al. 2017; Igrejas et al. 1999; Kiszonas and Morris 2018; Nazco et al. 2014; Nieto-Taladriz et al. 1997; Raciti et al. 2003; Sissons et al. 2005). The main reason for this popularity is that compared to other methods, SDS-PAGE is fast, cost-effective, and can be done with as little as one-half of a kernel, allowing the simultaneous identification of the allelic variants at all four glutenin loci. Even though SDS-PAGE is now a well-established technique, it also has several limitations. For example, high quality gels and robust interpretation of allele assignment requires expert personnel, and not all possible alleles can be discriminated, so LMW-GS may not always be classified correctly (Liu et al. 2010).

3.1.2 Two-Dimensional Gel Electrophoresis

Arguably the most informative gel-based method for the discrimination of different LMW-GS alleles (Yahata et al. 2005), 2DE was first introduced in 1975 and separates proteins first on the basis of isoelectric point, and then on molecular mass. With 2DE thousands of protein species can be readily resolved and visualized on a single gel, so isomorphisms, polymorphisms, and structural changes such as post-translational modifications can be detected (Chevalier et al. 2004). Gene products, visualized as precise protein spots, are de facto genetic and physiological markers, which can be useful in assessing genetic variability, establishing genetic distances and phylogenetic relationships between organism lines, species and genus, and detecting post-translational effects induced by the environment (Thiellement et al. 1999).

It is significant that this methodology was first developed in relation to the separation of cereal grain proteins (Wrigley 1968, 1970), with further developments of high-resolution 2DE by Klose (1975), O'Farrell (1975), and Scheele (1975), demonstrating the enormous potential of this analytical technique for separating thousands of proteins in parallel. Early on, using aneuploid lines Wrigley (1968, 1970) proved it was possible to allocate chromosomal locations to genes for over 50 gliadin proteins. Much more recently, 2DE has been used to allocate chromosomal locations for a wider range of wheat grain protein genes (Islam et al. 2002, 2003a, b).

The accuracy of 2DE compared with SDS-PAGE is much greater and allows the identification of specific allelic variants that otherwise could not be detected, especially for the LMW-GS. However, different bands in SDS-PAGE separations are not always distinguishable in 2DE separations and compared to other gel-based methods, 2DE is not efficient either in cost or time. In fact, on average only one sample per day can be analyzed with this method making 2DE ineffective as a routine screening tool in breeding programs for quality selection. Also, like SDS-PAGE, it requires a degree of expertise.

3.1.3 Reversed-Phase High-Performance Liquid Chromatography

RP-HPLC is a type of high-performance liquid chromatography that allows the separation of proteins based on their hydrophobicity. Relative to the previously described gel-based methods, RP-HPLC is easy to perform, highly reproducible, quantitative, and rapid. However, it has not been widely used to identify different glutenin and gliadin alleles, mainly because of poor resolution and the high cost of the apparatus and reagents (Gao et al. 2010). In durum wheat, this technique has been used to discriminate the γ gliadins, γ-42 and γ-45 (Burnouf and Bietz 1984; Taha 1997), but has had limited use for differentiating glutenin alleles (Dong et al. 2009; Gao et al. 2010).

In recent years, reversed-phase ultraperformance liquid chromatography (RP-UPLC), a new technology based on RP-HPLC, has been developed (Wu 2006). Compared to RP-HPLC, more rapid separation, higher resolution, and greater sensitivity can be obtained by RP-UPLC analysis (Swartz 2005). RP-UPLC has received more and more attention and it has been successfully applied in food, chemical, and medicine analyses (Fekete et al. 2009; Zhou et al. 2007). Recently, a rapid and valid RP-UPLC method for water-soluble protein analysis in wheat grains has been established (Yu et al. 2012).

3.1.4 Matrix Assisted Laser Desorption/Ionization Time of Flight Mass Spectrometry (MALDI-TOF-MS)

MALDI-TOF-MS is a method of mass spectrometry that allows the accurate identification of the mass of a protein. In addition to its high accuracy and sensitivity, only a small sample is required (less than 1 pmol), and it is faster to perform (one minute per sample) than other common separation methods (Kussmann et al. 1997; Zheng et al. 2011). The analysis can be automated which makes it suitable for dealing with a large number of samples in a short time, ideal for wheat breeding programs or wheat grain trading, for instance. The use of MALDI-MS for the structural characterization of glutenin subunits appears particularly appropriate, because the presence of a large domain constituted of repeating sequence motifs makes the application of conventional sequencing procedures, such as Edman degradation, difficult. In addition, MALDI-MS provides an opportunity to determine whether post-translational modifications have occurred. Currently, the genome sequencing of many species and establishment of corresponding databases is facilitating the development of MS-based protein identification.

One of the first studies to apply this technique to the analysis of the gliadins and glutenins was by Dworschak et al. (1998), who demonstrated its efficiency in characterizing the HMW-GS profiles of both common and durum wheat. This suggested that MALDI-TOF-MS could be used in breeding programs for the rapid characterization of this class of prolamins. Successively, many more studies have been performed showing the reliability of this technique not only for the analysis of the HMW-GS (Liu et al. 2009; Zheng et al. 2011), but also for the identification of most of the gliadin and LMW-GS allelic variants (Cunsolo et al. 2004; Garozzo et al. 1999; Muccilli et al. 2005; Muccilli et al. 2011; Zhang et al. 2008). However; the high cost of the equipment and the fact that not all LMW-GS can be resolved due to their molecular weight similarity, methods like SDS-PAGE are still the preferred option for the routine analysis of prolamin allelic variation.

While MALDI-TOF-MS is often used for direct identification of (simple) protein mixtures, or in combination with 2DE, electrospray ionization (ESI) MS is a liquid-based method and is compatible with typical chromatographic separations of biosamples. ESI produces a range of charged species for each biomolecule, which increases the precision of mass assignments (Fenn et al. 1989; Kelleher et al. 1999). For glutenins and other proteins, most applications for ESI-MS involve protein identification by proteolytic digestion, followed by liquid chromatography and tandem MS of individual ions from the resulting peptide mixture. This 'bottom-up' or 'shotgun' approach has been used to verify HMW-GS sequences against their gene sequences (Cunsolo et al. 2003, 2004), wheat gluten composition (Qian et al. 2008), and even gluten structures (Lutz et al. 2012). This approach becomes challenging with increasing complexity of the protein mixture and post-translational modifications (Capriotti et al. 2011; Kelleher et al. 1999; Stephenson et al. 2002).

3.2 *Wheat Seed Proteome*

Seed tissues contain a large number of diverse proteins with different chemical characteristics, which necessitate special extraction and purification approaches to retain the quality of proteins for analysis on 2DE gels or by MS (Branlard and Bancel 2007). For cereal grains, the major storage components are starch and different carbohydrates, which have a detrimental effect on the extractability of proteins that are aggregated or coupled to other compounds, such as starch granules or cell wall elements. The solubility of proteins is therefore dependent on particle size and the homogeneity in size distribution achieved by grinding seed material. There are protocols based on water, salt, dilute alcohol, and weak acid/base soluble fractions for targeting specific cereal proteins (Görg et al. 2007).

The structure of the wheat grain can be separated into the outer layers, the germ, and the endosperm. The endosperm is the most important from a human consumption point of view, so not surprisingly is the fraction studied in most detail (Amiour et al. 2002; Dunbar et al. 1985; Dupont et al. 2011; Islam et al. 2002; Skylas et al. 2001; Tasleem-Tahir et al. 2012). The full spectrum of wheat endosperm proteins has been visualized in proteome studies involving the 2DE fractionation of the polypeptides (after rupture of disulfide bonds) followed by the identification of individual components. At least 1300 polypeptides have been counted, over 300 of which have been identified by N-terminal amino acid sequencing and matched to established protein database information (Skylas et al. 2000; Vensel et al. 2005).

3.2.1 HMW-GS

Techniques such as HPLC and lab-on-a-chip capillary electrophoresisas as well as mass spectrometry (MS) have been employed to identify and characterize the molecular weights of different HMW-GS with reasonable accuracy (Bean and Lookhart 2000; Gao et al. 2010; Zhang et al. 2008). Wheat prolamins, and especially HMW-GS, are large proteins and have long repetitive sequences with few tryptic cleavage sites, leading to peptide pools after digestion that are not ideal for tandem MS (Di Stefano et al. 2012). Hence, a 'top-down' approach in which intact proteins are measured and (partially) sequenced can be advantageous for primary structure determination and the detection of specific protein modifications (Kelleher et al. 1999). For top-down proteomics ESI-MS is preferred over MALDI as the ion source, because mono-charged protein ions, such as those generated by a MALDI source, cannot be detected at high resolution (Capriotti et al. 2011). A top-down approach using ESI-MS can reveal the transcriptome protein structure including the positioning of post-translational modifications (McLafferty 2011).

3.2.2 LMW-GS and Gliadins

Despite the abundance of LMW-GSs, they have received much less attention than the HMW-GSs, probably due to their complexity, heterogeneity and co-migration with gliadins in SDS-PAGE. In the SDS-PAGE system, gliadins have been used as markers, providing an indirect way to define LMW-GS alleles (Jackson et al. 1996). It has become apparent that the proteomics characterization of LMW-GS extracts is very challenging. Indeed, these subunits are encoded by many genes with similar structural characteristics. Most of these genes have not been isolated and characterized yet, making the application of standard proteomic techniques and the determination of a direct correspondence between a gene sequence and its encoded protein difficult. Furthermore, it should be noted that in LMW-GS characterization, even though cleavage enzymes such as chymotrypsin and thermolysin have been tested (Vensel et al. 2007), the use of trypsin remains widespread. Therefore, as the amino acid sequences of LMW-GSs have a long repeating motif and a scarcity of cleavable tryptic sites, the number of the resulting tryptic peptides suitable for MS analysis is scant. A possible way to identify LMW-GSs is to perform MS/MS experiments on the few available tryptic peptides and try to interpret the results obtained in the light of the peculiar primary structure of this class of proteins. Specifically, it should be considered that: (i) sequence coverage will obviously be low owing to the long repeating motif lacking tryptic cleavage sites, (ii) the few peptides with masses suitable for MS/MS fragmentation derived from the N-terminal and C-terminal domains would not be able to be used to discriminate the single subunits because they are common to LMW-GS groups, and (iii) a single 2DE gel spot may frequently contain more proteins belonging to the same group of LMW-GS, but differing by point substitutions.

Despite these intrinsic limitations, several studies on LMW-GSs by both gel-free and gel-based methodologies have been performed. RP-HPLC has proven to be a highly efficient tool for the qualitative and quantitative investigation and isolation of intact gliadins and LMW-GS (Huebner and Bietz 1985; Marchylo et al. 1989; Wieser et al. 1998), sometimes in combination with ESI-MS (Mamone et al. 2000; Muccilli et al. 2005). Comparative analysis of the alkylated and non-alkylated B- and C-type LMW-GS fractions allowed the detection of 150 different components and a reliable determination of the number of cysteine residues present in 42 proteins, which were then tentatively classified based on their molecular mass and number of cysteine residues. Interestingly the results showed the extensive micro-heterogeneity of LMW-GSs, as several components were detected with minor differences in the relative molecular masses co-eluting in the same HPLC peak, indicating a higher complexity than expected from genomic analysis (Muccilli et al. 2005).

Direct MALDI-MS analysis of unfractionated LMW extracts usually shows a complex pattern of proteins in the 30–40 kDa range. The observed pattern may be suitable only for differentiating between wheat varieties, but the complexity of the mass spectra precludes the use of MALDI-MS as a stand-alone technique for the identification of individual components (Dworschak et al. 1998). In this respect, MALDI-TOF-MS has often been employed after the chromatographic isolation of LMW-GS extracts as a tool to determine the molecular mass of new subunits (An et al. 2006; Masci et al. 1995) and the number of cysteine residues (Masci et al. 1998).

An off-line combination of HPLC, MALDI-MS, tryptic digestion, MS/MS analysis, and database searching was performed on the entire gluten extract of a Canadian hard red spring wheat (Qian et al. 2008). The data obtained by this approach demonstrated that while the HMW-GSs were easily identified by means of their intact masses and their tryptic fragments according to the published cDNA sequences, the identification of the LMW-GS was limited for two reasons. (i) Several subunits have intact molecular masses too similar to be resolved by linear MALDI-TOF MS, and (ii) the DNA sequences of several LMW subunits are very similar (Ikeda et al. 2002) so most of the resulting tryptic peptides are identical. As a result, the data obtained do not provide definitive evidence that all of the expected subunits are indeed present.

All the proteomic studies performed on LMW-GSs by means of gel-based techniques encompass both the problems of achieving good electrophoretic separation for a class of proteins with similar electric charges and molecular masses, and the difficulties identifying proteins. The aim of these studies is not restricted to cultivar characterization (Mamone et al. 2009), but also includes efforts to better understand the role of LMW-GSs in the gluten matrix (Lindsay and Skerritt 1999) and seed maturation by comparing common wheat cultivars and translocated or transgenic ones (Scossa et al. 2008).

Recent studies on LMW-GS have focused on the role of these proteins in the gluten macropolymer and, consequently, on flour and semolina functionality by means of comparative proteomic analysis (2D–PAGE, in-gel digestion, and MS/MS analysis) on the LMW fraction of the durum wheat cultivar Svevo and of its derived 1BL.1RS translocated line (Gobaa et al. 2007, Muccilli et al. 2010). This translocation involves the substitution of the LMW and gliadin components associated with the *Glu-B3/Gli- B1* loci with monomeric v- and 40 kDa g-secalins, which are encoded by closely linked genes present at the *Sec-1* locus on the short arm of the chromosome 1R (Carrillo et al. 1990). No proteins homologous to the wheat LMW-GSs are present on this chromosome arm. The peptide fragment fingerprint approach, combining database searching of the MS/MS data and manual interpretation of tandem mass spectra, permitted the detection and characterization of almost all the N-terminal tryptic peptides of the B-type LMW-GSs identified. The conclusion was that all three types of typical LMW-GS, LMW-s, LMW-m, and LMW-I, are present in Svevo, whereas the latter group is the only one present when the *Glu-B3* locus is replaced by the short arm of the chromosome 1R (Muccilli et al. 2010). On the other hand, a comparative proteomic analysis performed on 16 doubled haploid lines, with or without the 1BL.1RS translocation, revealed quantitative and qualitative variations in prolamins and other endosperm proteins (Gobaa et al. 2007). Of particular interest was the identification of a γ-gliadin carrying nine cysteine residues, which was highly over-expressed in the 1BL.1RS translocated genotypes. This evidence suggests that the lack of LMW-GS is counterbalanced by an over-expression of a relatively similar prolamin. Indeed, the finding of a gliadin with an odd number of cysteines suggests that at least one residue is not involved in an intra-chain disulfide bond, but is available for covalent bonding to other polymeric glutenin subunits. The up-regulation of a γ-gliadin carrying nine cysteine residues upon the loss of an LMW-GS locus would mean that the glutenin polymerization is a regulated process. This process controls the amount of polymeric gluten subunits in order to maintain a certain level of polymerization in response to chromosomal rearrangement at their corresponding loci. A better understanding of these mechanisms may lead to further improvements in the rheology of the dough and in the quality of wheat.

4 Molecular Markers

Among the different factors that influence semolina quality, gluten strength is one of the most important, affecting pasta firmness and texture (Subira et al. 2014). Gluten strength is determined by both the quantity and quality of prolamins. Prolamin quantity is highly influenced by the environment and management practices, has low heritability, and is therefore difficult to breed for (Clarke et al. 2009). Conversely, prolamin 'quality' is determined by the allelic variation in glutenins and gliadins. To investigate potential impacts on rheology and end-product quality, several methods for the identification of specific prolamin alleles associated with strong or weak gluten have been developed and exploited for the manipulation of gluten composition (Liu et al. 2010; Rasheed et al. 2014). In general, prolamin allelic variation can be detected either on the basis of their relative differences at the protein level (molecular weight, isoelectric point, relative abundance, etc.) or on the basis of differences at the genic level (gene sequence polymorphisms).

4.1 Biochemical Markers

The first molecular markers used to discriminate between weak and strong gluten in durum wheat were the two γ gliadins γ-42 and γ-45, characterized on the basis of their relative mobility in A-PAGE (Bushuk and Zillman 1978). The associations between strong gluten and γ-45 and between weak gluten and γ-42 was first identified in pioneering work by Damidaux et al. (1978). This discovery greatly accelerated the process of durum wheat breeding for quality and brought an improvement in the overall technological quality of durum wheat varieties released in the years that followed. Indeed, all durum wheat varieties released into the Canada Western Amber Durum (CWAD) class after 1978 exhibited the strong gluten molecular marker γ-45 (Dexter 2008). Thanks to the improved quality of the CWAD released varieties, Canada became one of the major suppliers of durum wheat in high quality markets and resulted in a total annual production of around two million metric tons by the late 1970s. Similarly, characterization of varieties with these two markers was introduced in European breeding programs after the release of the French high-yielding semi-dwarf variety Durtal. Even though Durtal was viewed as a success from an agronomic standpoint, its end-use quality characteristics were so poor that the processing industries did not want to use it. Soon it was shown that Durtal possessed the γ-42 gliadin associated with weak gluten and later, breeding material with this marker was slowly eliminated from nursery stock.

As explained in Sect. 2, the genes encoding the γ gliadins are tightly linked with the genes encoding the ω gliadins and the LMW-GSs. For this reason, the γ-42 gliadin was found to be linked with the ω-33, 35 and 38 gliadins, and with the LMW-1 group. Similarly, the γ-45 gliadin was linked with the ω-35 gliadin and with the LMW-2 group (Payne et al. 1984b). After further studies it was found that the two γ gliadins pinpointed by Damidaux et al. (1978) were not the actual reason for the observed differences in gluten strength and that these differences were instead caused by different LMW-GS protein profiles (Payne et al. 1984b; Pogna et al. 1988; Ruiz and Carrillo 1994). Also, three additional LMW-GS patterns associated with the γ-45 gliadin (LMW-2), the γ-44 gliadin (LMW-2*), and with the gliadin γ-42 (LMW-1) were identified and associated with different degrees of gluten strength (Carrillo et al. 1990), further supporting the consensus that the LMW-GS were the major factors responsible for changes in gluten strength in durum wheat, and not the γ gliadins. For this reason, in the intervening years, even though the analysis of the γ gliadins remained an effective marker-based selection system, most studies were focused on the LMW-GS protein profiles of different durum wheat varieties and on the best ways to characterize and classify them (Aguiriano et al. 2008; Babay et al. 2015; Fois et al. 2011; Nieto-Taladriz et al. 1997).

4.2 DNA-Based Molecular Markers

Although biochemical markers are currently most frequently employed for prolamin analysis, DNA-based markers are preferred because they have several advantages. They are easy to use, inexpensive, give rapid results, require a minimum amount of DNA, and can be applied at any stage of the plant life cycle (Kiszonas and Morris 2018; Kumar 1999).

4.2.1 Hybridization-Based Markers

Hybridization-based markers are a class of DNA markers able to detect sequence polymorphism on the basis of the selective hybridization of a labelled DNA fragment (probe) to the genomic DNA of interest. The two main classes of hybridization-based markers are the restriction fragment length polymorphism (RFLP) markers and Diversity Array Technology (DArT) markers. RFLP markers can be considered the first DNA-based markers and consist of genomic DNA fragments produced by the cleavage of sequence specific restriction endonucleases, which are size-separated by gel electrophoresis followed by transfer to a solid support matrix, and hybridization to a labelled probe of known sequence. Even though RFLP markers associated with different *Glu-1* and *Glu-3* alleles have been identified in several QTL studies (Manifesto et al. 1998; Blanco et al. 1998; Campbell et al. 2001), they have not been successfully used for routine screening of the prolamin profile. RFLP markers are relatively expensive, time-consuming, and require expert personnel.

DArT is a method developed for high-throughput characterization of a given genotype based on the hybridization of DNA samples to solid-state platforms, and is able to detect sequence polymorphisms by recording the presence or absence of DNA fragments in genomic DNA. Because of its efficiency and its high genome coverage, DArT has been applied to durum wheat in numerous studies (Colasuonno et al. 2013; Laidò et al. 2013; Mantovani et al. 2008). However, as DArT exclusively generates dominant markers, other genotyping methods are often preferable.

4.2.2 PCR-Based Markers

PCR is a technique that allows the selective amplification of specific segments of DNA. Since its development in 1987 by Mullis, it has revolutionized the field of genetics and has underpinned the development of most of the DNA-based molecular markers typically employed for marker-assisted selection. Compared to protein-based markers, PCR-based markers have several advantages. They are easy and rapid to use including in automated systems, and they are accurate, economical, and can be applied at any stage of the plant life cycle (Kiszonas and Morris 2018; Kumar 1999). Also, with the advancement in molecular biology and analytical technologies, several systems have been developed in order to improve the accuracy and reduce the cost of gene polymorphism detection, such as PCR capillary electrophoresis, especially useful for the discrimination of DNA sequences based on their length polymorphism, and Kompetitive Allele Specific PCR genotyping system (KASP), mainly used for the detection of single nucleotide polymorphisms (SNPs).

Difficulties in genomic and molecular research on prolamins are due to (i) wheat being a polyploid with a large and complex genome, (ii) the high degree of homology among all glutenin genes and with gliadins, (iii) the high frequency of pseudogenes and repeat sequences, and (iv) the lack of a complete genomic reference sequence. Nevertheless, based on the nucleotide sequence of cloned prolamin genes, several gene-specific markers targeting the most common HMW-GS allelic variants have been developed (Table 2). Even though together these markers can detect the most common *Glu-A1* and *Glu-B1* alleles present in durum wheat varieties, only a few studies have used them for HMW-GS allelic characterization (Henkrar et al. 2017; Janni et al. 2017). The primary reason for this limited uptake is that several PCR reactions are often needed in order to generate a complete haplotype of the HMW-GS alleles. Consequently, SDS-PAGE is still more time-efficient and thus preferred for the HMW-GS allelic characterization in durum wheat.

Table 2 PCR markers targeting the most common HMW-GS allelic variants

Locus	Allele identified	Reference
Glu-A1	AxNull	Lafiandra et al. 1997
	Ax2*	Ma et al. 2003
	Ax2··	Giraldo et al. 2010
	Ax2* vs either the AxNull or the Ax1	Liu et al. 2008
Glu-B1	Bx20, Bx7oe, and Bx7	Butow et al. 2003
	Bx17 and Bx7*	Ma et al. 2003
	Bx7* and Bx7oe	Radovanovic and Cloutier 2003
	Bx17, Bx7, and Bx7oe	Butow et al. 2004
	Bx6	Schwarz et al. 2004
	Bx7oe	Ragupathy et al. 2008
	Bx14 and Bx17	Xu et al. 2008
	Bx7 vs Bx7*	Espí et al. 2012
	By8, By9, By16, and By20	Lei et al. 2006
	By8	Yan et al. 2009
	By8 and By18	Liang et al. 2015
	By8, By8*	Uthayakumaran et al. 2006

In the case of LMW-GS, allele specific markers are also available for most of the *Glu-A3* (Wang et al. 2010; Zhang et al. 2004) and *Glu-B3* alleles (Wang et al. 2009). These markers have been developed based on SNPs identified in specific LMW-GS gene sequences of common wheat and have been proven to efficiently detect most of the *Glu-A3* and *Glu-B3* alleles in durum wheat varieties (Henkrar et al. 2017; Igrejas et al. 2009). However, these sets of markers only captured the polymorphism of a few of the LMW-GS genes and they are not time-efficient since they require seven (Zhang et al. 2004), ten (Wang et al. 2009), or four (Wang et al. 2010) PCR reactions to characterize the LMW-GS allelic profile of a durum variety. More recently, different sets of LMW-GS markers have been developed based on the length polymorphism related to the repetitive region of each LMW-GS gene (Zhang et al. 2011). By using these markers, most of the LMW-GS genes of a common wheat variety could be amplified and differentiated thus allowing not only a better understanding of the genic complexity of the *Glu-3* loci, but also to associate specific LMW-GS gene profiles with the observed *Glu-3* allelic variants. These discoveries led to the development of a set of molecular markers able to effectively characterize most of the LMW-GS genes at each haplotype and to effectively identify the majority of the *Glu-A3* and *Glu-B3* alleles by using as few as one PCR reaction per locus (Ibba et al. 2018). However, it should be noted that the usefulness of these molecular markers has not yet been tested on durum wheat varieties.

4.2.3 Sequencing-Based Markers

Due to the development of next-generation sequencing technologies, genomic research and the selection of superior genotypes have been completely revolutionized. In general, next generation sequencing methods simultaneously sequence thousands to millions of DNA molecules allowing the genetic characterization of a wheat variety based on the SNPs across its genome in a cost and time-effective way. At present, different wheat germplasm collections are most commonly genotyped either through high density SNP arrays ('SNP chips') or through genotyping-by-sequencing (GBS). For wheat, three different gene-based SNP arrays are available, each differing in the number of SNP-based markers assembled (9 K, 90 K or 820 K) (Cavanagh et al. 2013; Wang et al. 2014; Winfield et al. 2016). Apart from being a high-throughput and reliable technique for wheat genotyping, the main advantages of using this technology are that the SNPs detected in the arrays are evenly distributed across the genome and are located in known genes. All of the wheat SNP chips have been used for genetic mapping (Maccaferri et al. 2015), QTL analysis (Colasuonno et al. 2014) and, more recently, for genomic selection in durum wheat (Haile et al. 2018). In the latter study, specifically, by using the Infinium iSelect 90 K SNP chip it was possible to predict gluten quality with up to 80% accuracy, indicating that this method could be a valid alternative to glutenin allele-specific markers for early screening of gluten quality. However, it should be noted that the cost of these assays is still high. Unless there is a significant decrease in the cost per sample, SNP chips are unlikely to be routinely used in wheat breeding programs.

GBS, on the other hand, is a de novo SNP discovery technique that begins by reducing the complexity of the genome through the use of specific restriction endonucleases, followed by sequencing of the resulting DNA fragments. As in the case of SNP chips, GBS has been used for both genetic mapping and genomic prediction (Fiedler et al. 2017; van Poecke et al. 2013). Specifically, Fiedler et al. (2017) evaluated the potential of using GBS for genomic prediction of gluten quality in a durum wheat breeding program, and with a total of 1699 GBS-derived markers they predicted gluten quality (as indicated by sedimentation volume) with 66% accuracy. Compared to the SNP chip array, the cost of GBS is much lower and allows the potential discovery of new markers. However, GBS has a high percentage of missing data, both in terms of individual wheat lines and specific loci, which can affect the robustness of the analysis.

An alternative to these methods that combines the accuracy of allele-specific markers with the high throughput of next-generation sequencing is tagged amplicon sequencing. This technique consists of a multiplexed PCR where multiple markers of interest are amplified and 'barcoded', followed by sequencing (Onda et al. 2018; Rife et al. 2015; Schnable Lab 2015). This method has been effectively applied in common wheat by Bernardo et al. (2015) who were able to multiplex 27 different markers targeting genes associated with both agronomic and economically important traits. Among these markers, two HMW-GS allele specific markers were included and successfully characterized. Even though tagged amplicon sequencing is still in its infancy for wheat, recent results (Bernardo et al. 2015) show that this method could effectively be used for marker assisted selection and, more specifically, could provide a more rapid and less expensive alternative for the simultaneous characterization of the prolamin allelic variants in durum wheat.

5 Contribution of Wheat Protein Composition to Quality

Improving the technological quality of durum wheat semolina has always been a central objective in durum wheat breeding programs (Clarke et al. 1998). Grain quality of durum wheat is a complex characteristic that includes several components. For pasta production, good cooking quality is assured by the formation of a continuous and strong network of denatured gluten proteins, which entraps the starch granules, limiting how much they swell and solubilise in boiling water. According to Shewry (1999) the impact of the environment on the synthesis of gluten proteins is essentially due to the prevalent temperature and availability of water during grain development. It is clear that the technological properties of wheat flour giving rise to the different end uses results from an association between genetic factors and environmental factors. However, different studies showed that if a certain wheat cultivar possesses some certain prolamin allele combinations at crucial loci, then it may exhibit valuable qualitative traits in terms of the end-product.

Testing for durum wheat quality is expensive and time-consuming. For this reason, typically only a limited number of lines are actually phenotyped for end-use quality mostly in later generations, so some lines of unacceptable quality are likely to be advanced (Fiedler et al. 2017). The effects of glutenin alleles on quality can be evaluated in unrelated cultivars or in segregating populations. In the former case, a sufficiently large number of cultivars with the same allele constitution should be compared to avoid confounding effects from other loci. Segregating lines from a cross with random recombinant inbred lines (RILs) is a better approach to estimate the genetic effects of particular loci, but more effort is required to produce this type of material.

5.1 Contribution of Wheat Protein Composition to Gluten Strength

The rheological properties of pasta depend on the viscoelastic gluten properties, with gluten strength being the main factor in pasta quality (Kovacs et al. 1994; Marchylo et al. 1998). The relationships between glutenin alleles and gluten strength, especially as determined by the SDS- sedimentation test (SDSST), has been analysed in different studies. This quality test is an indirect measure of gluten strength based on the extent of aggregation/precipitation of the gluten polymer. It is especially useful when evaluating breeding populations where grain quantity is limiting. Some studies have shown that the SDSST values are highly correlated with mixing development time measured with the Mixograph (MDT) (Babay et al. 2015; Brites and Carrillo 2001; Martínez et al. 2005; Ruiz and Carrillo 1995) and strength measured with the Alveograph (Brites and Carrillo 2001).

5.1.1 Contribution of HMW Glutenin Alleles

Studies of the relationships between HMW-GS coded by *Glu-1* and durum wheat quality led to the conclusion that the effects of HMW-GS were less significant than those of LMW-GS on gluten strength (Aguiriano et al. 2009; Ruiz and Carrillo 1995; Vázquez et al. 1996). In general, the associations between *Glu-1* alleles and gluten strength were either weak or contradictory, although the results may have been hampered by the reduced genetic variability at the *Glu-1* loci in modern durum wheat cultivars (Sissons et al. 2005). Some divergent results may also be due to interactions among HMW-GS and LMW-GS, so the effects of a given allele depend on the overall glutenin combination with which the given gene product interacts (Martínez et al. 2005; Ruiz and Carrillo 1995). Usually, LMW-GS have a dominant influence over the HMW-GS, making any main HMW-GS effect highly dependent on the LMW-GS background (Peña and Pfeiffer 2005). However, *Glu-1* allelic variants should also be considered when analysing the *Glu-3* contribution to gluten strength because some *Glu-1* alleles can significantly affect the average value measured for that *Glu-3* allele.

Glu-A1 Locus

The most studied alleles coded by *Glu-A1* are allele *a* (subunit 1), allele *b* (subunit 2∗) and allele *c* (Null subunit) (Fig. 1). Du Cros (1987) and Nazco et al. (2014) showed that none of these three alleles had a significant effect on gluten strength, but other authors found a positive relationship between *Glu-A1* alleles and gluten strength (Brites and Carrillo 2001; Ciaffi et al. 1995). Most of the studies which reported a significant effect of the *Glu-A1* alleles agreed that allele *c* has a negative influence (Ciaffi et al. 1991; du Cros 1987; Kaan et al. 1993; Porceddu et al. 1998; Turchetta et al. 1995), which was associated with lower values of gluten strength than alleles *a* and *b* (Brites and Carrillo 2001; Oak et al. 2004; Raciti et al. 2003; Ruiz and Carrillo 1995). Oak et al. (2004) reported that allele *b* was associated with higher gluten strength than allele *a.*

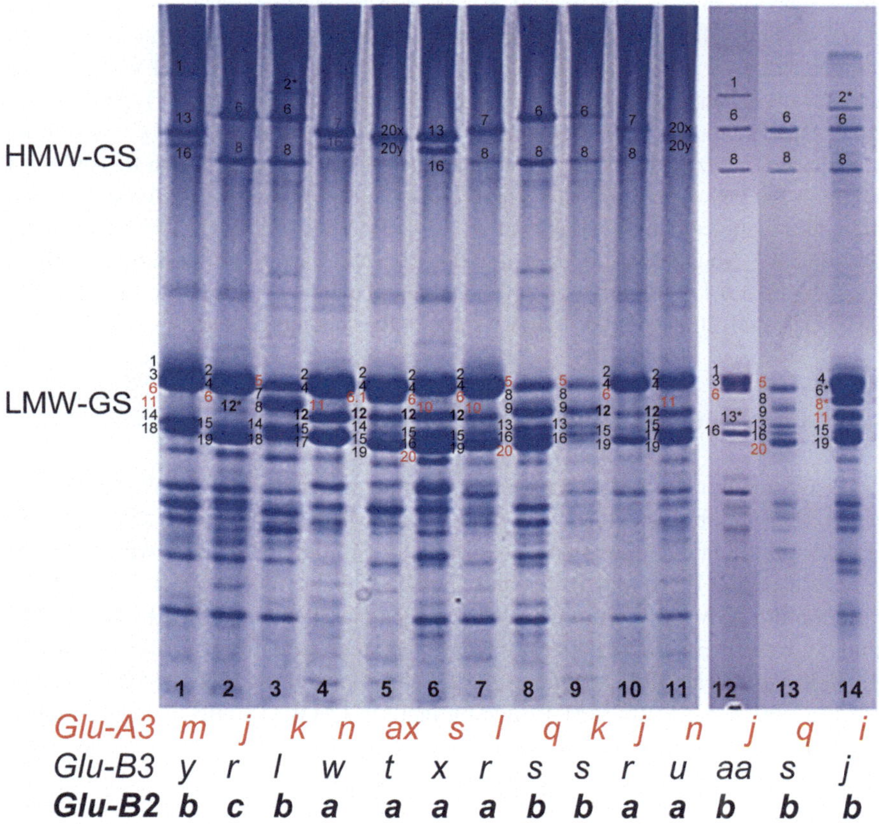

Fig. 1 SDS-PAGE of high-molecular-weight (HMW) and low-molecular weight (LMW) glutenin subunits of some wheat varieties. From left to right the varieties are Alaga, Alcala la Real, Andalucia-344, Ardente, Buck Cristal, Claro de Balazote, Cocorit, Fanfarrón, Langdon, Mexicali, Mundial, Blancal de Nules, Fanfarrón, and Mourisco Fino. The LMW glutenin alleles and encoded subunits are indicated for each variety with red, black and bold black symbols for *Glu-A3*, *Glu-B3,* and *Glu-B2* loci, respectively

Glu-B1 Locus

According to Boggini and Pogna (1989) the *Glu-B1* genotype is more relevant for gluten strength than active *Glu-A1* alleles. Turchetta et al. (1995) and Porceddu et al. (1998) found that HMW-GSs encoded by *Glu-B1* were responsible for 8% of the variation in gluten strength, while other studies have shown no influence of these proteins (Carrillo et al. 1990; Vázquez et al. 1996). Several studies agreed about the lower gluten strength values associated with allele *e* (subunits 20x + 20y) as opposed to other alleles such as allele *b* (subunits 7 + 8), *d* (subunits 6 + 8), *h* (subunits 14 + 15), and *z* (subunits 7 + 15) (Babay et al. 2015; Boggini and Pogna 1989; Brites and Carrillo 2001; Kaan et al. 1993; Magallanes-López et al. 2017; Oak et al. 2004; Raciti et al. 2003; Ruiz and Carrillo 1995; Sapirstein et al. 2007; Trad et al. 2014). The low gluten strength associated with HMW-GS 20x + 20y could be explained by the low proportion of unextractable polymeric protein (Edwards et al. 2007), a fraction positively associated with gluten strength (Gianibelli et al. 1995; Sissons and Batey 2003; Sissons et al. 2005). In contrast, Nazco et al. (2014) found a positive effect of allele *e* on SDSST values. Other alleles with negative effects on gluten strength were allele *a* (subunit 7) (Nazco et al. 2014) and f (subunits 13 + 16) (Babay et al. 2015; Oak et al. 2004). Conversely, Magallanes-López et al. (2017) found that allele f was associated with the strongest gluten in the genotypes analysed. The *Glu-B1* alleles with positive effects on gluten strength are alleles *b* and *d* (Babay et al. 2015; Boggini and Pogna 1989; Brites and Carrillo 2001; Magallanes-López et al. 2017; Ruiz and Carrillo 1995; Trad et al. 2014), *bc* (subunits 6 + 17) (Raciti et al. 2003) and *h* (subunits 14 + 15) (Brites and Carrillo 2001; Oak et al. 2004). Figure 1 shows some of the subunits encoded by *Glu-B1*.

5.1.2 Contribution of LMW Glutenin Alleles

LMW-GS are more important contributors to gluten strength than HMW-GS. Sissons et al. (2005) reported that the LMW-GS and HMW-GS accounted for 55% and 27% of the variation in the gluten index, respectively. Similar conclusions were made by Porceddu et al. (1998) and Brites and Carrillo (2001). With respect to the allelic variation in LMW-GS, durum wheats have been classified according to their LMW glutenin patterns and these patterns have been related to quality (Carrillo et al. 1990, 1991; Kovacs et al. 1995; Payne et al. 1984b; Pogna et al. 1988, 1990; Ruiz and Carrillo 1995). Since the LMW models include LMW-GS encoded by chromosomes 1A and 1B, the effect of the allelic variants at the *Glu-A3, Glu-B3,* and *Glu-B2* loci on pasta quality has been studied separately.

Glu-A3 Locus

Different studies have shown that *Glu-A3* alleles have a low influence on gluten strength (Aguiriano et al. 2009; Ruiz and Carrillo 1995; Vázquez et al. 1996). Magallanes-López et al. (2017) reported a significant effect of *Glu-A3* alleles on gluten quality, but in all cases smaller than effects of the environment and *Glu-B3* and *Glu-B1* alleles. Based on the results obtained in several studies, it was possible to divide the *Glu-A3* alleles present in wheat varieties into two groups (Table 3), the alleles related to higher values in the SDSST and MDT (positive effect on gluten strength), and the alleles related to lower values (negative effect on gluten strength). Some of these alleles are shown in Fig. 1. Some discrepancies regarding allele rank have arisen. The positive effect of allele *j* contrasts with the results reported by Martínez et al. (2005) and Aguiriano et al. (2009), who found weaker gluten to be associated with this allele. Also, Sissons et al. (2005) recommended avoiding the *l* allele if high gluten strength is required. Allele *s* is associated with better MDT values than allele *m* according to the results of Ruiz and Carrillo (1995). The significant negative effect of allele *k*, however, was a consistent result found in several studies.

Glu-B3 Locus

Different studies have reported that allelic variants at *Glu-B3* had a much greater effect on gluten quality than the variants at *Glu-A3* and *Glu-B2* (Brites and Carrillo 2001; Carrillo et al. 2000; Magallanes-López et al. 2017; Martínez et al. 2005; Ruiz and Carrillo 1995; Vázquez et al. 1996). Based on the results of several studies, *Glu-B3* alleles were classified into three groups according to the values measured in the SDSST and MDT (Table 4), the alleles associated with higher values (positive effect on gluten strength), the alleles related to lower values (negative effect on gluten strength), and those related to intermediate values. Most of these alleles are shown

Table 3 *Glu-A3* allele classification based on their positive or negative effect on gluten strength evaluated with the SDS sedimentation test or the Mixograph mixing development time. Alleles are ranked from stronger to weaker effects according to the results cited. The LMW-GS controlled by the *Glu-A3* allele in each case are also indicated

Effect	Allele 2013	Allele	LMW-GS	References
Positive	*Ax*	*ax*	6.1	Babay et al. 2015
	p	*h*	Null	Carrillo et al. 2000; Magallanes-López et al. 2017
	l	*c*	6 + 10	Carrillo et al. 2000
	j	*a*	6	Carrillo et al. 2000; Nazco et al. 2014; Magallanes-López et al. 2017
	m	*d*	6 + 11	Carrillo et al. 2000; Nazco et al. 2014; Magallanes-López et al. 2017
		new1	5* + 11 + 20	Aguiriano et al. 2009
Negative	*n*	*e*	11	Carrillo et al. 2000; Aguiriano et al. 2009
	o	*f*	6 + 11 + 20	Carrillo et al. 2000
	s	*g*	6 + 10 + 20	Carrillo et al. 2000
	k	*b*	5	Carrillo et al. 2000; Aguiriano et al. 2009; Sissons et al. 2005; Nazco et al. 2014; Magallanes-López et al. 2017

Table 4 *Glu-B3* allele classification based on their positive, intermediate or negative effect on gluten strength evaluated with the SDS sedimentation test or the Mixograph mixing development time. Alleles are ordered from stronger to weaker effects according to the results cited. The LMW-GS controlled by the *Glu-B3* allele in each case are also indicated

Effect	Allele 2013	Allele	LMW-GS	References
Positive	*t*	*c*	2 + 4 + 14 + 15 + 19	Vázquez et al. 1996; Carrillo et al. 2000; Brites and Carrillo 2001; Magallanes-López et al. 2017
	j	*j*	4 + 6* + 15 + 19	Brites and Carrillo 2001
	r	*a*	2 + 4 + 15 + 19	Ruiz and Carrillo 1995; Vázquez et al. 1996; Carrillo et al. 2000; Martínez et al. 2005; Nazco et al. 2014; Magallanes-López et al. 2017
Intermediate	*x*	*g*	2 + 4 + 15 + 16	Ruiz and Carrillo 1995; Carrillo et al. 2000
	u	*d*	2 + 4 + 15 + 17 + 19	Carrillo et al. 2000
	y	*h*	1 + 3 + 14 + 18	Carrillo et al. 2000; Ruiz et al. 2018
	w	*f*	2 + 4 + 15 + 17	Ruiz and Carrillo 1995; Carrillo et al. 2000; Magallanes-López et al. 2017
Negative	*s*	*b*	8 + 9 + 13 + 16	Ruiz and Carrillo 1995; Vázquez et al. 1996; Carrillo et al. 2000; Brites and Carrillo 2001; Martínez et al. 2005; Sissons et al. 2005; Magallanes-López et al. 2017
		i	7 + 8 + 14 + 18	Vázquez et al. 1996; Carrillo et al. 2000; Sissons et al. 2005
	Aa	*l*	1 + 3 + 13* + 16	Martínez et al. 2005
	k	*k*	8 + 9 + 13 + 16 + 19	Brites and Carrillo 2001
		new1	1 + 3 + 13* + 19	Aguiriano et al. 2009
	v	*e*	2 + 4 + 15 + 16 + 18	Carrillo et al. 2000

in the protein profiles in Fig. 1. Some discrepancies regarding allele rank were noted. For example, Magallanes-López et al. (2017) reported better values associated with allele *r* than with allele *t*, in contrast to the results obtained by Brites and Carrillo 2001, who found that allele *r* had worse values than alleles *t* and *j*. Sissons et al. (2005) found that allele *w* had a negative effect on gluten strength.

The main difference between the LMW-1 and LMW-2 patterns, associated with low and high gluten strength values respectively, is the presence of a strongly expressed *Glu-B3* coded protein band in the latter. This slowly migrating LMW-GS corresponds to the 42K LMW-GS (Masci et al. 2000) and is consistently the most abundant LMW-GS polypeptide (Masci et al. 1995). When the LMW-2 protein pattern is fractionated by the one-step SDS-PAGE procedure, this band is separated into the *Glu-B3* encoded subunits 2 and 4. These subunits are present in germplasm with the *Glu-B3* alleles *r, t, u, v, w,* and *x* (Nieto-Taladriz et al. 1997) and in other allelic variants recently reported (Ruiz et al. 2018). It is known that the presence of subunit 2 usually has a positive influence on gluten strength. However, alleles coding for this subunit can be associated with low gluten strength, such as allele *v*, whereas other alleles, such as allele *j* also linked to γ-45 but not coding for subunit 2, can have a positive effect on gluten strength (Carrillo et al. 2000; Brites and Carrillo 2001).

Glu-B2 Alleles

Three alleles at *Glu-B2* have been described so far, allele *a* (subunit 12), allele *b* (Null subunit) and allele *c* (subunit 12∗). No significant differences between alleles *a* and *b* were found (Babay et al. 2015; Brites and Carrillo 2001; Magallanes-López et al. 2017; Martínez et al. 2005; Ruiz and Carrillo 1996). In contrast, different studies have reported that allele *a* was associated with stronger gluten than allele *b* (Aguiriano et al. 2009, Martínez et al. 2005, Nazco et al. 2014) and allele *c* (Martínez et al. 2005). In contrast, Sissons et al. (2005) found that cultivars having allele *a* produce weak gluten.

5.2 Contribution of Wheat Protein Composition to Bread Quality

An appropriate balance between resistance to extension and extensibility, in conjunction with increased dough strength, is needed to breed durum wheat cultivars with loaf volume equivalent to that of bread wheat. However, a deficiency in dough extensibility is noted in durum wheat for breadmaking (Abecassis et al. 2012). Magallanes-López et al. (2017) reported that alleles coded by *Glu-B1* were the most significant for gluten extensibility evaluated by loaf volume and relevant Alveograph parameters. For this locus, several studies concur on the negative effect of allele *Glu-B1e* (Ammar et al. 2000; Edwards et al. 2007; Magallanes-López et al. 2017; Sapirstein et al. 2007) and the positive effect of allele *Glu-B1d* (Ammar et al. 2000; Magallanes-López et al. 2017). Magallanes-López et al. (2017) also reported that allele *Glu-B1f* was associated with the best breadmaking properties in durum wheat. It also seems that the presence of any subunit encoded by *Glu-A1* was favourable for bread making (Boggini et al. 1994). Considering LMW-GS composition, the LMW-2 pattern was associated with higher dough strength and higher loaf volumes than the LMW-1 (Ciaffi et al. 1995; Edwards et al. 2003).

Different studies have analysed the effects of added protein fractions to reconstituted flours and to a base semolina. Gliadin addition to dough resulted in weaker mixing curves but it increased the viscous nature of dough (Edwards et al. 2003, Sissons et al. 2005). Addition of glutenin to the base semolina increased the overall dough strength properties. Also, the higher expression of B-type LMW-GS was correlated with a higher gluten index (De Santis 2017), while increasing proportions of HMW-GS related to LMW-GS were generally associated with weak dough and gluten (Edwards et al. 2007).

5.3 Contribution of Wheat Protein Composition to Protein Content

Different studies have agreed that glutenin composition only weakly influences protein content (Brites and Carrillo 2001; Martínez et al. 2005; Ruiz and Carrillo 1995; Vázquez et al. 1996). Only a few studies found some *Glu-3* alleles to have significant effects on protein content. For *Glu-A3*, Aguiriano et al. (2009) reported a positive effect of the *new1* (5∗ + 11 + 20) allele relative to the *e* allele. For *Glu-B3*, Martínez et al. (2005) reported a positive effect of allele *s* relative to allele *r*. Contradictory results have been reported for HMW-GS, as Kaan et al. (1993) found a negative effect of *Glu-A1c* and no significant effect of *Glu-A1a* and *b*, whereas Aguiriano et al. (2009) found a negative influence of *Glu-A1b* and a positive influence of *Glu-A1a* and c. Both studies found no significant effects between the *Glu-B1b, d, e,* and *f* alleles.

6 Future Perspectives

The recently released bread and durum wheat reference genomes (Appels et al. 2018. Maccaferri et al. 2019) will facilitate molecular and genomic research on prolamins, especially on the complex LMW-GS family. Knowledge of the complete wheat genome will be an important asset to breeders, as specific traits can be linked with sequence polymorphisms, and new quality markers can be identified. However, it is unlikely that the manipulation of individual genes will be sufficient to study wheat gluten. The full complement of genomics and proteomics appears to be essential to understanding the genes, their products, and how those products interact to confer rheological properties. In the genomic era, proteomics is a powerful tool to investigate the expression, diversity, and interactions of gluten proteins, major determinants of the technological quality of wheat. The resulting knowledge will contribute to the strategic conservation of wheat genetic resources and improve and accelerate wheat breeding to meet the challenges of the twenty-first century (Ribeiro et al. 2013).

References

Abecassis J, Cuq B, Boggini G, Namoune H (2012) Other traditional durum-derived products. In: Sissons MJ, Carcea M, Marchylo M, Abecassis J (eds) Durum wheat chemistry and technology. 2nd edn, AACC International, St. Paul, MN, p. 177–199.

Aguiriano E, Ruiz M, Fité R, Carrillo JM (2008) Genetic variation for glutenin and gliadins associated with quality in durum wheat (*Triticum turgidum* L. ssp. *turgidum*) landraces from Spain. Spanish Journal of Agricultural Research 6: 599–609.

Aguiriano E. Ruiz M, Fité R, Carrillo JM (2009) Effects of N fertilisation, year and prolamin alleles on gluten quality in durum wheat (*Triticum turgidum* L. ssp. *turgidum*) landraces from Spain. Spanish Journal of Agricultural Research 7: 342–348.

Amiour N, Merlino M, Leroy P, Branlard G (2002) Proteomic analysis of amphiphilic proteins of hexaploid wheat kernels. Proteomics 2: 632–641.

Ammar K, Kronstad WE, Morris CF (2000) Breadmaking quality of selected durum wheat genotypes and its relationship with high molecular weight glutenin subunits allelic variation and gluten protein polymeric composition. Cereal Chemistry 77: 230–236.

An X, Zhang Q, Yan Y, Li Q, Zhang Y, Wang A et al.(2006) Cloning and molecular characterization of three novel LMW-i glutenin subunit genes from cultivated einkorn (*Triticum monococcum* L.). Theoretical and Applied Genetics 113: 383–395.

Appels R, Eversole K, Feuillet C, Keller B, Rogers J, Stein N et al. (2018) Shifting the limits in wheat research and breeding using a fully annotated reference genome. Science 361(6403):eaar7191.

Babay E, Hanana M, Mzid R, Slim-Amara H, Carrillo JM, Rodríguez-Quijano, M (2015) Influence of allelic prolamin variation and localities on durum wheat quality. Journal of Cereal Science 63: 27–34.

Bean SR, Lookhart GL (2000) Electrophoresis of cereal storage proteins. Journal of chromatography A 88: 23–36.

Bellil I, Hamdi O, Khelifi D (2014) Diversity of five glutenin loci within durum wheat (*Triticum turgidum* L. ssp. *durum* (Desf.) Husn.) germplasm grown in Algeria. Plant Breeding 133: 179–183.

Benmoussa M, Vezina L, Pagé M, Yelle S, Laberge S (2000) Genetic polymorphism in low-molecular-weight glutenin genes from *Triticum aestivum*, variety Chinese Spring. Theoretical and Applied Genetics 100: 789–793.

Beom H, Kim JS, Jang Y, Lim SH, Kim CK, Lee CK et al. (2018) Proteomic analysis of low-molecular-weight glutenin subunits and relationship with their genes in a common wheat variety. 3 Biotech 8: 56.

Bernardo A, Wang S, Amand PS, Bai G (2015) Using next generation sequencing for multiplexed trait-linked markers in wheat. PloS one 10: e0143890.

Bietz JA, Wall JS (1972) Wheat gluten subunits: Molecular weights determined by sodium dodecyl sulfate-polyacrylamide gel electrophoresis. Cereal Chemi stry 49: 416–430.

Blanco A, Bellomo MP, Lotti C, Maniglio T, Pasqualone A, Simeone R et al. (1998) Genetic mapping of sedimentation volume across environments using recombinant inbred lines of durum wheat. Plant Breed 117: 413–417.

Boggini G, Pogna NE (1989) The breadmaking quality and storage protein composition of Italian durum wheat. Journal of Cereal Science 9: 131–138.

Boggini G, Tusa P, Pogna NE (1994) Bread-making quality of durum wheat genotypes with atypical protein compositions. Tecnica molitoria, 42: 825–835.

Branlard G, Khelifi D, Lookhart G (1992) Identification of some wheat proteins separated by a two-step acid polyacrylamide gel electrophoresis and sodium dodecyl sulfate-polyacrylamide gel electrophoresis technique. Cereal chemistry (USA).

Branlard G, Bancel E (2007) Protein extraction from cereal seeds. In: Anonymous Plant Proteomics Springer, p. 15–25.

Branlard G, Picard B, Courvoisier C (1990) Electrophoresis of gliadins in long acrylamide gels: method and nomenclature. Electrophoresis 11: 310–314.

Branlard G, Rousset M (1987) Intérêt de l'analyse des protéines de réserve dans l'amélioration génétique du blé. Le Sélectionneur Français, 39:19–30.

Brites C, Carrillo JM (2000) Inheritance of gliadin and glutenin proteins in four durum wheat crosses. Cereal Research Communications 239–246.

Brites C, Carrillo JM (2001) Influence of high molecular weight (HMW) and low molecular weight (LMW) glutenin subunits controlled by *Glu-1* and *Glu-3* loci on durum wheat quality. Cereal Chemistry 78(1): 59–63.

Burnouf T, Bietz JA (1984) Reversed-phase high-performance liquid chromatography of durum wheat gliadins: Relationships to durum wheat quality. Journal of Cereal Science 2:.

Bushuk W (1981) Utilization of cereal proteins. Utilization of Protein Resources: 208–226.

Bushuk W, Zillman RR (1978) Wheat cultivar identification by gliadin electrophoregrams. I. Apparatus, method and nomenclature. Canadian Journal of Plant Science 58: 505–515.

Butow BJ, Gale KR, Ikea J, Juhasz A, Bedö Z, Tamas L et al. (2004) Dissemination of the highly expressed Bx7 glutenin subunit (*Glu-B1al* allele) in wheat as revealed by novel PCR markers and RP-HPLC. Theoretical and Applied Genetics 109: 1525–1535.

Butow BJ, Ma W, Gale KR, Cornish GB, Rampling L, Larroque O et al. (2003) Molecular discrimination of Bx7 alleles demonstrates that over expression has a major impact on wheat flour dough strength. Theoretical and Applied Genetics 107: 1524–1532.

Campbell KG, Finney PL, Bergman CJ, Gualberto DG, Anderson JA, Giroux MJ et al. (2001) Quantitative trait loci associated with milling and baking quality in a soft× hard wheat cross. Crop Science 41: 1275–1285.

Capriotti AL, Cavaliere C, Foglia P, Samperi R, Lagana A (2011) Intact protein separation by chromatographic and/or electrophoretic techniques for top-down proteomics. Journal of Chromatography A 1218: 8760–8776.

Carrillo JM, Martínez MC, Brites C, Nieto-Taladriz MT, Vázquez JF (2000) Relationship between endosperm proteins and quality in durum wheat (*Triticum turgidum* L. var. *durum*). Options mediterraneennes 40: 463–467.

Carrillo JM, Vazquez JF, Orellana J (1990) Relationship between gluten strength and glutenin proteins in durum wheat cultivars. Plant Breeding 104: 325–333.

Carrillo JM, Vazquez JF, Ruiz M, Albuquerque MM (1991) Relationships between gluten strength and gluten components in Spanish durum wheat landraces. In Proc. IV Intnl. Workshop on Gluten proteins. St Paul, Minnesota, USA, June (pp. 27–29).

Cassidy BG, Dvorak J, Anderson OD (1998) The wheat low-molecular-weight glutenin genes: characterization of six new genes and progress in understanding gene family structure. Theoretical and Applied Genetics 96: 743–750.

Cassidy BG, Dvorak J (1991) Molecular characterization of a low-molecular-weight glutenin cDNA clone from *Triticum durum*. Theoretical and Applied Genetics 81: 653–660.

Cavanagh CR, Chao S, Wang S, Huang BE, Stephen S, Kiani S et al. (2013) Genome-wide comparative diversity uncovers multiple targets of selection for improvement in hexaploid wheat landraces and cultivars. Proceedings of the national academy of sciences 110: 8057–8062.

Cheftel JC, Cuq, J-L, Lorient D (1985) Les protéines de blé. In: Protéines alimentaires. Biochimie, propriétés fonctionnelles, valeur nutriotionelle- Modifications chimiques. Technique et Documentation Lavoisier, Paris, France, p. 204–222.

Chevalier F, Martin O, Rofidal V Devauchelle AD, Barteau S, Sommerer N et al. (2004) Proteomic investigation of natural variation between Arabidopsis ecotypes. Proteomics 4: 1372–1381.

Ciaffi M, Benedettelli S, Giorgi B, Porceddu E, Lafiandra D (1991) Seed storage proteins of *Triticum turgidum* ssp. *dicoccoides* and their effect on the technological quality in durum wheat. Plant Breeding 107: 309–319.

Ciaffi M, Lafiandra D, Turchetta T, Ravaglia S, Barian H, Gupta R et al. (1995) Breadbaking potential of durum wheat lines expressing both x-and y-type subunits at the *Glu-A1* locus. Cereal Chemistry (USA).

Ciaffi M, Lee YK, Tamás L, Gupta R, Skerritt J, Appels R (1999) The low-molecular-weight glutenin subunit proteins of primitive wheats. III. The genes from D-genome species. Theoretical and Applied Genetics 98: 135–148.

Clarke FR, Clarke JM, Pozniak CJ, Knox RE, McCaig TN (2009) Protein concentration inheritance and selection in durum wheat. Canadian journal of plant science 89: 601–612.

Clarke JM, Marchylo BA, Kovacs M, Noll JS, McCaig TN, Howes NK (1998) Breeding durum wheat for pasta quality in Canada. Euphytica 100: 163–170.

Colasuonno P, Gadaleta A, Giancaspro A, Nigro D, Giove S, Incerti O et al. (2014) Development of a high-density SNP-based linkage map and detection of yellow pigment content QTLs in durum wheat. Molecular Breeding 34: 1563–1578.

Colasuonno P, Maria MA, Blanco A, Gadaleta A (2013) Description of durum wheat linkage map and comparative sequence analysis of wheat mapped DArT markers with rice and Brachypodium genomes. BMC genetics 14: 114.

Cunsolo V, Foti S, Saletti R, Gilbert S, Tatham AS, Shewry PR (2003) Structural studies of glutenin subunits 1Dy10 and 1Dy12 by matrix-assisted laser desorption/ionisation mass spectrometry and high-performance liquid chromatography/electrospray ionisation mass spectrometry. Rapid Communication Mass Spectrometry 17: 442–454.

Cunsolo V, Foti S, Saletti R, Gilbert S, Tatham AS, Shewry PR (2004) Structural studies of the allelic wheat glutenin subunits 1Bx7 and 1Bx20 by matrix-assisted laser desorption/ionization mass spectrometry and high-performance liquid chromatography/electrospray ionization mass spectrometry. Journal of Mass Spectrometry 39: 66–78.

Damidaux R, Autran JC, Grignac P, Feillet, P (1978) Mise en évidence de relations applicable en sélection entre électrophorégramme des gliadines et les propriétés viscoélastiques du gluten du Triticum durum Desf. CR Acad Sc Paris, Série D 287:701–704.

De Santis MA, Giuliani MM, Giuzio L, De Vita P, Lovegrove A, Shewry PR et al. (2017) Differences in gluten protein composition between old and modern durum wheat genotypes in relation to 20th century breeding in Italy. European Journal of Agronomy 87: 19–29.

De Vita P, Nicosia O, Nigro F, Platani C, Riefolo C, Di Fonzo N et al. (2007) Breeding progress in morpho-physiological, agronomical and qualitative traits of durum wheat cultivars released in Italy during the 20th century. European Journal of Agronomy, 26: 39–53.

Dexter JE (2008) The history of durum wheat breeding in Canada and summaries of recent research at the Canadian Grain Commission on factors associated with durum wheat processing. In Bosphorus 2008 ICC (International Cereal Congress).

Di Stefano V, Avellone G, Bongiorno D, Cunsolo V, Muccilli V, Sforza S et al. (2012) Applications of liquid chromatography-mass spectrometry for food analysis. Journal of Chromatography A 1259: 74–8.

Dong K, Hao CY, Wang AL, Cai M, Yan Y (2009) Characterization of HMW glutenin subunits in bread and tetraploid wheats by reversed-phase high-performance liquid chromatography. Cereal Research Communications 37: 65–73.

Dong L, Zhang X, Liu D, Fan H, Sun J, Zhang Z et al. (2010) New insights into the organization, recombination, expression and functional mechanism of low molecular weight glutenin subunit genes in bread wheat. PLoS One 5: e13548.

D'Ovidio R, Lafiandra D, Porceddu E (1996) Identification and molecular characterization of a large insertion within the repetitive domain of a high-molecular-weight glutenin subunit gene from hexaploid wheat. Theoretical and Applied Genetics 93: 1048–1053.

D'Ovidio R, Masci S, Porceddu E (1995) Development of a set of oligonucleotide primers specific for genes at the *Glu-1* complex loci of wheat. Theoretical and Applied Genetics 91: 189–194.

D'Ovidio R, Porceddu E, Lafiandra D (1994) PCR analysis of genes encoding allelic variants of high-molecular-weight glutenin subunits at the *Glu-D1* locus. Theoretical and Applied Genetics 88: 175–180.

Dreisigacker S, Sehgal D, Reyes Jaimez AE, Luna-Garrido B, Muñoz-Zavala S, Núñez-RíosC et al. (eds) (2016) CIMMYT Wheat Molecular Genetics: Laboratory Protocols and Applications to Wheat Breeding. Mexico, DF: CIMMYT.

du Cros DL, Joppa LR, Wrigley CW (1983) Two-dimensional analysis of gliadin proteinsassociated with quality in durum wheat: chromosomal location of genes for their synthesis. Theoretical and Applied Genetics 66: 297–302.

Du Cros DL (1987) Glutenin proteins and gluten strength in durum wheat. Journal of Cereal Science 5: 3–12.

Dunbar BD, Bundman DS, Dunbar BS (1985) Identification of cultivar specific proteins of winter wheat (*T. aestivum* L.) by high resolution two dimensional polyacrylamide gel electrophoresis and color-based silver stain. Electrophoresis 6: 39–43.

Dunn MJ (2000) From Genome to Proteome. Wiley, London.

Dupont FM, Vensel WH, Tanaka CK, Hurkman WJ, Altenbach SB (2011) Deciphering the complexities of the wheat flour proteome using quantitative two-dimensional electrophoresis, three proteases and tandem mass spectrometry. Proteome Science 9: 10.

Dworschak RG, Ens W, Standing KG, Preston KR, Marchylo BA, Nightingale MJ et al. (1998) Analysis of Wheat Gluten Proteins by Matrix-assisted Laser Desorption/Ionization Mass Spectrometry. Journal of Mass Spectrometry, 33: 429–435.

Edwards NM, Mulvaney SJ, Scanlon MG, Dexter JE (2003) Role of gluten and its components in determining durum semolina dough viscoelastic properties. Cereal Chemistry 80: 755–763.

Edwards NM, Gianibelli MC, McCaig TN, Dexter JE (2007) Relationships between dough strength, polymeric protein quantity and composition for diverse durum wheat genotypes. Journal of Cereal Science 45: 140–149.

Espí A, Giraldo P, Rodríguez-Quijano M, Carrillo JM (2012) A PCR-based method for discriminating between high molecular weight glutenin subunits Bx7 and Bx7* in *Triticum aestivum* L. Plant Breeding 131: 571–573.

Espí A, Rodríguez-Quijano M, Vázquez JF, Carrilo JM (2014) Molecular characterization of *Glu-B3* locus in wheat cultivars and segregating populations. Journal of Cereal Science 60: 374–381.

Fekete S, Fekete J, Ganzler K (2009) Validated UPLC method for the fast and sensitive determination of steroid residues in support of cleaning validation in formulation area. Journal of Pharmacology and Biomedical Analysis 49: 833–838.

Fenn JB, Mann M, Meng CK, Wong SF, Whitehouse CM (1989) Electrospray ionization for mass spectrometry of large biomolecules. Science 246: 64–71.

Fiedler JD, Salsman E, Liu Y, Michalak de Jimenez M, Hegstad JB, Chen B et al. (2017) Genome-Wide Association and Prediction of Grain and Semolina Quality Traits in Durum Wheat Breeding Populations. Plant Genome 10: 10.

Fois S, Schlichting L, Marchylo B, Dexter J, Motzo R, Giunta F (2011) Environmental conditions affect semolina quality in durum wheat (*Triticum turgidum* ssp. *durum* L.) cultivars with different gluten strength and gluten protein composition. Journal of the Science of Food and Agriculture 91: 2664–2673.

Gao L, Ma W, Chen J, Wang K, Li J, Wang S et al. (2010) Characterization and comparative analysis of wheat high molecular weight glutenin subunits by SDS-PAGE, RP-HPLC, HPCE, and MALDI-TOF-MS. Journal of Agricultural and Food Chemistry 58: 2777–2786.

Garozzo D, Cozzolino R, Giorgi SD, Fisichella S, Lafiandra D (1999) Use of hydroxyacetophenones as matrices for the analysis of high molecular weight glutenin mixtures by matrix-assisted laser desorption/ionization mass spectrometry. Rapid Communications in Mass Spectrometry 13: 2084–2089.

Gianibelli MC, Ruiz M, Carrillo JM, MacRitchie F (1995) Relationships between biochemical parameters and quality characteristics of durum wheats. In: Schofield JD (ed) Wheat Structure, Biochemistry and Functionality. Royal Society of Chemistry, Cambridge, p. 146–152.

Giraldo P, Rodríguez-Quijano M, Simon C, Carrillo JM (2010) Allelic variation in HMW glutenins in Spanish wheat landraces and their relationship with bread quality. Spanish Journal of Agricultural Research 8: 1012–1023.

Gobaa S, Bancel E, Kleijer G, Stamp P, Branlard G (2007) Effect of the 1BL. 1RS translocation on the wheat endosperm, as revealed by proteomic analysis. Proteomics 7:4349–4357.

Görg A, Klaus A, Lück C et al. (2007) Two-dimensional electrophoresis with immobilized pH gradients for proteome analysis. A laboratory manual. http://www.wzw.tum.de/proteomik

Gupta RB, Shepherd KW (1988) Low-molecular-weight glutenin subunits in wheat: their variation, inheritance and association with bread-making quality. In Proc. 7th International Wheat Genetics Symposium. T.E. Miller e R.M.D. Koebner (eds.). Cambridge, Vol. II, p. 943–949.

Gupta RB, Shepherd KW (1993) Production of multiple wheat-rye 1RS translocation stocks and genetic analysis of LMW subunits of glutenin and gliadins in wheats using these stocks. Theoretical and Applied Genetics 85: 719–728.

Haile JK, N'Diaye A, Clarke F, Clarke J, Knox R, Rutkoski J et al. (2018) Genomic selection for grain yield and quality traits in durum wheat. Molecular Breeding 38: 75.

Hailegiorgis D, Lee CA, Yun SJ (2017) Allelic composition and associated quality traits of the *Glu-1* and *Glu-3* loci in selected modern Ethiopian durum wheat varieties. Journal of Crop Science and Biotechnology 20: 387–392.

Henkrar F, El-Haddoury J, Iraqi D, Bendaou N, Udupa SM (2017) Allelic variation at high-molecular weight and low-molecular weight glutenin subunit genes in Moroccan bread wheat and durum wheat cultivars. 3 Biotech 7:1.

Huebner FR, Bietz JA (1985) Detection of quality differences among wheats by high-performance liquid-chromatography. Journal of Chromatography 327: 333–342.

Humphery-Smith I, Cordwell SJ, Blackstock WP (1997) Proteome research: complementarity and limitations with respect to the RNA and DNA worlds. Electrophoresis 18: 1217–1242.

Ibba MI, Kiszonas AM, Morris CF (2017) Evidence of intralocus recombination at the *Glu-3* loci in bread wheat (*Triticum aestivum* L.). Theoretical and Applied Genetics 130: 891–902.

Ibba MI, Kiszonas AM, Morris CF (2018) Development of haplotype-specific molecular markers for the low-molecular-weight glutenin subunits. Molecular Breeding 38: 68.

Igrejas G, Guedes-Pinto H, Carnide V, Branlard G(1999) The high and low molecular weight glutenin subunits and ω-gliadin composition of bread and durum wheats commonly grown in Portugal. Plant Breeding 118: 297–302.

Igrejas G, Juhász A, Gianibelli MC, Gale KR, Rahman S (2009) Low-molecular-weight glutenins in durum wheat: analysis of *Glu-A3* alleles using PCR markers. Plant Breeding 129: 574–577.

Ikeda TM, Nagamine T, Fukuoka H, Yano H (2002) Identification of new low-molecular-weight glutenin subunit genes in wheat. Theoretical and Applied Genetics 104: 680–687.

Islam N, Tsujimoto H, Hirano H (2003a) Wheat proteomics: relationship between fine chromosome deletion and protein expression. Proteomics 3: 307–316.

Islam N, Tsujimoto H, Hirano H (2003b) Proteome analysis of diploid, tetraploid and hexaploid wheat: towards understanding genome interaction in protein expression. Proteomics 3: 549–557.

Islam N, Woo SH, Tsujimoto H, Kawasaki H, Hirano H (2002) Proteome approaches to characterize seed storage proteins related to ditelocentric chromosomes in common wheat. Proteomics 2: 1146–1155.

Jackson EA, Holt LM, Payne PI (1983). Characterisation of high molecular weight gliadin and low-molecular-weight glutenin subunits of wheat endosperm by two-dimensional electrophoresis and the chromosomal localisation of their controlling genes. Theoretical and Applied Genetics 66: 29–37.

Jackson EA, Morel MH, Sontag-Strohm T, Branlard G, Metakovsky EV, Redaelli R (1996) Proposal for combining the classification systems of alleles of *Gli-1* and *Glu-3* loci in bread wheat (*Triticum aestivum* L.). Journal of Genetics and Breeding 50: 321–336.

Janni M, Cadonici S, Pignone D, Marmiroli N (2017) Survey and new insights in the application of PCR based molecular markers for identification of HMW-GS at the *Glu-B1* locus in durum and bread wheat. Plant Breeding 136: 467–473.

Jones RW, Taylor NW, Senti FR (1959) Electrophoresis and fractionation of wheat gluten. Archives. Biochemistry Biophysics 84: 363–376.

Joppa LR, Khan K, Williams ND (1983) Chromosomal location of genes for gliadin polypeptides in durum wheat *Triticum turgidum* L. Theoreticaland Applied Genetics 64: 289–293.

Kaan F, Branlard G, Chihab B, Borries C (1993) Relations between genes coding for grain storage protein and two pasta cooking quality criteria among durum wheat (*T. durum* Desf.) genetic resources. Journal of Genetics Breeding, 47: 151–156.

Kelleher NL, Lin HY, Valaskovic GA, Aaserud DJ, Fridriksson EK, McLafferty FW (1999) Top down versus bottom up protein characterization by tandem high-resolution mass spectrometry. Journal of the American Chemical Society 121: 806–812.

Khelifi D, Branlard G (1991) A new two-step electrophoresis method for analysing gliadin polypeptides and high and low molecular weight subunits of glutenin of wheat. Journal of Cereal Science 13: 41–47.

Kiszonas AM, Morris CF (2018) Wheat breeding for quality: A historical review. Cereal Chemistry 95: 17–34.

Klose J (1975) Protein mapping by combined isoelectric focussing and electrophoresis of mouse tissues. A novel approach to testing for induced point mutations in mammals. Humangenetik 26:231–243.

Kovacs MIP, Dahlke G, Noll JS (1994) Gluten viscoelasticity: its usefulness in the Canadian durum wheat breeding program. Journal of Cereal Science 19: 251–257.

Kovacs MIP, Howes NK, Leisle D, Zawistowski J (1995). Effect of two different low molecular weight glutenin subunits on durum wheat pasta quality parameters. Cereal Chemistry 72: 85–87.

Kreis M, Shewry PR, Forde BG, Miflin BJ (1985) Structure and evolution of seed storage proteins and their genes with particular reference to those of wheat, barley and rye. In Oxford Surveys of Plant Cell and Molecular Biology. B.J. Miflin (ed.). Oxford Univ. Press, Oxford, Vol. II, p. 253–317.

Kumar LS (1999) DNA markers in plant improvement: an overview. Biotechnology Advances 17: 143 182.

Kussmann M, Nordhoff E, Rahbek-Nielsen H, Haebel S, Rossel-Larsen M, Jakobsen L (1997) Matrix assisted laser desorption/ionization mass spectrometry sample preparation techniques designed for various peptide and protein analytes. Journal of Mass Spectrometry 32: 593–601.

Laemmli UK (1970) Cleavage of structural proteins during the assembly of the head of bacteriophage T4. Nature 227: 680–685.

Lafiandra D, Tucci GF, Pavoni A, Turchetta T, Margiotta B (1997) PCR analysis of x-and y-type genes present at the complex *Glu-A1* locus in durum and bread wheat. Theoretical and Applied Genetics 94: 235–240.

Laidò G, Mangini G, Taranto F, Gadaleta A, Blanco A, Cattivelli L, et al. (2013) Genetic diversity and population structure of tetraploid wheats (*Triticum turgidum* L.) estimated by SSR, DArT and pedigree data. PLoS ONE 8:e67280.

Lei ZS, Gale KR, He ZH, Gianibelli C, Larroque O, Xia XC (2006) Y-type gene specific markers for enhanced discrimination of high molecular weight glutenin alleles at the *Glu-B1* locus in hexaploid wheat. Journal of Cereal Science 43: 94–101.

Lerner SE, Cogliatti M, Ponzio NR, Seghezzo ML, Molfes ER, Rogers WJ (2004) Genetic variation for grain protein components and industrial quality of durum wheat cultivars sown in Argentina. Journal of Cereal Science 40: 161–166.

Levy AA, Zaccai M, Millet E, Feldman M (1988) Utilization of wild emmer for the improvement of grain protein percentage of cultivated wheat. In Proceedings of the Seventh International Wheat Genetics Symposium/edited by TE Miller and RMD Koebner. Cambridge: Published by the Institute of Plant Science Research, Cambridge Laboratory, c1988.

Liang X, Zhen S, Han C, Wang C, Li X, Ma W et al. (2015) Molecular characterization and marker development for hexaploid wheat-specific HMW glutenin subunit 1By18 gene. Molecular Breeding 35: 1–16.

Lindsay MP, Skerritt, JH (1999) The glutenin macropolymer of wheat flour doughs: structure–function perspectives. Trends in Food Science & Technology 10: 247–253.

Liu C-Y (1995) Identification of a new low Mr glutenin subunit locus on chromosome 1B of durum wheat. Journal of Cereal Science 21: 209–213.

Liu L, Ikeda TM, Branlard G, Peña RJ, Rogers WJ, Lerner SE et al. (2010) Comparison of low molecular weight glutenin subunits identified by SDS-PAGE, 2-DE, MALDI-TOF-MS and PCR in common wheat. BMC Plant Biology 10: 124.

Liu C-Y, Shepherd KW (1995) Inheritance of B subunits of glutenin and ω- and γ-gliadins in tetraploid wheats. Theoretical and Applied Genetics 90: 149–1157.

Liu L, Wang AL, Appels A (2009) A MALDI-TOF based analysis of high molecular weight glutenin subunits for wheat breeding. Journal of Cereal Science 50(2): 295–301.

Liu S, Chao S, Anderson JA (2008) New DNA markers for high molecular weight glutenin subunits in wheat. Theoretical and Applied Genetics 118: 177–183.

Lutz E, Wieser H, Koehler P (2012) Identification of disulfide bonds in wheat gluten proteins by means of mass spectrometry/electron transfer dissociation. Journal of Agricultural and Food Chemistry 60: 3708–3716.

Ma W, Zhang W, Gale KR (2003) Multiplex-PCR typing of high molecular weight glutenin alleles in wheat. Euphytica 134: 51–60.

Maccaferri M, Harris NS, Twardziok SO, Pasam RK, Gundlach H, Spannagl M et al (2019) Durum wheat genome highlights past domestication signatures and future improvement targets. Nature genetics 51: 885–895.

Maccaferri M, Ricci A, Salvi S, Milner SG, Noli E, Martelli PL et al. (2015) A high-density, SNP-based consensus map of tetraploid wheat as a bridge to integrate durum and bread wheat genomics and breeding. Plant Biotechnology Journal 13: 648–663.

MacKey J (2005) Wheat: its concept, evolution, and taxonomy. In: Royo C, Nachit MM, Di Fonzo N, Araus JL, Pfeiffer WH, Slafer GA (eds) Durum wheat breeding: current approaches and future strategies. Food Products Press, New York, p. 3–61.

McLafferty FW (2011) A Century of Progress in Molecular Mass Spectrometry. Annual Review of Analytical Chemistry 44: –22.

MacRitchie F (1992) Physicochemical properties of wheat proteins in relation to functionality. Advances in Food and Nutrition Research 36: 1–87.

MacRitchie F, Du Cros DL, Wrigley CW (1990) Flour polypeptides related to wheat quality. In Advances in Cereal Science and Technology. Y. Pomeranz (ed.). American Association Cereal Chemists, St. Paul, Minnesota, USA, Vol. X, p. 79–145.

Magallanes-López AM, Ammar K, Morales-Dorantes A, González-Santoyo H. Crossa J, Guzmán C (2017) Grain quality traits of commercial durum wheat varieties and their relationships with drought stress and glutenins composition. Journal of Cereal Science 75: 1–9.

Mamone G, Ferranti P, Chianese L, Scafuri L, Addeo F (2000) Qualitative and quantitative analysis of wheat gluten proteins by liquid chromatography and electrospray mass spectrometry. Rapid Communications in Mass Spectrometry 14: 897–904.

Mamone G, Picariello G, Caira S, Addeo F, Ferranti P (2009) Analysis of food proteins and peptides by mass spectrometry-based techniques. Journal of Chromatography 1216: 7130–7142.

Manifesto MM, Feingold S, Hopp HE, Schlatter AR, Dubcovsky J (1998) Molecular markers associated with differences in bread-making quality in a cross between bread wheat cultivars with the same high Mr glutenins. Journal of Cereal Science 27: 217–227.

Mantovani P, Maccaferri M, Sanguineti MC, Tuberosa R, Catizone I, Wenzl P (2008) An integrated DArT-SSR linkage map of durum wheat. Molecular Breeding 22: 629–648.

Marchylo BA, Dexter JE, Clarke JM, Ames N (1998) Effects of protein content on CWAD quality. In: Fowler DB, Geddes WE, Johnston AM, Preston KR (eds) Proc. Symp. on Wheat Protein and Marketing. University Extension Press, University of Saskatchewan, Saskatoon, p 53–62.

Marchylo BA, Kruger JE, Hatcher DW (1989) Quantitative reversed-phase high-performance liquid chromatographic analysis of wheat storage proteins as a potential quality prediction tool. Journal of Cereal Science 9: 113–130.

Martínez MC, Ruiz M, Carrillo JM (2004) New B low Mr glutenin subunit alleles at the *Glu-A3, Glu-B2* and *Glu-B3* loci and their relationship with gluten strength in durum wheat. Journal of Cereal Science 40: 101–107.

Martínez MC, Ruiz M, Carrillo JM (2005) Effects of different prolamin alleles on durum wheat quality properties. Journal of Cereal Science 41: 123–131.

Masci S, Lew EJL, Lafiandra D, Porceddu E, Kasarda DD (1995) Characterization of low-molecular-weight glutenin subunits in durum wheat by RP-HPLC and N-terminal sequencing. Cereal Chemistry 72: 100–104.

Masci S, D'Ovidio R, Lafiandra D, Kasarda DD (1998) Characterization of a low-molecular-weight glutenin subunit gene from bread wheat and the corresponding protein that represents a major subunit of the glutenin polymer. Plant Physiology 118: 1147–1158.

Masci S, D'Ovidio R, Lafiandra D, Kasarda DD (2000) A 1B-coded low-molecular-weight glutenin subunit associated with quality in durum wheats shows strong similarity to a subunit present in some bread wheat cultivars. Theoretical and Applied Genetics 100: 396–400.

Mazza M, Lori A, Pasquini M, Pogna NE (1996) Evidence for omega-gliadins encoded by the Gli-B5 locus in durum wheat (*Triticum turgidum* spp. *durum*). Journal of Genetics & Breeding 50: 197–201.

McIntosh RA, Yamazaki Y, Dubcovsky J, Morris CF, Appels R, Xia X et al. (2013) Catalogue of gene symbols for wheat. https://shigen.nig.ac.jp/wheat/komugi/genes/symbolClassList.jsp

Mecham DK, Kasarda DD, Qualset CO (1978). Genetic aspects of wheat gliadin proteins. Biochemical Genetics 16: 831–853.

Melas V, Morel MH, Feille P (1993). Les sous-unites glutenines du ble de faible poids moleculaire: des proteines d'avenir? Industries des Cereales 3–3.

Melas V, Morel MH, Autran J-C, Feillet P (1994) Simple and rapid method for purifying low molecular weight subunits of glutenin from wheat. Cereal Chemistry 71: 234–237.

Metakovsky EV, Knežević D, Javornik B (1991) Gliadin allele composition of Yugoslav winter wheat cultivars. Euphytica 54: 285–295.

Moragues M, Moralejo M, Sorrells ME, Royo C (2007) Dispersal of durum wheat landraces across the Mediterranean basin assessed by AFLPs and microsatellites. Genetic Resources and Crop Evolution 54: 1133–1144.

Moragues M, Zarco-Hernández J, Moralejo MA, Royo C (2006) Genetic diversity of glutenin protein subunits composition in durum wheat landraces [*Triticum turgidum* ssp. *turgidum* convar. *durum* (Desf.) MacKey] from the Mediterranean Basin. Genetic Resources and Crop Evolution 53: 993–1002.

Muccilli V, Cunsolo V, Saletti R, Foti S, Masci S, Lafiandra D (2005) Characterization of B- and C-type low molecular weight glutenin subunits by electrospray ionization mass spectrometry and matrix assisted laser desorption/ionization mass spectrometry. Proteomics 5: 719–728.

Muccilli V, Cunsolo V, Saletti R, Foti S, Masci S, Lafiandra D (2010). Characterisation of a specific class of typical low molecular weight glutenin subunits of durum wheat by a proteomic approach. Journal of Cereal Science 51: 134–139.

Muccilli V, Lo Bianco M, Cunsolo V, Saletti R, Gallo G, Foti S (2011) High molecular weight gluten in subunits in some durum wheat cultivars investigated by means of mass spectrometric techniques. Journal of Agriculture and Food Chemistry 59: 12226–12237.

Mullis KB, Erlich HA, Arnheim N, Horn GT, Saiki RK, Scharf S J (1987) Process for amplifying, detecting, and/or cloning nucleic acid sequences. U.S. Patent 4,683,195.

Nazco R, Peña RJ, Ammar K, Villegas D, Crossa J, Royo C (2014) Durum wheat (*Triticum durum* desf.) Mediterranean landraces as sources of variability for allelic combinations at *Glu-1/Glu-3* loci affecting gluten strength and pasta cooking quality. Genetic Resources and Crop Evolution 61: 1219–1236.

Nieto-Taladriz MT, Ruiz M, Martínez MC, Vazquez JF, Carrillo JM (1997) Variation and classification of B low-molecular-weight glutenin subunit alleles in durum wheat. Theoretical and Applied Genetics 95: 1155–1160.

Oak M, Tamhankar S, Rao V, Bhosale S (2004) Relationship of HMW, LMW glutenin subunits and γ-gliadins with gluten strength in Indian durum wheats. Journal of Plant Biochemistry and Biotechnology 13: 51–55.

O'Farrell PH (1975) High-resolution two-dimensional electrophoresis of proteins. Journal of Biological Chemistry 250: 4007–4021.

Onda Y, Takahagi K, Shimizu, Inoue K, Mochida K (2018) Multiplex PCR targeted amplicon sequencing (MTA-Seq): simple, flexible, and versatile SNP genotyping by highly multiplexed PCR amplicon sequencing. Frontiers in Plant Science 9: 201.

Osborne TB (1907) The proteins of the wheat Kernel. Carnegie Inst. Washington. Washington D.C. Publ 84: 1–119.

Osborne TB (1924) The Vegetable Proteins, 2nd edn. London.

Park OK (2004) Proteomic studies in plants. Journal of Biochemistry and Molecular Biology 37: 133–138.

Payne PI, Law CN, Mudd EE (1980) Control by homoeologous group 1 chromosomes of the high-molecular-weight subunits of glutenin, a major protein of wheat endosperm. Theoretical and Applied Genetics 58: 113–120.

Payne PI (1987) Genetics of wheat storage protein and the effect of allelic variation on breadmaking quality. Annual Review of Plant Physiology 38: 141–153.

Payne PI, Corfield KG (1979) Subunit composition of wheat glutenin proteins, isolated by gel filtration in a dissociating medium. Planta 145: 83–88.

Payne PI, Corfield KG, Blackman JA (1979) Identification of a high molecular weight subunit of glutenin whose presence correlates with breadmaking quality in six crosses of bread wheat. Theoretical and Applied Genetics 55: 153–159.

Payne PI, Holt LM, Jackson EA, Law CN (1984a). Wheat storage proteins: their genetics and their potential for manipulation by plant breeding. Philos. Trans. R. Soc., London, Ser. B, 304:359–371.

Payne PI, Holt LM, Law CN (1981) Structural and genetical studies on the high-molecular weight subunits of wheat glutenin. Theoretical and Applied Genetics 60: 229–236.

Payne PI, Jackson EA, Holt LM (1984b) The association between gamma-gliadin 45 and gluten strength in durum wheat varieties: a direct causal effect or the result of genetic linkage? Journal of Cereal Science 2: 73–81.

Pennington SR, Dunn MJ (eds.) (2001) Proteomics: From Protein Sequence to Function. BIOS Scientific Publishers, Oxford, UK, p 1–309.

Peña RJ, Pfeiffer WH (2005) Breeding methodologies and strategies for durum wheat quality improvement. Durum Wheat Breeding: Current Approaches and Future Strategies. Food Product Press. The Haworth Press Inc., New York, 663–772.

Pogna NE, Autran JC, Mellini F, Lafiandra D, Feillet P (1990) Chromosome 1B-encoded gliadins and glutenin subunits in durum wheat: genetics and relationship to gluten strength. Journal of Cereal Science 11: 15–34.

Pogna NE, Lafiandra D, Feillet P, Autran JC (1988) Evidence for a direct causal effect of low molecular weight subunits of glutenins on gluten viscoelasticity in durum wheats. Journal of Cereal Science 7: 11–214.

Porceddu E, Turchetta T, Masci S, D'Ovidio R., Lafiandra D, Kasarda DD et al. (1998) Variation in endosperm protein composition and technological quality properties in durum wheat. Euphytica 100: 197–205.

Qi PF, Wei YM, Ouellet T, Chen Q, Tan X, Zheng YL (2009) The γ-gliadin multigene family in common wheat (Triticum aestivum) and its closely related species. BMC Genomics 10: 168.

Qian YW, Preston K, Krokhin O, Mellish J, Ens W (2008) Characterization of Wheat Gluten Proteins by HPLC and MALDI TOF Mass Spectrometry. Journal of the American Society for Mass Spectrometry 19: 1542–1550.

Raciti CN, Doust MA, Lombardo GM, Boggini G, Pecetti L (2003) Characterization of durum wheat Mediterranean germplasm for high and low molecular weight glutenin subunits in relation with quality. European Journal of Agronomy 19: 373–382.

Radovanovic N, Cloutier S (2003) Gene-assisted selection for high molecular weight glutenin subunits in wheat doubled haploid breeding programs. Molecular Breeding 12: 51–59.

Ragupathy R, Naeem HA, Reimer E, Lukow OM, Sapirstein HD, Cloutier S (2008) Evolutionary origin of the segmental duplication encompassing the wheat *Glu-B1* locus encoding the overexpressed Bx7 (Bx7OE) high molecular weight glutenin subunit. Theoretical and Applied Genetics 116: 283–296.

Rasheed A, Xia X, Yan Y, Appels R, Mahmood T, He Z (2014) Wheat seed storage proteins: advances in molecular genetics, diversity and breeding applications. Journal of Cereal Science 60: 11–24.

Ribeiro M, Nunes-Miranda JD, Branlard G, Carrilllo JM, Rodriguez-Quijano M, Igrejas G (2013) One hundred years of grain omics: identifying the glutens that feed the world. Journal of Proteome Research 12: 4702–4716.

Rife TW, Wu S, Bowden R et al. (2015) Spiked GBS: a unified, open platform for single marker genotyping and whole-genome profiling. BMC Genomics 16: 248.

Rodríguez-Quijano M, Lucas R, Ruiz M, Giraldo P, Espi A, Carrillo JM (2010) Allelic variation and geographical patterns of prolamins in the USDA-ARS Khorasan wheat germplasm collection. Crop Science 50: 2383–2391.

Ruiz M, Carrillo JM (1993) Linkage relationships between prolamin genes on chromosomes 1A and 1B of durum wheat. Theoretical and Applied Genetics 87: 353–360.

Ruiz M, Carrillo JM (1994) Separate effects on gluten strength of *Gli-1* and *Glu-3* prolamin genes on chromosomes 1A and 1B in durum wheat. Journal of Cereal Science 21:137–144.

Ruiz M, Carrillo JM (1995) Relationships between different prolamin proteins and some quality properties in durum wheat. Plant Breeding 114: 40–44.

Ruiz M, Carrillo JM (1996) Gli-B3/Glu-B2 encoded prolamins do not affect selected quality properties in the durum wheat cross 'Abadía'×'Mexicali 75'. Plant Breeding 115: 410–412.

Ruiz M, Bernal G, Giraldo P (2018) An update of low molecular weight glutenin subunits in durum wheat relevant to breeding for quality. Journal of Cereal Science 83: 236–244.

Sapirstein HD, David P, Preston KR, Dexter JE (2007) Durum wheat breadmaking quality: effects of gluten strength, protein composition, semolina particle size and fermentation time. Journal of Cereal Science 45: 150–161.

Scheele GA (1975) Two-dimensional gel analysis of soluble proteins. Characterisation of guinea pig exocrine pancreatic proteins. Journal of Biological Chemistry 250: 5375–5385.

Schnable Lab (2015) Genotyping by multiplexing amplicon sequencing (GBMAS), c2001 2015 [updated 2015 Jul 6, cited 2015 Jul 29] Available: http://schnablelab.plantgenomics.iastate.edu/resources/protocols/

Schwarz G, Felsenstein FG, Wenzel G (2004). Development and validation of a PCR-based marker assay for negative selection of the HMW glutenin allele *Glu-B1-1d* (Bx-6) in wheat. Theoretical and Applied Genetics 109: 1064–1069.

Scossa F, Laudencia-Chingcuanco D, Anderson OD, Vensel WH, Lafiandra D, D'Ovidio R (2008). Comparative proteomic and transcriptional profiling of a bread wheat cultivar and its derived transgenic line overexpressing a low molecular weight glutenin subunit gene in the endosperm. Proteomics 8: 2948–2966.

Shewry PR (1999) The synthesis, processing, and deposition of gluten proteins in the developing wheat grain. Cereal Foods World 44: 587–589.

Shewry PR, Miflin BJ, Lew EJL, Kasarda DD (1983) The preparation and characterization of an aggregated gliadin fraction from wheat. Journal of Experimental Botany 34: 1403.

Shewry PR, Napier JA, Tatham AS (1995) Seed storage proteins: structures and biosynthesis. The Plant Cell 7: 945–956.

Shewry PR, Tatham AS (1990) The prolamin storage proteins of cereal seeds: structure and evolution. Biochemical Journal 267: 1–12.

Shewry PR, Tatham AS (1997) Disulphide bonds in wheat gluten proteins. Journal of Cereal Science 25: 207–227.

Shewry PR, Tatham AS, Forde J, Kreis M, Miflin BJ (1986) The classification and nomenclature of wheat gluten proteins: a reassessment. Journal of Cereal Science 4: 97–106.

Shulka TP (1975) Cereal proteins: chemistry and food applications. Critical Reviews in Food Science and Nutrition 6: 1–77.

Sissons MJ, Ames NP, Hare RA, Clarke JM (2005) Relationship between glutenin subunit composition and gluten strength measurements in durum wheat. Journal of Science Food and Agriculture 85: 2445–2452.

Sissons MJ and Batey IL (2003). Protein and starch properties of some tetraploid wheats. Cereal Chemistry 80: 468–475.

Skylas DJ, Copeland L, Rathmell W, Wrigley CW (2001) The wheat-grain proteome as a basis for more efficient cultivar identification. Proteomics 1: 1542–1546.

Skylas DJ, Mackintosh JA, Cordwell ST et al. (2000) Proteome approach to the characterisation of protein composition in the developing and mature wheat-grain endosperm. J Cereal Sci 32:169–188.

Stephenson JL, McLuckey SA, Reid GE, Wells JM, Bundy JL (2002) Ion/ion chemistry as a top-down approach for protein analysis. Curren Opinion in Biotechnology 13: 57–64.

Subira J, Pena R, Álvaro F, Ammar K, Ramdani A, Royo C (2014) Breeding progress in the pasta-making quality of durum wheat cultivars released in Italy and Spain during the 20th century. Crop Pasture and Science 65: 16–26.

Swartz ME (2005) UPLC: an introduction and review. Journal of Liquid Chromatography and Related Technologies 28: 1253–1263.

Taha SA (1997) Prediction of durum wheat quality from electrophoretic and high-performance liquid chromatography pattern of gliadins. Acta Alimentaria 26: 117–130.

Tao HP, Kasarda DD (1989) Two-dimensional gel mapping and N-terminal sequencing of LMW-glutenin subunits. Journal of Experimental Botany 40: 1015–1020.

Tasleem-Tahir A, Nadaud I, Chambon C, Branlard G (2012) Expression profiling of starchy endosperm metabolic proteins at 21 stages of wheat grain development. Journal of Proteome Research 11: 2754–277.
Tatham AS, Shewry PR (1995) The S-poor prolamins of wheat, barley and rye. Journal of Cereal Science 22: 1–16.
Thiellement H, Bahrman N, Damerval C, Plomion C, Rossignol M, Santoni V, et al. (1999) Proteomics for genetic and physiological studies in plants. Electrophoresis 20: 2013–2026.
Trad H, Ayed S, Rhazi L, Slim A, da Silva JAT, Hellal R, et al. (2014) Comparative quality analysis of gluten strength and the relationship with high molecular weight glutenin subunits of 6 tunisian durum wheat genotypes. Food Science and Biotechnology 23: 1363–1370.
Turchetta T, Ciaffi M, Porceddu E, Lafiandra D (1995) Relationship between electrophoretic pattern of storage proteins and gluten strength in durum wheat landraces from Turkey. Plant Breeding 114: 406–412.
Uthayakumaran S, Listiohadi Y, Baratta M, Batey IL, Wrigley CW (2006) Rapid identification and quantitation of high-molecular-weight glutenin subunits. Journal of Cereal Science 44: 34–39.
van Poecke RM, Maccaferri M, Tang J, Truong HT, Janssen A, van Orsouw N, et al. (2013) Sequence-based SNP genotyping in durum wheat. Plant Biotechnology Journal 11: 809–817.
Vázquez JF, Ruiz M, Nieto-Taladriz MT, Albuquerque MM (1996) Effects on gluten strength of low Mr glutenin submits coded by alleles at *Glu-A3* and *Glu-B3* loci in durum wheat. Journal of Cereal Science 24: 125–130.
Vensel WH, Dupont FM, Chan R, Hurkman WJ (2007) Mass spectrometry based identifications of LMW glutenin subunits. In: Lookhart GL, Ng PKW, editors. Gluten proteins 2006. Minneapolis, MN: AACCI. P. 347–351.
Vensel WH, Tanaka CK, Cai N, Wong JH, Buchanan BB, Hurkman WJ (2005) Developmental changes in the metabolic protein profiles of wheat endosperm. Proteomics 5: 1594–1611.
Waines JG and Payne PI 1987 Electrophoretic analysis of the high-molecular-weight glutenin subunits of *Triticum monococcum, T. urartu*, and the A genome of bread wheat (*T. aestivum*). Theoretical and Applied Genetics 74: 71–76.
Wang LH, Li GY, Peña RJ, Xia XC, He ZH (2010) Development of STS markers and establishment of multiplex PCR for *Glu-A3* alleles in common wheat (*Triticum aestivum* L.). Journal of Cereal Science 51: 305–312.
Wang S, Wong D, Forrest K, Allen A, Chao S, Huan BE, et al. (2014) Characterization of polyploid wheat genomic diversity using a high-density 90 00 single nucleotide polymorphism array. Plant Biotechnology Journal 12: 787–796.
Wang LH, Zhao XL, He ZH, Ma W, Appels R, Peña RJ, (2009) Characterization of low-molecular-weight glutenin subunit *Glu-B3* genes and development of STS markers in common wheat (*Triticum aestivum* L.). Theoretical and Applied Genetics 118: 525–539.
Wasinger VC, Cordwell SJ, Cerpa-Poljak A, Yan JX, Gooley AA, Wilkins MR, et al. (1995) Progress with gene-product mapping of the Mollicutes: *Mycoplasma genitalium*. Electrophoresis 16: 1090–1094.
Wieser H (2007) Chemistry of gluten proteins. Food Microbiology 24: 115–119.
Wieser H, Antes S, Seilmeier W (1998) Quantitative determination of gluten protein types in wheat flour by reversed-phase high-performance liquid chromatography. Cereal Chemistry 75: 644–650.
Williams KL (1999) Genomes and proteomes: towards a multidimensional view of biology. Electrophoresis 20: 678–688.
Winfield MO, Allen AM, Burridge AJ, Barker GL, Benbow HR, Wilkinson PA, et al. (2016) High-density SNP genotyping array for hexaploid wheat and its secondary and tertiary gene pool. Plant Biotechnology Journal 14: 1195–1206.
Woychik JH, Boundy JA, Dimler RJ (1961) Starch gel electrophoresis of wheat gluten proteins with concentrated urea. Archives of Biochemistry and Biophysics 94: 477–482.
Wrigley CW (1968) Gel electrofocusing-a technique for analysing multiple protein samples by isoelectric focusing. Science Tools 15: 17–23.
Wrigley CW (1970) Protein mapping by combined gel electrofocusing and electrophoresis: application to the study of genotypic variations in wheat gliadins. Biochemical Genetics 4:509–516.

Wu Y, Engen JR, Hobbins WB (2006) Ultra performance liquid chromatography (UPLC) further improves hydrogen/deuterium exchange mass spectrometry. Journal of the American Society for Mass Spectrometry 17: 163–167.
Xu Q, Xu J, Liu CL et al. (2008) PCR-based markers for identification of HMW-GS at *Glu-B1x* loci in common wheat. Journal of Cereal Science 47: 394–398.
Yahata E, Maruyama-Funatsuki W, Nishio Z, Tabiki T, Takata K, Yamamoto Y et al. (2005) Wheat cultivar-specific proteins in grain revealed by 2-DE and their application to cultivar identification of flour. Proteomics 5: 3942–3953.
Yan YM, Jiang Y, An XN, Pei YH, Pei XH, Zhang YZ, et al. (2009) Cloning, expression and functional analysis of HMW glutenin subunit 1By8 gene from Italy pasta wheat (*Triticum turgidum* L. ssp. *durum*). Journal of Cereal Science 50: 398–406.
Yu Z, Han C, Wang S, Lv D, Chen G, Li X et al. (2012) Fast separation and characterization of water-soluble proteins in wheat grains by reversed-phase ultra performance liquid chromatography (RP-UPLC). Journal of Cereal Science 57: 288–294.
Zhang Q, Dong YM, An XL, Wang AL, Zhang YZ et al. (2008) Characterization of HMW glutenin subunits in common wheat and related species by matrix-assisted laser desorption/ionization time-of-flight mass spectrometry (MALDI-TOF-MS). Journal of Cereal Science 47: 252–261.
Zhang W, Gianibelli MC, Rampling LR, Gale KR (2004) Characterisation and marker development for low molecular weight glutenin genes from *Glu-A3* alleles of bread wheat (*Triticum aestivum* L.). Theoretical and Applied Genetics 108: 1409–1419.
Zhang X, Liu D, Yang W, Liu K, Sun J, Guo X, et al. (2011) Development of a new marker system for identifying the complex members of the low molecular- weight glutenin subunit gene family in bread wheat (*Triticum aestivum* L.). Theoretical and Applied Genetics 122: 1503–1516.
Zheng W, Peng YC, Ma JH, Appels R, Sun DF, Ma WJ (2011) High frequency of abnormal high molecular weight glutenin alleles in Chinese wheat landraces of the Yangtze-River region. Journal of Cereal Science 54: 401–418.
Zhou JH, Zhao J, Yuan HC, Meng Y, Li Y, Wu L et al. (2007) Comparison of UPLC and HPLC for determination of trans-10-hydroxy-2-decenoicacid content in royal jelly by ultrasound-assisted extraction with internal standard. Chromatographia 66: 185–190.

Gluten Analysis

Réka Haraszi, Tatsuya M. Ikeda, Roberto Javier Peña, and Gérard Branlard

Abstract Wheat, barley and rye are sources of gluten and diverse food products are made from the grains of these cereals. Despite some species-specific differences, the molecular properties of the gluten proteins show similar characteristics in forming a unique protein network that has been extensively described in terms of its subunits and composition, its function in bakery products and its implications for human health.
There are many reasons for analysing gluten to serve purposes as diverse as assessing flour quality, selecting and breeding suitable cereal varieties, identifying varieties, identifying the source of gluten in a product, and quantifying gluten in food and drink, especially to protect gluten intolerant consumers.

The level of gluten in food and drink deemed to be safe for people with coeliac disease or non-coeliac gluten sensitivity is set in legislations. Various systems are in place worldwide to regulate food labelling and various testing methods are used or are available to quantify gluten, but cross-border standardisation to harmonize the quantification of gluten in food products has yet to be agreed.

To analyse gluten, the proteins must be separated from other possibly interfering food components. As it is difficult to solubilize gluten, gluten extraction is a critical part of the process. The level of gluten, the type of food matrix and the available technology impose further limitations and challenges.

In this chapter, we describe a strategy to select the most suitable gluten analysis approach according to the purpose, sample type, gluten level and performance characteristics required. The advantages and disadvantages of qualitative and quantitative

R. Haraszi (✉)
Campden BRI, Chipping Campden, UK
e-mail: reka.haraszi@campdenbri.co.uk

T. M. Ikeda
Western Region Agricultural research Center, NARO, Fukuyama, Japan

R. J. Peña
International Maize and Wheat Improvement Center (CIMMYT), Texcoco, Mexico

G. Branlard
INRAE, UCA UMR1095 GDEC, Clermont-Ferrand, France

G. Igrejas et al. (eds.), *Wheat Quality For Improving Processing And Human Health*, https://doi.org/10.1007/978-3-030-34163-3_6

gluten analysis techniques, such as gel electrophoresis, immunoassays, asymmetric field flow fractionation multi-angle laser light scattering, chromatography and chromatography coupled methods are covered.

1 The Purpose of the Gluten Analysis Defines the Right Technique

Reasons to analyse gluten may be to comply with food labelling legislation, to ensure food safety, to assess food quality (protein composition and functionality) or to identify and track varieties in breeding programs. There are already several qualitative and quantitative methods to serve the spectrum of needs for gluten analysis (Table 1).

Table 1 Overview of the most frequent purposes of gluten analysis in cereals and foods. Gluten levels vary from high (>100 mg/kg) to low (<100 mg/kg). RP-HPLC, reverse-phase high-performance liquid chromatography; SE-HPLC, size exclusion HPLC; PCR, polymerase chain reaction; LC-MS, liquid chromatography mass spectrometry, AFFFF-MALLS, asymmetrical flow field flow fractionation multi-angle laser light scattering; HMW, high molecular weight; LMW, low molecular weight; ELISA, enzyme-linked immunosorbent assay

Purpose		Gluten level in sample	Methods
Qualitative			
Breeding or quality assessments	Breeding new varieties	High	Gel electrophoresis, RP-HPLC
	Variety identification based on gluten composition		
Food safety	Presence/absence of gluten source, gluten speciation for labelling e.g. 'contains wheat'	Any	Gel electrophoresis, LC-MS, PCR
Methodological	Sample preparation for LC-MS analysis		Gel electrophoresis
	Checking enzymatic digestion efficiency prior to LC-MS analysis		
Clinical	Testing the immune response to gluten in serums	Low	Western blot
Quantitative			
Breeding or quality assessments	Quantification of total gluten	High	ELISA
	Gliadin to glutenin ratio		SE-HPLC
	HMW to LMW glutenin ratio		RP-HPLC
	Unextractable polymeric protein		SE-HPLC
	Size and mass of polymers	Any	AFFFF-MALLS
	Quantification of gluten in non-gluten containing food		ELISA, LC-MS
Food safety and labelling	Testing for gluten when used as a processing aid	Any	ELISA, LC-MS
	Testing for gluten in non-food materials		ELISA, LC-MS
	Food labelling – "Gluten free" (<20 mg/kg) and "very low gluten" (21–100 mg/kg)	Low	ELISA, LC-MS
	Quantification of partly or fully hydrated gluten		Competitive ELISA, LC-MS
	Validation of production line cleaning (swabs)		ELISA

Some of the methods are routinely used, while some require optimisation for use with gluten proteins. The purpose of the gluten analysis is an essential aspect to consider when first selecting an appropriate testing method. Other major determinants are the sample type and the expected or estimated level of gluten proteins in the samples. The diversity of food matrices that need to be dealt with is in itself a challenge and requires specific consideration for food analytics. The extraction of the target analyte, the gluten proteins, is more critical when the aim is quantification. Often a method is effective for samples with high levels of gluten, but less so for samples with only trace levels.

2 The Importance of Sample Type in Gluten Analysis

The grain quality of wheat, barley, rye and their cross varieties is often tested in cereal science and breeding. In these genetic materials, the level of gluten is high and the protein composition is the main characteristic of interest. Food products may contain gluten or gluten-containing cereals that have been added intentionally or may contain gluten due to unintentional contamination of raw materials during processing or product handling. Risk assessments of possible sources of contamination can be a way of estimating the expected level of gluten and gluten source before testing (Table 2).

Table 2 Overview of food sample types most often tested for gluten

Sample type	Gluten level	Foods	References (examples)
Gluten-containing cereals and cereal-based products (not heated)	High	Grains, flour, whole meal and other milling products from wheat, spelt, kamut, barley and rye	Batey et al. 1991; Bönick et al. 2017; Bromilow et al. 2017a; Colgrave et al. 2015; Cornec et al. 1994; DuPont et al. 2005; García-Molina and Barro 2017; Han et al. 2015; Labuschagne and Aucamp 2004; Larroque et al. 2007; Lexhaller et al. 2017; Lookhart et al. 1986, 1995, 2003; Martínez-Esteso et al. 2016, 2017; Peña et al. 2004; Qian et al. 2008; Schalk et al. 2017a, 2017b, 2018; Manfredi et al. 2015; Altenbach et al. 2010; Aghagholizadeh et al. 2017; Singh et al. 1991; Tatham et al. 2000; Zilic et al. 2011; Tanner et al. 2016; Guo et al. 2016
Cereal-based products (heat processed)		Bakery products, bread, pasta, noodles	García et al. 2005; Khamis 2014
Hydrolysed gluten content		Fermented products (e.g. beer, soy sauce, vinegar)	Allred et al. 2014; Cao et al. 2017; Sajic et al. 2017; Li et al. 2018

(continued)

Table 2 (continued)

Sample type	Gluten level	Foods	References (examples)
Gluten-free or low-gluten foods	Low	Non-gluten cereals and pseudo cereals (e.g. oat, maize, rice, sorghum, buckwheat, quinoa, amaranth and chia)	Real et al. 2012
		Soy, legumes, pulses	Melini et al. 2017
		Spices	
		Non-gluten containing foods and drinks	Taylor et al. 2018
		Fermented non-gluten products (e.g. soy sauce, vinegar, gluten-free beer, malt extracts and processed oats)	Panda et al. 2015
Process validation		Cleaning validation samples (swabs) Non-intentional gluten containing foods and drinks (when gluten is used as a processing aid)	
Non-food materials	Any	Binding agents or fillers, cosmetics, medicines, pet foods, children's toys	Hlywiak 2008

3 Gluten Analysis Methods

Many authors have published test protocols for gluten identification and quantification. Some methods are widely used, but often require optimisation for particular situations. Exact protocols for immunoassays, enzyme-linked immunosorbent assay (ELISAs) and Western blotting are not discussed here, as the manufacturer's instructions must be followed for each reagent kit. An overview of commercially available gluten ELISA kits was published recently with their specifications (Melini and Melini 2018). Rapid methods and convenient formats developed by some of the main ELISA manufacturers such as dip sticks (Glutentox from Biomedal (Biomedal 2017; Bromilow et al. 2017b), Rida-quick from R-bio-pharm, Veratox R5 from Neogen, etc.) and handheld devices (Taylor et al. 2018) are also available but are not discussed here in detail. The principle behind these methods is usually an immune reaction and they are less sensitive than standard ELISAs because the LOD is higher, but the assays are much faster to do.

3.1 Chromatography and Coupled Techniques for Gluten Analysis

Molecular profiling using reversed-phase (RP) or size exclusion (SE) high performance liquid chromatography (HPLC) has been widely used since the 1990s not only in wheat quality characterisation (Lookhart et al. 1986, 1995; Batey et al. 1991) but also for other applications after optimising the protocols (Table 3).

There are no standard methods for gluten analysis using liquid chromatography (LC) and LC coupled with mass spectrometry (MS) techniques. LC-MS or LC-MS/MS is considered to be a powerful and highly sensitive proteomics technique that is in high demand for food testing required for gluten-free labelling (Haraszi et al. 2011). There are several LC-MS platforms that differ in the technologies used for ionisation (e.g. electrospray ionisation (ESI) or matrix assisted laser desorption ionization (MALDI), fragmentation (e.g. triplequadrupole, quadrupole time of flight (QTOF) or Orbitrap), detection (e.g. collision induced dissociation, higher-energy collisional dissociation), acquisition modes (e.g. data dependent analysis, data independent analysis, multiple reaction monitoring) and data analysis tools (vendor specific search engines, databases and other bioinformatics packages) (Table 4). There is a definite need to standardise the different data analysis platforms and several researchers advise using multiple platforms to ensure the comparability of results (e.g. Fiedler et al. 2014; Bromilow et al. 2017b; Martínez-Esteso et al. 2016).

Several instruments and a range of extraction and digestion methods are used to identify proteins by LC-MS. A general workflow (e.g. Juhász et al. 2015a; Martínez-Esteso et al. 2016) and a table of published wheat gluten markers (Table 4) are evidence of the feasibility of using LC-MS or LC-MS/MS techniques for gluten analysis but to date these methods are not used routinely. As well as the need for expertise, the costs of instrumentation and maintenance are still limiting factors. Gluten quantification using LC-MS/MS requires that a set of peptide markers can

Table 3 Applications for which high performance liquid chromatography (HPLC) methods are used for gluten analysis. SE, size exclusion; RP, reversed phase; MALLS, multi-angle laser light scattering; MALDI-TOF, matrix assisted laser desorption ionisation - time of flight; LC-MS, liquid chromatography mass spectrometry, ELISA, enzyme-linked immunosorbent assay

Purpose	Technique	References (examples)
Characterisation of gluten sub-fractions	SE-HPLC	Cornec et al. 1994
Prediction of wheat quality	SE-HPLC	Labuschagne and Aucamp 2004
Molecular weight distribution of gluten proteins	RP-HPLC coupled with MALLS	Larroque et al. 2007
Characterisation and identification of wheat gluten proteins	RP-HPLC coupled with MALDI-TOF	Qian et al. 2008
Variety identification	RP-HPLC	Han et al. 2015
Fractionation of gluten proteins for subsequent LC-MS or ELISA testing to characterise the immunogenic fractions	RP-HPLC	Martínez-Esteso et al. 2017; Scherf 2016; Schalk et al. 2017b
Gluten quantification for studying the effect of N fertilisation	RP-HPLC	García-Molina and Barro 2017

Table 4 Published sequences of wheat gluten peptides obtained from LC-MS/MS studies that may be used for identification or quantification. Peptides were obtained with trypsin, chymotrypsin or thermolysin

Protein	Peptide sequence	Equipment	Bioinformatics	References
Alpha-gliadin	ALQTLPAMCNVY	Waters Synapt G2 nanoLC-ESI-QTOF, Xevo-TQS	PLGS, Skyline	Martínez-Esteso et al. 2016
	CQAIHNVVHAIIL	Thermo Scientific LTQ XL linear ion trap	Bioworks 3.3, Xcalibur	Manfredi et al. 2015
	DVVLQQHNIAHGR	Agilent 6530 LC-ESI-QTOF	Spectrum Mill	Liao et al. 2017
	DVVLQQPNIAHASSK	Thermo Scientific nanoLC-ESI Q-Exactive hybrid quadrupole-orbitrap	PEAKS	Aghagholizadeh et al. 2017
	FQPSQQNPQAQGF	Bruker HCT-Ultra PTM iontrap MS, Thermo Scientific TSQ Vantage	DataAnalysis 3.4, Mascot	Schalk et al. 2017a
	LQLQPFPQPQLPY	Waters Synapt G1 nanoLC-ESI-QTOF & Xevo TQS	PLGS, Skyline	van den Broeck et al. 2015
		Waters nanoLC -Thermo Scientific LTQ XL	Mascot, Scaffold, Skyline, Xcalibur	Fiedler et al. 2014
	LQLQPFPQPQLPYPQPHLPYPQPQPF	Waters Synapt G1 nanoLC-ESI-QTOF & Xevo TQS	PLGS, Skyline	van den Broeck et al. 2015
	LQLQPFPQPQLPYPQPQLPYPQPQ LPYPQPQPF			
	LQLQPFPQPQLPYPQPQLPYPQPQPF			
	LQLQPFPQPQLPYPQPQPF	Waters Synapt G1 nanoLC-ESI-QTOF & Xevo TQS	PLGS, Skyline	van den Broeck et al. 2015
		Bruker HCT-Ultra PTM iontrap MS Thermo Scientific TSQ Vantage	DataAnalysis 3.4, Mascot	Schalk et al. 2017a
	LWQIPEQSR	Agilent 6530 LC-ESI-QTOF	Spectrum Mill	Liao et al. 2017
	NLALQTLPAMCNVYIPPYCTIVPFGIFGTN	Waters Synapt G2 nanoLC-ESI-QTOF, Xevo-TQS	PLGS, Skyline	Martínez-Esteso et al. 2016
	QIPEQSR	Thermo Scientific LTQ XL linear ion trap	Bioworks 3.3, Xcalibur	Manfredi et al. 2015
	QQILQQQLIPCRDVVL	Waters Synapt G1 nanoLC-ESI-QTOF & Xevo TQS	PLGS, Skyline	van den Broeck et al. 2015
	QQQLIPCRDVVL			

	RPQQPYPQPQPQY	Waters Synapt G1 nanoLC-ESI-QTOF & Xevo TQS	PLGS, Skyline	van den Broeck et al. 2015
		Bruker HCT-Ultra PTM iontrap MS, Thermo Scientific TSQ Vantage	DataAnalysis 3.4, Mascot	Schalk et al. 2017a
		Waters nanoLC - Thermo Scientific LTQ XL	Mascot, Scaffold, Skyline, Xcalibur	Fiedler et al. 2014
	RPQQPYPQPQPQYSQPQHPIS QQQQQQQQQQQEQQIL	Waters Synapt G2 nanoLC-ESI-QTOF, Xevo-TQS	PLGS, Skyline	Martínez-Esteso et al. 2016
	RPQQPYPQSQPQY	Waters Synapt G1 nanoLC-ESI-QTOF, Xevo TQS	PLGS, Skyline	van den Broeck et al. 2015
	VSQQSYQLLQQLCCLQLWQTPEQSR	Waters Synapt G2 nanoLC-ESI-QTOF, Xevo-TQS	PLGS, Skyline	Martínez-Esteso et al. 2016
		Thermo Scientific nanoLC-ESI- Q-Exactive hybrid quadrupole-orbitrap	PEAKS	Aghagholizadeh et al. 2017
Avenin-LIKE	SAWEPQHPSSPEHQPTPQPQEHPVPHQK	Waters Synapt G2 nanoLC-ESI-QTOF, Xevo-TQS	PLGS, Skyline	Martínez-Esteso et al. 2016
	TAWEPHHPSSPEQQPTPQPQEQPVPHQK			
Gamma-gliadin	LQCQAIHNVVHAIILHQQQK	Waters Synapt G2 nanoLC-ESI-QTOF, Xevo-TQS	PLGS, Skyline	Martínez-Esteso et al. 2016
	NYLLQQCDPVSLVSSLVSMILPR			
	RPLFQLIQGQGIIRPQQPAQLEVIR			
	AFPQPQQTFPHQPQQQVPQPQQPQQPF	Applied Biosystems QSTAR Pulsar i MS	X!Tandem, Mascot Daemon, Scaffold, Protein Prophet	Altenbach et al. 2010
	ALRTLPTMCNVY	Waters Synapt G2 nanoLC-ESI-QTOF, Xevo-TQS	PLGS, Skyline	Martínez-Esteso et al. 2016
	ANIDAGIGGQ	Applied Biosystems QSTAR Pulsar i MS	X!Tandem, Mascot Daemon, Scaffold, Protein Prophet	Altenbach et al. 2010
	APFASIVADIGGQ	Agilent 6530 LC-ESI-QTOF	Spectrum Mill	Liao et al. 2017
		Applied Biosystems QSTAR Pulsar i MS	X!Tandem, Mascot Daemon, Scaffold, Protein Prophet	Altenbach et al. 2010

(continued)

Table 4 (continued)

Protein	Peptide sequence	Equipment	Bioinformatics	References
	APFASIVAGIGGQ	Agilent 6530 LC-ESI-QTOF	Spectrum Mill	Liao et al. 2017
		No MS	No MS	Srinivasan et al. 2015
		Applied Biosystems QSTAR Pulsar i MS	X!Tandem, Mascot Daemon, Scaffold, Protein Prophet	Altenbach et al. 2010
	ASIVADIGGQ	Applied Biosystems QSTAR Pulsar i MS	X!Tandem, Mascot Daemon, Scaffold, Protein Prophet	Altenbach et al. 2010
	ASIVAGIGGQ	Bruker HCT-Ultra PTM iontrap MS Thermo Scientific TSQ Vantage	DataAnalysis 3.4, Mascot	Schalk et al. 2017a
				Schalk et al. 2018
	ASIVAGISGQ	Applied Biosystems QSTAR Pulsar i MS	X!Tandem, Mascot Daemon, Scaffold, Protein Prophet	Altenbach et al. 2010
	ASIVASIGGQ			
	CQQPQQTIPQPHQTF			
	FHQPQQQFPQPQQPQQ			
	FHQPQQQFPQPQQPQQSFPQQQRP			
	FRQPQQPFY			
	FRQPQQPFYQQPQQTFPQPQQ			
	FYQQPQQTFPQPQQ			
	GIIQPQQPAQLEGIRSLVL			
	HQPQQQFPQPQQPQQSFPQ			
	HQPQQQFPQPQQPQQSFPQQQRPF			
	IIMQQEQRQG			
	IIMQQEQRQGVQ			
	IIQPQQPAQYE			
	IIQPQQPAQYEVIRS			
	ILLPLSQQQQL			
	ILLPLSQQQQLGQGTL			

	INVPYANIDAGIGGQ			
	IQILRPLFQ			
	IQPSLQQR			
	IQPSLQQRL			
	KAPFASIVADIGGQ			
	LAQIPRQ			
	LPLSQQQQVGQGSLVQGQGIIQ PQQPAQL	Waters Synapt G2 nanoLC-ESI-QTOF, Xevo-TQS	PLGS, Skyline	Martínez-Esteso et al. 2016
	LQPHQIAQL	Thermo Scientific LTQ XL linear ion trap	Bioworks 3.3, Xcalibur	Manfredi et al. 2015
	LQPHQPF	Applied Biosystems QSTAR Pulsar i MS	X!Tandem, Mascot Daemon, Scaffold, Protein Prophet	Altenbach et al. 2010
		Thermo Scientific LTQ XL linear ion trap	Bioworks 3.3, Xcalibur	Manfredi et al. 2015
	LQPHQPFSQQPQQ	Applied Biosystems QSTAR Pulsar i MS	X!Tandem, Mascot Daemon, Scaffold, Protein Prophet	Altenbach et al. 2010
	LQPQQPQQSFPQQQQPL	Applied Biosystems QSTAR Pulsar i MS	X!Tandem, Mascot Daemon, Scaffold, Protein Prophet	Altenbach et al. 2010
		Bruker HCT-Ultra PTM iontrap MS Thermo Scientific TSQ Vantage	DataAnalysis 3.4, Mascot	Schalk et al. 2017a
				Schalk et al. 2018
	LQPQQPQQSFPQQQQPLIQ	Applied Biosystems QSTAR Oulsar i MS	X!Tandem, Mascot Daemon, Scaffold, Protein Prophet	Altenbach et al. 2010
	LQPQQPQQSFPQQQQPLIQL			
	LQPQQPQQSFPQQQQPLIQLSL			
	LQQPQQPFPQPQQQLPQPQQPQQ			
	LQQQCSPVAMPQR	Thermo Scientific LTQ XL linear ion trap	Bioworks 3.3, Xcalibur	Manfredi et al. 2015
	NFLLQQCNHVSLVSSLVSIILPR	Applied Biosystems QSTAR Pulsar i MS	X!Tandem, Mascot Daemon, Scaffold, Protein Prophet	Altenbach et al. 2010
	NFLLQQCNPVSLVSSLISMILPR			

(continued)

Table 4 (continued)

Protein	Peptide sequence	Equipment	Bioinformatics	References
	NIQVDPSGQVQW	Bruker HCT-Ultra PTM iontrap MS	DataAnalysis 3.4, Mascot	Schalk et al. 2017a
		Thermo Scientific TSQ Vantage		Schalk et al. 2018
	NIQVDPSGQVQWLQQQLVPQLQQPL	Applied Biosystems QSTAR Pulsar i MS	X!Tandem, Mascot Daemon, Scaffold, Protein Prophet	Altenbach et al. 2010
	PFIQPSLQQR			
	QGVQILVPL			
	QLAQLEAIR			
	QLVQGQGIIQPQQPAQY	Waters nanoLC -Thermo Scientific LTQ XL	Mascot, Scaffold, Skyline, Xcalibur	Fiedler et al. 2014
	QPFPQQPQQPYPQQPQQPFPQT QQPQQPFPQSK	Applied Biosystems QSTAR Pulsar i MS	X!Tandem, Mascot Daemon, Scaffold, Protein Prophet	Altenbach et al. 2010
	QQCCQQLAR	Shimadzu nano HPLC, SCIEX 5600 TripleTOF MS	Protein Pilot 4.0 (SCIEX)	Li et al. 2018
	QQLPQPQQPQQSFPQQQR	Applied Biosystems QSTAR Pulsar i MS	X!Tandem, Mascot Daemon, Scaffold, Protein Prophet	Altenbach et al. 2010
	QSFPQQQRPF			
	QSFPQQQRPFIQPSLQQR			
	RQPQQPF			
	SDCQNMQQQCCQQLAQIPR	Thermo Scientific nanoLC-ESI- Q-Exactive hybrid quadrupole-orbitrap	PEAKS	Aghagholizadeh et al. 2017
	SDCQVMQQQCCQQLAQIPR	Agilent 6530 LC-ESI-QTOF	Spectrum Mill	Liao et al. 2017
				Simonato et al. 2011
	SFPQQQPPF	Applied Biosystems QSTAR Pulsar i MS	X!Tandem, Mascot Daemon, Scaffold, Protein Prophet	Altenbach et al. 2010
	SIIMQQEQRQGVQIRRPL			
	SQQPQQAFPQPQQTFPHQPQQQVPQ PQQPQQPF			
	SQQQQLGQGTL			
	SQQQQLGQGTLVQGQGIIQPQQL			

	SQQQLGQGTLVQGQGIIQPQQLAQL			
	SQQQLGQGTLVQGQGIIQPQQ LAQLEAIRSL			
	SQQQVGQGIL			
	SQQQVGQGILVQGQGIIQPQQPAQL			
	SQQQVGQGSL			
	SQQQVGQGSLVQGQGIIQPQQPAQL			
	TQQPQQPFPQFQQPHQPF			
	VDPGYQVHWPQQQPFPQPQQP			
	VHWPQQQPFPQPQQP			
	VPPECSIIRAPF			
	VPPECSIMR	Thermo Scientific LTQ XL linear ion trap	Bioworks 3.3, Xcalibur	Manfredi et al. 2015
	VPPNCSTINVPY	Applied Biosystems QSTAR Pulsar i MS	X!Tandem, Mascot Daemon, Scaffold, Protein Prophet	Altenbach et al. 2010
	VPPNCSTINVPYANIDAGIGGQ6			
	VQGQGIIQPQQL	Waters nanoLC -Thermo Scientific LTQ XL	Mascot, Scaffold, Skyline, Xcalibur	Fiedler et al. 2014
	VQGQGIIQPQQLAQLEAIRSL	Applied Biosystems QSTAR Pulsar i MS	X!Tandem, Mascot Daemon, Scaffold, Protein Prophet	Altenbach et al. 2010
	VQQQIPIVQPSVL	Thermo Scientific LTQ XL linear ion trap	Bioworks 3.3, Xcalibur	Manfredi et al. 2015
	VSPDCSTINAPF	Applied Biosystems QSTAR Pulsar i MS	X!Tandem, Pascot Paemon, Pcaffold, Protein Prophet	Altenbach et al. 2010
	VSPDCSTINAPFASIVVGIGGQ			
	VTILRPLFQ			
	VYVPPYCST			
	YQQPQQTFPQPQ			
	YQQPQQTFPQPQQ			
	YQQQQVGQGTLVQGQGIIQPQQPAQL			
Gliadin/ avenin-like	TTISPSSDVTTDMGGY	Waters Synapt G2 nanoLC-ESI-QTOF, Xevo-TQS	PLGS, Skyline	Martínez-Esteso et al. 2016

(continued)

Table 4 (continued)

Protein	Peptide sequence	Equipment	Bioinformatics	References
HMW-GS	HVSVEHQAASL	Bruker HCT-Ultra PTM iontrap MS, Thermo Scientific TSQ Vantage	DataAnalysis 3.4, Mascot	Schalk et al. 2018
				Schalk et al. 2017a
	LQPGQGQQGY	Bruker HCT-Ultra PTM iontrap MS, Thermo Scientific TSQ Vantage	DataAnalysis 3.4, Mascot	Schalk et al. 2018
				Schalk et al. 2017a
	QGQGPQGKQGY	Waters Synapt G2 nanoLC-ESI-QTOF, Xevo-TQS	PLGS, Skyline	Martínez-Esteso et al. 2016
	QQPGQGQHPEQGK	Shimadzu nano HPLC, SCIEX 5600 TripleTOF MS	Protein Pilot 4.0 (SCIEX)	Li et al. 2018
	QQPGQGQQPEQGQQPGQGQQGYYP TFPQQPGQGK	Waters Synapt G2 nanoLC-ESI-QTOF, Xevo-TQS	PLGS, Skyline	Martínez-Esteso et al. 2016
	QQPGQGQQTR	Shimadzu nano HPLC, SCIEX 5600 TripleTOF MS	Protein Pilot 4.0 (SCIEX)	Li et al. 2018
	QVVDQQLAGR			
	TASLQQPGQGQQGHYPASL	Bruker HCT-Ultra PTM iontrap MS, Thermo Scientific TSQ Vantage	DataAnalysis 3.4, Mascot	Schalk et al. 2018
				Schalk et al. 2017a
	TTSLQQSGQGQQGY	Waters Synapt G2 nanoLC-ESI-QTOF, Xevo-TQS	PLGS, Skyline	Martínez-Esteso et al. 2016
	VAKNQQLAAQLPAMCR			
HMW-GS Ax2	QQDQQSGQGQQPGQR	Shimadzu nano HPLC, SCIEX 5600 TripleTOF MS	Protein Pilot 4.0 (SCIEX)	Li et al. 2018
	QQPGQGQQLR			
	YYPTSPQQPGQEQQPR			
HMW-GS Bx14	DVSPGCRPITVSPGTR	Shimadzu nano HPLC, SCIEX 5600 TripleTOF MS	Protein Pilot 4.0 (SCIEX)	Li et al. 2018
HMW-GS Bx17	AQQLAAQLPAMCRLEGSDALSTR LEGSDALSTR	Thermo Scientific nanoLC-ESI-Q-Exactive hybrid quadrupole-orbitrap	PEAKS	Aghagholizadeh et al. 2017
	DVSPGCRPITVSPGTR			
	QDQQPGQR			
	QQGYYPTSPQQPGQGQQLGQGQPG YYPTSQQPGQK			

	QQSGQGQQPGQGQQSGQGQQPGQG QQAYYPTSSQQSR			
	QYEQQPVVPSK			
	RYYPSVTSSQQGSYYPGQASPQQSGQG QQPGQEQQPGQGQ			
HMW-GS By15	CCQQLR	Thermo Scientific nanoLC-ESI- Q-Exactive hybrid quadrupole-orbitrap	PEAKS	Aghagholizadeh et al. 2017
	ELQESSLEACRQVVDQQLAGRLPWST GLQMR			
	ELQESSLEQCR			
	LPWSTGLQMR			
	QLQCERELQESSLEACR			
	QVVDQQLAGR			
	VQQPATQLPIMCR			
HMW-GS Dx5	ACQQVMDQQLR	Thermo Scientific nanoLC-ESI- Q-Exactive hybrid quadrupole-orbitrap	PEAKS	Aghagholizadeh et al. 2017
	AQQLAAQLPAMCR			
	DISPECHPVVVSPVAGQYEQQIVVPPK			
	ELQELQER			
	GGSFYPGETTPPQQLQQR			
	IFWGIPALLK	AB SCIEX QTRAP® 4500	MIDAS™ workflow	Lock 2014
	IFWGIPALLKR	Thermo Scientific nanoLC-ESI- Q-Exactive hybrid quadrupole-orbitrap	PEAKS	Aghagholizadeh et al. 2017
	PQQPGQWQQPEQGQPR			
	YYPSVTCPQQVSYYPGQASPQRPGQGQQ PGQGQQGYYPTS			

(continued)

Table 4 (continued)

Protein	Peptide sequence	Equipment	Bioinformatics	References
HMW-GS Dy10	AQQPATQLPTVCR	Thermo Scientific nanoLC-ESI-Q-Exactive hybrid quadrupole-orbitrap	PEAKS	Aghagholizadeh et al. 2017
	CCQQLR			
	ELQESSLEACR			
	LPWSTGLQMR	AB SCIEX QTRAP® 4500	MIDAS™ workflow	Lock 2014
		Thermo Scientific nanoLC-ESI Q-Exactive hybrid quadrupole-orbitrap	PEAKS	Aghagholizadeh et al. 2017
	QLQCERELQESSLEACR	Thermo Scientific nanoLC-ESI Q-Exactive hybrid quadrupole-orbitrap	PEAKS	Aghagholizadeh et al. 2017
	QQPVQGQQPEQGQQPGQWQQGYYPTS PQQLGQGQQPR			
	QVVDQQLAGR	AB SCIEX QTRAP® 4500	MIDAS™ workflow	Lock 2014
	QVVDQQLAGRLPWSTGLQMR	Thermo Scientific nanoLC-ESI Q-Exactive hybrid quadrupole-orbitrap	PEAKS	Aghagholizadeh et al. 2017
	SVAVSQVAR	AB SCIEX QTRAP® 4500	MIDAS™ workflow	Lock 2014
		Thermo Scientific nanoLC-ESI Q-Exactive hybrid quadrupole-orbitrap	PEAKS	Aghagholizadeh et al. 2017
HMW-GS Dy3	ACQQVMDQQLR	Thermo Scientific nanoLC-ESI Q-Exactive hybrid quadrupole-orbitrap	PEAKS	Aghagholizadeh et al. 2017
	AQQPATQLPTVCR			
	DISPECHPVVVSPVAGQYEQQIVVPPK GGSFYPGETTPPQQLQQR			
	ELKACQQVMDQQLR			
	GGSFYPGETTPPQQLQQR			
	IFWGIPALLKP			
LMW-d	QQHQIPQQPQQFPQQQQF	Shimadzu LC - SCIEX QTRAP 5500	Analyst 1.6.1, Skyline	Bönick et al. 2017
	SQQQISQQPQQLPQQQQIPQQPQQF			

LMW-GS	GQQPQQQQL	Bruker HCT-Ultra PTM iontrap MS, Thermo Scientific TSQ Vantage	DataAnalysis 3.4, Mascot	Schalk et al. 2018
				Schalk et al. 2017a
	GSSLTSIDGQ	Waters Synapt G2 nanoLC-ESIQTOF, Xevo-TQS	PLGS, Skyline	Martínez-Esteso et al. 2016
	GVGTRVGAY			
	LQQCSQTAY			
	QLPQIPEQSR	Shimadzu nano HPLC, SCIEX 5600 TripleTOF MS	Protein Pilot 4.0 (SCIEX)	Li et al. 2018
	QQQPLPPQQTFPQQPL	Bruker HCT-Ultra PTM iontrap MS Thermo Scientific TSQ Vantage	DataAnalysis 3.4, Mascot	Schalk et al. 2018
				Schalk et al. 2017a
	SHHQQQQPIQQQPQPF	Shimadzu nano HPLC SCIEX 5600 TripleTOF MS	Protein Pilot 4.0 (SCIEX)	Li et al. 2018
	SIILQEQQQGF	Bruker HCT-Ultra PTM iontrap MS Thermo Scientific TSQ Vantage	DataAnalysis 3.4, Mascot	Schalk et al. 2018
				Schalk et al. 2017a
	SQMLQQSICHVMQQQCRQQLR	Waters Synapt G2 nanoLC-ESI-QTOF, Xevo-TQS	PLGS, Skyline	Martínez-Esteso et al. 2016
	VQQQIPVVQPSIL	Bruker HCT-Ultra PTM iontrap MS Thermo Scientific TSQ Vantage	DataAnalysis 3.4, Mascot	Schalk et al. 2018
	VQQQLPVVQPSIL	Bruker HCT-Ultra PTM iontrap MS Thermo Scientific TSQ Vantage	DataAnalysis 3.4, Mascot	Schalk et al. 2017a
LMW-i	ILPTMCSVNNPLYR	Thermo Scientific nanoLC-ESI Q-Exactive hybrid quadrupole-orbitrap	PEAKS	Aghagholizadeh et al. 2017
	SQMLQQSICHVMQQQCCQQLPQIP QQSRYEAIR			
	TTTSVPFGVGTGVGAY`			
	VFLQQQCIPVAMQR			
LMW-m	QLPQIPEQSRYDAIR	Thermo Scientific nanoLC-ESI Q-Exactive hybrid quadrupole-orbitrap	PEAKS	Aghagholizadeh et al. 2017
	SQTLWQSSCHVMQQQCCR			
	VFLQQQCSPVAMPQSLAR			

(continued)

Table 4 (continued)

Protein	Peptide sequence	Equipment	Bioinformatics	References
LMW-s	SQMLQQSSCHVMQQQCCQQLPQIPQQSRYEAIR	Thermo Scientific nanoLC-ESI Q-Exactive hybrid quadrupole-orbitrap	PEAKS	Aghagholizadeh et al. 2017
	TTTSVPFDVGTGVGAY			
	VFLQQQCSPVAMPQSLAR			
Omega-gliadin	HQQPEQIISQQPQQPF	Waters Synapt G2 nanoLC-ESI-QTOF, Xevo-TQS	PLGS, Skyline	Martínez-Esteso et al. 2016
	ILQPQQPLPQQPQQPF			
	SPHQPQQPFPQPQRPTPL			
	HQQQLPQQQFPQQQFPQQQFPQQQQFPQQQQL			
	FPHQSQQPF	Bruker HCT-Ultra PTM iontrap MS Thermo Scientific TSQ Vantage	DataAnalysis 3.4, Mascot	Schalk et al. 2017a
	GSSLTSIGGQ			
	QPHQPQQPYPQQ	Shimadzu LC - SCIEX QTRAP 5500	Analyst 1.6.1, Skyline	Bönick et al. 2017
	QQYPQQQPSGSDVISISGL	Bruker HCT-Ultra PTM iontrap MS Thermo Scientific TSQ Vantage	DataAnalysis 3.4, Mascot	Schalk et al. 2017a

be targeted and based on peptide fragmentation the amount of gluten can be detected. It is clear from Table 4 that only very few wheat gluten peptides have been identified in the multiple studies using different LC-MS platforms. LQLQPFPQPQLPY, LQLQPFPQPQLPYPQPQPF, RPQQPYPQPQPQY and VSQQSYQLLQQLCCLQLWQTPEQSR from alpha-gliadin, APFASIVADIGGQ, APFASIVAGIGGQ, LQPHQPF and LQPQQPQQSFPQQQQPL from gamma-gliadin, LPWSTGLQMR and SVAVSQVAR from HMW-GS Dy10 were each found in at least two studies. The abundance of certain proteins or peptides may be as low as zero or below a detectable limit. Even if their abundance is sufficiently high to be detected, these peptides are obtained via enzymatic digestion after protein extraction. The use of chymotrypsin, as opposed to trypsin, proved to be more successful for gluten digestion due to the particular amino acid composition of gluten proteins and the limited number of trypsin specific cleavage sites (e.g. Sealey-Voyksner et al. 2010; Martínez-Esteso et al. 2016).

It is thus more feasible to use chromatography and coupled techniques as confirmatory or identification approaches in gluten analysis. Relative quantification of the different gluten protein types is routinely done by SE- and RP-HPLC, but absolute quantification of gluten components is not yet fully achievable using LC-MS platforms.

As well as SE-HPLC, it is highly advisable to use another analytical technique, asymmetrical flow field-flow fractionation multi-angle laser light scattering (AFFFF-MALLS) to fully characterise storage protein polymers that have accumulated in cereal grains and are present in flour. In this case, the molecular screening is performed in a trapezoidal shaped cell where polymers are subjected to a double cross-flow gradient followed by multi-angle detection of deviation of a laser beam (Lemelin et al. 2002). This technique does not involve a stationary phase so the absence of protein shearing forces offers the possibility of measuring several polymer parameters (such as molecular mass and radius of gyration) and hence knowing the distribution of these molecular characteristics within the sample analysed and calculating the polydispersity index. Such measurements are not possible with SE-HPLC separation, which often has a cut-off of about 1000 kDa. AFFFF-MALLS has proved useful for characterizing polymer masses in flour, for example, by showing which of their properties explain the environmental stability of bread making quality (Lemelin et al. 2005), that they are highly influenced when wheat grain is subjected to ozone treatment (Goze et al. 2017), and that they undergo the unfolded protein response caused by environmental stresses during protein accumulation (Branlard et al. 2015). AFFFF-MALLS is likely to be the tool of choice for further research especially that aimed at reducing the polymer masses to render the gluten better for consumer health.

3.2 Electrophoresis Techniques

Electrophoresis techniques for gluten analysis are very specific and widely used. Detailed protocols and highlighted applications are provided here for the most frequently used techniques (Table 5 and Annexes).

Table 5 Gel electrophoresis methods to study gluten proteins

Method	Purpose	Advantage	Disadvantage	References
Sodium dodecyl sulfate polyacrylamide gel electrophoresis	To separate proteins based mainly on their molecular weights	Detecting small variations in protein size	Difficult to separate many proteins having similar molecular weights. Measured molecular weights are not accurate.	Singh et al. 1991; Peña et al. 2004; Ikeda et al. 2008
Acid polyacrylamide gel electrophoresis	To separate proteins based on their molecular weights and charges.	Useful to separate gliadins.	Gel handling is difficult. Interpretation of the banding patterns is difficult.	Metakovsky and Novoselskaya 2001; Branlard et al. 1990
Two-dimensional gel electrophoresis	To separate more proteins based on their isoelectric point and molecular weights	Separate more proteins. Usable for protein sequencing and MS analysis.	Time consuming. Immobilised IEF gels are more expensive.	Dumur et al. 2004; Ikeda et al. 2006; Liu et al. 2010

3.2.1 Sodium Dodecyl Sulphate Polyacrylamide Gel-Electrophoresis (SDS-PAGE)

SDS-PAGE is the most frequently used technique for HMW-GS analysis and is partly used for LMW-GS analysis. This technique has the advantage of allowing the detection of small size variations in HMW-GSs, but it is less useful for separating LMW-GSs and gliadins because they include many proteins with similar molecular weights. When using SDS-PAGE for gluten analysis, it is necessary to first block the free SH residues of the component proteins by alkylation with 4-vinylpyridine. The concentration of the bis-acrylamide cross-linker and the pH of the separation gel are also important aspects to optimise when separating gluten proteins. Better separation is obtained using a lower bis-acrylamide concentration (1.3%C) and lower pH (pH 8.5). A standard protocol used at the Wheat Chemistry and End–Use Quality Laboratory of CIMMYT is shown at the end of this chapter in Appendix I with methods for selective extraction then electrophoresis of gliadins and glutenins for SDS-PAGE.

3.2.2 Acid Polyacrylamide Gel-Electrophoresis (A-PAGE)

A-PAGE is currently only used for the advantages it offers for analysing gliadins due to the difficulty of handling the gels. A-PAGE separates gliadins better than SDS-PAGE, because it separates them based on their molecular weights and charges. There is a huge diversity of gliadins. Although it is difficult to interpret the banding patterns, the catalog by Metakovsky et al. (2018) lists 182 alleles at the six *Gli* loci of common wheat that may be useful for genomic analysis of gliadin gene families. A protocol used at INRA (France) is shown at the end of this chapter in Appendix II.

3.2.3 Two-Dimensional Gel Electrophoresis (2-DE)

The 2-DE technique separates more proteins based on their isoelectric point and molecular weights. It has been used for gluten protein sequencing (Ikeda et al. 2006) and MS analysis (Liu et al. 2010). The cost of immobilised isolelectric focusing (IEF) gels and instruments for IEF is nevertheless a limiting factor.

3.3 Challenges in Gluten Analysis

Gluten testing is undoubtedly a challenge and has been recently reviewed (Melini and Melini 2018). Due to the unique properties of gluten proteins, routine methods that are suitable for general protein analysis have often been found to be unsuccessful or have required prior protocol modification. While immunoassays have been shown to be suitable for routine gluten analysis in relation to compliance with food legislation and labelling, the limitations and challenges of other methods such as LC-MS are apparent. Critical factors like the complexity of the food matrix, the type of antibody in immunoassays, gluten extraction procedures and lack of reference material can all impact the reliability of immune-detection of gluten proteins and the need for harmonisation has been clearly highlighted.

3.3.1 Definition of Gluten

One of the challenges of gluten analysis is the ambiguity with which it is defined. In bread-making, the gluten is obtained when flour is added to water then mixed and washed with salt solution until other flour compounds, particularly starch and soluble proteins, are removed. The remaining viscoelastic portion is classically called gluten. By contrast, the legislative definition of gluten in Europe encompasses oat as a gluten source and defines gluten proteins according to their insolubility in 0.5 M NaCl (Codex 2008). The properties of oat avenins are however distinctly different from the properties of wheat gluten, barley hordeins and rye secalins, especially from the point of view of their toxicity (Real et al. 2012). Wheat research most frequently refers to the Osborne definition of gluten. Historically, wheat proteins were classified as water-soluble albumins, salt-soluble globulins, alcohol-soluble gliadin (prolamins) and insoluble glutenin (glutelins) (Osborne 1924). There is now a need in food labelling to display the gluten source, whether wheat, barley or rye, as some consumers may suffer from food allergy. Differentiating between wheat, barley and rye gluten is difficult, especially with certain methods (e.g. immunoassays) due to the similar sequence characteristics and solubility of gluten proteins. The definition of gluten is therefore specific for the selected extraction and analysis method so it is very important to state this especially in food safety applications. Conversely, if the purpose of gluten testing is for legislative labelling purposes, the legislative definition of gluten may determine what extraction and testing methods need to be used.

3.3.2 Solubility and Extractability of Gluten

The solubility of gluten proteins depends on the extraction solvents used (pH, ion strength, polarity) and the composition of the surrounding matrix. Gluten is most often extracted with either 60% or 80% ethanol (van den Broeck et al. 2009; Mena et al. 2012), 55% isopropanol (Colgrave et al. 2015), isopropanol and NaI (DuPont et al. 2005), or multi-step protocols based on the Osborne fractionation using a series of extraction solvents (Lookhart and Bean 1995; Zilic et al. 2011; Fallahbaghery et al. 2017).

The extraction efficiency of gluten proteins also depends on the fat and carbohydrate content of the matrix. In the future, it may be easier to design a standardised protocol for extracting gluten from wheat and other cereals than from food, especially processed food. For example, the presence of lipids and polyphenols influence protein solubility and the molecules can interfere with protein detection and identification when present in protein fractions.

Gluten solubility can be aided by converting the disulfide bonds into sulfhydryl groups using reducing agents such as dithiothreitol or beta-mercaptoethanol. In the presence of urea, proteins can be denatured and SDS can mask the surface charges of peptides and proteins. The use of polyvinylpyrrolidone was shown to aid gluten extraction from chocolate or cacao containing samples (Mena et al. 2012; Satsuki-Murakami et al. 2018). Fish gelatin, a reducing agent (Tris (2-carboxyethyl)-phosphine) and an anionic surfactant (N-lauroylsarcosine) are used in the universal prolamin and glutelin extractant solution (UPEX) before extraction with 80% ethanol, which is claimed to be suitable for all types of subsequent analysis techniques such as ELISA and LC-MS (Mena et al. 2012). Recently, a rapid, simple, and reproducible protocol for extraction and digestion of gluten proteins was published that is suitable for LC-MS quantification (Li et al. 2019).

The different extraction methods target various proportions of the different gluten protein types. The purity of the obtained gluten fractions can vary not only due to the presence of non-protein compounds but also of other non-target proteins. For example, the glutenin fraction contains gliadins while the LMW-GS fraction may contain omega-gliadins. The sequence homology between gliadins and LMW-GS means they have similar affinity for extraction buffers and is the main reason for their co-extraction.

3.3.3 Gluten Protein Sequences and Structure

Gluten proteins have a great amount of sequence homology within and between species. The secondary structure and conformation of the gluten protein chains differ however due to the presence of S-containing amino acids and the various polypeptide chain lengths. The S content of proteins makes them prone to disulfide bridge formation, which is a dynamic chemical bonding between the S-S and the reduced SH-SH forms. The sequence characteristics determine the physical and chemical properties of the proteins, which are very similar for the corresponding gluten protein subgroups of different species (e.g. HMW-GS in wheat and D-hordeins in barley). This homologue behavior can be advantageous and disadvantageous depending on the purpose of testing. When total gluten content is analysed, extraction is easier if the compounds of interest have similar properties. When the aim is to define the source of gluten (e.g. whether it is from wheat, barley or rye) or to characterise or quantify the different subgroups or even to target certain sequences, sequence homology is a major problem.

Accessibility of enzymes and antibodies to the target protein/peptide/epitope sequence is a substantial limiting factor in gluten detection methods. The use of different mono- or polyclonal antibodies in immunoassays, the specificity of antibodies, and the abundance of the immune-responsive protein sites are often the reason for variation in the performance of ELISA kits (Schopf and Scherf 2018). Enzymes can only cleave proteins if they have physical access to their specific cleavage sites on the relevant section of the polypeptide. Enzyme accessibility is therefore a major factor when producing peptides for LC-MS detection. Unfolding of the three-dimensional and secondary structure of the protein chain for digestion is a crucial step.

In MS-based proteomics, the identification of protein/peptide sequence is based on using a protein sequence database and comparing it to the detected mass of an ionised peptide fragment. Identification is based on known amino acid residue masses, cleavage rules of the applied enzyme(s) and allowed missed cleavage(s). The proteins may have post-translational modifications that could themselves be modified during processing (e.g. deamidation). Modifications can be fixed or variable and can affect all or just some of the amino acid residues. Consequently, identification is limited by the number and completeness of sequences available in the database. It is only recently that the wheat genome sequencing project was completed, and a reference genome became available (International Wheat Genome Sequence Consortium 2018). Once the contents of the genome database are converted into searchable expressed protein sequences, then the capabilities of MS based protein identification methods will improve. Correct annotations are also important when identifying proteins or the plant source. Annotations of gluten proteins in the current databases (e.g. www.uniprot.org) are often incorrect but a manually curated prolamin sequence database (including gluten) has now been created (www.propepper.net, Juhász et al. 2015b). A similar database dedicated to gluten has been developed as a tool for proteomic studies (Glu.Pro V1.0, Bromilow et al. 2017b).

The quantities of individual gluten proteins in a sample might be low and the peptide quantities even lower. The expression level of individual proteins are species and variety specific and will differ depending on the growing/environmental conditions. Biotic and abiotic stresses have an impact on the expression levels of proteins and protein groups (see 'Effects of environmental changes on the allergen content of wheat grain' chapter). In any gluten analysis method that relies on sequence data for identification or quantification, it is crucial to select abundant target peptides/proteins that are unique for the species, the total gluten content or a particular variety independent of the possible effects of stresses.

3.3.4 Method Performance Characteristics

The aim of gluten analysis determines the required sensitivity of a method. The surrounding food matrix is often the limiting factor in the achievable LOD or lower limit of quantification.

In immunoassays, the antibodies selected to target gluten peptides may cross-react with other non-target proteins creating false positive results or have more affinity for certain proteins perhaps from other species. In gluten ELISAs, overestimation and underestimation of gluten from one or other species are known issues. Indeed the antibodies in certain kits (e.g. R5) were developed against peptides/proteins of a particular species (e.g. barley hordein) and therefore the assay overestimates the quantity of proteins from that species. Continuing with the example of the R5 ELISA, test results are an underestimation of the actual level of wheat gluten because glutenin detection is not accounted for (e.g. Dostalek et al. 2006). Recent developments to detect total gluten content in oat by using a multiplex assay showed that it is possible to overcome this issue by selecting a better set of antibodies raised against gliadins and glutenins (Boison et al. 2018).

In MS-based gluten identification, if the target peptide for quantification is selected carefully and is unique for the gluten or its specific fraction, the possibility of cross reactivity can be excluded. In LC-MS/MS methods, the difficulty is to achieve limits of detection that are similar or lower than those for ELISAs.

The performance of commercially available ELISA kits was investigated by some researchers who mostly concurred on the need to improve gluten extraction, gluten peptide detection and calibrants, while debating the use of a suitable reference sample in the assays (Sharma 2012; Diaz-Amigo and Popping 2012; Bruins Slot et al. 2015; Bugyi et al. 2012; Torok et al. 2015; Panda et al. 2015; Martínez-Esteso et al. 2017; Rzychon et al. 2017; Lexhaller et al. 2017).

3.3.5 Standardisation and Harmonisation of Gluten Analysis

There is a lack of agreement on the level of performance necessary for gluten detection methods employed to comply with food safety legislation. Standardisation would bring gluten testing results into conformity with a standard. To arrive at an agreement would need acceptance of the use of a certified reference material (CRM) or a specific calibrant, not only for immunoassays but for any other suitable methodologies such as LC-MS/MS protocols.

Harmonisation of gluten detection would have to involve consideration of any processes that could contribute to making the results of different measurement procedures comparable by recognizing, understanding and explaining any disparities to generate uniform data or reliably convert it. The analytics community is well aware of the need for harmonisation and standardisation, but it is acknowledged not to be a straightforward exercise.

The outcome of various gluten analysis techniques (immunoassays, chromatography or MS) may be based on detection of a single peptide (e.g. a 33-mer), an individual protein (e.g. P18573 alpha-gliadin), a protein group (e.g. alpha-gliadins) or total gluten. It is difficult to determine accurate or meaningful conversion factors between

- peptides and individual proteins
- peptides and gluten
- individual proteins and gluten or
- gluten and the plant species of its origin.

For example, in ELISA methods gliadin is often measured and the data converted to represent gluten by using a conversion factor of 2. It is well known that the variation of the gliadin to glutenin ratio is variety dependent and also influenced by the environmental stresses. Although, it is not accurate, the use of a single conversion factor is currently the best approximation. Standardisation efforts may help to overcome the inaccuracies caused by using this factor of 2 when measuring gliadins and expressing gluten levels (Wieser and Koehler 2009; Diaz-Amigo and Popping 2013; Koerner et al. 2013; Bruins Slot et al. 2015).

Regardless of what aspect of gluten is measured, there should be a single agreed compound, a robust marker, that any method refers or converts to when expressing gluten analysis results. Options include the use of the same calibrants or standards or reference materials, although the ultimate solution may be the use of multiple techniques or more than one standard. Publications that compare method performances, reviewed the status of standardisation and harmonisation efforts of gluten analysis providing a high-resolution picture of the state of the art (Haraszi et al. 2011; Bugyi et al. 2013; Mena and Sousa 2015; Bruins Slot et al. 2015, 2016; Martínez-Esteso et al. 2017; Rzychon et al. 2017; Deora 2018; Alves et al. 2017; Melini and Melini 2018).

To date, standardisation and harmonisation of gluten detection remains unresolved, but two priorities are clear. An agreement on the specific analyte(s)/target(s)/set of markers is required to improve and make gluten measurements comparable (Martínez-Esteso et al. 2016). Well-characterised reference materials representative of all the different subgroups of gluten proteins are required (Martínez-Esteso et al. 2017).

Standardisation and harmonisation of analysis methods in gluten detection would also trigger a smoother implementation of the various food safety legislations world-wide helping people to consume gluten-free or low gluten foods safely. Last but not least, harmonisation would allow the food industry to better deal with gluten risk assessment, allergen management and communication of the associated issues (Melini and Melini 2018).

Appendix I

SDS-PAGE protocol used by the Wheat Chemistry and End–Use Quality Laboratory of CIMMYT

Reagents and main steps of the procedure were published in Peña et al. (2004, revised in 2018).

PREPARATION OF REAGENTS
Reagent 1. 1 M Tris pH 8.5

Dissolve 30.3 g of Tris in 220 ml of distilled water. Adjust pH to 8.5 with concentrated HCl. Bring the total volume up to 250 ml with distilled water. Store in the refrigerator.

Reagent 2. 1 M Tris pH 6.8
Dissolve 12.1 g of Tris in 64 ml of distilled water. Adjust pH to 6.8 with concentrated HCl. Bring the total volume up to 100 ml with distilled water. Store in the refrigerator.

Reagent 3. 10% sodium dodecyl sulfate (SDS)
Dissolve 5 g of SDS in 40 ml distilled water and bring the total volume up to 50 ml with distilled water. Store at room temperature (20–25 °C).

Reagent 4. Stock acrylamide solution (40%T and 1.3%C) for running and stacking gels
Weigh 0.52 g of bis-acrylamide and 39.5 g of acrylamide. Dissolve in approximately 70 ml of distilled water and bring the total volume up to 100 ml. Homogenize and filter the solution before use. Store in a dark (translucent) container in the refrigerator.
CAUTION: use gloves and mask when working with acrylamide.

Reagent 5. Stock sample buffer for total protein extracts
Dissolve 12 g of glycerol in 36 ml of distilled water. Add 0.76 g of Tris, 4.0 g of SDS, 750 mg of dithiothreitol (1.5% DTT) and 6.0 mg of bromophenol blue. Adjust to pH 6.8 with concentrated HCl and bring the total volume up to 50 ml with distilled water.

Reagent 6. Tris-glycine stock for running buffer
Put 30.0 g of Tris in a 1-L beaker. Add 887 ml of distilled water and stir. While stirring the solution first add 144 g of glycine then 10 g of SDS. The pH of the solution should be 8.3. **If the pH is higher, do not adjust it with HCl as it would alter the desired ionic strength.**

Reagent 7. Staining and destaining solutions
Solution 1 is a 12% trichloroacetic acid solution. Dilute 120 g of trichloroacetic acid in distilled water and make up to 1 L.
Solution 2 is the staining solution. Mix 400 ml of methanol, 100 ml of glacial acetic acid, 0.1 g of Coomassie Brilliant Blue R-250, and 500 ml of distilled water (1 L in total).

Reagent 8. 1.5% ammonium persulfate
Dissolve 150 mg of ammonium persulfate in 10 ml of distilled water. Prepare fresh immediately before use.

Procedures

Extraction of total proteins
Weigh 20 mg of sample into a microcentrifuge tube. Add 300 μl of sample buffer (reagent 5) and incubate for 5 min in an Eppendorf Thermomixer comfort at 90 °C and 1400 rpm. Centrifuge for 5 min at 13000 rpm in an Eppendorf Microcentrifuge 5415C (Brinkmann Instruments Inc., NY).

Gels for total protein extracts from durum wheat and bread wheat
The final acrylamide concentration to be used in the gel will depend on the size and type of bands to be examined in detail. For the running gel, 9–10% acrylamide is the most common gel concentration when examining high molecular weight glutenins from whole grain protein extracts. However, for better separation of subunits 2 and 2*, gels of around 13% acrylamide are more appropriate.

Formulas to determine gel concentrations

$$\text{Concentration of acrylamide}(\%\text{T}) = \frac{\text{g of acrylamide} + \text{g of bis} - \text{acrylamide}}{\text{total volume of solution}} \quad 100$$

$$\text{Concentration of bis-acrylamide}(\%\text{C}) = \frac{\text{g of bis} - \text{acrylamide}}{\text{g of Acrylamide} - \text{g of bis} - \text{acrylamide}} \quad 100$$

Staining and destaining gels

Submerge the gels for 5 min in a 12% trichloroacetic acid solution (reagent 1), then place them in staining solution (solution 2) for 4 h, and finally for destaining place them in distilled water for 24 h.

Reagents for glutenin and gliadin extracts

Solution 1.1. 0.08 M Tris-HCl buffer pH 8.0
Weigh 4.8 g of Tris and 20 g of SDS and dissolve them in 500 ml of distilled water, adjusting the pH to 8.0 with HCl.
Solution 1.2. 50% propanol
Solution 1.3. Mix 500 ml of each of solution 1.1 and 1.2
Solution 1.4. For 10 ml of solution 1.3, mix in 200 mg of DTT. This reagent should be prepared the same day.
Solution 1.5. For 10 ml of solution 1.3, add 140 µl of 4-vinyl-pyridine and mix. This reagent should be prepared the same day.
Solution 1.6. Glutenin extraction buffer
2% SDS (w/v), 40% glycerol (w/v), 0.02% (w/v) bromophenol blue.
Dissolve 20 g of glycerol in 36 ml of distilled water. Add 0.76 g of Tris, 1.0 g of SDS, and 10.0 mg of Coomassie Brilliant Blue R. Adjust the pH to 6.8 with concentrated HCl. Make the volume up to 50 ml with distilled water.
Solution 1.7. Gliadin extraction buffer
2% SDS (w/v), 40% glycerol (w/v), 0.02% (w/v) bromophenol blue. Dissolve 20 g of glycerol in 36 ml of distilled water. Add 0.76 g of Tris, 1.0 g of SDS, and 10.0 mg of Coomassie Brilliant Blue R. Adjust the pH to 8.0 with concentrated HCl. Make the volume up to 50 ml with distilled water.

Running gel (For two 15%T gels, 17 cm high and 1 mm thick)

24.7 ml of Tris pH 8.5 (Reagent 1)
24.4 ml of Acrylamide Stock for Running Gel (Reagent 4)
650 µl of 10% SDS.
14.0 ml of Distilled Water

This mixture is deaerated with a Sonics Branson 5510 for 2–4 min. Immediately filter the mix through coarse filter paper. Add 1.3 ml of fresh 1.5% ammonium persulfate (reagent 8). Assemble frames for pouring gels then add 50 µl of TEMED to the mixture. Mix gently for 5 seconds and pour the gel solution into the frames. Immediately apply 1 ml of butyl alcohol to the top of the gel to prevent the formation of a meniscus. The alcohol should be applied slowly with a syringe.

Stacking gel (For two 4.8%T gels, 3.0 cm high and 1.0 mm thick)

Before adding the stacking gel solution, remove the butyl alcohol from the top of the running gel.

6.12 ml of Distilled Water
1.10 ml of Tris pH 6.8 (Reagent 2)
1.05 ml of acrylamide stock for stacking gel (reagent 4).
90 µl of 10% SDS.
380 µl of 1.5% ammonium persulfate

Mix reagents together gently, add 50 µl of TEMED, and mix again. Apply to the top of the running gel, being careful that there are no bubbles in the lanes of the Teflon lane former (comb). This can be achieved by inserting the Teflon comb at an angle of approximately 30°. Let the gel stand for 30 min to 1 hr. Very carefully remove the Teflon comb and with a syringe remove the excess solution which did not polymerise. Fill the lanes with running buffer (see below).

Running buffer

Mix 400 ml of stock for running buffer (reagent 6) and 3.6 L of distilled water. Use approximately 2.5 L of buffer per tank (e.g. Protean BIO-RAD equipment, Bio-Rad Laboratories, Richmond, California) for 2 gels (4.5 L for a buffer tank for 6 gels).

Running the gel

Load the samples in each lane (4–8 µl of extracted protein). Assemble the equipment, connect the hoses of the cooling system (15 °C) to running water and turn on the power. The running time depends on the current (mA) and the porosity (%T) of the gel. Electrophoretic separation can be run at 12.5 mA per gel for about 19 h (overnight). When using more than 25 mA per gel (runs of approx. 8 h), it is necessary to use a cooling apparatus to maintain temperature at around 15 °C. For a shorter running time of 4–5 h, run the electrophoresis at 35–40 mA per gel at 15 °C until the colored line arrives at the bottom edge of the gel. In order to maintain ionic strength, it is recommended to use a small pump connecting the lower to the upper buffer containers to recirculate the running buffer.

Preserving gels

Fresh gels may be kept for limited time if placed in polyethylene plastic bags in the refrigerator. Drying the gels with a gel drier is more convenient to preserve the gels for longer.

Selective extraction of gliadins and glutenins for SDS-PAGE

According to Singh et al. (1991) with modifications.

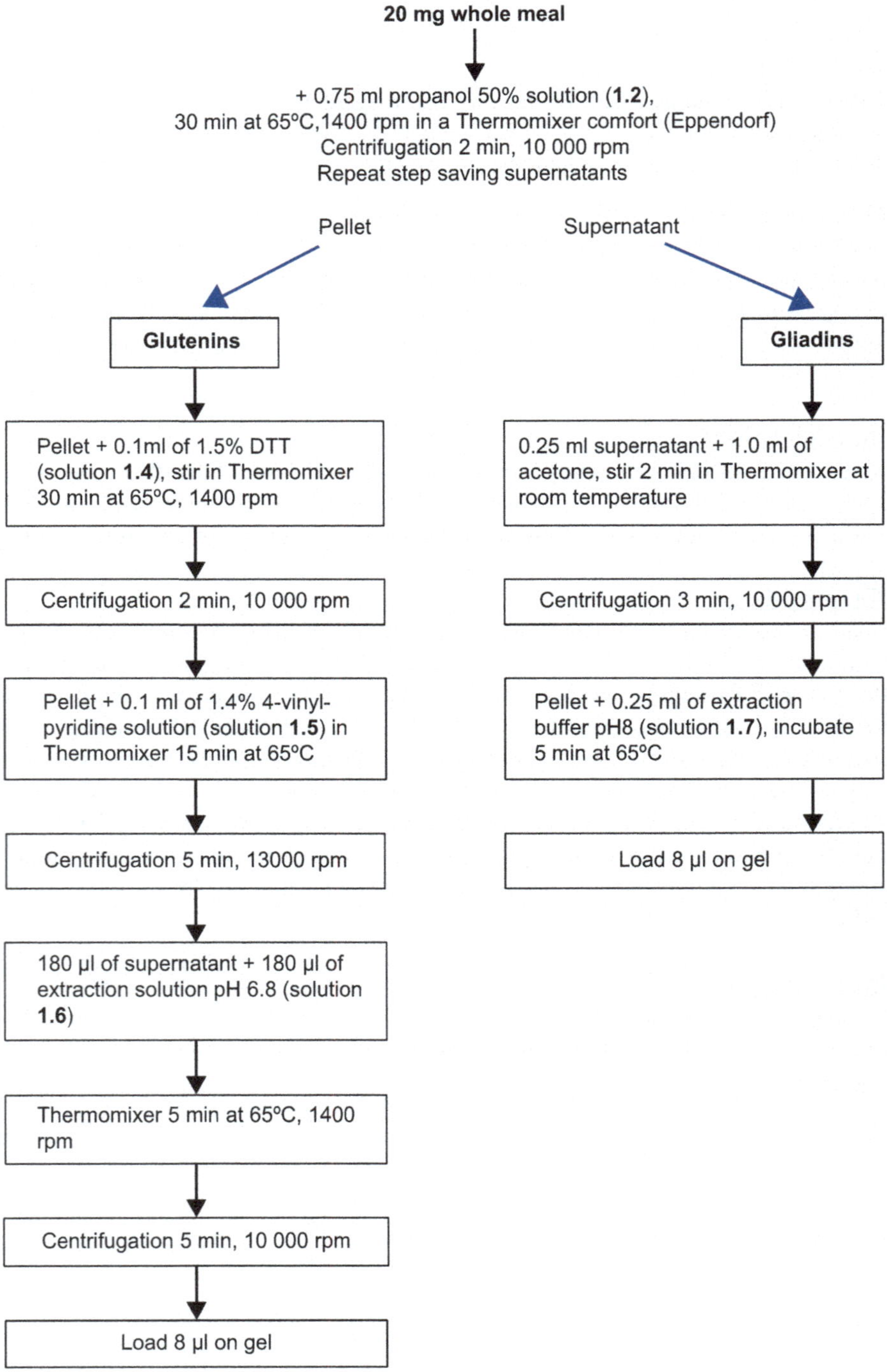

Appendix II

A-PAGE protocol for gliadin analysis

The main steps are described in Metakovsky and Novoselskaya (2001) for gels of standard size (18 × 20 cm). For long acrylamide gels (18 × 32 cm) it is advisable to use the protocol previously described by Branlard et al. (1990).

Reagents

1. 70% ethanol
2. 1.5% aluminium lactate pH 3.1

Dissolve 15 g of aluminium lactate in 800 ml of distilled water*. Adjust pH to 3.1 with lactic acid. Bring the total volume to 1 L with distilled water.

3. 1% ascorbic acid

Dissolve 0.1 g of ascorbic acid in 10 ml of distilled water. Prepare fresh for each use.

4. 1% $Fe(SO_4)_3$

Dissolve 0.1 g of $Fe(SO_4)_3$ in 10 ml of distilled water.

5. 40% acrylamide

It is highly advisable to purchase 40% acrylamide solution ready to use. Alternatively, dissolve 40 g of high quality acrylamide in 100 ml of distilled water. Store at 4 °C.

6. 2% bis-acrylamide.

It is highly advisable to purchase 2% bis-acrylamide solution ready to use.
Alternatively, dissolve 2 g of high quality bis-acrylamide in 100 ml of distilled water. Store at 4 °C.

7. 0.33% hydrogen peroxide

Purchase 3% hydrogen peroxide (10 volumes) from a pharmacy or store and keep it at 4 °C. Make 1.1 ml of 3% hydrogen peroxide (10 volumes) up to 10 ml with distilled water. Prepare fresh for each use.

8. Stock sample buffer

Dissolve 3 ml of glycerol in 1 ml of 1.5% aluminum lactate pH 3.1. Add a small portion (a few grains of powder) of pyronin or methyl green. Bring the total volume up to 10 ml with distilled water.

9. Stock running buffer (0.15% aluminium lactate pH 3.1)

Dilute 100 ml of 1.5% aluminum lactate with 900 ml of distilled water.

*Dr. Metakovsky consistently uses distilled water. The classic ultra-pure water influences the resolution of the gliadins and may alter the mobility of some gliadin alleles. Twice-deionized water (today referred to as ultra-pure water) was used by Branlard et al. (1990).

Procedures

Gel preparation

8.3% Acrylamide, 0.42% Bis-Acrylamide, 0.1% Ascorbic Acid, 0.00067% $Fe(SO_4)_3$, 0.15% Aluminum Lactate, pH 3.1

The glass plates (20 × 20 cm and 2 mm thickness for a normal gel or 30 × 20 cm and 2 mm thickness for a long gel for higher resolution) are cleaned with 70% ethanol then by spreading droplets of glycerol until a thin film forms and drying with KimWipes. This treatment makes it easier to remove gels from plates.

Mix 21 ml of 2% bis-acrylamide and 20.8 ml of 40% acrylamide. Add 67 µl of 1% $Fe(SO_4)_3$, 10 ml of 1% ascorbic acid, 10 ml of 1.5% aluminum lactate, and bring the total volume up to 100 ml, filter through filter paper and degas, then cool on ice before use. Add 0.4 ml of 0.33% hydrogen peroxide, mix well and pour rapidly into the glass plate assembly. The comb is inserted and the gel is polymerized for 10 min.

Gliadin extraction

Weigh about 20 mg of sample into a microcentrifuge tube. Add 150 µl of 70% ethanol and incubate for 30 min at room temperature. Centrifuge the sample for 5 min at 12500 rpm. Transfer the supernatant to a microcentrifuge tube and mix with an equal volume of the sample buffer. Freshly prepared (within two days) samples should be used to obtain well resolved electrophoretic profiles.

EXTRACTION OF GLIADINS FOR A-PAGE

about 20 mg of whole meal

↓

+ 0.2 ml or 0.15 ml of 70% ethanol solution
30 min at 40°C or room temperature
centrifuge for 10 min at 10 000 rpm
keep the supernatant

↓

+ equal volume of sample buffer

↓

load 20 to 30 µl

Electrophoresis

After rinsing the wells with running buffer, the glass plates containing the gel are placed in the vertical slab gel apparatus. Then the wells are filled with the buffer and 20 to 30 µl of the samples are slowly loaded. Electrophoresis from the anode (the upper buffer) to the cathode (the lower buffer) is performed for 10–20 min at 220 V and then for 2.5–3 h at 550 V without buffer circulation. The lower tank of the vertical apparatus is cooled with tap water or in a cooling system at 10 °C.

Staining/destaining

The gels are kept on one of the glass plates to reduce the risk of damaging the gels. The staining procedure is the same as for the SDS-PAGE method.

References

Aghagholizadeh R, Kadivar M, Nazari M, Mousavi F, Azizi MH, Zahedi M, Rahiminezhad MR (2017) Characterization of wheat gluten subunits by liquid chromatography – mass spectrometry and their relationship to technological quality of wheat. Journal of Cereal Science 76: 229–235.

Allred LK, Sealey-Voyksner JA and Voyksner RD (2014) Evaluation of qualitative and quantitative immunoassays to detect barley contamination in gluten-free beer with confirmation using LC-MS/MS. Journal of AOAC International 97(6): 1615–1625.

Altenbach SB, Vensel WH, DuPont FM (2010) Analysis of expressed sequence tags from a single wheat cultivar facilitates interpretation of tandem mass spectrometry data and discrimination of gamma gliadin proteins that may play different functional roles in flour. BMC Plant Biol 10: 7.

Alves TO, D'Almeida CTS, Ferreira MSL (2017) Determination of Gluten Peptides Associated with Celiac Disease by Mass Spectrometry, In "Celiac Disease and Non-Celiac Gluten Sensitivity" Publisher: InTech DOI: https://doi.org/10.5772/67547

Batey IL, Gupta RB, MacRitchie F (1991) Use of size-exclusion high-performance liquid chromatography in the study of wheat flour proteins: an improved chromatographic procedure. Cereal Chemistry 68: 207–209.

Biomedal (2017) https://www.foodnavigator.com/Article/2017/11/24/Biomedal-targets-gluten-detection-and-quantification. Last accessed in Nov 2018.

Boison J, Allred L, Almy D, Anderson L, Baumert J, Bhandari S, Cebolla A, Chen Y, Crowley E, Diaz-Amigo C, Doi H, Don C, Downs M, Dubiel N, et al. (2018) Standard Method Performance Requirements (SMPRs®) 2017.021: Quantitation of Wheat, Rye, and Barley Gluten in Oats. J AOAC Int 101(4): 1238–1242.

Bönick J, Huschek G, Rawel HM (2017) Determination of wheat, rye and spelt authenticity in bread by targeted peptide biomarkers. Journal of Food Composition and Analysis 58, 82–91.

Branlard G, Picard B, Courvoisier C (1990) Electrophoresis of gliadins in long acrylamide gels: Method and nomenclature. Electrophoresis 11: 310–314.

Branlard G, Lesage VS, Bancel E, Martre P, Méleard B, Rhazi L (2015) Coping with wheat quality in a changing environment - Proteomics evidence for stress caused by environmental changes. In "Advances in Wheat Genetics: From Genome to Field. Proceedings of the 12th International Wheat Genetics Symposium" (Y. Ogihara, S. Takumi, H. Handa Eds), Yokohama, Japan, 8-14 September 2013. ISBN: 978-4-431-55674-9, 255-264.

Bromilow SNL, Gethings LA, Langridge J, Shewry PR, Buckley M, Bromley MJ, Mills ENC (2017a) Comprehensive Proteomic Profiling of Wheat Glutens using a Combination of Data-Independent and Data-Dependent Acquisition. Frontiers in Plant Science 7: 2020.

Bromilow S, Gethings LA, Buckley M, Bromley M, Shewry PR, Langridge JI, Clare Mills EN (2017b) A curated gluten protein sequence database to support development of proteomics methods for determination of gluten in gluten-free foods. Journal of Proteomics 163: 67–75.

Bruins Slot ID, Fels-Klerx HJ, Bremer MGE G, Hamer RJ (2015) Evaluating the performance of gluten ELISA test kits: The numbers do not tell the tale. Cereal Chemistry 92: 513–521.

Bruins Slot ID, Fels-Klerx HJ, Bremer MGE G, Hamer RJ (2016) Immunochemical Detection Methods for Gluten in Food Products: Where Do We Go from Here? Critical Reviews in Food Science and Nutrition 56: 2455–2466.

Bugyi Z, Török K, Hajas L, Adonyi Z, Poms RE, Popping B, Diaz-Amigo C, Kerbach S, Tömösközi S (2012) Development of incurred reference material for improving conditions of gluten quantification. Journal of AOAC International 95: 382–387.

Bugyi Z, Török K, Hajas L, Adonyi Z, Popping B, Tömösközi S. (2013) Comparative study of commercially available gluten ELISA kits using an incurred reference material. Qual. Assur. Saf. Crops Foods 5: 79–87.

Cao W, Watson D, Bakke M, Panda R, Bedford B, Kande PS, Jackson LS, Garber EAE (2017) Detection of Gluten during the Fermentation Process To Produce Soy Sauce. Journal of Food Protection 80(5): 799–808.

Codex Alimentarius Commission, Foods for Special Dietary Use for Persons Intolerant to Gluten. Rome, Italy, 2008.

Colgrave ML, Goswami H, Byrne K, Blundell M, Howitt CA, Tanner GJ (2015) Proteomic profiling of 16 cereal grains and the application of targeted proteomics to detect wheat contamination. Journal of Proteome Research 14: 2659–2668.

Cornec M, Popineau Y, Lefebvre J (1994) Characterisation of Gluten Subfractions by SE-HPLC and Dynamic Rheological Analysis in Shear. Journal of Cereal Science 19(2): 131–139.

Deora NS (2018) Current detection and quantification method for gluten to support the gluten-free claim: an insight about ELISA method. Journal of Nutritional Health & Food Engineering 8: 75–76.

Diaz-Amigo C, Popping B (2012) Gluten and gluten-free: issues and considerations of labeling regulations, detection methods, and assay validation. J. AOAC Int. 95: 337–348.

Diaz-Amigo C, Popping B (2013) Accuracy of ELISA detection methods for gluten and reference materials: a realistic assessment. Journal of Agricultural and Food Chemistry 61: 5681–5688.

Dostalek P, Hochel I, Mendez E, Hernando A, Gabrovska D (2006) Immunochemical determination of gluten in malts and beers. Food Addit Contam 23: 1074–1078.

Dumur J, Jahier J, Bancel E, Laurière M, Bernard M, Branlard G (2004) Proteomic analysis of aneuploid lines in the homeologous group 1 of the hexaploid wheat cultivar Courtot. Proteomics 4: 2685–2695.

DuPont FM, Chan R, Lopez R, Vensel WH (2005) Sequential extraction and quantitative recovery of gliadins, glutenins, and other proteins from small samples of wheat flour. J. Agric. Food Chem. 53: 1575–1584.

Fallahbaghery A, Zou W, Byrne K, Howitt CA, Colgrave ML (2017) A Comparison of Gluten Extraction Protocols Assessed by LC-MS/MS Analysis. Joural of Agricultural and Food Chemistry 65: 2857–2866.

Fiedler KL, McGrath SC, Callahan JH, Ross MM (2014) Characterization of Grain-Specific Peptide Markers for the Detection of Gluten by Mass Spectrometry. Journal of Agricultural and Food Chemistry, 62: 5835–5844.

García E, Llorente M, Hernando A, Kieffer R, Wieser H, Méndez E (2005) Development of a general procedure for complete extraction of gliadins for heat processed and unheated foods. Eur J Gastroenterol Hepatol. 17: 529–39.

García-Molina MD, Barro F (2017) Characterization of Changes in Gluten Proteins in Low-Gliadin Transgenic Wheat Lines in Response to Application of Different Nitrogen Regimes. Front. Plant Sci. 8: 257.

Goze P, Rhazi L, Lakhal L, Jacolot P, Pauss A, Aussenac T (2017) Effects of ozone treatment on the molecular properties of wheat grain. Journal of Cereal Science 75: 243–251.

Glutentox. http://www.keydiagnostics.com.au/images/PDF/new/GlutenTox%20Reader/GlutenTox_Reader_Brochure_KD_12-14.pdf Last accessed in Nov 2018.

Guo B, Luan H, Lin S et al. (2016) Comparative proteomic analysis of two barley cultivars (*Hordeum vulgare* L.) with contrasting grain protein content. Frontiers in Plant Science 7: 542.

Han C, Lu X, Yu Z, Li X, Ma W & Yan Y (2015) Rapid separation of seed gliadins by reversed-phase ultra-performance liquid chromatography (RP-UPLC) and its application in wheat cultivar and germplasm identification. Bioscience, Biotechnology and Biochemistry. 79: 808–815.

Haraszi R, Chassaigne H, Maquet A, Ulberth F (2011) Analytical methods for detection of gluten in food - method developments in support to the legislations on labelling of foodstuffs. Journal of AOAC International 94: 1006–25.

Hlywiak KH (2008) Hidden sources of gluten. Practical Gastroenterology, In: Celiac disease: a comprehensive review and update, series #2, Eds: Muralidhar Jatla, M.D., Ritu Verma, M.D., pp. 27-34, 39.

Ikeda TM, Araki E, Fujita Y, Yano H (2006) Characterization of low-molecular-weight glutenin subunit genes and their protein products in common wheats. Theoretical and Applied Genetics 112: 327–334.

Ikeda TM, Branlard G, Peña RJ, Takata K, Liu L, He ZH, Lerner SE, Kolman MA, Yoshida H, Rogers WJ (2008) International collaboration for unifying *Glu-3* nomenclature systems in common wheat. Proc 11th Int Wheat Genet Symp, Sydney University Press, Sydney, Australia O42.

International Wheat Genome Sequencing Consortium (2018) Shifting the limits in wheat research and breeding through a fully annotated and anchored reference genome sequence. Science https://doi.org/10.1126/science.aar7191.

Juhász A, Haraszi R, Békés F, Skylas DJ, Wrigley CW (2015a) Proteomic Analysis Relevant to Cereal Grains. In H. Corke, J. Faubion, K. Seetharaman and C. Wrigley (Ed.), Encyclopedia of food grains (pp. 146–153) Oxford, United Kingdom: Elsevier.

Juhász A, Haraszi R, Maulis CS (2015b) ProPepper: a curated database for identification and analysis of peptide and immune-responsive epitope composition of cereal grain protein families, Database, Volume 2015, 2015, bav100, https://doi.org/10.1093/database/bav100

Khamis M (2014) Characterization and evaluation of heat-treated wheat flours. PhD thesis, Department of Grain Science and Industry, College of Agriculture, Kansas State University, Manhattan, Kansas.

Koerner TB, Abbott M, Godefroy SB, Popping B, Yeung JM, Diaz-Amigo C, Roberts J, Taylor SL, Baumert JL, Ulberth F, Wehling P, Koehler P (2013) Validation procedures for quantitative gluten ELISA methods: AOAC allergen community guidance and best practices. Journal of AOAC International 96: 1033–1040.

Labuschagne MT, Aucamp U (2004) The use of size exclusion high performance liquid chromatography (SE-HPLC) for wheat quality prediction in South Africa. South African Journal of Plant and Soil, 21:1, 8–12.

Larroque OR, Sharp P, Bekes F (2007) Molecular Weight Distribution of Gluten Proteins. In: Buck HT, Nisi JE, Salomón N (eds) Wheat Production in Stressed Environments. Developments in Plant Breeding, vol 12. Springer, Dordrecht.

Lemelin E, Aussenac T, Salvo I, Quandalle C (2002) Size characterization of wheat proteins by asymmetrical flow field flow fractionation and multi-angle laser light scattering. AACC Annual meeting, Montreal, Québec 98–108.

Lemelin E, Branlard G, Salvo L, Lein V, Aussenac T, Daydé J (2005) Breadmaking stability of wheat flours: relation between mixing properties and molecular weight distribution of polymeric glutenins. Journal of Cereal Science 42: 317–326.

Lexhaller B, Tompos C, Scherf KA (2017) Fundamental study on reactivities of gluten protein types from wheat, rye and barley with five sandwich ELISA test kits. Food Chemistry 237: 320–330.

Li H, Byrne K, Galiamov R, Mendoza-Porras O, Bose U, Howitt CA, Colgrave ML (2018) Using LC-MS to examine the fermented food products vinegar and soy sauce for the presence of gluten. Food Chemistry 254: 302–308.

Li H, Byrne K., Howitt CA, Colgrave ML (2019) Efficient Extraction and Digestion of Gluten Proteins. In: Wang X, Kuruc M (eds) Functional Proteomics. Methods in Molecular Biology, vol 1871. Humana Press, New York, NY, 405–412.

Liao YS, Kuo JH, Chen BL, Tsuei HW, Lin CY, Hsu-Yang L, Hwei-Fang C (2017) Development and Validation of the Detection Method for Wheat and Barley Glutens Using Mass Spectrometry in Processed Foods. Food Anal. Methods. 10: 2839.

Liu L, Ikeda TM, Branlard G, Peña RJ, Rogers WJ, Lerner SE, Kolman MA, Xia X, Wang L, Ma W, Appels R, Yoshida H, Wang A, Yan Y, He Z (2010) Comparison of low molecular weight glutenin subunits identified by SDS-PAGE, 2-DE. MALDI-TOF-MS and PCR in common wheat. BMC Plant Biol 10: 124.

Lock S (2014) Gluten detection and speciation by Liquid chromatography mass spectrometry (LC-MS/MS). Foods 3: 13–29.

Lookhart GL, Albers LD, Bietz JA (1986) Comparison of polyacrylamide gel electrophoresis and high performance liquid chromatography analysis of gliadin polymorphism in the wheat cultivar 'Newton. Cereal Chemistry 63: 497–500.

Lookhart GL, Bean SR (1995) Separation and characterization of wheat protein fractions by high-performance capillary electrophoresis. Cereal Chemistry 72: 527–532.
Lookhart GL, Bean SR, Bietz JA (2003) Reversed-phase high-performance liquid chromatography in grain applications. Cereal Foods World 48: 9–16.
Manfredi A, Mattarozzi M, Giannetto M, Careri M (2015) Multiplex liquid chromatography-tandem mass spectrometry for the detection of wheat, oat, barley and rye prolamins towards the assessment of gluten-free product safety. Analytica Chimica Acta, 895: 62–70.
Martínez-Esteso MJ, Nørgaard J, Brohée M, Haraszi R, Maquet A, O'Connor G (2016) Defining the wheat gluten peptide fingerprint via a discovery and targeted proteomics approach. Journal of Proteomics 147: 156–168.
Martínez-Esteso MJ, Brohee M, Norgaard J, O'Connor G (2017) Label-free proteomic analysis of wheat gluten proteins and their immunoreactivity to ELISA antibodies. Cereal Chemistry 94: 1–7.
Melini F, Melini V, Luziatelli F, Ruzzi M (2017) Current and Forward-Looking Approaches to Technological and Nutritional Improvements of Gluten-Free Bread with Legume Flours: A Critical Review. Comprehensive reviews in food science and food safety. 16: 1101–1122.
Melini F, Melini V (2018) Immunological Methods in Gluten Risk Analysis: A Snapshot, Safety. 4: 56.
Mena MC, Lombardía M, Hernando A, Méndez E, Albar JP (2012) Comprehensive analysis of gluten in processed foods using a new extraction method and a competitive ELISA based on the R5 antibody. Talanta 91: 33–40.
Mena MC, Sousa C (2015) Analytical Tools for Gluten Detection. Policies and Regulation. In: Arranz E, Fernández-Bañares F, Rosell CM, Rodrigo L, Peña AS, editors. Advances in the Understanding of Gluten Related Pathology and the Evolution of Gluten-Free Foods. Barcelona, Spain: Omnia Science pp. 527–564.
Metakovsky EV, Novoselskaya AY (2001) Gliadin allele identification in common wheat. I. Methodological aspects of the analysis of gliadin pattern by one-dimensional polyacrylamide gel electrophoresis. Journal of Genetics & Breeding 45: 317–324.
Metakovsky E, Melnik V, Rodriguez-Quijano M, Upelniek V, Carrillo JM (2018) A catalog of gliadin alleles: Polymorphism of 20th-century common wheat germplasm. The Crop Journal 6: 628–641.
Osborne TB (1924) The Vegetable Proteins, 2nd Edition, London.
Panda R, Zoerb HF, Cho CY, Jackson LS, Garber EAE (2015) Detection and Quantification of Gluten during the Brewing and Fermentation of Beer Using Antibody-Based Technologies, Journal of Food Protection. 78: 1167–1177.
Peña RJ, Gonzalez-Santoyo H, Cervantes F (2004) Relationship between Glu-D1/Glu-B3 allelic combinations and bread-making quality-related parameters commonly used in wheat breeding. In: Masci S., Lafiandra, D. (eds.). Proceedings of the 8th Gluten Workshop. Viterbo, 156-157b.
Qian Y, Preston K, Krokhin O, Mellish J, Ens W (2008) Characterization of Wheat Gluten Proteins by HPLC and MALDI TOF Mass Spectrometry. Journal of The American Society of Mass Spectrometry 19: 1542–1550.
Real A, Comino I, de Lorenzo L, Merchán F, Gil-Humanes J, et al.. (2012) Molecular and Immunological Characterization of Gluten Proteins Isolated from Oat Cultivars That Differ in Toxicity for Celiac Disease. PLoS ONE 7: e48365.
Rzychon M, Brohée M, Cordeiro F, Haraszi R, Ulberth F, O'Connor G (2017) The feasibility of harmonizing gluten ELISA measurements. Food Chemistry 234: 144–154.
Sajic N, Oplatowska-Stachowiak M, Streppel L, Drijfhout JW, Salden M, Koning F (2017) Development and in-house validation of a competitive ELISA for the quantitative detection of gluten in food. Food Control, 80: 401–410.
Satsuki-Murakami T, Kudo A, Masayama A, Ki M, Yamano T (2018) An optimized extraction method for gluten analysis in cacao-containing products using an extraction buffer with polyvinylpyrrolidone. Food Control 84: 70–74.
Schalk K, Koehler P, Scherf KA (2018) Targeted liquid chromatography tandem mass spectrometry to quantitate wheat gluten using well-defined reference proteins. PLoS ONE 13: e0192804.

Schalk K, Lang C, Wieser H, Koehler P, Scherf KA (2017a) Quantitation of the immunodominant 33-mer peptide from α-gliadin in wheat flours by liquid chromatography tandem mass spectrometry. Sci Rep.; 7: 45092.

Schalk K, Lexhaller B, Koehler P, Scherf KA (2017b) Isolation and characterization of gluten protein types from wheat, rye, barley and oats for use as reference materials. PLoS ONE 12: e0172819.

Scherf KA (2016) Impact of the preparation procedure on gliadin, glutenin and gluten contents of wheat starches determined by RP-HPLC and ELISA. Eur Food Res Technol 242: 1837.

Schopf M, Scherf KA (2018) Wheat cultivar and species influence variability of gluten ELISA analyses based on polyclonal and monoclonal antibodies R5 and G12. Journal of Cereal Science 83: 32–41.

Sealey-Voyksner JA, Khosla C, Voyksner RD, Jorgenson JW (2010) Novel aspects of quantitation of immunogenic wheat gluten peptides by liquid chromatography-mass spectrometry/mass spectrometry. J Chromatogr A. 18;1217: 4167-83.

Sharma GM (2012) Immunoreactivity and detection of wheat proteins by commercial ELISA kits. J. AOAC Int. 95: 364–371.

Simonato B, Mainente F, Tolin S, Pasini G (2011) Immunochemical and mass spectrometry detection of residual proteins in gluten fined red wine. J Agric Food Chem 59: 3101–3110.

Singh NK, Shepherd KW, Cornish GB (1991) A simplified SDS-PAGE procedure for separating LMW subunits of glutenin. J. Cereal Sci 14: 203–208.

Srinivasan B, Focke-Tejkl M, Weber M et al. (2015) Usefulness of recombinant γ-gliadin 1 for identifying patients with celiac disease and monitoring adherence to a gluten-free diet. J Allergy Clin Immunol 136: 1607–1618.e3.

Tanner GJ, Blundell MJ, Colgrave ML, Howitt CA (2016) Creation of the first ultra-low gluten barley (Hordeum vulgare L.) for coeliac and gluten-intolerant populations. Plant Biotechnol J 14: 1139–1150.

Tatham AS, Gilbert SM, Fido RJ, Shewry PR (2000) Extraction, Separation, and Purification of Wheat Gluten Proteins and Related Proteins of Barley, Rye, and Oats, Methods in Molecular Medicine, Vol. 41: Celiac Disease: Methods and Protocols, Ed: M. N. Marsh © Humana Press Inc., Totowa, NJ pages 55-73.

Taylor SL, Nordlee JA, Jayasena S, Baumert JL (2018) Evaluation of a Handheld Gluten Detection Device. J Food Prot. 81: 1723–1728.

Torok K, Hajas L, Bugyi Z, Balazs G, Tomoskozi S (2015) Investigation of the effects of food processing and matrix componentson the analytical results of ELISA using an incurred gliadin reference material candidate. Acta Aliment. 44: 390–399.

van den Broeck HC, Cordewener JHG, Nessen MA, America AHP, van der Meer IM (2015) Label free targeted detection and quantification of celiac disease immunogenic epitopes by mass spectrometry. Journal of Chromatography A, 60-71.

van den Broeck HC, America AHP, Smulders MJM, Bosch D, Hamer RJ, Gilissen L, van der Meer IM (2009) A modified extraction protocol enables detection and quantification of celiac disease-related gluten proteins from wheat. J. Chromatogr. B: Anal. Technol. Biomed. Life Sci. 877: 975–982.

Wieser H, Koehler P (2009) Is the calculation of the gluten content by multiplying the prolamin content by a factor of 2 valid? Eur. Food Res. Technol. 229: 9–13.

Zilic S, Barac M, Pesic M, Dodig D, Ignjatovic-Micic D (2011) Characterization of proteins from grain of different bread and durum wheat genotypes. Int. J. Mol. Sci. 2011, 12: 5878–5894.

Proteomics as a Tool in Gluten Protein Research

Maryke T. Labuschagne and Gilberto Igrejas

Abstract Gluten proteins account for 80% of wheat grain protein and are the largest contributor to wheat quality. Proteomics tools can be deployed in a programme designed to manipulate gluten proteins to improve quality and functional properties, to understand gluten structure and interrelationships between its components, and potentially to reduce allergies. The aim of this chapter is to review developments in the proteomics of gluten proteins, mainly from the last decade. It is clear that the technology used for gluten proteomics has developed significantly in this period, and the publication of the first completely sequenced wheat genome in 2014 has facilitated the application of these techniques in cereal research. Proteomics was shown to be useful for studying the effects of various biotic and abiotic stress conditions on gluten proteins during grain development. Proteomics will be increasingly important in investigating genotype by environment interaction in terms of baking quality characteristics. Great strides have also been made in the use of proteomics to identify gluten peptides with allergenic or toxic sequences. The integration of functional genomics, proteomics, bioinformatics, breeding and genetic resources is contributing to our understanding of the genetic and biochemical bases of quality traits in wheat. Technology is continually being developed and applied to elucidate interactions between biological molecules at all stages of the flow of genetic information in biological systems, and proteomics in combination with genomics will continue to play an important role in gluten protein research.

M. T. Labuschagne (✉)
Department of Plant Sciences, University of the Free State, Bloemfontein, South Africa
e-mail: labuscm@ufs.ac.za

G. Igrejas
Department of Genetics and Biotechnology, University of Trás-os-Montes and Alto Douro, Vila Real, Portugal

G. Igrejas et al. (eds.), *Wheat Quality For Improving Processing And Human Health*, https://doi.org/10.1007/978-3-030-34163-3_7

1 Introduction

Wheat is adapted to a wide range of environments and is the most widely grown, processed and consumed cereal in the world, because of the unique functional properties of its flour. The functional and nutritional properties of cereals are largely determined by their grain protein composition (Cunsolo et al. 2012). Gluten proteins account for 80% of wheat grain protein and are the largest contributor to wheat quality (D'Ovidio and Masci 2004). Gluten consists of hundreds of proteins present as monomers, the gliadins, or as polymers, the glutenins. Glutenins are linked by inter and intrachain disulphide bonds (Wieser 2007). Duplication and translocation of ancestral genes probably led to the origin of the diverse gluten protein encoding genes, accompanied by amino acid substitutions and insertion and deletion of peptide sequences (Shewry et al. 1986). The result is a complex mixture of homologous proteins of different molecular mass and charge, which makes gluten characterization difficult (Bonomi et al. 2013).

Proteomics is one of the more modern approaches for analysing cereal grain protein (Finnie et al. 2011). According to Wilkins et al. (1995) who coined the word, 'proteomics', it is the study of proteins in terms of their structure, composition, functions and interactions that direct the activities of the cell. Proteomics gives a better understanding of an organism than genomics, because genomics only gives a rough estimate of protein expression. The genome is constant, but protein expression is influenced by many factors. The central dogma of molecular biology for all living biological systems describes the flow of genetic information from DNA to mRNA, and on to the synthesis of polypeptides, which in turn, are assembled into active protein molecules. Newly emerging technologies encompass genomics (DNA), transcriptomics (mRNA) and proteomics (proteins) as well as the rapidly expanding and exciting field of bioinformatics, which provides complementary tools and interactive sequence databases (Skylas et al. 2005). The value of the proteomic approach has been demonstrated in a range of biological systems (Hochstrasser 1998; Blackstock and Weir 1999; Thiellement et al. 1999; Komatsu et al. 1999). Proteomics is now a major research area and in wheat is proving to be a powerful tool to elucidate the expression of proteins and how they contribute to the value of the grain. The interest in cereal proteomics goes beyond the elucidation of structure and function relationships. Plant proteomics is beginning to make practical contributions to applied fields through the identification and characterization of allergens, through studies of the equivalence of transgenic crops, genotyping, and heterosis, and for applications in food science, quality control and traceability. Many proteomics studies are aimed at generating knowledge on how to improve crop quality when contending with biotic and abiotic stress and in terms of nutritional and processing quality (Salekdeh and Komatsu 2007). For example, how genotype by environment interaction influences wheat baking quality (Holman et al. 2013) or how quality is influenced by stress, pathogens and yield (Cunsolo et al. 2012). The huge amount of data generated in proteomic experiments has led many journals to recommend, and some to require, that original data be submitted to public repositories (Jorrín-Novo et al. 2009).

2 Gluten Proteins

Gluten proteins confer viscoelasticity to dough, which determines its suitability for specific end-use products. The relative amounts of glutenins and gliadins play a large role in determining dough characteristics and end-use quality (Altenbach et al. 2016). Glutenins consist of high molecular weight glutenin subunits (HMW-GS) and low molecular weight glutenin subunits (LMW-GS). Members of both these groups contain few arginine and lysine residues and are rich in proline and glutamine, defining features of prolamins (Bromilow et al. 2017a). The HMW-GS in hexaploid bread wheat account for 7–15% of gluten protein and are encoded by three homoeologous loci, *Glu-A1, Glu-B1* and *Glu*-D1, on the long arms of group 1 chromosomes. Each of these loci have two HMW-GS genes encoding the x and y type subunits, respectively (Delcour et al. 2012). HMW-GS have up to seven cysteine residues, mostly in the N and C terminal regions (Shewry and Tatham 1997). Molecular characteristics of HMW-GS genes and their amino acid structures can be used to predict protein structure and possible roles in defining dough properties (Rasheed et al. 2014).

LMW glutenin and gliadins are encoded by alleles at *Glu-3* and *Gli-1* loci, respectively, which are linked genetically, and are believed to be derived from the same ancestral group of genes as their amino acid sequences are closely related (Singh and Shepherd 1988). The *Glu-A3, Glu-B3* and *Glu-D3* loci encoding the LMW-GS are located on the short arms of group 1 chromosomes (Singh and Shepherd 1988). The LMW-GS make up 20–35% of gluten protein. Each of these loci has several LMW-GS genes, and each of these genes can have two or more alleles (Zhang et al. 2013). The *Glu-3* locus was shown to have a positive effect on loaf volume (Clarke et al. 2003). Gliadins consist of four subfamilies, α/β with seven or six cysteine residues respectively, γ with nine cysteine residues, and δ and ω with only one cysteine residue (Veraverbeke and Delcour 2002). The α and γ gliadins are sometimes called C-type LMW-GS and ω gliadins D-type LMW-GS (Altenbach et al. 2016). The gliadin genes are situated at six loci, *Gli-A1, Gli-B1 and GliD1* on the short arms of the group 1 chromosomes, and *Gli-A2, GliB2* and *Gli-D2* on the short arms of the group 6 chromosomes. Gliadin makes up 40–50% of gluten protein. Gliadins can modify the extensibility of gluten by acting as plasticizer (Barak et al. 2015).

3 The Glutenin Macropolymer

Glutenin macropolymers (GMP) are formed in the gluten matrix when HMW-GS and LMW-GS covalently interact by intermolecular disulphide bonds (Don et al. 2006). GMPs with a molecular mass of more than 250 kD are a key determinant of dough functionality and end-use quality (MacRitchie 2014; Juhász et al. 2015). The functional GMPs (with the largest molecular mass) are insoluble in protein extraction buffers in non-reducing conditions. Gliadins with an odd number of cysteine residues can also bind covalently with GMPs (Juhász et al. 2015). The polymeric fraction, which cannot be extracted with sodium dodecylsulphate (SDS), and the glutenin fraction, which is insoluble in 50% (v/v) 1-propanol, are good indicators of the presence of functional GMPs and are significantly related to end-use quality. In the extraction protocol using 50% (v/v) 1-propanol, the soluble glutenin has a lower molecular mass. As both the soluble and insoluble fractions consist of HMW-GS and LMW-GS, the amounts present in the two fractions reflect the behaviour of glutenin subunits during polymerization into GMPs (Sapirstein and Fu 1998). The three *Glu-1* loci interact and show an additive effect, with the strongest interaction being between *Glu-B1* and *Glu-D1*. The *Glu-1* loci in order of influence from the strongest to the weakest are *Glu-D1*, then *Glu-B1*, followed by *Glu-A1* (Yang et al. 2014). *Glu-D1* is functionally dominant over *Glu-A1* and *Glu-B1*. The effect of 1Dx2 is stronger than that of 1Dy12 in quality characteristics, due to its more prominent role in functional GMP formation. In general, the role of the x-type HMW-GS (83–88 kDa) is more important than that of the y-type subunits (67–74 kDa). It was speculated that the x-type HMW-GS interact with LMW-GS to form functional GMPs (Wang et al. 2017). Most x-type subunits have four conserved cysteine residues, while y-type subunits have seven cysteine residues, and this may be a reason for differences in baking quality. Likewise there is an extra cysteine residue in the middle of the Ax2*B subunit, which is a variant of Ax2*, that positively affects gluten properties (Juhász et al. 2001).

LMW-GS were first identified in gel-filtrated extracts of wheat flour as high-molecular-weight gliadins linked by disulphide bonds, distinguishing them from monomeric gliadins (Beckwith et al. 1966). Starch gel electrophoresis gave further evidence for their existence, but it was technically difficult to separate them from co-migrating gliadins (Elton and Ewart 1966). Later, Nielsen et al. (1968) designated this particular fraction low-molecular-weight glutenin, because the viscosity and electrophoretic mobility of its components differed from those of the gliadin fraction. In pioneering work, Payne and Corfield (1979) analysed reduced glutenin components by SDS polyacrylamide gel electrophoresis (SDS-PAGE) and showed that the glutenin subunits could be subdivided into A, B, and C groups, according to their mobility, the A group corresponding to HMW-GS and the B and C groups to LMW-GS. They also demonstrated that HMW gliadins, once reduced and separated by SDS-PAGE, had mobilities similar to those of B and C subunits of LMW-GS.

Reversed-phase ultraperformance liquid chromatography (RP-UPLC), a technology based on reversed-phase high performance liquid chromatography (RP-HPLC), was developed in the early 2000s (Wu et al. 2006). LMW-GS can aggregate to form smaller polymers which interact covalently with y-type HMW-GS via disulphide bonds. With mass spectrometry (MS) it was possible to prove that disulphide bonds form between glutenin subunits, demonstrating x-x and x-y HMW-GS interactions, and interactions between y-type HMW-GS and LMW-GS (Lutz et al. 2012). So according to the current model of GMP structure, the HMW-GS form the backbone and the LMW-GS form the branches through bonds with the y-type HMW-GS (Wieser 2007). The amount and polymerization characteristics of functional GMPs are highly influenced by the number and structural features of different HMW-GS and LMW-GS. The proteins encoded by the three *Glu-1* loci differ significantly in their ability to incorporate the HMW-GS and LMW-GS into functional GMPs, with Glu-D1 having the strongest potency (Wang et al. 2017).

LMW-GS are subdivided into B, C and D subunits, according to their structural and functional properties. B subunits have a typical LMW-GS structure encoded by genes on chromosome 1. The C and D subunits are more gliadin-like and are encoded by genes on chromosome 6. They are structurally similar to gliadins, but function like glutenins, as they can form intermolecular disulphide bonds through unpaired cysteine residues. LMW-GS are a complex group with as many as 30 proteins between 32–45 kDa. B-type LMW-GS consist of a short N-terminal region, the variable repetitive region and conserved C terminal region. LMW-GS are classified as LMW-m, LMW-s and LMW-i types based on their N-terminal amino acid, methionine, serine or isoleucine respectively (D'Ovidio and Masci 2004). They all have eight cysteine residues mostly in the C-terminal region, seven of which are conserved between different LMW-GS. Six of the cysteines are involved in intramolecular bonds. The two additional cysteine residues, unique to LMW-GS, are not able to form intrachain bonds (Wieser 2007). Gliadins are usually monomers, but those with extra cysteine residues can be incorporated into the gluten polymer. These proteins can act as chain terminators of the gluten polymer and thus limit its size. LMW-s type has a short N terminal region followed by a variable repetitive region and conserved C-terminal region. It has a single cysteine residue in the repetitive region and seven cysteine residues in the C-terminal region, and the first and seventh cysteine residues are involved in forming disulphide bonds. LMW-i type proteins lack the N-terminal region and start at the repetitive domain. They have eight cysteine residues in the C-terminal domain, with the third and seventh involved in intermolecular disulphide bonds (D'Ovidio and Masci 2004). There is significant interest in determining structure-function relationships in wheat gluten proteins to understand and predict end-use quality (Yahata et al. 2005).

4 Proteomics

In recent years there has been significant progress in protein separation techniques, including two dimensional gel electrophoresis (2DE), liquid chromatography, tryptic digest MS and database searching (Buts et al. 2014). Among the most commonly used methods in proteomics, 2DE can be readily used to resolve and visualize thousands of protein species on a single gel, so isomorphisms, polymorphisms and structural changes, such as post translational modification (PTM) can be detected (Chevalier et al. 2004; David et al. 1997). Gene products, visualized as precise protein spots, are de facto genetic and physiological markers (Damerval et al. 1988; de Vienne et al. 1996), which can be useful in assessing genetic variability and for establishing genetic distances and phylogenetic relationships between organism lines, species and genera. Peptides separated by 2DE are prepared for MS through matrix-assisted laser desorption/ionization (MALDI) or electrospray ionization (ESI). The mass to charge ratio of peptides are measured to generate MS spectra. Different labelling strategies have been used for comparing peaks, such as in vivo isotopic labelling and stable isotopic labelling with labelled amino acids (Yates et al. 2009).

Proteomics bridges the gap between studying DNA and proteins (Ribeiro et al. 2013). In plants, proteomics analysis is done on whole organisms, tissues, cells, and sub-cellular fractions to compare various processes. Global analysis of plant proteins should give a better understanding of gene function and regulation than analysis of genes alone (Buts et al. 2014). The genome remains unchanged, but the protein expression is modified as genes are turned on and off in response to environmental conditions. DNA sequence analysis cannot predict the active form of a protein, and the quantification of RNA does not always reflect the corresponding protein levels. Although DNA sequencing is the most efficient way to determine amino acid sequences, conventional methods of sequencing highly repetitive proteins, such as HMW subunits, are prone to error. In addition, it does not provide any information on PTM, which can have major effects on the properties of the proteins (Carr et al. 1991; Bahr et al. 1994). Expression proteomics is used to quantitatively and qualitatively study expression of proteins under different conditions (Beyene et al. 2016). Currently there are two complementary proteomic approaches, the so-called gel-based and gel-free approaches. The gel-based approach represents the cornerstone of proteome analysis and uses the unequalled resolving power of 2DE to separate complex protein mixtures. Recently-developed gel-free approaches employ chromatography separation technologies instead. The two approaches differ in how extraction, separation, and detection of proteins and peptides are done. As a consequence of the complex chemical nature of proteins and due to the large dynamic range of their concentrations, each approach is able to focus on only a sub-fraction of the full protein set (Cunsolo et al. 2011). The gel-based approach has been the most widely used so far, which gives information such as molecular weight, isoelectric point, presence or absence of proteins and PTM (Lodha et al. 2013). Both approaches are followed by an MS step, the basis for protein characterization and identification. Improved mass spectrometers together with gene and genomic sequence databases have made MS a critical tool in protein and proteome analysis. Mass analysers available include ion trap (IT), time of flight (TOF), quadrupole (Q) and Fourier transform ion cyclotron resonance (FTICR). TOF analysers combined with MALDI were found to be very sensitive with high resolution and accuracy though incompatible with MS/MS, but this limitation was overcome by TOF/TOF analysis, which can be used with MALDI and ESI sources (Cunsolo et al. 2012). MS strategies for the detection of PTM, such as the glycosylation and phosphorylation of proteins, in electrophoretically separated proteins have been reported in a number of studies (Mørtz et al. 1996; Qin and Chait 1997; Jensen et al. 1998; Packer and Harrison 1998; Larsen and Roepstorff 2000; Larsen et al. 2001).

Bottom-up protein analysis or shotgun proteomics (peptide-centric proteomics) refers to the characterization of peptides released from the protein through proteolysis. Identification of peptides is done by comparing mass spectra derived from peptide fragmentation with theoretical spectra generated from *in silico* digestion of a protein database. The top-down approach allows MS analysis of intact proteins that have not been cleaved. It preserves the protein structure, including PTM, instead of measuring peptides (Jorrín-Novo et al. 2015; Bansal et al. 2016).

The development of gel-free, label-free methodologies has led proteomics research into a new phase. MS-based approaches are the standard currently. The development of the Orbitrap allowed the characterization of peptides at sub parts-per-million measurement accuracy (Yates et al. 2009). Label-free MS makes precise quantitation difficult, but is sufficient for comparison of relative protein abundance in full proteomes (Boja and Rodriguez 2012). The first gel-free proteomic analysis was done with multidimensional protein identification technology (MudPIT). MudPIT uses a combination of liquid chromatography mass spectrometry (LC-MS/MS) and HPLC. Quantitative proteomics approaches, such as isotope-coded affinity tags (ICAT), targeted mass tags (TMT) and isobaric tags for relative and absolute quantitation (iTRAQ) have been widely used in descriptive and comparative proteomics studies of plant development and metabolic strategies in abiotic stress adaptation (Matros et al. 2011). Most peptide-based quantitative proteome analyses are comparative or relative, and are based on chemical labelling of peptides and are suitable for plant protein/peptide labelling. This includes iTRAQ and ICAT (Agrawal et al., 2013). Techniques such as iTRAQ and high-resolution LC-MS/MS have allowed precise comparison of differentially abundant proteins. iTRAQ is a good method for studies of proteomic changes in polyploids such as wheat because it can simultaneously examine many proteins, and is also sensitive enough to measure protein abundance in related species when coupled with MS. This technique can be used to study genome duplication, divergence and hybridization in polyploid species (Bansal et al. 2016).

5 Proteomics as a Tool in Gluten Protein Analysis

Knowledge of the genetics, structure and composition of storage proteins has significantly expanded in the last decades due to biochemical and molecular studies (Ma et al. 2014). Wheat storage proteins cause celiac disease, baker's asthma and wheat-dependent exercise-induced anaphylaxis. A key to understanding the mechanisms of these diseases is to study the wheat genetic variants that are better tolerated (Rasheed et al. 2014). The wheat genome is very large and complex (17 Gb) (Šafář et al. 2010). The first draft sequence of the wheat genome was released in 2014 (IWGSC 2014). Sequenced genomes of crops greatly aid in the process of obtaining proteomics results (Reddy et al. 2015). Wheat proteomics is complementary to genomics in the sense that proteomics can be used to understand how the genome regions are involved in determining grain protein composition, enzyme activity and the expression of specific genes in different growing conditions (León et al. 2009, 2010). Gluten proteomics is important to link genomic data and functional biology, providing data on PTM and changes in protein expression. Gluten protein has an abundance of different protein homologues with slightly different amino acid sequences. High proportions of proline and glutamine residues and low proportions of lysine and arginine in primary sequences can, however, make protein cleavage for MS analysis difficult (Bansal et al. 2016).

Integration of functional genomics, proteomics, bioinformatics, breeding and genetic resources is aiding the understanding of the genetic and biochemical bases of quality traits in wheat. This information must be incorporated into breeding programs together with high-throughput screening techniques to combine abundant yield and agronomic characteristics with good quality (Rasheed et al. 2014). Proteomics tools can be used to study the structures and interrelationships of gluten proteins to facilitate the improvement of quality and functional properties and the reduction of allergenicity (Ribeiro et al. 2013). Years of selection for alleles linked to good quality have led to a reduction in genetic variation. Proteomics can be used as a tool to identify novel genes in the larger wheat gene pool, which can be incorporated into breeding programs. This will also reduce the erosion of genetic variability that has been taking place. Proteomics techniques have been successfully used to characterize gluten proteins, especially with the use of MALDI-TOF-MS. Proteomic data can fill the data gaps related to genotype by environment interaction, climate adaptation and disease and pest resistance. The full complement of genomics and proteomics is essential to understand genes, their products, interactions and rheological properties (Ribeiro et al. 2013).

SDS-PAGE is still commonly used in breeding programmes to screen for superior subunits linked to good quality, but this technique confounds the identification of some subunits, and the quantification of subunits is very difficult (Yan et al. 2003). Analysis by 2DE combines isoelectric focusing based on isoelectric point, followed by separation according to molecular weight by SDS-PAGE. It is a powerful tool to identify protein polymorphisms in wheat flours. In the 1980s RP-HPLC was developed for gluten protein identification (Bietz 1983). RP-HPLC had several advantages over SDS-PAGE such as better resolution and automation. RP-HPLC separates proteins based on surface hydrophobicity, but sometimes subunits have the same elution times, where RP-HPLC cannot separate them. Later RP-HPLC was combined with LC-MS/MS for characterization of glutenin subunits (Dong et al. 2009). LC-MS development allowed a better understanding of gluten heterogeneity in different wheat varieties and also allowed discrimination of varieties in terms of quality (Lagrain et al. 2013). MALDI-TOF-MS became a very useful tool in analysing HMW glutenins especially (Chen et al. 2007; Dong et al. 2009). Storage protein allele analysis is effective for genotyping genetic resources.

HMW-GS genes are closely linked and are difficult to manipulate with conventional breeding. Qualitative and quantitative changes in HMW-GS to produce stronger dough, for example, can be induced by introduction and expression of additional and novel genes through genetic transformation (León et al. 2009). Proteomics tools have been used to check the effectiveness of transformation events by quantifying expression levels of each group of storage proteins in wheat endosperm. Modifying gliadin content is a strategy used for reducing allergenicity in celiac patients (Wieser 1996).

The close sequence similarity between different HMW-GS, including the low frequencies of arginine and lysine residues, makes MS identification complicated, but MALDI-TOF-MS results showed good agreement with predicted masses of intact subunits and published DNA sequences. It was suggested that unfractionated HMW-GS should be analysed using MALDI-TOF-MS to identify and screen for glutenins associated with good quality (Liu et al. 2009). When data from MALDI-TOF-MS and LC-MS/MS were combined for HMW-GS protein regions, coverage of more than 95% was seen for all the subunits that were investigated. This data generally confirmed the gene-derived sequences, although there were some point mutations, insertions and deletions in the repetitive domain. This study also confirmed a total lack of PTM in HMW-GS (Mamone et al. 2009). Cunsolo et al. (2012) suggested that the nomenclature system of HMW-GS should be changed to one based on amino acid sequence similarity rather than electrophoretic similarity, as the HMW-GS are so similar. The detection of intact HMW-GS with ESI-MS has yet to be reported. A direct, accurate and sensitive detection of HMW-GS would be very useful to monitor wheat contamination in foods, which is necessary for people on a gluten-free diet.

The LMW-GS and gliadins are more complicated than the HMW-GS. The length of the repetitive domain largely determines the variation in LMW-GS size (Bonomi et al. 2013). Cunsolo et al. (2012) reported that characterization of LMW-GS is very complicated, as they have a high number of genes with similar structural characteristics. As a consequence, MALDI mass spectra obtained from the tryptic digests of all LMW-GS present very few signals and all attempts to identify proteins by database searches using the experimental mass-to-charge ratios (m/z) have been unsuccessful. It is therefore difficult to directly identify which gene sequences correspond to the encoded proteins. Amino acid sequences of LMW-GS show a long repeating pattern with very few cleavable tryptic sites, resulting in very few tryptic peptides that are analysable in MS. Cunsolo et al. (2012) suggested using MS/MS to analyse the available tryptic peptides. Ribeiro et al. (2013) found UPLC to be effective for characterization of LMW-GS and water soluble proteins. Using fine particles (1.7 μm diameter) to pack small-diameter columns, this system can produce a column performance as high as up to 100,000–300,000 theoretical plates per meter. Compared to RP-HPLC, separation was more rapid, resolution was higher, and sensitivity greater with RP-UPLC (Swartz 2005). UPLC has received more and more attention, and it has been successfully applied to food, chemical, and medicine analyses (Zhou et al. 2007). MALDI-TOF-MS showed promise for analysing LMW-GS, but more effort is needed for its use to become routine (Rasheed et al. 2014).

6 Proteomics of Gluten Proteins

Originally, the core separation technology of proteomics was 2DE, a system that is well suited to the separation of complex mixtures of proteins. It is significant that this methodology was first developed to separate cereal grain proteins by Wrigley (1968, 1970), with further developments of high-resolution 2DE by O'Farrell

(1975), Klose (1975), and Scheele (1975), demonstrating the enormous potential of this analytical technique.

Quantitative 2DE was reported to be a precise way to calculate HMW-GS to LMW-GS ratios and identify components in the gluten polymer (Vensel et al. 2014). This method can be coupled with MS/MS to identify individual proteins that can contribute to glutenin polymer formation and flour quality. The HMW-GS and ω gliadins appear in distinct areas of the gel as discrete spots. By extracting gluten with 0.5% SDS the extractable polymeric protein (EPP) and unextractable polymeric protein (UPP), which can only be extracted with sonication, were analysed. UPP were separated again using steric exclusion HPLC. The polymeric proteins were eluted in the first peak and monomeric proteins in the last part of chromatogram. Proteins corresponding to each elution peak were then analysed with 2DE. The HMW-GS and LMW-GS were the principle proteins in peak 1. The ω, α, and γ gliadins were predominant in peak 2 from EPP and UPP. Some LMW-GS proteins were visible in both fractions. There was a third peak for EPP and UPP which contained α amylase/trypsin inhibitors and purinins, but also some α and γ gliadins. Different fractions with the same pI and molecular weight gave different MS/MS results, showing the wheat proteome to be complex. Gliadins with an odd number of cysteine residues, known to be polymer chain terminators, could be distinguished from monomeric α, γ, and ω gliadins with six, eight and no cysteine residues. Some gliadins with an odd number of cysteines were found in both the UPP and EPP fractions. Non-gluten proteins (triticins, globulins and serpins) were more highly represented in EPP peak 1 than in other fractions (Vensel et al. 2014). The accumulation of serpins in wheat grain has a potential role in flour functionality. A glutamine rich motif in serpin is similar to those in glutamine-rich storage proteins that can impact dough rheology (Wu et al. 2012). It is possible that certain serpins can covalently link to glutenin polymers and serve as chain terminators (Vensel et al. 2014).

The development in the late 1980s of two "soft" desorption/ionization MS techniques capable of producing molecular ions of intact biomolecules, ESI (Fenn et al. 1989) and MALDI (Karas and Hillenkamp 1988; Tanaka et al. 1988), has revolutionized the field of protein investigation by rendering polypeptides analysable by MS. Moreover, the use of MALDI-MS for the structural characterization of glutenin subunits appears particularly appropriate, because the presence of a large domain constituted of repeating sequence motifs makes the application of conventional sequencing procedures, such as Edman degradation, difficult. In addition, MS provides a way of determining whether PTM of the proteins has occurred. MALDI-MS and RP-HPLC/ESI-MS combined with enzymatic digestion were used for the direct characterization of the gene-derived sequences of some HMW subunits isolated from different cultivars of bread wheat (Foti et al. 2000). ESI-MS was found to produce a range of charged species, which meant that more mass assignments could be made. It involves a process of proteolytic digestion for protein identification followed by LC and MS/MS. This bottom-up or shotgun approach was used to verify HMW-GS sequences against gene sequences (Cunsolo et al. 2004), wheat gluten composition (Qian et al. 2008), and gluten structure (Lutz et al. 2012). Considering the limitations of trypsin digestion of the repetitive prolamin sequences (Di Stefano et al. 2012), a top-down approach may be better to measure intact and partially sequenced proteins. This can be useful for determining primary structure and detection of protein modification. For top-down analysis, ESI is preferred over MALDI. Lagrain et al. (2013) used SDS-PAGE and RP-HPLC combined with ESI-quadrupole time of flight (QTOF) for LC-MS. They found very little PTM of HMW-GS. They concluded that HMW-GS can be identified by ESI-MS after isolating the proteins from flour, dissolving them in water/acetonitrile and separating them by RP-HPLC. RP-HPLC has been used successfully in combination with ESI-MS for gliadin and LMW-GS determination (Muccilli et al. 2010). Wheat proteomics, such as detection of HMW-GS combined with full transcriptome analysis, can be useful for genetic improvement of wheat (Mamone et al. 2009). A top-down approach using LC-ESI-MS can generate information on transcriptome protein structure, including details of PTM (McLafferty 2011).

Lee et al. (2016) studied the link between proteins expressed in a wheat variety to LMW-GS genes and haplotypes identified with *Glu-A3*, *Glu-B3* and *Glu-D3* gene-specific primers. To associate LMW-GS gene sequences with specific flour proteins they used 2D SDS-PAGE and identified 19 spots by N-terminal sequencing and LC-MS/MS. Seventeen LMW-GS spots were in-gel digested with chymotrypsin followed by MS/MS analysis. Six of these proteins matched proteins encoded at *Glu-B3* loci. They found that the relationship between *Glu-A3*, *Glu-B3* and *Glu-D3* mobility alleles by SDS-PAGE and their LMW-GS haplotypes were conserved among different wheat cultivars. Chymotryptic peptides of individual 2DE spots from the glutenin protein fraction were well matched with their gene haplotypes. The authors used ESI-MS/MS to analyse chymotryptic peptides in spots rather than MALDI-TOF, which was also done by Dupont et al. (2011) and Vensel et al. (2014). They identified 33 LMW-GS genes in one cultivar, but only 17 spots in 2DE. It could be that some 2DE spots are made up of multiple proteins with small differences in size and charge. Two spots were associated with the *Glu-A3* locus, six with *Glu-B3* locus, and nine with the *Glu-D3* locus. *Glu-D3* encodes the most abundant LMW-GS but plays only a minor role in quality variation between cultivars (Zhang et al. 2012).

In a study by Bonilla et al. (2018), HMW-GS and LMW-GS were extracted then separated by SDS-PAGE. The excised protein bands were analysed with MALDI-TOF-MS. Antibodies were developed against LMW and HMW glutenins using a proteomics approach. For glutenin subunit identification, gel bands were digested with trypsin, then digests were placed on a MALDI plate and resultant spectra compared with protein databases such as those curated by the National Center for Biotechnology Information, UniProt and ExPASy. The same was done for gliadin. Antibodies specific for each subunit were thus pinpointed. The antibodies were conjugated with quantum dots, and then applied to dough samples to label glutenin subunits. MALDI-TOF MS technology represents a powerful tool to quickly and accurately analyse glutenin composition for breeding purposes (Dworschak et al. 1998).

7 Grain Development Proteomics

Before reporting on the proteomic analyses carried out on glutenins, it must be recalled that a very large study was performed on albumins and globulins expressed in developing endosperm of wheat (*Triticum aestivum*) grain. These proteins were extracted in starchy endosperm at 21 stages (every 50°Cd) from 0 to 1003 °Cd allowing to identify 487 proteins classified in 17 biochemical functions and nine major expression profiles were found (Tasleem-Tahir et al. 2012). Many of these enzymes were involved in metabolic functions including starch and storage protein accumulation.

7.1 Development of Glutenin

Quality in wheat is largely determined by grain development processes. Therefore protein synthesis and regulation during grain development is very important to wheat quality. Starch and storage proteins show similar dynamics of accumulation as they are deposited during grain filling, about 14–28 days post-anthesis (DPA). Using iTRAQ, the HMW-GS were found to accumulate from 14 to 21 DPA, with peak expression at 28 DPA. Two LMW-GS and some gliadin/avenin like proteins were dramatically up-regulated at 28 DPA. Five serpins were significantly increased at 28 DPA. On the whole, iTRAC was found to be very useful to investigate differentially expressed proteins during grain development (Ma et al. 2014).

Mazzeo et al. (2017) harvested grains of durum cultivar Svevo at six stages of grain development between 3 DPA (early development) to 30 DPA (maturity/dessication). The gliadin proteome was then characterized using LC-MS/MS of chymotryptic digests of HPLC isolated subunits. Due to poor detection of the gliadin fraction, Western blotting was done using anti-gliadin antibodies. Glutenins were separated using SDS-PAGE and HPLC. In this case SDS-PAGE was more effective than HPLC for separation. The gluten protein subsets found at different stages of plant development were not related to the cultivar used. The α and γ gliadins were synthesized the earliest. Gliadins and glutenins steadily increased during grain development. HMW glutenins cumulated before LMW glutenins, suggesting the HMW glutenins form an initial polymeric backbone from which LMW-GS branches form. By the milk phase of grain development, the gluten proteins were similar to those in mature grain (Shewry et al. 2009).

7.2 Effects of Abiotic Stress on the Grain Proteome

Abiotic stress causes protein dysfunction. Proteins may be newly synthesized, or increase or decrease in quantity due to stress. Affected proteins may be involved in signalling, translation, host defence mechanisms, or carbohydrate or amino acid metabolism. In response to stress proteins may mediate these functions directly or indirectly, for example, by synthesising osmolytes or regulating the genome, which can influence physical features or enzymatic activity, for example, of antioxidant enzymes. Therefore elucidating mechanisms in plant stress responses is pertinent from both pure and applied viewpoints (Wang et al. 2004). Genome sequence information cannot explain gene function and developmental and regulatory biology or stress response mechanisms. To investigate this, comprehensive approaches are necessary for quantitative and qualitative analyses of the products of gene expression at transcriptome, proteome and metabolome levels (Timperio et al. 2008). Messenger RNA and protein levels cannot be correlated through transcriptome analysis alone, so proteomics is the most suitable method to analyse biochemical pathways and complex responses of plants to environmental stimuli. Proteomics is also an essential link between the transcriptome and the metabolome (Gray and Heath 2005). The current trend of rising global temperature is predicted to continue in the future. Wheat production is significantly affected by abiotic stress, especially high temperatures at grain filling. The influence of a short heat stress (HS: 4h at 38°C 4 consecutive days) on wheat grain, during the accumulation phase, was studied using proteomics analysis of total proteins and non prolamin fraction: The abundance of several proteins were revealed affected 26 days after the end of HS (Majoul et al. 2003, 2004). Heat stress affects grain protein synthesis, cell metabolism including that of carbohydrates, and enzymes involved in transcription and translation, and thus disrupts grain development. In a study using iTRAQ, 256 proteins were differentially expressed between normal and high temperatures. LMW-GS were significantly reduced at 15 and 20 DPA under heat stress (Zhang et al. 2017).

Yang et al. (2011) used 2DE and MS to investigate the effect of heat and drought stress applied at two stages of grain filling. The type and time of stresses affected the synthesis of individual protein fractions. Albumins were significantly reduced by stress. Glutenin accumulation increased by 85–159% during grain filling in response to stress. Combined stress events caused a significantly larger effect than single stress events. The α and γ gliadin fractions decreased due to drought stress.

Zhang et al. (2014) subjected three wheat cultivars to drought stress during grain filling, then analysed the mature grain proteomes by 2DE and MALDI-TOF-TOF. Significant increases in albumin and gliadin were seen under drought stress. Fourteen differentially expressed proteins were identified, eight of which are potential complex-forming proteins.

7.3 *Effects of Fertilizer Application on the Grain Proteome*

Post-anthesis fertilizer application was found to double protein content of flour (Altenbach et al. 2011). When no fertilizer was applied, 74% of protein was EPP while 26% was UPP. When post-anthesis fertilizer was applied, 67% of protein was EPP and 33% was UPP. Soluble and insoluble proteins were separated with 2DE (Altenbach et al. 2011). The polymers in each fraction represent a range of sizes rather than distinct size classes, therefore EPP and UPP had many of the same proteins. It is very difficult to quantify the proteins in each fraction because of the complexity of gluten proteins and the overlap between some LMW-GS and α and γ gliadin in 2DE (Altenbach et al. 2016). The proportions of some proteins differ in the EPP and UPP fractions, when analysed by MS/MS, indicating that there may be differences in the way the subunits link (Vensel et al. 2014). LMW-i proteins are more frequently found in EPP and LMW-s more frequently in UPP, showing that they play different roles in the gluten polymer. The proportion of chain terminators were very different in EPP compared to UPP (Vensel et al. 2014). Only α and ω chain terminators increased in response to post-anthesis fertilizer. The increases were more pronounced in UPP, indicating that sizes of the largest polymers may be limited by fertilizer application. Serpins and triticins (non-gluten proteins) were more abundant in EPP than UPP, but with fertilizer serpins were increased in both fractions and triticins were increased in UPP. On the whole, the HMW-GS to LMW-GS ratio increased in both polymer fractions due to post-anthesis fertilizer increases. While this could be positive for quality, the proportions of chain terminators and of several non-gluten proteins also increased, which may explain why a large increase in protein content was not accompanied by the same increase in UPP (Altenbach et al. 2016). The largest changes due to post-anthesis fertilizer in both EPP and UPP fractions were increases in ω chain terminators, serpins and α chain terminators. Triticin also significantly increased in UPP polymers. In the EPP polymers there were increases of about 10% in HMW-GS and 18% in the UPP due to fertilizer. There were small but significant decreases in LMW-s proteins, γ gliadin and farinins in UPP polymers. The total amount of glutenin was much higher in large polymers than small polymers, but did not change with fertilizer. The HMW-GS to LMW-GS ratio was higher in UPP than EPP and increased somewhat in response to fertilizer. This study showed the importance of LMW-GS structure. MS/MS identification of LMW-GS showed that its composition can influence polymer structure and size. LMW-s made up most of UPP and LMW-i most of EPP.

Hurkman et al. (2013) compared the effects of temperature and fertilizer on the gluten proteome of developing endosperm in the wheat cultivar Butte 86. The ratio of gliadins to glutenins increased in response to increased fertilizer and high temperatures. This was due to an increase in ω gliadins and some α gliadins. The ratio of HMW-GS to LMW-GS also increased in response to heat and increased fertilizer. Of the gliadins with an odd number of cysteines, which probably function as LMW-GS, only cys-type ω gliadins showed a response to the two treatments. Omega gliadins encoded by *Gli-B3* are implicated in wheat dependent exercise-induced anaphylaxis. This fraction was significantly increased in response to the two treatments. Several of the α gliadins, seen as the most immunogenic factors in celiac disease, increased in response to the two treatments. The two treatments elicited a similar response in terms of gluten proteins, such that most of the HMW-GS and ω gliadins and some α gliadins increased, while several LMW-GS and a minor γ gliadin decreased. This knowledge could be useful when designing transgenic approaches to determine the exact roles of proteins and how the proteome responds to fertilizer and high temperature stress.

In a study where gluten protein extracts were analysed with LC-MS using two different platforms, a total of 2736 gluten peptides were identified, 1548 with QTOF and 1031 with linear trap quadrupole, but only 157 peptides were common to both discovery platforms due to the different physical principles applied, such as the difference in ionization. This allowed the identification of a complementary range of peptides by the different platforms. In total, 127 and 63 gluten proteins were found with one or three unique peptides, respectively. Of the 63 with three unique peptides, 26 were gliadins (4 ω, 14 α and 8 γ) and 37 glutenins (29 LMW GS and 8 HMW GS). Sample preparation was not a limiting factor in gluten protein profiling (Bromilow et al., 2017a).

Gliadins and glutenins respond differently to N fertilization with gliadins accumulating more in grain, affecting the extensibility of dough made from the derived flour (Kindred et al. 2008). Wan et al. (2013) found that γ gliadin varied under different N treatments and different growth stages, being significantly higher under high N levels. Using iTRAQ technology, Yu et al. (2017) identified 16 and 12 differentially expressed proteins at respectively 7 and 18 DPA after N treatments. At 7 DPA γ gliadin, LMW-GS and HMW-GS were up-regulated. At 18 DPA gliadin and glutenin increased significantly after N treatment, and more protein bodies accumulated in the endosperm after N treatment. The number and size of protein bodies were significantly higher after N treatment. Specific proteins that were more abundant were γ gliadin, HMW-GS PW212, DX5 and y, and LMW-GS group 4 type II. These storage proteins are sometimes the main components of protein bodies. Chope et al. (2014) found that high level N treatment increased HMW-GS and gliadin content and their relative gene expression, but the amount of LMW-GS was reduced.

7.4 *High CO_2 Concentration*

A predicted increase of CO_2 concentration due to climate change is expected to increase yield in crops, particularly in C3 plants like wheat and rice, but to the detriment of quality (Leakey et al. 2009). Högy et al. (2009) reported changes in the wheat proteome due to increased CO_2 levels. Fernando et al. (2015) did a study on spring wheat with a maximum of 550 μm mol^{-1} CO_2. They used 2DE for gluten protein analysis, followed by MALDI-TOF/TOF (MS/MS). After analysis, Mascot was used for database searches against the Plant section of Swiss-Prot. Grain protein content was reduced by 9% in response to increased CO_2. A total of 39 protein spots were differentially expressed with 17 spots changing by more than two fold, and another 22 spots more than 1.5 fold. With MALDI-TOF/TOF MS, three spots were identified as being up-regulated and three down regulated by a factor of 1.5 to 2 under increased CO_2. There were fewer HMW-GS but more serpin Z1C and 1-Cys peroxiredoxin. Serpins are chymotrypsin-like serine protease inhibitors that protect storage proteins in the mature grain from being digested by insects or pathogens. Therefore the CO_2 increase could induce protection mechanisms to prevent premature proteolysis of seed storage proteins. Dough mixing time was longer for flour from wheat grown under high CO_2, but peak height was reduced by 7%. There was an 11% decrease in bread loaf volume. LMW-GS, which are synthesized in early grain filling, did not change under higher CO_2 concentrations (Shewry et al. 2009).

7.5 *Biotic Stress*

Li et al. (2018) analysed the proteome of wheat grain infested with powdery mildew by 2DE of proteins extracted from milled flour. Spots digested with trypsin were analysed by MALDI-TOF-TOF MS. Total starch was significantly reduced in infected plants, with a rise in the amylose to amylopectin ratio, and increases in total protein (+5.8%), glutenins (+12%) and gliadins (+15.6%) in infected seeds, but the glutenin to giadin ratio did not change. Of the 36 differentially expressed spots, 29 were upregulated and seven were downregulated in diseased grains. MS results showed 10 of the spots were storage proteins, while seven others were linked to protein synthesis and 11 to carbohydrate metabolism. The significant increase in triticin, serpin and HMW-GS in the infected seeds may relate to improved wheat quality.

8 Celiac Disease

Celiac disease is a human autoimmune enteropathy that arises when genetically predisposed people ingest gluten or related proteins such as hordeins and secalins (Kagnoff 2007). When the gluten is digested, due to the high proline content of prolamin, some peptides resist digestion (Hausch et al. 2002). The immune system reacts to the resistant gluten-derived peptides by making lymphocytes and antibodies against them that attack the lining of the small intestine. The intestinal villi are destroyed and the person cannot absorb nutrients (Tye-Din et al. 2010).

Proteomics potentially offers an alternative method to monitor gluten proteins in food. Central to MS is that a functionally annotated genome of the species is required or a curated set of sequences such as UniProt. However although the wheat genome has been sequenced, it is not in a form that is suitable for proteomics analyses. With the purpose of identifying as many proteins as possible at as high a certainty as possible, Bromilow et al. (2017b) developed a curated database of wheat gluten protein sequences, GluPro V1.0, which can be used expressly for identifying sequences responsible for allergies. Based on 630 full length cDNA sequences from bread wheat, the database is tailored for proteomics data mining. The final database comprised 55 HMW-GS sequences, 224 LMW-GS sequences and 185 α, 154 γ and 12 ω gliadin sequences, which is consistent with the types of gluten proteins observed in 2DE. LMW-GS were further classified into seven groups according to N-terminal amino acid sequences. There are 30 LMW-I group sequences, which lack the non-repetitive N-terminal sequence and start with the repetitive domain, and 31 LMW-s group sequences. The other 163 sequences are homologous sequences often differing by one amino acid at the start of the N-terminal domain, the LMW-m1, m2, m3, m4 and m5 groups. This grouping differs from the phylogenetic groups. The phylogenetic analysis of γ gliadin sequences showed four major and two minor groups. The α gliadins showed three groups based on the presence or absence of four immunodominant celiac toxic peptides (Bromilow et al. 2017b).

Proteomic profiling is the basis for quantification of gluten in food products and flour. Wheat, rye, barley and oats were screened with LC-MS/MS for cereal-specific peptide markers. Fewer proteins were identified in unsequenced cereal species. Four selective and sensitive peptide markers were identified which could detect a gluten concentration as low as 15 mg kg^{-1}, and contamination was found in commercial rye, millet, oats, sorghum, buckwheat and some soy varieties (Colgrave et al. 2015). Harmonized regulations on gluten-free labelling have been implemented since July 2016 through regulation (EU) No. 828/2014 in the Europe Union. Very low gluten is defined as less than 100 mg kg^{-1}, and gluten free as less than 20 mg kg^{-1} (Martínez-Esteso et al. 2016). MS is an attractive alternative to the enzyme linked immunosorbent assay (ELISA) which was proven to be inconsistent and not comparable between studies. In the study of Martínez-Esteso et al. (2016) RP-HPLC was followed by multi-enzymatic digestion with subsequent MS analysis. They detected specific peptide sequences with a set of peptide markers selected to include unique protein sequences and peptides with known immunogenic/toxic sequences in celiac disease. Specific peptide sequences from all gluten protein sub-groups were thus identified.

In proteomics, trypsin is generally used for protein identification and quantification, but gluten has few of the amino acids lysine and arginine needed for tryptic digestion, so pepsin and chymotrypsin should be used instead (Sealey-Voyksner et al. 2010). Bromilow et al. (2017a) also suggested that chymotrypsin should be used for protease digestion of gluten protein, although longer peptides are then formed compared to when trypsin is used, which could complicate MS data acquisition. They found QTOF to have better sequence coverage than other methods. Missed cleavages could enhance identification levels when using platforms that can sequence longer peptides. They found that proteomic profiling of plant proteins, such as gluten, is not limited by the methodology but rather by lack of genetic data in a form that can be handled by proteomics informatics pipelines designed to deal with highly polymorphic proteins with amino acid substitutions and both repeating sequences and deletions.

It has been hypothesised that modern wheat breeding practices may have contributed to the increase in celiac disease prevalence during the latter half of the twentieth century. Results presented by Ribeiro et al. (2016) do not support this hypothesis as *Triticum aestivum* spp. *vulgare* landraces, which were not subjected to breeding practices, presented higher amounts of potential immunostimulatory epitopes for celiac disease when compared to modern varieties (Ribeiro et al. 2016). Other results showed a high degree of heterogeneity between varieties. For example, the variety Amon had the most immunostimulatory epitopes (330.72 ± 4.09 g kg^{-1}) and Fiuza the least (79.92 ± 1.65 g kg^{-1}). The fourfold difference in toxic epitope content between Amon and Fiuza reveals the enormous potential that wheat genetic diversity represents for the development of celiac safe, or at least low-toxicity wheat-based products (Ribeiro et al. 2017).

Old wheat lines are more genetically diverse. Prandi et al. (2017) tested whether old wheat varieties developed before the first world war and modern varieties developed since then (with different ploidy levels) contained immunogenic and toxic sequences. All the wheats tested contained immunogenic and toxic sequences, and none were found to be safe for celiac disease sufferers. The γ gliadin had peptides with sequences that elicit an adaptive response in celiac patients. The toxic sequences occurred in α and γ gliadins and LMW-GS. Epitopes are not recognized equally by all patients. The study showed that old *Triticum* varieties had more immunogenic and toxic peptides than modern ones. Therefore wheat breeding has not been responsible for the increase in celiac disease. An increase in wheat consumption could play a role, as well as modification of the human immune system (Prandi et al. 2017). These results are in agreement with those of Ribeiro et al. (2016) who found higher levels of toxic epitopes in wheat landraces than in modern wheat varieties.

9 Conclusions

The purpose of proteome research is to recognize and identify all proteins and their expression patterns in a single cell or tissue in particular physiological conditions. The genome sequencing of many species and the establishment of corresponding open-access databases facilitates the development of MS-based protein identification. MS is widely applied to identify and quantify proteins including those subject to PTM or involved in interactions with other proteins, and this in turn has accelerated the development of proteomics (Ruan et al. 2008). Proteomics of wheat, including the reliable detection of HMW-GS, LMW-GS and gliadins, combined with a full transcriptome analysis would offer an effective approach for guiding and monitoring the genetic improvement of wheat. Wheat proteomics can be used to determine genotype with environment interaction, to define the genomic regions important for grain protein composition, the enzymes involved, and which genes are expressed under stress conditions. A combination of genomics and proteomics can assist in learning more about the genes, gene products, interactions and rheological properties of gluten proteins. The most important application of proteomics will probably be to discover, then score marker proteins associated with genotype by environment interactions. Proteomics provides a powerful set of tools with which to investigate the expression, diversity and interaction of gluten proteins.

References

Agrawal GK, Sarkar A, Righetti PG, Pedreschi R, Carpentier S, Wang T, Barkla BJ, Kohli A, Ndimba BK, Bykova NV, Rampitsch C, Zolla L, Rafudeen MS, Cramer R, Bindschedler LV, Tsakirpaloglou N, Ndimba RJ, Farrant JM, Renaut J, Job D, Kikuchi S, Rakwal R (2013) A decade of plant proteomics and mass spectrometry: Translation of technical advancements to food security and safety issues. Mass Spec Rev 32: 335–365.

Altenbach SB, Tanaka CK, Hurkman WJ, Whitehand LC, Vensel WH, Dupont FM (2011) Differential effects of a post-anthesis fertilizer regimen on the wheat flour proteome determined by quantitative 2-DE. Proteome Sci 9:46.
Altenbach SB, Tanaka CK, Whitehand LC, Vensel WH (2016) Effects of post-anthesis fertilizer on the protein composition of the gluten polymer in a US bread wheat. J Cer Sci 68: 66–73.
Bahr U, Karas M, Hillekamp F (1994) Analysis of biopolymers by matrix-assisted laser desorption/ionization (MALDI) mass spectrometry. Fresenius J Anal Chem 348: 783–791.
Bansal M, Sharma M, Kanwar P, Goyal A (2016) Recent advances in proteomics of cereals. Biotechnology and Genet Eng Rev 32: 1–17.
Barak S, Mudgil D, Khatkar BS (2015) Biochemical and functional properties of wheat gliadins: a review. Crit Rev Food Sci Nutr 55: 357–368.
Beckwith AC, Nielsen HC, Wall JS, Huebner FR (1966) Isolation and characterization of a high-molecular-weight protein from wheat gliadin. Cereal Chem 43: 14–28.
Beyene B, Haile G, Matiwos T, Deribe H (2016) Review on proteomics technologies and its application for crop improvement. Innov Sys Design Eng 7: 7–15.
Bietz JA (1983) Separation of cereal proteins by reverse-phase high performance liquid chromatography. J Chromat 255: 219–238.
Blackstock WP, Weir MP (1999) Proteomics: quantitative and physical mapping of cellular proteins. Trends Biotech 17: 121–127.
Boja ES, Rodriguez H (2012) Mass-spectrometry based targeted quantitative proteomics: achieving sensitive and reproducible detection of proteins. Proteomics 12: 1093–1110.
Bonilla JC, Ryan V, Yazar G, Kokini JL, Bhunia AK (2018) Conjugation of specifically developed antibodies for high- and low-molecular weight glutenins with fluorescent quantum dots as a tool for their detection in wheat flour dough. J Agric Food Chem 66: 4259–4266.
Bonomi F, Iametti S, Mamone G, Ferranti P (2013) The performing protein: beyond wheat proteomics? Cer Chem 90: 358–366.
Bromilow SNL, Gethings LA, Langridge JI, Shewry PR, Buckley M, Bromley MJ, Mills ENC (2017a) Comprehensive proteomic profiling of wheat gluten using a combination of data-independent and data-dependent acquisition. Front Plant Sci 7: 2020.
Bromilow S, Gethings LA, Buckley M, Bromley M, Shewry PR, Langridge JI, Clare Mills EN (2017b) A curated gluten protein sequence database to support development of proteomics methods for determination of gluten in gluten-free foods. J Proteomics 163: 67–75.
Buts K, Michielssens S, Hertog MLATM, Hayakawa E, Cordewener J, America AHP, Nicolai BM, Carpentier SC (2014) Improving the identification rate of data independent label-free quantitative proteomics experiments on non-model crops: A case study on apple fruit. J Proteomics 105: 31–45.
Carr SA, Hemling ME, Bean MF, Roberts GD (1991) Integration of mass spectrometry in analytical biotechnology. Anal Chem 63: 2802–2824.
Chen J, Lan P, Tarr A, Yan YM, Francki M, Appels R, Ma W (2007) Matrix-assisted laser desorption/ionization time-of-flight based wheat gliadin protein peaks are useful molecular markers for wheat genetic study. Rapid Commun Mass Spectrom 21: 2913–2917.
Chevalier F, Martin O, Rofidal V, Devauchelle AD, Barteau S, Sommerer N, Rossignol M (2004) Proteomic investigation of natural variation between Arabidopsis ecotypes. Proteomics 4: 1372–1381.
Chope GA, Wan Y, Penson SP, Bhandari DG, Powers SJ, Shewry PR, Hawkesford MJ (2014) Effects of genotype, season, and nitrogen nutrition on gene expression and protein accumulation in wheat grain. J Agric Food Chem 62: 4399–4407.
Clarke BC, Phongkham T, Gianibelli M, Beasly H, Bekes F (2003) The characterisation and mapping of a family of LMW-gliadin genes: effects on dough properties and bread volume. Theor Appl Genet 106: 629.
Colgrave ML, Goswami H, Byrne K, Blundell M, Howitt CA, Tanner GJ (2015) Proteomic profiling of 16 cereal grains and the application of targeted proteomics to detect wheat contamination. J Proteome Res 14: 2659–2668.

Cunsolo V, Foti S, Saletti R, Gilbert S, Tatham AS, Shewry PR (2004) Structural studies of the allelic wheat glutenin subunits 1Bx7 and 1Bx20 by matrix-assisted laser desorption/ionization mass spectrometry and high-performance liquid chromatography/electrospray ionization mass spectrometry. J Mass Spectrom 39: 66–78.

Cunsolo V, Muccilli V, Saletti R, Foti S (2011) Mass spectrometry in the proteome analysis of mature cereal kernels. Mass Spec Rev 31: 448–465.

Cunsolo V, Muccilli V, Saletti R, Foti S (2012) Mass spectrometry in the proteome analysis of mature cereal grains. Mass Spec Rev 31: 448–465.

Damerval C, Vartanian N, de Vienne D (1988) Differential two-dimensional protein patterns as related to tissue specificity and water conditions in *Brassica napus* var *oleifera* root system. Plant Phys 86:1304–1309.

David JL, Zivy M, Cardin ML, Brabant P (1997) Protein evolution in dynamically managed populations of wheat: adaptive responses to macro-environmental conditions. Theor Appl Genet 95: 932–941.

de Vienne D, Burstin J, Gerber S, Leonardi A, Le Guilloux M, Murigneux A, Beckert M, Bahrman N, Damerval C, Zivy M (1996) Two-dimensional electrophoresis of proteins as a source of monogenic and codominant markers for population genetics and mapping the expressed genome. Heredity 76: 166–177.

Delcour JA, Joye IJ, Pareyt B, Wilderjans E, Brijs K, Lagrain B (2012) wheat gluten functionality as a quality determinant in cereal-based food products. Ann Rev Food Sci Techn 3: 469–492.

Di Stefano V, Avellone G, Bongiorno D, Cunsolo V, Muccilli V, Sforza S, Dossena A, Drahos L, Vekey K (2012) Applications of HPLC-MS for food analysis. J Chromatogr A 1259: 74–85.

Don C, Mann G, Bekes F, Hamer RJ (2006) HMW-GS affect the properties of glutenin particles in GMP and thusGl flour quality. J Cer Sci 44: 127–136.

Dong K, Hao C, Wang A, Cai M, Yan Y (2009) Characterization of HMW glutenin subunits in bread and tetraploid wheats by reversed-phase high-performance liquid chromatography. Cer Res Com 37: 65–73.

D'Ovidio R, Masci S (2004) The wheat low-molecular weigh glutenin subunits. J Cer Sci 39: 321–339.

DuPont FM, Vensel WH, Tanaka CK, Hurkman WJ, Altenbach SB (2011) Deciphering the complexities of the wheat flour proteome using quantitative two-dimensional electrophoresis, three proteases and tandem mass spectrometry. Proteome Sci 9: 10.

Dworschak RG, Ens W, Standing KG, Preston KR, Marchylo BA, Nightingale MJ, Stevenson SG, Hatcher DW (1998) Analysis of wheat gluten proteins by matrix-assisted laser desorption/ionization mass spectrometry. J Mass Spectr 33:429–435.

Elton GAH, Ewart JAD (1966) Glutenins and gliadins electrophoretic studies. J Sci Food Agric 17: 34–38.

Fenn JB, Mann M, Meng CK, Wong SF, Whitehouse CM (1989) Electrospray ionization for mass spectrometry of large biomolecules. Science 246: 64–71.

Fernando N, Panozzo J, Tausz M, Norton R, Fitzgerald G, Khan A, Seneweera S (2015) Rising CO_2 concentration altered wheat grain proteome and flour rheological characteristics. Food Chem 170: 448–454.

Finnie C, Sultan A, Grasser KD (2011) From protein catalogues towards targeted proteomics approaches in cereal grains. Phytochem 72: 1145–1153.

Foti S, Maccarrone G, Saletti R, Roepstorff P, Gilbert S, Tatham AS, Shewry PR (2000) Verification of the cDNA deduced sequence of glutenin subunit 1Dx5 and an Mr 58 000 repetitive peptide by matrix-assisted laser desorption/ionisation mass spectrometry (MALDI-MS). J Cer Sci 31: 173–183.

Gray GR, Heath D (2005) A global reorganization of the metabolome in Arabidopsis during cold acclimation is revealed by metabolome fingerprinting. Physiol Plant 124: 236–248.

Hausch, F., Shan, L., Santiago, N. A., Gray, G.M., Khosla, C. (2002). Intestinal digestive resistance of immunodominant gliadin peptides. Am J Phys Gastrointest Liver Phys 283: G996–G1003.

Hochstrasser DF (1998) Proteome in perspective. Clin Chem Lab Medicine 36: 825–836.

Högy P, Zorb C, Langenkamper G, Betsche T, Fangmeier A (2009) Atmospheric CO2 enrichment changes the wheat grain proteome. J Cer Sci 50: 248–254.

Holman JD, Dasari S, Tabb DL (2013) Informatics of protein and posttranslational modification detection via shotgun proteomics. Meth Mol Biol 1002: 167–179.

Hurkman WJ, Tanaka CT, Vensel WH, Thilmony R, Altenbach SB (2013) Comparative proteomic analysis of the effect of temperature and fertilizer on gliadin and glutenin accumulation in the developing endosperm and flour from *Triticum aestivum* L. cv. Butte 86. Proteome Sci 11: 8.

International Wheat Genome Sequencing Consortium (IWGSC) (2014) A chromosome-based draft sequence of the hexaploid bread wheat (*Triticum aestivum*) genome. Science 345: 1251788.

Jensen ON, Larsen MR, Roepstorff P (1998) Mass spectrometric identification and microcharacterisation of proteins from electrophoretic gels: strategies and applications. Proteins 2: 74–89.

Jorrín-Novo JV, Maldonado AM, Echevarría-Zomeño S, Valledor L, Castillejo MA, Curto, M, José Valero S, Sghaier B, Donoso G, Redondo I (2009) Plant proteomics update (2007–2008): Second-generation proteomic techniques, an appropriate experimental design, and data analysis to fulfill MIAPE standards, increase plant proteome coverage and expand biological knowledge. J Proteom 72: 285–314.

Jorrín-Novo JV, Pascual J, Sánchez-Lucas R, Romero-Rodríguez MC, Rodríguez-Ortega MJ, Lenz C, Valledor L (2015) Fourteen years of plant proteomics reflected in *Proteomics*: Moving from model species and 2DE-based approaches to orphan species and gel-free platforms. Proteomics 15: 1089–1112

Juhász A, Tamás L, Karsai I, Vida G, Láng L, Bedö Z (2001) Identification, cloning and characterisation of a HMW-glutenin gene from an old Hungarian wheat variety, Bánkúti 1201. Euphytica 119: 75e79.

Juhász A, Békés F, Wrigley CW (2015) Wheat proteins. In: Z. Ustunol (ed.), Applied Food Protein. John Wiley & Sons, Ltd.

Kagnoff MF (2007) Celiac disease: pathogenesis of a model immunogenetic disease. J Clin Invest 117: 41–49.

Karas M, Hillenkamp F (1988) Laser desorption ionisation of proteins with molecular masses exceeding 10,000 Da. Anal Chem 60: 2299–2301.

Kindred DR, Verhoeven TM, Weightman RM, Swanston JS, Agu RC, Brosnan JM, Sylvester-Bradley R (2008) Effects of variety and fertiliser nitrogen on alcohol yield, grain yield, starch and protein content, and protein composition of winter wheat. J Cer Sci 48: 46–57.

Klose J (1975) Protein mapping by combined isoelectric focussing and electrophoresis of mouse tissues. A novel approach to testing for induced point mutations in mammals. Humangenetik 26: 231–243.

Komatsu S, Mahammad A, Rakwal R (1999) Separation and characterisation of proteins from green and etiolated shoots of rice (*Oryza sativa* L.): towards a rice proteome. Electrophoresis 20: 630–636.

Lagrain B, Brunnbauer M, Rombouts I, Koehler P (2013) Identification of intact high molecular weight glutenin subunits from the wheat proteome using combined liquid chromatography-electrospray ionization mass spectrometry. PLoS ONE 8(3): e58682.

Larsen MR, Roepstorff P (2000) Mass spectrometric identification of proteins and characterisation of their post-translational modifications in proteome analysis. Fresenius J Anal Chem 366: 677–690.

Larsen MR, Sorensen GL, Fey SJ, Larsen PM, Roepstorff P (2001) Phospho-proteomics: evaluation of the use of enzymatic de-phosphorylation and differential mass spectrometric peptide mass mapping for site specific phosphorylation assignment in proteins separated by gel electrophoresis. Electrophoresis 22: 223–238.

Leakey ADB, Ainsworth EA, Bernacchi CJ, Rogers A, Long SP, Ort DR (2009) Elevated CO2 effects on plant carbon, nitrogen, and water relations: Six important lessons from FACE. J Exp Bot 60: 2859–2876.

Lee JY, Beom HR, Altenbach SB, Lim YK, Kang C, Yoon U, Gupta R, Kim S, Ahn SA, Kim Y (2016) Comprehensive identification of LMW-GS genes and their protein products in a common wheat variety. Funct Integr Genom 16: 269.

León E, Marín S, Giménez MJ, Piston F, Rodríguez-Quijano M, Shewry PR, Barro F (2009) Mixing properties and dough functionality of transgenic lines of a commercial wheat cultivar expressing the 1Ax1, 1Dx5 and 1Dy10 HMW glutenin subunit genes. J Cer Sci 49: 148–156.
León E, Piston F, Rodríguez-Quijano M, Shewry PR, Barro F (2010) Stacking HMW-GS transgenes in bread wheat: Combining subunit 1Dy10 gives improved mixing properties and dough functionality. J Cer Sci 51: 13–20.
Li J, Liu X, Yang X, Li Y, Wang C, He D (2018) Proteomic analysis of the impacts of powdery mildew on wheat grain. Food Chem 261: 30–35.
Liu L,Wang A, Appels R, Ma J, Xia X, Lan P, He Z, Bekes F, Yan Y, Ma W (2009) A MALDI-TOF based analysis of high molecular weight glutenin subunits for wheat breeding. J Cer Sci 50: 295–301.
Lodha TD, Hembram P, Basak N (2013) A Successful approach to understand the from hylloclade of rice leaves free from cytosolic proteins: Application to study rice-Magnaporthe Oryzae interactions. Physiol Mol Plant Pathol 88: 28–35.
Lutz E, Wieser H, Koehler P (2012) Identification of disulphide bonds in wheat gluten proteins by means of mass spectrometry/electron transfer dissociation. J Agr Food Chem 60: 3708–3716.
Ma C, Zhou J, Chen G, Bian Y, Lv D, Li X, Wang Z, Yan Y (2014) iTRAQ-based quantitative proteome and phosphorylation characterization reveals the central metabolism changes involved in wheat grain development. Genomics 14: 1029.
MacRitchie F (2014) Theories of glutenin/dough systems. J Cer Sci 60: 4–6.
Majoul T, Bancel E, Triboi E, Ben Hamida J, Branlard G (2003) Proteomic analysis of the effect of heat stress on hexaploid wheat grain : characterization of heat-responsive proteins from total endosperm. Proteomics 3: 175–183.
Majoul T, Bancel E, Triboï E, Ben Hamida J, Branlard G (2004) Proteomic analysis of the effect of heat stress on hexaploid wheat grain : Characterization of heat-responsive proteins from non-prolamins fraction. Proteomics 4: 505–513.
Mamone G, Picariello G, Caira S, Addeo F, Ferranti P (2009) Analysis of food proteins and peptides by mass spectrometry-based techniques. J Chromatogr A 1216: 7130–7142.
Martínez-Esteso MJ, Nørgaard J, Brohée M, Haraszi R, Maquet A, O'Connor G (2016) Defining the wheat gluten peptide fingerprint via a discovery and targeted proteomics approach. J Proteomics 147: 156–168.
Matros A, Kaspar S, Witzel K, Mock HP. 2011. Recent progress in liquid chromatography-based separation and label-free quantitative plant proteomics. Phytochem 72:963–974.
Mazzeo MF, Di Stasio L, D'Ambrosio C, Arena S, Scaloni A, Corneti S, Ceriotti A, Tuberosa R, Siciliano RA, Picariello G, Mamone G (2017) Identification of early represented gluten proteins during durum wheat grain development. *J Agric Food Chem 65:* 3242–3250.
McLafferty FW (2011) A Century of Progress in Molecular Mass Spectrometry. Ann Rev Anal Chem 4: 1–22.
Mørtz E, Sareneva T, Julkunen I, Roepstorff P (1996) Does matrix-assisted laser desorption/ionisation mass spectrometry allow analysis of carbohydrate heterogeneity in glycoproteins? A study of natural human interferon-γ. J Mass Spectr 31: 1109–1118.
Muccilli V, Cunsolo V, Saletti R, Foti S, Margiotta B, Scossa F, Masci S, Lafiandra D (2010) Characterisation of a specific class of typical low molecular weight glutenin subunits of durum wheat by a proteomic approach. J Cer Sci 51: 134–139.
Nielsen HC, Beckwith AC, Wall JS (1968) Effect of disulphide-bond cleavage on wheat gliadin fractions obtained by gel filtration. Cer Chem 45: 37–47.
O'Farrell PH (1975) High-resolution two-dimensional electrophoresis of proteins. J Biol Chem 250: 4007–4021.
Packer NH, Harrison MJ (1998) Glycobiology and proteomics: is mass spectrometry the holy grail? Electrophoresis 19: 1872–1882.
Payne PI, Corfield KG (1979) Subunit composition of wheat glutenin proteins, isolated by gel filtration in a dissociating medium. Planta 145: 83–88.
Prandi P, Tedeschi T, Folloni S, Galaverna G, Sforza S (2017) Peptides from gluten digestion: A comparison between old and modern wheat varieties. Food Res Int 91: 92–102.

Qian YW, Preston K, Krokhin O, Mellish J, Ens W (2008) Characterization of wheat gluten proteins by HPLC and MALDI TOF mass spectrometry. J Am Soc Mass Spectr 19: 1542–1550.
Qin J, Chait BT (1997) Identification and characterisation of post-translational modifications of proteins by MALDI ion trap mass spectrometry. Anal Chem 69: 4002–4009.
Rasheed A, Xia X, Yan Y, Appels R, Mahmood T, He Z (2014) Wheat seed storage proteins: Advances in molecular genetics, diversity and breeding applications. J Cer Sci 60: 11–24.
Reddy TBK, Thomas AD, Stamatis D, Bertsch J, Isbandi M, Jansson J, Mallajosyula J, Pagani I, Lobos EA, Kyrpides NC (2015) The Genomes OnLine Database (GOLD) v.5: a metadata management system based on a four level (meta) genome project classification. Nucl Acids Res 43: D1099–1106.
Ribeiro M, Nunes-Miranda JD, Branlard G, Carillo JM, Rodriguez-Quijano M, Igrejas G (2013) One hundred years of grain omics: identifying the glutens that feed the world. J Proteome Res 12: 4702–4716.
Ribeiro M, Rodriguez-Quijano M, Nunes F, Carrillo J-M, Branlard G, Igrejas G (2016) New insights into wheat toxicity: Breeding did not seem to contribute to a prevalence of potential celiac disease's immunostimulatory epitopes. Food Chem 213: 8–18.
Ribeiro M, Rodríguez-Quijano M, Giraldo P, Pinto, L, Vázquez JF, Carrillo JM, Igrejas G (2017) Effect of allelic variation at glutenin and puroindoline loci on bread-making quality: favorable combinations occur in less toxic varieties of wheat for celiac patients. Eur Food Res Techn 243: 743–752.
Ruan S, Ma H, Wang S, Xin Y, Qian L, Tong J, Wang J (2008) Advances in plant proteomics-key techniques of proteome. Front Biol China 3: 245–258.
Šafář J, Šimková H, Kubaláková M, Číhalíková J, Suchánková P, Bartoš J, Doležel J (2010) Development of chromosome-specific BAC resources for genomics of bread wheat. Cytogenet Genome Res 129: 211–223.
Salekdeh GH, Komatsu S (2007) Crop proteomics. Aim at sustainable agriculture for tomorrow. Proteomics 7: 2976–29996.
Sapirstein HD, Fu BX (1998) Intercultivar variation in the quantity of monomeric proteins, soluble and insoluble glutenin, and residual protein in wheat flour and relationship to breadmaking quality. Cer Chem 75: 500–507.
Scheele GA (1975) Two-dimensional gel analysis of soluble proteins. Characterisation of guinea pig exocrine pancreatic proteins. J Biol Chem 250: 5375–5385
Sealey-Voyksner JA, Khosla C, Voyksner RD, Jorgenson JW (2010) Novel aspects of quantitation of immunogenic wheat gluten peptides by liquid chromatography–mass spectrometry/mass spectrometry. J Chrom A 1217: 4167–4183.
Shewry PR, Tatham AS, Forde J, Kreis M, Miflin BJ (1986) The classification and nomenclature of wheat gluten proteins: A reassessment. J Cer Sci 44:97–106.
Shewry PR, Tatham AS (1997) Biotechnology of wheat quality. J Sci Food Agric 73: 397–406.
Shewry PR, Underwood C, Wan Y, Lovegrove A, Bhandari D, Toole G, Mills ENC, Denyer K, Mitchell RAC (2009) Storage product synthesis and accumulation in developing grains of wheat. J Cer Sci 50: 106–112.
Singh NK, Shepherd KW (1988) Linkage mapping of genes controlling endosperm storage protein in wheat. Theor Appl Genet 75: 628–641.
Skylas DJ, Van Dyk D, Wrigley CW (2005) Proteomics of wheat grain. J Cer Sci 41: 165–179.
Swartz ME (2005) UPLC™: An introduction and review. J Liq Chrom & Rel Tech 28:1253–1263.
Tanaka K, Waki H, Ido Y, Akita S, Yoshida Y, Yoshida T (1988) Protein and polymer analysis of up to m/z 100,000 by laser ionisation time-of-flight mass spectrometry. Rapid Com Mass Spectr 2: 151–153.
Tasleem-Tahir A, Nadaud I, Chambon C, Branlard G (2012) Expression profiling of starchy endosperm metabolic proteins at 21 stages of wheat grain development. J Proteome Res 11(5): 2754–2773.
Thiellement H, Bahrman N, Damerval C, Plomion C, Rossignol M, Santoni V, de Vienne D, Zivy M (1999) Proteomics for genetic and physiological studies in plants. Electrophoresis 20: 2013–2026.

Timperio AM, Egidi MG, Zolla L (2008). Proteomics applied on plant abiotic stresses: role of heat shock proteins (HSP). J Proteomics 71: 391–411.

Tye-Din JA, Stewart JA, Dromey JA, Beissbarth T, van Heel DA, Tatham A, Henderson K, Mannering SI, Gianfrani C, Jewell DP, Hill AVS, McCluskey J, Rossjohn J, Anderson RP (2010). Comprehensive, quantitative mapping of T cell epitopes in gluten in celiac disease. *Sci Transl Med* **2:** 41ra51.

Vensel WH, Tanaka CK, Altenbach SB (2014). Protein composition of wheat gluten polymer fractions determined by quantitative two-dimensional gel electrophoresis and tandem mass spectrometry. Proteome Sci 12: 8.

Veraverbeke WS, Delcour JA (2002) Wheat protein composition and properties of wheat glutenin in relation to breadmaking functionality. Crit Rev Food Sci Nutr 42: 179–208.

Wan Y, Shewry PR, Hawkesford MJ (2013) A novel family of γ-gliadin genes are highly regulated by nitrogen supply in developing wheat grain. J Exp Bot 64: 161–168.

Wang W, Vinocur B, Shoseyov O, Altman A (2004) Role of plant heat-shock proteins and molecular chaperones in the abiotic stress response. Trends Plant Sci 9: 244–252.

Wang Z, Li Y, Yang Y, Liu X, Qin H, Dong Z, Zheng S, Zhang K, Wang D (2017) New insight into the function of wheat gluten proteins as investigated with two series of genetic mutants. Scientific Reports 7:3428.

Wieser H (1996) Relation between gliadin structure and coeliac toxicity. Acta Paed 85: 3–9.

Wieser H (2007) Chemistry of gluten proteins. Food Microbiol 24: 115–119.

Wilkins MR, Sanchez J, Gooley AA, Appel RD, Humphery-Smith I, Hochstrasser DF, Williams KL (1995) Progress with proteome projects: why all proteins expressed by a genome should be identified and how to do it. Biotechn Gen Eng Rev 13: 19–50.

Wrigley CW (1968) Gel electrofocusing-a technique for analysing multiple protein samples by isoelectric focusing. Science Tools 15: 17–23.

Wrigley CW (1970) Protein mapping by combined gel electrofocusing and electrophoresis: application to the study of genotypic variations in wheat gliadins. Bioc Gen 4: 509–516.

Wu Y, Engen JR, Hobbins WB (2006) Ultra performance liquid chromatography (UPLC) further improves hydrogen/deuterium exchange mass spectrometry. J Am Soc Mass Spectr 17: 163–167.

Wu MJ, Mckay S, Howes N, Chin J, Hegedus E (2012) Identification of novel serpin isoforms and serpin polymorphisms among Australian wheat cultivars. J Cer Sci 55: 202–209.

Yahata E, Maruyama-Funatsuki W, Nishio Z, Tabiki T, Takata K, Yamamoto Y, Tanida M, Saruyama H (2005) Wheat cultivar-specific proteins in grain revealed by 2-DE and their application to cultivar identification of flour. Proteomics 5: 3942–53.

Yan Y, Yu J, Jiang Y, Hu Y, Cai M, Hsam SL, Zeller FJ (2003) Capillary electrophoresis separation of high molecular weight glutenin subunits in bread wheat (*Triticum aestivum* L.) and related species with phosphate-based buffers. Electrophoresis 24:1429–36.

Yang F, Jørgensen AD, Li H, Søndergard I, Finnie C, Svensson B, Jiang D, Wollenweber B, Jacobsen S (2011) Implications of high-temperature events and water deficits on protein profiles in wheat (*Triticum aestivum* L. cv. Vinjett) grain. Proteomics 11: 1684–1695.

Yang Y, Li S, Zhang K, Dong Z, Li Y, An X, Chen J, Chen Q, Jiao Z, Liu X, Qin H, Wang D (2014) Efficient isolation of ion beam-induced mutants for homoeologous loci in common wheat and comparison of the contributions of *Glu-1* loci to gluten functionality. Theor Appl Genet 127: 359–372.

Yates JR, Ruse CI, Nakorchevsky A (2009) Proteomics by mass spectrometry: approaches, advances, and applications. Ann Rev Biomed Eng 11: 49–79.

Yu X, Chen X, Wang L, Yang Y, Zhu X, Shao S, Cui W, Xiong W (2017). Novel insights into the effect of nitrogen on storage protein biosynthesis and protein body development in wheat caryopsis. *J Exp Bot* 68: 2259–2274.

Zhang X, Liu D, Zhang J, Jiang W, Luo G, Yang W, Sun J, Tong Y, Cui D, Zhang A (2013) Novel insights into the composition, variation, organization, and expression of the low-molecular-weight glutenin subunit gene family in common wheat. J Exp Bot 64: 2027–2040.

Zhang X, Jin H, Zhang Y, Liu D, Li G, Xia X, He Z, Zhang A (2012) Composition and functional analysis of low-molecular-weight glutenin alleles with Aroona near-isogenic lines of bread wheat. BMC Plant Biol 12:243.

Zhang Y, Huang X, Wang L, Wei L, Wu Z, You M, Li B (2014). Proteomics analysis of wheat seed in response to drought stress. J Integr Agric 13: 919–925.

Zhang Y, Pan J, Huang X, Guo D, Lou H, Hou Z, Su M, Liang R, Xie C, You M, Li B (2017) Differential effects of a post-anthesis heat stress on wheat (*Triticum aestivum* L.) grain proteome determined by iTRAQ. Scientific Reports7: 3468.

Zhou J, Zhao J, Yuan H, Meng Y, Li Y, Wu L, Xue X (2007) Comparison of UPLC and HPLC for determination of trans-10-Hydroxy-2-Decenoic Acid content in royal jelly by Ultrasound-Assisted Extraction with Internal Standard. Chromatographia 66: 185–190.

Genotypic and Environmental Effects on Wheat Technological and Nutritional Quality

Eva Johansson, Gérard Branlard, Marta Cuniberti, Zina Flagella, Alexandra Hüsken, Eric Nurit, Roberto Javier Peña, Mike Sissons, and Daniel Vazquez

Abstract Technological (processing performance and end-product) and nutritional quality of wheat is in principle determined by a number of compounds within the wheat grain, including proteins, polysaccharides, lipids, minerals, heavy metals, vitamins and phytochemicals, effecting these characters. The genotype and environment is of similar importance for the determination of the content and composition of these compounds. Furthermore, the interaction between genotypes and the

E. Johansson (✉)
Department of Plant Breeding, The Swedish University of Agricultural Sciences, Alnarp, Sweden
e-mail: Eva.johansson@slu.se

G. Branlard
INRAE, UCA UMR1095 GDEC, Clermont-Ferrand, France

M. Cuniberti
Wheat and Soybean Quality Lab, National Institute of Agriculture Technology (INTA) Marcos Juárez, CC 21, Córdoba, Argentina

Z. Flagella
Department of Agricultural, Food and Environmental Sciences, University of Foggia, Foggia, Italy

A. Hüsken
Department of Safety and Quality of Cereals, Max Rubner-Institut, Federal Research Institute of Nutrition and Food, Detmold, Germany

E. Nurit
Mazan, France

R. J. Peña
International Maize and Wheat Improvement Center (CIMMYT), Texcoco, Mexico

M. Sissons
NSW Department of Primary Industries, Tamworth Centre for Crop Improvement, Calala, NSW, Australia

D. Vazquez
National Institute of Agriculture Research (INIA), La Estanzuela, Uruguay

G. Igrejas et al. (eds.), *Wheat Quality For Improving Processing And Human Health*, https://doi.org/10.1007/978-3-030-34163-3_8

cultivation environment may play a significant role. Many studies have evaluated whether the genotype or the environment plays the major role in determining the content of the mentioned compounds. An overall conclusion of these studies is that except for compounds encoded by single major genes, importance of certain factors mainly depend on how wide environments and how diverse cultivars are within these comparative studies. Comparing environments all over, e.g. across Latin America, ends up with a high significance of the environment while large studies including genotypes of wide genetic background result in a significant role for the genotype. In addition, for some technological properties and components, genotype has a higher effect (e.g. grain hardness and gluten proteins), while environment influences stronger on others (e.g. protein and mineral content).

Content and concentration of proteins, but also to some extent of starch, some non-starch polysaccharides and lipids, are essential in determining the technological quality of a wheat flour. For nutritional quality of the flour, the majority of the compounds are together the important determinant. Thus an increased understanding of environmental effects is essential. As to how the environment is influencing the content of the compounds, there are some differences. The protein content and composition is strongly affected by environmental factors influencing nitrogen availability and cultivar development time. However, these two factors are impacted by a range of environmental (temperature, precipitation, humidity/sun hours, etc.) and agronomic (soil properties, crop management practices such as seeding density, nitrogen fertilizer application timing and amount, etc.) components. Thus, to understand the interplay between the various environmental and agronomic factors impacting the technological quality of a wheat flour, modeling is a useful tool. Several other compounds, including minerals and heavy metals, are to a higher extent determined by site specific variation, resulting in similar rankings of entries across locations, although the total content is varying among years. The bioactive compounds and vitamins are a part of the defense mechanisms of plants and thus there is a variation in these compounds depending on prevailing biotic and abiotic stresses (heat, drought, excess rainfall, nutrition, diseases and pests). Thus, even for nutritional quality of wheat, incorporating all compounds of relevance in the evaluation would benefit from modeling tools.

1 Introduction

Wheat is the most important crop cultivated worldwide. In fact, wheat is among the three major crops (together with rice and maize) in terms of cultivated area and total yield. Furthermore, it is the major crop contributing to trade and with the highest variety of food products for human consumption (FAO 2017; Peña-Bautista et al. 2017). Products of consumption, originating from wheat, include e.g. bread, pasta, noodles, biscuits and other types of confectionary products (Kumar et al. 2011), each product having specific technological quality requirements (Johansson et al.

2013; Peña-Bautista et al. 2017). Due to the high amount of human consumption of wheat products, the nutritional value of wheat is a key component (Hussain 2012; Peña-Bautista et al. 2017) together with its functional properties. Wheat products contribute more calories and nutrients to the human food than any other crop (Peña-Bautista et al. 2017), and is an important source of proteins, minerals, carbohydrates and vitamins to the human body (Fardet 2010).

Both technological and nutritional quality of wheat is to a major extent determined by presence, content and composition of different compounds in the wheat grain (Hussain 2012; Johansson 1995). Technological quality is a result of the balance between different components in the wheat grain and flour, including proteins, starch, lipids, water, etc., and the interaction within and between these components (Johansson 1995; Johansson et al. 2013; Kasarda 1989). The wheat grain contains 54–72% of starch which is thereby the main component of the grain, although proteins (normally 8–20% of the wheat grain) are considered as the components of highest importance for the technological quality of wheat (Kasarda 1989; Pomeranz 1988; Wall 1979). The wheat proteins are mainly known to contribute structure and function to wheat derived food products for human consumption.

Nutritionally, wheat contributes substantially to the human diet primarily with protein, dietary fiber, starch, minerals and vitamins (Fardet 2010; Shewry 2007; Simmonds 1989). However, the amino acid composition of the wheat proteins is not the most beneficial as a food source for humans, due to the relatively low content of essential amino acids in the wheat proteins (Jiang et al. 2008). The dominating dietary fibers in the wheat grain are beta glucan and fructan present in the starchy endosperm and arabinoxylan and cellulose from the bran (Andersson et al. 2013). Important minerals in the wheat grain are calcium, magnesium, potassium, sodium, phosphorus, copper, iron, selenium and zinc (Hussain 2012). Among the vitamins, wheat contains vitamin D and vitamin K, while the most common carotenoid in wheat is lutein (Hussain et al. 2015), and the most common tocol is β-tocotrienol (Hussain et al. 2012), both of significant impact on human health.

The content of all compounds in plants are impacted by genotype and environment as well as their interactions (Johansson et al. 2014). However, the level of influence on these different parameters depends on selection of plant material, cultivation environment and cultivation practices (Johansson et al. 2014). Therefore, a small to large impact is seen from genotypic and environmental factors on technological and nutritional quality in wheat.

This chapter aims at a review and compilation, as well as a discussion and concluding remarks as regards to how and when cultivar, natural environmental factors and cultivation practices are influencing technological and nutritional quality in wheat. Furthermore, the background reasons for the variation in quality will be sought for and discussed in order to formulate concluding predicting models of how genotypes and environments can be combined in order to fine tune technological and nutritional characters in wheat. For a broader and basic understanding, this chapter is compiling results from a range of different environments across the globe.

2 Genotypic Effects on Technological and Nutritional Quality

The genotype influences in particular all components involved in determining the technological and nutritional quality of wheat and encoding genes have been determined for the majority of the compounds although with somewhat different results for those compounds determined by quantitative traits loci (QTL) (Table 1).

Table 1 Quality related traits in wheat, their relations to quality and their encoding genes

Trait	Compound	Encoding gene/s	Product	References (examples)
Gluten proteins	HMW-GS LMW-GS Gliadins	1AL, 1BL, 1DL 1AS, 1BS, 1DS 1AS,1BS,1DS,6A,6B,6D	Bread, Pasta, Noodle, Biscuits, Pan bread etc	Payne & Lawrence 1983; Payne et al. 1984, 1987; Shewry et al. 1987; Margiottta et al. 1996; Johansson 1996; Liang et al. 2010; He et al. 2005; Johansson et al. 2013; Schmid et al. 2016
Protein polymerization	Gluten proteins	-	Bread, Pasta Noodle, Biscuits Pan bread etc	Johansson et al. 2001; 2013
Protein concentration	Gluten proteins	1, 2, 3A, 3D, 4BS, 5AL, 6, 7	Bread, Pasta Noodle, Biscuits Pan bread etc	Finney & Barmore 1948 Johansson et al. 2001 2004, 2013 Groos et al. 2003 Prasasd et al. 2003 Blanco et al. 2006

(continued)

Table 1 (continued)

Trait	Compound	Encoding gene/s	Product	References (examples)
Plant development	Gluten protein concentration polymerization	-	Bread, Pasta, Noodle, Biscuits Pan bread etc	Malik et al. 2013 Johansson et al. 2013
Grain hardness	Puroindolines Lysophospholipids	5D	Bread, Pasta Noodle, Biscuits Pan bread etc	Pasha et al. 2010 Liu et al. 2017
Waxy wheat	Low amylose starch	7A, 7D, 4A	Noodle	Graybosch 1998 Kim et al. 2003
Dietary fiber	Arabinoxylan Beta glucan	6B 1B, 3A, 5B, 6D	Nutrition	Charmet et al. 2009 Nemet et al. 2010 Andersson et al. 2013 De Santis et al. 2018
Minerals content	Iron, Zinc	2A, 5A, 6B, 7A	Nutrition	Peleg et al. 2009 Tiwari et al. 2009
Vitamin B	B1, B2	-	Nutrition	Batifoulier et al. 2006 Shewry et al. 2011 Nurit et al. 2016 Li et al. 2018
Phytochemicals	Tocotrienols Lutein Ferulic acid	- - -	Nutrition	Shewry et al., 2010 Hussain et al. 2012 Fratianni et al. 2013

2.1 *Technological Quality*

The gluten proteins are well known being a major contributor determining the technological quality of wheat. Both protein content and composition is to various parts (from less and up to 100%) determined by the wheat genotype (Johansson et al. 2013). Therefore, the gluten proteins, their content and composition and their genotypically determined variation and relations to quality are among the most thoroughly studied characters of the wheat grain (Johansson et al. 2013).

Among the gluten proteins, the high molecular weight-glutenin subunits (HMW-GS) are known for making the highest impact on dough viscoelastic properties and technological quality of wheat flour. The HMW-GS are also the most easily determined among the gluten proteins, as each wheat genotype only contain three-to-six subunits encoded by genes on the long arm of the group 1 chromosomes (Payne et al. 1984; Shewry et al. 1987). However, due to the relatively large variation of HMW-GS, with a total number of 30 different subunits described in early publications (Margiotta et al. 1996; Payne et al. 1984; Shewry et al. 1987) and several hundred summarized in more recent versions of the wheat gene catalogue (MacIntosh et al. 2008), impact and variation among genotypes is substantial. Both the total number of HMW-GS (e.g. 3 vs 6 subunits totally) and the composition of specific subunits (e.g. 2 + 12 vs 5 + 10) plays a role in determining dough strength and extensibility, where studies have shown e.g. HMW-GS 21* + 21*y, 14 + 15 and 5 + 10 attributed to stronger dough properties than the null allele, 6 + 8 and 2 + 12 (Johansson et al. 1993; Margiotta et al. 1996; Payne & Lawrence 1983; Payne et al. 1987; Shewry et al. 1992). One of the major reasons for differences in the impact on quality from different HMW-GS is in their number of cysteine residues, where e.g. 5 + 10 contains 12 cysteine residues compared to 2 + 12 having 11 cysteine residues (Shewry et al. 1992). A higher amount of cysteine residues is suggested to contribute to a more complex/or tightly linked protein polymer (Blechl and Anderson 1996; Blechl et al. 1998; Johansson et al. 2013).

Studies during later years have proven also the impact of low molecular weight (LMW)-GS on technological quality (Liang et al. 2010). Although, it is thought that specific LMW-GS contribute to enhance the size and viscoelasticity of the glutenin macropolymer, the molecular explanation for these differences is less well understood. Recent findings have indicated LMW-GS to act as polymer chain terminators (Schmid et al. 2016). Besides the glutenins, also the gliadins are participating in building the protein polymers within processed products of wheat and several studies have proven the impact of various gliadins on technological quality (Branlard and Dardevet 1985; Johansson 1996). The specific protein composition of HMW-GS, LMW-GS and gliadins are basically genetically determined (Brites and Carrillo 2001; He et al. 2001; Johansson et al. 2013).

Grain hardness, total grain protein concentration, cultivar determined development time (i.e. the differences in development time of a cultivar from sowing to anthesis and maturity) and polymerization of the proteins (%UPP) are additional gluten protein related characteristics that are influencing the technological quality of the wheat flour (Finney and Barmore 1948; Johansson et al. 2001; Johansson et al. 2013; Malik et al. 2013; Pasha et al. 2010). Grain hardness is strongly genetically determined by the presence or absence of certain puroindolines encoded on genes present on chromosome 5D (Pasha et al. 2010). The puroindolines are affecting the adhesion of storage proteins to the starch in the wheat grain (Presinzka et al. 2016). Variation in grain protein concentration is partly genetically determined by a number of quantitative genes and mostly positively correlated to bread volume but also to less polymerized proteins and weaker gluten (Finney and Barmore 1948;

Johansson et al. 2001, 2004, 2013). Plant development time to anthesis, which is partly genetically determined, has been shown to correlate with gluten protein polymerization, affecting dough strength in a way similar to that related to HMW-GS composition (Johansson et al. 2013; Malik et al. 2011).

Genotypically determined variation in wheat grain starch and lipids, effecting technological quality of wheat, is also well known and has been described in the literature. Waxy starch is one such character, where the production of starch amylose is genetically determined through three genes on chromosomes 7A, 7D and 4A (Graybosch 1998), resulting in higher viscosity of the starch which is beneficial in noodle production (Kim et al. 2003). Furthermore, content of lysophospholipids, the most abundant among the lipids in the wheat grain, forming complexes with amylose, are mostly genetically determined (Liu et al. 2017).

2.2 Nutritional Quality

A number of components in wheat contribute substantially in the human diet to health benefits; protein, dietary fibers, minerals, vitamins (e.g. B vitamins) and phytochemicals, where e.g. wheat contribute 20% of the dietary fibers in the human diet in the UK (Shewry and Hey 2015). Composition of genetically determined nutritional components in wheat have been widely evaluated. Most of the nutritional traits are determined by QTLs at various chromosomes, making their exact definition and use in breeding challenging (Michell and Shewry 2015).

Grain protein concentration is a typical trait of nutritional quality in wheat determined by QTLs (Blanco et al. 2006; Groos et al. 2003; Prasad et al. 2003). Quantitative genes encoding grain protein concentration in wheat have been identified on chromosomes 2A, 3A, 4D and 7D, each being responsible for 10% of the phenotypic variation (Groos et al. 2003). Another study defined QTLs on chromosomes 2B, 2D, 3D and 7A as stable encoding genes for grain protein concentration in wheat (Prasad et al. 2003). In durum wheat, QTLs with a major impact on grain protein concentration has been detected on chromosomes 2A, 6A and 7B (Blanco et al. 2006).

In principle all dietary fiber components have been shown having a genetic component of determination (Andersson et al. 2013; De Santis et al. 2018). Arabinoxylans, the major component of the cell walls, of which the water extractable part in particular is studied are determined by two major quantitative traits loci, of which the one located on chromosome 6B explained 59% of the phenotypic variation (Charmet et al. 2009). Additionally a QTL encoding water extractable arabinoxylans was found located on chromosome 1BL explaining 32–37% of the variation in relative viscosity (Martinant et al. 1998). Beta glucans, contribute about 20% of the cell walls (Nemeth et al. 2010) and the beta glucans have been found encoded by QTLs on chromosomes 1B, 3A, 5B and 6D (Manickavelu et al. 2011).

Genetic determination have been evaluated of the grain mineral content in wheat, and the main focus has been on Fe and Zn, being the two minerals of highest significance for human health (Peleg et al. 2009). Studies have indicated a *Gpc-B1* locus, being involved in coding for flag leaf senescence and grain protein concentration, also being involved in determining the Zn and Fe content in wheat grain (Distelfeld et al. 2007). In wild emmer wheat, consistently having higher minerals content than cultivated bread and durum wheat, QTLs for minerals were located in similar chromosomal regions as QTLs for grain protein concentration, i.e. 2A, 5A, 6B and 7A (Peleg et al. 2009). QTLs for Fe and Zn have been mapped to the chromosomes 2A and 7A in diploid wheat (Tiwari et al. 2009). Most studies in modern wheat have focused on the *Gpc-B1* locus (Tabbita et al. 2017) although modern genomic tools open new opportunities to further understand possible improvements in grain mineral content in wheat (Borrill et al. 2014).

Genotypic determination of the vitamin B complex has received less attention. Recent studies have indicated a large variation among genotypes in content of different types of vitamin B, as well as of changes in content and their digestibility while processed (Andersson et al. 2010; Batifoulier et al. 2006; Lampi et al. 2008; Nurit et al. 2016; Sampson et al. 1996; Shewry et al. 2011). Genome-wide association mapping and development of single nucleotide polymorphism (SNP) markers has been carried out recently to allow improvements of vitamin B1 and B2 in wheat (Li et al. 2018).

Among the tocols and carotenoids, wheat in general shows high contents of tocotrienols and lutein, two compounds contributing highly to human health and more limitedly in other foods (Fratianni et al. 2013; Hussain et al. 2012; Husain et al. 2015). Ferulic acid is the most commonly found phenolic acid in wheat (Vaher et al. 2010). Clear genotypic variation in content of various phytochemicals have been demonstrated in various studies (Shewry et al. 2010), most likely being the results of various QTLs (similarly as the case in e.g. maize; Wong et al. 2003).

Thus, for all compounds influencing technological and nutritional quality of wheat, there is a strong to very strong component of genetical influence, for most nutritional and part of the technological quality in the form of a number of QTL, while for specific composition of storage proteins, they are each determined by specific genes.

3 Environmental Effects on Technological and Nutritional Quality

Environmental effects on quality aspects involves every parameter that influences the quality except those being completely genotypically determined, summarized in Table 2. The environmental factors affecting quality can be divided into those mainly not possible to control by the farmer and those clearly possible to influence. Factors possible to effect by the farmer are those included in agronomic-crop management practices (which also includes the selection of the crop and genotype to grow). Climatic factors are environmental factors less easily to influence by the farmer. Below, we have divided the environmental effects into climatic effects (weather and locality = where the crop is grown), agronomic and crop management effects (how the crop is grown), and effects from biotic and abiotic stresses.

Table 2 Summary of environmental (including agronomic) factors influencing quality-related compounds of wheat and their effects

Environment	Factor	Event	Compound	Quality	Reference (some selected)
Weather	Temperature Precipitation	Plant development	Biomass	Technological quality	Randall & Moss 1990
			Protein		Johansson et al. 2002, 2013
			Starch		Malik 2012
			Prot. polymer.		
		-	Dietary fiber	Nutrition	Lampi et al. 2010
		-	Vitamin Bs		Shewry et al. 2010, 2011
		-	Phytochemicals		Hussain 2012
Locality	Soil	Plant development	Biomass	Technological quality	Vazquéz et al 2012
			Protein		Johansson et al. 2013
			Starch		Uhlen et al. 2015
			Prot. polymer.		
		-	Minerals	Nutrition	Distelfeld et al. 2007
		-	Heavy metals		Yilmaz et al. 2008
		-	Vitamin Bs		Hussain 2012
		-	Phytochemicals		Nurit et al. 2015
	Nitrogen	Plant development	Biomass	Technological quality	Malik 2012
			Protein		Vázquez et al. 2012
			Starch		Johansson et al. 2013
			Prot. polymer.		Uhlen et al. 2015
Irrigation	Water	Plant development	Protein	Technological quality	Oweis et al. 1999 Rharrabti et al. 2001
Fertilizers	Nitrogen	Plant development	Biomass	Technological quality	Johansson et al. 2001, 2005

(continued)

Table 2 (continued)

			Protein		Malik et al. 2011
			Starch		Giuliani et al. 2011a, b
			Prot. polymer.		Zhong et al. 2018
	Sulphur	-	Prot. Polymer.	Technological Quality	Zhao et al. 1997
					Flaete et al. 2005
Cropping systems	Nitrogen	Plant development	Biomass	Technological quality	Hussain et al. 2012
			Protein		Hussain 2012
			Starch		Grahmann et al. 2014
			Prot. polymer.		
		-	Minerals	Nutrition	Ryan et al. 2004
Biotic stress	Pests/ pesticides	Plant development	Biomass	Technological quality	Egli 1998
			Protein		Husenov 2018
			Starch		
			Prot. polymer.		
		-	Phytochemicals	Nutrition	Johansson et al. 2014
Abiotic stress	CO_2	Plant development	Biomass	Technological quality	Wieser et al. 2008
			Protein		Myers et al. 2014
	Heat		Starch		Li et al. 2013a, b
			Prot.polymer		Guzmán et al. 2016
	Drought				Magallanes-Lopez et al. 2017
					De Santis et al. 2017

3.1 *Effect of the Cultivation Environment (Weather and Locality) on Technological and Nutritional Quality*

The cultivation environment with its specific weather and soil conditions influences considerably the quality of the wheat. Weather effects and their influences on the technological and nutritional quality are discussed below, while heat and drought stress is discussed under the section of biotic and abiotic stresses.

3.1.1 Technological Quality

The weather largely influence the crop development from sowing and until harvest, not least for germination and development of the green biomass of the plant. The temperature and precipitation, both during early plant development and the grain-filling period, are also influencing the technological quality of wheat (Johansson et al. 2013; Malik 2012). One major reason for the effect of weather during early crop development on technological quality is the negative correlation between the plant development time and the polymerization of the proteins in the wheat grain (Johansson et al. 2013). A prolonged plant maturation time until anthesis due to the weather conditions, mainly contributes to increases in green biomass and carbohydrate accumulation in the wheat plant (Malik 2012). The increase in biomass accumulation results in later plant stages to an increased accumulation of carbohydrates in the grains, diluting the grain protein concentration (Johansson et al. 2013). Thus, early plant development is one explanation for differences in grain protein concentration contributing to differences in technological quality. Similarly, weather parameters during the latter part of the wheat development time, i.e. after anthesis or during grain filling, can be explained through a number of molecular mechanisms (Johansson et al. 2013). In general, technological quality of wheat benefits from an increase in temperature during the grain-filling period up to a mean daily temperature of 30 °C but thereafter a temperature increase has a negative impact due to production of heat-stress proteins (Johansson et al. 2002; Randall and Moss 1990). However, variation among genotypes do exist (Blumenthal et al. 1995) and there are also genotypes showing improved quality under heat stress due to increased grain protein concentration (Hernández-Espinosa et al. 2018). In general the weather related effects during the grain-filling period on technological quality can be divided into three parts; (i) temperature effects on grain accumulation where a higher temperature results in a faster maturation of the wheat grain with less starch accumulation, high protein content and an increased gliadin/glutenin ratio (Randall and Moss 1990; Kumar et al. 2011), (ii) temperature effects on enzymatic activity, including alpha-amylase activity, and starch-hydrolyzing enzymes involved in pre-harvest sprouting and the protein disulfide isomerase involved in formation of disulfide bonds among gluten proteins and subunits (Every et al. 2003), (iii) humidity or water content in the grain, influenced by precipitation and temperature, affecting hydrogen and electrostatic bonds among proteins (Johansson et al. 2008, 2013).

The effect of locality on the technological quality of wheat is partly explained by the differences in weather among various locations. However, also soil parameters, such as soil type, nitrogen availability, water holding capacity and content of microorganisms are important characters influencing technological quality of wheat (Malik 2012). A limited number of studies have evaluated wheat materials over a broad range of localities, and part of the explanation of this is the genotypic adaptation to its respective cultivation conditions and applications, with differences in temperature, requirements of winter hardness, long/short day adaptations, etc. However, studies are also available comparing technological quality of wheat in a broad range of environments, proving the importance of the cultivation location for technological quality (Vázquez et al. 2012; Uhlen et al. 2015).

3.1.2 Nutritional Quality

Weather is also to a large extent influencing the nutritional quality of wheat, as the weather and cultivation location is influencing uptake, mobilization and accumulation of essentially all nutritional compounds in the wheat grain (Hussain 2012).

Grain protein concentration in the wheat grain is dependent on basically all environmental factors influencing N uptake, biomass storage, reallocation, transport and grain accumulation of proteins and carbohydrates (Bhullar and Jenner 1985; Johansson et al. 2013; Malik 2012). Grain protein and starch accumulation is known to be two different processes not influencing each other (Jenner et al. 1991), although increases and decreases of starch accumulation in the wheat grain is influencing the dilution rate of the protein at similar protein accumulation (Jenner et al. 1991; Johansson et al. 2013; Malik 2012;). Thus, the wheat plant has to be cultivated in a temperature that promotes growth to accumulate protein and starch at all, but at increased temperatures plant growth and thereby starch accumulation is reduced with higher protein concentration as a result. Similarly, drought results in decreased starch accumulation and higher grain protein concentrations in wheat (Dupont and Altenbach 2003).

The impact on dietary fiber from environmental factors e.g. weather and locality of cultivation, has not been as rigorously investigated as has the environmental impact on grain protein content. A major contribution to the understanding of the impact of weather on dietary fiber content was carried out within the HEALTHGRAIN project. This study evaluated the impact of genotype and environment on dietary fibers and phytochemical components, and found a clear impact of cultivation temperature and precipitation on dietary fiber, although the weather impact was larger on the phytochemical components. Furthermore, the genetic influence was in that project found higher on dietary fibers than the environmental factor which was opposite to what was found for the phytochemical compounds (Gebruers et al. 2010).

Minerals content in the wheat grain is known to be effected by the mineral content in the soil (Hussain 2012). Thus, locality of cultivation plays a considerable role for mineral and also for heavy metal accumulation in the wheat grain (Hussain 2012). However, more recent studies are also available reporting a relationship with grain protein and Fe and Zn concentration in the grain in a changing climate (Gao et al. 2012, Magallanes-López et al. 2017a, Velu et al. 2016). Thus, due to the fact that grain protein concentration and Fe and Zn accumulation are encoded on genes in the same chromosomal region (Distelfeld et al. 2007), environmental effects influencing the transcription of these genes might influence both grain protein and Fe and Zn accumulation.

Reports of environmental effects on the vitamin B complex are scarce. However, a study using diverse wheat genotypes from the UK, Poland, France and Hungary showed that 48–70% of the variations in vitamins B1, B3 and B6 was determined by environmental differences (Shewry et al. 2011). Similarly, clear differences in vitamin Bs were reported among wheat from two localities in France (Nurit et al. 2016). A positive correlation was found for the content of these B vitamins and temperature during grain filling, while for vitamin B2 a positive correlation was found with precipitation during the three-month period before heading (Shewry et al. 2011).

The HEALTHGRAIN project is also one of few studies that has evaluated the effects of weather and cultivation location on phytochemical composition in the wheat grain (Shewry et al. 2010). Content of tocols and phenols was found to vary considerably in the diverse cultivation environment with localities spread across Europe and the variation was found different for different cultivars (Lampi et al. 2010). Temperature and precipitation were found as major contributors to the variation in phytochemical content of grains (Lampi et al. 2010; Shewry et al. 2010).

To conclude, weather and in particular temperature and precipitation, but also cultivation location effect the content and composition of in principle all compounds influencing technological and nutritional quality, although the extent of influence from various weather and locality related parameters and in relation to genotypic influences vary dependent on compound and to genetic background.

3.2 Agronomy-Crop Management Effects on Technological and Nutritional Quality

Cropping includes the opportunity to influence the quality of the produced crop by the use of various cultivation variables. Selection of superior cultivars can promote optimum crop yield and quality, and the genotypic influence has been described above. Similarly, selection of locality for the cultivation influences the quality, as also described above, partly due to the weather conditions at that particular locality. However, also the soil conditions at the locality, including soil type, nitrogen and minerals content, water holding capacity, etc., have a substantial impact on quality of wheat. Besides the soil conditions, also inputs such as applications of fertilizers and pesticides as well as the cropping system are influencing the quality in wheat (Malik 2012). Below, available information as to impact of soil, agronomic practices including irrigation, application of fertilizers and pesticides and the use of cropping systems on technological and nutritional quality is compiled.

3.2.1 Technological Quality

Soil conditions influence the technological quality of wheat. However, differences in soil includes a range of parameters e.g. nutrient, nitrogen and sulphur availability, water holding capacity, soil type, microorganism content, etc., and their various impact and interactions are difficult to relate to the technological quality parameters (Holm et al. 2018; Malik 2012). A few studies have tried to explain differences in wheat quality with soil properties, e.g. coarse soil texture combined with low carbon dioxide content were reported to lead to water stress during grain filling resulting in higher grain protein levels (Stewart et al. 2002). A recent study reported low soil temperature combined with adequate precipitation in a sandy soil to result in the best malting quality in barley (Holm et al. 2018). In wheat, interactions of soil and weather conditions have been shown to have an impact on the bread-making quality (Erekul & Köhn 2006).

Irrigation provides an opportunity for greatly increased crop yield (Rharrabti et al. 2001). Increasing the crop yield potential with supplemental irrigation increases the amount of N that is required to optimize crop yield and quality (Oweis et al. 1999). Under irrigation, protein content tends to be decreased, due to the increased yield, as compared to at rainfed conditions (Rharrabti et al. 2001) and hard vitreous kernels can also decrease (Oweis et al. 1999).

Quite a number of studies have evaluated impact of fertilizers and especially nitrogen application on technological quality of wheat. Generally, the quality is impacted by total amount of nitrogen applied, when the nitrogen is applied (timing) and nitrogen availability in the soil and combined with the fertilizer (Grahmann et al. 2014; Johansson et al. 2013). Increased nitrogen application normally increases yield and also when applied at later stages of plant development it increases the grain protein concentration (Giuliani et al. 2011a, b; Grahmann et al. 2014; Johansson et al. 2013). Also the polymerization of the proteins (%UPP) is effected by nitrogen fertilization, where late applications decrease %UPP (Johansson et al. 2001, 2005; Malik et al. 2011). However, a recent study indicated that top dressings of N at the emergence of the flag leaf resulted in higher content of glutenin and glutenin macropolymers than top dressings at emergence of the fifth leaf from the top (Zhong et al. 2018). Early nitrogen applications results in increased biomass and delayed plant development time decreases the gliadins to glutenins ratio, thereby effecting the technological quality (Johansson et al. 2013). Besides nitrogen from the fertilizer treatment, also Sulphur applications have shown a positive correlation to technological quality. Lack of Sulphur reduces the production of cysteine residues in the storage proteins, thereby also reducing the number of disulphide bonds and building of the polymeric network in the dough (Flaete et al. 2005; Zhao et al. 1997). Pesticide treatments might also influence the technological quality (Husenov 2018), which might be the results of a healthier plant with increased biomass production and possibly also a longer grain filling period as compared to non-treated plants.

Also, the cropping system is known to effect the technological quality of wheat (Grahmann et al. 2014; Hussain 2012). The major reason for wheat obtaining different technological quality when cultivated organically as compared to conventionally can be attributed to the nitrogen availability that normally differs between the systems (Hussain et al. 2012). In general, nitrogen availability is lower and delayed in organic farming as compared to conventional farming, resulting in lower grain protein concentrations and decreased %UPP (Hussain 2012).

3.2.2 Nutritional Quality

The locality and the soil type of the locality greatly impact the nutritional quality of wheat. It is well known that heavy metals in the soil are correlated to heavy metals in the wheat grain, thereby negatively impacting the nutritional quality (Hussain et al. 2012; Nan et al. 2002). Also, minerals content in the wheat grain is known to mirroring the contents of minerals in the soil (Hussain et al. 2010). At low level of e.g. Zn in the soil, grain content can be increased by soil or leaf fertilization with Zn (Yilmaz et al. 2008). Zn fertilization not only affects the Zn content in the grain but also impact yield and crop stand (Cakmak 2008). Similarly, Fe and Mn fertilization have been shown effective to increase the content of these elements in the wheat grain (Zeidan et al. 2010), while Se fertilization did not show a similar effect (Stroud et al. 2010). Studies on the effect of soil properties and fertilization regimes on the contents and composition of bioactive compounds are limited.

The cropping system has in several studies been shown to influence the nutrient content and nutrient density of the wheat grain, with higher values for organic than conventional wheat (Moreira-Ascarrunz et al. 2016; Ryan et al. 2004). Furthermore, intercropping with e.g. chick pea has shown positive effects on the mineral content in the wheat grain (Gunes et al. 2007). Cropping systems were not found to influence contents of bioactive compounds, such as tocols, carotenoids and phenolic compounds, as to the limited number of studies carried out (Konopka et al. 2012).

3.3 *Effects of Biotic and Abiotic Stresses*

Biotic and abiotic stresses, in terms of diseases and pests attacking the wheat versus e.g. heat, drought and saline stress are all known to influence the yield of wheat although they may also influence the quality of the wheat. The climate change is predicting an increase in heat, water and CO_2 stress, including water logging, frost, disease and pest dynamics (Porter et al. 2014; Reynolds et al. 2016), and the maintenance of grain quality under the climate change is an important goal for human nutrition and end-use functional properties (Nuttal et al. 2017).

3.3.1 Technological Quality

Reports on the effects of biotic stress on technological quality are limited. In general, diseases and pests are known to affect the yield of the wheat plant, often through mechanisms contributing to a reduced filling of the wheat grain to its optimal capacity (Egli 1998). A decrease in starch accumulation due to the attack of diseases and pests may often result in a decrease in flour milling quality of the grain (Husenov 2018) but in an increase in grain protein concentration which might result in positive effects on the technological quality of the wheat flour (Dimmock and Gooding 2002).

Average CO_2 concentration in the atmosphere is projected to increase from the current value of 400 ppm to 550 ppm by 2050 (Carter et al. 2007). Wheat productivity will most likely increase under CO_2 enrichment, resulting in changes of both the chemical composition of vegetative plant parts and grain quality (Högy and Fangmeier 2008). Increasing concentration of atmospheric CO_2 consistently reduces grain protein percentage and leaf N of wheat (Blumenthal et al. 1996; Myers et al.

2014; Panozzo et al. 2014; Weiser et al. 2008), while C/N ratio increases (Fangmeier et al. 1999; Wieser et al. 2008). Suggested mechanisms for the reduced leaf N are increased carbon assimilation, CO_2-effects on N uptake, allocation, or biochemical assimilation (Buchner et al. 2015; Tausz-Posch et al. 2014). Wheat gluten content has been found to decrease while gluten index increased, resulting in reduced pasta quality, with elevated CO_2 in durum wheat (Buchner et al. 2015).

Heat stress from either high temperatures (up to 30 °C) or heat shock (>30 °C) may alter amylose/amylopectin ratio, size distribution of starch granules, as well as grain protein composition, adversely affecting dough elasticity, strength and end-use properties (Hurkman et al. 2003; Li et al. 2013a, b). Generally, heat stress has been found to result in a weakening of the dough properties for both bread wheat and durum wheat (Cuniberti 2000; Guzmán et al. 2016; Magallanes-Lopez et al. 2017b), although some studies have reported instead an increase in gluten strength (Flagella et al. 2010; Hernández-Espinosa et al. 2018; Panozzo and Eagles 2000). A number of reasons have been reported for the weakening in gluten strength by heat stress; (i) a reduction in the glutenin/gliadin ratio (Cuniberti 2000), (ii) an increase in the amount of polymeric proteins up to 30 °C, where an optimum is reached, while the amount of polymeric proteins are instead decreasing with higher temperatures (Sampson et al. 1996). Cultivars has been found to synthesize heat shock proteins at heat stress (Blumenthal et al. 1998), and heat-tolerant cultivars were found to exhibit a more diverse and stronger production of heat shock proteins (Skylas et al. 2002). However, few specific studies have also reported an increased dough strength by high temperature in Mediterranean durum wheat, which may be explained by an impact of the heat shock proteins, resulting in an increase in glutenin macropolymer content (Flagella et al. 2010). A range of studies have evaluated up- and down-regulation of various proteins as a result of heat stress, showing changes in contents of HMW-GS, LMW-GS as well as in various gliadins (Altenbach 2012; Hurkan et al. 2013; Majoul-Haddad et al. 2013; Pompa et al. 2013).

Drought stress effects on technological quality of wheat has been evaluated to a lesser degree than has heat stress. However, the majority of the studies carried out imply that drought stress results in higher grain protein content (Flagella et al. 2010; Li et al. 2013a, b), with no change in gliadin/glutenin ratio (Daniel and Triboï 2002; Panozzo et al. 2001), but increase in polymeric protein and gluten strength (Blumenthal et al. 1998; Guzmán et al. 2016; Hernández-Espinosa et al. 2018; Li et al. 2013a; Magallanes-Lopez et al. 2017b). Some few studies have evaluated up- and down-regulation of proteins in relation to drought stress and have mainly reported shifts in some LMW-GS and gliadins (Giuliani et al. 2015; De Santis et al. 2017; Yang et al. 2011).

An effect of both heat and drought stress is also a shortened plant development time, which influences starch accumulation and thereby starch/protein quota, which might favor technological quality. Saline stresses might influence plant growth in similar way as drought stress thereby affecting the quota of starch and protein accumulation which might influence the protein accumulation and technological quality (Francois et al. 1986; Johansson et al. 2013).

3.3.2 Nutritional Quality

The nutritional quality of plants is known to be partly affected by biotic stresses. In particular polyphenols are described as defense molecules against pathogens and pests (Johansson et al. 2014). Polyphenolic compounds are also known to contribute to human health (Johansson et al. 2014). However, pests and diseases may also make a direct impact on the major components of the wheat grain, e.g. the starch and proteins, by e.g. feeding (Bardner and Fletcher 1974), thereby impacting the nutritional quality of the wheat grain.

Also, the content and composition of phytochemicals, such as polyphenols, carotenoids, tocols, etc., can be influenced by abiotic stresses e.g. drought, heat and saline stress, which have been reported in particular for polyphenolic compounds (Akula and Ravishankar 2011). Furthermore the abiotic stresses influence the quota of starch to proteins, thereby affecting the protein concentration in the wheat grain (Bardner and Fletcher 1974).

4 Importance of Genotype x Environment Interactions on Quality and Quality Stability

Variation in genotype response under different environmental conditions is known as genotype by environment interaction. Genotype-environment interactions (G x E) are important in evaluating cultivar adaptation, selecting parents, and developing improved genotypes. If the ranking of genotypes differs between environments this makes it more difficult to identify superior breeding lines since the measured parameter values are affected more by environmental variation than genetic differences. So it is common to do a G x E study using multi-environment trials to select genotypes with wide adaptability across environments and identify those performing best in high input systems (water and nitrogen fully available) and in resource limited ones (low water and fertility). An ideal stable genotype is one that performs for agronomic and quality across a wide range of environments.

It is important that wheat breeders know the heritability of agronomical and quality traits to improve them. However, these traits are influenced by the genotype and environment and also if correlations exist among traits, this will make progress more difficult (Barnard et al. 2002). Heritability is a parameter which is widely used in the establishment of breeding programs and formation of selection indexes (Falconer 1985). In general, heritability is low for the characteristics with agronomical importance since these are influenced by a large number of genes. Heritability alone is not sufficient and genetic advance indicates the magnitude of the expected genetic gain from one cycle of selection (Hamdi et al. 2003). High genetic advance coupled with high heritability estimates offers the most effective condition for selection (Larik et al. 2000).

There are several statistical approaches to the analysis and interpretation of G x E studies, including (i) Combined analysis of variance (ANOVA) (Zobel et al. 1988), (ii) Broad sense heritability calculated as the ratio the genotypic to the phenotypic variance (Falconer and Mackay 1996), (iii) Expected genetic advance (Allard 1960), (iv) Expected genetic advance as percent of mean (GAM), (v) Stability analysis (Eberhart and Russell 1996), (vi) additive main effects and multiplicative interaction (AMMI) model (Altay 2012; Akcura et al. 2009; Gauch and Zobel 1997;; Ilker et al. 2011). The AMMI model often displays the results as a biplot that allows simultaneous display of both samples and variables from a two-way data matrix and allows visualization of the interrelationship among environments, genotypes, and G x E interaction (Gabriel 1971). Biplots have been constructed by plotting the first principal component (PC1) scores of the genotypes and the traits against their respective scores for the second principal component (PC2) that resulted from singular-value decomposition (SVD) of trait-standardized data in each environment (Yan and Rajcan 2002). The biplot is often useful for visualization of genetic correlation among traits and in identifying genotypes that are superior in desired traits and hence could be candidates for use as parents in a breeding program or could be directly released for commercial production.

4.1 Technological Quality

Effects of GxE interactions have been examined in a range of studies focused on various environments and different wheat types for a range of technological purposes. Fewer studies have been carried out on e.g. durum wheat for pasta making than on bread wheat for bread-making quality. In general, many studies are reporting some parameters e.g. protein content to be determined by the environment to a higher extent than more quality related traits as e.g. specific protein content (Ames et al. 1999; Boggini et al. 1997; Johansson et al. 2013; Mariani et al. 1995; Novaro et al. 1997). From all studies comparing genotype and environmental interactions and their relative influence on technological quality parameters, it is clearly shown that their impact is effected by the broadness and ranges of selected genotypes and environments. As the specific environment can greatly influence the outcomes, a G x E study should be conducted by all breeding programs for the environments of relevance.

Genotype by environment interactions have proved to be highly significant in most genotype x environment studies (Hristov et al. 2010; Johansson et al. 2000; Koppel and Ingver 2010; Malik et al. 2011; Malik 2012; Rozbicki et al. 2015; Thomason and Phillips 2006; Williams et al. 2008). Although, in some cases the interaction was neglected in comparison to both genetic and environmental effects (Laidig et al. 2017). These interactions are explained by the differential and specific response of different genotypes to each environmental factor, such as growing temperatures, soil, diseases, etc. The studies on grain quality genotype x environment interactions are limited (Hatfield and Walthall 2015). This is in particular relevant as the importance of the subject is expected to grow in the future (Koppel and Ingver 2010) due to global climate change (Steffen et al. 2011). For such studies, the choice of environments and genotypes to be studied have a major impact on the results (Williams et al. 2008). The importance of the selection of genotypes was evidenced when several researches proved that the interaction with the environment was higher for groups of genotypes containing only HMW-GS 5 + 10 than those containing 2 + 12 (Hristov et al. 2010; Johansson et al. 2000).

Interactions between genotype and environment depends on the factors studied, e.g. genotype has been found interacting more with nitrogen fertilization than with temperature (Malik et al. 2011). However, other authors have instead observed higher interaction of location by year than the interaction of genotype with environmental factors (Rozbicki et al. 2015). Interactions become even more important when it is considered that in addition to reach a proper value for each parameter, the value should be stable (Steffen et al. 2011). Therefore, stability is important to select proper genotypes, which depends on genotype x environment interactions (Johansson et al. 2003). Moreover, considering stability, specific protocols have been proposed to select cultivars in order to properly consider the genotype by environment interactions (Rozbicki et al. 2015). Although the importance of quality stability is unquestionable (Koppel and Ingver 2010), there is no consensus how to define or measure it. Several concepts have been applied, and several statistics have been proposed to assess quality stability (Lin et al. 1986; Robert 2002). The most intuitive perception is that a genotype is stable if the variance or coefficient of variability among environments is small, which is basically "homeostasis". However, this concept usually leads to low quality genotypes, since the superior quality ones will have a better response to better quality environments (Vázquez and Castro 2018). A second concept is that a genotype is stable when the response to environments is parallel to the mean of all studied genotypes. In another concept of stability, it depends on unpredictable irregularities in the response to an environment: it is less stable the more it deviates from the expected behavior.

Several researchers who worked on quality stability observed that the genotypes that are more stable for one parameter, are less stable for others (Barić et al. 2004; Koppel and Ingver 2010; Lemelin et al. 2005; Mut et al. 2010). However, other studies have shown possibilities to find cultivars stable for most quality characteristics (Grausgruber et al. 2000).

4.2 Nutritional Quality

Most of the nutritional compounds in wheat are determined by genotype x environment interactions to various degrees. However, in general, the effect of genotype x environmental interactions has been evaluated and determined to a less extent in wheat for nutritional characters as compared to the studies on technologically important parameters.

Protein concentration in wheat is one of the major compounds related to nutrition in wheat. Wheat, being a staple food for one third of the world population, contributes more protein to the human diet than any other cereal crop (Kumar et al. 2011). The grain protein concentration is known to be largely determined by genotype x environment interactions (Gomez-Becerra et al. 2010). The studies on the impact of the genotype x environment interactions on nutritious protein from wheat for human consumption are limited.

The diversity screen within the HEALTHGRAIN project clearly showed the impact of genotype x environment interactions on the dietary fiber content on wheat grains (Gebruers et al. 2010). Different dietary fiber related compounds were found to be determined to a similar degree by the genotype and the environment and interactions were found to have a relatively high impact indicating limited breeding opportunities for these compounds.

For minerals, the genotype x environmental interactions were found to account for 17 to 58% of the variation, and the effect varied as related to which mineral was evaluated (Joshi et al. 2010). Among the minerals, iron and zinc have been given the largest attention due to the high number of humans with deficit amounts of these minerals through their diet. For iron and zinc content in the wheat grain, high genetic x environment interactions have been reported and the need for increased understanding of relationships between the genetic and the environmental effects while breeding for high content varieties have been pointed out (Joshi et al. 2010). Nutritional yield from certain environments and from certain farming systems can be calculated (Moreira-Ascarrunz et al. 2016) and is impacted by the genotype x environmental interactions.

Also, the majority of the phytochemicals in the wheat grain has been found strongly influenced by the genotype x environment interactions (Lv et al. 2013). However, some studies have indicated considerably stronger effects by the genotype and the locations than for interactions between the two, making breeding for high contents of antioxidants and phytochemicals possible, especially for certain environments (Mpofu et al. 2006).

To conclude, all compounds affecting the nutritional quality of wheat are influenced by genetic x environment interactions, although to different extent depending on selection of compound, genotypes and environments for comparison.

5 Opportunities to Modulate Technological and Nutritional Quality

As shown above, basically all compounds affecting the technological and nutritional quality in wheat are impacted by both genotypes and environment and their interactions although to various extents. In breeding and cultivation of wheat, the breeders and the farmers have a desire to produce wheat with the most optimal quality for the produce of interest. Therefore, it is highly important to understand how and when various compounds are influenced by genotypic and environmental parameters and opportunities to modulate the quality towards optimal performance for various end-uses. Below is an attempt to describe opportunities to use the genotypic and environmental variation to govern production of the various compounds in wheat thereby contributing to the tailoring of the end-use quality of wheat.

5.1 By the Uses of Genotypes

Tailoring quality by the use of genotypes is easiest for characteristics determined by single genes, i.e. when a specific gene is encoding a specific compound, and the presence and absence of that specific compound is influencing the quality. For the compounds determining technological and nutritional quality of wheat, the only ones known as singly encoded on specific genes are the gluten proteins, i.e. the glutenins and gliadins (Payne and Lawrence 1983; Payne et al. 1984, 1987; Shewry et al. 1987; He et al. 2005), and the grain hardness determined by certain puroindolines (Pasha et al. 2010). Presence and absence as well as amounts of all other compounds are determined by multiple genes or QTLs and their expression is not determined only by the genotype but also by the environment, as described above. Thereby, tailoring quality in wheat is most easily done for bread-making and pasta making qualities by selection on glutenin and gliadin composition and for grain hardness by selection of puroindolines. Among the storage proteins, the HMW-GS have been found having the highest impact on bread-making quality, being the core of the polymers formed during dough formation and bread baking (Johansson et al. 1993; Payne et al. 1984, 1987). Here, the HMW-GS 5 + 10 are most significantly correlated with high gluten strength while 2 + 12 with low gluten strength (Johansson et al. 1993; Payne et al. 1984, 1987). The rest of the HMW-GS contribute in various degrees to gluten strength (Johansson 1995). Selecting for a suitable gluten protein composition is thereby possible by breeding, to breed strong wheat suitable for e.g. French bread or baguettes and weaker wheat for e.g. home baking or biscuits. During later years, methods have also been developed for selection within breeding programs on LMW-GS to select for suitable bread-making quality in bread wheat (Johansson 1995). Similarly as for the gluten proteins, the puroindolines are used for selection in breeding programs of grain hardness of common wheat, where medium hard to hard grains are specifically needed for bread making while soft wheat are specifically needed for biscuits production. Thus, variation in puroindolines is therefore a tool for the breeder in selection for end-use quality. Combining storage protein composition with puroindolines composition contribute to opportunities to tailor gluten strength and grain hardness and thereby end-use properties.

However, despite the fact that various genotypes have similar gluten protein and puroindoline composition, the end-use quality of wheat might vary based on genetical influence. The reason for this is other variations in protein content and composition, and also in other components of the wheat grain. Plant development time of the wheat crop, grain protein concentration, polymerization behavior of the gluten proteins, grain starch and lipid content and composition are all factors known to influence the technological quality of the wheat grain and its flour. By governing these factors, end-use quality can be tailored. Thus, breeding cultivars just to attain higher grain protein concentration may result in weak gluten and eventually a sticky dough (Finney and Barmore 1948; Johansson et al. 2001, 2004, 2013). Breeding cultivars to attain high content of %UPP instead would result in a stronger dough (Finney and Barmore; Johansson et al. 2001, 2013) and plant development time of the wheat might play a role for this character (Pasha et al. 2010).

Similarly, as for compounds determining technological quality, those determining nutritional quality have a strong genetic component mainly originating from QTLs in the wheat genome. Thus, genotypes with high contents can be selected in breeding programs to target nutritional composition in future wheat varieties. Thereby, genotypic variation and novel breeding technologies such as speed breeding and genomic selection or other novel genomic methods can be adopted to tailor highly nutritious wheat for target human populations. Fig. 1 summarizes the opportunities to use genotypes to tailor both technological and nutritional quality in wheat.

5.2 *By the Use of Environment*

The environment influences content and composition of essentially all components affecting both the technological and the nutritional quality of wheat, with the exception of those determined by single genes as described above. The proportion of the effects of the environment in relation to those of the genotype will differ depending on the outlay of the study, the compounds evaluated, cultivars included in the study and environmental factors selected. However, a couple of environmental factors have been ascribed larger impact for the quality compared to others, determining at least similar ranges of end-use quality as does the genotype. As for technological quality, the most documented environmental factors described to have highest impact are (i) everything that influences plant development time, (ii) nitrogen availability at certain growth stages and (iii) temperature and drought constraints during the majority of the time for plant development and grain maturation (Johansson et al. 2013). Similarly for nutritional quality, factors attributed to plant development, nitrogen availability, temperature and drought constraints have been determined as the main effectors for several of the nutritional components. Furthermore, biotic stresses, i.e. attacks from pathogens and insects, have been described as a major contributor to increased contents of some of the phytochemicals in wheat, in particular of the polyphenols as these compounds are a part of the plant defense mechanisms (Johansson et al. 2014).

Using environmental factors to modulate yield has actually two components, (i) first the selection of site for the cultivation, which determines both the soil status for the crop production and partly such environmental factors as temperature and precipitation being at least partly determined by selection of the cultivation location, (ii) and thereafter the inputs in terms of crop management (cultivation methods, cultivation system adopted and inputs in terms of fertilizers and pesticides applied).

Fig. 1 Impact of genotype and environment at early and late plant development on wheat quality

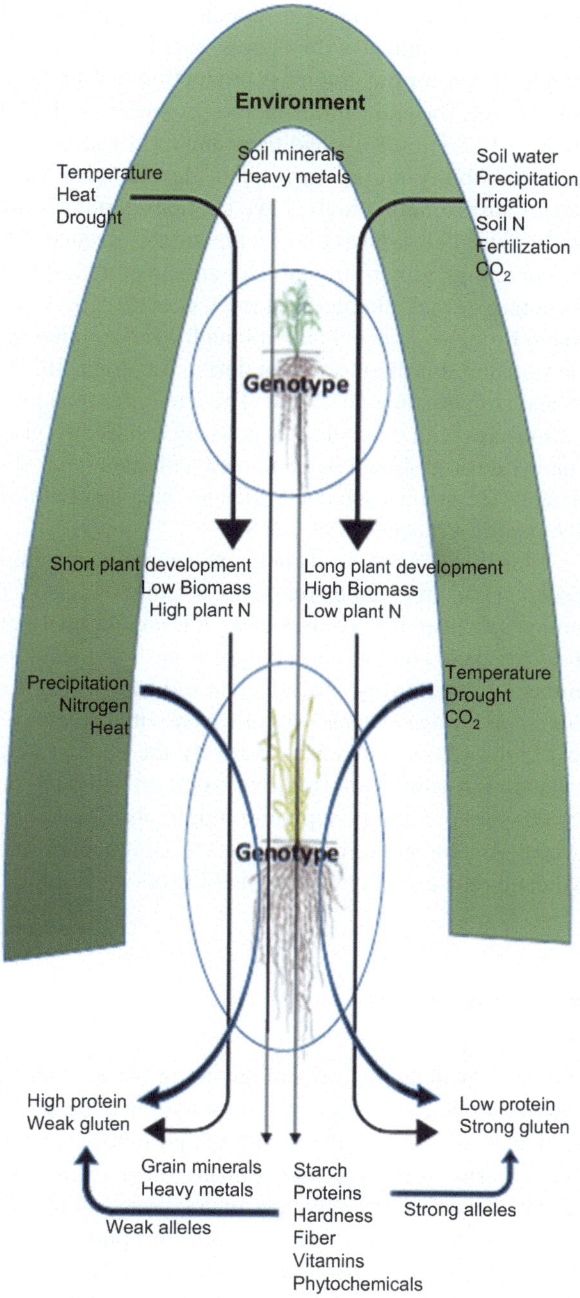

To obtain a strong gluten in dough, genotypes with high-quality gluten proteins should be combined with factors contributing to a long plant development and opportunities for high biomass production during the early plant growth, and thereafter, during the grain-filling factors promoting a short plant development (Johansson et al. 2013). Thus, soil conditions and fertilization strategies resulting in high nitrogen availability during early plant development and low availability during late stages are promoting high gluten strength. Similarly, low temperature ranges during early plant development and temperatures around 30 °C during grain filling also promote high gluten strength (Johansson et al. 2013). Opposite conditions prevails to obtain a weak gluten (Johansson et al. 2013). As to precipitation, high levels to avoid droughts is beneficial although oversaturated soil conditions or flooding are detrimental. Furthermore, rain during the grain-filling decreases gluten strength in wheat (Johansson et al. 2008). Thus, to tailor technological quality in wheat, cultivation sites with most suitable growing conditions on an average bases for the wheat quality desired should be combined with genotypes showing interacting characters. Thereafter, inputs such as fertilization at suitable times can be used to further tailor the quality (Vázquez et al. 2018).

Nutritional components in the wheat grain are composed of a variety of different compounds, and studies on environmental effects of their respective contents and composition are limited. Basically, it is well known that the environment plays a key role for their content and composition. Specific influences known are that i) the minerals content vary dependent on soil content of the minerals, ii) temperature during grain filling correlates significantly with content of vitamins B1, B3 and B6, and iii) biotic effects influence positively the content of polyphenols in plants. Thus, selection of cultivation location, based on soil conditions and daily mean averages of temperature and precipitation might also be used to tailor nutritional quality of wheat. A summary of opportunities to tailor technological and nutritional quality in wheat by the use of environmental factors is presented in Fig. 1.

6 Conclusions

Technological and nutritional quality in wheat is basically determined by the interplay of a range of compounds present in the wheat grain. With the exception of the gluten proteins, starch type and the puroindolines responsible for grain hardness, the majority of these compounds are impacted to a similar extent of the genotype and the environment and clear interactions among these factors are also seen. The gluten proteins, the starch type and the puroindolines are basically encoded by major single genes resulting in various allelic proteins correlating differently to wheat quality. As for the rest of the compounds influencing quality characteristics in wheat, these are encoded by multiple genes or QTLs, thereby being more vulnerable to environmental effects for their translation and modification. For several of these characters, enzymatic actions are also involved, e.g. protein disulphide isomerase is the enzyme involved in forming the disulphide bonds in the gluten polymer

(Johansson et al. 2002), and enzyme activities are known to be effected by temperature and humidity. Generally, the genotypes and environments can be seen as having similar ranges of effects on content and composition of quality related compounds encoded on multiple genes or QTLs, and the range of influence of genotypic and environmental variation is more limited to the selection of genotypes and environments for evaluation. The opportunities to tune the wheat quality is obvious by selection of wheat genotypes positively interacting with the environment of choice and thereafter fine-tuning with the use of inputs as can be seen from Fig. 1. However, the high impact of the environment for many of the quality characteristics in wheat and the changing climate across the world with increasing temperature and bursts of extreme weather, calls for development of novel selections tools for the breeders to cope with environmental fluctuations.

References

Akcura M, Kaya Y, Taner S (2009) Evaluation of durum wheat genotypes using parametric and nonparametric stability statistics. Turkish Journal of Field Crops 14:111–122.

Akula R, Ravishankar GA (2011) Influence of abiotic stress signals on secondary metabolites in plants. Plant Signal Behaviour 6:1720–1731.

Allard RW (1960) Principles of Plant Breeding. John Wiley and Sons. Inc., New York.

Altay F (2012) Yield stability of some Turkısh winter wheat (Triticum aestivum L.) genotypes in the western transitional zone of Turkey. Turkish Journal of Field Crops 17:129–134.

Altenbach S (2012) New insights into the effects of high temperature, drought and post-anthesis fertilizer on wheat grain development. Journal of Cereal Science 56:39–50.

Ames NP, Clarke JM, Marchylo BA, Dexter JE, Woods SM (1999) Effect of environment and genotype on durum wheat gluten strength and pasta viscoelasticity. Cereal Chemistry 76:582–586.

Andersson AAM, Kamal-Eldin A, Aman P (2010) Effects of environmental and variety on alkyresorcinols in wheat in the HEALTHGRAIN diversity screen. Journal of Agriculture and Food Chemistry 58:9299–9305.

Andersson AAM, Andersson R, Piironen V, Lampi A-M, Nyström L, Boros D, Fra's A, Gebruers K, Courtin CM, Delcour JA, Rakszegi M, Bedo Z, Ward JL, Shewry PR, Åman P (2013) Contents of dietary fiber components and their relation to associated bioactive components in whole grain wheat samples form the HEALTHGRAIN diversity screen. Food Chemistry 136:143–1248.

Bardner R, Fletcher KE (1974) Insect infestations and their effects on teh growth and yield of field crops: a review. Bulletin of Entomological Research 64:141–160.

Barić M, Pecina M, Šarčević H, Kereša S (2004) Stability of four Croatian bread winter wheat (Triticum aestivum L.) cultivars for quality traits. Plant Soil and Environment 50:402–408.

Barnard AD, Labuschagne MT, Van Niekerk HA (2002). Heritability estimates of bread wheat quality traits in The Western Cape Province of South Africa. Euphytica. 127:115–122.

Batifoulier F, Verny MA, Chanliad E, Remesy C, Demingné C (2006) Variability of B vitamin concentrations in wheat grain, milling fractions and bread products. European Journal of Agronomy 25:163–169.

Bhullar SS, Jenner CF (1985) Differential responses to high temperatures of starch and nitrogen accumulation in the grain of four cultivars of wheat. Australian Journal of Plant Physiology 12:363–375.

Blanco A, Simeone R, Gadaleta A (2006) Detection of QTLs for grain protein content in durum wheat. Theoretical and Applied Genetics 112:1195–1204.

Blechl AE, Anderson OD (1996) Expression of a novel high-molecular-weight glutenin subunit gene in transgenic wheat. Nature Biotechnology 14:875–879.

Blechl AE, Hung QL, Anderson OD (1998) Engineering changes in wheat flour by genetic transformation. Journal of Plant Physiology 152:703–707.

Blumenthal C, Bekes F, Gras PW, Barlow EWR, Wrigley CW (1995) Identification of wheat genotypes tolerant to the effects of heat stress on grain quality. Cereal Chemistry 72:539–544.

Blumenthal CS, Rawson HM, McKenzie E, Gras PW, Barlow EWR, Wrigley CW (1996) Changes in wheat grain quality due to doubling the level of atmospheric CO2. Cereal Chemistry 73:762–766.

Blumenthal CS, Stone PJ, Gras PW, Bekes F, Clarke B, Barlow EWR, Appels R, Wrigley CW (1998) Heat-shock protein 70 and dough-quality changes resulting from heat stress during grain filling in wheat. Cereal Chemistry 75:43–50.

Boggini G, Doust MA, Annicchiarico P, Pecetti L (1997) Yielding ability, yield stability, and quality of exotic durum wheat germplasm in Sicily. Plant Breeding 116:541–545.

Borrill P, Connorton JM, Balk J, Miller AJ, Sanders D, Uauy C (2014) Biofortification of wheat grain with iron and zinc: integrating novel genomic resources and knowledge from model crops. Frontiers in Plant Science 5:1–8.

Branlard G, Dardevet M (1985) Diversity of grain proteins and bread wheat quality: I. Correlation between gliadin bands and flour quality characteristics. Journal of Cereal Science 3:329–343.

Brites C, Carrillo JM (2001) Influence of high molecular weight (HMW) adn low molecular weight (LMW) glutenin subunits controlled by *Glu-1* and *Glu-3* loci on durum wheat quality. Cereal Chemistry 78:59–63.

Buchner P, Tausz M, Ford R, Leo A, Fitzgerald GJ, Hawkesford MJ, Tausz-Posch S (2015) Expression patterns of C- and N- metabolism related genes in wheat are changed during senescence under elevated CO2 in dry-land agriculture. Plant Science 236:239–249.

Cakmak I (2008) Enrichment of cereal grains with zinc: agronomic or genetic biofortification? Plant and Soil 302:1–17.

Carter TR, Jones RN, Lu X, Bhadwal S, Conde C, Mearns LO, O'Neill BC, Rounsvell MDA, Zurek MB (2007). New assessment methods and the characterization of future conditions. In: Parry ML, Canziani OF, Palutikof PJ, van der Linden PJ, Hanson CE (eds.) Climate Change: Impacts, Adaptation and Vulnerability. Contribution of working group II to the fourth assessment report of the Intergovernmental Panel on Climate Change, pp.133–171.

Charmet G, Masood-Quraishi U, Ravel C, Romeuf I, Balfourier F, Perretant MR, Joseph JL, Rakszegi M, Guillon F, Sado PE, Bedo Z, Saulnier L (2009) Genetics of dietary fiber in bread wheat. Euphytia 170:155.

Cuniberti MB (2000) Caloric stress and their effect in the argentine wheat quality. In: 11th Cereal and Bread Congress-Cereals, Health and Life-Surfers Paradise, Gold Coast, Australia. 11-14 September, pp. 144.

Daniel C, Triboï E (2002) Changes in wheat protein aggregation during grain development: effects of temperatures and water stress. European Journal of Agronomy 16:1–12.

De Santis MA, Giuliani MM, Giuzio L, De Vita P, Lovegrove A, Shewry PR, Flagella Z (2017) Differences in gluten protein composition between old and modern durum wheat genotypes in relation to 20th century breeding in Italy. European Journal of Agronomy 87:19–29.

De Santis MA, Kosik O, Passmore D, Flagella Z, Shewry P, Lovegrove A (2018) Comparison of the dietary fiber composition of old and modern durum wheat (Triticum turgidum spp. Durum) genotypes. Food Chemistry 244:304–310.

Dimmock JPRE, Gooding MJ (2002) The influences of foliar diseases, and their control by fungicides, on the protein concetration in wheat grain: a review. Journal of Agricultural Science 138:349–366.

Distelfeld A, Cakmak I, Peleg Z, Ozturk L, Yazici AM, Budak H, Saranga Y, Fahima T (2007) Multiple QTL-effects of wheat *Gpc-B1* locus on grain protein and micronutrient concentrations. Physiologia Plantarum 129:635–643.

Dupont FM, Altenbach SB (2003) Molecular and biochemical impacts of environmental factors on wheat grain development and protein synthesis. Journal of Cereal Science 38:133–146.

Eberhart SA, Russell WA (1966) Stability parameters for comparing varieties. Crop Sci 6:36–40.
Egli D (1998) Seed biology and the yield of grain crops. CAB International, New York, NY, USA. pp. 178.
Erekul O, Köhn W (2006) Effect of weather and soil conditions on yield components and bread-making quality of winter wheat (Triticum aestivum L.) and winter triticale (Triticosecale Wittm.) varieties on North-east Germany. Journal of Agronomy and Crop Science 192:452–464.
Every D, Griffin WB, Wilson PE (2003) Ascorbate oxidase, protein disulfide isomerase, ascorbic acid, dehydroascorbic acid and protein levels in developing wheat kernels and their relationship to protein disulfide bond formation. Cereal Chemistry 80:35–39.
Falconer DS (1985) Introductions to Quantitative Genetics. Longman, London.
Falconer DS, Mackay TFC (1996) Introduction to quantitative genetics. 4th Ed. Longman, New York, p 580.
Fangmeier A, De Temmerman L, Mortensen L, Kemp K, Burke J, Mitchell R, van Oijen M, Weigel HJ (1999) Effects on nutrients and on grain quality in spring wheat crops grown under elevated CO2 concentrations and stress conditions in the European, multiple-site experiment 'ESPACE-wheat'. European Journal of Agronomy 10:215–229.
FAO (2017) Food and Agriculture Organization of the United Nations. FAOSTAT Database. Rome, Italy: FAO.
Fardet A (2010) New hypotheses for the health protective mechanisms of whole grain cereals: What is beyond fiber? Nutritional Research Reviews 23:65–134.
Finney KF, Barmore (1948) Loaf volume and protein content of hard winter and spring wheats. Cereal Chemistry 25:291.
Flaete NES, Hollung K, Ruud L, Sogn T, Faergestad EM, Skarpeid HJ, Magnus EM, Uhlen AK (2005) Combined nitrogen and sulphur fertilisation and its effect on wheat quality and protein composition measured by SE-FPLC and proteomics. Journal of Cereal Science 41:357–369.
Flagella Z, Giuliani MM, Giuzio L, Volpi C, Masci S (2010) Influence of water deficit on durum wheat storage protein composition and technological quality. European Journal of Agronomy 33:197–207.
Francois LE, Maas EV, Donovan TJ, Youngs VL (1986) Effect of salinity on grain yield and quality, vegetative growth, and germination of semi-dwarf and durum wheat. Agronomy Journal 78:1053–1058.
Fratianni A, Giuzio L, Di Criscio T, Flagella Z, Panfili G (2013) Response of carotenoids and tocols of durum wheat in relation to water stress and sulfur fertilization. Journal of Agriculture and Food Chemistry 61:2583–2590.
Gabriel KR (1971) The biplot graphic display of matrices with application to principal component analysis. Biometrika 58:453–467.
Gao XP, Lukow OM, Grant CA (2012) Grain concentration of protein, iron and zinc and bread-making quality in spring wheat as affected by seeding date and nitrogen fertilizer management. Journal of Geochemical Exploration 121:36–44.
Gauch HG, Zobel RW (1997) Identifying mega-environments and targeting genotypes. Crop Science 37: 311–326.
Gebruers K, Dornez E, Bedö Z, Rakszegi M, Frás A, Boros D, Courtin CM, Delcour JA (2010) Environment and genotype effects on the content of dietary fiber and its components in wheat in the HEALTHGRAIN diversity screen. Journal of Agriculture and Food Chemistry 58:9353–9361.
Giuliani MM, Giuzio L, De Caro A, Flagella Z (2011a) Relationships between nitrogen utilization and grain technological quality in durum wheat. I. Nitrogen translocation and nitrogen use efficiency for protein. Agronomy Journal 103:1487–1494.
Giuliani MM, Giuzio L, De Caro A, Flagella Z (2011b) Relationships between nitrogen utilization and grain technological quality in durum wheat. II. Grain yield and quality. Agronomy Journal 103:1668–1675.
Giuliani MM, Palermo C, De Santis MA, Mentana A, Pompa M, Giuzio L, Masci S, Centonze D, Flagella Z, (2015) Differential expression of durum wheat gluten proteome under water stress during grain filling. Journal of Agriculture and Food Chemistry 63:6501–6512.

Gomez-Becerra HF, Erdem H, Yazici A, Tutus Y, Torun B, Ozturk L, Cakmak I, (2010) Grain concentrations of protein and mineral nutrients in a large collection of spelt wheat grown under different environments. Journal of Cereal Science 52:342–349.

Grahmann K, Verhulst N, Peña RJ, Buerkert A, Vargas-Rojas L, Govaerts B (2014) Durum wheat (Triticum durum L.) quality and yield affected by tillage-straw management and nitrogen fertilization practice under furrow-irrigated conditions. Field Crops Research 164:166–177.

Grausgruber H, Oberforster M, Werteker M, Ruckenbauer P, Vollmann J (2000) Stability of quality traits in Austrian-grown winter wheats. Field Crops Research 66:257–267.

Graybosch RA (1998) Waxy wheat: Origin, properties and prospects. Trends Food Sci Technol 9:135–142.

Groos C, Robert N, Bervas E, Charmet G (2003) Genetic analysis of grain protein-content, grain yield and thousand-kernel weight in bread wheat. Theoretical and Applied Genetics 106:1032–1040.

Gunes A, Inal A, Adak MS, Alpaslan M, Bagci EG, Erol T, Pilbeam DJ (2007) Mineral nutrition of wheat, chickpea and lentil as affected by mixed cropping and soil moisture. Nutrient Cycling in Agroecosystems 78:83–96.

Guzmán C, Autrique JE, Mondal S, Singh RP, Govindan V, Morales-Dorantes A, Posadas-Romano G, Crossa J, Ammar K, Peňa RJ (2016) Response to drought and heat stress on wheat quality, with special emphasis on bread-making quality, in durum wheat. Field Crops Research 186:157–165.

Hamdi A, El-Ghareib AA, Shafey SA, Ibrahim MAM (2003) Genetic variability, heritability and expected genetic advance for earliness and seed yield from selection in lentil. Egypt Journal of Agricultural Research 81:125–137.

Hatfield JL, Walthall CL (2015) Meeting global food needs: realizing the potential via genetics × environment × management interactions. Agronomy Journal 107:1215–1226.

He ZH, Liu L, Xia C, Liu JJ, Peña RJ (2005) Composition of HMW and LMW glutenin subunits and their effects on dough properties, pan bread, and noodle quality of Chinese bread wheats. Cereal Chemistry 82:345–350.

Hernández-Espinosa N, Mondal S, Autrique E, Gonzalez-Santoyo H, Crossa J, Huerta-Espino J, Singh RP, Guzmán C (2018) Milling, processing and end-use quality traits of CIMMYT spring bread wheat germplasm under drought and heat stress. Field Crops Research 215:104–112.

Holm L, Malik AH, Johansson E (2018) Optimizing yield and quality in malting barley by the governance of field cultivation conditions. Journal of Cereal Science 82:230–242.

Hristov N, Mladenov N, Djuric V, Konkic-Spika A, Marjanovic-Jeromela A, Simic D (2010) Genotype by environment interactions in wheat quality breeding programs in southeast Europe. Euphytica 174:315–324.

Hurkman WJ, McCue KF, Altenbach SB, Korn A, Tanaka CK, Kothari KM, Johnson EL, Bechtel DB, Wilson JD, Anderson OD, DuPont FM (2003) Effect of temperature on expression of genes encoding enzymes for starch biosynthesis in developing wheat endosperm. Plant Science 164:873–881.

Husenov B (2018) Towards an impact on food security for Tajikistan. Acta Universitatis Agriculturae Sueciae 2018:5.

Hussain A (2012) Quality of organically produced wheat from diverse origin. Thesis. Acta Universitatis Agriculturae Sueciae 2012:18.

Hussain A, Larsson H, Kuktaite R, Johansson E (2010) Mineral composition of organically grown wheat genotypes: contribution to daily minerals intake. International Journal of Environmental Research and Public Health 7:3442–3456.

Hussain A, Larsson H, Kuktaite R, Johansson E (2012) Healthy food from organic wheat: Choice of genotypes for production and breeding. Journal of the Science of Food and Agriculture 92:2826–2832.

Hussain A, Larsson H, Kuktaite R, Olsson ME, Johansson E (2015) Carotenoid content in organically produced wheat: relevance for human nutritional health on consumption. International Journal of Environmental Research and Public Health 12:14068–14083.

Högy P, Fangmeier A (2008) Effects of elevated atmospheric CO2 on grain quality of wheat. Journal of Cereal Science 48:580–591.
Ilker E, Geren H, Unsal R, Sevim I, Tonk FA, Tosun M, (2011) AMMI biplot analysis of yield performances of bread wheat cultivars grown at different locations. Turkish Journal of Field Crops 16:64–68.
Jenner CF, Ugalde TD, Aspinall D (1991) The physiology of starch and protein deposition in the endosperm of wheat. Australian Journal of Plant Physiology 18:211–226.
Jiang XL, Tian JC, Hao Z, Zhang WD (2008) Protein content and amino acid composition in grains of wheat-related species. Agricultural Science China 7:272–279.
Johansson E (1995) Wheat grain proteins: accumulation and composition in breeding for improved bread-making quality. Thesis. The Swedish University of Agricultural Sciences. Svalöv, Sweden.
Johansson E (1996) Quality evaluation of D zone omega gliadins. Plant Breeding 115:57–62.
Johansson E, Henriksson P, Svensson G, Heneen WK (1993) Detection, chromosomal location and evaluation of the functional value of a novel high Mr glutenin subunit found in Swedish wheats. Journal of Cereal Science 17:237–245.
Johansson E, Prieto-Linde ML, Jönsson J (2001) Effects of wheat cultivar and nitrogen application on storage protein composition and bread-making quality. Cereal Chemistry 78:19–25.
Johansson E, Svensson G, Tsegaye S (2000) Genotype and environment effects on bread-making quality of Swedish grown wheat cultivars containing HMW glutenin subunits 2+12 or 5+10. Acta Agriculturae Scandinavica 49:225–233.
Johansson E, Nilsson H, Mazhar H, Skerritt J, MacRitchie F, Svensson G (2002) Seasonal effects on storage proteins and gluten strength in four Swedish wheat cultivars. Journal of the Science of Food Agriculture 82:1305–1311.
Johansson E, Prieto-Linde M-L, Kuktaite R, Andersson A, Jönsson JÖ (2003) Stability in wheat quality: a utopia or a future possibility. In: EA Pogna, M Romanò, NE Pogna, G Galterio (Eds.) Proceedings of the Tenth International Wheat Genetics Symposium (pp. 443-446). Istituto Sperimentale per la Cearicoltura, Roma, Italy.
Johansson E, Prieto-Linde ML, Svensson G (2004) Influence on nitrogen application rate and timing on grain protein composition and gluten strength in Swedish wheat cultivars. Journal of Plant Nutrition and Soil Science 167:345–350.
Johansson E, Kuktaite R, Andersson A, Prieto-Linde ML (2005) Protein polymer built-up during wheat development: influences of temperature and nitrogen timing. Journal of the Science of Food and Agriculture 85:473–479.
Johansson, E., Prieto-Linde, M.-L., and Gissén, C. (2008) Influences of weather, cultivar and fertilizer rate on grain protein accumulation in field-grown wheat, and relations to grain water content and falling number. Journal of the Science of Food and Agriculture 88:2011–2018.
Johansson E, Malik AH, Hussain A, Rasheed F, Newson WR, Plivelic T, Hedenqvist MS, Gällstedt M, Kuktaite R (2013) Wheat gluten polymer structures: The impact of genotype, environment and processing on their functionality in various applications. Cereal Chemistry 90:367–376.
Johansson E, Hussain A, Kuktaite R, Andersson SC, Olsson ME (2014) Contribution of organically grown crops to human health. International Journal of Environmental Research and Public Health 11:3870–3893.
Joshi AK, Crossa J, Arun B, Chand R, Trethovan R, Vargas M, Ortiz-Monasterio I (2010) Gentype x environment interaction for zinc and iron concentration of wheat grain in eastern Gangetic plains of India. Field Crops Research 116:268–277.
Kasarda DD (1989) Glutenin structure in relation to wheat quality. In: Wheat is Unique. American Association of Cereal Chemistry, St Paul, MN, pp. 277–302.
Kim W, Johnson JW, Graybosch RA, Gaines CS (2003) Physicochemical properties and end-use quality of wheat starch as a function of waxy protein alleles. J Cereal Science 37:195–204.
Konopka I, Tanska M, Faron A, Stepien A, Wojtkowiak K (2012) Comparison of the phenolic compounds, carotenoids and tocochromanols content in wheat grain under organic and mineral fertilization regimes. Molecules 17:12341–12356.

Koppel R, Ingver A (2010). Stability and predictability of baking quality of winter wheat. Agronomy Research 8:637–644.

Kumar P, Yadava RK, Gollen B, Kumar S, Verma RK, Yadav S (2011) Nutritional contents and medicinal properties of wheat: a review. Life Sciences and Medicine Research 22: 1–10.

Laidig F, Piepho HP, Rentel D, Drobek T, Meyer U, Huesken A (2017) Breeding progress, environmental variation and correlation of winter wheat yield and quality traits in German official variety trials and on-farm during 1983–2014. Theoretical and Applied Genetics 130:223–245.

Lampi AM, Nurmi T, Ollilainen V, Piironen V (2008) Tocopherols and tocotrienols in wheat genotypes in the HEALTHGRAIN diversity screen. Journal of Agriculture and Food Chemistry 56:9716–9721.

Lampi A-M, Nurmi T, Piironen V (2010) Effects of the environment and genotype on tocopherols and tocotrienols in wheat in the HEALTHGRAIN diversity screen. Journal of Agriculture and Food Chemistry 58:9306–9313.

Larik AS, Malik SI, Kakar AA, Naz MA (2000) Assessment of heritability and genetic advance for yield components in G. hirsutum. Science Khyber 13:39–44.

Lemelin E, Branlard G, Salvo L, Lein V, Aussenac T, Dayde J (2005) Breadmaking stability of wheat flours: Relation between mixing properties and molecular weight distribution of polymeric glutenins. Journal of Cereal Science 42:317–326.

Li Y, Wu Y, Hernandez-Espinosa N, Peña RJ (2013a) The influence of drought and heat stress on the expression of end-use quality parameters of common wheat. Journal of Cereal Science 57:73–78.

Li Y, Wu Y, Hernandez-Espinosa N, Peña RJ (2013b) Heat and drought stress on durum wheat: responses of genotypes, yeild, and quality parameters. Journal of Cereal Science 57:398–404.

Li JY, Liu JD, Wen W, Zhang PZ, Wan YX, Xia XC, Zhang Y, He ZH (2018) Genome-wide association mapping of vitamins B1 and B2 in common wheat. Crop Journal 6:263–270.

Liang D, Tang J, Peña RJ, Singh R, He X, Shen X, Yao D, Xia X, He Z (2010) Characterization of CIMMYT bread wheat for high- and low-molecular weight glutenin subunits and other quality-related genes with SDS-PAGE, RP-HPLC and molecular markers. Euphytica 172:235–250.

Lin CS, Binns MR, Leftovitch LP (1986) Stability analysis: where do we stand? Crop Science 26:894–900.

Liu L, Guo Q, He ZH, X, XC, Waters DLE, Raymond CA, King GJ (2017) Genotypic variation in wheat flour lysophospholipids. Molecules 22:909.

Lv J, Lu Y, Niu Y, Whent M, Ramadan MF, Costa J, Yu L (2013) Effect of genotype, environment, and their interaction on phytochemical compositions and antioxidant properties of soft winter wheat flour. Food Chemistry 138:454–462.

Magallanes-López AM, Hernandez-Espinosa N, Velu G, Posadas-Romano G (2017a) Variability in iron, zinc and phytic acid content in a worldwide collection of commercial durum wheat cultivars and the effect of reduced irrigation on these traits. Food Chemistry 237:499–505.

Magallanes-Lopez AM, Ammar K, Morales-Dorantes A, Gonzalez-Santoyo H, Crossa J, Guzman C (2017b) Grain quality traits of commercial durum wheat varieties and their relationships with drought stress and glutenins composition. Journal of Cereal Science 75:1–9.

Majoul-Haddad T, Bancel E, Martre P, Triboi E, Branlard G (2013) Effect of short heat shocks applied during grain development on wheat (Triticum aestivum L.) grain proteome. Journal of Cereal Science 57:486–495.

Malik AH (2012) Governing grain protein concentration and composition in wheat and barley. Acta Universitatis Agriculturae Sueciae 2012:55.

Malik AH, Prieto-Linde ML, Kuktaite R, Andersson A, Johansson E (2011) Individual and interactive effects of cultivar maturation time, nitrogen regime and temperature level on accumulation of wheat grain proteins. Journal of the Science of Food and Agriculture 91:2192–2200.

Malik AH, Kuktaite R, Johansson E (2013) Combined effect of genetic and environmental factors on the accumulation of proteins in the wheat grain and their relationship to bread-making quality. Journal of Cereal Science 57:170–174.

Manickavelu A, Kawaura K, Imamura H, Mori M, Ogihara Y (2011) Molecular mapping of quantitative trait loci for domestication traits and b-glucan content in a wheat recombinant inbred line population. Euphytica 177:179–190.

Margiotta B, Urbano M, Colaprico G, Johansson E, Buonocore F, D'Ovidio R, Lafiandra D (1996) Detection of Y-type subunit at the GLU-A1 locus in some Swedish bread wheat lines. Journal of Cereal Science 23:203–211.

Mariani BM, D'Egidio MG, Novaro P (1995) Durum wheat quality evaluation: Influence of genotype and environment. Cereal Chemistry 72:194–197.

Martinant JP, Cadalen T, Billot A, Chartier S, Leroy P, Bernard M, Saulnier L, Branlard G (1998) Genetic analysis of water extractable arabinoxylans in bread wheat endosperm. Theoretical and Applied Genetics 97:1069–1075.

Michell RAC, Shewry PR (2015) Dietary fiber: wheat genes for enhanced human health. Adv Wheat Genetics: from Genome to Field 411-419.

Moreira-Ascarrunz SG, Larsson H, Prieto-Linde ML, Johansson E (2016) Mineral nutritional yield and nutrient density of locally adapted wheat genotypes under organic production. Foods 5:89.

Mpofu A, Sapirstein HD, Beta T (2006) Genotype and environmental variation in phenolic content, phenolic acid composition, and antioxidant activity of hard spring wheat. Journal of Agriculture and Food Chemistry 54:1265–1270.

Mut, Z, Aydin N, Bayramoglu HO, Ozcan H (2010) Stability of some quality traits in bread wheat (Triticum aestivum) genotypes. Journal of Environmental Biology 31:489–495.

Myers SS, Zanobetti A, Kloog I, Huybers P, Leakey ADB, Bloom AJ, Carlisle E, Dietterich LH, Fitzgerald G, Hasegawa T, Holbrook NM, Nelson RL, Ottman MJ, Raboy V, Sakai H, Sartor KA, Schwartz J, Seneweera S, Tausz M, Usui Y (2014) Increasing CO2 threatens human nutrition. Nature 510:139–142.

Nan Z, Zhao C, Li J, Chen F, Sun W (2002) Relations between soil properties and selected heavy metal concetrations in spring wheat (Triticum aestivum L.) grown in contaminated soils. Water Air and Soil Pollution 133:205–213.

Nemeth C, Freeman J, Hones HD, Sparks C, Pellny TK, Wilkinson MD, Dunwell J, Andersson AM, Åman P, Guillon F, Saulnier L, Mitchell RAC, Shewry PR 2010) Down-regulation of the CSLF6 gene results in decreased (1,3:1,4)-β-D-glucan in endosperm of wheat. Plant Physiology 152:1209–1218.

Novaro P, D'Egidio MG, Bacci L, Mariani BM (1997) Genotype and environment: their effect on some durum wheat quality characteristic. Journal of Genetics and Breeding 51:247–252.

Nurit E, Lyan B, Pujos-Guillot E, Branlard G, Piquet A (2016) Change in B and E vitamin and lutein, β-sitosterol contents in industrial milling fractions and during toasted bread production. Journal of Cereal Science 69:290–296.

Nuttal JG, O'Leary GJ, Panozzo JF, Wlaker CK, Barlow KM, Fitzgerald GJ (2017) Models of grain quality in wheat - A review. Journal of Field Crops Research 202:136–145.

Oweis T, Pala M, Ryan J (1999) Managment alternatives for improved durum wheat production under supplemental irrigation in Syria. European Journal of Agronomy 11:255–266.

Panozzo J, Eagles HA (2000) Cultivar and environmental effects on quality characters in wheat. II protein. Australian Journal of Agricultural Research 51:629–636.

Panozzo JF, Eagles HA, Wootton M (2001) Changes in protein composition during grain development in wheat. Australian Journal of Agricultural Research 52:485–493.

Panozzo J, Walker CK, Partington DL, Neumann NC, Tausz M, Seneweera S, Fitzgerald GL (2014) Elevated carbon dioxide changes grain protein concentration and composition and compromises baking quality. A FACE study. Journal of Cereal Science 60:461–470.

Pasha I, Anjum FM, Morris CF (2010) Grain hardness: A major determinant of wheat quality. Food Science and Technology 16:511–522.

Payne PI, Lawrence GJ (1983) Catalogue of alleles for the complex loci, Glu-A1, Glu-B1 and Glu-D1 which code for high-molecular-weight subunits of glutenin in hexaploid wheat. Cereal Research Communications 11:29.

Payne PI, Holt LM, Jackson EA, Law CN, Damaina AB (1984) Wheat storage proteins: their genetics and their potential for manipulation by plant breeding. Philosophical Transactions of the Royal Society of London Series B Biological Sciences 304:359–371.

Payne PI, Nightingale MA, Krattiger AF, Holt LM (1987) The relationship between HMW glutenin subunit composition and the bread-making quality of British-grown wheat varieties. Journal of the Science of Food and Agriculture 40:51–65.

Peleg Z, Cakmak I, Ozturk L, Yazici A, Jun Y, Budak H, Korol AB, Fahima, T, Saranga Y (2009) Quantitative trait loci conferring grain mineral nutrient concentrations in durum wheat x wild emmer wheat RIL population. Theoretical and Applied Genetics 119:353–369.

Peña-Bautista RJ, Hernandez-Espinosa N, Jones JM, Guzman C, Braun HJ (2017) Wheat-based foods: Their global and regional importance in the food supply, nutrition and health. Cereal Foods World 62:231–249.

Pomeranz Y (1988) Chemical composition of kernel structures. In: Pomeranz Y (ed) Wheat: Chemistry and Technology. Y Pomeranz vol I. AACC, St Paul, MN, USA, pp. 91–158.

Pompa M, Giuliani MM, Palermo C, Agriesti F, Centonze D, Flagella Z (2013) Comparative analysis of gluten proteins in three durum wheat cultivars by a proteomic approach. Journal of Agriculture and Food Chemistry 61:2606–2617.

Porter JR, Xie L, Challinor AJ, Cochrane K, Howden SM, Iqbal MM, Lobell DB, Travasso MI, Netra Chhetri N, Garrett K, Ingram J, Lipper L, McCarthy N, McGrath J, Smith D, Thornton P, Watson J, Ziska L (2014) Food security and food production systems, In: Climate change 2014: impacts, adaptation and vulnerability. Part A: global and sectoral aspects. Cambridge University Press. New York, USA, pp. 485–533.

Prasad M, Kumar N, Kulwal P, Röder M, Balyan H, Dhaliwal H, Gupta P (2003) QTL analysis for grain protein content using SSR markers and validation studies using NILs in bread wheat. Theoretical and Applied Genetics 106:659–667.

Randall PJ, Moss HJ (1990) Some effects of temperature regime during grain filling on wheat quality. Australian Journal of Agricultural Research 41:603–617.

Reynolds MP, Quilligan E, Aggarwal PK, Bansal CK, Cavalieri AJ, Chapman SC, Chapotin SM, Datta SK, Duveiller E, Gill KS, Jagadish KSV, Joshi AK, Koehler AK, Kosina P, Krishnan S, Lafitte R, Mahala RS, Muthurajan R, Paterson AH, Boddupalli Prasanna BM, Rakshitt S, Rosegrant MW, Sharma I, Ravi RS, Sivasankar S, Vadez V, Valluru R, Prasad VPV, Yadav OP (2016) An integrated approach to maintaining cereal productivity under climate change. Global Food Security 8:9–18.

Rharrabti Y, Villegas D, Garcia del Moral LF, Aparicio N, Elhani S, Royo C (2001) Environmental and genetic deterimantion of protein content and grain yield in durum wheat under Mediterranean conditions. Plant Breeding 120:381–388.

Robert N (2002) Comparison of stability statistics for yield and quality traits in bread wheat. Euphytica 128: 333–341.

Rozbicki J, Ceglińska A, Gozdowski D, Jakubczak M, Cacak-Pietrzak G, Mądry W, Golba J, Piechociński M, Sobczyński G, Studnicki M, Drzazga T (2015). Influence of the cultivar, environment and management on the grain yield and bread-making quality in winter wheat. Journal of Cereal Science 61:126–132.

Ryan MH, Derrick JW, Dann PR (2004) Grain mineral concentrations and yield of wheat grown underorganic and conventional management. Journal of the Science of Food and Agriculture 84:207–216.

Sampson DA, Wen QB, Lorenz K (1996) Vitamin B6 and pyridoxine glucoside content of wheat and wheat flours. Cereal Chemistry 73:770–774.

Schmid M, Wieser H, Koehler P (2016) Isolation and characterization of high-molecular-weight (HMW) gliadins from wheat flour. Cereal Chemistry 93:536–542.

Shewry PR (2007) Improving the protein content and composition of cereal grain. Journal of Cereal Science 46:239–250.

Shewry PR, Hey SJ (2015) The contribution of wheat to human diet and health. Food Energy and Security 4:178–202.

Shewry PR, Tatham AS, Forde J, Kreis M, Miflin BJ (1987) The classification and nomenclature of wheat gluten proteins: A reassessment. Journal of Cereal Science 4:97–106.

Shewry PR, Halford NG, Tatham AS (1992) High molecular weight subunits of wheat glutenin. Journal of Cereal Science 15:105–120.

Shewry PR, Piironen V, Lampi A-M, Edelmann M, Kariluoto S, Nurmi T, Fernandez-Orozco R, Ravel C, Charmet G, Andersson AAM, Åman P, Boros D, Gebruers K, Dornez E, Courtin CM, Delcour JA, Rakszegi M, Bedo Z, Ward JL (2010) The HEALTHGRAIN wheat diversity

screen: Effects of genotype and environment on phytochemicals and dietary fiber components. Journal of Agriculture and Food Chemistry 58:9291–9298.
Shewry PR, Van Schaik F, Ravel C, Charmet G, Rakszegi M, Bedo Z, Ward JL (2011) Genotype and environmental effects on the contents of vitamin B1, B2, B3 and B6 in wheat grain. Journal of Agriculture and Food Chemistry 59:10564–10571.
Simmonds DH (1989) Wheat and wheat quality in Australia. Australian Wheat Board. CSIRO, Australia.
Skylas DJ, Cordwell SJ, Hains PG, Larsen MR, Basseal DJ, Walsh BJ, Blumenthal C, Rathmell W, Copeland L, Wrigley CW (2002) Heat shock of wheat during grain filling: Proteins associated with heat tolerance. Journal of Cereal Science 35:175–188.
Steffen W, Persson Å, Deutsch L, Zalasiewicz J, Williams M, Richardson K, Crumley C, Crutzen P, Folke C, Gordon L, Molina M (2011). The Anthropocene: From global change to planetary stewardship. Ambio 40: 739–761.
Stewart CM, McBratney AB, Skerritt JH (2002) Site-specific durum wheat quality and its relationship to soil properties in a single field in Northen New South Wales. Precision Agriculture 3:155–168.
Stroud JL, Broadley MR, Foot I, Fairweather-Tait SJ, Hart DJ, Hurst R, Knott P, Mowat H, Norman K, Scott P, Tucker M, White PJ, McGrath SP, Zhao FJ (2010) Soil factors affecting selenium concetration in wheat grain and the fate and speciation of Se fertilizers applied to soil. Plant and Soil 332:19–30.
Tabbita F, Pearce S, Barneix AJ (2017) Breding for increase grain protein and micronutrient content in wheat: Ten years of the *GPC-B1* gene. Journal of Cereal Science 73:183–191.
Tausz-Posch S, Armstrong RD, Tausz M (2014) Nutrient use and nutrient use efficiency of crops in a high CO2 atmosphere. In: Hawkesford MJ (ed.) Nutrient Use Efficiency in Plants, Plant Ecophysiology 10. Springer International Publishing, Switzerland.
Thomason WE, Phillips SB (2006) Methods to evaluate wheat cultivar testing environments and improve cultivar selection protocols. Field Crops Research 99:87–95.
Tiwari VK, Rawat N, Chhuneja P, Neelam K, Aggarwal R, Randhawa GS, Dhaliwal HS, Keller B, Singh K (2009) Mapping of quantitative trait loci for grain iron and zink concentration in diploid A genom wheat. Journal of Heredity 6:771–776.
Uhlen AK, Dieseth JA, Koga S, Böcker U, Hoel B, Anderson JA, Moldestad A (2015) Variation in gluten quality parameters of spring wheat varieties of different origin grown in contrasting environments. Journal of Cereal Science 62:110–116.
Vaher M, Matso K, Levandi T, Helmja K, Kaljurand M (2010) Phenolic compounds and the antioxidant activity of the bran, flour and whole grain of different wheat varieties. Proceedings of the Chemical Society 2:76–82.
Wall JS (1979) The role of wheat proteins in determining baking quality. In: Laidman DL, Wyn Jones RG (eds) Recent advances in the biochemistry of cereals. London/New York: Academic. pp. 275–311.
Vázquez D, Berger AG, Cuniberti M, Bainotti C, de Miranda MZ, Scheeren PL, Jobet C, Zúniga J, Cabrera G, Verges R, Peña RJ (2012) Influences of cultivar and environment on quality of Latin American wheats. J Cereal Science 56:196–203.
Vázquez D, Berger A, Prieto-Linde ML, Johansson E (2018) Can nitrogen fertilization be used to modulate yield, protein content and bread-making quality in Uruguayan wheat? Journal of Cereal Science 85:153–161.
Vázquez D, Castro M (2018) Mejoramiento para calidad estable en ambientes variables. In: S Germán, M Quincke, D Vázquez, M Castro, S Pereyra, P Silva, A García (Eds.), Seminario internacional "1914-2014, un siglo de mejoramiento de trigo en La Estanzuela" (pp. 212–220). UCTT/INIA, Montevideo, Uruguay.
Velu G, Guzman C, Mondal S, Autrique JE, Huerta J, Singh RP (2016) Effect of drought and elevated temperature on gran zinc and iron concentrations in CIMMYT spring wheat. Journal of Cereal Science 69:182–186.
Wieser H, Manderscheid R, Martin E, Weigel HJ (2008) Effects of elevated atmospheric CO2 concentrations on the quantitative protein composition of wheat grain. Journal of Agriculture and Food Chemistry 56:6531–6535.

Williams RM, O'Brien L, Eagles HA, Solah VA, Jayasena V (2008). The influences of genotype, environment, and genotype x environment interaction on wheat quality. Australian Journal of Agriculture Research 59:95–111.

Wong JC, Lambert RJ, Tadmore Y, Rocheford TR (2003) QTL associated with accumulation of tocopherols in maize. Crop Science 43:2257–2266.

Yan W, Rajcan I (2002). Biplot evaluation of test sites and trait relations of soybean in Ontario. Crop Science 42:11–20.

Yang F, Jorgensen AD, Li H, Sondergaard I, Finnie C, Svensson B, Jiang D, Wollenweber B, Jacobsen S (2011) Implications of high-temperature events and water deficits on protein profiles in wheat (Triticum aestivum L. cv. Vinjett) grain. Proteomics 11:1684–1695.

Yilmaz A, Ekiz H, Torun B, Gultekin I, Karanlik S, Bagci SA, Cakmak I (2008) Effect of different zinc application methods on grain yield and zinc concentration in wheat cultivars grown on zinc-deficient calcareous soils. Journal of Plant Nutrition 20:461–471.

Zeidan MS, Mohamed MF, Hamouda HA (2010) Effect of foliar fertilization of Fe, Mn and Zn on wheat yield and quality in low sandy soils fertility. World Journal of Agricultural Science 6;696–699.

Zhao FJ, Withers PJA, Evans EJ, Monaghan J, Salmon SE, Shewry PR, McGrath SP (1997) Sulphur nutrition: An important factor for the quality of wheat and rapeseed. Soil Science and Plant Nutrition 43:1137–1142.

Zhong YX, Xu D, Hebelstrup KH, Yang DL, Cai J, Wang X, Zhou Q, Cao WX, Dai T, Jiang D (2018) Nitrogen topdressing timing modifies free amino acids profiles and genes expression of storage proteins in wheat grain. BMC Plant Biology 18: 353.

Zobel R., Wright MJ, Gauch HG (1988) Statistical analysis of yield trial. Agron. Journal 80: 388–393.

Improving Wheat Nutritional Quality through Biofortification

Sewa Ram and Velu Govindan

Abstract Wheat is an important cereal crop that contributes significantly to the human diet. Different parts of the wheat grain provide different nutrients. Wheat germ is rich in vitamins B and E, protein, unsaturated fats, minerals, and carbohydrates, while the bran consists mostly of insoluble carbohydrates, protein, traces of B vitamins and minerals, and some anti-nutritional factors such as phytic acid. The endosperm is the largest part of the grain and consists mainly of starch and protein. There are increasing concerns about the deficiency of vitamins and minerals in the human diet, a condition commonly referred to as "hidden hunger" that has serious and widespread consequences in developing countries where cereals are the main source of food and nutrition. The low bioavailability of essential micronutrients, especially iron and zinc in humans and some farm animals, contributes not only to micronutrient deficiency but also to phosphorus pollution. Existing interventions to provide micronutrients such as with pharmaceutical supplements or industrial fortification of food products are effective yet have some limitations particularly in rural settings. Biofortification, the production of new food crops with higher micronutrient densities, may be a more apt approach. For example, enhancing wheat micronutrient density and bioavailability could lead to both improved human health and more sustainable agriculture. This can be accomplished by understanding the genetic diversity of wheat iron and zinc content and the genetic and molecular factors underlying these traits. Fertilizer application to crops has the potential to complement the gains made through genetic biofortification. Progress made in both genetic and agronomic strategies for wheat iron and zinc biofortification including the enhancement of bioavailability will be reviewed in this chapter.

S. Ram (✉)
ICAR-Indian Institute of Wheat and Barley Research, Karnal, India
e-mail: sewa.ram@icar.gov.in

V. Govindan
CIMMYT, Mexico City, Mexico

G. Igrejas et al. (eds.), *Wheat Quality For Improving Processing And Human Health*, https://doi.org/10.1007/978-3-030-34163-3_9

1 Introduction

Wheat is an important cereal crop that contributes significant quantities of protein, carbohydrate, fats, minerals and vitamins to the human diet. Wheat also has a high amount of dietary fibre that contributes additional health benefits associated with the consumption of whole grain products, including reduced risks of obesity, type 2 diabetes, cardiovascular disease and colorectal cancer. The distribution of nutrients within the wheat kernel is typical of many cereals. The germ is rich in vitamins B and E, high quality protein, unsaturated fats, minerals, and carbohydrates, while the bran consists mostly of the insoluble carbohydrates such as cellulose, protein, traces of B vitamins and minerals, as well as anti-nutritional factors such as phytic acid. The endosperm is the largest part of the grain and consists mainly of starch and protein, and trace amounts of vitamins and minerals. Though wheat has many nutritional qualities, it lacks adequate amounts of several essential nutrients such as vitamins A, B_{12} and C, most fats, micronutrients and the essential amino acid lysine.

There are increasing concerns about a lack of vitamins and minerals in the human diet, a condition commonly referred to as "hidden hunger". Such deficiencies occur when intake and absorption of vitamins and minerals are too low to sustain good health and development. According to the World Health Organization, at least 2 billion individuals around the world, or one in three people, suffer from hidden hunger, with South Asia and sub-Saharan Africa being the most affected (WHO 2017). The problem is particularly serious in developing countries where cereals are the main source of food and nutrition. The consequences of hidden hunger in terms of mortality, impaired physical and cognitive development, and eye problems are staggering. Malnutrition is estimated to contribute to more than 45 percent of all child deaths. Every year iron (Fe) and zinc (Zn) deficiencies cause the deaths of about 800000 children. Disability-adjusted life years (DALYs) are calculated as the sum of years of life lost and the years lived with disability based on 291 causes and 20 age groups of both sexes. Fe deficiency is responsible for 2.4% of global worldwide DALYs, while the corresponding value for Zn deficiency is 2.9% (Mutangadura 2004).

The widespread consumption of white flour, which is made predominantly from the endosperm of wheat grain discarding the bran in the milling process, has worsened the degree of Fe and Zn malnutrition. This is because Fe and Zn accumulate in higher concentrations in the embryo and aleurone layer than in endosperm of wheat grain (Cakmak et al. 2010). For this reason, the consumption of whole grain rather than white wheat flour has been advocated as a way to increase daily Fe and Zn intake. Current ways of ensuring sufficient micronutrient intake, such as taking pharmaceutical supplements or eating foods which have been fortified during industrial production, are effective but have some limitations particularly in rural settings. An alternative approach is to develop new food crops with higher micronutrient densities, often referred to as "biofortification". In the long term, increasing the production of micronutrient-rich foods and improving dietary diversity will substantially reduce micronutrient deficiencies. In the near term, consuming biofortified crops can help address micronutrient deficiencies by increasing the daily adequacy of intakes of an individual over a lifetime (Bouis et al. 2011).

Cereals have genetically low concentrations of Zn and Fe compounded by low bioavailability. About half of world cereal production comes from soils low in plant-available Zn, so the quality of the grain produced is poor with respect to Zn content. The situation of Fe deficiency in cereals is similar. Independent of the total amount or concentration of an essential micronutrient in a foodstuff, if it is in a form which cannot be assimilated and used by the body it not only contributes to micronutrient deficiency, but may have other negative effects. As will be explained, the low bioavailability of cereal Fe and Zn in humans and monogastric animals has led to phosphorus (P) pollution of the environment. Enhancing the micronutrient density and bioavailability in wheat eaten by humans and monogastric animals could therefore lead to improved human health, less pollution of water bodies, and more sustainable agriculture.

As a starting point, this can first be addressed by understanding more about the genetic diversity in wheat Fe and Zn accumulation and the genetics and molecular mechanisms responsible for the diverse traits observed. Large amounts of information have recently been generated on various cereal genes related to starch quality, mineral nutrition, and anti-nutritional factors. Additionally, agronomic interventions through fertilizer application have shown potential to complement gains made through genetic biofortification.

2 Comparative Advantages of Biofortified Grain

Over the last 50 years, agricultural research for developing countries has increased the production and availability of calorically dense staple crops, but the production of micronutrient-rich non-staples, such as vegetables, pulses and animal products, has not increased in equal measure. Non-staple food prices have increased steadily and substantially, making it more difficult for the poor to afford a good quality diet (Bouis et al. 2011). There is no single solution to alleviate micronutrient deficiencies and a multifold approach is needed. Biofortification complements existing interventions while having two key comparative advantages. It can be more cost-effective in the long term and reach underserved rural populations who may have limited access to diverse diets or other micronutrient interventions. Unlike the continual financial outlays required for supplements and commercial fortification processes, an upfront investment in plant breeding yields micronutrient-rich biofortified planting material for farmers to grow at virtually zero additional cost. Target micronutrient levels for biofortified crops can be set to meet the specific dietary needs of women and children based on current consumption habits. Once developed, nutritionally improved crops can be evaluated and adapted to new environments and geographies, multiplying the benefits of the initial investment. Once the micronutrient trait has been mainstreamed into the core breeding objectives of national and international crop development programs, recurrent expenditures by agricultural research institutes for monitoring and maintenance are minimal. Biofortification empowers farmers by combining the micronutrient trait with other agronomic and consumption traits that are preferred. After fulfilling the household's food needs, surplus biofortified crops make their way into rural and urban retail outlets. Recently in a community based study of 6005 participants in Delhi, India, pairs of women of child bearing age (WCBA) or children were randomly allocated to receive either high zinc biofortified wheat flour (HZn, 30 mg/kg zinc daily) or low zinc biofortified wheat flour (LZn, 20 mg/kg zinc daily) for 6 months, 360 g of flour for WCBA and 120 g for children daily (Sazawal et al. 2018). Baseline and endline blood samples were obtained to assess the presence of hematological markers and the zinc status, and data on compliance and morbidity were collected. Compared to children in LZn group, children in the HZn group had 17% (95% CI, 6 to 31%, $p = 0.05$) and 40% (95% CI, 16 to 57%, $p = 0.0019$) fewer days with pneumonia and vomiting respectively. WCBA in the HZn group also had 9% fewer days with fever compared to the LZn group, a statistically significant result.

3 Mineral Content of Wheat Grain

Successful crop improvement through plant breeding depends on the extent of genetic variation in the target traits present in the gene pool of the available germplasm. When breeding for higher Fe and Zn concentrations in the grain, the task is further complicated by the fact that the grain micronutrient concentrations depend largely on environmental conditions, particularly soil composition. Despite advances in breeding for efficiency of uptake or mobilization to the grain, the Fe and Zn concentrations in mature grain are limited by how much of the minerals are available in soil. It would be difficult, if not impossible, to develop varieties that produce grain with nutritionally meaningful concentrations of these minerals when the plants are grown in severely deficient soils. Crops grown in the field with application of manure tend to produce grain with higher Fe and Zn values. For example, Welch et al. (2005) reported that an Indian wheat cultivar C306 accumulated 130 mg/kg of Zn and 220 mg/kg of Fe in grain when grown under hydroponic conditions with high mineral content. However, the cultivar had Fe and Zn contents in the range of 35–45 mg/kg and 30–40 mg/kg, respectively when grown under field conditions in India.

The range of values for Fe concentration in grain among hexaploid wheat, *Triticum dicoccon*, and landraces grown under field conditions was 25–56 mg/kg, with a mean of 37 mg/kg, while the range for Zn concentration was 25–65 mg/kg, with a mean of 35 mg/kg. It was noted that the genotypes with the highest levels of Fe and Zn were low-yielding, unadapted genotypes (Monasterio and Graham 2000). The search for germplasm that accumulated higher levels of Fe and Zn led to a more in-depth evaluation of landraces and, finally, the definition of a secondary gene pool including tetraploid and diploid progenitors of hexaploid wheat with enhanced micronutrient status (Monasterio and Graham 2000). *Triticum dicoccoides*, *Aegilops tauschii*, *Triticum monococcum*, and *Triticum boeticum* were found to be among the most promising sources of high Fe and Zn levels in the grain. Some of these genotypes showed values as high as 142 mg/kg of Zn, but such high values may have been due to past application of manure in some locations. A recent study evaluated a set of high-yielding lines in field conditions. The Zn concentrations generally ranged between 15 and 35 mg/kg but were as high as 43 mg/kg in some genotypes, while Fe concentrations ranged from 20 to 60 mg/kg (Oury et al. 2006).

4 Grain Fe and Zn Targets as per Recommended Dietary Allowance Requirements

Taking into account the bioavailability, the daily intake and the estimated average requirements of a balanced human diet, some general estimations were made by the interdisciplinary HarvestPlus initiative, which was launched in 2003, to set tentative breeding targets for wheat. In Pakistan and northern India the target is to increase Fe and Zn levels respectively 25 and 10 mg/kg above the baseline, the mean of all genotypes grown in the region. This translates on average into total Zn and Fe levels in the grain of 45 and 60 mg/kg, respectively. In our opinion, there is sufficient genetic variability to develop wheat varieties with increased Zn levels in the grain. While there is also promising genetic variability in Fe content, the lower bioavailability of Fe compared to Zn, means that target levels for Fe must be significantly higher and are therefore more challenging. Recommended dietary allowance (RDA) requirements for Fe and Zn consumption differ for different age groups, but typical requirements for a healthy person are given in Table 1.

5 Phytic Acid and Micronutrient Bioavailability

Phytic acid, also known as inositol hexaphosphate (IP6) (Fig. 1) or phytate, is the principal storage form of phosphorus in many plant tissues, especially in the grass (wheat, rice, rye, barley, etc.) and bean families, and is a major antinutritional factor. Phosphorus consumed in this form is generally not bioavailable to humans because they lack phytase, the digestive enzyme required to separate phosphorus from the

Table 1 Mineral content of wheat and wheat fractions (mg/kg) and RDA requirements (mg)

Wheat sample	Fe	Zn	Cu	Mn
Wheat	18–31	21–63	1.8–6.2	24–37
Bran	74–103	56–141	8.4–16.2	72–144
Germ	41–58	<100–144	7.2–11.8	101–129
Flour	3.5–9.1	3.4–10.5	0.62–0.63	2.1–3.5
Maximum adult RDA	15	15	1.5–3.0	2–5

RDA, recommended dietary allowances are the daily levels of intake of essential nutrients judged to be adequate to meet the known nutritional needs of most healthy persons. RDA values are taken from Grusak and Penna (1999) and other values from Betschart (1988)

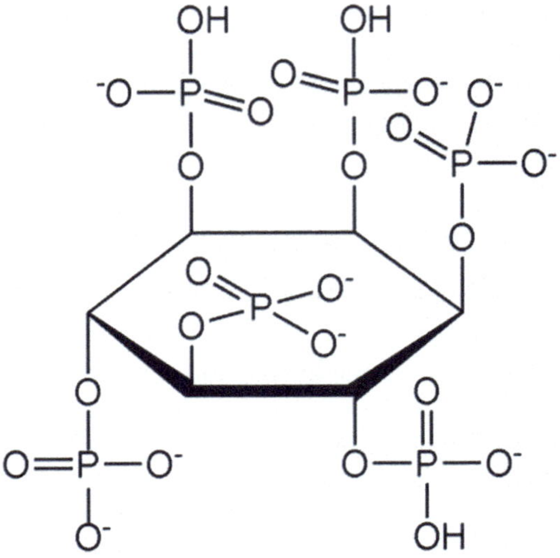

Fig. 1 Structure of phytic acid

phytate molecule. Phytic acid binds strongly to metallic cations such as those of K, Mg, Mn, Fe, Ca and Zn to form a mixed salt called phytin making these minerals less bioavailable to humans. Phytin accumulates in seed protein bodies that are either dispersed or in dense inclusions called globoids. The stable complexes of these minerals with phytic acid can be the cause of micronutrient deficiencies (e.g. of Fe and Zn), particularly in poor countries where diets are primarily seed based. Monogastric animals such as pigs, poultry and fish cannot digest phytic acid either, so animal feeds are often supplemented with P. As a result a large proportion of phytic acid is excreted leading to the accumulation of polluting amounts of P in soil and water.

The bioavailability of these nutrients can be improved by reducing phytic acid content as well as increasing phytase activity in the grain. For breeding to improve the nutritional quality of wheat, it is therefore necessary to identify genotypes having low phytic acid content and high phytase activity along with higher Fe and Zn content. Phytic acid accounts for up to 85% of the total P stored in the seed, so developing low phytate crops may improve P availability in animal feeds and reduce phytic acid excretion, thus lessening the negative effect of animal waste on the environment. While increasing the availability of iron and zinc will significantly benefit human nutrition, recently phytic acid has been studied for its potential anti-carcinogenic properties. Phytic acid may have some preventive effects in prostate, breast, pancreatic and colon cancer. The mechanism, however, is not yet understood. Therefore, it may not be advisable to completely eliminate phytic acid from wheat grain but to reduce it to a minimum level which is sufficient as a source of minerals during seed germination and beneficial to human beings.

In addition to phytic acid content, phytase activity is important in increasing micronutrient availability. Higher phytase activity in the grain may result in more extensive phytate degradation in the human stomach. Significant positive correlations have been reported between native phytase activity and respectively P utilization (Oloffs et al. 2000) and micronutrient bioavailability (Lopez et al. 2003). Higher phytase levels may contribute substantially to the gastrointestinal hydrolysis of phytate in non-ruminant animals (Steiner et al. 2009). Phytase levels have been increased in transgenic wheat expressing microbial genes (Brinch-Pedersen et al. 2006). However, the transgenic approach has the drawbacks of unstable gene expression and low acceptability of plant products genetically modified with microbial genes. Therefore, wheat may be an alternative source of phytases for applications in food processing, because they are more accepted by consumers and the allergenic potential is assumed to be lower (Greiner and Konietzny 2006), because cereals and legumes (and hence their phytases) are already part of the human diet. By contrast, phytase from *Aspergillus niger* is assumed to be a high risk factor for occupational asthma and rhinitis (Baur et al. 2002). In summary, raising phytase levels in wheat can be beneficial both economically and environmentally.

5.1 Variation in Phytate and Phytase Levels among the Wheat Genotypes

Screening a lot of germplasm including synthetic hexaploids showed that the phytic acid contents of wheat varied about threefold and ranged from 9.0 mg/g to 26.3 mg/g. Similarly in varieties developed in India differences in phytic acid contents were up to 3.4 fold, while differences were up to 5.9 fold in synthetic hexaploids (Fig. 3). The phytase unit (FTU) is a measure of the activity of the enzyme releasing inorganic phosphate from excess substrate, sodium phytate. Phytase activities in released varieties ranged from 284 FTU/kg to 962 FTU/kg with a mean of 516 FTU/kg, while in synthetic hexaploids activities varied from 255 FTU/kg to 1518 FTU/kg with a mean of 634 FTU/kg. It is interesting to note that greater variability was observed in this set of synthetic hexaploids. Recently we have developed mutants with very high levels of phytase (>2400 FTU/kg) in the wheat variety PBW 502 developed in India (unpublished data). Synthetic hexaploids with higher phytase levels and mutants can be used to enhance the diversity in enzyme levels in bread and durum wheats. More synthetic hexaploids can be used for enzyme studies to explore the full extent of variation available. The large diversity in phytase levels in synthetic hexaploids may be because the synthetics were developed by crossing different species of tetraploids with different accessions of *Ae. squarrosa* (diploids). Possibly all the possible variations were not tapped during the evolution of wheat by natural crossing between diploids and tetraploids. Fewer hybridization events in a restricted geographic region might have led to the narrow genetic diversity of bread wheat. Among cereals, higher phytase potential has been reported in rye and triticale (Greiner and Konietzny 2006). In addition, mutation breeding has been used to develop high phytase lines and low phytic acid lines in the background of high yielding varieties in India (unpublished data).

5.2 Environmental Influence on Phytic Acid and Phytase Levels

Both environment and genotype can have significant effects on phytase levels (Ram et al. 2010) with genotypic effects being greater. Other studies also indicated environmental and genotypic effects on phytase levels (Okot-Kotber et al. 2003; Liu et al. 2007; Steiner et al. 2009). The large genotypic influence on phytase level was shown by high heritability ($h^2 = 98.38$). Phytate levels were also influenced more by genetic factors than by environmental factors with high heritability ($h^2 = 82.58$). Varieties with higher levels of phytase in one location tended to have higher values in other locations as well. The high heritability of phytase levels indicates that the trait is controlled by one or two major genes. Traits with higher genotypic influence are much more useful in breeding programmes. As there is a larger genotypic effect for phytase levels and a larger variation in activity, there is greater scope for manipulating phytase levels through breeding in wheat and thus may be very useful in improving bioavailability of Fe and Zn to human beings.

6 Breeding Strategies for Higher Fe and Zn Concentrations in Wheat Grain

An added difficulty in breeding wheat for improved micronutrient status is that grain yield and micronutrient levels must be increased simultaneously if farmers are to accept the new Fe and Zn enriched cultivars. This is because increased Fe and Zn concentrations are not visible traits, and tangible agronomic advantages are critical when convincing farmers to adopt such cultivars. As with grain protein, the concentrations of Fe and Zn decrease when yield increases. However, the phytate concentration also decreases (Monasterio and Graham 2000) so the overall effect on the final biovailability of Fe and Zn is difficult to predict. On the other hand, there is a positive correlation between Fe and Zn which allows both nutrients to be improved simultaneously (Monasterio and Graham 2000). Various methods are used to improve self-pollinated crops for quantitatively inherited traits. As our understanding of the underlying genetic control of Fe and Zn concentration is poor, breeding has focused on crossing materials of unrelated parentage and intermediate micronutrient status with the aim of identifying transgressive segregants. Provided sufficiently large F2 and F3 population sizes are maintained and genetic drift minimized, the F4 and later generations can be screened for Fe and Zn concentration once a higher level of homozygosity has been achieved. This approach works well when the target genes are distributed in adapted wheat backgrounds. However, insufficient variability in Fe and Zn concentration in adapted wheat cultivars and breeding lines has led to the search for new genetic variability in the secondary wheat gene pool.

The following breeding method relies on the production of very large populations and significant investment in inductively couple plasma optical emission spectroscopy (ICP-OES) to analyse micronutrient status. New primary hexaploid synthetic wheats with significantly higher Fe and Zn concentrations were used as nonrecurrent parents in double backcrosses (BC) to adapted wheat at CIMMYT, Mexico. It was necessary to identify BC1F1 progeny high in Fe and Zn with which the second backcross could be made. However, evaluating Fe and Zn concentrations in single plant is difficult as variation in the soil in which they are grown often gives misleading results. The process was time and labour consuming. Many more BC were made than were kept because the Zn and Fe status could not be determined by testing the BC plants with ICP-OES until they were mature. For this reason, only BC progeny from plants high in Fe and Zn concentration were advanced. This approach increased the probability of identifying progeny rich in Fe and Zn with superior agronomic type knowing that the percentage of the donor parent decreases in the cross. The BC2F2 populations were then grown over a large area and intense selection pressure was applied for agronomic type and disease resistance. Single plants selected from the BC2F2 population were advanced to F4 using the modified pedigree method. In the F4 generation, Fe and Zn were once again measured in two environments. Lines which accumulated high levels of Fe and Zn at both locations were advanced to F6 and head rows were selected from which near homozygous advanced lines were derived. The advanced lines were increased under disease pressure and selected lines subsequently tested in a replicated trial for grain yield and Fe and Zn concentration at one location in North Western Mexico. Rather than test the materials further in Mexico, those with the highest yield and highest micronutrient concentration were sent to South Asia for replicated yield trials for more extensive evaluation in the target region. A major drawback is that selection pressure for Fe and Zn cannot be applied in early generations grown on the CIMMYT site in Mexico as the evaluation of single plants does not give an accurate measure of micronutrient status.

Double haploids (DHs) have also been used to rapidly advance promising genetic materials. Given the complexity of inheritance and the need to simultaneously increase both grain yield and Fe and Zn content, DHs should not be made on unselected F1 progeny. They should only be used judiciously and produced on BC progeny confirmed as being high in Fe and Zn, for example, the BC2F4 or equivalent generation. Experience at CIMMYT using these methodologies has been mixed. Lines developed using primary hexaploid *T. dicoccon*-based synthetics, backcrossed twice to the adapted recurrent parent, selected for yield and Fe and Zn status in Mexico, then tested widely in Pakistan, showed good yield and improved Fe concentration but little significant improvement for Zn. Soil and grain analysis in systematic checks at the CIMMYT research station at Ciudad Obregón showed that soil Zn concentration may have been much more heterogeneous than Fe. Large variations in soil Zn can confound or mask genetic differences among lines, and hamper the identification of lines with genetically superior concentrations of grain Zn. One strategy to reduce this problem is to use a systematic check, alpha lattice designs, and spatial analyses of segregating and advanced populations.

Another potential strategy that needs further study is the use of Zn-containing fertilizer, whether applied to leaves or soil, to try to homogenize soil Zn concentration. Oury et al. (2006) concluded that given the generally strong genotype-by-environment interaction for Fe and Zn concentrations, screening for these traits would be highly unreliable in breeding for enhanced micronutrient concentration. The development of cheaper and more rapid screening assays for Fe and Zn based on colorimetric tests or near infra-red spectroscopy (NIRS) may allow plant breeders to phenotype plants more efficiently and accurately so greater selection pressure can be applied in early generations, minimizing the effect of "misclassified" lines on eventual outcomes. The complex inheritance of Fe and Zn concentration will also complicate the development and use of molecular markers, given that many genes are likely to be involved possibly with additive effects. However, if quantitative trait loci (QTLs) with large effects can be identified and tagged, cultivar development will be greatly facilitated as selection for these QTLs in early generations will improve the probability of identifying high-yielding lines that can have an impact on human nutrition. Recently a link was found between a gene regulating senescence, GpcB1 6BS, and high levels of protein, Fe, and Zn in the grain. Incorporating GpcB1 6BS into wheat could potentially result in grain with improved levels of Fe and Zn (Uauy et al. 2006, Distelfeld et al. 2007).

The 'hotspot' genomic regions and associated molecular markers for grain Zn concentration were identified through traditional QTL mapping and association mapping studies, which will be useful for marker assisted breeding (Velu et al. 2016, 2018). According to preliminary genomic prediction analysis the correlation between observed and predicted values for grain Zn and Fe were about 50% across environments suggesting genomic selection models can be applied to accelerate breeding efficiency (Velu et al. 2016).

6.1 CIMMYT Experience in Wheat Breeding for High Zn

Biofortification work at CIMMYT is done by a team of researchers including plant breeders, pathologists, cereal chemists and human nutrition specialists with the funding support from the HarvestPlus program. It is now part of the Agriculture for Nutrition and Health program managed by the CGIAR Systems Organization. Breeding for enhanced Fe and Zn concentrations was initially quite challenging, due to (i) the limited genetic variation in micronutrient contents in the adapted varieties and elite breeding germplasm and (ii) the complexity of genetic and metabolic networks controlling the homeostasis of Fe and Zn in wheat grain.

In the early 2000s, scientists conducted a large-scale screen of Zn and Fe content in wheat landraces (local, indigenous varieties) and wild relatives conserved in CIMMYT's germplasm bank near its Mexico City headquarters. Among them, some of the tetraploid emmer wheats (*T. dicoccum*), progenitors of hexaploid wheat such as *Ae. tauschii*, *T. monococcum* and *T. boeticum,* and landraces were seen to be the most promising sources for high grain Fe and Zn concentration, followed by *T. spelta*, and *T. polonicum* (Ortiz-Monasterio et al. 2007; Monasterio and Graham 2000). Subsequent field evaluation of a set of Mexican and Iranian landraces screened in Zn-enriched soil conditions at Cd. Obregón in Northern Mexico showed more than twofold variation in both grain Zn (40–96 mg/kg) and Fe (27–56 mg/kg) (Mondal et al. 2016; Vikram et al. 2016). The rich genetic diversity for Zn and Fe in wheat in different wild species and landraces provides novel alleles for genetic enhancement of Zn and Fe.

CIMMYT initiated biofortification breeding in 2007 by crossing these high micronutrient carrying synthetic wheats, *T. spelta,* and landraces with high-yielding adapted wheat germplams and then selecting plants with agronomic and disease resistance traits in segregating populations and selecting plants with high Zn and Fe content in advanced generations. The continued conventional breeding approach has resulted in the incorporation of several novel alleles for high grain Zn and Fe in elite, high-yielding germplasm (Velu et al. 2014). The decision to focus on breeding for Zn instead of Fe was based largely on the assumption that Zn is more bioavailable than Fe in human diets in terms of the percentage absorption, so increasing Zn would have a greater nutritional impact. To date, seven biofortified wheat varieties have been released with higher Fe and Zn content: Zincol 2016 in Pakistan; Zinc Shakti (Chitra), WB 02, HPBW 01 and Ankur Shiva in India; BARI Gom 33 in Bangladesh; and recently Nohely F2018 in Mexico. Zincol 2016 has 20% more Zn (+6 mg/kg Zn) while Zinc Shakti (Chitra) possesses 40% more Zn (+14 mg/kg Zn). Zincol 2016 was also released in Afghanistan, Iran and Egypt as Muqawim-09, Mehrgan and Misr1, respectively. In addition, the new biofortified wheat varieties WB 02 and HPBW 01 currently occupy a considerable area in the North Western Plains Zone of India and contain about 20% more Zn than commonly grown non-biofortified varieties.

Since 2007, over 5000 simple and top/back crosses were made between parents with high micronutrient contents and those with high-yield potential followed by early-generation (F2-F4) selection for agronomic traits, and selection for grain Zn, grain yield, and yield stability in advanced generations (F5-F7). The non-destructive X-ray fluorescence based high-throughput technique has been developed and routinely used to screen a large number of breeding lines for grain Zn and Fe contents (Paltridge et al. 2012). This progress was made by focusing strongly on phenotyping plants growing in Zn-homogenized fields at Ciudad Obregón, Mexico. Wheat lines with higher grain Zn also had high Zn when grown in trials in South Asia, as part of the Harvest Plus Yield Trial (HPYT). This multi-site analysis revealed high heritability and high genetic correlations between locations including Ciudad Obregón and target environments (Velu et al. 2012). National partners have shown an increasing interest in growing the HPYT in the intervening years (Fig. 2). HPYT was initiated with 10 sets of trials for the 2010–11 crop season and 78 sets were tested during the 2017–18 season.

The number of crosses with high Zn lines were gradually increased by around 10% each year to retain essential core traits such as yield potential and stability, heat and drought tolerance, resistance to the three rusts *Fusarium* head blight, *Septoria tritici* blotch, and spot blotch, and end-use quality. The addition of Zn as a core trait will therefore require a significant expansion of population sizes, phenotyping for Zn, yield testing and expansion of land use, phenotyping, genotyping, molecular

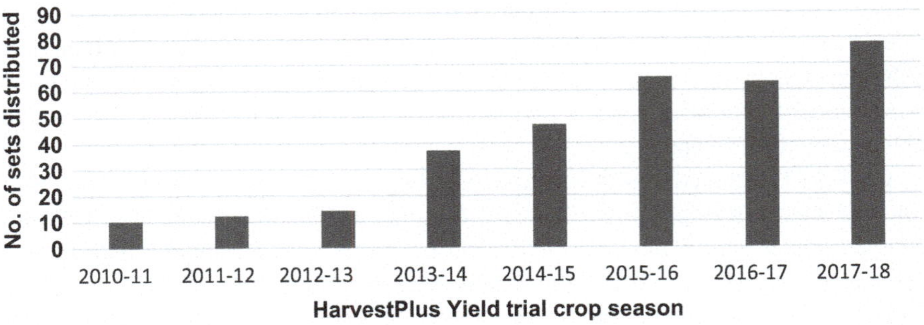

Fig. 2 Number of sets of HarvestPlus Yield Trials distributed globally each season

assisted selection (both the development and use), and genomic selection. It is anticipated that molecular breeding tools will make selection more efficient, but without an increase in breeding activities, the goal will not be achievable. In addition, antinutrients that limit bioavailability such as phytates can be reduced to a certain threshold, and factors that enhance bioavailability (high phytase activity, inulin-type fructans) can be increased to ensure higher Zn bioavailability in humans as complementary strategies.

7 Agronomic Biofortification

To complement genetic biofortification efforts, Zn fertilization is an agronomic biofortification strategy where foliar application of Zn-containing fertilizers during the early grain filling stage of wheat enhances grain Zn concentration by 20–30% (Cakmak, 2008). Cakmak et al. (2010) suggested that Zn and Fe fertilizers applied to soil and leaves can have a positive role by increasing the respective metal concentrations in durum wheat grain. The same authors also claimed that increased activity of Zn and Fe in source tissues (flag leaf and stem) during grain filling could be increased by additional soil or foliar application of Zn and Fe (Cakmak et al. 2010). Habib (2012) showed that simultaneous Zn and Fe application increased their respective concentrations in the grain more than separate applications. However, the final concentrations of Zn and Fe depend on the size of wheat grains (Velu et al. 2011) and number of grains per spike (Nowack et al. 2008). In stem girdling experiments using the radioactive isotope ^{65}Zn, Timsina (2014) demonstrated the role of phloem transport of Zn in wheat plants. The zinc tracer supplied on the upper leaf was transported to lower leaves and root tips. Ciccolini et al. (2017) reported the influence of Fe and Zn biofortification by foliar spray on the concentrations and potential bioavailability of Fe and Zn, and the amount of health-promoting compounds in wholemeal flour of two common wheat varieties, an old one and a modern one. The effect of this biofortification on milling and bread making was also studied. The concentration of Zn (+78%) and its bioavailability (+48%) increased in

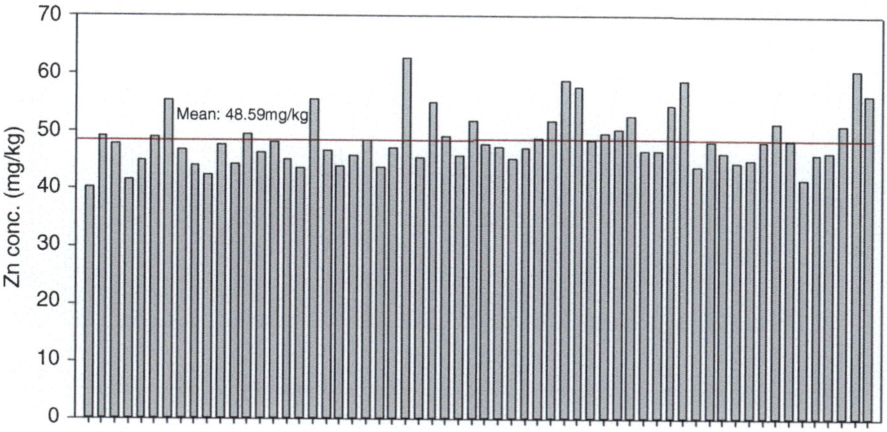

Fig. 3 Genetic variability of grain Zn concentration in genotypes in a low Zn environment (-Zn). The red line indicates the mean Zn concentration (n = 60)

the flour of the old variety, but it was ineffective in increasing Fe concentration in either variety. However, the old variety showed a higher concentration (+41%) and bioavailability (+26%) of Fe than the modern one. Wholemeal flour had more Fe, more Zn and more health-promoting compounds compared to white flour. Bread making only slightly changed Fe and Zn concentrations but greatly increased their bioavailability (77% and 70%, respectively). All these results will be of great interest when developing a production chain for enriched functional bread that has a protective role against chronic cardio-vascular diseases.

Field trials conducted in the CIMMYT experimental station at Ciudad Obregón, Mexico using a set of 60 biofortified and commercial varieties grown under field conditions with (+Zn) and without (-Zn) Zn treatments showed interesting results (Velu et al. unpublished). The grain Zn concentration of genotypes under -Zn conditions showed a large variation from 40 to 63 mg/kg, with a mean value of about 49 mg/kg (Fig. 3). This demonstrates the wide variability present in grain Zn concentrations in this set of germplasm. Figure 4 illustrates the differences in grain Zn concentrations of the same genotypes in +Zn conditions, that is with foliar application of Zn fertilizer. The average increase in grain Zn was about 9 mg/kg and varying responses of different genotypes were noted. In the most responsive genotype, the grain Zn concentration increased by as much as 19 mg/kg (Fig. 4). A strong positive association was observed between -Zn and + Zn with an R^2 value of 0.46 ($P < 0.01$) showing the high heritability of grain Zn concentrations with and without Zn fertilization.

Depending on the soil Zn status, application of Zn-containing fertilizers to the soil may also contribute to improved grain Zn levels. Field trials revealed that fertilizing the soils with a combination of Zn and nitrogen (N), in addition to foliar fertilization, further enhanced the Zn levels in the plants and the grain (Cakmak 2009). These results indicate that soil fertilization strengthens root growth and the root uptake system which improves the plant's ability to retain zinc and supports

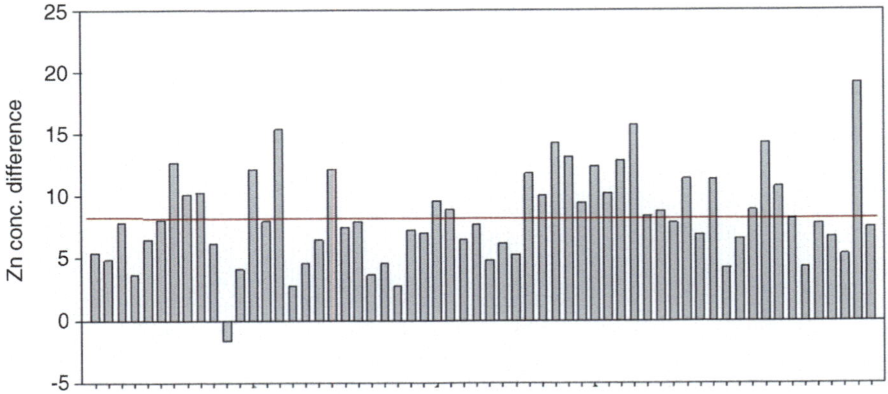

Fig. 4 Differential responses of genotypes to Zn fertilization. The difference in grain Zn concentration was measured for each variety grown in a Zn-rich environment (+Zn) and in a Zn-poor environment (-Zn). The bars represent the varieties in the same order as in Fig. 3. The red line indicates the mean Zn concentration difference (n = 60)

increases in yields. Increased atmospheric CO_2 may cause a decline in the nutrient concentration of crops, including Zn. Properly applied Zn fertilizer is a rapid, effective solution to reduce Zn deficiency, especially for small-scale farmers who grow most of their own food. Combining plant breeding technologies with agronomic biofortification techniques would amplify the benefits of both approaches and holds significant promise for reducing zinc deficiency in human nutrition while improving crop yield and resilience.

7.1 N Fertilization Influences Fe and Zn Content in Grain

N fertilization is known not only to increase wheat grain yield but also to facilitate the uptake of Fe and Zn in wheat grain (Cakmak et al. 2010). The uptake and transport of Fe and Zn to grain is probably facilitated by metal chelating compounds, such as 2-deoxymugineic acid (DMA) mainly for the translocation of Fe and Zn from flag leaves to grain in wheat (Kutman et al. 2010). The latter authors reported that N nutrition is critical in both the uptake and translocation of Zn and Fe to wheat grain and also showed that at high N rates, nearly 80% and 60% of total shoot Zn and Fe, respectively, were harvested with the grain. Improving N status of plants from low to sufficient resulted in a threefold increase in the shoot Fe content of wheat plants (Aciksoz et al. 2011). This demonstrated that N is a critical player in the uptake and accumulation of Zn in plants and thus deserves special attention in Zn biofortification strategies for food crops. Depending on the N supply, Zn remobilization from pre-anthesis sources provided almost all the grain Zn when the Zn supply was withheld at anthesis. Cakmak et al. (2010) found co-localization of protein, Fe and Zn in the embryo and aleurone layer of wheat grain, which indicates that protein-rich grains accumulate higher amounts of Zn and Fe. Increasing Zn and N supply had a major impact on Zn accumulation in the endosperm, which reached concentrations higher than the current breeding targets (Persson et al., 2016). Protein-rich wheat grain accumulated 57 mg kg^{-1} of Zn and 71 mg kg^{-1} of Fe compared to 30 mg kg^{-1} of Zn and 36 mg kg^{-1} of Fe in protein-poor grain. This showed that higher protein or nitrogen content favours the accumulation of Zn and Fe in wheat grain. Singh et al. (2018) investigated the interactive effect of N, Zn, and Fe on grain yield, protein content and nutrient concentrations in a pot experiment conducted in an environmentally controlled growth chamber. They concluded that split application of 160 kg ha^{-1} N at sowing and then at stem elongation in combination with soil and foliar application of Zn and Fe, could be a viable agricultural practice to enhance protein content and Zn and Fe concentrations in grain.

8 Transgenic Approaches

Low flour extraction rates (75–80%) result in the removal of significant proportions of the aleurone layer and a lesser amount of micronutrients. In many areas of the developing world, such as Central Asia and China, where low flour extraction rates are common, a transgenic approach may be needed to develop Fe-rich flour by introgressing an additional gene encoding the iron-binding protein ferritin into wheat. Since most Fe and Zn accumulates in the aleurone layer, if more of these micronutrients are deposited in the starchy endosperm, they are less likely to be lost from low-extraction-rate flour. To put this in context though, a transgenic approach for developing wheat with high provitamin A carotenoids is not envisioned in the short term because the yellow-to-orange colour of the resulting flour would have significant acceptability problems in most regions where wheat is consumed and because most provitamin A carotenoids would probably be oxidized and lost during processing to produce flour. Further oxidizing events during cooking, boiling, or baking may result in foods with lower than desirable levels of provitamin A.

9 Fe/Zn Uptake, Transportation and Remobilization

Much work has been carried out to understand the distinct routes Fe and Zn take to reach the grain in diploid crop species such as rice, maize, and barley. Conservation of these pathways between species allows predictions to be made about Fe and Zn transport in wheat where less is known. Recent reviews have comprehensively covered the pathways in model crops (Curie et al. 2009; Waters and Sankaran 2011; Borg et al. 2012; Lee et al. 2012; Schroeder et al. 2013) so here an outline of the putative pathways in wheat is given with discussion of the key steps to target for crop improvement as proposed by Borrill et al. 2014 (Fig. 5).

The uptake of Fe and Zn from the soil occurs via two processes in plants: direct uptake of Fe^{2+} and Zn^{2+} by ZRT and IRT-like proteins (ZIPs), or via secretion of

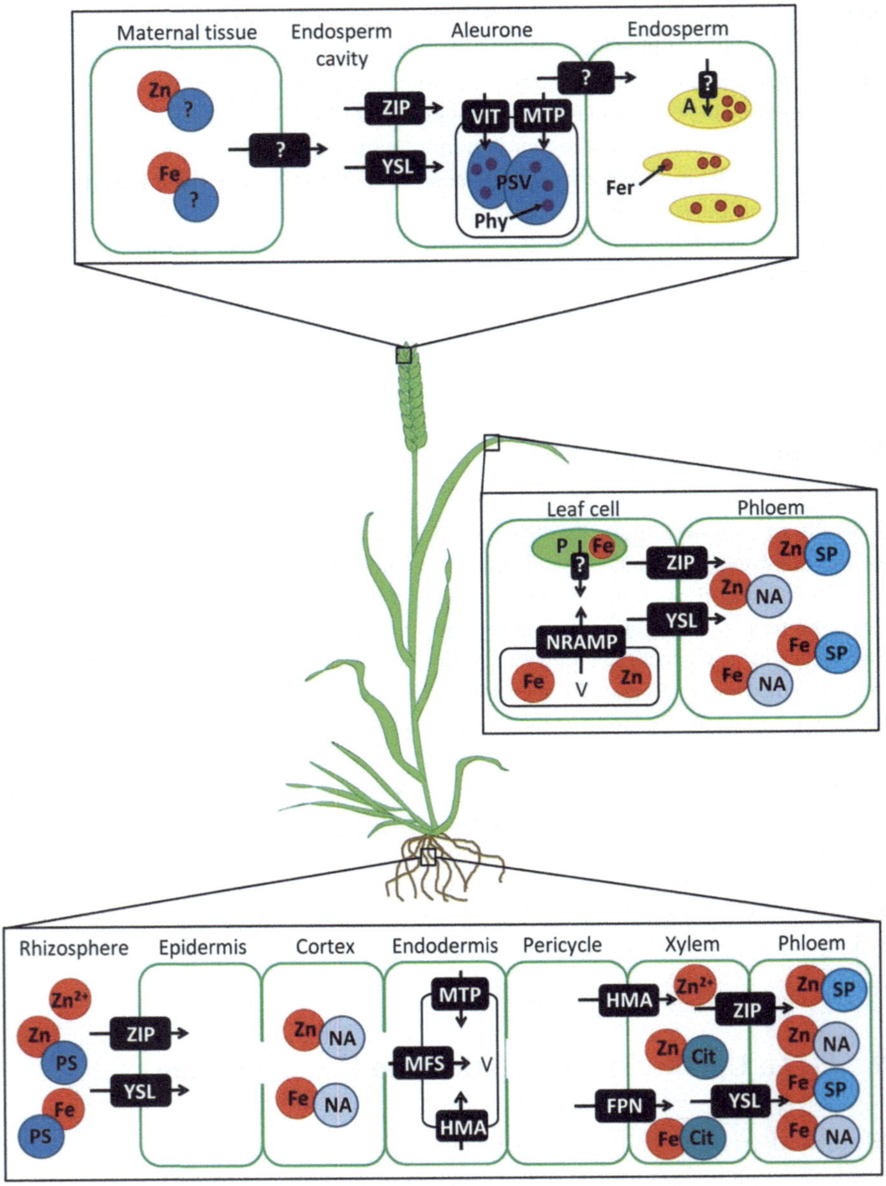

Fig. 5 Simplified proposed pathway for Fe and Zn uptake and translocation to the grain in wheat. Putative classes of transport proteins are shown as white text in black boxes and are based on evidence from other species, while question marks indicate unidentified transporters. Uptake from soil (lower panel): Fe and Zn bound to phytosiderophore (PS) and free Zn^{2+} are absorbed from the soil by root epidermal cells. Fe and Zn move via the apoplast and symplast to the pericycle, but may be sequestered into vacuoles *en route*. Fe and Zn are loaded into the xylem and transferred into the phloem in the root, basal shoot or leaf tissues (not shown). Translocation from leaves (middle panel): Fe and Zn are remobilized from leaf cell plastids (P) and vacuoles (V) and loaded into the phloem for transport to the ear. Loading into ear (upper panel): Fe and Zn are exported

phytosiderophores (PSs) which chelate Fe cations and are then taken up by yellow stripe like (YSL) transporters. The chelation route is generally used for Fe uptake in monocots such as wheat. PSs are exuded into the root medium to increase mobilization of Zn and Fe when the latter are deficient in the soil. The type of PSs wheat roots predominantly exude, like DMA, is the same under Zn and Fe deficiency (Rengel and Römheld 2000). The relationship between Fe transport to shoots and differential exudation of PSs by wheat genotypes has been proposed to be a physiological mechanism behind differential genotypic tolerance to Zn deficiency. Greater tolerance to Zn deficiency among wheat genotypes is associated with the increased exudation of PSs, increased Fe uptake rate, and less transport of Fe to shoots (Rengel and Römheld 2000).

In many steps of Fe and Zn transport the same families of proteins are involved, but the two metals are treated separately by plants often with the involvement of different members of multigene families. Metal chelators such as nicotianamine (NA) are important for radial movement of Fe and Zn through the root (Deinlein et al. 2012) and the transport of Zn into the vacuole affects overall Zn transport through the roots into the shoot (Haydon et al. 2012). Fe and Zn are loaded into the xylem where Zn can move as a cation or in a complex with organic acids such as citrate (Lu et al. 2013), and Fe is chelated by citrate. Transfer from xylem to phloem can occur in the root or basal part of the shoot or during remobilization from the leaves during grain filling and is facilitated by ZIP and YSL family proteins. In wheat all nutrients enter the grain from the phloem because the xylem is discontinuous. In the phloem Fe and Zn are transported as complexes with NA or small proteins. Certain transporters carrying the cations from the maternal tissue into the endosperm cavity and into the aleurone and embryo have been proposed, for example, members of the ZIP, YSL, and metal tolerance protein (MTP) families. In addition, vacuolar transporters are also associated with enhanced Fe and Zn accumulation in the endosperm. Connorton et al. (2017) identified two vacuolar iron transporter paralogs, VIT1 and VIT2, in wheat. Greater than twofold increases in iron were found in white flour fractions when *TaVIT2* was overexpressed under the control of an endosperm-specific promoter. In wheat grain most Fe and Zn is located in the aleurone layer which is lost during milling. This problem is further compounded by the fact that Fe in these tissues is deposited mainly in protein storage vacuoles where it is bound to phytate, which makes it poorly bioavailable to humans (Borg et al. 2009). Ferritin, which forms large Fe-rich nanoparticles, is generally regarded as a more bioavailable storage form and is present in endosperm amyloplasts (Balmer et al. 2006). Thus it is important to not only consider the total content of Fe and Zn in grain, but also the tissue localization and speciation (chelates, protein particles or other) which affect their bioavailability.

Fig.5 (continued) from the maternal tissue into the endosperm cavity. After uptake into the aleurone layer most Fe and Zn are sequestered in protein storage vacuoles (PSVs) bound to phytate (Phy). A small proportion of Fe and Zn may enter the endosperm and be stored bound to ferritin (Fer) in amyloplasts (A). Abbreviations: Cit, citrate; FPN, ferroportin; HMA, heavy metal ATPase; MFS, major facilitator superfamily transporter; MTP, metal tolerance protein; NA, nicotianamine; NRAMP, natural resistance-associated macrophage protein; SP, small proteins; VIT, vacuolar iron transporter; YSL, yellow stripe like transporter; ZIP, ZRT and IRT like protein. From Borrill et al. (2014)

10 Conclusions

Development of micronutrient-rich wheat grains using biofortification rationale and techniques can substantially reduce micronutrient deficiencies in the human diet. Large variations in Fe and Zn content in wheat germplasm have now been described which has identified an important genetic resource for improving micronutrient content in high-yielding backgrounds. Both genetic and agronomic improvement in wheat Fe and Zn content can lead to improved health of human beings. High phytase and low phytic acid wheat genotypes have been developed which can enhance the bioavailability of micronutrients to human beings and some farm animals. Our understanding of the molecular basis of micronutrient absorption and translocation to grains has increased and this will help in improving the efficiency of Fe and Zn absorption and accumulation in grains.

References

Aciksoz BS, Yazicici A, Ozturk L, Cakmak I (2011) Biofortification of wheat with iron through soil and foliar application of nitrogen and iron fertilizers. *Plant and Soil* 349: 215–225.

Balmer Y, Vensel WH, Dupont FM, Buchanan BB, Hurkman WJ (2006) Proteome of amyloplasts isolated from developing wheat endosperm presents evidence of broad metabolic capability. Journal of Experimental Botany 57: 1591–1602.

Baur X, Melching-Kollmuss S, Koops F, Strasburger K, Zober A (2002) IgE-mediated allergy to phytase – A new animal feed additive. Allergy 57: 943–945.

Betschart AA (1988) Nutritional quality of wheat and wheat foods. In: Wheat Chemistry and Technology. Ed. Y. Pomeranz. St. Paul, Minnesota, USA. pp. 91–132.

Borg S, Brinch-Pedersen H, Tauris B, Holm P (2009) Iron transport, deposition and bioavailability in the wheat and barley grain. Plant and Soil 325: 15–24.

Borg S, Brinch-Pedersen H, Tauris B, Madsen LH, Darbani B, Noeparvar S, et al. (2012) Wheat ferritins: improving the iron content of the wheat grain. Journal of Cereal Science 56: 204–213.

Borrill P, Connorton JM, Balk J, Miller A, Sanders D, Uauy C (2014) Biofortification of wheat grain with iron and zinc: integrating novel genomic resources and knowledge from model crops. Frontiers in Plant Science doi: https://doi.org/10.3389/fpls.2014.00053.

Bouis HE, Hotz C, McClafferty B, Meenakshi JV, Pfeiffer WH (2011) Biofortification: A new tool to reduce micronutrient malnutrition. Food and Nutrition Bulletin 32: 31S–40S.

Brinch-Pedersen H, Hatzack F, Stoger E, Arcalis E, Pontopidan K., Holm PB (2006) Heat-stable phytases in transgenic wheat (*Triticum aestivum* L.): Deposition pattern, thermostability and phytate hydrolysis. Journal of Agricultural and Food Chemistry 54 (13): 4624–4632.

Cakmak I (2008) Enrichment of cereal grains with zinc: agronomic or genetic biofortification?. Plant and Soil 302: 1–17.

Cakmak I (2009) Enrichment of fertilizers with zinc: an excellent investment for humanity and crop production in India. Journal of Trace Elements in Medicine and Biology 23: 281–289.

Cakmak I, Pfeiffer, WH, Mcclafferty B (2010) Biofortification of durum wheat with zinc andiron. Cereal Chemistry 87: 10–20.

Ciccolini V, Pellegrino E, Coccina A, Fiaschi AI, Cerretani D, Sgherri C, Quartacci MF, Ercoli L (2017) Biofortification with iron and zinc improves nutritional and nutraceutical properties of common wheat flour and bread. *Journal of Agricultural and Food Chemistry* 65 (27): 5443–5452.

Connorton JM, Jones ER, Rodríguez-Ramiro I, Fairweather-Tait S, Uauy C, Balka J (2017) Wheat vacuolar iron transporter TaVIT2 transports Fe and Mn and is effective for biofortification. Plant Physiology 174: 2434–2444.

Curie C, Cassin G, Couch D, Divol F, Higuchi K, Jean M, et al. (2009) Metal movement within the plant: contribution of nicotianamine and yellow stripe 1-like transporters. Annals of Botany 103: 1–11.

Deinlein U, Weber M, Schmidt H, Rensch S, Trampczynska A, Hansen TH, et al. (2012) Elevated nicotianamine levels in *Arabidopsis halleri* roots play a key role in zinc hyper accumulation. Plant Cell 24: 708–723.

Distelfeld A, Cakmak I, Peleg Z, Ozturk L, Yazici AM, Budak H, et al. (2007) Multiple QTL-effects of wheat *Gpc-B1* locus on grain protein and micronutrient concentrations. Physiologia Plantarum 129: 635–643.

Greiner R, Konietzny U (2006) Phytase for food application. Food Technology and Biotechnology 44 (2): 125–140.

Grusak MA, Penna DD (1999) Improving the nutrient composition of plants to enhance human nutrition and health. Annual Review Plant Physiology & Plant Molecular Biology 50: 133–161.

Habib M (2012) Effect of supplementary nutrition with Fe, Zn chelates and urea on wheat quality and quantity. African Journal of Biotechnology 11: 2661–2665.

Haydon MJ, Kawachi M, Wirtz M, Hillmer S, Hell R, Kramer U (2012) Vacuolar nicotianamine has critical and distinct roles under iron deficiency and for zinc sequestration in *Arabidopsis*. Plant Cell 24: 724–737.

Kutman UB, Yildiz B, Cakmak I (2010) Improved nitrogen status enhances zinc and iron concentrations both in the whole grain and the endosperm fraction of wheat. *Journal of Cereal Science* 53: 118–125.

Lee S, Jeon JS, An G (2012) Iron homeostasis and fortification in rice. Journal of Plant Biology 55: 261–267.

Liu ZH, Wang HY, Wang XE, Zhang GP, Chen PD, Liu DJ (2007) Phytase activity, phytate, iron, and zinc contents in wheat pearling fractions and their variation across production locations. Journal of Cereal Science 45: 319–326.

Lopez HW, Krespine V, Lemaire A, Coudray C, Coudray CF, Messager A, Demigne C, Remesy C (2003) Wheat variety has a major influence on mineral bioavailability; studies in rats. Journal of Cereal Science 37: 257–266.

Lu L, Tian S, Zhang J, Yang X, Labavitch JM, Webb SM, et al. (2013) Effi-cient xylem transport and phloem remobilization of Znin the hyper accumulator plant species *Sedumal fredii*. New Phytologist 198: 721–731.

Monasterio I, Graham R (2000) Breeding for trace minerals in wheat. Food Nutrition Bulletin 21(4): 392–396.

Mondal S, Rutkoski JE, Velu G, Singh PK, Crespo-Herrera LA, Guzmán C, et al. (2016). Harnessing diversity in wheat to enhance grain yield, climate resilience, disease and insect pest resistance and nutrition through conventional and modern breeding approaches. Frontiers in Plant Science 7: 991. doi: https://doi.org/10.3389/fpls.2016.00991.

Mutangadura GB (2004) World Health Report 2002: Reducing risks, promoting healthy life. Geneva: World Health Organization.

Nowack B, Schwyzer I, Schulin R (2008) Uptake of Zn and Fe by wheat (*Triticum aestivum* var. Greina) and transfer to the grain in the presence of chelating agents (Ethylene diamine disuccinic acid and Ethylene diamine tetra acetic acid). *Journal of Agricultural and Food Chemistry* 56: 4643–4649.

Okot-Kotber, M, Yong KJ, Bagorogoza K, Liavoga A (2003) Phytase activity in extracts of flour and bran from wheat cultivars: enhanced extractability with ß- glucanase and endo-xylanase. Journal of Cereal Science 38: 307–315.

Oloffs K, Cossa J, Jeroch H (2000) The importance of native phytase activity in wheat on the phosphorus utilization in broilers and laying hens. Archiv fur Geflugelkunde 64(4): 157–161.

Ortiz-Monasterio I, Palacios-Rojas N, Meng E, Pixley K, Trethowan R, Pena RJ (2007) Enhancing the mineral and vitamin content of wheat and maize through plant breeding. Journal of Cereal Science 46: 293–307.

Oury FX, Leenhardt F, Rémésy C, Chanliaud E, Duperrier B, Balfourier F, Charmet G (2006) Genetic variability and stability of grain magnesium, zinc and iron concentrations in bread wheat. European Journal of Agronomy 25: 177–185.

Paltridge NG, Milham PJ, Ortiz-Monasterio JI, Velu G, Yasmin Z, Palmer LJ, Guild GE, Stangoulis JCR (2012) Energy-dispersive X-ray fluorescence spectrometry as a tool for zinc, iron and selenium analysis in whole grain wheat. Plant and Soil 361: 251–260.

Persson DP, deBang TC, Pedas PR, Kutman UB, Cakmak I, Andersen B (2016) Molecular speciation and tissue compartmentation of zinc in durum wheat grains with contrasting nutritional status. *New Phytologist* 211: 1255–1263.

Ram S, Verma A, Sharma S (2010) Large variability exists in phytase levels among Indian wheat varieties and synthetic hexaploids. Journal of Cereal Science 52: 486–490.

Rengel Z, Römheld V (2000) Root exudation and Fe uptake and transport in wheat genotypes differing in tolerance to Zn deficiency. Plant and Soil 222: 25–34.

Sazawal S, Dhingra U, Dhingra P, Dutta A, Deb S, Kumar J, Devi P, Prakash A (2018) Efficacy of high zinc biofortified wheat in improvement of micronutrient status, and prevention of morbidity among preschool children and women - a double masked, randomized, controlled trial. Nutrition Journal 17: 86.

Schroeder JI, Delhaize E, Frommer WB, Guerinot ML, Harrison MJ, Herrera-Estrella L, et al. (2013) Using membrane transporters to improve crops for sustainable food production. Nature 497: 60–66.

Singh BR, Timsina YN, Lind OC, Cagno S, Janssens K (2018) Zinc and iron concentration as affected by nitrogen fertilization and their localization in wheat grain. Frontiers in Plant Science 9: 307. doi: https://doi.org/10.3389/fpls.2018.00307.

Steiner T, Mosenthin R, Zimmermann B, Greiner R, Roth S (2009) Distribution of phytase activity, total phosphorus and phytate phosphorus in legume seeds, cereals and cereal by-products as influenced by harvest year and cultivar. *Animal Feed Science and Technology* 133 (3): 320–334.

Timsina YN (2014) Effect of nitrogen fertilization on zinc and iron uptake and yield components of wheat. Master thesis, Norwegian University of Life Sciences, As, Akershus, 80.

Uauy C, Distelfeld A, Fahima T, Blechl A, Dubcovsky J (2006) A NAC gene regulating senescence improves grain protein, zinc, and iron content in wheat. Science 314: 1298–1301.

Velu G, Ortiz-Monasterio I, Cakmak I, Hao Y, Singh RP (2014) Biofortification strategies to increase grain zinc and iron concentrations in wheat. Journal of Cereal Science 59: 365–372.

Velu G, Singh RP, Huerta-Espino J, Peña-Bautista RJ, Arun B, Mahendru-Singh A, Yaqub-Mujahid M, Sohu VS, Mavi GS, Crossa J, Alvarado G, Joshi AK, Pfeiffer WH (2012) Performance of biofortified spring wheat genotypes in target environments for grain zinc and iron concentrations. Field Crops Research 137: 261–267.

Velu G, Ortiz-Monasterio I, Singh RP, Payne T (2011) Variation for grain micronutrients concentration in wheat core-collection accessions of diverse origin. *Asian Journal of Crop Science* 3: 43–48.

Velu, G, Singh RP, Crespo-Herrera L, Juliana P, Dreisigacker S, Valluru R (2018) Genetic dissection of grain zinc concentration in spring wheat for mainstreaming biofortification in CIMMYT wheat breeding. Scientific Reports 8: 13526. (https://www.nature.com/articles/s41598-018-31951-z).

Velu G, Tutus Y, Gomez-Becerra HF, Hao Y, Demir L, Kara R, et al. (2016) QTL mapping for grain zinc and iron concentrations and zinc efficiency in a tetraploid and hexaploid wheat mapping populations. Plant and Soil 411: 81–99. doi:https://doi.org/10.1007/s11104-016-3025-8.

Vikram P, Franco J, Burgueño-Ferreira J, Li H, Sehgal D, Saint-Pierre C, et al. (2016) Unlocking the genetic diversity of Creole wheats. Scientific Reports 6: 23092.

Waters BM, Sankaran RP (2011) Moving micronutrients from the soil to the seeds: Genes and physiological processes from a biofortification perspective. Plant Science 180: 562–574.

Welch RM, House WA, ortiz-monasterio I, Cheng Z (2005) Potential for improving bioavailable zinc in wheat grain (*Triticum* species) through plant breeding. Journal of Agricultural and Food Chemistry 53: 2176–2180.

WHO (2017) The world health report. World Health Organization, Geneva, Switzerland (accessed online June 1, 2017).

Phenolic Compounds in Wheat Kernels: Genetic and Genomic Studies of Biosynthesis and Regulations

Domenica Nigro, Heinrich Grausgruber, Carlos Guzmán, and Barbara Laddomada

Abstract Whole wheat grains are an important source of bioactive components, particularly of phenolic acids and flavonoids. Due to the health-promoting effects of these phenolics, nowadays, the increase of their content in mature kernels is of great interest and a potential target for wheat breeding programs. The biogenesis of phenolics occurs through the general phenylpropanoid pathway, which is ubiquitous in plant cell walls and leads to the synthesis of secondary metabolites that are involved in plant defence and structural support. This chapter reviews the current knowledge in phenylpropanoid chemistry, and the genetic and molecular basis for the biosynthesis of phenolic acids and anthocyanins in wheat grains. Also, advances in assessing genetic variation in the content and composition of these components in wheat germplasm are reviewed, including the effects of different environmental conditions on their accumulation in mature kernels. The recent, ongoing genomic studies are reviewed providing updates on quantitative trait loci and genes involved in the synthesis and accumulation of phenolics in wheat kernels. Finally, the promise and limitations of breeding programs to potentially develop wheat cultivars rich in phenolic components are discussed.

D. Nigro
Department of Soil, Plant & Food Sciences, Plant Breeding Section, University of Bari, Bari, Italy

H. Grausgruber
Department of Crop Sciences, University of Natural Resources and Life Sciences, Tulln an der Donau, Austria

C. Guzmán
Departamento de Genética, Escuela Técnica Superior de Ingeniería Agronómica y de Montes, Edificio Gregor Mendel, Campus de Rabanales, Universidad de Córdoba, Córdoba, Spain

B. Laddomada (✉)
Institute of Sciences of Food Production, Lecce, Italy
e-mail: barbara.laddomada@ispa.cnr.it

G. Igrejas et al. (eds.), *Wheat Quality For Improving Processing And Human Health*, https://doi.org/10.1007/978-3-030-34163-3_10

1 Chemistry and Biosynthesis of Phenolic Acids and Anthocyanins in Wheat

Phenolic compounds are characterized as having at least one benzene ring with one or more substituted hydroxyl groups. Ubiquitous secondary metabolites in plants, phenolics are classified according to their molecular mass as melanins, suberin, tannins, and lignins. Most wheat phenolics reported are phenolic acids and flavonoids (Liu et al. 2010). Based on their chemistry, phenolic acids are classified as hydroxy derivatives of either cinnamic or benzoic acid (BA). Flavonoids are derivatives of benzo-γ-pyrone and are classified into anthocyanidins, flavonols, flavans, flavanones, flavones, isoflavones and hydrolysable tannins based on their heterocycle oxidation states and aromatic ring positions.

1.1 Chemical Structure

The hydroxy derivatives of cinnamic acid are the most common phenolic acids in wheat kernels and include some major components such as the ferulic, sinapic and *p*-coumaric acids (Li et al. 2008; Laddomada et al. 2015a). Ferulic acid alone accounts for about 90% of the total phenolic acids in mature wheat kernels (Lempereur et al. 1997). The hydroxy derivatives of BA include some minor components, such as the frequently found vanillic, syringic and *p*-hydroxybenzoic acids (Li et al. 2008; Laddomada et al. 2015a).

Phenolic acids are also classified as C_6-C_3, C_6-C_2 or C_6-C_1 depending on the length of the carbon side chains. They share the C_6 moiety, the benzene ring. The C_6-C_3 compounds include cinnamic acid and its derivatives, C_6-C_2 compounds are phenylacetic acid and its derivatives, while C_6-C_1 compounds include BA and its derivatives. Mechanisms have been proposed to explain the antioxidant activity of these compounds, and most likely depend on the number of hydroxyl groups on the benzene ring and ortho-substitution with the electron donor methoxy group (Kikuzaki et al. 2002).

Flavonoids generally have a 15-carbon skeleton, consisting of two phenyl rings connected by a three-carbon bridge that usually forms a third ring (C_6-C_3-C_6). By far the most common coloured flavonoids are the anthocyanins, which have a core pigment structure, the aglycone anthocyanidin, with a sugar moiety bound at different hydroxylated positions. Around 90% of all anthocyanins are based on cyanidin, delphinidin, pelargonidin, and their methylated derivatives (Schwinn and Davies 2004).

1.2 Biosynthetic Pathway

Phenolic acids are synthesized along the phenylpropanoid pathway, a ubiquitous pathway in plant cell walls that is responsible for the synthesis of a wide range of other secondary metabolites, such as flavonoids, coumarins, lignin and lignans, all involved in plant defence, structural support or survival of higher plants (Vogt 2010). Not surprisingly, light, temperature, hormones, biotic and abiotic stresses and mechanical damage may influence the biosynthesis of phenolic acids (Qin et al. 2014).

1.2.1 Biogenesis of Hydroxy Derivatives of Cinnamic Acid

Hydroxycinnamic acids are synthesized during the very initial stages of the phenylpropanoid pathway (Fig. 1). In the first step, phenylalanine is deaminated to trans-cinnamic acid by L-phenylalanine ammonia-lyase (PAL). In turn, trans-cinnamic acid is converted to *p*-coumaric acid by trans-cinnamate 4-monooxygenase, also known as cinnamic acid 4-hydroxylase (C4H), then *p*-coumaric acid is converted to caffeic acid by *p*-coumaric acid hydroxylase (C3H). Methylation of caffeic acid, catalysed by caffeic acid 3-*O*-methyltransferase (COMT), leads to the biosynthesis of ferulic acid, which is converted to sinapic acid by the subsequent actions of the enzymes ferulate 5-hydroxylase (F5H) and COMT.

1.2.2 Biogenesis of Hydroxy Derivatives of Benzoic Acid

The biogenesis of the hydroxy derivatives of BA has been recently reviewed (Widhalm and Dudareva 2015). Hydroxybenzoic acids are mainly synthesized from *p*-coumaric acid and its hydroxy- and methoxy derivatives by removal of a 2-carbon fragment from the C_6-C_3 precursor (Geissman and Hinreiner 1952), as shown in Fig. 2 for the case of *p*-hydroxybenzoic acid. In *Oryza sativa* L. and *Hordeum vulgare* L., it was found that *p*-coumaric acid is converted to *p*-hydroxybenzoic acid, while syringic acid is formed from sinapic acid. Tracer experiments showed that U-^{14}C-Phe and CA-β-^{14}C labels are incorporated into BAs (El-Basyouni et al. 1964). When wheat leaves were fed with radiolabeled *p*-coumaric acid, the $C_{1'}$ and $C_{2'}$ carbons of the precursor side chain were released as acetate/acetyl-CoA (Vollmer et al. 1965). Nevertheless, different routes based on reactions such as hydroxylation, *O*-methylation and demethoxylation are also possible for the biogenesis of these types of phenolic acids (El-Basyouni et al. 1964).

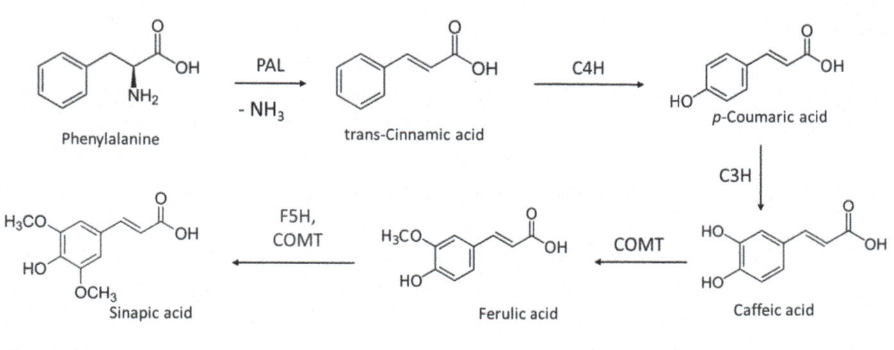

Fig. 1 Phenylpropanoid pathway leading to the biosynthesis of hydroxycinnamic acids

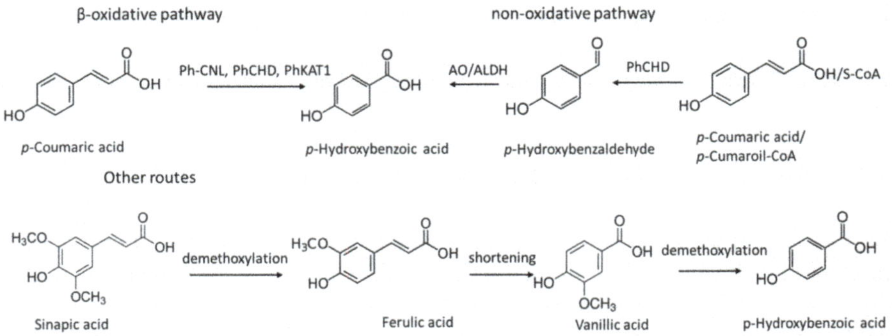

Fig. 2 Metabolic pathways leading to the biosynthesis of hydroxybenzoic acids

Side chain shortening can occur via three pathways: the CoA-dependent β-oxidative route, the CoA-dependent non-oxidative route, and the CoA-independent non-oxidative route. In *Petunia*, the β-oxidative route begins with the activation of *p*-coumaric acid to the corresponding CoA-ester by cinnamoyl-CoA ligase (Ph-CNL). Next, the C3' carbon of the cinnamoyl-CoA (CA-CoA) propyl side chain is hydrated to make 3-hydroxy-3-phenylpropanoyl-CoA, followed by oxidation of the newly formed hydroxyl group to produce 3-oxo-3-phenylpropanoyl-CoA, both reactions being are catalyzed by the bifunctional CA-CoA hydratase/dehydrogenase (PhCHD). The final step of the β-oxidative route is cleavage of the β-keto thioester intermediate to produce BA-CoA, with the release of acetyl-CoA. This reaction is catalyzed by 3-ketoacyl thiolase 1 (PhKAT1).

The characteristic feature of the non-oxidative pathways is that benzaldehydes (BDs) are key metabolic intermediates in the formation of BAs. The first step in the CoA-independent non-oxidative pathway is the hydration of free *p*-coumaric acid to 3-hydroxy-3-phenylpropanoic acid intermediates, which are then subjected to side chain degradation via a reverse aldol reaction that releases acetate to produce BDs. Unlike the CoA-independent non-oxidative pathway, the CoA-dependent route starts from CoA esters that are converted to 3-hydroxy-3-phenylpropanoyl-CoA intermediates, which undergo side chain shortening via a reverse aldol reaction with the release of acetyl-CoA to form BDs. The final common step of the non-oxidative pathways is the oxidation of BDs to BAs by aldehyde oxidases and/or aldehyde dehydrogenases.

Another route leading to hydroxybenzoic acids was suggested to be based on hydroxylation and *O*-methylation of C_6-C_1 precursors (El-Basyouni et al. 1964). For instance, carboxyl-labelled BA was converted to ortho and *p*-hydroxybenzoic acid in leaf disks of *Gaultheria* and *Primula* species. A further route based on demethoxylation was found in *Triticum* species (El-Basyouni et al. 1964), where it was found that ^{14}C-labelled phenylalanine and ^{14}C-labelled phenolic acids administered to shoots were converted to *p*-hydroxybenzoic, vanillic and syringic acids. In particular, sinapic acid was found to be demethoxylated to ferulic acid, which was converted by side chain shortening to vanillic acid, which was in turn demethoxylated to *p*-hydroxybenzoic acid (Fig. 2).

1.2.3 Biogenesis of Anthocyanins

The flavonoid pathway is also part of the larger phenylpropanoid pathway. The precursors *p*-coumaroyl-CoA and malonyl-CoA are derived from phenylalanine and citrate, respectively, and are synthesized to naringenin chalcone by chalcone synthase (CHS), and further catalysed by chalcone isomerase (CHI) to colorless naringenin. Naringenin is converted to dihydrokaempferol by flavanone 3-hydroxylase (F3H). Further enzymatic conversions transform the colorless dihydroflavanols to anthocyanins. First, dihydroflavanols are reduced to leucoanthocyanidins, followed by oxidation, dehydration, and glycosylation of leucocyanidin, leucodelphinidin, and leucopelargonidin to cyanidin-3-glucoside, delphinidin-3-glucoside and pelargonidin-3-glucoside, respectively. The great diversity of anthocyanin pigments is due to further secondary modification of anthocyanidin 3-glucosides by glycosylation, methylation, and acylation (Holton and Cornish 1995; Schwinn and Davies 2004).

2 Exploring the Genes Involved in the Biosynthesis of Phenolic Compounds

2.1 Genes Encoding Phenolic Pathway Enzymes

Enzymes involved in the biosynthesis of other phenylpropanoids and lignin have been widely characterized in a number of plants, but several steps remain unknown, so our understanding of the pathway is still under revision (Vanholme et al. 2013; Wang et al. 2015). Nevertheless some metabolic engineering of the pathway has been possible. As mentioned previously, the biosynthetic steps of hydroxycinnamic derived phenolic acid have been well described, but hydroxybenzoic derived phenolic acid biosynthesis is still to be fully defined.

In most plants, the general phenylpropanoid pathway is a collection of the first two or three enzymatic steps, generating intermediates subsequently channelled into specific branch pathways to produce flavonoids, stilbenes, monolignols, phenolic acids, and coumarins. Specifically, reactions carried out by phenylalanine ammonia lyase (PAL), cinnamate 4-hydroxylase (C4H) and 4-coumaroyl CoA ligase (4CL) catalyse reaction characteristic of this pathway (Vogt 2010). Then 3-*O*-methyltransferase (COMT) and ferulate 5-hydroxylase (F5H) complete the suite of enzymes in the hydroxycinnamic pathway.

In a recent study, the *Arabidopsis thaliana* phenylpropanoid pathway enzyme sequences were used to identify the main enzymes involved in the biosynthesis of *p*-coumaric, ferulic and sinapic acids in wheat (Nigro et al. 2017). The *A. thaliana* gene sequences were used as a query to retrieve orthologous gene sequences from the monocot model *Brachypodium distachyon*, and from *Oryza sativa*, *Hordeum vulgare*, *Zea mays*, *Triticum urartu*, *Aegilops tauschii* and *Triticum aestivum* from the Ensembl Plant database (http://plants.ensembl.org/). For each species, a single gene was found for C4H, C3H and F5H, whereas two copies were identified for the COMT and PAL gene families (Table 1). The putative genes were reported as COMT1 and COMT2 and PAL1 and PAL2 based on published data (Ma et al. 2016) and known enzymatic pathways (Vogt 2010).

In a similar approach, five putative genes associated with hydroxycinnamic derived acids were identified in the Arabidopsis database. Based on information reported by Widhalm and Dudareva (2015), gene sequences coding for cinnamate-CoA ligase (CNL), *4*-coumarate-CoA ligase (4CL), cinnamoyl-CoA hydratase-dehydrogenase (CHD), 3-ketoacyl-CoA thiolase (KAT) and 1,4-dihydroxy-2-naphthoyl-CoA thioesterase (DHNAT) were retrieved from the Arabidopsis genome and used as queries to identify orthologous sequences in the same range of grass species (Table 1).

In order to define the genetic distance between the genes within the two biosynthetic pathways, the retrieved protein coding sequences from different species were aligned by using the ClustalW method with Mega7 software. Phylogenetic analysis was carried out using the neighbor-joining method and a 1000-replication bootstrap test for significance. The phylogenetic tree was generated with Mega7 (http://www.ebi.ac.uk/Tools/phylogeny/) and modified with FigTree software (http://tree.bio.ed.ac.uk/software/figtree/). Figure 3a shows the results for enzymes involved in the biosynthesis of hydroxy derivatives of cinnamic acid. All the orthologues clustered in the same clades of the phylogenetic tree and shared common conserved motifs in the cDNA sequences. The phylogenetic analysis revealed very high similarity among the C4H, C3H and F5H orthologue cDNAs, which were closer than those of the COMT and PAL gene families. The tree has a common branch that underwent functional diversification, first with two branches differentiating F5H and, more recently, a second diversification between C4H and C3H.

The evolutionary relationship among enzymes involved in hydroxybenzoic acid biosynthesis is represented in Fig. 3b. The genes involved in this pathway are more diverse among themselves, than are the hydroxycinnamic acid biosynthetic genes. Even though the 4CL enzyme seems to have a common branch with DHNAT, they probably underwent subsequent diversification, so the 4CL is really phylogenetically closer to the CNL enzyme, which also has ligase activity.

Also, KAT and CHD originated from a common branch, which functionally diversified into two different groups, having thiolase and dehydrogenase activity, respectively, with both Arabidopsis sequences being distantly related to their grass counterparts.

Anthocyanin pigmentation of the wheat pericarp is encoded by the *Pp1* and *Pp3* loci on chromosome 2AL and the short arms of the homeologous group 7 chromosomes, respectively (Khlestkina et al. 2010). Complementary dominant genes were described for alien gene introgressions (Tereschenko et al. 2012). *Ba1* and *Ba2* loci on chromosomes 4B and 4A, respectively, are responsible for anthocyanin accumulation in the aleurone layer (the aleurone is blue) and were introgressed as chromosome translocations, additions or substitutions from *Elytrigia pontica* (syn. *Lophopyrum ponticum*), *Triticum boeoticum* or *Triticum monococcum* (Zeller et al. 1991, Metzger and Sebesta 2004, Qualset et al. 2005, Buresova et al. 2015).

Table. 1 Ensembl entries of phenolic acid metabolism genes retrieved from *Arabidopsis thaliana, Brachypodium distachyon, Oryza sativa, Hordeum vulgare, Zea mays, Triticum aestivum, Aegilops tauschii* and Triticum urartu. (EnsemblPlants website: http://plants.ensembl.org/)

Gene	Enzyme	*A. thaliana*	*B. distachyon*	*O. sativa*	*H. vulgare*	*Z. mays*	*T. aestivum*	*Ae. Tauschii*	*T. urartu*
Hydroxycinnic acid biosynthesis									
PAL1	Phenylalanine ammonia-lyase	AT2G37040	BRADI_5g 15830v3	Os02g0 626400	HORVU6Hr1 G058820	Zm00001 d003015	Traes_2BL_ C051606EA,1	F775_06188	TRIUR3_ 22522
PAL2		AT3G53260	BRADI_3g 49260v3	OS04G 0518400	HORVU0Hr1 G016330	GRMZM2G 029048_T01	Traes_1BS_ BD86C90A7.1	F775_06189	TRIUR3_ 02596
C4H	Trans-cinnamate 4-monooxygenase	AT2G30490	BRADI_2g 53470v3	OS05G 0320700	HORVU3Hr1 G080830	Zm00001 d009858	TRAES3BF00 6600010CFD	F775_29972	TRIUR3_ 18982-T1
C3H	p-coumarate 3-hydroxylase	AT2G40890	BRADI_2g 21300v3	OS05G 0494000	HORVU3Hr1 G078660	Zm00001 d038555	Traes_1AL_ A0B81FF76.1	F775_27986	TRIUR3_ 23576
COMT1	Caffeic acid 3-O-methyl transferase	AT5G54160	BRADI_2g 02390v3	OS08T0 157500-01	HORVU3Hr1 G116770	Zm00001 d049541	Traes_6BS_ 881DA479E	F775_31276	TRIUR3_ 02449
COMT2		AT1G33030	–	OS08G 0157500	HORVU7Hr1 G082280	GRMZM5 G814904	TRAES3BF06 5400030CFD	F775_32449	TRIUR3_ 32612-T1
F5H	Ferulate-5-hydroxylase	AT4G36220	BRADI_3g 30590v3	OS10G0 512400	HORVU1Hr1 G047220	Zm00001 d013862	TRAES3BF057 900080CFD	F775_13391	TRIUR3_ 24298
Hydroxybenzoic acid biosynthesis									
CNL	Cinnamate-CoA ligase	AT1G65880	BRADI_4g 37570v3	BGIOSG A029291	HORVU3Hr1 G115740	Zm00001 d048446	TraesCS5A01 G356800	F775_18414	TRIUR3_ 25561
4CL	4-coumarate-CoA ligase	At1g51680	BRADI_3g 37300v3	BGIOSG A020757	HORVU6Hr1 G030390	Zm00001 d015459	TraesCS6A01 G151700	F775_32463	TRIUR3_ 22492
CHD	Cinnamoyl-CoA hydratase-dehydrogenase	AT4G29010	BRADI_3g 10180v3	BGIOSG A007953	HORVU3Hr1 G013880	Zm00001 d053308	TraesCS6A01 G125800	F775_26789	TRIUR3_ 15771
KAT	3-ketoacyl-CoA thiolase	AT5G48880	BRADI_3g 27960v3	BGIOSG A033076	HORVU1Hr1 G038330	Zm00001 d014093	TraesCS1A01 G143900	–	TRIUR3_ 28769
DHNAT	1,4-dihydroxy-2-naphthoyl-CoA thioesterase	AT5G48950	BRADI_1g 12080v3	BGIOSG A010016	HORVU3Hr1 G013400	Zm00001 d024770	TraesCS4A01 G282200	F775_15757	TRIUR3_ 27108

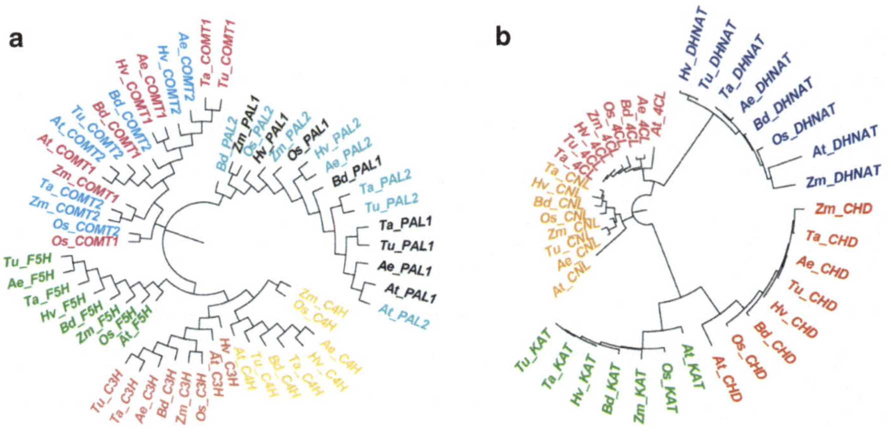

Fig. 3 Phylogenetic trees of the (**a**) hydroxycinnamic and (**b**) hydroxybenzoic acid derivative genes from *Arabidopsis thaliana* (At), *Brachypodium distachyon* (Bd), *Zea mays* (Zm), *Hordeum vulgare* (Hv), *Oryza sativa* (Os), *Aegilops tauchii* (Ae), *Triticum uraru* (Tu) and *Triticum aestivum* (Ta). Gene abbreviations: *PAL, phenylalanine ammonia-lyase*; *C4H, trans-cinnamate 4-monooxygenase*; *C3H, p-coumarate 3-hydroxylase*; *COMT, caffeic acid 3-O-methyltransferase*; *F5H, ferulate-5-hydroxylase*; *CNL, cinnamate-CoA ligase*; *4CL, 4-coumarate-CoA ligase*; *CHD, cinnamoyl-CoA hydratase-dehydrogenase*; *KAT, 3-ketoacyl-CoA thiolase*; and *DHNAT, 1,4-dihydroxy-2- naphthoyl-CoA thioesterase*

2.2 Genes Regulating Phenylpropanoid Biosynthesis

Besides identifying and characterizing the enzymatic genes involved in phenolic compound pathways, a number of studies have focused on how they are regulated. It has been shown that the transcriptional regulation of the structural genes is controlled by different transcription factors (TFs), such as v-myb myeloblastosis viral oncogene homologs (MYB), basic helix-loop-helix protein (bHLH), WD-repeat protein (WDR), NAC and WRKY, which can regulate the expression of enzymatic genes at the transcription level in several ways during plant development. In addition to canonical TFs, plant microRNAs (miRNAs), such as miR828 and miR858, and small interfering RNAs (siRNAs) have been shown to play critical regulatory roles in phenylpropanoid biosynthesis.

MYB proteins form one of the largest plant TF families involved in regulating the phenylpropanoid biosynthetic pathway, and thus the synthesis of phenylpropanoid-derived compounds (Liu et al. 2015). The N-terminal region of these TFs harbors the conserved MYB domain that is required for DNA binding, while the C-terminal modulator region is more variable and responsible for the regulatory activity. To date, four MYB classes have been identified based on the number and position of the MYB domain(s), 1R (R1/2, R3-MYB), 2R (R2R3-MYB), 3R (R1R2R3-MYB), and 4 (Du et al. 2012). The 2R class is the largest MYB class in plants (Jin and Martin 1999; Dubos et al. 2010) and the members regulate developmental processes, responses to biotic and abiotic stresses, and primary and secondary metabolism, including phenylpropanoid biosynthesis. MYB-regulated flavonoid and monolignol pathways are largely conserved in plants, although most studies have been carried out in model plants or dicots (Hichri et al. 2011; Craven-Bartle et al. 2013; Liu et al. 2015; Xu et al. 2015; Zhou et al. 2015, Mu et al. 2015). Knowledge of MYBs regulating the metabolism of other phenylpropanoids, such as phenolic acids, is limited, especially for monocots.

Phenolic acids have been extensively studied in *Salvia miltiorrhiza*, an important medicinal plant of great economic and medicinal value because of its bioactive components. MYB TFs, which may act as activators or repressors in the biosynthetic pathway of phenolic acids, have been well characterized in this species (Wang et al. 2013; Zhang et al. 2013; Li and Lu, 2014). Heterologous expression in *S. miltiorrhiza* of *A. thaliana* PAP1 increased the level of both rosmarinic and salvianolic acids (Zhang et al. 2010; Wang et al. 2013), and the constitutive expression of AtMYB12 increased chlorogenic acid content in different species (Luo et al. 2008; Rommens et al. 2008; Payyavula et al. 2013; Qiu et al. 2013).

While most MYBs function as activators in phenylpropanoid biosynthesis, some act as repressors, including subgroup 4 R2R3-MYBs and some R3 or R3-related MYBs. R2R3-MYB repressors include AtMYB3, AtMYB4, AtMYB7 and AtMYB32 (Jin et al. 2000; Preston et al. 2004; Dubos et al. 2010; Fornalé et al. 2014; Zhou et al. 2017). AtMYB3 binds to the promoter of AtC4H to repress its transcription (Zhou et al. 2017). AtMYB4 can repress AtC4H transcription to limit sinapate ester production in Arabidopsis, and AtMYB4 overexpression represses AtC4H, AtCHS and At4CL3 expression in transgenic plants (Jin et al. 2000).

Similarly, in *Salvia miltiorrhiza,* SmMYB39, a R2R3-MYB protein, represses phenolic acid biosynthesis by negatively regulating the expression of SmC4H and the tyrosine aminotransferase gene (SmTAT) (Zhang et al. 2013). SmMYB39 overexpression caused a decrease in phenolic acid content, while downregulating it caused a dramatic increase in 4-coumaric acid, rosmarinic acid, salvianolic acid B, and salvianolic acid. Fornalé et al. (2014) demonstrated that AtMYB7 is a TF that specifically represses flavonol biosynthesis. Disruption of AtMYB7 increases flavonol accumulation and expression of PAL, C4H, and 4CL genes. On the contrary, when AtMYB7 is overexpressed, flavonol content decreases concurrently with the downregulation of other flavonoid biosynthetic pathway genes. Another *A. thaliana* MYB protein, AtMYB32, represses lignin biosynthesis specifically in pollen by repressing AtCOMT expression (Preston et al. 2004).

AtMYB75 is an anthocyanin biosynthesis activator that negatively regulates monolignol biosynthesis. Expression of AtPAL1, AtC4H, At4CL1, AtC3H, AtCCoAOMT and AtF5H genes is higher in plants with mutant MYB75 (Bhargava et al. 2010). Zhang et al. (2010) also found that in *S. miltiorrhiza* several phenolic acid biosynthetic pathway genes, such as SmPAL2, SmC4H, Sm4CL2, hydroxyphenylpyruvate reductase gene and rosmarinic acid synthase-like gene, are strongly upregulated in AtMYB75 transgenic plants. The biosynthesis of anthocyanin and phenolic acids in Arabidopsis are also negatively regulated by AtMYBL2, an R3-MYB-related protein (Matsui et al. 2008; Dubos et al. 2008).

A few maize TFs that participate in the regulation of specific pathway genes have been characterized. One of the first to be isolated is ZmMYB-IF35 which promotes the accumulation of ferulic and chlorogenic acids when constitutively expressed in maize (Dias and Grotewold 2003). Other TFs were reported, such as the ZmMYB40/ZmMYB95 paralogs (Dias et al. 2003; Heine et al. 2007). By searching Ensembl Plants (http://plants.ensembl.org/index.html) and updated genome project websites we found that ZmMYB-IF35 and ZmMYB40 correspond to the same locus and are synonymous. Other reported maize TFs include ZmMYB31/ZmMYB42/ZmMYB11, which belong to subgroup 4 of the R2R3-MYB family and function as pathway repressors (Sonbol et al. 2009; Fornalé et al. 2010; Vélez-Bermúdez et al. 2015; Agarwal et al. 2016), and ZmMYB111 and ZmMYB148, which were recently proposed to control the expression of maize PAL genes (Zhang et al. 2016).

Recent studies by Agarwal et al. (2016) led to the identification of ZmMYB31 and ZmMYB42 syntelogs in rice and sorghum. Syntelogs of MYB31 and MYB42 bind to phenylpropanoid genes that function in the early, mid and late stages of the pathway in three monocot grasses, even though some genes have been duplicated and have diverged. Yang et al. (2017) identified 11 TFs recognizing 10 or more phenolic gene promoters, including two R2R3-MYBs. MYB65 and MYB19 are both suggested to be involved in PAL and other phenolic gene expression. By considering synteny, we aimed to define whether orthologous genes of these transcription factors were present in *A. thaliana, B. distachyon, O. sativa, H. vulgare, Z. mays, T. aestivum, Ae. tauschii* and *T. urartu*. Of the known Arabidopsis and maize TFs, 11 have orthologous genes in the species examined. All the MYB protein sequences retrieved in this way were aligned using the ClustalW method with Mega7 software. Phylogenetic analysis was carried out using the neighbor-joining method and a 1,000-replication bootstrap test for significance. The tree was generated with Mega7 (http://www.ebi.ac.uk/Tools/phylogeny/).

As expected, the phylogenetic tree showed two clades, one grouping all MYB proteins, and a smaller one corresponding to MYBLs proteins, which do not have the same structure as proteins in the MYB 2R group but is a R3-MYB related protein. All the MYBs investigated are derived from diversification from three main branches. MYB65 and MYB19 share the same root with MYB111 which subsequently diversified from them. Similarly, MYB4, MYB11, MYB32 and MYB42 share common motifs in their amino acid sequences, which are more diverse from the latter clade. Among them, a third branch shows that MYB40 and MYB12 proteins have a common origin. Further studies will be needed to determine the evolutionary background of this important gene family, which could be a starting point for studying TF regulation of phenolic acid gene expression in wheat.

MicroRNAs (miRNAs) are a class of endogenous small noncoding RNAs, 20–24 nucleotides in length, which regulate complementary mRNAs (Iwakawa and Tomari (2015). Plant miRNAs are involved in different aspects of growth, development, primary metabolism, secondary metabolism and responses to biotic and abiotic stresses by inducing translational repression or degrading target mRNAs at the post-transcriptional level (Zhang and Wang 2015). Endogenous trans acting small interfering RNAs (tasiRNAs) are 21 nucleotides in length and can repress gene expression through post-transcriptional gene silencing in plants (Peragine et al. 2004; Vazquez et al. 2004; Allen et al. 2005).

Deng and Lu (2017) extensively reviewed the regulation of phenylpropanoid biosynthesis by miR828-MYB and miR828-tasiRNA-MYB cascades, citing a number of studies showing that miRNAs are involved in targeting R2R3-MYB proteins in a number of dicot species (e.g. in *Arabidopsis*, *Vitis*, *Populus*) (Fig. 4).

The evolutionary relationships were obtained by using the neighbor-joining method on 76 amino acid retrieved sequences for ten MYB TFs and one MYBL. Protein sequences were reported for *Arabidopsis thaliana* (At), *Brachypodium distachyon* (Bd), *Zea mays* (Zm), *Hordeum vulgare* (Hv), *Oryza sativa* (Os), *Aegilops tauchii* (Ae), *Triticum uraru* (Tu) and *Triticum aestivum* (Ta).

Interestingly, they found MIR828 loci in gymnosperms and basal monocots, but not in rice, sorghum or maize. In a bioinformatics approach, we tried to identify syntenic regions for MIR828 and MI858 in *B. distachyon, O. sativa, H. vulgare, Z. mays, T. urartu, Ae. tauschii* and *T. aestivum*, but none was found. This suggests different mechanisms control MYB expression in dicots and monocots or possibly the involvement of miRNAs yet to be found.

The genes encoding proteins for anthocyanin biosynthesis are spatially and temporally regulated by transcriptional factors belonging to the following families: (i) R2R3-MYB; (ii) MYC, encoding proteins with sequence homology to the bHLH DNA binding/dimerization domain found in the Myc oncoproteins; and (iii) WD40 proteins. One member of each family must be functional for the transcriptional activation of structural genes (Mol et al. 1998; Hichri et al. 2011). Candidate genes *TaMYC1* (*TaPpb1*) and *ThMYC4E* were identified for *Pp3* and *Ba1*, respectively (Shoeva et al. 2014, Li et al. 2017, Jiang et al. 2018, Shoeva 2018). A gene cluster

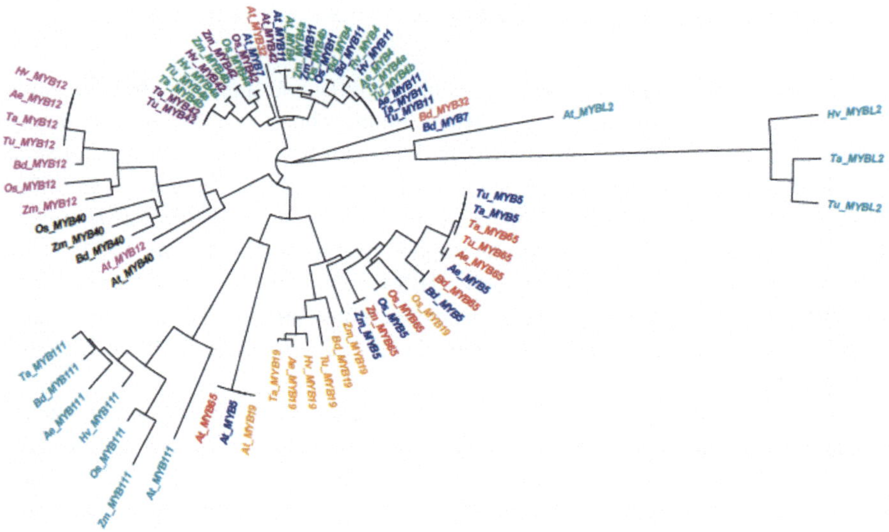

Fig. 4 Phylogenetic tree of the identified transcription factors

on chromosome 7D is responsible for the anthocyanin coloration of various wheat organs (Khlestkina et al. 2010, 2015) which seems to be associated with structural gene *TaPpm1* (Jiang et al. 2018).

3 Occurrence of Phenolic Compounds in Wheat Grain

In mature wheat grains, 75–80% of phenolic acids occur as insoluble bound forms being esterified to cell wall polymers, 20–25% are esterified to sugars and other low molecular mass compounds, and only 0.5–2% are soluble and free (Žilić 2016). Indeed, phenolics are structural components of plant cell walls. Ferulic acid and *p*-coumaric acid especially form ester bonds with arabinose units of arabinoxylans (Fulcher 1982). Phenolic acids also play a key role in the formation of secondary walls as they are precursors of monolignols in the biosynthesis of lignin. For these reasons, the outer layers of wheat grains are the compartments that contain the most phenolic acids. The aleurone layer has the highest concentration of total phenolic acids, especially ferulic and trans-sinapic acid, whereas the pericarp has abundant ferulic acid dehydrodimers, and the embryo abundant ferulic acid (Ragaee et al. 2014, Lempereur et al. 1997). Conversely, the starchy endosperm contains the least phenolic acid (Liyana-Pathirana and Shahidi 2006). The bran fraction also contains the most anthocyanins as these flavonoid pigments are located either in the pericarp or aleurone layer (Siebenhandl et al. 2007).

3.1 Analytical Methods for Extraction, Identification and Quantification of Phenolic Acids and Anthocyanins

Several analytical assays are used to obtain, identify and quantify phenolic acids (see the review by Dwivedi et al. 2016). Soluble phenolic acids are commonly extracted using polar solvents (i.e. methanol, ethanol or acetone), whereas alkaline or acidic hydrolysis is needed to release insoluble components (Adom et al. 2003; Li et al. 2008; Parker et al. 2005). Subsequent quantification is performed by Folin-Ciocalteu assays and visible (VIS) spectrophotometry (Singleton et al. 1999), or by high performance liquid chromatography (HPLC) coupled with a diode array detector, HPLC/mass spectrometry (MS), or reverse phase (RP)-HPLC to identify and quantify individual components (Laddomada et al. 2015a; Li et al. 2008; Nicoletti et al. 2013).

Current methods to analyse phenolic acids are complex, time consuming, laborious and inadequate for routine screening of large collections. These constraints are still the bottleneck for association mapping studies (Furbank and Tester 2011). Other analytical approaches that are faster, cheaper and simpler than chromatographic methods, such as those based on near infrared (NIR) spectroscopy, should be tested, developed and calibrated (Dwivedi et al. 2016).

Anthocyanins are usually extracted from wholemeal flour or the bran fraction with a methanol-hydrochloric acid solution or other polar solvent (Abdel-Aal and Hucl 1999; Tyl and Bunzel 2012). Syed Jaafar et al. (2013) reported the use of an accelerated solvent extractor for anthocyanin extraction. The total anthocyanin content can be easily determined by ultraviolet (UV)/VIS spectrometry (Abdel-Aal and Hucl 1999). Individual anthocyanins were identified by various spectrometric and chromatographic methods such as HPLC, liquid chromatography and MS (LC-MS), nuclear magnetic resonance spectrometry and MS, or high performance thin layer chromatography (Abdel-Aal et al. 2006; Tyl and Bunzel 2012; Syed Jaafar et al. 2013; Oberlerchner et al. 2018).

3.2 Role of Phenolic Acids and Anthocyanins in Human Health and Quality of Wheat Based Products

Phenolic acids are the major group of phytochemicals in the wheat grain and even though they are secondary metabolites and non-nutrient components, they contribute to wheat quality because they have a range of health-related antioxidant activities. Several in vitro and in vivo studies demonstrated that phenolic acids function as free-radical scavengers, reducing agents and quenchers of singlet oxygen formation, some of the main causes of oxidative damage to DNA and lipids (Sevgi et al. 2015; Graf 1992). Phenolic acids also protect low density lipoproteins from oxidation by reactive oxygen species, which is associated with the initial steps of atherosclerosis (Yu et al. 2005).

Besides their antioxidant properties, phenolic acids defend against carcinogenesis by scavenging carcinogenic agents and inducing the apoptosis of cancer cells or inhibition of aspects of angiogenesis (Thomasset et al. 2007; Ramos 2008). Also, phenolic compounds may exert anti-hypertensive, anti-microbial, anti-inflammatory or photo-protective activities (Bravo 1998; Whent et al. 2012; Laddomada et al. 2015b). Some studies showed that phenolic acids are able to activate specific endogenous antioxidant mechanisms (e.g. the Nrf2 pathway) that reduce the risks of a number of processes in endothelial dysfunction and inflammation (Juurlink et al. 2014). Of the phenolic acids, ferulic acid has the highest antioxidant activity due to the presence of three distinctive molecular motifs capable of free radical scavenging (Itagaki et al. 2009).

Ferulic acid in its free form is efficiently absorbed by the intestine (Bourne and Rice-Evans, 1998; Bourne et al. 2000). On the contrary, the bound forms of phenolic acids are only partially metabolized in the stomach and small intestine. Some recent investigations have pointed to gut microbiota having an influence on the bioavailability and biological activity of wheat phenolics (Marín et al. 2015; Cardona et al. 2013). Ester- and ether-linked phenolic acids can be released in the intestine by the action of intestinal microbes (Andreasen et al. 2001). Indeed, by reaching the colon mostly undigested, bound phenolics can exert their unique antioxidant, anti-inflammatory and anticancer activities locally (Andreasen et al. 2001; Drankham et al. 2003; Vitaglione et al. 2008). More studies are needed for the health claims about phenolic acids to gain the approval of the regulatory authorities such as the United States Food and Drug Administration and European Food Safety Authority.

Phenolic acids can actually influence some end-product quality parameters such as colour, aroma, and taste. For example, phenolic acids are subject to the activity of polyphenol oxidases (PPO) (E.C. 1.14.18.1) so they may cause undesired discoloration of noodles and dough browning (Taranto et al. 2012). Colour is one of the main factors influencing pasta quality, of both dried and fresh types, because some consumers prefer yellow to amber pasta. Browning can be caused by PPO activity that mostly occur in the aleurone layer of the kernel. Indeed, PPO activity increases at higher flour extraction rates (Hatcher and Kruger 1993; Okot-Kotber et al. 2001). However, oxidative enzymes can be inactivated by the high temperatures used for pasta drying, or by other thermal treatments, such as pasteurization, that are generally practised in the production of industrially produced fresh pasta.

Anthocyanin-pigmented wheat varieties are traditionally used for food and beverages in Ethiopia (Geleta et al. 2009). The first commercial purple wheat variety was released in New Zealand for providing colour and texture in wholemeal bread (Griffin 1987). In Europe, wholemeal bakery products made from purple wheat were introduced in 2006 by Backaldrin®, an initiative of The Kornspitz Company, Asten, Austria, and marketed under the brand PurPur®. In Canada, purple wheat products are marketed by InfraReady Products, Saskatoon, under the registered trademark AnthoGrain™ (Grausgruber et al. 2018). In China, coloured products such as instant noodles and soy sauce are already marketed (Li and Beta 2011). As the anthocyanins of purple and blue wheat are located in the outer layers of the grain, wholemeal or bran-enriched flour has to be used to process such food products. It is necessary to take into account that heat treatments during processing can degrade anthocyanins (Li et al. 2007; Hiemori et al. 2009).

3.3 Processing Technologies to Enhance Phenolic Content in Wheat-Based Products

Numerous factors affect the content, composition, and stability of phenolic acids that can be found in derived wheat products. These may depend on the type of flour used, on the presence of other ingredients, and on processing (Laddomada et al. 2015b). Phenolic acids are mainly concentrated in the outer layers of mature grains, so the use of those fractions has been considered to enhance phenolic content and antioxidant properties of wheat-based products. Indeed, due to the negative impact of wheat bran on bread and pasta quality (Edwards et al. 1995; Hemdane et al. 2016), a number of debranning processes have been developed to overcome these problems (Hemery et al. 2007; Blandino et al. 2013; Fares et al. 2010).

Debranning, or pearling, is the progressive removal of bran layers by consecutive abrasion of kernels, that lead to different by product classes, namely first, second and third debranning fractions (Hemery et al. 2011). The addition of low levels of the second debranning fraction result in higher phenolic content and antioxidant activity of end-products, without affecting their physical properties (Blandino et al. 2013). Also, the use of residuals of the second and third debranning steps can reduce mycotoxin contamination compared to the first debranning fraction (Brouns et al. 2012; Rizzello et al. 2012). Micronization of debranning fractions was also proposed to improve health features of bread products without altering the rheological properties of the dough (Rizzello et al. 2012). The addition of second and third debranning micronized fractions to re-milled semolina did not affect bread volume and crumb hardness, but improved its functional properties by increasing the content of dietary fiber and phenolics with antioxidant activity (Pasqualone et al. 2017).

Natural phenolic extracts can also serve to improve the antioxidant properties of wheat based products, such as pasta (Pasqualone et al. 2015) and bread (Sivam et al. 2011). Ultrasound-assisted technology, an environmentally-friendly system for extracting bioactive compounds from natural sources, was used to recover phenolic compounds from wheat bran or other plant sources (Wang et al. 2008). Pasta supplemented with bran extracts displayed a higher antioxidant activity and phenolic content compared to conventional products, and had good sensory properties (Pasqualone et al. 2015).

Sourdough fermentation was shown to improve functional, textural and sensory properties of bread supplemented with bran fractions (Rizzello et al. 2012). A number of bran pre-treatments (i.e. fermentation, and enzyme or heat treatments) were also proposed to enhance the bio-accessibility of phenolic acids and minimize the negative effects of bran on the quality of end-products (De Kock et al. 1999; Salmenkallio-Marttila et al. 2001). The use of yeast, lactic acid bacteria or enzymes for bran fermentation improved the bioavailability of phenolic compounds, and the loaf volume, crumb firmness and shelf life of bread products (Katina et al. 2007; Mateo Anson et al. 2011; Salmenkallio-Marttila et al. 2001).

4 Genetic Variability, Genomic Studies and Breeding Perspectives

4.1 Variability in Phenolic Acids and Anthocyanins in Bread and Durum Wheat Germplasm

Genetic variation in the phenolic acid profiles of wheat germplasm has been explored over the past 20 years, using several analytical methods (Adom et al. 2003; Dinelli et al. 2009; Wu et al. 1999; Kyung-Hee et al. 2006; Nicoletti et al. 2013). However, only a few works have tackled the evaluation of large sets of samples across multiple environments to estimate trait heritability (Menga et al. 2010;

Martini et al. 2014; Mpofu et al. 2006; Brandolini et al. 2013; Heimler et al. 2010; Serpen et al. 2008; Li et al. 2008; Fernandez-Orozco et al. 2010; Laddomada et al. 2017).

Overall it has been confirmed that hydroxy derivatives of cinnamic acids, especially ferulic, sinapic and *p*-coumaric acids are the most common phenolic acids in whole wheat grains. Conversely, hydroxy derivatives of BA (i.e. vanillic, syringic, and 2,4-dihydroxybenzoic acids) are minor components both in common and durum wheat genotypes. Definitively, the bound fraction is the most abundant, representing about 80% of total phenolic acids (Li et al. 2008; Fernandez-Orozco et al. 2010; Laddomada et al. 2015a).

Up to 3.5 fold variation in phenolic acid content was noted in common and durum wheat collections (Li et al. 2008; Fernandez-Orozco et al. 2010; Laddomada et al. 2017). These levels of variation were significantly lower than those observed in other cereals (Dwivedi et al. 2016). However studies of large wheat collections revealed that phenolic acid contents of old wheat varieties differed only slightly from those of modern wheats apart from some minor components (Dinelli et al. 2009; Laus et al. 2015; Shewry and Hey 2015; Laddomada et al. 2015a). Among tetraploid subspecies, durum cultivars contain more total phenolic acids than other tetraploid subspecies (Giambanelli et al. 2013; Brandolini et al. 2013; Laddomada et al. 2017).

Winter and bread wheat cultivars were the most extensively studied for variations in phenolic acids (Shewry et al. 2010). Though a wide variation was observed for phenolic acids among different common wheat lines, low heritability levels were reported, being less than 0.09 for free and conjugated fractions, 0.26 for bound phenolic acids, and about 0.3 for total phenolic acids (Shewry et al. 2010). Indeed, a strong influence of environmental factors and genotype × environment interactions were responsible for the low heritability (Shewry et al. 2010).

In tetraploid wheats significant effects of genotype, year, and genotype by year interaction were also found for individual and total phenolic acids (Laddomada et al. 2017; Menga et al. 2010). Heritability values varied for individual phenolic acids, ranging from 0.48 for syringic acid to 0.70 for ferulic acid, while for total phenolic acid content the heritability value was 0.63 in a study carried out in one location in southern Italy over 2 years (Laddomada et al. 2017). Nevertheless both in bread and in durum wheat some precise genotype × environment combinations result in more stable and higher phenolic acid content (Fernandez-Orozco et al. 2010).

Positive correlations were observed between the contents of total phenolic acids and other bioactive components (Shewry et al. 2010). Slight positive correlations were found between bound phenolic acids and water-extractable arabinoxylans from bran (Shewry et al. 2010), but have not necessarily been confirmed (Shewry et al. 2010; Laddomada et al. 2017), probably because any factor affecting grain size will also affect the concentration of bran components in wholemeal flour (Shewry et al. 2010).

Reported values of total anthocyanin content vary due to the different environmental conditions and different genetic material used in the various studies. Generally, blue aleurone types contain more concentrated anthocyanin than purple pericarp types (Abdel-Aal et al. 2006; Iriki et al. 2007; Žofajová et al. 2012; Syed Jaafar et al.

2013). Abdel-Aal et al. (2006) argued that anthocyanins in purple wheat are more prone to environmental effects due to the location of the pigment in the pericarp. Bustos et al. (2012) demonstrated that anthocyanin concentration increased in wheat that was fertilized and harvested early. More than 20 anthocyanins were detected in purple and blue wheat (for review see Lachman et al. 2017). Generally, the anthocyanin profile of purple wheat is more complex than that of blue wheat. The dominant aglycone in purple wheat is cyanidin, whereas delphinidin predominates in blue wheat (Abdel-Aal et al. 2006; Syed Jaafar et al. 2013; Böhmdorfer et al. 2018).

4.2 Abiotic and Biotic Factors Influencing Phenolics Accumulation in Wheat Grain

The effect of abiotic and biotic factors on phenolic acid accumulation in mature wheat kernels has been the focus of a few studies but more are needed to underpin the initial evidence (Dwivedi et al. 2016). Adaptation to abiotic stresses requires different protective mechanisms based on structural or biochemical responses. Phenolic acids and flavonoids are two important groups of secondary metabolites which have crucial roles in defence against adverse conditions.

Biosynthesis of phenolic acids and flavonoids may increase under extreme temperatures and drought conditions due to their importance in modelling cell wall structure and their antioxidant properties that counteract the effects of reactive oxygen species.

In a recent study, Shamloo et al. (2017) studied the effects of genotype and temperature on accumulation of plant secondary metabolites in three Canadian and Australian wheat cultivars grown under controlled environments. Phenolic acid and flavonoid contents increased in mature grains with the increase in the growing temperature from 20 °C to 30 °C. A strong effect of genotype was observed on the response to growing temperatures with some showing a higher increase in total phenolic acids and total flavonoid contents compared to others. However, the study had some limitations because the only variable parameter was the temperature, while other environmental factors such as water deficit, elevated CO_2 or UV light could have contributed to the physiological response. When the effect of terminal heat stress was evaluated in bread wheat genotypes, it had a negative influence on several secondary metabolites, including phenolic acids (Shahid et al. 2017).

The effect of water stress on durum wheat genotypes was recently evaluated (Liu et al. 2018). Phenolic acids accumulated differently in mature grains of different genotypes, independent of whether they were resistant, tolerant or sensitive to stress (Liu et al. 2018). Some lines did not exhibit any significant change in phenolic acids under the stress, while some others had higher concentrations of these metabolites compared to the controls (Liu et al. 2018). So far, the accumulation of phenolic acids in response to heat stress and drought seems to depend on genotype, environment and their interaction, but more studies are needed to clarify the relative significance of the different factors. Also, it is not clear if modifications in grain phenolic content caused by heat or drought are indirectly influenced by bran to endosperm ratios or by grain filling and grain size.

Nitrogen fertilization may also have the effect of increasing total free phenolic acids, and decreasing conjugated soluble phenolic acids in bread wheat grain (Stumpf et al. 2015). High levels of solar UV radiation result in an increase in phenolic acids and flavonoids in the grain of red and white Canadian wheat (Lukow et al. 2012). As constituents of cell walls, phenolic acids may be associated with resistance to pests and pathogens (Santiago et al. 2013). For example, in bread wheat, the phenolic acids that accumulated during grain development contributed positively to *Fusarium* resistance (McKeehen et al. 1999).

There are numerous reports that abiotic stress, such as that caused by heavy metals, or osmotic or oxidative conditions, can affect anthocyanin synthesis in purple wheat. While most reports deal with plant organs other than the grain, Hosseinian et al. (2008) and de Leonardis et al. (2015) observed increases in anthocyanin accumulation in wheat grains after heat stress. Overall, a better understanding of abiotic and biotic stress on phenolic secondary metabolites would inform decisions on which genotypes with enhanced phenolic acid content could be selected to suit particular environments.

4.3 Quantitative Trait Loci and Genes Associated With Phenolic Accumulation in Wheat Grain

To date, only a few studies have focused on the genetic control of phenolic acid content and these were on different species such as soybean (Li et al. 2016), eggplant (Prohens et al. 2013; Plazas et al. 2013), tomato (Sacco et al. 2013), and apple (Chagné et al. 2012; Verdu et al. 2014).

Phenolic acids in cereals have been investigated in rice, barley and sorghum, but most studies were limited to determining the total phenolic acid content and not considering individual phenolic compounds (Cai et al. 2015; Jin et al. 2009; Mohammadi et al. 2014; Rhodes et al. 2014). There are recent reports of studies on single phenolic acids, such as cinnamic-derived phenolic acids (ferulic, chlorogenic and caffeic acids), in relation to their role in wheat resistance to pathogens like *Fusarium graminearum* (Kazan et al. 2012; Gauthier et al. 2016; Atanasova-Penichon et al. 2016). The genetic variability of phenolic compounds in durum wheat was investigated in a collection of 111 tetraploid wheat (*T. turgidum* L.) genotypes, about half of which were durum cultivars and the remainder landraces and wild types (Pasqualone et al. 2014; Laddomada et al. 2017).

Traditional QTL approaches have been extensively used to investigate the most important agronomic traits in wheat, such as grain protein content and yield (Maccaferri et al. 2008; Blanco et al. 2012). For this approach it is necessary to develop specific segregating populations of at least 120–150 lines (usually bi-parental mapping populations generated ad hoc for the chosen trait, such as recombinant inbred lines), to be phenotyped for the desired trait and genotyped with a high number of polymorphic DNA-based molecular markers. However, QTL detection is then limited to loci segregating between crosses. As the detected QTL cover many centimorgans, additional steps are required to narrow down the QTL region and/or clone the genes.

The first study of QTL and genes involved in individual phenolic acid accumulation in wheat grain was carried out by Nigro et al. (2017). A tetraploid wheat collection was genotyped with 81,587 gene-associated single nucleotide polymorphisms (SNPs) in a genome wide association study (GWAS), an approach generally used to identify the genomic regions controlling a quantitative trait. Linkage disequilibrium-based association mapping is a recent, alternative approach that uses a set of genotypes (germplasm accessions, breeding lines, cultivars) representing the products of hundreds of recombination cycles, thus providing higher resolution QTL mapping (Rafalski 2010). The limitation of association mapping (GWAS and candidate gene approaches) is the high frequencies of false-positive and false-negative associations, which depend on population structure, relative kinship among individuals, and on multiple testing of thousands of markers. By taking a comparative genomics approach, Nigro et al. (2017) identified six candidate genes, *PAL1, PAL2, C4H, C3H, COMT1* and *COMT2*, involved in the biosynthesis of hydroxycinnamic acid in wheat, and about 50 SNPs were found in the coding sequences. For each candidate gene the chromosomal location and accurate map position based on two reference consensus linkage maps were determined. In the absence of information on the genetic basis of phenolic acid metabolism in wheat, identifying genes and QTL for phenolic acid composition and content in wheat grain is important. The collection was phenotyped for the accumulation of individual phenolic compounds as well as total soluble phenolic components. The GWAS detected 22 QTL distributed on almost all durum wheat chromosomes. Two QTL for *p*-coumaric acid coincide with the PAL2 and C3H genes on chromosome arms 2AL and 1AL, respectively. This first study made it possible to identify candidate gene-based markers, useful for elucidating the mechanism of phenolic acid accumulation in wheat kernels and exploiting the genetic variability of phenolic acid content for the nutritional improvement of wheat end-products. Similar studies might be carried out focusing on hydroxybenzoic derived phenolic acid, as well as TFs that might regulate the whole phenolic acid biosynthesis pathway. The identification of functional markers and precise map positions can be particularly useful for breeders in marker-assisted selection programs.

4.4 Breeding Perspectives

There is a considerable amount of literature documenting the variation in phenolic acids in wheat germplasm that is potentially available for breeding purposes. However, some of this variability depends on environmental factors (i.e. temperature, water availability, soil features, agronomy) and genotype × environment interactions. If the trait has low heritability, the variation that does exist would be maniable by breeders. However, there are different viewpoints, one more pessimistic because of the low stability of phenolic acid contents across very diverse growing environments (Shewry et al. 2010), while another is more optimistic about the possibility of selecting stable genotypes with higher amounts of phenolic acids for certain growing sites (Fernandez-Orozco et al. 2010; Menga et al. 2010; Laddomada et al. 2017). Indeed, bound phenolic acids are less influenced by the environment. In addition, different studies converged on the fact that the extent of variation related to environmental conditions may vary from genotype to genotype, with some genotypes being more stable over several growing seasons. Based on such evidence, several authors suggested that genotypes with higher and more stable contents of phenolic acids could be selected for cultivation and for breeding targeted to specific wheat growing areas (Fernandez-Orozco et al. 2010; Menga et al. 2010; Laddomada et al. 2017). Also, grain producers might exploit the impact the environment has on phenolic acid accumulation by selecting growing locations that result in higher contents of phenolic acids in mature grains.

An interesting aspect that might be considered is to estimate diferulate content. Dimerization of ferulic acid is essential for the formation of cross-links between arabinoxylans in the cell wall. For over 20 years, 5–5'-coupled diferulic acid was the only ferulic acid dehydrodimer known, but recently other dehydrodiferulic acid forms were identified that are involved in the cross-linking of cell wall polysaccharides in wheat. The total amount of different forms of dehydrodiferulic acids associated to soluble and insoluble dietary fibre can make up a large proportion of the total phenolic acids in wheat (Lempereur et al. 1997; Bunzel et al. 2001). However, diferulates have not been evaluated in extensive genetic studies because it is complex and time-consuming to identify and quantify them in soluble and especially in insoluble dietary fibre. Estimating such an important group of phenolic acid components might add to our understanding of the heritability and health potential of phenolic acids in wheat.

The different inheritance of the purple pericarp and blue aleurone traits may be readily exploited by targeted cross combinations of respective germplasm. In this way and by selecting for adaptation to the prevalent environmental conditions, wheat lines with both increased total anthocyanin content and increased antioxidant activity were created (Syed Jaafar et al. 2013; Varga et al. 2013; Böhmdorfer et al. 2018; Grausgruber et al. 2018).

5 Conclusions

Phenolics are an important class of components arising from the phenylpropanoid pathway that contribute to the health benefits of whole wheat grains. Several approaches have been taken to improve the phenolic contents of wheat-based foods. The use of specific wheat debranning fractions or bran extracts as ingredients to make pasta or baking products has been proposed, notwithstanding possible undesirable effects on dough rheology and the sensory properties of end-products.

There is a large genetic variation in the phenolic components of wheat germplasm that could be used to select wheat genotypes with higher contents of phenolic acids in mature grains. Though phenolic acid content is influenced by environmental factors and genotype × environment interactions, several common and durum wheat genotypes were found to have stable and elevated contents of phenolic acids in certain wheat-growing areas over several years.

More effort is needed to advance association studies to identify genes and QTL controlling phenolic acids. A major constraint is the complex, time-consuming, and extremely laborious analytical methodologies used to analyze phenolic acid profiles. For the GWAS screening of large wheat collections, more simple approaches have been suggested, such as NIR spectroscopy, that will need to be validated or adjusted to analyze all phenolic components.

Increasing knowledge of the phenylpropanoid pathway is leading to the identification of further structural and regulatory genes involved in the biosynthetic pathway, making it more likely that key genetic factors will be found that may contribute to the design of wheat lines with more phenolic compounds.

References

Abdel-Aal ESM, Hucl P (1999) A rapid method for quantifying total anthocyanins in blue aleurone and purple pericarp wheats. Cereal Chem 76: 350–354.

Abdel-Aal ESM, Young JC, Rabalski I (2006) Anthocyanin composition in black, blue, pink, purple, and red cereal grains. J Agric Food Chem 54: 4696–4704.

Adom KK, Sorrells EM, Liu RH (2003) Phytochemical profiles and antioxidant activity of wheat varieties. J Agric Food Chem 51: 7825–7834.

Agarwal T, Grotewold E, Doseff AI et al. (2016) MYB31/MYB42 syntelogs exhibit divergent regulation of phenylpropanoid genes in maize, sorghum and rice. Sci Rep 6:28502.

Allen E, Xie Z, Gustafson AM et al. (2005) microRNA-directed phasing during trans-acting siRNA biogenesis in plants. Cell 121: 207–221.

Andreasen MF, Kroon PA, Williamson G et al. (2001) Intestinal release and uptake of phenolic antioxidant diferulic acids. Free Radic Biol Med 31:304–314.

Atanasova-Penichon V, Barreau C, Richard-Forget F (2016) Antioxidant secondary metabolites in cereals: Potential involvement in resistance to Fusarium and mycotoxin accumulation. Front Microbiol 7, 566.

Bhargava A, Mansfield SD, Hall HC et al. (2010) MYB75 functions in regulation of secondary cell wall formation in the Arabidopsis inflorescence stem. Plant Physiol 154: 1428–1438.

Blanco A, Mangini G, Giancaspro A et al. (2012). Relationships between grain protein content and grain yield components through quantitative trait locus analyses in a recombinant inbred line population derived from two elite durum wheat cultivars. Mol Breed 30: 79–92.

Blandino M, Sovrani V, Marinaccio F et al. (2013) Nutritional and technological quality of bread enriched with an intermediated pearled wheat fraction. Food Chem 141: 2549–2557.

Böhmdorfer S, Oberlerchner JT, Fuchs C, Rosenau T, Grausgruber H (2018) Profiling and quantification of grain anthocyanins in purple pericarp× blue aleurone wheat crosses by high-performance thin-layer chromatography and densitometry. Plant Methods 14: 29.

Bourne L, Paganga G, Baxter D et al. (2000) Absorption of ferulic acid from low-alcohol beer. Free Radic Res 32: 273–280.

Bourne LC, Rice-Evans C (1998) Bioavailability of ferulic acid. Biochem Biophys Res Commun 253: 222–227.

Brandolini A, Castoldi P, Plizzari L et al. (2013) Phenolic acids composition, total polyphenols content and antioxidant activity of *Triticum monococcum*, *Triticum turgidum* and *Triticum aestivum*: a two-year evaluation. J Cereal Sci 58: 123–131.
Bravo L (1998) Polyphenols: chemistry, dietary sources, metabolism, and nutritional significance. Nutr Rev 56: 317–333.
Brouns F, Hemery Y, Price R et al. (2012) Wheat aleurone: Separation, composition, health aspects, and potential food use. Crit Rev Food Sci Nutr 52: 553–568.
Bunzel M, Ralph J, Marita JM et al. (2001) Diferulates as structural components in soluble and insoluble cereal dietary fibre. J Sci Food Agric 81: 653–660.
Buresova V, Kopecky D, Bartos J, Martinek P, Watanabe N, Vyhnanek T, Dolezel J. (2015) Variation in genome composition of blue-aleuron wheat. Theor Appl Genet 128: 278–282.
Bustos DV, Riegel R, Calderini DF (2012) Anthocyanin content of grains in purple wheat is affected by grain position, assimilate availability and agronomic management. J Cereal Sci 55: 257–264.
Cai SG, Han ZG, Huang YQ, Chen ZH, Zhang GP, Dai F. (2015) Genetic diversity of individual phenolic acids in barley and their correlation with barley malt quality. J Agric Food Chem 63: 7051–7057.
Cardona F, Andrés-Lacueva C, Tulipani S et al. (2013) Benefits of polyphenols on gut microbiota and implications in human health. J Nutr Biochem 24: 1415–1422.
Chagné D, Krieger C, Rassam M, et al. (2012) QTL and candidate gene mapping for polyphenolic composition in apple fruit. BMC Plant Biol 12: 12.
Craven-Bartle B, Pascual MB, Cánovas FM et al. (2013) A Myb transcription factor regulates genes of the phenylalanine pathway in maritime pine. Plant J 74: 755–766.
De Kock S, Taylor J, Taylor JRN (1999) Effect of heat treatment and particle size of different brans on loaf volume of brown bread. LWT-Food Sci Technol 32: 349–356.
de Leonardis AM, Fragasso M, Beleggia R et al. (2015) Effects of heat stress on metabolite accumulation and composition, and nutritional properties of durum wheat grain. Int J Mol Sci 16: 30382–30404.
Deng Y, Lu S (2017). Biosynthesis and regulation of phenylpropanoids in plants. Crit Rev Plant Sci 36: 257–290.
Dias AP, Grotewold E (2003) Manipulating the accumulation of phenolics in maize cultured cells using transcription factors. Biochem Eng J 14: 207–216.
Dias AP, Braun EL, McMullen MD, Grotewold E (2003) Recently duplicated maize R2R3 Myb genes provide evidence for distinct mechanisms of evolutionary divergence after duplication. Plant Physiol 131: 610–620.
Dinelli G, Carretero AS, Di Silvestro R et al. (2009) Determination of phenolic compounds in modern and old varieties of durum wheat using liquid chromatography coupled with time-of-flight mass spectrometry. J Chromatogr A 1216: 7229–7240.
Drankham K, Carter J, Madl R et al. (2003) Antitumor activity of wheats with high orthophenolic content. Nutr Cancer 47:188–194.
Du H, Feng BR, Yang SS et al. (2012) The R2R3-MYB transcription factor gene family in maize. PLoS One 7:e37463.
Dubos C, Le Gourrierec J, Baudry A et al. (2008) MYBL2 is a new regulator of flavonoid biosynthesis in *Arabidopsis thaliana*. Plant J 55: 940–953.
Dubos C, Stracke R, Grotewold E et al. (2010) MYB transcription factors in Arabidopsis. Trends Plant Sci 15: 573–581.
Dwivedi SL, Upadhyaya HD, Chung IM et al. (2016) Exploiting phenylpropanoid derivatives to enhance the nutraceutical values of cereals and legumes. Front Plant Sci 7: 763.
Edwards NM, Biliaderis CG, Dexter JE (1995) Textural characteristics of wholewheat pasta and pasta containing non-starch polysaccharides. J Food Sci 60: 1321–1324.
El-Basyouni SZ, Chen D, Ibrahim RK et al. (1964) The biosynthesis of hydroxybenzoic acids in higher plants. Phytochemistry 3: 485–492.

Fares C, Platani C, Baiano A et al. (2010) Effect of processing and cooking on phenolic acid profile and antioxidant capacity of durum wheat pasta enriched with debranning fractions of wheat. Food Chem 119: 1023–1029.

Fernandez-Orozco R, Li L, Harflett C, Shewry PR, Ward JL (2010) Effects of environment and genotype on phenolic acids in wheat in the HEALTHGRAIN diversity screen. J Agric Food Chem 58: 9341–9352.

Fornalé S, Lopez E, Salazar-Henao JE et al. (2014) AtMYB7, a new player in the regulation of UV-sunscreens in *Arabidopsis thaliana*. Plant Cell Physiol. 55: 507–516.

Fornalé S, Shi XH, Chai CL et al. (2010) ZmMYB31 directly represses maize lignin genes and redirects the phenylpropanoid metabolic flux. Plant J 64: 633–644.

Fulcher RG (1982) Fluorescence microscopy of cereals. Food Microstructure 1: 167–175.

Furbank RT, Tester M (2011) Phenomics technologies to relieve the phenotyping bottleneck. Trends Plant Sci 16: 635–644.

Gauthier L, Bonnin-Verdal MN, Marchegay G et al. (2016) Fungal biotransformation of chlorogenic and caffeic acids by *Fusarium graminearum*: new insights in the contribution of phenolic acids to resistance to deoxynivalenol accumulation in cereals. Int J Food Microbiol 221: 61–68.

Geissman TA, Hinreiner E (1952) Theories of the biogenesis of flavonoid compounds. Bot Rev 18: 77–164.

Geleta N, Eticha F, Grausgruber H (2009) Preservation of tetraploid wheat landraces in the West Central Highlands of Ethiopia. In: Splechtna BE (ed), Proc Int Symp Preservation of biocultural diversity – a global issue, 6–8 May 2008, Vienna, Austria, pp. 91–98. BOKU-University of Natural Resources and Life Sciences, Vienna.

Giambanelli E, Ferioli F, Koc,aoglu B, Jorjadze M, Alexieva I,Darbinyan N, D'Antuono LF (2013) A comparative studyof bioactive compounds in primitive wheat populationsfrom Italy, Turkey, Georgia, Bulgaria and Armenia. J Sci Food Agric 93: 3490–3501.

Graf E (1992) Antioxidant potential of ferulic acid. Free Rad Biol Med 13: 435–448.

Grausgruber H, Atzgersdorfer K, Böhmdorfer S (2018) Purple and blue wheat - Health-promoting grains with increased antioxidant activity. Cereal Foods World 63: 217–220.

Griffin WB (1987) Outcrossing in New Zealand wheats measured by occurrence of purple grain. NZ J Agric Res 30: 287–290.

Hatcher DW, Kruger JE (1993) Distribution of polyphenol oxidase in flour millstreams of Canadian common wheat classes milled to three extraction rates. Cereal Chem 70: 51–55.

Heimler D, Vignolini P, Isolani L et al. (2010) Polyphenol content of modern and old varieties of *Triticum aestivum* L. and *T. durum* Desf. grains in two years of production. J Agric Food Chem 58: 7329–7334.

Heine GF, Malik V, Dias AP et al. (2007) Expression and molecular characterization of ZmMYB-IF35 and related R2R3-MYB transcription factors. Mol Biotechnol 37: 155–164.

Hemdane S, Jacobs PJ, Dornez E et al. (2016) Wheat (*Triticum aestivum* L.) bran in bread making: A critical review. Compr Rev Food Sci Food Safety 15: 28–42.

Hemery Y, Rouau X, Lullien-Pellerin V et al. (2007) Dry processes to develop wheat fractions and products with enhanced nutritional quality. J Cereal Sci 46: 327–347.

Hemery Y, Chaurand M, Holopainen U, Lampi AM, Lenthinen P, Piironen V, Sadaoudi A, Rouau X (2011) Potential of dry fractionation of what bran for the development of food ingredients, part I: Influence of ultra-fine grinding. J Cereal Sci 53: 1–8.

Hichri I, Barrieu F, Bogs J, et al. (2011) Recent advances in the transcriptional regulation of the flavonoid biosynthetic pathway. J Exp Bot 62: 2465–2483.

Hiemori M, Koh E, Mitchell AE (2009) Influence of cooking on anthocyanins in black rice (*Oryza sativa* L. *japonica* var. SBR). J Agric Food Chem 57: 1908–1914.

Holton TA, Cornish EC (1995) Genetics and biochemistry of anthocyanin biosynthesis. Plant Cell 7: 1071–1083.

Hosseinian FS, Li W, Beta T (2008) Measurement of anthocyanins and other phytochemicals in purple wheat. Food Chem 109: 916–924.

Iriki N, Ishii G, Kuwabara T et al. (2007) Effects of anthocyanin on antioxidative activity and seed dormancy in wheat grain. Jpn J Crop Sci 76:569-575 (in Japanese, English abstract).

Itagaki S, Kurokawa T, Nakata C et al. (2009) *In vitro* and *in vivo* antioxidant properties of ferulic acid: A comparative study with other natural oxidation inhibitors. Food Chem 114: 466–471.

Iwakawa HO, Tomari Y (2015) The functions of micro- RNAs: mRNA decay and translational repression. Trends Cell Biol 25: 651–665.

Jiang W, Liu T, Nan W et al. (2018) Two transcription factors TaPpm1 and TaPpb1 co-regulate anthocyanin biosynthesis in purple pericarps of wheat. J Exp Bot 69: 2555–2567.

Jin H, Martin C (1999) Multifunctionality and diversity within the plant MYB-gene family. Plant Mol Biol 41: 577–585.

Jin H, Cominelli E, Bailey P et al. (2000) Transcriptional repression by AtMYB4 controls production of UV-protecting sunscreens in Arabidopsis. EMBO J 19: 6150–6161.

Jin L, Xiao P, Lu Y, Shao Y, Shen Y, Bao J (2009) Quantitative trait loci for brown rice color, phenolics, flavonoid contents, and antioxidant capacity in rice grain. Cereal Chem 86: 609–615.

Juurlink BHJ, Azouz HJ, Aldalati AMZ et al. (2014) Hydroxybenzoic acid isomers and the cardiovascular system. Nutr J 13: 63.

Katina K, Laitila A, Juvonen R et al. (2007) Bran fermentation as a means to enhance technological properties and bioactivity of rye. Food Microbial 24: 175–186.

Kazan K, Gardiner DM, Manners JM (2012) On the trail of a cereal killer: recent advances in *Fusarium graminearum* pathogenomics and host resistance. Mol Plant Pathol 13: 399–413.

Khlestkina EK, Pshenichnikova TA, Röder MS et al. (2010) Clustering of anthocyanin pigmentation genes in wheat group 7 chromosomes. Cereal Res Commun 37: 391–398.

Khlestkina EK, Shoeva OY, Gordeeva EI (2015) Flavonoid biosynthesis genes in wheat. Russ J Genet Appl Res 5: 268–278.

Kikuzaki H, Hisamoto M, Hirose K, Akiyadm K, Taniguchi H (2002). Antioxidant properties of ferulic acid and its related compounds. J Agric Food Chem 50: 2161–2168.

Kyung-Hee K, Tsao R, Yang R et al. (2006) Phenolic acid profiles and antioxidant activities of wheat bran extracts and the effect of hydrolysis conditions. Food Chem 95: 466–473.

Lachman J, Martinek P, Kotíková Z et al. (2017) Genetics and chemistry of pigments in wheat grain - A review. J Cereal Sci 74: 145–154.

Laddomada B, Caretto S, Mita G (2015a) Wheat bran phenolic acids: bioavailability and stability in whole wheat-based foods. Molecules 20: 15666–15685.

Laddomada B, Durante M, Mangini G et al. (2017). Genetic variation for phenolic acids concentration and composition in a tetraploid wheat (*Triticum turgidum* L.) collection. Genet Resour Crop Evol 64: 587–597.

Laddomada B, Durante M, Minervini F et al. (2015b) Phytochemical characterization and anti-inflammatory activity of extracts from the whole-meal flour of Italian durum wheat cultivars. Int J Mol Sci 16: 3512–3527.

Laus MN, Di Benedetto NA, Caporizzi R, Pastore D (2015) Evaluation of phenolic antioxidant capacity in grains of modern and old durum wheat genotypes by the novel QUENCHERABTS approach. Plant Foods Hum Nutr 70: 207–214.

Lempereur I, Rouau X, Abecassis J (1997) Genetic and agronomic variation in arabinoxylan and ferulic acid content of durum wheat (*Triticum durum* L.) grains and its milling fractions. J Cereal Sci 25: 103–110.

Li C, Lu S (2014) Genome-wide characterization and comparative analysis of R2R3-MYB transcription factors shows the complexity of MYB-associated regulatory networks in Salvia miltiorrhiza. BMC Genom 15: 277.

Li L, Shewry PR, Ward JL (2008) Phenolic acids in wheat varieties in the HEALTHGRAIN diversity screen. J Agric Food Chem 56: 9732–9739.

Li MW, Muñoz NB, Wong CF et al. (2016). QTLs regulating the contents of antioxidants, phenolics, and flavonoids in soybean seeds share a common genomic region. Front Plant Sci 7: 854.

Li N, Li S, Zhang K et al. (2017) *ThMYC4E*, candidate *Blue aleurone 1* gene controlling the associated trait in *Triticum aestivum*. PLoS One 127:e1081116.

Li W, Beta T (2011) Flour and bread from black-, purple-, and blue-colored wheats. In: Preedy VR, Watson RR, Patel VB (eds), Flour and breads and their fortification in health and disease prevention, pp. 59–67. Academic Press, London.

Li W, Pickard MD, Beta T (2007) Effect of thermal processing on antioxidant properties of purple wheat bran. Food Chem 104: 1080–1086.

Liu Q, Qiu Y, Beta T (2010) Comparison of antioxidant activities of different colored wheat grains and analysis of phenolic compounds. J Agric Food Chem 58: 9235–9241.

Liu J, Osbourn A, Ma P (2015) MYB transcription factors as regulators of phenylpropanoid metabolism in plants. Mol Plant 8: 689–708.

Liu H, Bruce DR, Sissons M, Able AJ, Able JA (2018) Genotype-dependent changes in the phenolic content of durum under water-deficit stress. Cereal Chem 95: 59–78.

Liyana-Pathirana CM, Shahidi F (2006) Importance of insoluble-bound phenolics to antioxidant properties of wheat. J Agric Food Chem 54: 1256–1264.

Lukow O, Suchy J, Adams K et al. (2012) Effect of solar radiation, plant maturity and post-harvest treatment on the color and phenolic and carotenoid contents in seed of red and white Canadian wheat. J Plant Cell Sci 3: 1–13.

Luo J, Butelli E, Hill L et al. (2008) AtMYB12 regulates caffeoyl quinic acid and flavonol synthesis in tomato: expression in fruit results in very high levels of both types of polyphenol. Plant J 56: 316–326.

Ma D, Li Y, Zhang J et al. (2016) Accumulation of phenolic compounds and expression profiles of phenolic acid biosynthesis related genes in developing grains of white, purple, and red wheat. Front Plant Sci 7: 528.

Maccaferri M, Sanguineti M C, Corneti S et al. (2008) Quantitative trait loci for grain yield and adaptation of durum wheat (*Triticum durum* Desf.) across a wide range of water availability. Genetics 178: 489–511.

Marín L, Miguélez EM, Villar CJ et al. (2015) Bioavailability of dietary polyphenols and gut microbiota metabolism: Antimicrobial properties. BioMed Res Int 2015: 905215.

Martini D, Taddei F, Nicoletti I et al. (2014) Effects of Genotype and Environment on Phenolic Acids Content and Total Antioxidant Capacity in Durum Wheat. Cereal Chem 91: 310–317.

Mateo Anson N, Aura AM, Selinheimo E et al. (2011) Bioprocessing of wheat bran in whole wheat bread increases the bioavailability of phenolic acids in men and exerts antiinflammatory effects *ex vivo*. J Nutr 141: 137–143.

Matsui K, Umemura Y, Ohme-Takagi M (2008) AtMYBL2, a protein with a single MYB domain, acts as a negative regulator of anthocyanin biosynthesis in Arabidopsis. Plant J 55: 954–967.

McKeehen JD, Busch RH, Fulcher RG (1999) Evaluation of wheat (*Triticum aestivum* L.) phenolic acids during grain development and their contribution to *Fusarium* resistance. J Agric Food Chem 47: 1476–1482.

Menga V, Fares C, Troccoli A, Cattivelli L, Baiano A (2010) Effects of genotype, location and baking on the phenolic content and some antioxidant properties of cereal species. Int J Food Sci Tech 45: 7–16.

Metzger RJ, Sebesta E (2004) Registration of three blue-seeded wheat genetic stocks exhibiting xenia. Crop Sci 44: 2281–2283.

Mohammadi M, Endelman J, Nair N, Shiaoman C, Jones S, Muehlbauer G, Ullrich S, Baik B-K, Wise M, Smith K (2014) Association mapping of grain hardness, polyphenol oxidase, total phenolics, amylose content, and β-glucan in US barley breeding germplasm. Mol Breed 34: 1229–1243.

Mol J, Grotewold E, Koes R (1998) How genes paint flowers and seeds. Trends Plant Sci 3:212–217.

Mpofu A, Sapirstein HD, Beta T (2006) Genotype and environmental variation in phenolic content, phenolic acid composition, and antioxidant activity of hard spring wheat. J Agric Food Chem 54: 1265–1270.

Mu HM, Du XJ, Zhang XS (2015) Study on plants MYB transcription factors regulate biological synthesis of phenylpropanoid metabolism. North Garden 24: 171–174.

Nicoletti I, Daniela MD, De Rossi A et al. (2013) Identification and quantification of soluble free, soluble conjugated, and in soluble bound phenolic acids in durum wheat (*Triticum turgidum* L

vardurum) and derived products by RP-HPLC on a semimicro separation scale. J Agric Food Chem 61: 11800–11807.
Nigro D, Laddomada B, Mita G et al. (2017) Genome-wide association mapping of phenolic acids in durum wheat. J Cereal Sci 75: 25–34.
Oberlerchner JT, Fuchs C, Grausgruber H et al. (2018) À côté calibration - Making optimal use of time and space in quantitative high performance thin layer chromatography. J Chromatogr A 1533: 193–198.
Okot-Kotber M A, Liavoga A, Yong KJ, Bagorogoza K (2001) Activity and inhibition of polyphenol oxidase in extracts of bran and other milling fractions from a variety of wheat cultivars. Cereal Chem 78:514–520.
Parker ML, Ng A, Waldron KW (2005) The phenolic acid and polysaccharide composition of cell walls of bran layers of mature wheat (*Triticum aestivum* L. cv. *Avalon*) grains. J Sci Food Agric 85: 2539–2547.
Pasqualone A, Delvecchio LN, Gambacorta G et al. (2015) Effect of supplementation with wheat bran aqueous extracts obtained by ultrasound-assisted technologies on the sensory properties and the antioxidant activity of dry pasta. Nat Prod Commun 10: 1739–1742.
Pasqualone A, Laddomada B, Centomani I et al. (2017) Bread making aptitude of mixtures of re-milled semolina and selected durum wheat milling by-products. LWT-Food Sci Technol 78: 151–159.
Pasqualone A, Delvecchio LN, Mangini G, Taranto F, Blanco A (2014). Variability of total soluble phenolic compounds and antioxidant activity in a collection of tetraploid wheat. Agr Food Sci 23: 307–316.
Payyavula RS, Singh RK, Navarre DA (2013) Transcription factors, sucrose, and sucrose metabolic genes interact to regulate potato phenylpropanoid metabolism. J Exp Bot 64: 5115–5131.
Peragine A, Yoshikawa M, Wu G, et al. (2004) SGS3 and SGS2/SDE1/RDR6 are required for juvenile development and the production of trans-acting siRNAs in Arabidopsis. Genes Dev 18: 2368–2379.
Plazas M, Andújar I, Vilanova S et al. (2013) Breeding for chlorogenic acid content in eggplant: interest and prospects. Not Bot Horti Agrobot 41: 26–35.
Preston J, Wheeler J, Heazlewood J, Li SF, and Parish RW (2004) AtMYB32 is required for normal pollen development in *Arabidopsis thaliana*. Plant J 40: 979–995.
Prohens J, Whitaker BD, Plazas M et al. (2013) Genetic diversity in morphological characters and phenolic acids content resulting from an interspecific cross between eggplant, *Solanum melongena*, and its wild ancestor (*S. incanum*). Ann Appl Biol 162: 242–257.
Qin YZ, Tai G, Xie KY et al. (2014). Ambient light alters gene expression pattern of enzymes and transcription factors involved in phenylpropanoid metabolic pathway in potato under chilling stress. Agric Sci Technol 15: 1899–1904.
Qiu J, Gao F, Shen G et al. (2013) Metabolic engineering of the phenylpropanoid pathway enhances the antioxidant capacity of *Saussurea involucrata*. PLoS One 8: e70665.
Qualset CO, Soliman KM, Jan CC, Dvořák J, McGuire PE, Vogt HE (2005) Registration of UC66049 *Triticum aestivum* blue aleurone genetic stock. Crop Sci 45: 432.
Rafalski JA 2010 Association genetics in crop improvement. Curr Opin Plant Biol 13: 174–180.
Ragaee S, Seetharaman K, Abdel-Aal ESM (2014) The impact of milling and thermal processing on phenolic compounds in cereal grains. Crit Rev Food Sci Nutr 54: 837–849.
Ramos S (2008) Cancer chemoprevention and chemotherapy: dietary polyphenols and signalling pathways. Mol Nutr Food Res 52: 507–526.
Rhodes DH, Hoffmann JrL, Rooney WL, Ramu P, Morris GP, Kresovich S (2014) Genome-wide association study of grain polyphenol concentrations in global sorghum [*Sorghum bicolor* (L.) Moench] germplasm. J Agric Food Chem 62: 10916–10927.
Rizzello CG, Coda R, Mazzacane F (2012) Micronized by-products from debranned durum wheat and sourdough fermentation enhanced the nutritional, textural and sensory features of bread. Food Res Int 46: 304–313.
Rommens CM, Richael CM, Yan H (2008) Engineered native pathways for high kaempferol and caffeoylquinate production in potato. Plant Biotechnol J 6: 870–886.

Sacco A, Di Matteo A, Lombardi N et al. (2013). Quantitative trait loci pyramiding for fruit quality traits in tomato. Mol Breed 31: 217–222.

Salmenkallio-Marttila, M, Katina K, Autio K (2001) Effects of bran fermentation on quality and microstructure of high-fiber wheat bread. Cereal Chem 78: 429–435.

Santiago R, Barros-Rios J, Malvar RA (2013) Impact of cell wall composition on maize resistance to pests and diseases. Int J Mol Sci 14: 6960–6980.

Schwinn KE, Davies KM (2004) Flavonoids. In: Davies KM (ed), Plant pigments and their manipulation, Blackwell Publishing Ltd, Oxford, UK, p 92–149.

Serpen A, Gökmen V, Karagöz A et al. (2008) Phytochemical quantification and total antioxidant capacities of emmer (*Triticum dicoccon* Schrank) and einkorn (*Triticum monococcum* L.) wheat landraces. J Agric Food Chem 56: 7285–7292.

Sevgi K, Tepe B, Sarikurkcu C (2015) Antioxidant and DNA damage protection potentials of selected phenolic acids. Food Chem Toxicol 77: 12–21

Shahid M, Saleem MF, Anjum SA, Shahid M, Afzal I (2017) Effect of terminal heat stress on proline, secondary metabolites and yield components of wheat (*Triticum aestivum* L.) genotypes. Philipp Agric Sci 100: 278–286.

Shamloo M, Babawale EA, Furtado A et al. (2017) Effects of genotype and temperature on accumulation of plant secondary metabolites in Canadian and Australian wheat grown under controlled environments. Sci Rep 7: 9133.

Shewry PR, Hey S (2015) Do "ancient" wheat species differ from modern bread wheat in their contents of bioactive components? J Cereal Sci 65: 236–243.

Shewry PR, Piironen V, Lampi AM, Edelmann M, Kariluoto S, Nurmi T, Fernandez-Orozco R, Ravel C, Charmet G, Andersson AAM et al. (2010) The HEALTHGRAIN wheat diversity screen: Effects of genotype and environment on phytochemicals and dietary fiber components. J Agric Food Chem 58: 9291–9298.

Shoeva OY (2018) Complex regulation of the *TaMyc1* gene expression in wheat grain synthesizing anthocyanin pigments. Mol Biol Rep 45: 327–334.

Shoeva OY, Gordeeva EI, Khlestkina EK (2014) The regulation of anthocyanin synthesis in the wheat pericarp. Molecules 19: 20266–20279.

Siebenhandl S, Grausgruber H, Pellegrini N et al. (2007) Phytochemical profile of main antioxidants in different fractions of purple and blue wheat, and black barley. J Agric Food Chem 55: 8541–8547.

Singleton VL, Orthofer R, Lamuela-Raventos RM (1999) Analysis of total phenols and other oxidation substrates and antioxidants by means of Folin-Ciocalteu reagent. Method Enzymol 299: 152–178.

Sivam AS, Sun, Waterhouse D et al. (2011) Physicochemical properties of bread dough and finished bread with added pectin fiber and phenolic antioxidants. J Food Sci 76: H97–H107.

Sonbol FM, Fornale S, Capellades M et al. (2009) The maize ZmMYB42 represses the phenylpropanoid pathway and affects the cell wall structure, composition and degradability in Arabidopsis thaliana. Plant Mol Biol 70: 283–296.

Stumpf B, Yan F, Honermeier B (2015) Nitrogen fertilization and maturity influence the phenolic concentration of wheat grain (*Triticum aestivum*). J Plant Nutr Soil Sci 178: 118–125.

Syed Jaafar SN, Baron J, Siebenhandl-Ehn S et al. (2013) Increased anthocyanin content in purple pericarp × blue aleurone wheat crosses. Plant Breed 132: 546–552.

Taranto F, Delvecchio LN, Mangini G et al. (2012)Molecular and physic-chemical evaluation of enzymatic browning of whole meal and dough in a collection of tetraploid wheats. J Cereal Sci 55: 405–414.

Tereschenko OY, Pshenichnikova TA, Salina EA et al. (2012) Development and molecular characterization of a novel wheat genotype having purple grain colour. Cereal Res Commun 40: 210–214.

Thomasset SC, Berry DP, Garcea G et al. (2007) Dietary polyphenolic phytochemicals - promising cancer chemopreventive agents in humans? A review of their clinical properties. Int J Cancer 120: 451–458.

Tyl CE, Bunzel M (2012) Antioxidant activity-guided fractionation of blue wheat (UC66049 *Triticum aestivum* L.). J Agric Food Chem 60: 731–739.

Vanholme R, Cesarino I, Rataj K et al. (2013) Caffeoyl shikimate esterase (CSE) is an enzyme in the lignin biosynthetic pathway in Arabidopsis. Science 341: 1103–1106.

Varga M, Bánhidy J, Czeuz Let al. (2013) The anthocyanin content of blue and purple coloured wheat cultivars and their hybrid generations. Cereal Res Commun 41: 284–292.

Vazquez F, Vaucheret H, Rajagopalan R et al. (2004). Endogenous trans-acting siRNAs regulate the accumulation of Arabidopsis mRNAs. Mol Cell 16: 69–79.

Vélez-Bermúdez IC, Salazar-Henao JE, Fornale S et al. (2015) A MYB/ZML complex regulates wound-induced lignin genes in maize. Plant Cell 27: 3245–3259.

Verdu CF, Guyot S, Childebrand N et al. (2014). QTL analysis and candidate gene mapping for the polyphenol content in cider apple. PLoS One 9: e107103.

Vitaglione P, Napolitano A, Fogliano V (2008) Cereal dietary fibre: a natural functional ingredient to deliver phenolic compounds into the gut. Trends Food Sci Tech 19: 451–463.

Vogt T (2010) Phenylpropanoid biosynthesis. Mol Plant 3: 2–20.

Vollmer KO, Reisener HJ, Grisebach H (1965) The formation of acetic acid from p-hydroxycinnamic acid during its degradation to p-hydroxybenzoic acid in wheat shoots. Biochem Biophys Res Commun 21: 221–225.

Wang J, Sun B. Cao Y et al. (2008) Optimisation of ultrasound-assisted extraction of phenolic compounds from wheat bran. Food Chem 106: 804–810.

Wang D, Song Y, Chen Y et al. (2013) Metabolic pools of phenolic acids in Salvia miltiorrhiza are enhanced by co-expression of *Antirrhinum majus* Delila and Rosea1 transcription factors. Biochem Eng J 74: 115–120.

Wang P, Dudareva N, Morgan JA et al. (2015) Genetic manipulation of lignocellulosic biomass for bioenergy. Curr Opin Chem Biol 29: 32–39.

Whent M, Huang H, Xie Z et al. (2012) Phytochemical composition, anti-inflammatory, and anti-proliferative activity of whole wheat flour. J Agric Food Chem 60: 2129–2135.

Widhalm JR, Dudareva N (2015) A familiar ring to it: biosynthesis of plant benzoic acids. Mol Plant 8: 83–97.

Wu H, Haig T, Pratley J (1999) Simultaneous determination of phenolic acids and 2,4-dihydroxy-7-methoxy-1,4-benzoxazin-3-one in wheat (*Triticum aestivum* L.) by gas chromatography–tandem mass spectrometry. J Chromatogr A 864: 315–321.

Xu W, Dubos C, Lepiniec L (2015) Transcriptional control of flavonoid biosynthesis by MYB-bHLH-WDR complexes. Trends Plant Sci 20: 176–185.

Yang F, Li W, Jiang N et al. (2017) A maize gene regulatory network for phenolic metabolism. Mol Plant 10: 498–515.

Yu L, Zhou K, Parry JW (2005) Inhibitory effects of wheat bran extracts on human LDL oxidation and free radicals. LWT-Food Sci Technol 38: 463–470.

Zeller FJ, Cermeno MC, Miller TE (1991) Cytological analysis on the distribution and origin of the alien chromosome pair conferring blue aleurone color in several European common wheat (*Triticum aestivum* L.) strains. Theor Appl Genet 81: 551–558.

Zhang B, Wang Q (2015) MicroRNA-based biotechnology for plant improvement. J Cell Physiol 230: 1–15.

Zhang J, Wang M, Cheng F et al. (2016). Identification of microRNAs correlated with citrus granulation based on bioinformatics and molecular biology analysis. Postharvest Biol Technol 118: 59–67.

Zhang S, Ma P, Yang D et al. (2013) Cloning and characterization of a putative R2R3 MYB transcriptional repressor of the rosmarinic acid biosynthetic pathway from *Salvia miltiorrhiza.* PLoS One 8: e73259.

Zhang YA, Yan YP, Wang ZZ (2010) The Arabidopsis PAP1 transcription factor plays an important role in the enrichment of phenolic acids in *Salvia miltiorrhiza*. J Agric Food Chem 58: 12168–12175.

Zhou M, Li W, Sun Z et al. (2015) Production and transcriptional regulation of proanthocyanidin biosynthesis in forage legumes. Appl Microbiol Biotechnol 99: 3797–3806.

Zhou M, Zhang K, Sun Z et al. (2017) LNK1 and LNK2 co-repressors interact with the MYB3 transcription factor in phenylpropanoid biosynthesis. Plant Physiol 174: 1348–1358.

Žilić S (2016) Phenolic compounds of wheat. Their content, antioxidant capacity and bioaccessibility. MOJ Food Process Technol 2: 85–89.

Žofajová A, Pšenáková I, Havrlentová M et al. (2012) Accumulation of total anthocyanins in wheat grain. Agriculture (Pol'nohospodárstvo) 58: 50–56.

Wheat Cell Wall Polysaccharides (Dietary Fibre)

Peter Shewry, Ondrej Kosik, Till Pellny, and Alison Lovegrove

Abstract Wheat is a major source of dietary fibre in the human diet, with whole grain containing about 11–15% fibre/g dry wt. However, in most countries wheat is most widely consumed after milling to give white flour, reducing the fibre content to less than 5%. The major dietary fibre components in white flour are the cell wall polysaccharides arabinoxylan and β-glucan. This chapter therefore focuses on these components, reviewing their structures and properties, biosynthesis, variation in amount and composition and genetic control. This provides a basis for increasing the content of wheat fibre and manipulating its properties to optimise the health benefits of wheat-based foods.

1 Introduction

Cereals, and wheat in particular, are major sources of dietary fibre. For example, in the UK, which is relatively prosperous with a highly varied diet, bread contributes about 20% of the total daily fibre intake in adults (slightly less in children), with a further 6–9% coming from breakfast cereals (which include wheat products) (Bates et al. 2014a, b; Steer et al. 2008). Although white bread contains substantially less fibre than wholemeal (as discussed below), it nevertheless accounts for 11% of the daily fibre intake in UK adults (Steer et al. 2008). This contribution can be expected to be greater in countries where higher proportions of wholegrain are consumed. For example, in Finland total cereals (which comprise between 48% and 68% wheat, depending on age and gender) account for 37% (women) and 45% (men) of the total fibre intake (Helldán et al. 2012). Fibre intake is also likely to be higher if wheat contributes a higher proportion of diet (such as in North Africa and Central Asia, where wheat can contribute 50% or more of calorific intake) (USAID 2011).

P. Shewry (✉) · O. Kosik · T. Pellny · A. Lovegrove
Department of Plant Sciences, Rothamsted Research, Harpenden, Hertfordshire, AL5 2JQ, UK
e-mail: peter.shewry@rothamsted.ac.uk

G. Igrejas et al. (eds.), *Wheat Quality For Improving Processing And Human Health*, https://doi.org/10.1007/978-3-030-34163-3_11

2 Content and Composition of Dietary Fibre in Whole Grain

Although definitions of dietary fibre vary, the most widely accepted is that proposed by the EU (Commission Directive 2008/100/EC, 28 October 2018): "carbohydrate polymers with three or more monomeric units (to exclude mono- and disaccharides, simple sugars of one or two molecules) which are neither digested nor absorbed in the small intestine". It additionally recognises that "fibre has been traditionally consumed as plant material and has one or more beneficial physiological effects…" and that it "may be closely associated in the plant with lignin or other non-carbohydrate components" which "when extracted with the carbohydrate polymers for analysis of fibre may be considered as fibre".

Whole wheat grain contains between about 11% and 15% dietary fibre, with the major components being cell wall polysaccharides, lignin, fructans and resistant starch (Table 1). However, resistant starch and fructans are considered elsewhere in this volume (Chapters 4 and 17, respectively) and will not be discussed further here.

The major cell wall polysaccharides of whole wheat grain are arabinoxylan (AX), cellulose ((1 → 4)-β-D-glucan) and (1 → 3) (1 → 4)-β-D-glucan (β-glucan) and this article therefore focuses on these components, and in particular on AX. Furthermore, our intention is to provide a broad overview, and the reader is referred to Saulnier et al. (2007) and Stone and Morell (2009) for more detailed reviews of the structures and properties of AX and of wheat carbohydrates, respectively.

AX comprises a backbone of β-D-xylopyranosyl (xylose) residues linked through (1 → 4) glycosidic linkages with some residues being substituted with α-L-arabinofuranosyl (arabinose) residues at either position 3 or positions 2 and 3 (Fig. 1a). Some arabinose residues present as single substitutions at position 3 may themselves be substituted with ferulic acid at the 5 position. Oxidation of ferulate present on adjacent AX chains may occur to give dehydrodimers (diferulate cross-links). Diferulate cross-linking is important because it affects the physio-chemical properties (notably solubility and viscosity) of AX and hence the behaviour in food processing and the gastro-intestinal (GI) tract. AX is therefore often divided into water-extractable (WE-AX) and water-unextractable (WU-AX) forms.

Cellulose ((1 → 4)-β-D-glucan) comprises glucose residues joined by (1 → 4) linkages. Individual cellulose molecules have extended ribbon-like conformations which facilitate their packing into microfibrillar arrays stabilised by hydrogen

Table 1 Contents of total dietary fibre and dietary fibre components in 129 winter wheat varieties (taken from data in Andersson et al. 2013)

	Range	Mean
Total dietary fibre (%)	11.5–15.5	13.4
Klason lignin (%)	0.74–2.03	1.33
Arabinoxylan (%)	5.53–7.42	6.49
Cellulose (%)	1.67–3.05	2.11
β-Glucan (%)	0.51–0.96	0.73
Fructan (%)	0.84–1.85	1.28

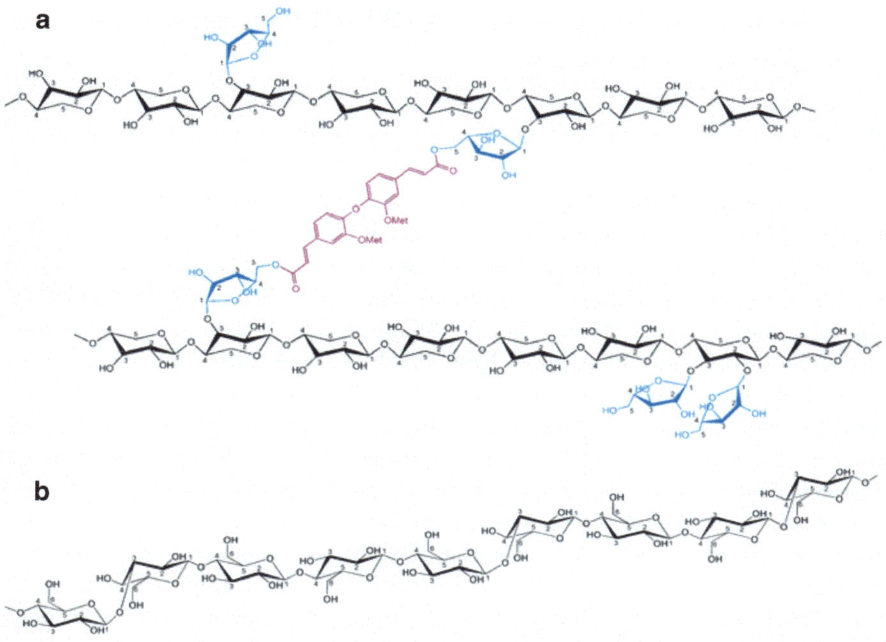

Fig. 1 Schematic structures of the major cell wall polysaccharides of wheat grain. (**a**), Arabinoxylan (AX). The xylan backbone is shown in black, arabinose linked to the 3 and 2 + 3 positions of the xylose residues in blue, and diferulate linked to the 5 positions of adjacent arabinose residues in red. (**b**), (1 → 3)(1 → 4)-β-D-Glucan (β-glucan), showing (1 → 3) and (1 → 4) linkages

bonding throughout their length. This results in a highly crystalline structure and cellulose is very difficult to solubilise.

B-glucan ((1 → 3) (1 → 4)- β -D-glucan) comprises glucose residues joined by (1 → 3) and (1 → 4) linkages (Fig. 1b). Single (1 → 3) linkages are usually separated by two or three (1 → 4) linkages, but longer stretches of (1 → 4) linked glucan units (sometimes referred to as "cellulose-like" regions) have been reported for wheat bran β-glucan (Li et al. 2006). The molar ratio of tri- to tetrasaccharides (DP3/DP4) of β-glucan extracted from wheat bran was found to be 3.7–4.5, trisaccharides accounting for 67.1–72.3%, tetrasaccharides for 21–24.2% and oligosaccharides of 5 and above residues for 6.7–8.7% (Lazaridou and Biliaderis 2007). The longest oligosaccharide found in mixed-linkage glucan from wheat bran was 12 residues (Lazaridou et al. 2004). However, detailed studies of the structure of β-glucan from wheat starchy endosperm have not been reported. Wheat β-glucan shows low solubility, with about 10–15% of the total in wholemeal samples being soluble in hot water (Nemeth et al. 2010).

Lignin is only present in the pericarp/seed coat of wheat (Stone and Morell 2009) and hence recovered on milling in the bran but not white flour. It is a complex polymer of aromatic alcohols and the detailed structure of cereal grain lignin has been described by Bunzel et al. (2004).

3 Contents and Composition of Cell Wall Polysaccharides in Wheat Grain Tissues

The amounts and proportions of cell wall polysaccharides vary between tissues, as summarized in Table 2.

The cell walls of the starchy endosperm account for about 2–3% of the dry weight. Although they have been studied in detail for many years, only two studies have reported the relative proportions of individual polysaccharides. The widely accepted composition, dating from analyses reported by Mares and Stone (1973), states that they comprise about 70% AX, 20% β-glucan, 2% cellulose and 7% glucomannan (a polymer comprising chains of (1 → 4)- β -linked D-mannose and (1 → 4)- β -linked D-glucose units) (Mares and Stone 1973). However, a recently published study reports a substantially higher proportion of cellulose (28.4%) and lower proportions of AX (60.9%) and β-glucan (6.3%), with 4% glucomannan (Gartaula et al. 2018). These data are therefore also included in Table 2.

Gartaula et al. (2018) also reported 0.3% arabinan and 0.3% xyloglucan, and immunolabelling of developing tissues has also shown the presence of xyloglucan, together with callose ((1 → 3)-β-D-glucan) and pectin (Chateigner-Boutin et al. 2014; Palmer et al. 2015; Pellny et al. 2012). Starchy endosperm AX contains only low levels of ferulic acid: 0.2–0.4% (w/w) of WE-AX and 0.6–0.9% (w/w) of WU-AX (Bonnin et al. 1998). The structural features that determine AX solubility are still not fully understood, but are considered to include the degree of arabinosylation (A:X ratio), the extent of diferulate cross-linking and the molecular weight (as discussed by Saulnier et al. 2007).

The aleurone cells have thick cell walls which account for about 35–40% of the dry weight (Barron et al. 2007). These comprise 29% β-glucan, 65% arabinoxylan and 2% each of cellulose and glucomannan (Bacic and Stone 1981). However, the

Table 2 Contents and compositions of cell walls in wheat grain tissues (% dry weight) (revised based on Shewry et al. 2010b)

Tissue	Cell walls (% dry weight)	Components				
		Cellulose	Lignin	Xylan	β-glucan	Glucomannan
Starchy endosperm Mares and Stone, 1973)	2–3	2	0	70	20	7
Starchy endosperm (Gartaula et al. 2018)	Not determined	28.4	Not determined	60.9	6.3	4.0
Bran		29	8	64	6	Not determined
Aleurone	40	2–4	0	62–65	29–34	2
Outer pericarp (beeswing)		30	12	60	Not determined	Not determined

ratio of arabinose to xylose is lower than that of starchy endosperm AX, 0.41 and 0.47 for two cultivars compared to 0.81 and 0.87, respectively (Barron et al. 2007). The aleurone AX are highly esterified and cross-linked with about 3.2% of the AX dry weight being ferulic acid and 0.45% being diferulic acid (Antoine et al. 2003; Parker et al. 2005). Additional esterification with *p*-coumaric acid and acetyl groups also occurs (Antoine et al. 2004; Rhodes and Stone 2002).

The outer layers together comprise about 45–50% cell wall material (Barron et al. 2007). The major tissue is the pericarp which comprises about 30% cellulose, 60% arabinoxylan and 12% lignin (Stone and Morell 2009). The pericarp AX has a complex structure, being highly branched with galactose and glucuronic acid residues, and is often termed glucuronoarabinoxylan (GAX). It also has high contents of ferulic acid and diferulic acids (Antoine et al. 2003; Parker et al. 2005; Saulnier and Thibault 1999), ferulic acid trimer (Barron et al. 2007) and acetylation (Mandalari et al. 2005). Immunolabelling shows that the aleurone and pericarp cell walls also contain pectic polysaccharides (Chateigner-Boutin et al. 2014; Palmer et al. 2015).

The scutellum and embryonic axis of the germ contain about 12% and 25% of neutral carbohydrate, respectively, with arabinose and xylose accounting for about 65% of the total sugars released on hydrolysis (Barron et al. 2007). Other sugars released were glucose, galactose and (from the embryonic axis only) mannose.

4 Biosynthesis of Cell Wall Polysaccharides

Carbohydrate-Active en-Zymes (CAZymes) are responsible for the synthesis, modification and degradation of carbohydrate polymers in plants (Pinard et al. 2015). They are organised into classes based on their amino acid sequence similarity, integrating both structural and mechanistic features (Henrissat and Davies 2000). The glycosyl transferase (GT) enzymes are capable of transferring an activated sugar nucleotide (NDP-sugar) onto a specific acceptor and can catalyse formation of glycosidic bonds. The enzymes responsible for the synthesis of cell wall polysaccharides, and their encoding genes, have been largely identified over the past two decades.

The xylan backbone of AX is synthesised by three proteins encoded by the IRX9, IRX10 and IRX14 genes, with IRX9 and IRX14 being members of the glycosyltransferase 43 (GT43) family and IRX10 a member of the GT47 family (Lovegrove et al. 2013). It is proposed that these three proteins form a single xylan synthase complex (XSC) anchored inside the Golgi apparatus (Zeng et al. 2016). The addition of arabinose units to the xylan backbone is more complex and less well-understood. It appears that at least four specific enzymes are required: to add arabinose to the 3 position of monosubstituted xylose and to the 2 and 3 positions of disubstituted xylose, and to add feruloylated arabinose to the 3 position of monosubstituted xylose. Only one of these, which is encoded by the XAT1 gene (a member of GT61 family) has been identified so far; responsible for adding arabinose residues at the 3 position to give mono-substituted xylose in wheat (Anders et al. 2012). It has also been shown that the glucuronic acid substitution present in GUX is controlled by GUX1 and GUX2 (both GT8) (Bromley et al. 2013).

The synthesis of β-glucan, unlike other hemicelluloses, takes place predominantly at the plasma membrane, close to the cell wall (Lockhart 2015). Two major families of enzymes are involved, which are members of the cellulose synthase-like (Csl) GT2 family called CslF and CslH. CSLF6 and CSLH1 are the dominant genes involved in β-glucan synthesis in the walls of grasses including wheat grain endosperm (Wilson et al. 2015). Expression of a CSLF6 RNAi constructs resulted in a 30–52% decrease in β-glucan in wheat grain (Nemeth et al. 2010). The CSLF6 enzyme catalyses the formation of both $(1 \rightarrow 3)$ and $(1 \rightarrow 4)$ linkages, with the position and flexibility of the catalytic site determining the fine structure of the polymer (Dimitroff et al. 2016).

Cellulose chains are also synthesised at the plasma membrane, by cellulose synthase (CesA) complexes (Schneider et al. 2016) which themselves are assembled in the Golgi apparatus or endoplasmatic reticulum with the aid of STELLO proteins (Zhang et al. 2016) and then transported to and inserted into the plasma membrane via the trans-Golgi network and small vesicles (Crowell et al. 2009). Although it has been generally thought that the CesA complex comprises 36 subunits, recent work suggests they may comprise only 18 or 24 subunits (Kumar et al. 2017). The cellulose synthase complex (rosette) responsible for synthesising the primary cell wall (as in the wheat starchy endosperm) is composed of CesA1, CesA3 and CesA6 subunits, all of which are part of the GT2 family (Gonneau et al. 2014).

5 Genetic and Environmental Impacts on the Content and Extractability of AX and β-Glucan

The studies in Table 1 showed substantial variation in the contents of cell wall polysaccharides in whole grain of wheat. These proportions will be expected to vary with seed size and shape, which will affect the proportions of bran and starchy endosperm, as well as variation in the compositions of the individual tissues.

A number of quantitative studies on the contents of AX and β-glucan have been reported, some of which are summarised in Table 3. This includes data only from studies where 10 or more genotypes have been analysed, and includes data only for wholemeal and white flour. Data for bran have been omitted, as the use of small scale laboratory mills in some studies will result in low and variable flour yields (and hence variation in the purity of brans).

These data show wide variation in content, from 0.26–1.40% WE-AX, 1.35–2.87% TOT-AX and 0.25–0.63% β-glucan in white flour, and from 0.34–0.92% WE-AX, 3.10–10.74% TOT-AX and 0.18–1.18% β-glucan in whole grain. The data of Gebruers et al. (2008) are shown in Fig. 2, which also shows that the proportion of WE-AX varies from about 20–50% of TOT-AX; this may affect the processing properties of flours derived from these wheats.

Table 3 Variation in the contents (% dry wt) of TOT-AX, WE-AX and total β-glucan in whole grain and white flour from comparative studies of wheat genotypes

Number of Lines	Tissue	Total β-glucan			WE-AX			Total AX			
		min	max	mean	min	max	mean	min	max	mean	Ref[(1)]
18	whole grain				0.34	0.83	0.59	4.52	6.79	5.89	1
49	flour	0.25	0.63	0.44	0.48	1.23	0.81				2
20	flour				0.36	0.78	0.56				3
22	whole grain				0.36	0.83	0.56	5.53	7.79	6.36	4
22	flour				0.26	0.91	0.49				5
20	flour				0.26	0.75	0.51	1.66	2.87	2.18	6
20	whole grain							4.79	6.92	5.76	
151	flour				0.30	1.40	0.50	1.35	2.75	1.90	7
	whole grain	0.50	0.95	0.75							
23/6	flour				0.30	0.75	0.50	1.65	2.75	2.00	8
	whole grain	0.50	0.65	0.60							
50	whole grain				0.39	0.92	0.67	3.10	4.70	3.99	9
338	whole grain	0.22	1.18					3.75	8.30		10
27	Whole grain	0.18	0.65					4.34	10.74		

[(1)]1. Hong et al. (1989) (analyses of 7 hard red, 7 hard white and 4 club wheats grown on two sites. Values reported for pentosans, recalculated assuming 10% water content) 2. Andersson et al. (1992) (determined by sugar analysis); 3. Andersson et al. (1994); 4. Saulnier et al. (1995); 5. Martinant et al. (1999) (22 lines grown at 3 locations for 2 years); 6. Ordaz-Ortiz and Saulnier (2005); 7. Gebruers et al. (2008) (131 winter and 20 spring lines grown on one site); 8. Gebruers et al. (2010a, b) (23 lines grown on 6 sites and 3 lines on 5 sites); Li et al. (2006) (25 spring and 25 winter wheats each grown in 3 environments); 10. Pritchard et al. (2011) (338 Australian winter wheats grown at various times and locations in Australia, 27 Chinese wheats grown in China).

In most studies it is not possible to partition this variation between the effects of genotype, environment and interactions between these factors. However, several studies have estimated the heritability of AX fractions based on analyses of samples grown in multiple environments, in most cases these have shown heritabilities of above 50%. For example, Hong et al. (1989) calculated that the genotypic variance was 1.6 times greater than the environmental variance for WE-AX (pentosans) and 2.4 times for TOT-AX (pentosans) in wholemeal flours of 18 wheat lines grown on two sites in the USA, while Martinant et al. (1999) calculated the broad sense heritability (genotypic variance/phenotypic variance) for WE-AX in white flour of 19 cultivars grown on three locations in France as 0.75. Similarly, Dornez et al. (2008) calculated broad sense heritabilities of 0.53 for TOT-AX and 0.96 for WE-AX in wholemeal flour of 14 cultivars grown in Belgium for three years while Shewry et al. (2010a), using material reported by Gebruers et al. (2010a), reported 60% heritability for WE-AX and 72% heritability for TOT-AX in white flour, based on multisite analyses of 26 lines. Yang et al. (2016) similarly reported broad sense heritabilities of 0.50 for TOT-AX, 0.38 for WU-AX and 0.71 for WE-AX in a population of 240 recombinant inbred lines.

Finnie et al. (2006) also determined the effects of genotype and environment on AX in 7 spring and 20 winter wheats grown in 10 and 12 environments,

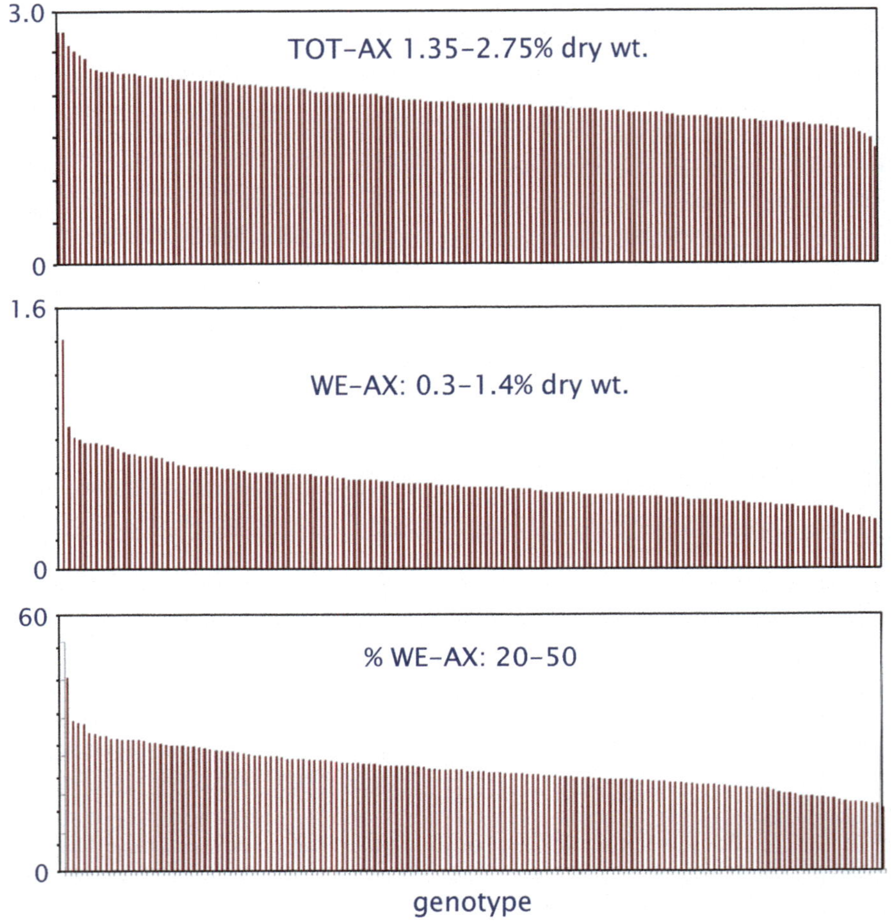

Fig. 2 Contents of TOT-AX and WE-AX and % of WE-AX in white flour of 151 wheat genotypes grown on the same site. Plotted from data reported by Gebruers et al. (2008)

respectively. Genotype was the primary determinant of variation in the amounts of TOT-AX and WE-AX, but environment had a greater effect on the amount of water-unextractable AX. Even greater effects of environment were reported by Li et al. (2006), with environment having much greater effects on both TOT-AX and WE-AX than the genotype.

Comparisons of durum wheat cultivars grown in different environments also showed effects of genotype and genotype x environment on AX (Ciccoritti et al. 2011) and on AX and β-glucan (De Santis et al. 2018).

Shewry et al. (2010a) compared weather data with the analyses of AX reported by Gebruers et al. (2010a) to show a significant negative correlation between the %WE-AX in white flour and the mean temperature between heading and harvest, and a significant positive correlation with the total precipitation over the same period.

The differences in AX solubility discussed above may be determined by differences in AX structure, including the degree of arabinose substitution (with high substitution expected to increase solubility) ferulic acid cross-linking (expected to decrease solubility) and chain length, but will also be affected by the activities of xylanases and xylanase inhibitors, as discussed by Gebruers et al. (2010b). In particular, xylanases are produced during pre-harvest sprouting.

6 Variation in AX and β-glucan Structure

The simplest approach to compare the structure of AX and β-glucan in genotypes is by "enzyme fingerprinting". In this procedure the AX is hydrolysed by a specific endoxylanase and the oligosaccharides released (called AXOS) then separated by high performance ion-exchange chromatography (HP-AEC). The structures of the AXOS produced by specific endoxylanases have been determined by mass spectroscopy (Ordaz-Ortiz and Saulnier 2005), allowing their proportions to be used as a fingerprint for variation in structure (and particularly the degree of arabinosylation). Similarly, digestion of β-glucan ((1 → 3,1 → 4)-β-D-glucan) with lichenase (endo-β-glucanase) releases mainly oligosaccharides (GOS) of three (G3) and 4 (G4), reflecting the relative distributions of the (1 → 3) and (1 → 4) linkages.

Enzyme fingerprinting was therefore used to compare AX structure in white flour of 150 lines grown on a single site, with the datasets being compared by Principal Component Analysis (PCA) (Shewry et al. 2010c) (Fig. 3). The major separation is in dimension 1, which reflects differences in the degree of arabinosylation of the AXOS released by digestion.

Variation in the structure of AX in different parts of the starchy endosperm has also been demonstrated, using fingerprinting and spectroscopic analyses of tissue samples, and direct microspectroscopic imaging of sections of tissue. Fourier-transform infra-red (FT-IR) microspectroscopy has been used to determine differences in the degree of arabinosylation of AX, showing that this varies across the starchy endosperm, with the AX in the central cells being more highly substituted than that in the outer cells (Barron et al. 2005). Furthermore, this gradient varies between cultivars (Saulnier et al. 2009; Toole et al. 2011), as shown for six cultivars grown together on the same site in Fig. 4 (Toole et al. 2011), and is affected by the growth conditions, with the degree of substitution being lower in grain grown under hot dry conditions (Toole et al. 2007). Velickovic et al. (2016) used a different approach, mass spectroscopy imaging (MALDI-MS), to determine two oligosaccharide fractions released from AX, a pentasaccharide called AX5 (containing a single arabinose substitution) and a hexasaccharide called AX6 (containing one disubstituted xylose). The ratio of these two oligosaccharides was taken to represent differences in AX substitution, as measured by FT-IR above, with similar results in terms of distribution within the grain (AX6 being enriched in the centre of the grain) and differences between cultivars. Variation also occurs in the ratio of β-glucan to AX within different parts of the endosperm (Saulnier et al. 2009) and in the

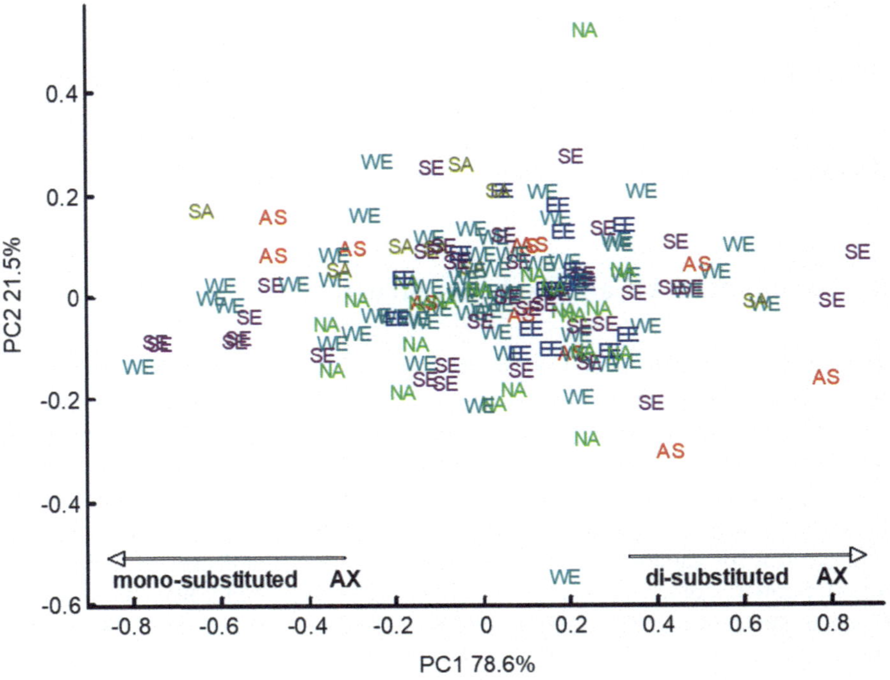

Fig. 3 Principal Component Analysis of enzyme fingerprinting of AX from lines in the HEALTHGRAIN diversity screen (150 lines). PC1 and PC2 are the first two principal components in PCA. WE: cultivars originating from Western Europe (58), Austria (4), France (23), Germany (12), United Kingdom (15), Switzerland (3), Netherlands (1); SE: Southern Europe and Mediterranean countries (23): Italy (15), Bulgaria 93), Romania (5), Turkey (4), Israel (1); EE: Eastern and Central Europe (25): Hungary (8), Yugoslavia (6), Croatia (1), Czech Republic (1), Kazakhstan (1), Russia (5), Poland (3); AS: Asia and Oceania (14): Australia (8), China (3), Korea (2), New Zealand (1); NA: North America (21): Canada (6), USA (15); SA: South America(8): Argentina (2), Mexico (6). Taken from Shewry et al. (2010c)

structure of β-glucan (shown by the ratio of G3 and G4 fragments determined by MALDI-MS imaging) (Velickovic et al. 2016).

These differences in AX and β-glucan amount and structure across the starchy endosperm indicate that similar differences will be observed between milling fractions as well as between grain samples of different genotypes and when grown under different conditions. This poses a challenge for delivering defined amounts of fibre in wheat-based foods.

7 Genetic Control of AX Amount

A number of quantitative trait loci (QTLs) for AX have been reported, based on classical Mendelian crosses and association genetics. The earliest report, by Martinant et al. (1999) used two mapping populations and determined WE-AX, extract viscosity (which is largely determined by WE-AX) and the ratio of

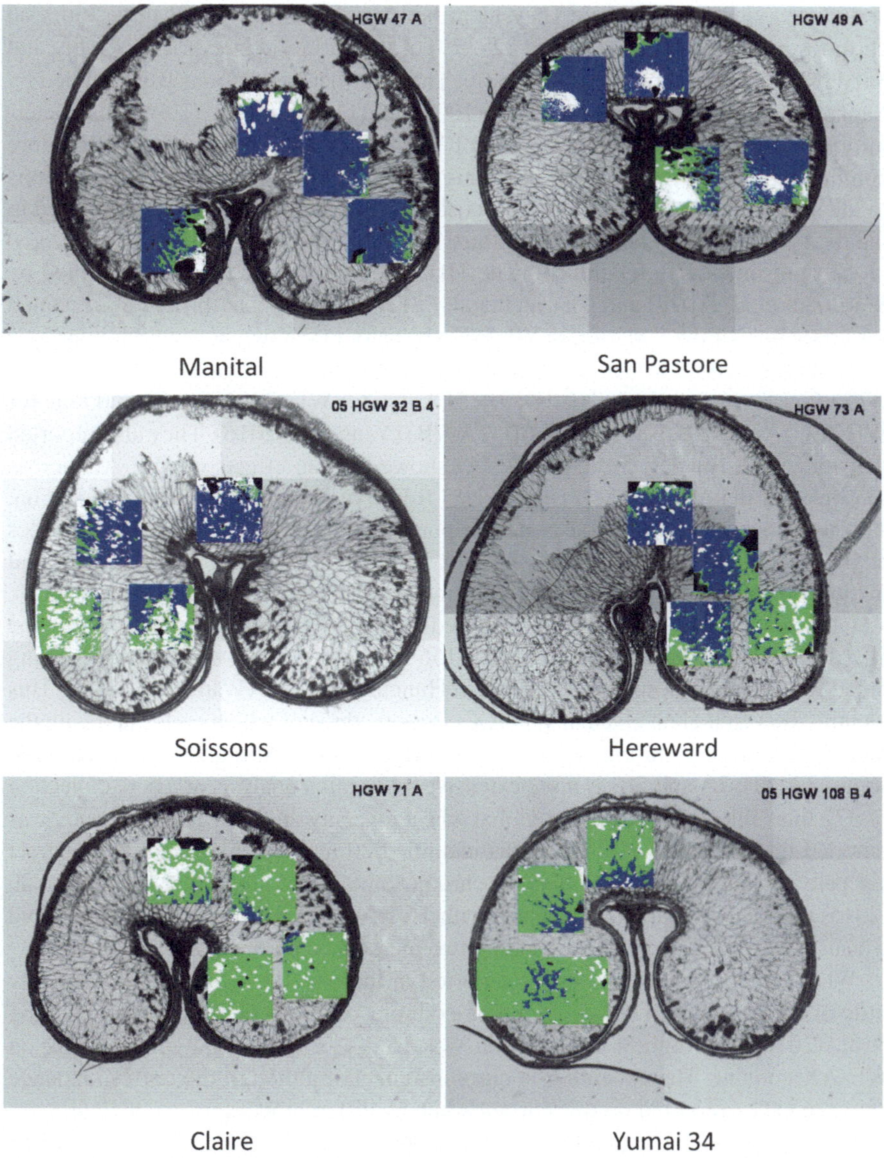

Fig. 4 FT-IR microspectroscopy images of areas of cell-wall only sections of grain of six wheat genotypes FT-IR overlaid onto the corresponding visible microscope images (prepared as described by Barron et al., 2005; Toole et al., 2007). Images are coloured to represent differences in the degree of substitution of the AX with arabinose, with areas of high arabinosylation being shown in blue and low arabinosylation in green. Taken from Toole et al. (2011)

arabinose:xylose in WE-AX. They identified a major QTL for all three traits on chromosome 1B, which explained 32–37% of the variation in extract viscosity and 35–42% of the variation in the A:X ratio. Three further populations were studied by Laperche et al. (2007), Perretant et al. (2000), and Quraishi et al. (2009), also determining extract viscosity as a measure of WE-AX. Quraishi et al. (2011) therefore combined data from these three populations with analyses of two new populations to identify "meta-QTL" for extract viscosity. This allowed the 12 QTL identified in the five populations to be reduced to three meta-QTL for WE-AX viscosity, located on chromosomes 1B, 3D and 6B. The 1B QTL corresponded to that identified by Martinant et al. (1999) and Charmet et al. (2009) reported that this QTL accounted for up to 59% of the variation in WE-AX viscosity in the two new populations.

However, a more recent analysis of 240 inbred lines reported a larger number of QTL: four for WE-AX (on 1B, 1D, 3B, 5B), two for WU-AX (1B, 1D) and nine for WE-AX 1A, 1B, 2B, 3B, 5A, 5B, 6B, 7A, 7B) (Yang et al. 2016). They also reported that four QTL (on 1B, 5A, 5B and 7A) showed stable effects on WE-AX across seasons, but that the 1B QTL was derived from rye on the 1BL/1RS translocation. This translocation has previously been reported to result in higher WE-AX (Bordes et al. 2011; Selanere and Andersson 2002; Yang et al. 2014) but is not now present in most breadmaking wheats due to detrimental effects on quality (Graybosch 2001).

Quraishi et al. (2011) also reported association genetic analysis of the HEALTHGRAIN diversity collection of 156 wheat lines (131 winter and 20 spring bread wheats and five spelt lines grown in Hungary in 2005) (Ward et al. 2008). This identified seven loci involved in WE-AX viscosity, three of which co-located with the meta-QTLs on chromosomes 1B, 3D and 6B and four additional loci on chromosomes 3A, 5B, 7A and 7B. A more extensive association analysis of a core collection of 372 lines selected to represent global wheat diversity and grown in two years was reported by Bordes et al. (2011). This identified 20 markers associated with extract viscosity, including a strong effect on chromosome 1BL. Similarly, Marcotuli et al. (2015) reported 19 markers associated with TOT-AX in a collection of 104 tetraploid wheats grown on a single site, including two markers on chromosome 1B.

With few exceptions, the QTLs reported in these studies account for relatively little of the variation in AX content. For example, of the 15 QTLs reported by Yang et al. (2016), the highest contribution was 15.2% of the phenotypic variance in WE-AX content. The exception is chromosome 1B, although the correspondence between QTLs reported on this chromosome by different workers is not clear.

Our studies have focused on the Chinese bread wheat cultivar Yumai 34, which was shown by Gebruers et al. (2008) to have the highest contents of both TOT-AX and WE-AX out of 150 genotypes grown on the same site. We have analysed lines from four crosses, with the cultivars Ukrainka, Claire, Altai and Valoris. The latter cross is particularly interesting, as Valoris also has a high content of AX in flour (Gebruers et al., 2008) and was therefore used as a parent by Charmet et al. (2009). The lines from these crosses were mapped using the 35 K Axiom array, which comprises 35,000 markers randomly distributed on the wheat genome. This facilitated the identification of two major QTLs, on chromosome 1B from Yumai 34 and chromosome 6B from Valoris, and the development of a molecular marker for the 1B QTL which was validated by analysis of high AX lines selected using biochemical screening (Lovegrove et al. 2020).

8 Future Prospects

The existence of wide variation in the AX content of white flour, and the high heritability, indicate that it should be amenable to selection in plant breeding programmes. Thus, WE-AX and TOT-AX in flour are particularly attractive targets for selection in plant breeding programmes. However, there have been two major limitations to progress.

Firstly, although Charmet et al. (2009) suggested that their 1B QTL accounted for up to 59% of the variance in AX amount in crosses, this QTL was not precisely mapped and was not identified as having a similar major effect in other studies. Secondly, the lack of tightly linked markers or simple chemical screens (such as NIR spectroscopy) means that selection requires lengthy and expensive chemical analyses. Nevertheless, Tremmel-Bede et al. (2017) showed that high AX lines with competitive yields and good agronomic performance can be developed, using chemical analysis for selection. These lines exploited the strong 1B QTL in Yumai 34 and the availability of molecular markers for this and for the 6B QTL from Valoris should allow these two QTLs to be stacked to achieve even higher contents of AX in flour.

The identification of genes for AX synthesis, and the demonstration that they can be down-regulated in transgenic wheat with effects on AX amount, structure and properties (Anders et al. 2012; Freeman et al. 2016, 2017; Lovegrove et al. 2013), indicates that GM and mutagenesis (TILLING) (Targeting Induced Local Lesions in Genomes) (McCallum et al. 2000) may be used in the future to fine tune the amount and properties of AX for specific end uses.

There is also a much more serious barrier to progress, which is a lack of demand for grain and flour with high levels of endogenous fibre from consumers or grain processors. Until this demand is generated it is unlikely that breeders will embark on lengthy and costly breeding programmes for high fibre wheat. However, there is now a large body of literature detailing the many health benefits of cereal fibre consumption including reducing the risk of many chronic diseases such as coronary heart disease, stroke, diabetes, some types of cancer (Anderson et al. 2009; Buttriss and Stokes 2008; SACN 2005). It is to be hoped that the provision of molecular tools to breeders to enable them to select wheat lines with increased dietary fibre content easily with no yield penalty will, in the near future provide the pull for the breeding of healthier wheat cultivars.

Acknowledgements Rothamsted Research receives grant-aided support from the Biotechnology and Biological Sciences Research Council (BBSRC) of the UK and the work forms part of the Designing Future Wheat strategic programme (BB/P016855/1).

References

Anders N, Wilkinson MD, Lovegrove A, Freeman J, Tryfona T, Pellny TK, Weimar T, Mortimer JC, Stott K, Baker JM, Defoin-Platel M, Shewry PR, Dupree P, Mitchell RAC (2012) Glycosyl transferases in family 61 mediate arabinofuranosyl transfer onto xylan in grasses. Proceedings of the National Academy of Sciences *USA* 109: 989–993.

Andersson R, Westerlund E, Aman P (1994) Natural variations in the contents of structural elements of water-extractable non-starch polysaccharides in white flour. Journal of Cereal Science 19: 77–82.

Andersson R, Westerlund E, Tilly A-C, Aman P (1992) Natural variations in the chemical composition of white flour. Journal of Cereal Science 17: 183–189.

Andersson AAM, Andersson R, Piironen V, Lampi A-M, Nystrom L, Boros D, Fras A, Gebruers K, Courtin CM, Delcour JA, Raskzegi M, Bedo Z, Ward JL, Shewry PR, Aman P (2013) Contents of dietary fibre components and their relation to associated bioactive components in whole grain wheat samples from the HEALTHGRAIN diversity screen. Food Chemistry 136: 1243–1248.

Anderson JW, Baird P, Davis Jr RH, Ferreri S, Knudtson M, Koraym A, Waters V, Williams C.L (2009) Health benefits of dietary fiber. Nutrition Reviews 67: 188–205.

Antoine C, Peyron S, Lullien-Pellerin V, Abecassis J, Rouau X (2004), Wheat bran tissue fractionation using biochemical markers, Journal of Cereal Science 39: 387–393.

Antoine C, Peyron S, Mabille F, Lapierre C, Bouchet B, Abecassis J, Rouau X (2003) Individual contribution of grain outer layers and their cell wall structure to the mechanical properties of wheat bran. Journal of Agriculture and Food Chemistry 51: 2026–2033.

Bacic A, Stone BA (1981) Chemistry and organisation of aleurone cell wall components from wheat and barley. Australian Journal of Plant Physiology 8: 475–495.

Barron C, Parker ML, Mills ENC, Rouau X, Wilson RH (2005) FT-IR imaging of wheat endosperm cell walls in situ reveals compositional and architectural heterogeneity related to grain hardness. Planta 220: 667–677.

Barron C, Surget A, Rouau X (2007) Relative amounts of tissues in mature wheat (*Triticum aestivum* L.) grain and their carbohydrate and phenolic acid composition. Journal of Cereal Science 45: 88–96.

Bates B, Lennox A, Prentice A, Bates C, Page P, Nicholson S, Swan G (Eds) (2014a) National Diet and Nutrition Survey: Results from Years 1–4 (combined) of the Rolling Programme (2008/2009–2011/2012). Executive Summary, Public Health England, London, UK.

Bates B, Lennox A, Prentice A, Bates C, Page P, Nicholson S, Swan G (Eds) (2014b) National Diet and Nutrition Survey: Results from Years 1–4 (combined) of the Rolling Programme (2008/2009–2011/2012). Public Health England, London.

Bonnin E, Le Goff A, Saulnier L, Charand M, Thibault JF (1998) Preliminary characterisation of endogenous wheat arabinoxylan-degrading enzymic extracts. Journal of Cereal Science 28: 53–62.

Bordes J, Ravel C, Le Gouis J, Lapierre A, Charmet G, Balfourier F (2011) Use of a global wheat core collection for association analysis of flour and quality traits. Journal of Cereal Science 54: 137–147.

Bromley JR, Busse-Wicher M, Tryfona T, Mortimer JC, Zhang Z, Brown DM, Dupree P (2013) GUX1 and GUX2 glucuronyltransferases decorate distinct domains of glucuronoxylan with different substitution patterns. Plant Journal 74: 423–434.

Bunzel M, Ralph J, Lu F, Hatfield R, Steinhart H (2004) Lignins and ferulate-coniferyl alcohol cross-coupling products in cereal grains. Journal of Agricultural and Food Chemistry 52: 6496–6502.

Buttriss JL, CS Stokes (2008) Dietary fibre and health, an overview. Nutrition Bulletin 33: 186–200.

Charmet G, Masood-Quraishi U, Ravel C, Romeuf I, Balfourier F, Perretant MR, Joseph JL, Rakszegi M, Guillon F, Sado PE, Bedo Z, Saulnier L (2009) Genetics of dietary fibre in bread wheat, Euphytica 170: 155–168.

Chateigner-Boutin AL, Bouchet B, Alvarado C, Bakan B, Guillon F (2014) The wheat grain contains pectic domains exhibiting specific spatial and development-associated distribution. PLoS One PMID: 24586916.

Ciccoritti R., Scalfati G., Cammerata A., Sgrulletta, D (2011) Variations in content and extractability of durum wheat (*Triticum turgidum* L. var *durum*) arabinoxylans associated with genetic and environmental factors. International Journal of Molecular Sciences 12: 4536–4549.

Crowell EFF, Bischoff V, Desprez T, Rolland A, Stierhof Y-D, Schumacher K, Gonneau M, Höfte H, Vernhettes S (2009) Pausing of Golgi Bodies on Microtubules Regulates Secretion of Cellulose Synthase Complexes in *Arabidopsis*. The Plant Cell 21: 1141–1154.

De Santis MA, Kosik O, Passmore D, Flagella Z, Shewry PR (2018) Comparison of the dietary fibre composition of old and modern durum wheat (*Triticium turgidum* spp. *Durum)* genotypes. Food Chemistry 244: 304–310.

Dimitroff G, Little A, Lahnstein J, Schwerdt JG, Srivastava V, Bulone V, Burton RA, Fincher GB (2016) (1,3;1,4)-β-Glucan Biosynthesis by the CSLF6 Enzyme: Position and Flexibility of Catalytic Residues Influence Product Fine Structure. Biochemistry 55: 2054–2061.

Dornez E, Gebruers K, Joye IJ, de Ketelaere B, Lenartz J, MassauxC, Bodson B, Delcour JA, Courtin CM (2008) Effects of genotype, harvest year and genotype-by-harvest year interactions on arabinoxylan, endoxylanase activity and endoxylanase inhibitor levels in wheat kernels. Journal of Cereal Science 47: 180–189.

Finnie SM, Bettge AD, Morris CF (2006) Influence of cultivar and environment on water-soluble and water-insoluble arabinoxylans in soft wheat. Cereal Chemistry 83: 617–623.

Freeman J, Lovegrove A, Wilkinson MD, Saulnier L, Shewry PR, Mitchell RAC (2016) Effect of suppression of arabinoxylan synthetic genes in wheat endosperm on chain length of arabinoxylan and extract viscosity. Plant Biotechnology Journal 14: 109–116.

Freeman J, Ward JL, Kosik O, Lovegrove A, Wilkinson MD, Shewry PR, Mitchell RA C (2017) Feruloylation and structure of arabinoxylan in wheat endosperm cell walls from RNAi lines with suppression of genes responsible for backbone synthesis and decoration. Plant Biotechnology Journal 15: 1429–1438.

Gartaula G, Dhital S, Netzel G, Flanagan BM, Yakubov GE, Beahan CT, Collins HM, Burton RA, Bacic A, Gidley MJ (2018) Quantitative structural organisation model for wheat endosperm cell walls: Cellulose as an important constituent. Carbohydrate Polymers 196: 199–208.

Gebruers K, Dornez E, Boros D, Frás A, Dynkowska W, Bedő Z, Rakszegi M, Delcour JA, Courtin CM (2008) Variation in the content of dietary fiber and components thereof in wheats in the HEALTHGRAIN Diversity screen. Journal of Agricultural and Food Chemistry 56: 9740–9749.

Gebruers K, Dornez E, Bedo Z, Rakszegi M, Frais A, Boros D, Courtin CM, Delcour JA (2010a) Environment and genotype effects on the content of dietary fiber and it's components in wheat in the HEALTHGRAIN diversity screen. Journal of Agricultural and Food Chemistry 58: 9353–9361.

Gebruers K, Dornez E, Bedo Z, Rakszegi M, Courtin CM, Delcour JA (2010b) Variability in xylanase and xylanase inhibitor activities in different cereals in the HEALTHGRAIN diversity screen and contribution of environment and genotype to this variability in common wheat. Journal of Agricultural and Food Chemistry 58: 9362–9371.

Gonneau M, Desprez T, Guillot A, Vernhettes S, Hofte H (2014) Catalytic Subunit Stoichiometry within the Cellulose Synthase Complex. Plant Physiology 166: 1709–1712.

Graybosch RA (2001) Uneasy unions: Quality effects of rye chromatin transfers to wheat. Journal of Cereal Science 33: 3–16.

Helldán A, Raulio S, Kosola M, Tapanainen H, Ovaskainen M-L, Virtanen S (2012) The National FINDIET 2012 Survey. National Institutes of Health and Welfare, Helsinki, Finland.

Henrissat B, Davies GJ (2000) Glycoside Hydrolases and Glycosyltransferases. Families, Modules, and Implications for Genomics. Plant Physiology 124: 1515–1519.

Hong BH, Rubenthaler GL, Allen RE (1989) Wheat pentosans. I. Cultivar variation and relationship to kernel hardness. Cereal Chemistry 66: 369–373.

Kumar M, Atanassov I, Turner S (2017) Functional Analysis of Cellulose Synthase (CESA) Protein Class Specificity. Plant Physiology 173: 970–983.

Laperche A, Brancourt-Hulmel M, Heumez E, Gardet O, Hanocq E, Devienne-Barret F, Le Gouis J (2007) Using genotype x nitrogen interaction variables to evaluate the QTL involved in wheat tolerance to nitrogen constraints. Theoretical and Applied Genetics 115: 399–415.

Lazaridou A, Biliaderis C G, Micha-Screttas M, Steele BR (2004) A comparative study on structure-function relations of mixed-linkage (1→3), (1→4) linear β-D-glucans. Food Hydrocolloids 18: 837–855.

Lazaridou A, Biliaderis CG (2007) Molecular aspects of cereal β-glucan functionality: Physical properties, technological applications and physiological effects. Journal of Cereal Science 46: 101–118.

Li W, Cui SW, Kakuda Y (2006) Extraction, fractionation, structural and physical characterization of wheat β-D-glucans. Carbohydrate Polymers 63: 404–408.

Lockhart J (2015) Uncovering the Unexpected Site of Biosynthesis of a Major Cell Wall Component in Grasses. The Plant Cell 27: 483.

Lovegrove A, Wilkinson MD, Freeman J, Pellny TK, Tosi P, Saulnier L, Shewry PR, Mitchell RAC (2013) RNAi suppression of genes in glycosly transferase families 43 and 47 in wheat starchy endosperm causes large decreases in arabinoxylan content. Plant Physiology 163: 95–107.

Lovegrove A, Wingen LU, Plummer A, Wood A, Passmore D, Kosik O, Freeman J, Mitchell RAC, Hassall K, Ulker M, Tremmel-Bede K, Rakszegi M, Bedo Z, Perretant M-R, Charmet G, Pont C, Salse J, Leverington Waite M, Orford S, Burridge A, Pellny TK, Shewry PR, Griffiths S. (2020) Identification of a major QTL and associated molecular marker for high arabinoxylan fibre in white flour. PLOS ONE https://doi.org/10.1371/journal.pone.0227826.

Mandalari G, Faulds CB, Sancho AI, Bisignano G, LoCurto R, Waldron KW (2005) Fractionation and characterisation of arabinoxylans from brewers' spent grain and wheat bran. Journal of Cereal Science 42: 205–212.

Marcotuli I, Houston K, Waugh R, Fincher GB, Burton RA, Blanco A, Gadaleta A (2015) Genome wide association mapping for arabinoxylan content in a collection of tetraploid wheats. PLOS One 10(7) e0132787.

Mares DJ, Stone BA (1973) Studies on wheat endosperm. I. Chemical composition and ultrastructure of the cell walls. Australian Journal of Biological Science 26: 793–812.

Martinant JP, Billot A, Bouguennec A, Charmet G, Saulnier L, Branlard G (1999) Genetic and environmental variations in water-extractable arabinoxylans content and flour extract viscosity. Journal of Cereal Science 30: 45–48.

McCallum CM, Comai L, Greene EA, Henikoff S (2000) Targeted screening for induced mutations. Nature Biotechnology 18: 455–7.

Nemeth C, Freeman J, Jones HD, Sparks C, Pellny TK, Wilkinson MD, Dunwell J, Andersson AAM, Åman P, Guillon F, Saulnier L, Mitchell RAC, Shewry PR (2010) Down-regulation of the CSLF6 gene results in decreased (1,3;1,4)-β-D-glucan in endosperm of wheat. Plant Physiology 152: 1209–1218.

Ordaz-Ortiz JJ, Saulnier L (2005) Structural variability of arabinoxylans from wheat flour. Comparison of water-extractable and xylanase-extractable arabinoxylans. Journal of Cereal Science 42: 119–125.

Palmer R, Cornuault V, Marcus SE, Knox P, Shewry PR, Tosi P (2015) Comparative in situ analyses of cell wall matrix polysaccharide dynamics in developing rice and wheat grain. Planta 241: 669–685.

Parker ML, Ng A, Waldron KW (2005) The phenolic acid and polysaccharide composition of cell walls of bran layers of mature wheat (*Triticum aestivum* L. cv. Avalon) grains. Journal of the Science of Agricultural and Food Chemistry 85: 2539–2547.

Pellny TK, Lovegrove A, Freeman J, Tosi P, Love CG, Knox P, Shewry PR, Mitchell RAC (2012) Cell walls of developing wheat starchy endosperm: comparison of composition and RNA-Seq transcriptome. Plant Physiology 158: 612–627.

Perretant MR, Cadalen T, Charmet G, Sourdille P, Nicolas P, Boeuf C, Tixier MH, Branlard G, Bernard S, Bernard M (2000) QTL analysis of bread-making quality in wheat using a doubled haploid population. Theoetrical and Applied Genetic, 100: 1167–1175.

Pinard D, Mizrachi E, Hefer CA, Kersting AR, Joubert F, Douglas CJ, Mansfield SD, Myburg AA (2015) Comparative analysis of plant carbohydrate active enZymes and their role in xylogenesis. BMC Genomics 16: 402.

Pritchard JR, Lawrence GJ, Larroque O, Li Z, Laidlaw HKC, Morell MK, Rahman S (2011) A survey of β-glucan and arabinoxylan content in wheat. Journal of Agricultural Food Chemistry 91: 1298–1303.

Quraishi UM, Abrouk M, Bolot S, Pont C, Throude M, Guilhot N, Confolent C, Bortolini F, Praud S, Murigneux A, Charmet G, Salse J (2009) Genomics in cereals: from genome-wide conserved orthologous set (COS) sequences to candidate genes for trait dissection. Functional and Integrative Genomics 9: 473–484.

Quraishi U-M, Murat F, Abrouk M, Pont C, Confolent C, Oury FX, Ward J, Boros D, Gebruers K, Delcour JA, Courtin CM, Bedő Z, Saulnier L, Guillon F, Balzergue S, Shewry PR, Feuillet C, Charmet G, Salse J (2011) Combined meta-genomics analyses unravel candidate genes for the grain dietary fibre content in bread wheat (*Triticum aestivum* L.). Functional and Integrative Genomics 11: 71–83.

Rhodes DI, Stone BA (2002) Proteins in walls of wheat aleurone cells. Journal of Cereal Science 36: 83–101.

SACN (2005). Carbohydrates and Health, UK Scientific Advisory Committee on Nutrition https://www.gov.uk/government/publications/sacn-carbohydrates-and-health-report

Saulnier L, Peneau N, Thibault J-F (1995) Variability in grain extract viscosity and water-soluble arabinoxylan content in wheat. Journal of Cereal Science 22: 259–264.

Saulnier L, Thibault JF (1999) Ferulic acid and diferulic acids as components of sugar beet pectins and maize bran heteroxylans. Journal of the Science of Food and Agriculture 79: 396–402.

Saulnier L, Sado P-E, Branlard G, Charmet G, Guillon F (2007) Wheat arabinoxylans; exploiting variation in amount and composition to develop enhanced varieties. Journal of Cereal Science 46: 261–281.

Saulnier L, Robert P, Grintchenko M, Jamme F, Bouchet B, Guillon F (2009). Wheat endosperm cell walls: Spatial heterogeneity of polysaccharide structure and composition using micro-scale enzymatic fingerprinting and FT-IR microspectroscopy. Journal of Cereal Science 50: 312–317.

Schneider R, Hanak T, Persson S, Voigt CA (2016) Cellulose and callose synthesis and organization in focus, what's new? Current Opinion in Plant Biology 34: 9–16.

Selanere ML, Andersson R (2002) Cell wall composition of 1B/1R translocation wheat grains. Journal of the Science of Food and Agriculture 82: 538–545.

Shewry PR, Piironen V, Lampi A-M, Edelmann M, KariluotoS, Nurmi T, Fernandez-Orozco R, Ravel C, Charmet G, Andersson AAM, Åman P, Boros D, Gebruers K, Dornez E, Courtin CM, Delcour JA, Rakszegi M, Bedő Z, Ward JL (2010a) The HEALTHGRAIN wheat diversity screen: effects of genotype and environment on phytochemicals and dietary fiber components. Journal of Agricultural and Food Chemistry 58: 9291–9298.

Shewry PR, Freeman J, Wilkinson M, Pellny T, Mitchell RAC (2010b) Challenges and opportunities for using wheat for biofuel production. In: Energy Crops. N Halford and A Karp, Eds. RSC, London, pp. 13–26.

Shewry PR, Saulnier L, Guillon F, Gebruers K, Courtin C, Delcour J, Toole G, Boros D, Salse J, Ravel C, Mills ENC, Ward JL, Charmet G. (2010c) In: Dietary Fibre: New Frontiers for Food and Health (eds. J.W. van der Kamp, J.M. Jones, B.V. McCleary and D.L. Topping). Wageningen Academic Publishers, Wageningen, Netherlands, pp. 65–78.

Steer T, Thane C, Stephen A, Jebb S (2008) Bread in the diet: consumption and contribution to nutrient intakes of British adults. Proceedings of the Nutrition Society 67: E363.

Stone B, Morell MK (2009) In Wheat Chemistry and Technology, 4th ed., K. Khan and P.R. Shewry Eds. American Association of Cereal Chemists, St Paul, MN.

Toole GA, Wilson RH, Parker ML, Wellner NK, Wheeler TR, Shewry PR, Mills ENC (2007) The effect of environment on endosperm cell-wall development in *Triticum aestivum* during grain filling: an infrared spectroscopic imaging study. Planta 225: 1393–1403.

Toole GA, Le Gall G, Colquhoun IJ, Bedo Z, Saulnier L, Shewry PR, Mills ECN (2011) Spectroscopic analysis of diversity of arabinoxylan structrues in endosperm cell walls of wheat cultivars (*Triticum aestivum*) in the HEALTHGRAIN diversity collection. Journal of the Science of Agricultural and Food Chemistry 59: 7075–7082.

Tremmel-Bede K, Lang OL, Torok K, Tomoskozi S, Vida G, Shewry PR, Bedo Z, Rakszegi M. (2017) Development and characterisation of wheat lines with increased levels of arabinoxylan. Euphytica 213: 291.

USAID (2011) Regional View of Wheat Markets and Food Security in Central Asia https://documents.wfp.org/stellent/groups/public/documents/ena/wfp238576.pdf

Velickovic D, Saulnier L, Lhomme M, Damond A, Guillon F (2016) Mass spectrometric imaging of wheat (*Triticum* spp.) and barley (*Hordeum vulgare* L.) cultivars: distribution of major cell wall polysaccharides according to their main structural features. Journal of the Science of Agricultural and Food Chemistry 64: 6249–6256.

Ward J, Poutanen K, Gebruers K, Piironen V, Lampi A-M, Nystrom L, Anderson A A M, Åman P, Boros D, Rakszegi M, Bedo Z, Shewry PR (2008) The HEALTHGRAIN cereal diversity screen: concept, results and prospects, J Agric Food Chem 56: 9699–9709.

Wilson SM, Ho YY, Lampugnani ER, Van de Meene AML, Bain MP, Bacic A. Doblin MS (2015) Determining the Subcellular Location of Synthesis and Assembly of the Cell Wall Polysaccharide (1,3; 1,4)-β-d-Glucan in Grasses. Plant Cell. 27: 754–771.

Yang L, Huang YL, Chang P, Yan J, Zhang YL, Xia XC, Tian YB, He ZH, Zhang Y (2014) QTL Mapping for arabinoxylans content and its relationship with processing quality in common wheat. Acta Agrononica Sinica 40: 1695–1701.

Yang L, Zhao D, Yan J, Zhang Y, Xia X, Tian Y, He Z, Zhang Y (2016) QTL mapping of grain arabinoxylan contents in common wheat using a recombinant inbred line population. Euphytica 208: 205–214.

Zeng W, Lampugnani ER, Picard KL, Song L, Wu A, Farion IM, Zhao J, Ford K, Doblin MS, Bacic A (2016) Asparagus IRX9, IRX10, and IRX14A Are Components of an Active Xylan Backbone Synthase Complex that Forms in the Golgi Apparatus. Plant Physiology 171: 93–109.

Zhang Y, Nikolovski N, Sorieul M, Vellosillo T, McFarlane HE, Dupree R, Kesten C, Schneider R, Driemeier C, Lathe R, Lampugnani E, Yu X, Ivakov A, Doblin MS, Mortimer JC, Brown SP, Persson S, Dupree P (2016) Golgi-localized STELLO proteins regulate the assembly and trafficking of cellulose synthase complexes in Arabidopsis. Nature Communications 7: 11656.

Grain Quality in Breeding

Marcelo Helguera, Aigul Abugalieva, Sarah Battenfield, Ferenc Békés, Gérard Branlard, Martha Cuniberti, Alexandra Hüsken, Eva Johansson, Craig F. Morris, Eric Nurit, Mike Sissons, and Daniel Vazquez

Abstract Grain characteristics (hardness, protein content/quality, starch properties, enzymatic activity, etc.) play an important role in the definition of end use quality for wheat-based products. Among them, gluten strength and extensibility, mostly determined by glutenin and gliadin composition, are two of the main factors that determine gluten quality. The complex inheritance of most quality traits has led to the development of indirect tests used in breeding for early and advanced generation selection. The main focus of breeders is adding resistance to biotic stress (fungi, insects, nematodes, etc.) and increasing grain yield while selection for quality often occurs in later generations. This often results in the propagation of poor quality lines that must be later discarded. Evaluation of quality in early generations requires suitable tests, preferably non-destructive. Increasing knowledge of the genes involved in quality will facilitate more precise and effective selection. Recent advances in wheat genome sequencing and the extensive genotyping of mapping populations has led to a precise molecular characterization of high molecular weight (HMW) and low molecular weight (LMW) glutenins, as well as the discovery of genes associated with quality traits like grain

M. Helguera (✉)
National Institute of Agricultural Technology (INTA), Marcos Juárez, Argentina
e-mail: helguera.marcelo@inta.gob.ar

A. Abugalieva
Kazakh Scientific Research Institute of Agriculture and Plant Growing, Almalybak, Kazakhstan

S. Battenfield
Syngenta, Junction City, KS, USA

F. Békés
FBFD PTY LTD, Sydney, NSW, Australia

G. Branlard
INRAE, UCA UMR1095 GDEC, Clermont-Ferrand, France

M. Cuniberti
Wheat and Soybean Quality Laboratory, National Institute of Agricultural Technology (INTA), Buenos Aires, Argentina

G. Igrejas et al. (eds.), *Wheat Quality For Improving Processing And Human Health*, https://doi.org/10.1007/978-3-030-34163-3_12

hardness, starch composition (e.g., waxy genes), etc. Massive genomic data will impact in breeding programs allowing quality fine tuning by precise selection of glutenins, starch, hardness and other traits, for specific end uses through marker assisted selection, genomic selection, etc. This chapter will describe different methods used for quality selection in breeding programs and research, and some examples of integration of local breeding programs with the extremely diverse end-uses of wheat based on a series of case-studies. Current and potential approaches to quality evaluation in durum wheat, wild relatives and synthetic wheat breeding programs will be also presented.

1 Methods for Quality Selection and Evaluation

In this section we discuss the efficiency of a diverse subset of methods used for predicting different aspects of wheat quality and its potential use in breeding programs. This subset includes commonly used near-infrared (NIR) spectroscopy and Payne Score, and more sophisticated methods such as the Wheat Simulator and Protein Quality Index (PQI). More specific and quality research methods are also discussed including, the Protein Scoring System (PSS), the High Performance Liquid Chromatography (HPLC) and variants SE-HPLC and RP-HPLC for %UPP determination, and the LC-MS/MS analytical method for screening of water-soluble vitamins. It should be mentioned that there are many methods extensively used for quality selection in breeding programs not included here (such as SDS-sedimentation, SRC, SKCS, among others) that have been already reviewed in other publications. The choice of methods discussed in this section is related with the expertise of co-authors taking part of this book chapter.

A. Hüsken
Department of Safety and Quality of Cereals, Max Rubner-Institut, Detmold, Germany

E. Johansson
Department of Plant Breeding, The Swedish University of Agricultural Sciences, Alnarp, Sweden

C. F. Morris
Western Wheat Quality Laboratory, USDA-ARS, Pullman, WA, USA

E. Nurit
Mazan, France

M. Sissons
NSW Department of Primary Industries, Tamworth Centre for Crop Improvement, Calala, NSW, Australia

D. Vazquez
Instituto Nacional de Investigación Agropecuaria (INIA), La Estanzuela, Uruguay

1.1 *Applying NIR Techniques in Quality Related Selection in Wheat Breeding*

Since the early work of Rubenthaler and Pomeranz (1987), achieving good correlations of water absorption, mixing time, and loaf volume of hard red winter (HRW) wheat to flour NIR spectra, near-infrared (NIR) spectroscopy has been used as a rapid, accurate, and non-destructive technique for measuring many wheat quality parameters. Screening large numbers of lines for several parameters shows that NIR methods are practicable. Based on previous results (Dowell et al. 2006) NIR shows the potential for predicting protein content, moisture content, and flour color b* values with accuracies suitable for process control ($r^2 > 0.97$). Many other parameters were predicted with accuracies suitable for rough screening including test weight, average single kernel diameter and moisture content, SDS sedimentation volume, color a* values, total gluten content, Mixograph, Farinograph, and Alveograph parameters, loaf volume, specific loaf volume, baking water absorption and mixing time, gliadin and glutenin content, flour particle size, and the percentage of dark hard and vitreous kernels. However when the influence of protein content was removed from the analyses, the only factors that could be predicted by NIR with $r^2 > 0.70$ were moisture content, test weight, flour color, free lipids, flour particle size, and the percentage of dark hard and vitreous kernels. Nowadays, NIR is widely applied to the measurement of cereal quality and cereal product composition. The technique enables the rapid assessment of protein, wet gluten, moisture, and ash, and with lower reliability also determination of Zeleny sedimentation and water absorption. In general, NIR can predict these parameters with a high degree of accuracy, as the relevant spectral regions show reasonably clear differences with changing sample composition. Some success was achieved even when modeling some rheological parameters, especially those being measured by the Farinograph (Hrušková et al. 2001), Extensograph (Delwiche et al. 1998) and Alveograph (Czuchajowska and Pomeranz 1991; Jirsa et al. 2008). Prediction of dough properties by NIR spectra analysis, however, is influenced by many factors, especially errors of reference methods and results dependent on the protein content of tested flours. Reliability of computed characteristics of dough varies according to the calibration sample set, and the extent and quality range of flour parameters (Hrušková and Šmejda 2003). Thus, NIRS can be used to predict many grain quality and functionality traits, but mainly because of the high correlations of these traits to protein content. Another way of utilising the NIR technique in quality related screening is to look for quantitative estimations of protein data related to quality attributes. Wesley et al. (2001) reported successful prediction of gliadin and glutenin content from NIRS. In another study (Scholz et al. 2007), partial least squares regression gave high r^2 values between many protein parameters and NIR/NIT (near-infrared transmittance) spectra of flours, while no such relationship was found for whole wheat grains. The highest correlations were found for the total amount of extractable and unextractable proteins and the monomer/polymer protein ratio. Some positive relationships were also found between the NIR/NIT spectra and the percentage of total unextractable polymeric protein in the total polymeric protein and the percentage of large unextractable polymeric protein in the total large polymeric protein. Predictive methods for high value traits are crucial to facilitate increased genetic gain in plant breeding programs. With small quantities of grain required (<200 g) and rapid turnaround time (<1 min of scanning time), high-end NIR predictions of grain quality traits provide informative data for weighed index selection decisions in a similar time frame to grain yield. Of course, appropriate NIR instrumentation does not come cheap, and the prediction calibrations are only of value when the training population is built from a large and robust dataset drawn from environments representative of the target breeding program and encompasses relevant genetic diversity. In Australian wheat breeding, elite grain quality is central for cultivar adoption, and thus, overall genetic merit. Calibration equations to predict compositional grain attributes (e.g. grain protein) have been common place for several years, even on relatively low cost NIR instruments. However, recent improvements in NIR instrumentation and dynamic biometrics to build dynamic calibration equations for high value flour traits (e.g. milling yield, extensibility, Rmax, colour and ash content) and more recently for derived end product testing (e.g. baking loaf volume) has increased the importance and relevance of NIR in wheat breeding programs that place a strong importance on high grain quality. At Dow Seeds, these robust NIR predictive correlations of physical grain attributes ($r^2 > 0.8$), flour traits ($r^2 = 0.4–0.8$), and end product testing ($r^2 > 0.7$) enable effective discrimination decisions within early stage breeding populations and fixed lines tested within standard crop-season. As the scanned seed is still viable with a rapid turn-around time of NIR, early stage selections can be made in time to allow contra season nursery for enriched germplasm to further increase the rate of overall genetic gain.

1.2 Predicting Dough Properties from Genetic and Biochemical Data

1.2.1 Dough Strength and Extensibility

Relating the protein composition to certain quality traits by statistical means is a frequently used methodology to relate structure/composition to functionality in cereal science. The classic tool used by breeders is the Payne score (Payne et al. 1987) providing a single number to estimate dough strength from the HMW glutenin allelic composition. Despite the large success in using the Payne protein marker score in breeding programs over the years, significant limitations of this the method have been realized. The Payne score in original form and even all later alternative calculation methods are not applicable to fully describe breadmaking quality: they are simple and very useful tools to estimate the most important rheology characteristics of the dough, namely dough strength and – in cases of the more up-to-date formulas – extensibility. These essential characteristics of the dough are directly related to the structure of proteins comprising gluten, and therefore it is meaningful to relate the composition of these proteins and the characteristics of the dough. Therefore it is not surprising that attempts to predict breadmaking quality simply with Payne score type models have failed (Hamer et al. 1992). Another type of limitation of the original Payne score is that it takes into account only the contribution of the HMW glutenin subunits in relation to dough strength. Several attempts have been introduced to involve the LMW glutenin alleles in similar mathematical formulas (Békés et al. 2006; Cornish et al. 2006; Gupta et al. 1991, 1994; Oury et al. 2010). By the application of sophisticated statistical approaches, the Wheat Simulator (Eagles et al. 2002), and the Protein Quality Index (PQI) (Békés et al. 2006) went one step further: they are capable of describing the effects of both HMW and LMW-GS alleles on dough strength and extensibility, individually and the pair-wise interactions among the alleles (Fig. 1). As it is shown in Fig. 1, besides estimating not only Rmax but also Ext, and besides considering the individual and interactive contributions of both HMW and LMW-GS alleles, the number of HMW-GS alleles have been increased in PQI from 13 to 17 with some very important alleles such as the *Glu-B1al* allele (with the overexpressing subunit 7). The quite large (33×33) matrixes for the estimation of both Rmax and Ext are available on the official web-site of AACCI where the allelic composition, Payne score and PQI of more 8500 wheat varieties, cultivated in more than 80 countries are tabulated (Békés and Wrigley 2013, 2017). The approach of relating allelic composition to quality attributes is possible with careful data selection and applying robust mathematical tools: the genetic potential of a line, with a certain combination of alleles on the six glutenin coding loci, can meaningfully be predicted where both the contribution of the individual alleles and their pair-wise interactions play equally important role (Baracskai et al. 2011; Békés et al. 2006; Cornish et al. 2006). The original version of the PQI model is to predict the genetic potential of dough strength and extensional properties of dough of a wheat flour with 12% protein content and with the ratios of glutenin to gliadin and HMW to LMW GS of 1.0 and 0.2, respectively. The further developed version of the model (Békés et al. 2006) is capable of considering the effects of the expression levels of the different storage proteins genes, so that the actual dough parameters can be predicted. The input to this model is the allelic composition and the quantitative protein composition (including UPP%) (Gupta et al. 1993), while the output provides a good estimate of the actual dough strength and extensibility of a given sample ($r^2 > 0.85$ and $r^2 > 0.75$, respectively). The application of the PQI model on different sample populations (Baracskai et al.

$$Q = \sum_{13}^{i=1} \alpha_i{}^*(q_H)_i \qquad Q = \sum_{17}^{i=1} \alpha_i{}^*(q_H)_i + \sum_{16}^{j=1} \alpha_i{}^*(q_L)_j + \sum_{17}^{i=1}\sum_{16}^{j=1} \beta_{i,j}{}^*(q_H)_j{}^*(q_L)_j \quad (1)$$

Payne score Payne, 1987 — **Protein Quality Index (PQI)** Békés et al, 2006

Fig. 1 The mathematical formula of Payne score and PQI. The presence or absence of an individual HMW-GS $(\mathbf{q_H})_i$ or LMW-GS $(\mathbf{q_L})_j$ allele in the sample is coded by **1** or **0**, respectively. The $(\boldsymbol{\alpha}_i)$ and $(\boldsymbol{\beta}_i)$ weighting factors describe the individual and interactive contribution of an allele, respectively, determining **Q** quality attribute (Rmax in Payne score, Rmax or Extensibility in PQI)

2011; Békés et al. 2006) importantly point to the impact of allelic interactions as a major variable determining dough properties (Fig. 2). The realisation of the existence of the large contribution of allele–allele interactions in wheat flour doughs may alter our way of utilising our knowledge of relating genetic/chemical information to quality attributes in wheat breeding, in the grain industry and in basic research. In breeding, the real value of a certain allele has to be investigated in several backgrounds to be able to realize its interaction potential. Consequently, different allelic combinations, rather than certain individual glutenin alleles should be targeted in breeding situations to develop new lines with certain quality attributes, especially to improve extensibility.

The interactive effects of the alleles present in commercial wheat flour blends are responsible for the well-known problem in the grain industry: dough properties, such as dough strength and extensibility are not simply additive characteristics. They usually show non-linear relationships with the blend formulation (Békés et al. 1998, 2001): applying the Fig. 1 equation to describe the dough strength (Rmax) of a blend with two (u and v) components, it is clear that the difference between the non-linear and linear model does contain only interactive elements (non-linearity is the function of the interactions of alleles of component u with those of component v). If the allelic composition of the components as well as the $\alpha_{i,j}$ and $\beta_{i,j}$ parameters are known, the quality attributes of blends can be estimated, providing an efficient way of developing non-linear optimization models for blend formulation. It is important to note that the quality attributes of hybrid wheats with their complex genetic makeup can be estimated using the quality attributes of sibling lines and the above mathematical model.

At the time of the introduction of Payne score only the sodium dodecyl sulphate polyacrylamide gel electrophoresis (SDS-PAGE) based methodology was available to identify the glutenin alleles in a sample. While in case of HMW-GS alleles it is a simple, routine task, the identification of LMW-GS alleles is rather complicated and requires special skills. Since the routine application of the above prediction methods requires knowledge of both HMW and LMW-GS alleles, sensitive, reproducible

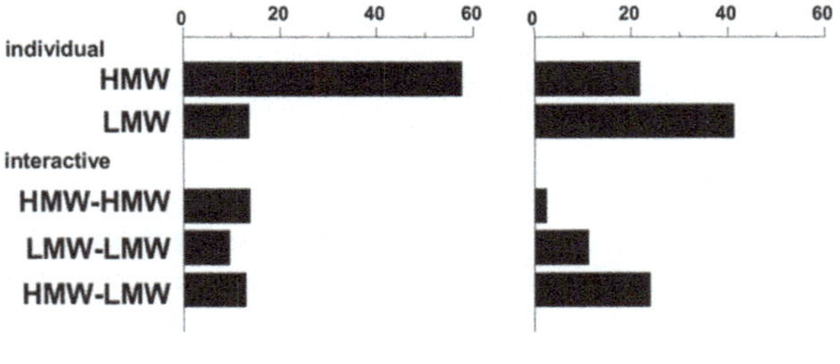

Fig. 2 Typical relative contributions of the individual and interactive effects of HMW and LMW glutenin alleles on dough strength (Rmax) and extensibility (Ext) in a set of wheat varieties (n = 107) using the data of Baracskai et al. (2011)

and high throughput methodologies had to be developed. The molecular marker (Abdel-Mawgood 2008; Howitt 2010; Howitt et al. 2006; Liu et al. 2012), and MALDI-TOF (Gao et al. 2010; Liu et al. 2009, 2010; Ma et al. 2009; Wang et al. 2015) technology developed for both HMW and LMW-GS alleles nowadays provide a solution for this task. In Australia, service companies, specialised for the identification of glutenin alleles, even for the application of Wheat Simulator (Eagles et al. 2002), or PQI (Békés et al. 2006) are available to wheat breeders. Illustrating the capabilities of NIR-based and PQI estimation of Rmax and Ext, the relationships between predicted and measured values of a small routine sample set is shown in Fig. 3. These levels of relationships provide a reliable, cost-effective solution in breeding situation to rank and select/omit breeder's lines with high throughput.

1.2.2 Estimation of Water Absorption

The above models however are only suitable for the estimation of certain dough properties. Other parameters such as water absorption (WA) cannot be predicted on the basis of gluten proteins only. Water absorption – the amount of water needed to hydrate flour components to produce a flour with optimum consistency – can be described as a function of protein content, the amounts of pentosans and the level of starch damage in the sample (Bushuk and Békés 2002). Experiments carried out with flours supplemented with different protein classes resulted in the observation that while mixing requirement, dough strength and extensibility significantly depends on the glutenin to gliadin ratio, WA is not sensitive to this ratio (Uthayakumaran et al. 1999). However, the ratio of the amount of gluten proteins to soluble proteins can significantly alter WA (Tömösközi et al. 2004). Supplementing

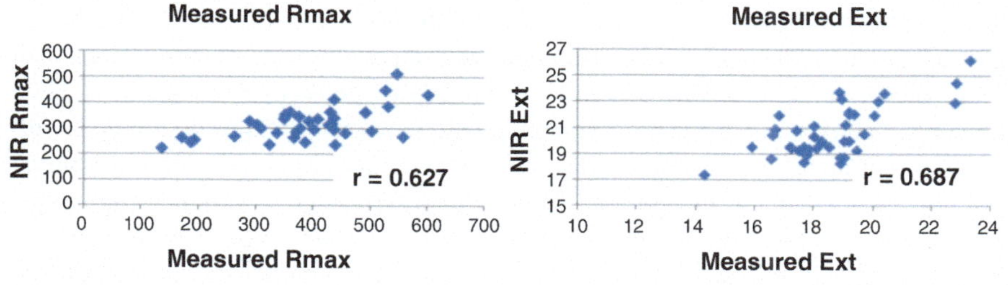

Fig. 3 Comparison of measured Rmax and extensibility data with predicted values derived from NIR calibration or PQI method

wheat flour with soluble proteins of different origin and polarity showed that polarity/hydrophobicity as well as the charge distribution of albumins and globulins are the key features changing the amount of water needed for hydration (Tömösközi et al. 2002). A significantly high positive correlation ($r^2 > 0.8$) was obtained in 63 straight run flours between WA and damaged starch content compared to a value of $r^2 = 0.5$ for protein content (Tara et al. 1972). Prediction equations using multiple linear models with protein content and starch damage have been developed and it was found that more than 90% of the variation in WA is covered by these two parameters. It was also concluded that the remaining variation is mostly related to the pentosan content of the flours. The relationships between the pentosan composition and water absorption of flours indicated that arabinoxylans (AX) play the major role and the largest effect is caused by the soluble – small and medium sized – arabinoxylans, while the large polymers do not have marked effects on WA (Primo-Martin and Martinez-Anaya 2003). A linear mathematical model has been developed to estimate WA from protein content, starch damage, AX content and the relative amount of soluble proteins with a reasonable correlation ($r^2 = 0.65$) between measured and estimated data. The introduction of a new parameter, related to the cultivar dependent quantitative composition of soluble proteins, determined by lab-on-a-chip (LOC) analysis, largely improved the predictability of WA. Based on the large variation among the level of AX and certain soluble protein components in wheat flour as well as their significant contribution to determine WA it was concluded that they can be target traits to alter during wheat breeding programs to improve WA. This predictive equation provides explanations for the relative importance of several components of the flour in relation to water absorption, however the direct measurement of water absorption using small scale equipment such as the DoughLab seems to be a simpler and less time-consuming option to determine this important parameter.

1.2.3 Predicting End-Product Quality

In the last 20 years numerous researchers have attempted to predict bread quality by combining measurements made from grain, flour, or dough and combining them into prediction models (Békés 2012a, b). Protein content and composition, falling number, ash content, water absorption, mixing and rheological parameters have been applied in these studies in different combination using different linear mathematical models. The published predictions showed a wide range of correlations with loaf volume ($r^2 = 0.39–0.78$). A new generation of predictive models have been developed (Békés 2012b). All of the previously mentioned models do not deal with the fact that most relationships between grain or flour parameters and loaf volume are not linear: there is an optimum level of energy needed to produce the largest loaf; below the level the dough is under-mixed, and above that it is over-mixed, resulting in smaller loafs. So, for a set of technological parameters and ingredients, there is an optimal dough strength and extensibility. The Protein Scoring System (PSS) (Békés 2012b) uses the Morup- Olesen transformation (Møup and Olesen

1976) (developed originally for describing the effects of individual essential amino acid levels in food or feed stuffs on the biological value) on certain dough parameters prior to multiple regression. Using only four parameters –protein content, water absorption, dough strength (Rmax), extensibility– a $r^2 > 0.85$ can be achieved with low standard error of prediction on loafs produced with commercial bread-making formulations. The model is in use in both breeding and quality control situations applying dough property parameters predicted based on NIR spectra, making the end-product quality estimation incomparably cheaper than carrying out baking tests, with stronger relationships between predicted and measured data than the direct loaf volume prediction from NIR spectra.

1.3 HPLC Methods to Determine Quality in Wheat

High performance liquid chromatography (HPLC) has been applied within research during the last 30 years for determining protein quality in wheat. One of the major steps towards developing HPLC methods for quality determination occurred in the lab of Prof. Finlay MacRitchie in Australia, from which the first publications about the topic came around 1990 (Singh et al. 1990). The work in this lab resulted in the current world-wide use of size exclusion (SE)-HPLC technique for determination of %UPP (Gupta et al. 1993). Simultaneously as the Australian group developed the SE-HPLC method, a reversed phase (RP) based HPLC method was developed in Canada for prediction of wheat quality (Marchylo et al. 1989). The RP-HPLC method was later refined in order to quantify the amount of certain protein groups and subunits (Wieser and Seilmeier 1998), and by applying various extraction buffers to understand how proteins in the polymer are bound (Kuktaite et al. 2004; Rasheed et al. 2016). Since the development of the SE-HPLC method to determine %UPP, this method has been used widely in research laboratories around the world. The method is based on a two step extraction (Gupta et al. 1993), where the first step extracts proteins soluble in SDS, while the second step extracts proteins soluble by sonication. In principle, the first step extracts polymers and monomers not bound to large polymeric proteins structures through disulphide bonds. In the second step, the sonication is primarily breaking the disulphide bonds, although sonication in too tough conditions is also able to break peptide bonds, while a too weak sonication does not break enough disulphide bonds to solubilize all the proteins in the wheat grain (Gupta et al. 1993). The %UPP is calculated combining information from both the chromatograms as %UPP = amount (area under the chromatogram) of unextractable (chromatogram from sonication) polymeric protein (PP)/(amount unextractable + extractable (chromatogram from SDS) PP) ×100. Thus, %UPP is describing how large part of the polymeric proteins, being extracted by the two-step extraction procedure adopted, that is not easily extractable (sonication is needed instead of SDS) and not bound with disulphide bonds. The method as described above and as developed by MacRitchie's lab has been applied as such in many labs; e.g. in Sweden (Johansson et al. 2001; Kuktaite et al. 2004; Rasheed et al. 2016), Italy (Pirozi et al. 2008), Norway (Tronsmo et al. 2003), the USA (Naeem et al. 2012), Canada (Edwards et al. 2007) and China (He et al. 2005). In the majority of these studies, the %UPP has been correlated to wheat quality, and primarily a positive correlation has been stressed towards other quality evaluations, measuring gluten strength. A number of studies have also used the total amount of SDS extractable proteins (proteins from extraction in step one) measured with SE-HPLC as a measurement of grain protein concentration as a close correlation prevails (Malik et al. 2011, 2013). Furthermore, some studies have divided the SE-HPLC chromatograms into different parts where the majority of the glutenins and gliadins are present, extrapolating that the glutenins are forming the PP while the gliadins are forming the monomeric proteins (MP). Thereafter the glutenin to gliadin ratio from the SE-HPLC chromatogram has been correlated to gluten quality (Park et al. 2006; Wang et al. 2014). In a recent study, alternative calculation methods were used for %UPP (Hu et al. 2017). Although a definition is presented in this publication, such alternative definitions of already existing abbreviations is a bit messy. The RP-HPLC methods have been used in a range of studies both to quantify different subunits, as e.g. HMW-GS can be distinguished as separate peaks with the appropriate running conditions and column for the analyses (Wieser and Seilmeier 1998; Wieser et al. 1998). However, RP-HPLC methods have also been used to quantify amounts of proteins extractable with a range of different extraction buffers (Johansson et al. 2001, 2002, 2003; Kuktaite et al. 2004, 2011; Rasheed et al. 2016). Thus, such analyses can contribute to an understanding of the relations of proteins that are cross-linked through different types of bonds including disulphide, sulphur and peptide bonds (Ceresino et al. 2019; Rasheed et al. 2018). Cross-links between storage proteins of wheat is a major contributor to the quality of wheat (Johansson et al. 2013) and their existence in the wheat grain and their interchange during the mixing and bread-making process are affecting the quality of the wheat dough (Hussain et al. 2012; Johansson et al. 2013). In general, the HPLC methods have contributed largely to an increased understanding of protein behaviour in the wheat grain and in flour during mixing as well as in dough during bread-making (Johansson et al. 2013). Still, the HPLC methods are not used to a large extent in plant breeding for improving bread-making quality. Several of the methods, especially SE-HPLC to determine %UPP, are high through-put, need only a small amount of material, and are relatively cheap. Consequently, at least SE-HPLC for %UPP determination should be of interest to implement as a selection method for plant breeders to select for wheat quality. However, despite the many positive characteristics of the method, it has also some clear drawbacks. The major drawback is the lack of consistency of the method over different labs using the method. This is due to differences in set up and running conditions but also to the fact that the type of equipment, column and even temperature during running are affecting the results. Also columns from different suppliers result in differences in chromatograms and even columns from the same supplier may not be consistent. Thus, to get comparable results, samples always need to be run on the same column, in the same batch and in the same lab. For comparisons over various columns, batches and even more so over different labs, standard samples need to be used and results recalculated according to these standards. Besides issues with repeatability, a correct interpretation of the HPLC results often require experience running this kind of analyses.

1.4 LC-MS/MS Method and Perspectives to Improve the Evaluation of Vitamins in Wheat

Wheat as one of the major world agricultural products, has so far received great attention regarding its technological quality attributes but little work focuses on its nutritional quality. Water-soluble vitamins are an important class of compounds that require quantification from food sources to monitor nutritional value. Measurement of vitamins in food is complicated and represents a complex analytical problem for several reasons. Firstly, vitamins are compounds having diverse chemical structures and properties. As a consequence, it is very difficult to develop a single method for their simultaneous determination. Secondly, wheat vitamins are present at relatively low concentrations that require sensitive methods for their analyses (Fardet 2010). Finally the susceptibility of vitamins to degradation by exposure to light, air, heat, alkaline pH and their diverse forms make their extraction from food matrices very challenging. An important amount of studies have been devoted to the development of analytical methods which could explore and monitor the nutrient composition of whole-grain wheat or end-use products. Nevertheless, in the objective of a large scale varietal screening, none of the reported methods were suitable for the analysis of wheat flour and wheat food products. The present study briefly describes the LC-MS/MS method which was recently developed. In addition some perspectives to improve the nutritional evaluation of vitamins in wheat will be discussed.

1.4.1 Evaluation of the LC-MS/MS Analytical Method for the Simultaneous Screening of Seven Water-Soluble Vitamins

Taking advantage of high sensitivity and selectivity of the MS/MS detection the extraction procedure was considerably simplified. Effectively one of the main goal was to find a procedure allowing the analysis of all vitamins present in food products in a single chromatographic run and to simplify as much as possible the extraction procedure in order to apply this method at high throughput. The analytical procedure was optimized by taking into account both the nature of the analyses and the nature of the matrix. An enzymatic mixture (Ndaw et al. 2000) was selected and optimized for the analysis of the free and some chemically bound forms of vitamins in foodstuffs. This extraction is simple, fast, accurate and can be extended to different wheat food sample (wholemeal grain, flour, dough, bread, toasted bread, biscuit). Due to the heterogeneous nature of the materials being studied, the important issue of the matrix effect was investigated to avoid any interferences which could induce bias during the analysis. This difficulty was overcome by the use of one isotope labelled internal standard for each class of compounds (the seven analyzed vitamins were classified into three classes of homologous compounds). The excellent precision and bias (Nurit et al. 2015) of the method within the different materials validated the choice of using three internal standards. The choice of a reverse column coupled with accurate chromatographic conditions have allowed the separation of the seven water-soluble vitamins in 15 min with excellent performances in terms of peak shape and peak intensity for most of the vitamins. The method was applied to the determination of water-soluble vitamins in manufacturing wheat-based food products (Nurit et al. 2016) and in a large set of 195 genetically diverse cultivars (Nurit 2015). The 195 accessions were chosen from the INRA worldwide bread wheat core collection of 372 accessions curated by the Clermont Ferrand genetic resources Center (Balfourier et al. 2007). This core collection, representative of the world's wheat diversity, was also studied for agronomic performances and quality traits (Bordes et al. 2008). The 195 accessions brought a huge phenotypic diversity as revealed particularly on the seven water soluble vitamins (see chapter "Environmental effects on wheat technological and nutritional quality" in this book). The simplicity of the extraction procedure, as well as the direct injections of the extract in the LC- MS/MS system make this method rapid and potentially high-throughput. As a consequence, this procedure is suitable for a fast screening of vitamin contents in wheat flour and wheat-based food products with the objective of a large scale varietal screening.

1.4.2 Perspectives about Improving Techniques

Water-soluble vitamins are an important class of compounds that require quantification from food to monitor nutritional value. Nevertheless, most of the vitamins exits as groups of chemically related compounds having similar biological activity capable of meeting a nutritional requirement (frequently called "vitamers"). For the case of water-soluble vitamins, niacin (nicotinic acid + nicotinamide) and vitamin B6 (pyridoxine, pyridoxal and pyridoxamine) constitute an interesting case, in which, the glycosylated forms of pyridoxine and nicotinic acid are prevalent in plant-derived foods (Gregory et al. 1991). In addition, it has been shown that in the milky kernel of maize, nicotinamide adenine dinucleotide (NAD) and nicotinamide adenine dinucleotide phosphate (NADP) were the predominant form in which niacin occurs (Wall and Carpenter 1988). Common extraction processes for these vitamers involve alkali or acid hydrolysis under heating condition (either in a boiling water bath or in an autoclave). Such extraction releases free vitamins from its bound forms. In the presented LC-MS/MS method such extraction which was destructive for an important part of vitamins led us to achieve a gentler extraction method which did not allow the complete release of the vitamins present in the sample as NAD, NADP, nicotinic acid glucoside and pyridoxine glucoside. In data previously reported (Sampson et al. 1996), there was a significant fraction (average of 68%) of vitamin B6 in wheat present as pyridoxine glucoside. An important additional finding was the changes in the distribution of niacin compounds in corn during its development. Immature sweet corn is an effective source of NAD and NADP, whereas niacin in mature field corn is largely present as glycosylated forms of niacin (Wall and Carpenter 1988). These glycosylated forms were reported as nutritionally unavailable (Gregory et al. 1991; Wall and Carpenter 1988). Thus, it will be of interest to develop a simple, reliable and high throughput analytical method which allows the measurement of all biologically active water–soluble vitamers. Therefore, it will be interesting to take full advantage of the high sensitivity and selectivity of the MS/MS detection. The challenge will be to optimize the MS/MS parameters for the identification and quantitation of the different binding forms of water-soluble vitamins and to find chromatographic conditions allowing the separation of the bound and free forms of vitamins in a short run time with a good performance in terms of peak shape and peak intensity. Such methodology should help to develop generalized approaches to account for differences in bioavailability of vitamers and to thoroughly enlighten the consumer about the nutritional labeling of cereal products. However, the current high costs of chemically analyzing large populations by LC-MS/MS may limit its use for breeding (particularly for early breeding generations which implies a large number of samples to be screened) and, consequently, LC-MS/MS will probably not become a standard selection tool, unless effort in developing analytical technique such as infrared spectroscopy are taken, then development of the calibration models for quantifying the wheat flour vitamins could be achieved using primary reference data obtained from LC-MS/MS method.

2 Quality Selection in Breeding Programs, Case Studies

In this section we review quality selection strategies considering bread wheat breeding in Germany, Uruguay, Argentina and USA (bread and soft wheats). Durum wheat quality selection strategies are also reviewed. The last part of this chapter covers quality evaluation in wheat wild relatives considering the experience of Kazakhstan.

2.1 Wheat Breeding for Improved Baking Quality in Germany

2.1.1 Introduction

Today, the German wheat classification system categorizes varieties according to their baking quality as part of the registration process. E- (elite) wheats have the highest quality, followed by A- (blending), B- (bread making) and C- (not usable for baking) wheats, the latter having the lowest quality. Assignment to a certain quality group is dependent on particular minimum requirements with respect to individual quality traits and on a comparison with a defined reference variety. New varieties are registered and protected after successfully passing 3 years of official testing in several environments for agronomic value, homogeneity and novelty. Direct and indirect quality parameters (loaf volume, dough elasticity and surface, crude protein content, falling number, sedimentation value, water absorption and milling yield) for baking quality are part of the official variety approval and registration process in

Germany (BSA 2017). Selection decisions for quality in early generations with limited grain and large number of samples are performed by markers linked to the known *Glu-1* and *Glu-3* glutenin loci, followed by indirect phenotypic selection based on correlated quality traits like crude protein content, sedimentation value, grain hardness and falling number. The development of alternative breeding strategies to increase the selection gain per year is a continuous challenge for plant breeders; as a consequence, genomic selection is already being applied as an efficient approach to support wheat quality improvement in German breeding programs.

2.1.2 Breeding Progress

From the historical perspective breeding for improved baking quality of winter wheat was very successful in Germany after World War II. The introduction of shorter varieties allowed higher levels of nitrogen application as well as late top dressing, and together with the release of varieties with better protein quality it was possible to produce winter wheat with acceptable baking quality. After returning to self-sufficiency after World War II, Germany still had to import about two million tonnes of high quality baking wheat from Canada every year until the 1970s. In the course of the 1970s, however, winter wheat production in Germany was able to cover the domestic demand of wheat with sufficient baking quality (Laidig et al. 2017). The aim of obtaining wheat of better baking quality has been accomplished over the decades by both the elimination of inferior quality varieties and the development of new varieties with superior quality. This was confirmed by trends in yield and baking quality of winter wheat varieties released during 1961–2008 and 1983–2014 in Germany (Hartl et al. 2010; Laidig et al. 2017). Hartl et al. (2010) observed a clear increase of grain and protein yield, sedimentation value and loaf volume based on genetic improvements, whereas protein content was only raised in magnitudes obligatory for reaching the criteria for high baking quality wheat. They assumed that selection decisions in early generations were mainly carried out on the basis of sedimentation value in German wheat breeding programs. As a side effect, application of this single selection criterion might have led to the increased occurrence of so-called correlation breakers, i.e. varieties that achieved, despite a high sedimentation value, only medium loaf volumes. However, on the positive side, Hartl et al. (2010) concluded that breeding progress in Germany has improved the specific loaf volume (loaf volume/% crude protein) progressively and that an increasing number of registered varieties reached high to very high loaf volumes with a medium protein content. Laidig et al. (2017) found a large significant gain in grain yield (24%), but a strong decline in protein concentration (−8.0%) between 1983 and 2014. They showed that both traits are strongly negatively related, which was consistent over all varieties and also within quality groups. However, the study indicated that losses in baking quality were mitigated in Germany between 1983 and 2014 or even partially improved. Improvement of indirect baking quality parameters were achieved for falling number (5.8%), sedimentation value (7.9%), hardness (13.4%), water absorption (1.2%) and milling yield (2.4%). The apparent gain in sedimentation value provided evidence that progress in baking quality was mainly due to improved protein quality as well as their interaction with other fractions like the puroindolines that correspond to grain hardness. The authors further reported that grain yield, falling number and protein concentration were highly influenced by environment, whereas for sedimentation value, hardness, water absorption and loaf volume it was stated that genotypes accounted for more than 60% of total variation. They concluded that the strong negative relation between grain yield and protein concentration on the one side and the strong association between protein concentration, sedimentation value and loaf volume on the other side makes it difficult to achieve breeding progress in traits related to end-use quality.

2.1.3 Conclusions

In summary, breeding for improved baking quality in wheat is determined largely by the common negative correlation between yield and crude protein content. Over recent decades, wheat breeding in Germany has concentrated on yield, so that newer varieties, generally, have higher yields and lower crude protein content. As compensation for lower crude protein content, there has been an efficient selection for higher gluten quality. For the future, selection of wheat varieties that combine high yield and sufficient baking quality in lower nitrogen-input and higher climate-variability cropping systems may require an advanced breeding approach. There is a need for varieties with a maximum ability for nitrogen uptake, high nitrogen remobilization and reallocation efficiency during grain filling and an efficient conversion of nitrogen into high-quality proteins. What is sought-after are not necessarily wheat varieties that consistently achieve 13% protein, but varieties that will be suitable for bread-making at 11–12% crude protein with stable quality under increasingly fluctuating environmental conditions.

2.2 Wheat Breeding for Improved Bread-Making Quality in Uruguay

Quality screening in Uruguay has been reported since 1925 (Boerger 1928), when test weight, flour extraction and baking performance were evaluated for advanced lines. Although Farinograph was applied since 1929, during the subsequent decades, several faster tests such as Pelshenke and Zeleny sedimentation proved to be useful. This was the practice until the early 90s, when the SDS sedimentation test (Amaya et al. 1991) was incorporated as a key element for early generation screenings. Due to markets requirements (Peña 2007), a more sophisticated approach was adopted since the 1990s. While gluten strength was still evaluated by the same SDS sedimentation test in early generation screening, selection for rheological properties in later steps became essential. The Mixograph proved to be more effective than the Farinograph to characterize mixing properties since it is faster and a smaller sample is required. Extensional properties were proven to be key for breadmaking properties, so the Alveograph was included in the protocols (Vázquez and Watts 2004). Protein/gluten content was evaluated by two independent methodologies. Wet gluten was included as a characterization parameter due to requirements from the milling industry. Traditional Kjeldahl analyses were used until near infrared spectrometry (NIRS) technology was developed specifically for the breeding program (Vázquez et al. 2007). Although protein and rheological properties are considered the most relevant characteristics, several independent parameters should be considered as well, such as milling behaviour and proper seed dormancy and resistance to preharvest sprouting through Falling Number. Grain hardness was traditionally evaluated by the pearling index method. More recently, Particle Size Index was used to develop a NIRS equation (Vázquez et al. 2007), which is extensively used. Once selected genotypes are almost ready to be released, breadmaking tests are performed, including both the standard AACC 10–10 and one specifically designed for local breads (Paulley et al. 2004). Further tests have been evaluated. For example, Mixolab proved to generate information that is complementary with that which is traditionally used (Vázquez and Veira 2015). Pentosans were also evaluated and proved to significantly affect rheological properties of the dough (Garófalo et al. 2011). However, these parameters have not been included in the routine characterization of the genotypes because of the cost-benefit ratio. All genotypes are evaluated in several environments, considering the importance of the genotype by environment interactions (Vázquez et al. 2012) and stability (Vázquez and Castro 2018). Several technologies are being considered for the near future. New NIRS calibrations are on the agenda, including equations for gluten strength (Vázquez et al. 2007). Genomic selection proved to be useful, mostly when combined with other tools (Lado et al. 2018).

2.3 Wheat Breeding for Improved Bread-Making Quality in Argentina

Records of wheat introduction in Argentina appeared in the early XVI century by Spanish conquerors, and for more than three centuries wheat was cultivated on a low scale, at a level close to mining communities in the central-east part of the country. We have to wait until the end of the XIX century to witness a rapid expansion of the crop boosted by the European immigration wave, the development of land (railway) and sea transportation systems and the increase of food demand by Europe. Wheat breeding started at the beginning of the XX century with efforts concentrated on improving disease resistance, bread-making quality, tolerance to the main abiotic stresses, and grain yield, mostly as a response to disease tolerance, lodging resistance, and management practices (Slafer and Andrade 1989). National and international markets demanded wheats with strong gluten pushing breeding programs to release high breadmaking quality wheats. However, the lack of a premium price prompted by quality segregation pressed farmers to sacrifice quality for yield, as wheat was exported as a commodity. This situation has changed over the last 10 years as more than 70% of the crop area are planted to good quality high yielding cultivars. In 2018 the total grain production was 18.4 million tonnes (average yield 3.2 t/ha) (trigoargentino 2018). In 2017, 12.6 million tonnes were exported to 47 different destinations; Brazil was the main buyer with 40% of total wheat exported (Indec 2018). Argentina is the main wheat producing and exporting country in Latin America. Wheat trading is based on Standard of Commercial Grades that includes in the hard wheat price a bonus of 2% by protein content when above 11% (13.5% moisture basis) if the test weight is greater than 75 kg/hl. When protein content is under 10.9% gradual and cumulative discounts are applied: 10.9–10% a discount of 2%, 9.9–9.0% a discount of 3% and less than 9% a discount of 4% (norma 2018). In Argentina grain and flour quality characterization have shown some logical modifications over time. For example, a standard quality characterization for varieties released during the 1950–70's considered (i) physical and chemical tests on grain including test weight, weight per 1000 kernels, flour yield, protein content and ash content; (ii) flour tests including ash content, protein content, wet gluten content and gas production (Elion 1933); (iii) flour baking tests, baking water absorption, baking mix time, baking loaf volume, crumb grain (0–100) and crumb whiteness (0–100); and (iv), the Alveograph (Chopin) dough testing device parameters P, L, G, W and P/L. During the 1970–80s with the support of CIMMYT, local breeding programs introduced Mexican germplasm with a positive impact on yield genetic gain (Lo Valvo et al. 2018). Also, Pelshenke test (early generation), flour (dough) testing using the Zeleny test, and the Farinograph (Brabender) dough testing device with parameters water absortion, development time and stability time, were included for grain and flour quality characterization, and during the 1980–90's falling number was included as an additional flour test. We have to wait until 1998 to find a significant contribution to grain and flour quality characterization for variety release: the creation of an internal quality classification system. The proposal defined three wheat Quality Groups (QG): QG 1 involved cultivars with extra strong gluten suitable for blending; QG 2, cultivars adapted to traditional baking in Argentina (fermentation time longer than 8 h); and QG 3, cultivars suitable for direct baking methods (fermentation time less than 8 h). The classification system was based on the information obtained from a set of quality tests including test weight, protein content, wet gluten content, flour yield/ash content ratio, Alveograph W, Farinograph stability, and baking loaf volume (Cuniberti and Otamendi 2004). The next relevant contribution to grain and flour quality characterization was the creation of a quality index based on weighted contributions of the same set of quality tests used in the above quality classification system, with 45% of the variation of the quality index accounted by the variation of W value from the Alveograph and baking loaf volume variables (Salomon and Miranda 2003). In 2007 grain and flour quality characterization for cultivar release included for first time grain color (Minolta, parameters $L*$, $a*$ and $b*$), HMW-GS, 1BL/1RS and 1AL/1RS wheat-rye translocations, determined by SDS-PAGE and A-PAGE, respectively (Bainotti et al. 2009). Lastly, in 2011, grain hardness (SKCS score) was included with no additional modifications in grain and flour quality characterization for cultivar release to the present day. In Argentina more than 98% of wheat production is bread wheat, mostly hard red spring with 120 cultivars in the market; and remaining 2% are durum, soft and waxy wheats. In general terms breeding programs select bread-making quality at different stages of the process as it follows: various technological micro-tests like Zeleny and SDS sedimentation are used to select strong gluten in segregating materials (individual plants) at early generations. Selected lines advanced to preliminary and regional yield trials are evaluated and selected based on grain properties using test weight, weight of 1000 kernels and protein content. Milling, flour, and dough properties of lines advanced to preliminary trials are evaluated considering flour extraction (>70%), wet gluten content and Mixograph (National Manufacturing) parameter mixing time. After that, lines advanced to regional yield trials are evaluated based on rheology properties including Alveograph (Chopin) parameters W and P/L, Farinograph (Brabender) parameter stability and baking loaf volume. Table 1 summarise typical quality test for bread wheat selection in a breeding program in Argentina. A putative new variety is evaluated based on official regional yield trial information considering 2 years of trials in three locations or 3 years in one location and QG is defined by comparison with three reference varieties (Bainotti et al. 2009, 2017).

A special mention should be given to the use of molecular markers in breeding for quality traits. First attempts of marker assisted breeding in Argentina come from the late 1990s with the INTA National Wheat Breeding Program and were focused on selection of superior HMW-GS in small segregating populations by SDS-PAGE. More recently, molecular information of breadmaking quality protein/traits HMW-GS (Gianibelli et al. 2002), LMW-GS (Lerner et al. 2009; Demichelis et al. 2018), wheat-rye translocations (Vanzetti et al. 2013), puroindolines (Moiraghi et al. 2013), *Gpc-B1* for grain protein content (Tabbita et al. 2013), and the GBSS I gene (Vanzetti et al. 2009), among others, was exploited by breeding programs for germplasm characterization (introductions, crossing blocks), and selection (RILs) with the aim of fixing valuable alleles (HMW-GS *Glu-B1al* allele, GPC, others) into adapted germplasm. Examples of marker assisted breeding are cultivars Biointa 2004 and MS INTA 416 with introgression of disease resistance genes *Lr47* and *Fhb1* (Bainotti et al. 2009, 2017); the release of commercial cultivars with marker assisted introgression of quality traits is still a pending task.

2.4 Hard Wheat Breeding for Improved Bread-Making Quality in USA

Wheat breeding crosses are made to align attributes of two varieties in order to make progeny which are genetically superior to both parents. Crosses are made and traditional selection occurs or the doubled haploid process is applied to make

Table 1 Quality tests used in Argentina for bread wheat selection in breeding programs

Test	Application	Sample type	Sample Size	Comments/protocol
Zeleny test	Measures the swelling of the proteins in a solution of lactic acid and propanol. Indicates the quality of the gluten	Flour	3.2 g	To select strong gluten in segregating materials at early generations (F_{2-3-4}). Strong correlation with protein content. AACC 56–61
Test weight	Measures the density of a grain and how well the endosperm has filled out	Grain	500 g	For characterization of lines advanced to preliminary trials (F_{6-7}) and regional yield trials (F_{8-9-10}). Values higher than 76 kg/hl
Weight of 1000 kernels	Grain characterization	Grain	1000 kernels	For characterization of lines advanced to preliminary trials (F_{6-7}) and regional yield trials (F_{8-9-10}). Values higher than 35 g
NIR	Predicts protein content	Grain	5 g	For characterization of lines advanced to preliminary trials (F_{6-7}) and regional yield trials (F_{8-9-10}). Values higher than 10.5%, 13.5% moisture basis. AACC 39–21
Wet gluten content	Gluten characterization	Flour	10 g	For characterization of lines advanced to Preliminary Trials (F_{6-7}). Values higher than 25%. IRAM 15864
Mixograph (National Manufacturing)	Indicator of gluten strength	Flour	10 g	For characterization of lines advanced to preliminary trials (F_{6-7}). Parameter mixing time > 3 min. AACC 54–40
Alveograph (Chopin)	Indicator of gluten strength and breadmaking quality	Flour	250 g	For characterization of lines advanced to regional yield trials (F_{8-9-10}). Parameters W > 240 and P/L = > 1. AACC 54–30 A
Farinograph (Brabender)	Measures flour water absorption and dough mixing characteristics; indicator of gluten strength	Flour	50 g	For characterization of lines advanced to Regional Yield Trials (F_{8-9-10}). Parameter stability should be between 10–40 min. IRAM 15855
Baking loaf volume	Indicator of dough suitability for breadmaking quality	Flour	100 g	For characterization of lines advanced to regional yield trials (F_{8-9-10}). AACC 10-10B
SDS-PAGE	HMW-GS characterization	Grain	3–5 grains	For characterization of lines advanced to Regional Yield Trials (F_{8-9-10}). Lawrence and Shepherd (1980)

segregating populations of wheat into fixed, true breeding lines. Through the time of segregation, generally only molecular markers or grain grading are used to make selections for wheat quality. Finally the line is fixed, or true breeding, but at this first stage, usually very little seed is available, and almost all the seed needs to be used in multiplication for the next year. Grain grading for kernel size, color, and vitreousness may occur at this stage. Typically, only enough seed is available for genotyping and planting of first year yield trial. Following the first yield trial, a limited amount of excess seed should be available for wheat quality testing. Wheat grain is assessed for pre-milling characteristics that impact marketing. These tests include kernel weight, test weight per volume, color, hardness, vitreousness of the kernel, and total protein content, and any other tests which may be conducted on NIRS systems. At this stage of heavy seed limitation and high entry numbers, small-scale tests are typically utilized, such as small scale milling (Brabender) along with smaller rheological testing such as Mixograph (National Manufacturing), Zeleny or SDS-sedimentation tests (AACC 2010). Though many international standards require Farinograph or Alveograph testing, the Mixograph is preferred by most US breeding programs. With these data, as well as field phenotypes, and any molecular data or GS prediction, typically one round of breeding selection is made, and entry numbers are highly reduced. After this stage, and into replicated yield trials, enough seed is typically available to conduct milling, Mixograph, and baking tests. Individual breeders determine their breeding pipeline allocations to early generation work, molecular markers, yield trials, disease screening, and quality testing. Many times this leads to wheat quality testing occurring later in the pipeline and on fewer materials which are already highly screened for yield and adaptation traits. The art of breeding is applied in creating a pipeline in which genetic gains are coming rapidly and fewer lines need to be discarded due to mandatory traits. In US breeding programs, no official recommendation system is in place for the release of varieties. Breeders utilize milling, dough rheology, and baking tests, as well as molecular markers, along their release decision making pipeline to insure they individually release high quality wheat varieties. Internal decision making committees which consider all attributes of a wheat variety and vote upon release decision are common. Wheat is segregated by market class in the United States. The predominant types are hard red winter, hard white, hard red spring, soft red winter, and soft white wheat. These marketing classes all have unique end-user standards. The Wheat Quality Council (Wheat Quality Council 2018) has targets approved for each market class in the USA, which can be found in their reports.

2.5 Quality Evaluation in Soft White Wheat

The U.S. Pacific Northwest states of Idaho, Oregon, and Washington produce approximately 5.4 mmt of soft white wheat annually. This production is apportioned about 6% club and 94% soft white 'common' (those varieties having a compact and lax spike, respectively). Both winter and spring varieties are grown, but winter types dominate due to the generally higher production potential of winter wheats in general. Soft white wheats are bred and selected to have superior and unique end-use quality. Much of the work involved in evaluating experimental wheat breeding lines and conducting wheat quality research occurs at the U.S.D.A. Western Wheat Quality Laboratory (WWQL). Following, is a brief description of the relevant testing that occurs. The WWQL also evaluates hard red and white 'bread' wheats. Naturally, grain yield is the primary consideration of wheat breeders, followed by yield protection in the form of resistance to abiotic and biotic stresses. From the farmer perspective, high bulk density (test weight), appropriate protein content (usually 8.5–11.0%), and high Falling Numbers are their measure of quality. But from a breeding perspective, milling performance, functionality of starch, components influencing water relations, color, dough strength, and cookie and cake quality are measured and selected for.

Milling Performance Milling performance is evaluated using a 'MicroMill' (30 g samples), a modified Quadrumat Sr. (600 g), Buhler MLU-202 (1500 g), and Miag Multomat (40–300 kg) flour mills. Desirable criteria include soft kernel texture, ease of endosperm-bran separation, mellow friable endosperm, high break flour and straight grade flour yields with low ash, bright white color, and low starch damage. The Quadrumat system performs the bulk of sample testing, usually around 4500 individual breeding lines per year.

Starch The primary selection criterion for starch functionality relates to 'normal' vs. 'partial waxy' genotypes, the latter are targeted for the Northeastern Asian noodle markets of Japan and South Korea. The partial waxy trait is conferred by a single null allele in granule bound starch synthase, and causes a slight decrease in amylose and an increase in starch swelling. This increased swelling creates a softer but resilient noodle texture. The majority of soft white varieties have normal starch. Testing for partial waxy lines is achieved by using the Flour Swelling Volume Test, the RapidVisco Analyzer, and the Amylograph.

Water Relations A number of different factors can contribute to water relations in doughs and batters. In general, soft white wheat has low water absorption. Selection for low water absorption involves low values for water solvent retention capacity (SRC), sodium carbonate SRC, sucrose SRC, and low dough water absorption in the 10-g Mixograph and 50-g Farinograph. These tests are aimed at capturing low starch damage, low arabinoxylan content, and are certainly influenced by protein and bran contamination.

Color As mentioned above, soft white wheat flour should be bright white and free of bran specks. Additionally, breeding lines are screened for polyphenol oxidase activity using the AACC International Approved Method that employs L-DOPA (L-3,4-dihydroxyphenylalanine) as the substrate in a whole-seed assay. Further, raw white salted and alkaline noodle sheets are evaluated for brightness after 24 h using the C.I.E. L^*, a^*, and b^* color space. High L^* values and good color stability (low ΔL^*) are desirable.

Dough Strength Dough strength is the primary trait that separates soft white common and club wheat sub-classes. Whereas soft white common varieties range from moderately weak to moderately strong, club wheats are selected for uniformly weak dough strength. The methods used include flour SDS micro- sedimentation and the 10-g Mixograph. The Farinograph and Alevograph (AlveoLab) are also suitable but cannot produce sufficient sample throughput in early generations.

Cookie and Cake Quality All soft white breeding samples receive a cookie bake. The cookie test captures a number of different flour attributes but is especially influenced by starch damage and arabinoxylans. Large diameter cookies that result from greater oven spread are desirable. More advanced generation soft white common lines and all club wheat lines additionally receive a sponge cake bake test. The sponge cake method involves gently folding the flour into a prepared egg-sugar-water foam. Large volume cakes with a symmetrical shape and soft interior crumb are desirable. The combination of a very low moisture cookie and high moisture cake batter capture a wide range of products typically made from soft white wheat. On an occasional basis, steamed breads and boiled white salted noodles are evaluated. Many of the world's foods are best made using soft wheat flour. For these, U.S. Pacific Northwest soft white wheat is particularly well suited. As in all wheat breeding endeavors, this is not by accident or chance. A considerable amount of resources and a large variety of individual tests are employed to provide breeders a detailed portrait of end-use quality. These data are used in a rigorous selection process to ensure that growers, millers, processors and consumers receive the highest quality varieties, grain, flour and foods.

2.6 Current and Potential Approaches to Quality Evaluation in Durum Wheat Breeding Programs

Durum wheat breeding programs are scattered around the globe and present in many of the durum wheat producing nations (Canada, France, Italy, Spain, USA, Algeria, Morocco, Syria, Mexico, India, Germany, Kazakhstan, Austria, Turkey, Tunisia, Argentina and Australia). The breeding objectives for quality targets and extent of testing depend on the specific countries focus to supply durum wheat for internal use or export. While customers have specific requirements, key quality features should be met to sell in the market to meet the requirements for good pasta, couscous or other durum derived product quality. Nevertheless, many breeding objectives for quality are common across programs with differences related to specific issues for the country, for example screening for low grain cadmium content is important in Canada due to its higher soil content.

The breeder and quality chemist must try to satisfy the requirements of the entire production chain consisting of grower, grain trader, miller, pasta maker and consumer with each having different requirements (Troccoli et al. 2000). The typical process in a breeding program is to decrease the large number of lines (hundreds to thousands) arising from a cross, selecting the best lines for yield, adaptation, quality performance and disease resistance. Lines are assessed against check varieties which are commercial varieties grown under identical conditions to the test lines. If a line is to be advanced it should perform equally or better than the checks. Characteristics may correlate but in opposition, for example the inverse relationship between yield and protein makes it a challenge to select for both and the breeding team needs to make decisions about what criteria take the highest priority in advancing a line. However, a Canadian variety Strongfield achieved a 13.5% yield increase over the check, Kyle, but also a 0.3% units higher protein content showing both targets can be achieved (Li et al. 2018). For quality evaluation, the test depends on the amount of sample and resources and methods available. In general, the earlier the stage of the breeding program, the less seed available, the larger number of samples to test and the more limited the testing possible due to resources available. Quite often predictive tests are used in early generation testing which results in less accurate prediction of quality. Table 2 summarises typical durum quality tests for pasta derived products made from durum wheat [the reader is referred to other texts for more details on these tests and those applicable to other durum derived products in the reference list (Abecassis et al. 2012; Sissons et al. 2012)] used in breeding programs. These tests are designed to ensure released varieties meet the various grading standards in each country and are related to uniform kernel size, high test weight, a high proportion of vitreous kernels, a low percentage of sprouted kernels, meeting protein, moisture, semolina, dough and pasta quality and consumer/buyer requirements of the durum production chain. While the above tests are commonly used in later generation testing of durum wheat when sufficient sample and resources are available, the choice of test for early generation testing is different. Some of the potential early generation tests are summarised in Table 3 and have been applied to varying extents in breeding for quality. There are several tests to measure dough properties directly (Mixograph, Glutomatic, Farinograph, SDS Sedimentation (SDSS)) that have found routine application in breeding programs while the Glutopeak is showing potential to replace some of these methods being faster and easier to operate and better at discriminating poor from medium-strong dough strength. Other instruments like the Kieffer rig and CSIRO extension tester have only been used for research. To create the semolina (flour) needed to conduct dough tests from limited grain quantities, small-scale mills can be used such as the falling number and Udy Cyclone mills (which produce wholemeal) or similar, Brabender Quadrumat Junior mill and Chopin mill are capable of producing semolina from as little as 20 g of grain.

Many samples can be milled to produce samples for further testing like SDSS. Where very limited seed is available (<5 g) specialised micro-mill such as FQC 2000 can be used to make a "semolina" or when many samples are to be evaluated, biochemical based tests have found some favour for prediction of dough strength such as SE-HPLC to determine %UPP, 96 well microplate reader to measure % insoluble glutenin (IP%) and the swelling index of glutenin (SIG). The latter has found application in durum breeding (Li et al. 2013). The most useful and commonly used instrument for assessment of grain and semolina is NIR. Using scan-

Table 2 Typical durum quality tests for pasta derived products

Test	Application	Industry sector	Sample type	Sample size	Desirable range	Reference
Test weight	A measure of the density of a grain and how well the endosperm has filled out	Grain trader, miller	Grain	500 g	>74 kh/hl	AACC 55–10.01
Falling number	Used to assess weather damaged grain, which lowers starch viscosity	Miller, pasta maker	Wholemeal flour	100 g grain	>250 s	AACCI 56–81.03
Vitreous kernel	Kernels with translucent, vitreous appearance	Trader, miller pasta maker	Grain	300 kernels	>70%	ICC method 129
Screenings	Grain with high percentage of undersized grains reduces semolina yield	Grain trader, miller	Grain	>150 g	<5%	Particles below 2.0 mm
Protein	Dry pasta made from high-protein semolina (12%) is physically stronger and more elastic than pasta from lower protein semolina	Grain trader, miller, pasta maker	Grain	>50 g	>13%	AACC 46–30.01; ICC method (NIR) 159
Yellow pigment	The main pigments in durum wheat responsible for the yellow color are carotenoids	Pasta maker	Grain, semolina	2–3 g	Exceed check varieties	AACC 14–50.01
Yellowness	A bright, yellow color in semolina ensures a good color in the pasta	Pasta maker	Semolina, pasta	>20 g	Minolta b* >26	AACC 14–22.01
Moisture	Post harvest low moisture content is expected, which is necessary for the storage life of the grain	Grain trader, miller	Grain	>50 g	Below 12%	AACC method 44–15.02
Ash	Mineral content remaining when all the organic content has been removed by combustion at very high temperatures	Miller	Grain, semolina	2–3 g	<0.9%, dmb	AACC 08–01.01 ash- basic method
Semolina yield	Millers aim to produce as much semolina from a given amount of grain with minimal flour	Miller	Grain	>20 g	>65%	AACC 26–41.01
Granularity	Affects the amount and uniformity of water absorption during mixing	Pasta maker	Semolina	100 g		AACC 66–20.01
Gluten quality	Gluten strength	Pasta maker	Semolina	>2-100 g	Medium- strong dough	AACC 54–40.02, 54–22.01, 54–30.02, 54–10.01; ICC 158
Pasta quality	Capacity of the product to maintain good texture after cooking	Consumer	Pasta	>5 g	Firm, good colour, taste, aroma	AACC 14–22.01

ning instruments internal calibrations can be set up in the laboratory to predict a range of characteristics. Machines that can handle grain are desirable as they are non-destructive and faster to process samples than semolina/flour. Examples of some uses are listed in Table 3. The most accurate measures are those with a high RPD such as protein and moisture while many of the other predictors are better suited to rough screening (Sissons et al. 2006) which is still useful to make changes to the direction of a population of plants in the desired way. More recent technology such as image analysis is more accurate at assessing grain defects such as hard vitreous kernels, black point percentage, identifying insect infestation than NIR and grain grading is being used in some countries at grain silo receival stands with potential for future applications (Saini et al. 2014). Starch properties can be readily measured using small samples of semolina or ground pasta using an RVA (Table 3). However, this has mainly been used for research as the role of starch in durum wheat quality evaluation has not been considered important enough to measure in a breeding program unless the RVA is being used to predict falling number. The RVA is particularly useful to discriminate waxy from high amylose durum where each has distinct RVA profiles (Lafiandra et al. 2012). Some equipment like the GRL extension tester and Viscoelastograph to measure pasta viscoelasticity are unique to a laboratory (Grain Research Laboratory) or no longer in production leaving texture analysers and cooking tests (cooking time, cooking loss, water absorption, total solids) as the main tools to assess pasta quality for breeding lines. Even small-scale pasta extruders can be used to prepare pasta for assessment (Table 3) but there are limitations with sufficient sample available for pasta analysis together with loss of appearance and colour leaving this approach more to research.

2.7 Quality Evaluation in Wheat Wild Relatives, the Case of Kazakhstan

Plant genetic diversity in improving agricultural production is a key factor providing adaptability to unpredictable environmental and climatic changes, maintain resilience in the face of variation of productions systems and meet the needs of the expanding human population (Esquinas-Alcazar 2005). However, in the search of high yield elite varieties, modern crops suffer the narrowing of the genetic base (Tester and Langridge 2010) to the point of virtually eliminating local germplasm generated during centuries of traditional agriculture. In the case of wheat, as a staple crop providing around 20% of human dietary energy, sustainable and steady increases in wheat yields, boosted by genetic diversity are vital for the food security of next generations. Fortunately, wild relatives and progenitors of bread wheat still preserve remarkable genetic diversity in terms of alleles that may contribute to adaptive processes which may be utilized to develop hardy high-yielding varieties combining also genetic variation for quality traits. For more details a recent review of the contribution of wild relatives as source of variation for wheat grain quality improvement has been published (Alvarez and Guzmán 2018).

Table 3 Small scale tests to evaluate durum quality

Test	Application	Sample, quantity	Comments/need, references
Swelling index of glutenin	Predict gluten strength; measures proportion of flour protein that consists of glutenin polymers of very large molecular weight	Semolina, flour, other (35–45 mg)	Samples with different glutenin swelling properties; application in breeding (Wang and Kovacs 2002; Sissons and Smit 2018; Uthayakumaran et al. 2007; Li et al. 2013)
SDS sedimentation volume	Measures the amount of sediment after mixing flour in an SDS–lactic acid solution for a fixed time; indicator of gluten strength	Flour, semolina, other (1 g)	Poor discrimination of moderately strong from strong gluten; most common early generation dough test (Dexter et al. 1980; Dick 1983)
Mixograph or similar	Measures dough mixing characteristics using a pin mixer; indicator of gluten strength	Flour, semolina (2-25 g)	Poor at discriminating moderately strong from strong genotypes; Extensively used in breeding programs (Dick and Youngs 1988)
Glutomatic system	Wet/dry gluten, gluten index; indicator of gluten strength	Flour, semolina (10 g)	Very weak samples or those with poor gluten development giving no results. Extensively used in breeding programs (Cubadda et al. 1992)
Glutopeak	Measures the aggregation behavior of gluten in flour samples; indicator of gluten strength	Flour, semolina, other (8–10 g)	Can predict GI, rapid; Gaining more interest (Sissons 2016; Sissons and Smit 2018; Marti et al. 2014; Wang et al. 2017)
Farinograph-E	Measures flour water absorption and dough mixing characteristics with a Z-arm mixer; indicator of gluten strength	Flour, semolina (10 g)	Slow with poor discrimination of moderately strong from strong types (ICC115/1; AACC54–21.02)
Small scale mills	Mill grain into flour or semolina	Grain (1–100 g)	Brabender Quadramat junior mill; Chopin CD2 mill; Extensively used (Varga et al. 2000; Békés et al. 2003)
NIR	Predict protein, moisture, ash, test weight, semolina yield, HVK, yellow pigment, b*, hardness, wet gluten	Grain, flour, semolina, pasta, other	Scanning or fixed wavelength NIR; extensively used in breeding programs (Sissons et al. 2006; McCraig et al. 1992; Wang et al. 2002; Osborne 2006; Wesley et al. 2005)
Image analysis	Predicts HVK, specks, semolina yield, blackpoint	Grain, flour, semolina, pasta	Image analyser with calibrations (Cervitec 1625, EyeFoss; Seedcount, Next Instruments). Good use in industry less so in breeding programs (Symons et al. 1996; Novaro et al. 2001; Gorretta et al. 2006; Wang et al. 2003)
RVA	Measures pasting viscosity of starch or semolina/pasta solution, falling number	Flour, starch, other (3–5 g)	Widely used (Batey and Curtin 2000; Grant et al. 2001; Aravind et al. 2013)
Texture analyser	Dry and cooked pasta texture	Dough, pasta	Texture analyser; extensively used AACC 66–50.01 (Sissons et al. 2008; Cubadda et al. 2007)

In Kazakhstan, interspecific crosses including winter bread wheat and wild relatives are being performed as method of creating fundamentally new plants that combine their hereditary basis of the most valuable features and characteristics of cultivated and wild plants (Kozhahmetov and Abugalieva 2014). Wild relatives used for interspecific crosses with bread wheat include tetraploids *Aegilops triaristata* (2n = 28 CUCUMM), *Triticum militinae* (2n = 28 A^tA^tGG), *Triticum timopheevii* (2n = 28 A^bA^bGG), *Aegylops cylindrica* (2n = 28 CCDD) and the hexaploid *Triticum kiharae* (2n = 42 A^bA^bGGDD), among others.

Advanced lines obtained from interspecific crosses are characterized considering agronomic and botanical descriptors, disease resistance and yield as previously described (Kozhahmetov and Abugalieva 2014). Also grain components together with their physico-chemical properties are used to define grain quality characteristics. Table 4 shows Farinograph and Alveograph parameters obtained from advanced synthetic lines selected in interspecific crosses. According to the physical flour and dough properties, synthetics varies according to the dilution test at a level of 80–170 farinograph units, with the best value for both liquefaction and valorimetric evaluation for genotypes 231 and 1712 with 90–42 farinograph units and 80–45 units, respectively (Table 4).

High molecular weight glutenins composition in synthetic forms has been also analyzed and is described in Table 5. Additional infomation including presence of 1BL/1RS rye translocation and Payne Score is also indicated. Eight genotypes showed no segregation for HMW-GS and rye translocation, two of them with a Payne Score of 10. Remaining genotypes were characterized by segregation of HMW-GS and rye translocation (Table 5). HMW-GS identification, and therefore quality prediction in wild relatives derived synthetics is difficult because of the important differences of these cultivars in terms of glutenins and other grain components important to define wheat quality (Abugalieva and Morgounov 2016).

Table 4 Physico-chemical properties obtained from transitional winter wheat forms. Farinograph and Alveograph parameters depending on the year of cultivation are described

		Dilution test[a]		Valorimetric test[a]			
Genotype	Cross	2016	2017	2016	2017	P/L	W
231	(Bezostaya1 × *Ae.triaristata*) × Karlygash	90	–	42	62	0.49	387
1712	Erythr.350 × *T.militinae*	80	100	45	50	0.34	221
1721–6	(Bezostaya 1 ×*T.m.*) × *T.m.*-6	130	100	30	43	0.27	279
1721–9	(Bezostaya 1 ×*T.m.*) × *T.m.*-9	150	70	28	56	0.39	177
1721–4	(Bezostaya 1 ×*T.m.*) × *T.m.*-4	150	120	21	40	0.36	174
1671	Zhetysu × *T.m.*	140	120	33	44	0.29	174
1727	Erythr.350 × *T.kiharae*	160	80	31	50	0.59	240
1676	Stekl.24 × *T.timopheevii*	–	160	–	35	0.21	157
1674	Zhetysu × *T.t.*	–	80	–	53	0.35	279
1718	Bezostaya1 × *Ae.cylindrica*	170	120	27	36	0.33	119
1825	Stekl.24 × *Ae.c.*	–	–	–	56	0.41	205
	Almaly	130	110	38	44	0.48	187

[a]Measured in Farinograph unit

Table 5 Protein marker composition including HMW-GS and wheat-rye translocations in the prediction and grain quality chracterization of winter and synthetic wheat forms

Genotype	Cross	*Glu-A1*	*Glu-B1*	*Glu-D1*	Status 1B/1R	Payne score
231	PEG 231 × Karlygash	2∗	7 + 9	5 + 10	1B/1B	9
1712	Erythr.350 × *T.militinae*	2∗	7 + 9	5 + 10	1B/1B	9
1721–6	PEG ×*T.m.*) (6)	0	7 + 9	2 + 12	1B/1B	5
1721–9	PEG ×*T.m.*) (9)	2∗	7 + 9	5 + 10	1B/1B	9
1721–4	PEG ×*T.m.*) (4)	0	7 + 9	2 + 12	1B/1B	5
1680	Steklovidnaya 24 × *T.m.*	2∗	7 + 8/7 + 9	5 + 10	1B/1B	10/9
1671	Zhetysu × *T.m.*	1	17 + 18/21 + 18	5 + 10	1B/1B/ 1B/1R	10
1671	Zhetysu × *T.m.*	2∗	7 + 9	5 + 10	1B/1B	9
1723	PEG × *T.kiharae*	1	7 + 9/6	5 + 10/4 + 10	1B/1B/ 1B/1R	8/6
1675	Zhetysu × *T.k.*	0	7∗	5 + 10/4 + 10	1B/1B	6
1675	Zhetysu × *T.k.*	2∗	7 + 8	5 + 10	1B/1B	10
1727	Erythrospermum350 × *T.k.*	2∗/0	7 + 9/6 + 8	5 + 10	1B/1B/ 1B/1R	9/6
1676	Steklov.24 × *T.timopheevii*	1	7 + 8	5 + 10	1B/1B	10
1674	Zhetysu × *T.t.*	0/1/2∗	7 + 9/7 + 8	5 + 10	1B/1B	7/10
1718	PEG × 1718	0	7 + 8/7∗	3 + 12	1B/1B	6/5
1825	Steklovidnaya 24 × *Ae. cylindrica*	0	21 + 8/7∗	5 + 10	1B/1B/ 1B/1R	6

Acknowledgements Jacques Bordes, François Balfourier who experimented the wheat accessions and Agnes Piquet for her assistance in E Nurit's PhD are greatly acknowledged. Certain parts of the work mentioned in this chapter were supported within project OTKA-K 16-119835, funded by the National Research, Development and Innovation Office, Hungary.

References

Abdel-Mawgood AL (2008) Molecular markers for predicting end-products quality of wheat (Triticum aestivum L.). African Journal of Biotechnology 7: 2324–2327.

Abecassis J, Cuq B, Boggini G, Namoune H (2012) Other Traditional Durum- Derived Products. In: Durum Wheat: Chemistry and Technology, AACC International Press, pp. 177–200.

Abugalieva AI, Morgounov AI (2016) Genetic potential of winter wheat grain quality in Central Asia. International J of Environmental and Science Education 11: 4869–4884.

AACC, American Association of Cereal Chemists. 2010. Approved methods of the AACC. St. Paul, MN, USA.

Alvarez JB, Guzmán C (2018) Interspecific and intergeneric hybridization as a source of variation for wheat grain quality improvement. Theoretical and Applied Genetics 131:225–251.

Amaya A, Peña RJ, Zarco-Hernández J, Rajaram S, A MK (1991) Quality (breadmaking) characteristic of normal (1B/1B) and 1B/1R translocation (1B/1R) wheats varying in dough stickiness character at two mixing speeds. Cereal Foods World 36: 701.

Aravind N, Sissons M, Fellows CM, Blazek J, Gilbert EP (2013) Optimisation of resistant starch ii and iii levels in durum wheat pasta to reduce in vitro digestibility while maintaining processing and sensory characteristics. Food Chemistry 136: 1100–1109.

Bainotti C, Fraschina J, Salines J, Nisi J, Dubcovsky J, Lewis S, et al. (2009) Registration of 'BIOINTA 2004' Wheat. Journal of Plant Registrations 3: 165–169

Bainotti CT, Lewis S, Campos P, Alberione E, Salines N, Gomez D, et al. (2017) MS INTA 416: A new Argentinean wheat cultivar carrying Fhb1 and Lr47 resistance genes. Crop Breeding and Applied Biotechnology 17: 280–286.

Balfourier F, Roussel V, Strelchenko P, Exbrayat-Vinson F, Sourdille P, Boutet G, et al. (2007) A worldwide bread wheat core collection arrayed in a 384-well plate. Theoret- ical and Applied Genetics 114: 1265–1275.

Baracskai I, Balázs G, Liu L, Ma W, Oszvald M, Newberry M, et al. (2011) A retrospective analysis of HMW and LMW glutenin alleles of cultivars bred in Martonvásár, Hungary. Cereal Research Communications 39: 225–236.

Batey IL, Curtin BM (2000) Effects on pasting viscosity of starch and flour from different operating conditions for the rapid visco analyser. Cereal Chemistry 77: 754–760.

Békés F (2012a) New aspects in quality related wheat research: 1. challenges and achievements. Cereal Research Communications 40: 159–184.

Békés F (2012b) New aspects in quality related wheat research: Ii. new methodologies for better quality wheat. Cereal Research Communications 40: 307–333.

Békés F, Wrigley CW (2013) Gluten alleles and predicted dough quality for wheat varieties worldwide: a great resource—free on the aacc international website. Cereal Foods World 58: 325–328.

Békés F, Wrigley CW (2017) Gluten alleles and predicted dietary intolerances: use of a database of wheats world-wide. EC Nutrition 10: 43–50.

Békés F, Larroque O, Hart P, O'Riordan B, Miskelly D, Baczynski M, et al. (1998) Non-linear behaviour of grain and flour blends made from dissimilar components. In: Proceedings 48th RACI conference, Eds.: Tarr A W, Ross A S, Wrigley C W, RACI conference, pp. 123–127.

Békés F, Larroque O, Hart P, O'Riordan B, Miskelly D, Wrigley CW (2001) Non-linear behaviour of grain and flour blends made from dissimilar com- ponents. In: Proc 11th Cereal and Bread Congress, Cereals, health and life, Surfers Paradise 2000, Eds: Wootton M, Batey IL, Wrigley CW, Cereal and Bread Congress, pp. 464–469.

Békés F, Lukow O, Uthayakumaran S, Mann G (2003) Small-scale dough testing methods. Wheat gluten protein analysis AACC, St Paul, USA pp. 173–198.

Békés F, Kemény S, Morell M (2006) An integrated approach to predicting end-product quality of wheat. European Journal of Agronomy 25: 155–162.

Boerger A (1928) Observaciones sobre agricultura, quince años de trabajos fitotécnicos en Uruguay [Agricultural notes: fifteen years of crop breeding in Uruguay].

Bordes J, Branlard G, Oury F, Charmet G, Balfourier F (2008) Agronomic characteristics, grain quality and flour rheology of 372 bread wheats in a worldwide core collection. Journal of Cereal Science 48: 569–579.

BSA (2017) Bundessortenamt, Beschreibende Sortenliste 2017. In: Getreide, Mais, Öl- und Faserpflanzen, Leguminosen, Rüben, Zwischen- früchte. Bundessortenamt, Hannover.

Bushuk W, Békés F (2002) Novel raw materials, technologies and products – new challenge for the quality control. In: Proceedings ICC (Budapest, Hungary, 26-29 May, 2002), pp. 14–19.

Ceresino EB, Kuktaite R, Sato HH, Hedenqvist MS, Johansson E (2019) Impact of gluten separation process and transglutaminase source on gluten based dough properties. Food Hydrocolloids 87: 661–669.

Cornish GB, Békés F, Eagles HA, Payne PI (2006) Prediction of Dough Properties for Bread Wheats. In: Gliadin and Glutenin: The Unique Balance of Wheat Quality, AACC International, Inc., pp. 243–280.

Cubadda R, Carcea M, Pasqui LA (1992) Suitability of the gluten index method for assessing gluten strength in durum wheat and semolina. Cereal foods world (USA).

Cubadda RE, Carcea M, Marconi E, Trivisonno MC (2007) Influence of gluten proteins and drying temperature on the cooking quality of durum wheat pasta. Cereal Chemistry 84:48–55.

Cuniberti M, Otamendi M (2004) Creating class distinction. World Grain 1.

Czuchajowska Z, Pomeranz Y (1991) Evaluation of Vital Dry Gluten Composition and Functionality in Breadmaking By Near-Infrared Reflectance Spectroscopy. Cereal Foods World 36: 439–446.

Delwiche SR, Graybosch RA, Peterson CJ (1998) Predicting protein composition, biochemical properties, and dough-handling properties of hard red winter wheat flour by near-infrared reflectance. Cereal Chemistry 75: 412–416.

Demichelis M, Vanzetti LS, Crescente JM, Nisi MM, Pflüger LA, Bainotti CT, et al. (2018) Significant Effects in Bread Making Quality Associated with the Gene Cluster Glu-D3 / Gli-D1 from the Bread Wheat Cultivar Prointa Guazú. Cereal Research Communications published online november 19, 2018.

Dexter JE, Matsuo RR, Kosmolak FG, Leisle D, Marchylo BA (1980) The suitability of the sds-sedimentation test for assessing gluten strength in durum wheat. Canadian Journal of Plant Science 60: 25–29.

Dick JW (1983) A modified screening test for rapid estimation gluten strength in early-generation durum wheat breeding lines. Cereal Chemistry 60: 315–318.

Dick JW, Youngs VL (1988) Evaluation of durum wheat, semolina and pasta in the united states. In: Durum Chemistry and Technology, Am. Assoc. Cereal Chem. St. Paul, MN. Dexter, J. E., pp. 237–248.

Dowell FE, Maghirang EB, Xie F, Lookhart GL, Pierce RO, Seabourn BW, et al. (2006) Predicting wheat quality characteristics and functionality using near-infrared spectroscopy. Cereal Chemistry 83:529–536.

Eagles HA, Hollamby GJ, Gororo N, Eastwood RF (2002) Estimation and utilisation of glutenin gene effects from the analysis of unbalanced data from wheat breeding programs. Australian Journal of Agricultural Research 53:367–377.

Edwards NM, Gianibelli MC, McCaig T, Clarke JM, Ames NP, Larroque OR, Dexter JE (2007) Relationships between dough strength, polymeric protein quantity and composition for diverse durum wheat genotypes. Journal of Cereal Science 45:140–149.

Elion E (1933) A simple volumetric method for measuring gas production during dough fermentation. Cereal Chemistry 10: 245–249.

Esquinas-Alcázar J (2005) Protecting crop genetic diversity for food security: Political, ethical and technical challenges. Nature Reviews Genetics 6: 946–953.

Fardet A (2010) New hypotheses for the health-protective mechanisms of whole-grain cereals: what is beyond fibre? Nutrition Research Reviews 23: 65–134.

Gao L, Ma W, Chen J, Wang K, Li J, Wang S, et al. (2010) Characterization and comparative analysis of wheat high molecular weight glutenin subunits by SDS-PAGE, RP-HPLC, HPCE, and MALDI-TOF-MS. Journal of Agricultural and Food Chemistry 58:2777–2786.

Garófalo L, Vázquez D, Ferreira F, Soule S (2011) Wheat flour non-starch polysaccharides and their effect on dough rheological properties. Industrial Crops and Products 34:1327–1331.

Gianibelli MC, Echaide M, Larroque OR, Carrillo JM, Dubcovsky J (2002) Biochemical and molecular characterisation of *Glu-1* loci in Argentinean wheat cultivars. Euphytica 128:61–73.

Gorretta N, Roger J, Aubert M, Bellon-Maurel V, Campan F, Roumet P (2006) Determining vitreousness of durum wheat kernels using near infrared hyperspectral imaging. Journal of Near Infrared Spectroscopy 14:231–239.

Grant L, Vignaux N, Doehlert D, McMullen M, Elias E, Kianian S (2001) Starch characteristics of waxy and nonwaxy tetraploid (triticum turgidum l. var. durum) wheats. Cereal Chemistry 78:590–595.

Gregory JF, Trumbo PR, Bailey LB, Toth JP, Baumgartner TG, Cerda JJ (1991) Bioavailability of Pyridoxine-5′-β-D-Glucoside Determined in Hu- mans by Stable-Isotopic Methods. The Journal of Nutrition 121:177–186.

Gupta RB, Békés F, Wrigley CW (1991) Prediction of physical dough proper- ties from glutenin subunit composition in bread wheats: correlation studies. Cereal Chemistry 68:328–333.

Gupta RB, Khan K, Macritchie F (1993) Biochemical basis of flour properties in bread wheats. i. effects of variation in the quantity and size distribution of polymeric protein. Journal of Cereal Science 18:23–41.

Gupta RB, Paul JG, Cornish GB, Palmer GA, Békés F, Rathjen AJ (1994) Allelic variation at glutenin subunit and gliadin loci, *glu-1*, *glu-3* and *gli-1*, of common wheats. i. its additive and interaction effects on dough properties. Journal of Cereal Science 19:9–17.

Hamer RJ, Weegels PL, Marseille JP (1992) Prediction of the breadmaking quality of wheat: the use of hmw glutenin-a subunit-based quality scoring systems. Journal of Cereal Science 15:91–102.

Hartl L, Mohler V, Henkelmann G (2010) Bread-making quality and grain yield in German winter wheat. I. History. Höhere Bundeslehr- und Forschungsanstalt für Landwirtschaft Raumberg-Gumpenstein, Irdning, Austria, pp. 25–28.

He GY, Jones HD, D'Ovidio R, Masci S, Chen M, West J, Butow B, Anderson OD, Lazzeri P, Fido R, Shewry PR (2005) Expression of an extended HMW subunit in transgenic wheat and the effect on dough mixing properties. Journal of Cereal Science 42(2):225–231.

Howitt CA (2010) Identification of grain variety and quality type. In: Cereal Grains, Elsevier, pp. 311–341.

Howitt CA, Gale KR, Juhász A (2006) Chapter 11 Diagnostic Markers for Quality. In: Gliadin and Glutenin: The Unique Balance of Wheat Quality, References Series, AACC International, Inc., pp. 333–361.

Hrušková M, Šmejda P (2003) Wheat flour dough alveograph characteristics predicted by NIR Systems 6500. Czech Journal of Food Sciences 21(No. 1):28–33.

Hrušková M, Bednarova M, Novotny F (2001) Wheat flour dough rheological characteristics predicted by nirsystems 6500. Czech Journal of Food Sciences 19(6):213–218.

Hu Y, Wang L, Li Z (2017) Modification of protein structure and dough rheological properties of wheat flour through superheated steam treatment. Journal of Cereal Science 76:222–228.

Hussain A, Larsson H, Kuktaite R, Prieto-Linde ML, Johansson E (2012) Towards the understanding of bread-making quality in organically grown wheat: Dough mixing behaviour, protein polymerisation and structural properties. Journal of Cereal Science 56:659–666.

INDEC (2018) https://www.indec.gob.ar/index.asp

Jirsa O, Hrušková M, Švec I (2008) Near-infrared prediction of milling and baking parameters of wheat varieties. Journal of Food Engineering 87:21–25.

Johansson E, Prieto-Linde ML, Jönsson J (2001) Effects of Wheat Cultivar and Nitrogen Application on Storage Protein Composition and Breadmaking Quality. Cereal Chemistry Journal 78:19–25.

Johansson E, Nilsson H, Mazhar H, Skerritt J, MacRitchie F, Svensson G (2002) Seasonal effects on storage proteins and gluten strength in four Swedish wheat cultivars. Journal of the Science of Food and Agriculture 82:1305–1311.

Johansson E, Prieto-Linde ML, Svensson G, Jönsson J (2003) Influences of cultivar, cultivation year and fertilizer rate on amount of protein groups and amount and size distribution of mono- and polymeric proteins in wheat. The Journal of Agricultural Science 140: 275–284.

Johansson E, Malik AH, Hussain A, Rasheed F, Newson WR, Plivelic T, et al. (2013) Wheat Gluten Polymer Structures: The Impact of Genotype, Environment, and Processing on Their Functionality in Various Applications. Cereal Chemistry 90:367–376.

Kozhahmetov KK, Abugalieva AI (2014) Using gene fund of wild relatives for common wheat improvement. International Journal of Biology and Chemistry 7:41–43.

Kuktaite R, Larsson H, Johansson E (2004) Variation in protein composition of wheat flour and its relationship to dough mixing behaviour. Journal of Cereal Science 40:31–39.

Kuktaite R, Plivelic TS, Cerenius Y, Hedenqvist MS, Gällstedt M, Marttila S, et al. (2011) Structure and Morphology of Wheat Gluten Films: From Polymeric Protein Aggregates toward Superstructure Arrangements. Biomacromolecules 12:1438–1448.

Lado B, Vázquez D, Quincke M, Silva P, Aguilar I, Gutiérrez L (2018) Resource allocation optimization with multi-trait genomic prediction for bread wheat (Triticum aestivum L.) baking quality. Theoretical and Applied Genetics 131: 2719–2731.

Lafiandra D, Masci S, Sissons M, Dornez E, Delcour J, Courtin C, Caboni MF (2012) Kernel components of technological value. In: Durum wheat chemistry and technology, 2nd edn, AACC International, Inc., pp. 85–124.
Laidig F, Piepho HP, Rentel D, Drobek T, Meyer U, Huesken A (2017) Breeding progress, environmental variation and correlation of winter wheat yield and quality traits in German official variety trials and on-farm during 1983–2014. Theoretical and Applied Genetics 130:223–245.
Lawrence GJ, Shepherd KW (1980) Variation in glutenin protein subunits of wheat. Australian Journal of Biological Sciences 33:221–233.
Lerner SE, Kolman MA, Rogers WJ (2009) Quality and endosperm storage protein variation in Argentinean grown bread wheat. I. Allelic diversity and discrimination between cultivars. Journal of Cereal Science 49:337–345.
Li L, Niu Y, Ruan Y, DePauw RM, Singh A, Gan Y (2018) Agronomic advancement in tillage, crop rotation, soil health and genetic gain in durum wheat cultivation: a 17-year canadian story. Preprints.
Li YF, Wu Y, Hernandez-Espinosa N, Peña RJ (2013) Heat and drought stress on durum wheat: Responses of genotypes, yield, and quality parameters. Journal of Cereal Science 57:398–404.
Liu L, Wang A, Appels R, Ma J, Xia X, Lan P, et al. (2009) A MALDI-TOF based analysis of high molecular weight glutenin subunits for wheat breeding. Journal of Cereal Science 50:295–301.
Liu L, Ikeda TM, Branlard G, Peña RJ, Rogers WJ, Lerner SE, et al. (2010) Comparison of low molecular weight glutenin subunits identified by SDS- PAGE, 2-DE, MALDI-TOF-MS and PCR in common wheat. BMC Plant Biology 10:124.
Liu Y, He Z, Appels R, Xia X (2012) Functional markers in wheat: current status and future prospects. Theoretical and Applied Genetics 125:1–10.
Lo Valvo PJ, Miralles DJ, Serrago RA (2018) Genetic progress in Argentine bread wheat varieties released between 1918 and 2011: Changes in physiolog- ical and numerical yield components. Field Crops Research 221:314–321.
Ma M, Wang A, Békés F, Newberry M, Gao L, Ma J, et al. (2009) High resolution identification of High and Low Molecular Weight Glutenin alleles by Matrix-Assisted Laser Desorption/Ionization Time- of-Flight Mass Spectrometry (MALDI-TOF-MS) in common wheat (Triticum aestivum L.). In: Gluten Proteins 2009. Proc. 10th Internat. Gluten Workshop Ed: Branlard G, pp. 271–275.
Malik AH, Prieto-Linde ML, Kuktaite R, Andersson A, Johansson E (2011) Individual and interactive effects of cultivar maturation time, nitrogen regime and temperature level on accumulation of wheat grain proteins. Journal of the Science of Food and Agriculture 91:2192–2200.
Malik AH, Kuktaite R, Johansson E (2013) Combined effect of genetic and environmental factors on the accumulation of proteins in the wheat grain and their relationship to bread-making quality. Journal of Cereal Science 57:170–174.
Marchylo BA, Kruger JE, Hatcher DW (1989) Quantitative reversed-phase high-performance liquid chromatographic analysis of wheat storage proteins as a potential quality prediction tool. Journal of Cereal Science 9:113–130.
Marti A, Cecchini C, D'Egidio MG, Dreisoerner J, Pagani MA (2014) Characterization of durum wheat semolina by means of a rapid shear-based method. Cereal Chemistry 91:542–547
McCraig TN, McLeod JG, Clarke JM, DePauw RM (1992) Measurement of durum pigment with a near-infrared instrument operating in the visible range. Cereal Chemistry 69:671–672.
Moiraghi M, Vanzetti L, Pflüger L, Helguera M, Pérez GT (2013) Effect of high molecular weight glutenins and rye translocations on soft wheat flour cookie quality. Journal of Cereal Science 58:424–430.
Møup ILK, Olesen ES (1976) New method for prediction of protein value from essential amino acid pattern [biological value of foods, nitrogen balance]. Nutrition Reports International 13:355–365.
Naeem H, Paulon D, Irmak S, MacRitchie F (2012) Developmental and envi- ronmental effects on the assembly of glutenin polymers and the impact on grain quality of wheat. Journal of Cereal Science 56:51–57.

Ndaw S, Bergaentzlé M, Aoudé-Werner D, Hasselmann C (2000) Extraction procedures for the liquid chromatographic determination of thiamin, ri- boflavin and vitamin B6 in foodstuffs. Food Chemistry 71:129–138.

norma (2018) https://www.bcr.com.ar/normas/normas/norma%20xx%20trigo%20pan.pdf

Novaro P, Colucci F, Venora G, D'Egidio M (2001) Image analysis of whole grains: a noninvasive method to predict semolina yield in durum wheat. Cereal Chemistry 78:217–221.

Nurit E (2015) Identification of genetic, environmental and technologic factors associated to the variability of vitamins in common wheat and wheat based food products. PhD thesis, Université Blaise Pascal-Clermont-Ferrand II.

Nurit E, Lyan B, Piquet A, Branlard G, Pujos-Guillot E (2015) Development of a LC-MS/MS method for the simultaneous screening of seven water- soluble vitamins in processing semi-coarse wheat flour products. Analytical and Bioanalytical Chemistry 407:3471–3479.

Nurit E, Lyan B, Pujos-Guillot E, Branlard G, Piquet A (2016) Change in B and E vitamin and lutein, β-sitosterol contents in industrial milling fractions and during toasted bread production. Journal of Cereal Science 69:290–296.

Osborne BG (2006) Applications of near infrared spectroscopy in quality screening of early-generation material in cereal breeding programmes. Journal of Near Infrared Spectroscopy 14:93–101.

Oury FX, Chiron H, Faye A, Gardet O, Giraud A, Heumez E, Rolland B, Rousset M, Trottet M, Charmet G, Branlard G (2010) The prediction of bread wheat quality: joint use of the phenotypic information brought by technological tests and the genetic information brought by HMW and LMW glutenin subunits. Euphytica 171:87–109.

Park SH, Bean SR, Chung OK, Seib PA (2006) Levels of Protein and Protein Composition in Hard Winter Wheat Flours and the Relationship to Breadmaking. Cereal Chemistry 83:418–423.

Paulley G, Vázquez D, Lysenko E, Preston KR (2004) Development of a laboratory baking test for Uruguayan French style hearth bread using Canadian wheat flour. Canadian Journal of Plant Science 84:949–954.

Payne PI, Nightingale MA, Krattiger AF, Holt LM (1987) The relationship between HMW glutenin subunit composition and the bread-making quality of British-grown wheat varieties. Journal of the Science of Food and Agriculture 40:51–65.

Peña RJ (2007) Current and Future Trends of Wheat Quality Needs. In: Buck HT, Nisi JE, Salomón N (eds) Wheat Production in Stressed Environments, Springer Netherlands, Dordrecht, pp. 411–424.

Pirozi M, Margiotta B, Lafiandra D, MacRitchie F (2008) Composition of polymeric proteins and bread-making quality of wheat lines with allelic HMW-GS differing in number of cysteines. Journal of Cereal Science 48:117–122.

Primo-Martin C, Martinez-Anaya M (2003) Influence of Pentosanase and Oxidases on Water-extractable Pentosans during a Straight Breadmaking Process. Journal of Food Science 68:31–41.

Rasheed F, Kuktaite R, Hedenqvist MS, Gällstedt M, Plivelic TS, Johansson E (2016) The use of plants as a "green factory" to produce high strength gluten-based materials. Green Chemistry 18:2782–2792.

Rasheed F, Plivelic TS, Kuktaite R, Hedenqvist MS, Johansson E (2018) Unraveling the Structural Puzzle of the Giant Glutenin Polymer—An Interplay between Protein Polymerization, Nanomorphology, and Functional Properties in Bioplastic Films. ACS Omega 3:5584–5592.

Rubenthaler G, Pomeranz Y (1987) Near-infrared reflectance spectra of hard red winter wheats varying widely in protein content and breadmaking potential. Cereal Chemistry 64:407–411.

Saini M, Singh J, Prakash N (2014) Analysis of wheat grain varieties using image processing-a review. International Journal of Science and Research 3:490–495.

Salomon N, Miranda R (2003) Indice de calidad industrial en trigo: una her- ramienta para determinar la aptitud de los materiales geneticos. In: Estrategias y metodologías utilizadas en el mejoramiento de trigo: un enfoque multidisciplinario, CIMMYT, pp. 163–174.

Sampson DA, Wen Q, Lorenz K (1996) Vitamin B6 and Pyridoxine Glucoside Content of Wheat and Wheat Flours. Cereal Chemistry 73:770–774.

Scholz E, Prieto-Linde ML, Gergely S, Salgó A, Johansson E (2007) Possibilities of using near infrared reflectance/transmittance spectroscopy for determination of polymeric protein in wheat. Journal of the Science of Food and Agriculture 87:1523–1532.

Singh NK, Donovan GR, Batey IL, MacRitchie F (1990) Use of sonication and size-exclusion high-performance liquid chromatography in the study of wheat flour proteins. I. Dissolution of total proteins in the absence of reducing agents. Cereal Chemistry 67:150–161.

Sissons M (2016) Glutopeak: A breeding tool for screening dough properties of durum wheat semolina. Cereal Chemistry 93:550–556.

Sissons M, Smit J (2018) Small-scale methods to assess the gluten properties of durum wheat. Cereal Chemistry 95:456–468.

Sissons M, Osborne B, Sissons S (2006) Application of near infrared reflectance spectroscopy to a durum wheat breeding programme. Journal of Near In- frared Spectroscopy 14:17–25.

Sissons M, Abecassis J, Marchylo B, Cubadda R (2012) Methods used to assess and predict quality of durum wheat, semolina, and pasta. In: Durum Wheat (Second Edition), Elsevier, pp. 213–234.

Sissons MJ, Schlichting LM, Egan N, Aarts WA, Harden S, Marchylo BA (2008) A standardized method for the instrumental determination of cooked spaghetti firmness. Cereal Chemistry 85:440–444.

Slafer GA, Andrade FH (1989) Genetic Improvement in Bread Wheat (Triticum aestivum) Yield in Argentina. Field Crops Research 21:289–296.

Symons SJ, Dexter JE, Matsuo RR, Marchylo BA (1996) Semolina speck counting using an automated imaging system. Cereal Chemistry 73:561–566.

Tabbita F, Lewis S, Vouilloz J, Ortega M, Kade M, Abbate P, Barneix A (2013) Effects of the *Gpc-B1* locus on high grain protein content introgressed into Argentinean wheat germplasm. Plant Breeding 132:48–52.

Tara KA, Bains GS, Finney PL (1972) Damaged Starch and Protein Contents in Relation to Water Absorption of Flours of Indian Wheats. Starch 24:342–345.

Tester M, Langridge P (2010) Breeding Technologies to Increase Crop Production in a Changing World. Science 327:818–822.

Tömösközi S, Békés F, Haraszi R, Gras PW, Varga J, Salgó A (2002) Application of Micro Z-arm dough mixer in wheat research - Effect of protein addition on mixing properties of wheat dough. Periodica Polytechnica Chemical Engineering 46:31–38.

Tömösközi S, Kindler Á, Varga J, Rakszegi M, Láng L, Bedo Z, et al. (2004) Determination of Breadmaking Quality of Wheat Flour Dough with Different Macro and Micro Mixers. In: The Gluten Proteins, The Royal Society of Chemistry, pp. 267–270.

trigoargentino (2018) http://www.trigoargentino.com.ar

Troccoli A, Borrelli GM, De Vita P, Fares C, Di Fonzo N (2000) Mini review: durum wheat quality: a multidisciplinary concept. Journal of Cereal Science 32:99–113.

Tronsmo KM, Magnus EM, Baardseth P, Schofield JD, Aamodt A, Færgestad EM (2003) Comparison of Small and Large Deformation Rheological Properties of Wheat Dough and Gluten. Cereal Chemistry 80:587–595.

Uthayakumaran S, Gras PW, Stoddard FL, Békés F (1999) Effect of Varying Protein Content and Glutenin-to-Gliadin Ratio on the Functional Properties of Wheat Dough. Cereal Chemistry 76:389–394.

Uthayakumaran S, Barker N, Batey IL, Dines J, Miskelly D, Wrigley CW (2007) Rapid methods to predict soft-wheat dough quality for specific food products. Cereal Chemistry 84:522–526.

Vanzetti LS, Pflüger LA, Rodríguez-Quijano M, Carrillo JM, Helguera M (2009) Genetic variability for waxy genes in Argentinean bread wheat germplasm. Electronic Journal of Biotechnology 12.

Vanzetti LS, Yerkovich N, Chialvo E, Lombardo L, Vaschetto L, Helguera M (2013) Genetic structure of Argentinean hexaploid wheat germplasm. Genetics and Molecular Biology 36:391–9.

Varga L, Foder D, Nánási J, Békés F, Southan M, Gras P, Rath C, Salgo A, Tömösközi S (2000) Laboratory mill for small-scale testing. In: Proceedings of the 7th International Workshop Gluten 2000, The Royal Society of Chemistry, Cambridge UK, pp. 317–320.

Vázquez D, Castro M (2018) Mejoramiento para calidad estable en ambientes variables. In: German S, Quincke M, Vázquez D, Castro M, Pereyra S, Silva P, García A (eds) Seminario internacional "1914–2014, un siglo de mejoramiento de trigo en La Estanzuela", INIA Serie Técnica 241: 212–220.

Vázquez D, Veira M (2015) Applicability of Mixolab test with local wheat flours. International Journal of Food Studies 4:78–87.

Vázquez D, Watts B (2004) Gluten Extensibility: A Key Factor in Uruguayan Wheat Quality. In: The Gluten Proteins, The Royal Society of Chemistry, pp. 279–282.

Vázquez D, Williams PC, Watts B (2007) NIR spectroscopy as a tool for quality screening. In: Buck HT, Nisi JE, Salomón N (eds) Wheat Production in Stressed Environments, Springer Netherlands, Dordrecht, pp. 527–533.

Vázquez D, Berger AG, Cuniberti M, Bainotti C, de Miranda MZ, Scheeren PL, Jobet C, Zúñiga J, Cabrera G, Verges R, Peña RJ (2012) Influence of cultivar and environment on quality of Latin American wheats. Journal of Cereal Science 56:196–203.

Wall J, Carpenter K (1988) Variation in availability of niacin in grain products. Food technology 42:198–204.

Wang A, Liu L, Peng Y, Islam S, Applebee M, Appels R, et al. (2015) Identification of Low Molecular Weight Glutenin Alleles by Matrix- Assisted Laser Desorption/Ionization Time-of-Flight Mass Spectrometry (MALDI-TOF-MS) in Common Wheat (Triticum aestivum L.). PLOS ONE 10:e0138981.

Wang C, Kovacs MIP (2002) Swelling index of glutenin test. i. method and comparison with sedimentation, gel-protein, and insoluble glutenin tests. Cereal Chemistry 79:183–189.

Wang D, Dowell FE, Dempster R (2002) Determining vitreous subclasses of hard red spring wheat using visible near-infrared spectroscopy. Cereal chemistry 79:418–422.

Wang K, Dupuis B, Fu BX (2017) Gluten aggregation behavior in high-shear- based glutopeak test: Impact of flour water absorption and strength. Cereal Chemistry 94:909–915.

Wang N, Dowell FE, Zhang N (2003) Determining wheat vitreousness using image processing and a neural network. Transactions of the ASAE 46:1143–1150.

Wang P, Chen H, Mohanad B, Xu L, Ning Y, Xu J, et al. (2014) Effect of frozen storage on physico-chemistry of wheat gluten proteins: Studies on gluten-, glutenin- and gliadin-rich fractions. Food Hydrocolloids 39:187–194.

Wesley IJ, Larroque O, Osborne BG, Azudin N, Allen H, Skerritt JH (2001) Measurement of Gliadin and Glutenin Content of Flour by NIR Spectroscopy. Journal of Cereal Science 34:125–133.

Wesley IJ, Ruggiero K, Osborne BG, Anderssen RS (2005) The challenge of single estimates in near infrared calibration and prediction: the measurement of durum vitreousness using receival instruments. Journal of Near Infrared Spectroscopy 13:333–338.

wheatqualitycouncil (2018) www.wheatqualitycouncil.org

Wieser H, Seilmeier W (1998) The influence of nitrogen fertilisation on quantities and proportions of different protein types in wheat flour. Journal of the Science of Food and Agriculture 76:49–55.

Wieser H, Antes S, Seilmeier W (1998) Quantitative Determination of Gluten Protein Types in Wheat Flour by Reversed-Phase High-Performance Liquid Chromatography. Cereal Chemistry 75:644–650.

High Throughput Testing of Key Wheat Quality Traits in Hard Red Spring Wheat Breeding Programs

Bin Xiao Fu, Kun Wang, Brigitte Dupuis, and Richard D. Cuthbert

Abstract Effective and efficient selection of key quality traits in early generations of a wheat breeding program is crucial when developing new wheat varieties with improved end-use quality. Here we propose and evaluate a screening protocol based on limited amounts of grain samples for early generation testing of wheat flour yield, flour water absorption and gluten strength. A modified protocol using the Quadrumat Junior (QJ) mill was developed to predict flour yield and compared to the standard protocol using the Bühler laboratory mill. The resulting flour samples (8 g) were tested with the GlutoPeak, a shear-based measuring device, to predict flour water absorption. Gluten strength was also assessed with the GlutoPeak and a rapid extensigraph method. Significant correlation ($r = 0.90$, $p < 0.001$) was found between QJ flour yield and the yield obtained in the Bühler mill. GlutoPeak torque was highly correlated with farinograph measurements of water absorption ($r = 0.91$, $p < 0.001$). Significant correlations ($r > 0.91$, $p < 0.001$) were found for GlutoPeak strength index and the maximum resistance to extension (R_{max}) of the modified extensigraph method. R_{max} that is conventionally used to evaluate lines in registration trials for dough properties in Canada. The mixing parameters obtained during dough preparation for the rapid extensigraph method provided additional information about dough strength and mixing requirements. With a four-fold increase in throughput, the proposed combination using QJ mill, GlutoPeak and rapid extensigraph methods requires as little as 200 g of wheat to predict milling performance, flour water absorption and gluten properties. This protocol can be widely adopted for screening key quality traits in wheat breeding programs.

B. X. Fu (✉) · K. Wang · B. Dupuis
Grain Research Laboratory, Canadian Grain Commission, Winnipeg, MB, Canada
e-mail: binxiao.fu@grainscanada.gc.ca

R. D. Cuthbert
Swift Current Research and Development Centre, Agriculture and Agri-Food Canada, Swift Current, SK, Canada

G. Igrejas et al. (eds.), *Wheat Quality For Improving Processing And Human Health*, https://doi.org/10.1007/978-3-030-34163-3_13

Abbreviations

CWRS Canada Western Red Spring
FAB farinograph absorption
QJ Quadrumat Junior
R_{max} maximum resistance to extension
T_{max} maximum torque; WA, water absorption.

1 Introduction

Wheat flour forms viscoelastic dough when mixed with water. The unique viscoelasticity derived directly from the gluten protein component allows wheat flour to be used for the production of many types of food products. The balance between the extensibility and elasticity in developed dough determines which applications the wheat flour is suited for and the quality of the end-product. Gluten governs the dough mixing requirements and viscoelasticity of doughs that are fermented, chemically leavened, sheeted, or extruded (Delcour et al. 2012). Other major factors determining whether a particular flour is suited for a chosen end use include protein content and starch properties. For optimum results, each flour product should fulfil functional specifications suitable for both processing and quality needs.

From a commercial perspective, wheat quality is a very complicated criterion, depending largely on the product type and the processing technology available (McFall and Fowler 2009). Four discerning quality factors determine the value of the wheat and its suitability for a product. These are protein content, soundness, milling performance, and dough properties (Uthayakumaran and Wrigley 2017).

Protein content, a wheat class-determining factor, is genotype dependent but is greatly influenced by growing conditions. Due largely to genetic selection, wheat grain with a high protein content tends to be hard and to have strong gluten ideal for bread making, whereas wheat grain with a low protein content tends to be soft and to have weak gluten more suitable for pastry making. Wheat is considered unsound when excessive moisture just prior to or during harvest causes the grain to sprout. Significant sprouting damage can have serious adverse effects on functionality because the elevated α-amylase activity induced on germination digests the starch stored in the grain. Wheat protein content and soundness can be rapidly measured by near infrared spectroscopy or effectively estimated with a falling number test (an indicator of α-amylase activity), respectively (Ross and Bettge 2009).

Milling quality is a very complex trait. The general physical properties, such as hardness, kernel weight, test weight, soundness, and vitreousness affect how readily wheat grain can be processed into flour. The key indicators of milling quality are yield, ash content and bran contamination in the flour. There is currently no alternative for predicting milling yield other than milling itself. Flour dough properties, which include water absorption (WA), mixing requirement, and viscoelasticity, are critical in determining wheat processing performance and end-product quality. Common practice involves milling enough flour to complete farinograph, mixograph, extensigraph, alveograph or other tests of the rheological properties of dough. While the farinograph and mixograph tests measure dough properties during mixing (absorption, mixing time, and stability), extensigraph or alveograph parameters are generally considered to be more relevant to dough handling, fermentation, and the baking process. However, the traditional protocol requires large amounts of wheat (3–5 kg) to be milled and throughput is minimal due to the time-consuming tests for dough properties. The combination of Bühler laboratory mill, farinograph and extensigraph protocols is of some use in the milling and baking industry where sample size and low throughput are not a major constraint. However, it is not applicable to screening the quality of breeding populations where sample numbers are usually high but the amount of grain in each sample is very limited (O'Brien and Cracknell 2018).

The purpose of this study was to develop a new test protocol for measuring wheat milling yield, flour WA, mixing requirement and dough viscoelasticity on small samples of wheat grain with a view to increasing the throughput of grain quality testing necessary for efficient wheat breeding.

2 Materials and Methods

2.1 *Wheat Samples Used for Developing Small-Scale Milling Protocol*

Three advanced breeding lines (set I) representing a wide range of milling quality were used to develop a small-scale milling protocol for flour yield prediction. To validate the new method, 20, 20, and 18 advanced breeding lines were selected from the 2015, 2016 and 2017 Canadian bread wheat variety registration trials, respectively. A composite sample of each line was made from wheat grown at multiple locations across Western Canada. All composite samples were graded as No. 2 Canada Western Red Spring (CWRS) or better. Wheat samples (2 kg each) were ground in a Bühler laboratory mill following the AACC International Method 26–21.02. Flour extraction rates were calculated as averages of duplicate millings and expressed as a proportion of the weight of the clean wheat. The lines selected in this study had flour yields ranging from 73% to 78%.

2.2 *Milling Protocol with Modified QJ Laboratory Mill*

Following the AACC International standard method 26–50.01 (Brabender Quadrumat Jr. Method), wheat was first tempered to a moisture content of 14% by mixing the conditioned wheat on a roller conveyor (Norpak Handling Ltd., Port Hope, Ontario) for 25 min at 50 Hz. The tempered wheat was left to rest overnight

(20 h) to allow for moisture penetration and equilibration prior to milling. If the initial moisture content of wheat was less than 11%, the wheat was tempered to 12% and then to 14% over two consecutive days.

The Quadrumat Junior II-G mill (C.W. Brabender Instruments, Inc., So. Hackensack, NJ) was pre-warmed for 30 min to about 4 °C above room temperature. The feed rate was controlled at 125 to 150 g of wheat per minute. Roll gaps were set at 0.75 mm between rolls 1 and 2, 0.04 mm between rolls 2 and 3 and 0.03 mm between rolls 3 and 4. To improve the milling efficiency and optimize the flour extraction rate, the reel sifter originally supplied with the QJ laboratory mill was removed, and the whole meal particles obtained were instead sifted through a Bühler MLUA GM sieve (Bühler AG, Uzwil) with an opening of 250 μm for 1 min at 260 rpm to separate bran from the ground whole meal to yield fine white flour. The flour extraction rate was expressed as the ratio of the amount of flour obtained after sieving to the weight of whole meal collected after milling.

2.3 *Wheat Samples Used for Predicting Water Absorption and Gluten Strength*

Thirty-two advanced breeding lines (set II) were chosen from the 2017 Canadian bread wheat variety registration trials. A composite sample of each line was made from wheat grown at multiple locations across Western Canada. All composite samples were graded as No. 1 CWRS and milled in a Bühler laboratory mill following the AACC International Method 26–21.02. Flour samples were prepared at a constant flour extraction rate of 74%. Farinograph test of Bühler milled flour was based on AACC International Method 54–21.02. Dough extension properties were examined by following a modified extensigraph protocol as described by Suchy et al. (2017). The flour samples used in this study showed a wide range of WA from 61.2 to 68.3% and gluten strength from R_{max} 311 to 774 Brabender units (BU). The baking quality of the flour samples was evaluated with a Lean No Time bake method (Dupuis and Fu 2017). Bread loaf volume was measured using a VolScan Profiler 300 (Stable Micro Systems, Surrey, UK) and expressed relative to 100 g of flour. Loaf top ratio, defined by subtracting pan height (mm) from loaf height (mm), then dividing by loaf width (mm), was used to provide an objective assessment of dough handling properties.

2.4 *GlutoPeak Test for Predicting Water Absorption and Gluten Properties*

Gluten aggregation properties of QJ milled flour were measured in the GlutoPeak (Brabender GmbH and Co KG, Duisburg, Germany) with a high-shear based method as previously described (Fu et al. 2017a; Wang et al. 2017). In a typical experiment, 8 g of flour (14% moisture basis) was dispersed in 10 mL of distilled water in a stainless steel sample cup. The speed of the rotating paddle was set at 2700 rpm. Temperature was controlled at 34 °C by circulating water through the jacketed sample cup. All measurements were performed in duplicate. During the GlutoPeak test, the flour-water slurry was subjected to the intense mechanical shearing action of the paddle rotating at high speed. The counter torque generated by flour hydration and gluten network formation upon mixing and the time required to reach peak resistance were recorded in a torque curve. The resulting GlutoPeak parameters include: peak time (PT), peak area (PA), maximum torque (T_{max}) which is the maximum resistance occurring during mixing, and GlutoPeak strength index (GSI) as defined by Wang et al. (2017).

2.5 *Rapid Extensigraph Test at Constant and Adjusted WA*

The rapid extensigraph test protocol proposed by Fu et al. (2017b) was used. To compare the rapid extensigraph test at fixed and variable WA, dough was prepared at either a fixed absorption of 67.5% or at an absorption level of the value predicted with the GlutoPeak +2% for each flour sample milled from CWRS wheat. After mixing, the dough was first placed in a sealed plastic container for 15 min, after which the dough was rounded, shaped, molded into a cylindrical shape and then allowed to rest in the humidified extensigraph chamber for 30 min before it was stretched. All samples were tested in duplicate from two separate mixings and the R_{max} value (BU), extensibility (cm) and extensigram area (cm^2) were reported.

2.6 *Statistical Analysis*

All data were analyzed with Microsoft Excel and SAS 9.4 software (SAS Institute, Gary, NC, U.S.A.). Tukey's test was done after the analysis of variance and significant differences are indicated at the level of $P < 0.05$.

3 Results and Discussion

3.1 *Prediction of Wheat Flour Yield with the Quadrumat Junior Mill*

Milling 2–3 kg of wheat in a Bühler laboratory mill equipped with three break rolls and three reduction rolls is the standard procedure (AACC International Method 26–21.02) for assessing the milling quality of candidate lines in wheat variety registration trials. However, this procedure is of limited value for phenotypic screening necessary for breeding programs or genetic mapping studies because sample numbers are usually large but amounts of each sample are very limited. The Brabender QJ laboratory mill has been widely adopted for preparing flour for flour and dough rheology analysis when only a limited amount of wheat is available for milling (Fu et al. 2017a, b; Wang et al. 2017). However, there is little evidence to show whether QJ milled flour is comparable to flour prepared with larger scale milling equipment such as a laboratory test mill, pilot mill or commercial mill.

Fu et al. (2017b) modified the original QJ milling procedure by replacing the reel sifter with a Bühler sifter for the separation of bran from flour. This modification improved milling efficiency and limited flour contamination between samples. In that procedure, wheat was tempered to 16% moisture before it was ground in the QJ mill with the internal sifter removed. The resulting whole meal was then sifted through a screen size of 315 μm to collect the flour. While efficient for the preparation of flour for quality analysis from a small amount of wheat sample, we recently found that the relationship of flour yield obtained from the above QJ milling protocol and the Bühler laboratory mill was not always reliable enough for screening grain milling performance.

The major difference between the QJ mill and the Bühler laboratory mill is that there is no provision for reduction passages in the QJ mill. Estimating flour yield from the QJ mill is based exclusively on the release of break flour from the three consecutive break passages and the additional sieving that follows. To improve the prediction of flour yield when reduction rolls are not used, the tempering moisture was lowered to 14% from 16% to enhance flour recovery from the outer layers of the grain. The sieve opening was changed from 315 μm to 200 μm to reduce bran contamination (Protocol 1 in Table 1). Even with these adaptations, flour yields for all three lines tested were significantly lower than with the standard Bühler laboratory mill. The ranking of the three lines for flour yield was different mainly due to the much lower yield from the BW406 line.

The large endosperm particles retained above the screen (200 μm) could have led to an underestimation of the flour yield. Therefore, an attempt was made to recover flour from these large endosperm particles. The whole meal was sifted through 500-μm (top) and 200-μm (bottom) sieves, then the middlings (the fraction collected between the 200-μm and 500-μm sieves) were ground with the QJ mill, and sifted again through the 200-μm opening to collect the flour (Protocol 2 in Table 1). Results showed that BW406 produced significantly more middlings than the

Table 1 Comparison of flour yields from Bühler laboratory and Quadrumat Junior mills for three selected Canadian Western red spring wheat lines

Milling protocol	Bühler mill	Quadrumat Junior mill Protocol 1	Protocol 2	Protocol 3
Sieving	---	200 μm, 1 min	Wholemeal, 200 and 500 μm, 1 min; Reduced middlings, 200 μm, 1 min	250 μm, 1 min
BW5022	77.6 ± 0.1[a]	75.5 ± 0.2[bc]	75.3 ± 0.1[bc]	77.5 ± 0.3[a]
BW406	75.0 ± 0.0[cd]	72.1 ± 0.1[f]	73.4 ± 0.1[e]	75.9 ± 0.2[b]
BW5021	74.4 ± 0.2[e]	73.1 ± 0.2[e]	73.2 ± 0.2[e]	75.4 ± 0.1[bc]

[a]Values labeled with the same superscript are not significantly different ($p < 0.05$) in Tukey's test

BW5022 and BW5021 lines, and subsequent reduction of the middlings increased the total flour yield from BW406 (Table 1). By adding the recovered flour from the middlings to the flour passed through the 200-μm screen, the total flour yield for each of the three lines was more in line with those obtained with the standard protocol for the Bühler laboratory mill.

These results indicate that the proportion of particles between 200 and 500 μm is an important aspect influencing the accuracy of flour yield prediction. However, additional milling of middlings is time consuming. To capture the flour particles in the middlings fraction without resort to additional milling, the opening of the screen was increased from 200 to 250 μm (Protocol 3). Total flour yields recovered with Protocol 2 and 3 were similar, showing that a sieve opening of 250 μm was effective in capturing flour from the middlings. More importantly, the ranking of the lines for flour yield with Protocol 3 was more in line with the standard Bühler milling test ranking.

To validate the modified QJ milling protocol with the 250 μm sieve opening (Protocol 3), 20, 20, and 18 advanced breeding lines were selected from the 2015, 2016 and 2017 Canadian bread wheat variety registration trials, respectively. Significant correlations ($r > 0.89$, $P < 0.001$) were found between observed flour yields for samples from the same lines processed with the QJ milling protocol 3 and with the Bühler laboratory mill for all three years. Therefore, a tempering moisture of 14% and sieving of the granular products from a single QJ mill with a 250-μm screen was adopted as the new QJ milling protocol for flour yield prediction and flour preparation for quality analysis.

3.2 *Prediction of Water Absorption with GlutoPeak*

One emerging technique that has been recently introduced by Brabender® (Brabender GmbH and Co KG, Duisburg, Germany) to evaluate wheat flour quality is the GlutoPeak apparatus (Melnyk et al. 2011). Differing from traditional dough rheological measurements, the GlutoPeak measures the aggregation behavior of wheat gluten proteins in a flour-water suspension under a high-speed shearing action (Melnyk et al. 2012). During the test, the counter torque that develops as the gluten network form upon mixing and the time required to reach peak resistance are registered as a torque curve. The GlutoPeak does not require large sample sizes (as little as 8 g of flour) or great technical skill and the test time is short (< 10 min).

The potential use of GlutoPeak as a rapid tool has been extensively explored to evaluate flour quality, particularly in relation to WA capacity (Marti et al. 2015; Fu et al. 2017a) and gluten strength (Marti et al. 2014; Sissons 2016; Wang et al. 2017). Fu et al. (2017a) evaluated the relationship between flour WA as measured by farinograph (FAB) and the GlutoPeak T_{max} for 83 advanced breeding lines selected from five Canadian bread wheat variety registration trials in 2015. Highly significant relationships were found for flour samples produced with the QJ mill and the Bühler laboratory mill according to values for GlutoPeak T_{max} and FAB. (Fig 1).

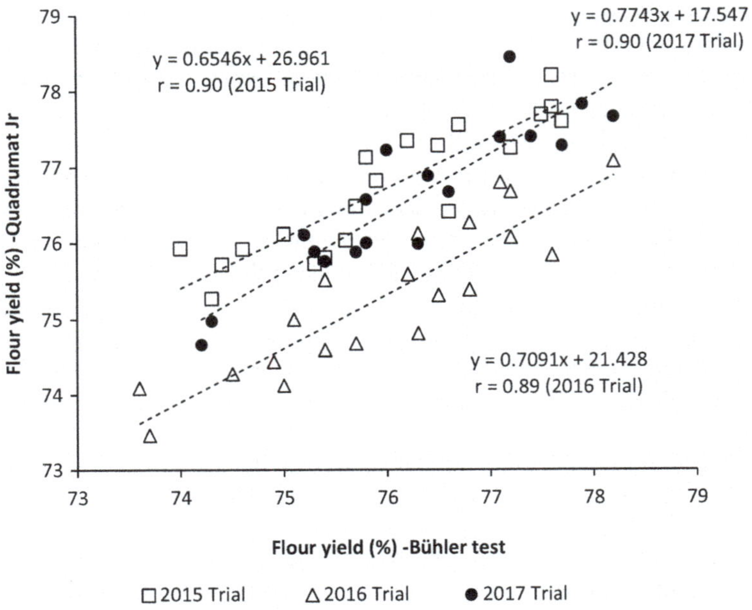

Fig. 1 Relationship between flour yield from the Bühler test and flour yield from the proposed Quadrumat Junior milling protocol for selected advanced breeding lines of hard red spring wheat

With our modified QJ milling protocol, the GlutoPeak T_{max} was still highly correlated ($r = 0.91$, $p < 0.001$) with FAB (Fig. 2). With the increase in WA from 61.2 to 68.3%, T_{max} increased from 44.5 to 54.0 BU with a standard error of estimate of 0.78% (Fig. 2). Among the 32 samples selected, residuals for 14 lines exhibited less than 0.5% difference between the predicted and measured absorption values, while residuals for the remaining 11 and 7 samples ranged between 0.6 to 1% and 1.1 to 1.8%, respectively. The accuracy of the proposed absorption prediction model based on flour milled from the modified QJ milling protocol was comparable to the previous model developed by Fu et al. (2017a) with a standard error of estimate of 0.58%.

3.3 Assessment of Gluten Strength with GlutoPeak

A systematic study was conducted by Wang et al. (2017) to elucidate the impact of WA and gluten strength on gluten aggregation behavior in GlutoPeak tests. They found that GlutoPeak peak time (PT) and peak area (PA) were positively associated with gluten strength, but negatively affected by WA. A new parameter, GlutoPeak strength index (GSI = PA × T_{max}), was introduced to account for the influence of WA on gluten strength prediction. This arithmetic product provided greater correlation with conventional dough strength parameters than with PA or PT based on analysis of 56 advanced breeding lines with a wide range of flour absorption from 55.4 to 68.4%.

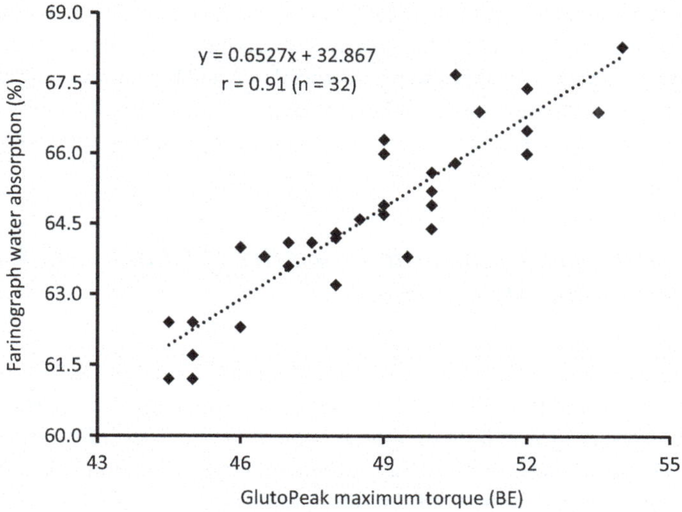

Fig. 2 Relationship between farinograph water absorption and GlutoPeak maximum torque for selected advanced breeding lines of hard red spring wheat

Table 2 Simple correlation coefficients (r) between GlutoPeak parameters and dough properties for flour milled from a selected set of advanced breeding lines of Canadian Western red spring wheat

Sample set II: FAB 61.2–68.3%, R_{max} 311–774 BU					
GlutoPeak parameters		T_{max}	PT	PA	GSI
Farinograph mixing properties	FAB	0.91***	−0.67***	−0.37*	-0.15^{NS}
	DDT	-0.10^{NS}	0.63***	0.72***	0.73***
	Stability	-0.02^{NS}	0.57***	0.76***	0.79***
Mixing and extensigraph parameters	MT	-0.16^{NS}	0.69***	0.89***	0.89***
	ME	-0.04^{NS}	0.53**	0.76***	0.78***
	R_{max}	-0.12^{NS}	0.70***	0.89***	0.91***
	Extensibility	0.20^{NS}	−0.45**	−0.55**	−0.52**
	EA	-0.06^{NS}	0.66***	0.85***	0.87***
Breadmaking performance	BA	0.89***	−0.59***	-0.26^{NS}	-0.03^{NS}
	MT	-0.20^{NS}	0.77***	0.95***	0.94***
	ME	-0.29^{NS}	0.78***	0.91***	0.87***
	BLV	-0.06^{NS}	0.36*	0.51**	0.51**
	LTR	-0.18^{NS}	0.63	0.79***	0.79***

*, **, ***, Significance at 5, 1, and 0.1% levels respectively in Tukey's test. ^{NS}, not significant ($p > 0.05$). BA, bake absorption; BLV, bread loaf volume; DDT, dough development time; EA, extensigraph area; FAB, farinograph absorption; GSI, GlutoPeak strength index (T_{max} × PA); LTR, loaf top ratio; ME, mixing energy; MT, mixing time; PA, peak area; PT, peak time; R_{max}, dough maximum resistance to extension; T_{max}, GlutoPeak maximum torque

Table 2 presents the relationship between GlutoPeak parameters of flour milled with the modified QJ protocol and dough strength properties of flour from the Bühler laboratory mill for the 32 selected advanced breeding lines used in this study. Among the GlutoPeak parameters, GSI, PA and PT were highly correlated with dough strength as measured by farinograph stability, pin mixer mixing time, mixing energy input, extensigraph maximum resistance to extension (R_{max}), and breadmaking performance as indicated by loaf top ratio and bread loaf volume. In the relatively narrow range of flour WA (61.2–68.3%), PA was comparable to GSI for predicting gluten strength.

3.4 Assessing the Viscoelastic Properties of Gluten with Rapid Extensigraph Measurements

The standard extensigraph method (AACC Method 54–10.01, ICC 114/1, ISO 5530-2) has been traditionally used to evaluate the rheological properties of flour dough by providing information on dough strength and extensibility. However, the need to mill large amounts of flour, the prerequisite for farinograph WA and the long dough resting time (135 min) limit the throughput capacity of testing to about 5–6 samples per day, which has deterred the wider adoption of the extensigraph as a rapid tool to measure dough properties of large populations of wheat samples.

A significant amount of work has been invested in developing small-scale dough extension tests to replace the standard extensigraph for early generation screening of dough properties (Abang Zaidel et al. 2008; Anderssen et al. 2004; Chen et al. 2009; Nash et al. 2006). Suchy et al. (2017) developed a modified dough preparation protocol with a 200-g Swanson-type pin mixer and a shorter dough resting time of 90 min down from 135 min in the extensigraph test. Dough developed with the pin mixer at higher WA (FAB plus 4%) appeared to be more functional in baking than dough prepared with the farinograph. Therefore, an extensigraph test based on dough prepared with a pin mixer can effectively discriminate dough strength when dough development is optimal. Fu et al. (2017b) further reduced the flour sample size requirement to 100 g and applied a short dough resting regime of 45 min, resulting in a ~three fold increase in throughput compared with the modified or standard extensigraph methods. However, a constant value of WA was adopted in the rapid extensigraph test due to the limited amounts flour samples and the time required to generate FAB values. As previously discussed by Fu et al. (2017b), variation in flour WA could affect dough strength measurements. The effective prediction of WA with the GlutoPeak as described above provides the possibility of adjusting WA when preparing dough for the rapid extensigraph test.

In this study, therefore, the rapid extensigraph test was conducted at constant WA of 67.5% (Fu et al. 2017b) and at 2% above GlutoPeak predicted absorption. The relationships presented in Table 3 are for the dough properties measured with the rapid extensigraph method at constant and adjusted WA and with the modified extensigraph method. Similar correlations were found for mixing time, extensigraph R_{max} and area between the two extensigraph methods and WA values. However, correlations were stronger for dough extensibility and mixing energy when WA was constant. Overall, the rapid extensigraph performed at constant WA is equally effective in measuring gluten properties as when WA is adjusted based on GlutoPeak predictions.(Table 4).

Table 3 Simple correlation coefficients (*r*) for dough properties measured by a rapid extensigraph method at constant and adjusted water absorptions compared to a modified extensigraph method for flour milled from a selected set of advanced breeding lines of hard red spring wheat

Sample set II: FAB 61.2–68.3%, R_{max} 311–774 BU						
Rapid extensigraph		Modified extensigraph				
		MT	ME	R_{max}	Extensibility	EA
Constant WA	MT	**0.89*****	0.87***	0.93***	−0.62**	0.86***
	ME	0.83***	**0.88*****	**0.88*****	−0.54**	0.82***
	R_{max}	0.90***	0.92***	**0.93*****	−0.54**	0.89***
	Extensibility	−0.27^{NS}	−0.36^{NS}	−0.25^{NS}	**0.51***	−0.12^{NS}
	EA	0.88***	0.87***	0.93***	−0.45**	**0.92*****
Predicted WA + 2%	MT	**0.92*****	0.85***	0.91***	−0.58**	0.85***
	ME	0.83***	**0.80*****	**0.83*****	−0.46**	0.80***
	R_{max}	0.94***	0.92***	**0.97*****	−0.58**	0.92***
	Extensibility	−0.19^{NS}	−0.16^{NS}	−0.16^{NS}	**0.37** ^{NS}	−0.07^{NS}
	EA	0.92***	0.90***	0.95***	−0.52**	**0.92*****

*, **, *** Significance at 5, 1, and 0.1% levels, respectively in Tukey's test. ^{NS}, not significant ($p > 0.05$). BU, Brabender unit; WA, water absorption; MT, mixing time; ME, mixing energy; R_{max}, dough maximum resistance to extension; EA, extensigraph area

Table 4 Comparison of standard quality testing methods used in Canadian wheat variety registration trials and proposed rapid testing protocol for screening

	Standard test methods	Proposed rapid protocol
Wheat milling quality		
Instrument	Bühler MLU 202 laboratory mill	Quadrumat junior mill
Sample size	3.0–5.0 kg	0.2–0.4 kg
Temper moisture	16.3%	14.0%
Milling protocol	AACCI 26–21.02	QJ mill protocol coupled with Bühler MLUA GM sieve (250 μm, 60 s)
Daily throughput	6–10	30–40
Flour water absorption		
Instrument	Farinograph	GlutoPeak
Sample size	50–150 g	8 g
Testing protocol	AACCI 54–21.02	Fu et al. (2017a); Wang et al. (2017)
Testing time	20–60 min per sample	8 min per sample
Testing condition	63 rpm, 30 °C	8 g flour/10 g H_2O (w/w) 2700 rpm, 34 °C
Daily throughput	6–9	40–50
Dough viscoelasticity		
Instrument	Extensigraph	Extensigraph
Flour requirement	200 g	100 g
Testing protocol	Suchy et al. (2017)	Fu et al. (2017b)

(continued)

Table 4 (continued)

	Standard test methods	Proposed rapid protocol
Water absorption	FAB +4%	Constant absorption or predicted FAB by GlutoPeak +2%
Mixing requirement	P2M software, 10% past peak time	
Testing time	90 min	45 min
Daily throughput	6–8	15–20

4 Conclusions

The standard quality evaluation protocol (based on a combination of Bühler laboratory mill, farinograph, and extensigraph methods) used in Canadian wheat variety registration trials, requires large amounts of wheat (3–5 kg) for milling and testing throughput is minimal due to the time-consuming tests for dough properties. To align quality selection with the requirements for registration, we therefore propose a rapid screening protocol for wheat flour yield, flour WA and dough properties of CWRS wheat based on limited grain samples in the early generations of wheat breeding programs. With almost a four-fold increase in throughput, the proposed protocol combining the use of QJ mill, GlutoPeak and rapid extensigraph methods requires as little as 200 g of wheat for predicting milling performance, flour WA and gluten properties. It can be widely adopted for screening these key quality traits in wheat breeding programs.

Acknowledgements The authors are grateful to Andrea Iverson, Altash Yirdaw, Ofelia Francisco-Pabalan, Yuming Chen and Shan Leng for conducting QJ milling, GlutoPeak and extensigraph analysis.

References

AACC International Approved Methods of Analysis. 11th Edition. Method 26-21.02 26-21.02 Experimental Milling—Bühler Method for Hard Wheat; 26-50.01 Brabender Quadrumat Jr. (Quadruplex) Method; 54-21.02 Rheological Behavior of Flour by Farinograph: Constant Flour Weight Procedure. AACC International, St. Paul, MN, U.S.A.

Abang Zaidel, D. N., Chin, N. L., Abdul Rahman, R., and Karim, R. 2008. Rheological characterization of gluten from extensibility measurement. J. Food Eng. 86 (4), 549–556.

Anderssen, R. S., Bekes, F., Gras, P. W., Nikolov, A., and Wood, J. T. 2004. Wheat-flour dough extensibility as a discriminator for wheat varieties. J. Cereal Sci. 39, 195–203.

Chen, R. Y., Seabourn, B. W., Xie, F., and Herald, T. J. 2009. A modified extensigraph method for evaluating dough properties of hard wheat breeding lines. Cereal Chem. 86, 582–589.

Delcour, J.A., Joye, I.J., Pareyt, B., Wilderjans, E., Brijs, K., and Lagrain, B. 2012. Wheat Gluten Functionality as a Quality Determinant in Cereal-Based Food Products. Annu. Rev. Food Sci. Technol. 3, 469–92.

Dupuis, B., and Fu, B. X. 2017. A new lean no time test baking method with improved discriminating power. J. Cereal Sci. 74, 112–120.

Fu, B. X., Wang, K., and Dupuis, B. 2017a. Predicting water absorption capacity of wheat flour using a high shear-based method with the GlutoPeak. J. Cereal Sci. 76, 116–121.
Fu, B.X., Wang, K., Dupuis, B., and Cuthbert, R. 2017b. A rapid extensigraph protocol for measuring dough viscoelasticity and mixing requirement. J. Cereal Sci. 76, 99–107.
Marti, A., Cecchini, C., D'Egidio, M. G., Dreisoerner, J., and Pagani, M. A. 2014. Characterization of durum wheat semolina by means of a rapid shear-based method. Cereal Chem. 91, 542–547.
Marti, A., Ulrici, A., Foca, G., Quaglia, L., and Pagani, M. A. 2015. Characterization of common wheat flours (Triticum aestivum L.) through multivariate analysis of conventional rheological parameters and gluten peak test indices. LWT-Food Sci. and Technol. 64, 95–103.
McFall, K.L., and Fowler, M.E. 2009. Overview of wheat classification and trade. pp. 439–454. In: Wheat Science and Trade. Ed. B.F. Carver, Wiley-Blackwell, Ames, Iowa, USA.
Melnyk, J. P. Dreisoerner, J., Bonomi, F., Marcone, M. F., and Seetharaman, K. 2011. Effect of the Hofmeister series on gluten aggregation measured using a high shear-based technique. Food Res. Intl. 44, 893–896.
Melnyk, J. P., Dreisoerner, J., Marcone, M. F., and Seetharaman, K. 2012. Using the gluten peak tester as a tool to measure physical properties of gluten. J. Cereal Sci. 56, 561–567.
Nash, D., Lanning, S. P., Fox, P., Martin, J. M., Blake, N. K., Souza, E., and Grayhosch, R. A. 2006. Relationship of dough extensibility to dough strength in a spring wheat cross. Cereal Chem. 83, 255–258.
O'Brien, L., and Cracknell, R.L. 2018. The application of testing methods in cereals breeding programs. pp. 15–39, In: The ICC Handbook of Cereals, Flour, Dough and Product Testing: Methods and Applications, Ed. S.P. Cauvain, DEStech Publications, Inc., USA.
Ross, A.S., and Bettge, A.D. 2009. Passing the test on wheat end-use quality. pp. 455–493. In: Wheat Science and Trade. Ed. B.F. Carver, Wiley-Blackwell, Ames, Iowa, USA.
Sissons, M. 2016. GlutoPeak: a breeding tool for screening dough properties of durum wheat semolina. Cereal Chem. 93, 550–556.
Suchy, J., Dupuis, B., Sakamoto, J., and Fu, B. 2017. Alternative dough preparation protocol for extensigraph test of dough strength. Cereal Chem. 94, 270–276.
Uthayakumaran, S., and Wrigley, C. 2017. Wheat: grain-quality characteristics and management of quality requirements. pp. 91–126, In: Cereal Grains: Assessing and managing Quality. Eds. C. Wrigley, I. Batey, and D. Miskelly, Woodhead Publishing Ltd., Cambridge, UK.
Wang, K., Dupuis, B., and Fu, B. X. 2017. Gluten aggregation behavior in high shear-based GlutoPeak test: impact of flour water absorption and strength. Cereal Chem. 94, 909–915.

Molecular Marker Development and Application for Improving Qualities in Bread Wheat

Zhonghu He, Awais Rasheed, Xianchun Xia, and Wujun Ma

Abstract Molecular marker technology has provided a novel and efficient tool for improving qualities in bread wheat. This chapter summarizes progress in gene cloning, gene specific marker (functional marker) development and validation, establishment of high-throughput platform in genotyping, as well as integration of molecular marker technology with conventional quality testing and traditional breeding since 2000. Comparative genomic approach was used to discover more than 20 loci controlling important quality traits, and to develop and validate around 66 gene-specific markers for quality traits such as high- and low-molecular-weight glutenin subunits, color associated traits including polyphenol oxidase (PPO) and yellow pigment, as well as starch parameters. Now the availability of reference wheat genome sequence and on-going efforts to sequence diverse wheat cultivars would offer new opportunities to identify loci responsible for various quality traits through improved genome-wide association study (GWAS) and analytical approaches. Development of high-throughput genotyping platform such as SNP arrays, genotyping-by-sequencing (GBS) and Kompetitive Allele-specific PCR (KASP) have been well-established and will accelerate molecular breeding progress for quality improvement. New cultivars carrying excellent breadmaking quality and outstanding agronomic performance such as Zhongmai 1062 and Jimai 23 were developed. Future strategies in using molecular markers in the context of gene-editing to fine tune allelic effects are also discussed.

Z. He (✉) · A. Rasheed
Institute of Crop Sciences, Chinese Academy of Agricultural Sciences (CAAS), Beijing, China

Global Wheat Program, International Maize and Wheat Improvement Center (CIMMYT), Beijing, China
e-mail: zhhecaas@163.com

X. Xia
Institute of Crop Sciences, Chinese Academy of Agricultural Sciences (CAAS), Beijing, China

W. Ma
State Agriculture Biotechnology Centre, School of Veterinary & Life Sciences, Murdoch University, Perth, West Australia, Australia

G. Igrejas et al. (eds.), *Wheat Quality For Improving Processing And Human Health*, https://doi.org/10.1007/978-3-030-34163-3_14

1 Methodology for Molecular Marker Development and Validation

The research area spanning wheat glutenins during the last 30 years is a classic example of evolution in diagnostic platforms used in wheat end-use quality improvement. The advances in molecular genetics and biochemistry have provided the basis for understanding the genetics, structure and composition of glutenins in wheat. The genes controlling HMW-GS are mapped to *Glu-1* loci on the long arms of homoeologous group 1 chromosomes named as *Glu-A1*, *Glu-B1* and *Glu-D1* (Payne 1987). SDS-PAGE was considered to be the simplest and commonly used technique to identify HMW-GS and LMW-GS alleles. Advances in molecular biology have overcome the low-resolution limitations of protein-based identification of HMW glutenin allele by application of specific PCR markers.

Development of these markers is based on DNA polymorphisms among the glutenin subunit genes and once available they can be considered as perfect or functional markers for HMW-GS alleles. The major advantages are high-throughput capability and genotyping at the vegetative stages (Liu et al. 2008). A total of 12 allele-specific markers at *Glu-1* loci have been reported, however only eight of them are frequently used in breeding (Table 1). Allele-specific PCR (AS-PCR) markers are available for the three most common x-type subunits at the *Glu-A1* locus i.e. *1Ax2∗, 1Ax1* and *1Ax Null* (Liu et al. 2008; Ma et al. 2003) (Table 1). At *Glu-B1*, allele-specific markers are available for x-type subunits *Bx7*, *Bx14*, and *Bx17* (Xu et al. 2008), *Bx6* (Schwarz et al. 2004), *Bx7^{OE}* (Butow et al. 2003; Ragupathy et al. 2008), and y-type subunits *By8*, *By16* and *By18* (Lei et al. 2006). At *Glu-D1*, markers are available to identify *1Dx2*, *1Dx5*, *1Dy10* and *1Dy12* (Liu et al. 2008). Out of six *Glu-1* genes, the gene *Glu-A1y* is usually not expressed in bread or durum wheat. Roy et al. (2018) introduced 1Ay gene in two Australian bread wheat cultivars, Livingston and Bonnie Rock, and three sister lines were developed. The introduction of 1Ay gene increased glutenin/gliadin ratio without affecting the relative amount of subunits, and increased gluten contents by 10%.

However, there is limited expertise in the world to diagnose LMW-GS alleles in bread and durum wheat. SDS-PAGE, 2-DE, MALDI-TOF-MS and PCR based markers were developed to detect the *Glu-3* allelic variation. Liu et al. (2010) compared the four techniques to assess their suitability for use in breeding programs. They indicated that PCR-based markers are the simplest, most accurate, lowest cost technique and therefore recommended them for the identification of *Glu-A3* and *Glu-B3* alleles in breeding programs. Seventeen allele-specific markers for *Glu-A3* and *Glu-B3* loci have been reported and used (Table 1), and multiplex PCR protocols have been developed to reduce costs of screening in practical breeding programs (Wang et al. 2010). Application of functional markers for identification of LMW-GS in various types of wheat germplasm has also been reported (Jin et al. 2011). Recently, the allelic differentiation of *Glu-3* loci was further differentiated using haplotype analysis and some more diagnostic markers were developed to identify *Glu-B3c* and *Glu-B3d* alleles (Ibba et al. 2017) (Table 1).

Table 1 List of all functional markers available in wheat along with their KASP counterpart and standard cultivars for allele identification

Trait	Gene	Marker	Allele	KASP[a]	Standard	Reference
Gluten elasticity	*Glu-A1*	UMN19	*Glu-A1* (Ax1, Ax2[a], AxNull)	gluA1.1_1594; gluA1.1_1883	Chinese Spring (CS), Opata 85,	Liu et al. 2008
	Glu-A1	Ax2[a]	*Glu-A1b* (Ax2[a])	As above	Pavon 76, Opata 85	Ma et al. 2003
	Glu-B1	TaBAC1215C06-F517/R964	*Glu-B1al* (Bx7^{OE})	Bx7^{OE}	Dorico, ProINTA Colibr 1, Klein Jabal	Ragupathy et al. 2008
	Glu-B1	cauBx642	*Glu-B1b* (7 + 8); *Glu-B1i* (17 + 18); *Glu-B1h* (14 + 15)	NA	CS, Jing771, Pm97034	Xu et al. 2008
	Glu-B1	ZSBy9F2/R2	*Glu-B1f* (13 + 16)	NA	Baxter	Lei et al. 2006
	Glu-B1	ZSBy8F5/By8R5	*Glu-B1* (By8)	NA	Sunco	Lei et al. 2006
	Glu-D1	UMN25F/25R	*Glu-D1* (Dx2, Dx5)	Glu-D1d_SNP	CS, Pavon 76	Liu et al. 2008
	Glu-D1	UMN26F/26R	*Glu-D1* (Dy10, Dy12)	Glu-D1d_SNP	CS, Pavon 76	Liu et al. 2008
	Glu-A3	LA1F/SA1R	*Glu-A3a*	NA	Neixiang 188, Chinese Spring	Wang et al. 2010
	Glu-A3	LA3F/SA2R	*Glu-A3b*	NA	Gabo, Pavon 76	Wang et al. 2010
	Glu-A3	LA1F/SA3R	*Glu-A3c*	NA	Pitic, Seri 82	Wang et al. 2010
	Glu-A3	LA3F/SA4R	*Glu-A3d*	NA	Nidera Baguette 10, Cappelle-Desprez	Wang et al. 2010
	Glu-A3	LA1F/SA5R	*Glu-A3e*	NA	Amadina, Marquis	Wang et al. 2010
	Glu-A3	LA1F/SA6R	*Glu-A3f*	NA	Kitanokaori, Renan	Wang et al. 2010
	Glu-A3	LA1F/SA7R	*Glu-Ag*	NA	Bluesky, Glenlea	Wang et al. 2010

(continued)

Table 1 (continued)

Trait	Gene	Marker	Allele	KASP[a]	Standard	Reference
	Glu-B3	SB1F/SB1R	*Glu-B3a*	NA	Chinese Spring	Wang et al. 2009
	Glu-B3	SB2F/SB2R	*Glu-B3b*	NA	Renan, Gabo	Wang et al. 2009
	Glu-B3	SB3F/SB4R	*Glu-B3c*	NA	Insignia, Halberd	Wang et al. 2009
	Glu-B3	SB4F/SB4R	*Glu-B3d*	NA	Pepital, Ernest	Wang et al. 2009
	Glu-B3	SB5F/SB5R	*Glu-B3e*	NA	Cheyenne	Wang et al. 2009
	Glu-B3	SB6F/SB6R	*Glu-B3fg*	NA	Fengmai 27	Wang et al. 2009
	Glu-B3	SB7F/SB7R	*Glu-B3g*	NA	Splendor, Cappelle-Desprez	Wang et al. 2009
	Glu-B3	SB8F/SB8R	*Glu-B3h*	NA	Aca 303, Pavon 76	Wang et al. 2009
	Glu-B3	SB9F/SB9R	*Glu-B3ad*	NA	Opata 85	Wang et al. 2009 Ikeda unpublished
	Glu-B3	SB10F/SB10R	*Glu-B3bef*	NA	Gawain	Wang et al. 2009
Grain texture	*Pina-D1*	Pina-N2	*Pina-D1a,b*	Pina-D1_INS	Chinese Spring, Zhongyou 9507	Chen et al. 2012
	Pinb-D1	Pinb-D1	*Pinb-D1a,b*	Pinb-D1_INS	Chinese Spring, Lorvin 10	Giroux and Morris 1997
	Pinb-D1	Pinb-DF/Pinb-DR	*Pinb-D1p*	No	Shannongyoumai 3	Li et al. 2008

	Pinb-B2	Pinb-B2v2	*Pinb-B2a, b*	Pinb2_IND	Chinese Spring, Zhongmai 175	Chen et al. 2010
Polyphenol oxidase	*Ppo-A1*	PPO18 and PPO33	*Ppo-A1a,b*	PPOA1_SNP	Zhengmai 9023, Jinmai 67	Sun et al.2005
	Ppo-D1	PPO16 and PPO29	*Ppo-D1a,b*	PPOD1_SNP	Chinese Spring, Mexipak-65	He et al.2007
Lipoxygenase	*TaLox-B1*	LOX16	*TaLox-B1a*	LoxB1_SNP	Chinese Spring	Geng et al. 2012
	TaLox-B1	LOX18	*TaLox-B1b*	LoxB1_SNP	Inqilab-91	Geng et al. 2012
	TaLox-B2	LOX-B23	*TaLox-B2a,b*	NA	Zhongmai 18, GC8901	Zhang et al. 2015
Phytoene synthase	*Psy-A1*	YP7A	*Psy-A1a,b*	PSY-A1_IND	Nongda 3291, Wanmai 33	He et al.2008
	Psy-B1	YP7B-1	*Psy-B1a,b*	NA	Jingdong 8, Jimai 38	He et al.2009
	Psy-B1	YP7B-2	*Psy-B1c*	PSY_B1c_SNP	Yannong 18	He et al. 2009
	Psy-D1	YP7D-1	*Psy-D1a,g*	Psy1Da-g	Chinese Spring, Zhou 8425B	Wang et al. 2009
Zeta-carotene desaturase	*TaZds-A1*	YP2A-1	*TaZds-A1a,b*	ZDS-A1_SNP	Chinese Spring, Zhongmai 175	Dong et al. 2012
	TaZds-D1	YP2D-1	*TaZds-D1a,b*	ZDS-D1_SNP	Chinese Spring, Sunstate	Zhang et al. 2011
Lycopene	*Lyce-B1*	NA	*TaLYC-B1a,b*	LYCE-B1_SNP	Zhoumai 8235B, Norin 61	Dong unpublihed
Phytoene desaturase	*PDS-B1*	NA	*TaPds-B1a,b*	PDS-B1_SNP	Zhou 8425B, Insignia	Dong unpublished
Peroxidase	*POD-A1*	POD-3A1,2	*TaPod-A1a,b*	PODA1_462_SNP	Norin 61, Norin 67	Wei et al.2015

(continued)

Table 1 (continued)

Trait	Gene	Marker	Allele	KASP[a]	Standard	Reference
Avenin-like protein	*ALPb-7A*	NA	*NA*	ALPb7A-3IND	Chinese Spring, Chara, Westonia	Chen et al. 2016
	ALPb-7A	NA	*NA*	ALPb7A_225SNP	Chinese Spring, Chara, Westonia	Chen et al. 2016
	ALPb-4A	TaALP-7A-F/R	*active-type ALP-7A*	ALPb4A_228_SNP	Chinese Spring, Chara, Westonia	Chen et al. 2016
	ALPb-4A	NA	*NA*	ALPb4A_773_SNP	Chinese Spring, Yitip	Chen et al. 2016
	ALPb-4A	NA	*NA*	ALPb4A_3IND	Chinese Spring, Chara, Westonia, Yitip	Chen et al. 2016
	ALPa-4A	NA	CS-type, Spitfire-type	ALPa4A_285_SNP	Spitfire	Chen et al. 2016
	ALPa-4A	NA	Wyalketchem-type	ALPa4A_184_SNP	Wyalketchem	Chen et al. 2016
Pre-harvest sprouting	*TaSdr-A1*	Sdr2A	*TaSdr-A1a,b*	SDRA1_643	Yangxiaomai, Zhongyou 9507	Zhang et al. 2014
	TaSdr-B1	Sdr2B	*TaSdr-B1a,b*	SDR_SNP	Yangxiaomai/Zhongyou 9507	Zhang et al., 2014
	Vp1-B1	Vp1B3	*Vp1-B1a,c*	Vp1B1-83_IND	Zhongyou 9507, Xinong 979	Yang et al. 2007
	Vp1-B1	Vp1B3	*Vp1-B1a,b*	Vp1B1-193_IND	Zhongyou 9507, Yongchuanbaimai	Yang et al. 2007
	Vp-1A	A17–19	*Vp-1Ab, c*	NA	Wanxianbaimaizi, Jing411	Chang et al. 2011
	TaDFR	DFR-F/R	TaDFR-Ba, b	NA	Taiyuan 566, Longmai 13	Bi et al. 2014
	Phs1	TaPHS1-SNP1	*Rio-type, NW-type*		RioBlanceo, NW97S186	Liu et al. 2013

Grain color	*Tamyb10-A1*	Tamyb10-A1	*R-A1a,b*	Tamby10-A1	Norin 10, Norin 61, Chinese Spring; Prina	Himi et al. 2011
	Tamyb10-A1	Tamyb10-A1	*R-A1a,* Norin-type	Tamby10-Nor17	Norin 10, Norin 61, Chinese Spring	Himi et al. 2011
	Tamyb10-B1	Tamyb10-B1	*R-B1a,b*	TamybR B1a-b	Norin 10, Norin 61,Chinese Spring	Himi et al. 2011
	Tamyb10-D1	Tamyb10-D1	*R-D1a,b*	TamybR D1a-b	Norin 10, Norin 61, Chinese Spring	Himi et al. 2011
	Tamyb10-D1	Tamyb10D	*Not designated*	NA	Yangxiaomai, Zhongyou 9507	Wang et al. 2016
Amylose content	*Wx-A1*	AFC/AR2	Null, Wild-type	NA	Norin 61, Kanton 107	Nakamura et al. 2002
	Wx-B1	BDFL/BRD	Null, Wild-type	WxB1_SNP	Norin 61, Kanton 107	Nakamura et al. 2002
	Wx-D1	BDFL/DRSL	Null, Wild-type	NA	Norin 61, California	Nakamura et al. 2002
Wheat bread-making quality	Wbm	NWPFor/Rev		Wbm_SNP	Mantol, Aca 601, Insignia	Furtado et al. 2015

[a]KASP markers are partially reported in Rasheed et al. (2016). Further information on other KASP markers can be obtained by personal communication to Zhonghu He or Awais Rasheed

A combination of different techniques was required to identify certain alleles of LMW-GS and these combinations are especially useful when characterizing new alleles. As more alleles are reported at *Glu-A3* and *Glu-B3* in bread wheat, more molecular markers will be needed to distinguish them in breeding germplasm. Liu et al. (2010) recommended a standard set of 30 cultivars to represent all known LMW-GS allelic variants for future studies. Among them, Chinese Spring, Opata 85, Seri 82 and Pavon 76 were recommended as a core set for use in SDS-PAGE gels. Use of the standard cultivar set was recommended to promote and facilitate information sharing on LMW-GS in order to ultimately enhance the global quality improvement efficiency in wheat.

Functional markers are developed from the polymorphisms within the coding sequences of functional genes which could be either single nucleotide polymorphisms (SNPs) or InDels. Fine mapping followed by map-based cloning is the most effective strategy to isolate the functional genes in plants (Yan et al. 2004). However, due to a large genome size, it had been very difficult to clone genes by map-based cloning in bread wheat. Alternatively, a significantly large number of genes, especially several genes related to wheat quality, have been cloned in wheat using comparative genomics approach. There is very high gene collinearity (synteny) among the grass genomes of maize, barley, rice, and *Brachypodium* which could facilitate gene discovery in wheat (El Baidouri et al. 2017; Valluru et al. 2014). A classic example is the cDNA sequence of maize *Psy1* gene (GenBank accession U32636) in that all wheat ESTs sharing high similarity with the reference gene were blasted and subjected to contig assembly (He et al. 2008). The wheat *Psy1* gene was cloned with PCR amplification, and a functional marker YP7A for discrimination of two alleles at *Psy-A1* locus was developed and validated using 217 Chinese cultivars and 240 $F_{2:6}$ lines from the cross of PH82–2/Neixiang 188. However, the recent reports of genome sequences of wheat and its immediate progenitors could facilitate the unprecedented discovery of functional genes and development of functional markers for use in wheat breeding (Rasheed et al. 2018).

Liu et al. (2012) documented 97 functional markers for detecting 93 alleles at 30 loci in bread wheat. This number has increased during the past 5 years due to rapid advancements in wheat genomics. Currently, there are 157 functional markers documented for more than 100 loci underpinning adaptability, grain yield, disease resistance, end-use quality and tolerance to abiotic stresses. Out of all these functional markers, almost 66 are related to end-use quality in wheat (Table 1).

2 Overview of Functional Markers Related to End-Use Quality in Wheat

While next generation sequencing (NGS) and SNP arrays are excellent choices for gene discovery and mapping, and for identifying linked markers for important traits. Such trait-associated markers, in addition to functional markers, are ideal for gene tagging and gene introgression in breeding. Functional markers for wheat end-use quality have been described along with tester germplasm for identification of alleles. Apart from gene-specific markers for *Glu-1* and *Glu-3* loci, there are several other newly identified genes underpinning bread-making quality in wheat (Table 1). Recently, a highly expressed bread-making gene (*wbm*) was identified in the transcriptome of developing wheat seed (Furtado et al. 2015). RNA-seq analysis revealed that the S-rich *wbm* gene was highly expressed consistently in all cultivars with good bread-making quality. Guzmán et al. (2016) later identified 8 of 56 CIMMYT cultivars carrying the *wbm* gene and concluded that the allele has a significant effect on overall gluten quality, gluten strength, gluten extensibility and bread-making quality. However, the effects were smaller than those associated with the *Glu-B1* and *Glu-D1* loci. Similarly, the wheat avenin-like protein (ALP) is important constituent of gluten and has shown positive effects on dough properties. Chen et al. (2016) isolated wheat ALP genes and developed a functional marker to identity active and silenced b-type ALP-7A gene, where the active type had significant effect on bread-making quality.

Polyphenol oxidase (PPO) activity responsible for brown discoloration of the wheat products especially Asian noodles, is an undesirable character. Several markers have been developed to identify PPO alleles on chromosomes 2A and 2D (He et al. 2007; Sun et al. 2005). The practical usage of these markers in wheat breeding for identification of genotypes with lower PPO activity is scientifically valid (Liang et al. 2010). Nevertheless, PPO gene located on chromosome 2B had limited polymorphism in Chinese wheat to develop a functional marker. Lipoxygenase activity is also a major determinant of color and processing quality of wheat products (Geng et al. 2012). A lipoxygenase (LOX) gene has been mapped to chromosome 4BS (*TaLox-B1*) and two allele-specific markers LOX16 and LOX18 amplify 489- and 791- bp PCR fragments in cultivars with higher and lower LOX activities, respectively (Geng et al. 2012). The gene, *TaLox-B1*, was sequenced and a SNP was identified in the third exon which helped in development of two markers for identifying alleles *TaLox-B1a* and *TaLox-B1b*. Zhang et al.(2015) reported two new loci for *TaLox* on chromosome 4BS and are designated as *TaLox-B2* and *TaLox-B3*. They also developed a functional marker, Lox-B23, to distinguish *TaLox-B2a, TaLox-B2b* and *TaLox-B3* alleles in bread wheat.

The color of wheat derived products is due to the yellow pigment content (YPC). Regional preference for color does exist, like bright white color is preferred for Chinese white salted noodles, whereas yellow alkaline noodles with bright yellow color are widely preferred in southeastern Asia and Japan (Parker et al. 1998). Carotenoids are responsible for yellow pigment (He et al. 2008) while phytoene synthase (PSY) and zeta-carotene desaturase (ZDS) are important enzymes in the biosynthetic pathway for carotenoid synthesis in wheat (Dong et al. 2012; Zhang et al. 2011). PSY genes are present on chromosomes 7AL, 7BL and 7DL and several allele-specific markers for PSY genes have been developed (He et al. 2008; He et al. 2009, Wang et al. 2009). The reverse genetics approaches using RNAi decreased the *Psy1* transcripts level by 54–76% and YPC was reduced by 26–35%. This indicated that PSY1 is the most important regulatory enzyme in carotenoid biosynthesis and a series of candidate genes involved in secondary metabolic pathways and core metabolic processes responded to *Psy1* down-regulation (Zhai et al. 2016). Similarly, markers for ZDS genes on chromosomes 2A and 2D can discriminate allelic difference in wheat (Dong et al. 2012; Zhang et al. 2011). More recently, a novel QTL for peroxidase (POD) activity was mapped and was annotated to be *TaPod-A1* gene determining flour color (Wei et al. 2015). Two functional markers were developed for two alleles amplifying 291- and 766-bp fragments in cultivars with lower and higher POD activities, respectively. Nigro et al. (2017) identified six candidate genes involved in the biosythesis of hydroxycinnamic acid in wheat.

Starch fractions account for almost 70% of the dry matter in wheat grain and greatly affect end-use quality especially Asian noodles. Waxy proteins are the products of granule bound starch synthase (GBSS I) genes on chromosomes 7A, 4A and 7D of wheat. Nakamura et al. (Nakamura et al. 2002) developed functional markers for waxy- and wild-type alleles and validated the alleles in a set of 30 lines using a single PCR reaction. Later, a high-throughput KASP marker is also developed for *Wx-B1* locus and further development of KASP markers for other *Wx* loci is in progress. A waxy mutant line carrying *Wx-D1d* allele has been identified and characterized at molecular level (Yi et al. 2017) and a KASP marker was developed for *Wx-D1d* allele which was tracked in backcross derived populations.

The tolerance to pre-harvest sprouting (PHS) is an important breeding objective in many countries, however the work on gene discovery is very limited since reliable phenotyping is a time-consuming activity. Liu et al. (2013) cloned a major QTL related to PHS and designed the KASP marker for *Phs1* allele. Another major gene, *Viviparous 1* as a regulator of late embryo development have shown significant effect on sprouting tolerance, was cloned and functional markers were developed for three different alleles (Yang et al. 2007). Chang et al. (2011) identified six alleles at *Vp-1A* locus, however no allelic variation was found at *Vp-1D* locus. Similarly, functional markers are also available for seed dormancy genes, *TaSdr-B1* and *TaSdr-A1* (Zhang et al. 2017) genes, which are major factors in tolerance to PHS. Red-grained wheat varieties are generally more tolerant to PHS as compared to white-grained varieties. The red pigment of grain coat is synthesized through the flavonoid biosynthesis pathway, in which the dihydroflavonol-4-reductase gene (DFR) is one of the gene involved in anthocyanin synthesis. Bi et al. (2014) cloned homeologous genes *TaDFR* in Chinese wheat, and no allelic variation was found at *TaDFR-A1* and *TaDFR-D1* genes on chromosomes 3A and 3D. However, two alleles were identified at *TaDFR-B1* locus characterized by 8-bp InDel. A CAPS marker was developed to differentiate red and white grain Chinese cultivars with distinct PHS resistance. Similarly, major grain color gene *Tamyb10* (transcription factor for *R-1* gene) on chromosomes 3A, 3B and 3D have been cloned and functional markers are available to identify the allelic variations (Himi et al. 2011). Wang et al. (2016) further developed a new STS marker to characterize *Tamyb10-D* gene in Chinese wheat cultivars differing in response to PHS. They concluded that wheat cultivars with 1629-bp fragment for *Tamyb10-D* were tolerant to PHS as compared to cultivras amplifying 1178- bp fragment. Rasheed et al. (2016) converted several of these markers including *TaSdr*, *TaVp1-B1* and *TaMFT-A1* genes into high-throughput KASP assays. All these functional markers for bread-making and processing quality provide a powerful toolkit to complement the phenotypic selection of wheat germplasm with desirable end-use quality features during breeding (Torada et al. 2016).

3 High-Throughput Genotyping for Wheat End-Use Quality

Almost all functional markers in wheat were gel-based markers, thus hinder the large-scale germplasm screening. Therefore, it is challenging to develop a high-throughput platform to use single markers in wheat breeding programs. Rasheed et al. (2017) highlighted six factors in developing such platforms; these included (i) number of data points that can be generated in a short time period, (ii) ease of use, (iii) data quality (sensitivity, reliability, reproducibility, and accuracy), (iv) flexibility (genotyping few samples with many SNPs or many samples with few SNPs), (v) assay development requirements, and (vi) genotyping cost per sample or data point. Sufficient recent reports indicate that LGC's KASP is an well-received global benchmark technology for such genotyping requirements in terms of both cost-effectiveness and high throughput (Semagn et al. 2014).

At a first step, several groups worked on converting gel-based functional markers to high-throughput KASP markers (http://www.cerealsdb.uk.net/cerealgenomics/CerealsDB/kasp_download.php?URL=). The numbers were increased to 72 after validation in a bread wheat diversity panel (Rasheed et al. 2016). This effort has continued in our group and we currently have more than 150 KASP markers for almost 100 functional genes (Rasheed et al. unpublished data). Currently, most gene mapping studies (both QTL and GWAS) use SNP arrays or NGS; the markers linked to QTL are SNPs and can be easily converted to KASP assays for further diagnosis or QTL introgression in breeding. Similarly, diagnostic KASP assay development is preferred due to the wide acceptance and usefulness of this technology during functional gene discovery. Several QTL linked to quality traits and SNPs in functional genes have been converted to KASP markers; examples include *Glu-1, Glu-3, Pin-D1* and *Ppo-A1* (Rasheed et al. 2016) . The functional genes for which KASP markers are available are listed in Table 1.

KASP provides the throughput required in breeding programs for gene tagging and gene introgression without compromising of flexibility. However, higher cost is still an issue because KASP mastermix is a commercial proprietary from LGC and there are no other competitors. Due to this limitation, several groups tried to develop other open source uni-plex SNP genotyping techniques like semi-thermal asymmetric reverse PCR (STARP) (Long et al. 2017) and Amplifluor-like (Jatayev et al. 2017) which can be used with any commercial mastermix, significantly reducing the per-data-point cost. More recently, other commercial alternatives of KASP assays were introduced, including PACE® mastermix from 3CR Biosciences (www.3crbio.com) and rhAmp from Integrated DNA Technologies® (https://www.idtdna.com/pages/products/qpcr-and-pcr/genotyping/rhamp-snp-genotyping). However, their acceptance in wheat breeding programs is yet to be seen.

4 QTL and GWAS for End-Use Quality in Wheat Using SNP Arrays

QTL mapping and GWAS have been exponentially increased in all major crops including wheat due to the introduction of high-throughput and cost-effective genotyping platforms (Rasheed et al. 2017). Due to the co-dominant nature and high abundance, SNPs are the ideal markers for QTL and GWAS studies. SNP arrays have become a cost-effective and high-throughput means for genotyping and currently several SNP arrays are available for wheat. A 90 K SNP array was developed and almost 3380 wheat accessions were characterized (Wang et al. 2014) and have been extensively used in QTL and GWAS experiments (Table 2). To overcome the several limitations in 90 K, Winfield et al. (2016) developed an 820 K Affymetrix Axiom SNP array from resequencing exomes of 43 bread wheat and wild species accessions representing the primary, secondary and tertiary gene pools. The 820 K SNP array was used to characterize 475 bread wheat and wheat relative accessions. A subset of SNPs from the 820 K array were then used to design a breeder-oriented Axiom 35 K SNP (Allen et al. 2017), which is effective in characterizing SNPs in wild relatives of wheat in a cost-effective manner (King et al. 2017). Recently, Rimbert et al. (2018) used whole-genome resequencing data from eight wheat accessions and discovered more than three million genome-wide SNPs from genic and intergenic regions that were mined for single-copy loci to design a 280 K SNP array. A 660 K SNP array developed at the Chinese Academy of Agricultural Sciences (CAAS) is currently in extensive use and have succeeded in identifying QTL for bread-making quality and kernel number (Jin et al. 2016) and constructing a high-density linkage map of *Agropyron cristatum* (Zhou et al. 2018). However, the features of this SNP array and criteria for selection of SNP markers were not revealed. More recently, we developed Wheat 50 K (Triticum TraitBreed array) and 15 K SNP arrays based on the most qualified SNPs selected from the Wheat 35 K, 90 K, and 660 K SNP chips. Around 135 and 150 functional markers, and 700 and 1000 SNPs tightly linked with known QTL are also included in the 50 K and 15 K SNP arrays, respectively. The new Wheat 50 K and 15 K SNP arrays are a significant step towards more uniform coverage of SNPs on all chromosomes, less frequency of redundant markers and cost-effective as compared to existing SNP arrays.

The above-mentioned arrays are useful tools for gene mapping. A brief summary of QTL and GWAS experiment for wheat end-use quality are presented in Table 2. Although genetic architecture of end-use quality in wheat has been reported using gene mapping strategies, the wide array of QTL identified have been rarely used in wheat breeding. Jin et al. (2016) identified QTL for processing quality in a recombinant inbred line (RIL) population from the Gaocheng 8901/Zhoumai 16 cross using Wheat 90 K and 660 K SNP arrays. Composite interval mapping identified 119 additive QTL on 20 chromosomes except 4D; among them, 15 accounted for more than 10% of the phenotypic variation across two or three environments. Twelve QTL for Mixograph parameters, 17 for RVA parameters and 55 for Mixolab parameters were new. Eleven QTL clusters were identified. Zhai et al. (2016) identi-

Table 2 A brief description of QTL mapping, genome-wide association studies and genomic selection in wheat for end-use quality

Study	Trait	Panel/population	Marker	QTL or selection accuracy	Reference
QTL mapping	End-use quality and mixograph	WCB414/SS RILs	DArT	19 QTL	Echeverry-Solarte et al. 2015
	Milling and baking quality	Several populations	DArT	75 QTL	Cabrera et al. 2015
	Mixograph and Mixolab	Gaocheng 8901/ Zhoumai 16 RILs	90 K and 660 K SNP array	119 QTL	Jin et al. 2016
	Arabinoxylan contents	PH82–2/Neixiang 188 RILs	SSRs/STS	15 QTL	Yang et al. 2016
	Processing quality	Ning7840/Clark RILs		41 QTL	Li et al. 2017
	Flour color	Gaocheng 8901/ Zhoumai 16 RILs	90 K SNP array	46 QTL	Zhai et al. 2016
	Processing quality	RAC875/Kukri	DArT		Maphosa et al. 2013
	Dough rheology	Drysdale/Gladius RILs		5 QTL	Maphosa et al. 2015
	Flour quality	Two RILs	DArT and SSRs	20 and 34 QTL	Deng et al. 2015
	Dough rheology	HI977/HD2329 RILs	SSRs	16 QTL	Prashant et al. 2015
GWAS	GPC and sedimentation volume	192 bread wheat lines	90 K SNP array	30 QTL	Liu et al. 2017a
	Flour related traits	469 bread wheat lines	90 K SNP array	105 QTL	Jernigan et al. 2017
	Flour color traits	166 bread wheat lines	90 K SNP array	32 QTL	Zhai et al. 2018
	Flour color traits	205 bread wheat cultivars	90 K SNP array	94 QTL	Jiang et al. 2018
	Vitamins B1 and B2	166 bread wheat lines	90 K SNP array	24 QTL	Li et al. 2018
	Strach granule size	166 bread wheat lines	90 K SNP array	48 QTL	Li et al. 2017
	Dough rheology and Alveograph	120 elite lines	DArT	20 QTL	Tadesse et al. 2015
Genomic selection	Bread-making quality	5520 advanced lines	GBS	0.32 to 0.62	Battenfield et al. 2016
	Bread-making quality	6095 advanced lines	GBS	0.52–0.93	Lado et al. 2018

(continued)

Table 2 (continued)

Study	Trait	Panel/population	Marker	QTL or selection accuracy	Reference
	Bread-making quality	Two bi-patental soft winter wheat populations	DArT and SSRs	0.42–0.66	Heffner et al. 2011
	Bread-making quality	840 winter wheats	DArTseq	0.38–0.63	Michel et al. 2018
	End-use quality	398 wheat lines	90 K SNP array	0 to 0.69	Hayes et al. 2017
	Grain yield and quality	170 cultivars and mapping population	90 K SNP array	0.-0.8	Haile et al. 2018
	End-use quality	635 winter wheat lines	15 K illumina array	0.50–0.79	Kristensen et al. 2018

fied 56 QTL for flour color-related traits and PPO activity from the same population. A GWAS experiment in 469 soft winter wheat cultivars identified 105 significant marker-trait associations for flour yield, lactic acid solvent retention capacity, flour SDS sedimentation and flour swelling volume using 90 K SNP array (Jernigan et al. 2017). QTL clusters were detected for grain quality on chromosomes 1B, 6B and 7B in a doubled haploid population CO940610/'Platte' (Dao et al. 2016). Maphosa et al. (2014) identified QTL for several bread-making quality traits including flour water absorption, protein content and dough rheology in a cross between Drysdale and Gladius. Genomic regions containing photoperiod sensitivity loci affected grain protein content while the *Ha* (puroindoline) locus on chromosome 5D was associated with loaf quality traits. Other QTL (on chromosomes 2B, 3B and 5A) were novel and not associated with any known quality or phenology genes. The new loci identified using GWAS approach need to be further validated in bi-parental populations before using in marker-assisted selection.

Arabinoxylans (AX) are major polymers of wheat grain cell walls and affect the end-use properties and nutritional quality. Yang et al.(2016) identified two pairs of epistatic QTL for AX in the RILs derived from the cross PH 82–2/Neixiang 188. Additionally, a QTL on chromosome 1B likely to be the 1B.1R translocation showed stable effects on AX contents across seasons. In addition, several GWAS studies have been conducted in Chinese wheats to underpinning the genetic basis of end-use quality (Marcotuli et al. 2015). More than 50 SNP markers were associated with grain protein contents and SDS sedimentation volume in 192 bread wheat lines from China genotyped with 90 K SNP array (Liu et al. 2017a). Similarly, 205 elite wheat cultivars genotyped with 90 K SNP array were used to identify 28, 30, 24 and 12 marker-traits associations for $L*$, $a*$, $b*$ traits and PPO activity. They found that a SNP within the *Pina-D1* was associated with all the color-related traits in wheat. Two GWAS experiments in 166 wheat cultivars and elite lines from China identified the marker-trait associations for starch granule size distribution (Li et al. 2017) and vitamins B1 and B2 using 90 K SNP array (Li et al. 2018).

Another significant quality limiting factor is the black point reaction in wheat. Black point is characterized by discoloration at the embryo end of kernels and downgrade end-use quality of the grain by discoloration. It is a serious problem in China, USA, Australia, Canada and Siberia. There is a huge knowledge gap on the actual cause of black point in wheat and the genetic basis of tolerance to black point reaction. Liu et al. (2016) identified 9 QTL for black point resistance in wheat using a RIL population derived from Linmai 2/Zhong 892 cross. Similarly, a GWAS for black point resistance was conducted in 166 diverse wheat cultivars mainly from China and 25 loci associated with black point resistance were identified (Liu et al. 2017b). These two studies provide novel insight into genetic architecture of black point resistance in wheat and the tightly linked SNP markers can be used in QTL and GWAS for black point resistance could be used for marker-assisted selection.

Although there are increasing numbers of reports to unravel the genetic architecture of wheat end-use quality using gene mapping approaches, the downstream translation of marker development to be used in wheat breeding is extremely slow.

5 Development of Wheat Cultivars With Desirable End-Use Quality

Development of the high-yield cultivars has always been the top priority worldwide, however quality has become an increasingly important objective over the years even in China and India. Gene-specific markers for quality traits as listed in Table 1, have been extensively used globally, including CIMMYT and CAAS, particularly for characterizing crossing parents and confirming the presence of targeted genes for advanced lines. Actually, improvement of quality and disease resistance such as for rusts and powdery mildew are the best examples of application of molecular markers in developing cultivars.

Improvement of dough strength and color related traits are major breeding objectives in China. Gene-specific markers and quality testing have been fully integrated into our breeding programs. Our strategies include (a) selection of crossing parents based on quality testing as well as genotype data from molecular markers, (b) limited backcross or single cross approach depending on the agronomic performance of both parents, (c) large population size such as 500–600 plants in BC_1(backcross 1), (d) selection of plants based on agronomic performance and molecular marker testing as well as quality testing including SDS sedimentation value or Mixograph data in segregating populations, and (e) yield and adaptation testing as well as quality evaluation and confirmation of presence of targeted genes by molecular markers. Seven quality cultivars such as Zhongmai 996, Zhongmai 998, Zhongmai 1062, Jimai 23, Zhongmai 29, and Zhongmai 578, have been released in various provinces of China by this approach. The molecular markers used included these for *PPO*, *PSY1*, *Glu-1* and *Glu-3* genes (He et al. 2014). Our experiences indicate that molecular markers can significantly improve breeding efficiency since only 5% of our crosses is targeted by molecular marker-assisted selection (MAS). Improvement of dough strength and color is relatively easy to achieve, however improvement of dough extensibility is more challenging, thus at least one crossing parent must confer outstanding extensibility. The other 95% of crosses are still managed by conventional breeding since no gene-specific marker is available for targeted traits such as yield and adaptation.

6 Role of Gene Editing for Quality Improvement

The recent advances in molecular genetics have made it possible to edit specific genes based on site-specific nucleases. This offers exciting potential to precisely edit targeted important genes with greater speed and accuracy. The extensive studies involving clustered regulatory interspaced short palindromic repeats (CRISPR)/CRISPR-associated protein 9 (Cas9) have been tested in several commercial crops (Lozano-Juste and Cutler 2014). The CRISPR/Cas9 is robust, affordable and easy to use and shown to be useful across a range of plant species (Jiang et al. 2013). Although, reports on the use of CRISPR/Cas9 are limited, it is expected that large-scale germplasm-characterization efforts in conjunction with CRISPR-based genome-editing technologies will herald a new era whereby crop plants can be precisely modified without necessarily use of physical seed samples that contain important traits. Towards wheat quality improvement, Sánchez-León et al. (2018) reported a successful modification of a gene related to end-use quality. They demonstrated that CRISPR/Cas9 modified the coeliac disease causing α-gliadin gene array and obtained non-transgenic, low–gluten wheat lines. Previously, traditional mutagenesis was failed to achieve low–gluten wheat due to the complexity of the *Gli*–2 locus and the high copy number of the α–gliadin genes. However, CRISPR/Cas9 efficiently and precisely targeted the conserved regions of the α–gliadin genes in bread and durum wheat, leading to high–frequency mutagenesis in most gene copies. Although, both CRISPR/Cas9 and RNAi are highly effective for obtaining wheat lines lacking coeliac disease epitopes, the CRISPR/Cas9 has the advantages of inducing stable and heritable mutations that do not involve the expression of a transgene, and also provide a phenotype that is independent of environmental conditions. It is expected the CRISPR/Cas9 and its various new variants would be an exceptional robust tool to precisely manipulate the functional genes underpinning wheat bread-making quality.

7 Genomic Selection for Quality Improvement

As mentioned above, past selection processes in wheat breeding relied on phenotypic traits that historically led to a non-steady rate of genetic gain in breeding. Genomic selection (also referred to genomic prediction) or genome-wide selection (GS) has emerged as a strategy extensively used in animal breeding to steadily achieve genetic gain. It has also shown significant outcomes in crop breeding (Bernardo 2016) in both pure line breeding and hybrid breeding (Crossa et al. 2017).

In GS, a test population representing the genetic diversity of a large breeding population is thoroughly genotyped and phenotyped to predict phenotypic performance based on genomically estimated breeding values (GEBVs). The large breeding population is then genotyped and the GEBVs are used to predict the phenotypes of lines in the population. According to Hickey et al. (2017), GS directly addresses four factors that affect the rate of genetic gain: (i) the speed of GS should be faster than phenotypic selection and breeders can recycle genotypes more quickly, (ii) selection intensity is greater than phenotypic selection and more individuals can be selected based on GEBVs, (iii) GEBVs are more accurate than estimated breeding values based on phenotype and pedigree alone, and (iv) GS can more efficiently integrate wide crossing and pre-breeding.

GS has emerged as a valuable tool for improving complex traits controlled by QTL with small effects. Various simulation models for predicting selection accuracy depend largely on marker density, marker type, size of training populations, and trait heritability. Due to its promise, GS has been practiced extensively in wheat breeding. GS has not only been applied to bread wheat cultivars to predict grain yield (Belamkar et al. 2018), disease resistance (Juliana et al. 2017), and end-use quality (Hayes et al. 2017), but also in wheat genetic resources to predict breeding value. GS has great potential for improving bread-making and end-use quality because most of the quality testing is laborious, time-consuming, costly, need large amount of seed and destructive in nature. Michel et al. (2018) tested more than 400 wheat accessions for protein content, dough viscoelastic and mixing properties related to baking quality, and predicted genomic selection accuracy between $r = 0.39–0.47$ for these traits. They postulated that GS can be applied 2–3 years earlier than direct phenotypic selection, and the estimated selection response was nearly twice as high in comparison with indirect selection by protein content for baking quality related traits. This considerable advantage of genomic selection could accordingly support breeders in their selection decisions and aid in efficiently combining superior baking quality with grain yield in newly developed wheat cultivars.

Previously, Heffner et al. (2011) conducted first genomic selection experiment in two soft winter wheat bi-parental populations. The prediction accuracy was greater than MAS for all the traits and the average ratio of GS accuracy to phenotypic selection accuracy was 0.66, 0.54, and 0.42 for training population sizes of 96, 48, and 24, respectively. These results provide further empirical evidence that GS could produce greater genetic gain per unit time and cost than both phenotypic selection and conventional MAS in plant breeding with use of year-round nurseries and inexpensive, high-throughput genotyping technology. Hayes et al. (2017) derived NIR and NMR predictions for 19 end-use quality traits in 398 wheat accessions and predicted selection accuracy in 2420 wheat accessions. The accuracy ranged from 0 to 0.47 before the addition of the NIR/NMR data, while after these data were added, it ranged from 0 to 0.69. Genomic predictions were reasonably robust across locations and years for most traits. Using NIR and NMR predictions of quality traits overcomes a major barrier for the application of genomic selection for grain end-use quality traits in wheat breeding.

The genomic selection prediction models were tested in CIMMYT bread wheat breeding program for end-use quality phenotypes. Battenfield et al. (2016) characterized 5520 breeding lines for basic quality parameters including flour yield, protein content, SDS-sedimentation and Mixograph and Alveograph performance. The prediction accuracy ranged from 0.32 (grain hardness) to 0.62 (mixing time). Similarly, two bread wheat populations, 1465 spring wheat lines from Uruguay and 6095 lines from CIMMYT, were used to predict the quality performance of progenies from single crosses. Overall, GS appeared to be a promising tool to facilitate the early generation selection for end-use quality in wheat and higher rates of genetic gain could be possible in bread wheat. Compared with QTL mapping and GWAS, GS has more promise in harnessing genetic gains from genetic resources for quantitative traits and is seen as a more reliable and useful approach (Bernardo 2016). However, the key challenges in successful practice of GS depend on cost-effectiveness and less biased approaches for genotyping, software for handling, quality control and joint analysis of genotypic, phenotypic and environment data, and a streamlined work flow for using GS within the overall breeding pipeline.

8 Conclusion and Future Prospects

Wheat end-use quality traits are difficult to breed because their phenotypic evaluation is costly, time-consuming and labor intensive. Furthermore, phenotyping for quality traits is only possible in the late breeding cycles due to the large amount of sample requirement and destructive nature of phenotyping assays. Therefore, these traits are ideal targets for marker-assisted selection or genome-wide selection. A major fraction of genes responsible for bread-making quality is known and their functional markers are available. Such genes could be easily deployed in breeding programs through MAS. Recently, a major barrier was overcome in practicing MAS by developing high-throughput KASP markers for several important wheat end-use quality traits and it is now possible to screen thousands of wheat accessions for major genes in a day. The successful use of CRISPR/Cas9 to desirably edit the functional genes indicated that future strategies can be designed in using molecular marker in the context of gene-editing to fine tune allelic effects of genes on major quality traits.

References

Allen AM, Winfield MO, Burridge AJ, Downie RC, Benbow HR, Barker GLA, et al. (2017) Characterization of a wheat breeders' array suitable for high-throughput SNP genotyping of global accessions of hexaploid bread wheat (*Triticum aestivum*). Plant Biotechnology Journal 15: 390–401.

Battenfield SD, Guzmán C, Gaynor RC, Singh RP, Peña RJ, Dreisigacker S, et al. (2016) Genomic Selection for Processing and End-Use Quality Traits in the CIMMYT Spring Bread Wheat Breeding Program. Plant Genome 9: 0005.

Belamkar V, Guttieri MJ, Hussain W, Jarquín D, El-Basyoni I, Poland J, et al. (2018) Genomic selection in preliminary yield trials in a winter wheat breeding program. G3: Genes, Genomes, Genetics 8: 2735–2747.

Bernardo R (2016) Bandwagons I, too, have known. Theoretical and Applied Genetics 129: 2323–2332.

Bi HH, Sun YW, Xiao YG and Xia LQ (2014) Characterization of DFR allelic variations and their associations with pre-harvest sprouting resistance in a set of red-grained Chinese wheat germplasm. Euphytica 195: 197–207.

Butow BJ, Ma WJ, Gale KR, Cornish GB, Rampling L, Larroque O, et al. (2003) Molecular discrimination of Bx7 alleles demonstrates that a highly expressed high-molecular-weight glutenin allele has a major impact on wheat flour dough strength. Theoretical and Applied Genetics 107: 1524–1532.

Cabrera A, Guttieri M, Smith N, Souza E, Sturbaum A, Hua D, et al. (2015) Identification of milling and baking quality QTL in multiple soft wheat mapping populations. Theoretical and Applied Genetics 128: 2227–2242.

Chang C, Zhang HP, Zhao QX, Feng JM, Si HQ, Lu J et al. (2011) Rich allelic variations of *Viviparous-1A* and their associations with seed dormancy/pre-harvest sprouting of common wheat. Euphytica 179: 343–353.

Chen F, Beecher B and Morris C (2010) Physical mapping and a new variant of *Puroindoline b-2* genes in wheat. Theoretical and Applied Genetics 120: 745–751.

Chen F, Zhang FY, Xia XC, Dong ZD and Cui DQ (2012) Distribution of puroindoline alleles in bread wheat cultivars of the Yellow and Huai valley of China and discovery of a novel puroindoline a allele without PINA protein. Molecular Breeding 29, 371–378.

Chen XY, Cao XY, Zhang YJ, Islam S, Zhang JJ, Yang RC, et al. (2016) Genetic characterization of cysteine-rich type-b avenin-like protein coding genes in common wheat. Science Reports 6: 30692.

Crossa J, Pérez-Rodríguez P, Cuevas J, Montesinos-López O, Jarquín D, de los Campos G, et al. (2017) Genomic selection in plant breeding: Methods, models, and perspectives. Trends in Plant Science 22: 961–975.

Dao HQ, Byrne PF, Reid SD and Haley SD (2016) Validation of quantitative trait loci for grain quality-related traits in a winter wheat mapping population. Euphytica 213: 5.

Deng Z, Tian J, Chen F, Li W, Zheng F, Chen J, et al (2015) Genetic dissection on wheat flour quality traits in two related populations. Euphytica 203: 221–235.

Dong CH, Ma ZY, Xia XC, Zhang LP and He ZH (2012) Allelic variation at the *TaZds-A1* locus on wheat chromosome 2A and development of a functional marker in common wheat. Journal of Integrative Agriculture 11: 1067–1074.

Echeverry-Solarte M, Kumar A, Kianian S, Mantovani EE, McClean PE, Deckard EL, et al. (2015) Genome-Wide mapping of spike-related and agronomic traits in a common wheat population derived from a supernumerary spikelet parent and an elite parent. The Plant Genome 8: 0089.

El Baidouri M, Murat F, Veyssiere M, Molinier M, Flores R, Burlot L, Alaet al. (2017) Reconciling the evolutionary origin of bread wheat (*Triticum aestivum*). New Phytologist 213: 1477–1486.

Furtado A, Bundock PC, Banks PM, Fox G, Yin X and Henry RJ (2015) A novel highly differentially expressed gene in wheat endosperm associated with bread quality. Science Reports 5: 10446.

Geng H, Xia X, Zhang L, Qu Y and He Z (2012) Development of functional markers for a lipoxygenase gene *TaLox-B1* on chromosome 4BS in common wheat. Crop Science 52: 568–576.

Giroux MJ and Morris CF (1997) A glycine to serine change in puroindoline b is associated with wheat grain hardness and low levels of starch-surface friabilin. Theoretical and Applied Genetics 95: 857–864.

Guzmán C, Xiao Y, Crossa J, González-Santoyo H, Huerta J, Singh R et al. (2016) Sources of the highly expressed wheat bread making (*wbm*) gene in CIMMYT spring wheat germplasm and its effect on processing and bread-making quality. Euphytica 209: 689–692.

Haile JK, N'Diaye A, Clarke F, Clarke J, Knox R, Rutkoski J, Bassi FM and Pozniak CJ (2018) Genomic selection for grain yield and quality traits in durum wheat. Molecular Breeding 38: 75.

Hayes BJ, Panozzo J, Walker CK, Choy AL, Kant S, Wong D, et al. (2017) Accelerating wheat breeding for end-use quality with multi-trait genomic predictions incorporating near infrared and nuclear magnetic resonance-derived phenotypes. Theoretical and Applied Genetics 130: 2505–2519.

He X, Zhang Y, He Z, Wu Y, Xiao Y, Ma C et al. (2008) Characterization of phytoene synthase 1 gene (*Psy1*) located on common wheat chromosome 7A and development of a functional marker. Theoretical and Applied Genetics 116: 213–221.

He XY, He ZH, Ma W, Appels R and Xia XC (2009) Allelic variants of *phytoene synthase 1* (*Psy1*) genes in Chinese and CIMMYT wheat cultivars and development of functional markers for flour colour. Molecular Breeding 23: 553–563.

He XY, He ZH, Zhang LP, Sun DJ, Morris CF, Fuerst EP et al. (2007) Allelic variation of polyphenol oxidase (PPO) genes located on chromosomes 2A and 2D and development of functional markers for the PPO genes in common wheat. Theoretical and Applied Genetics 115: 47–58.

He Z, Xia X, Peng S and Adam Lumpkin T (2014) Meeting demands for increased cereal production in China. Journal of Cereal Science 59: 235–244.

Heffner EL, Jannink J, Iwata H, Souza E and Sorrells ME (2011) Genomic selection accuracy for grain quality traits in biparental wheat populations. Crop Science 51: 2597–2606.

Hickey JM, Chiurugwi T, Mackay I and Powell W (2017) Genomic prediction unifies animal and plant breeding programs to form platforms for biological discovery. Nature Genetics 49: 1297–1298.

Himi E, Maekawa M, Miura H and Noda K (2011) Development of PCR markers for *Tamyb10* related to *R-1* red grain color gene in wheat. Theoretical and Applied Genetics 122: 1561–1576.

Ibba MI, Kiszonas AM, Guzmán C and Morris CF (2017) Definition of the low molecular weight glutenin subunit gene family members in a set of standard bread wheat (*Triticum aestivum* L.) varieties. Journal of Cereal Science 74: 263–271.

Jatayev S, Kurishbayev A, Zotova L, Khasanova G, Serikbay D, Zhubatkanov A, et al. (2017) Advantages of Amplifluor-like SNP markers over KASP in plant genotyping. BMC Plant Biology 17: 254.

Jernigan KL, Morris CF, Zemetra R, Chen J, Garland-Campbell K and Carter AH (2017) Genetic analysis of soft white wheat end-use quality traits in a club by common wheat cross. Journal of Cereal Science 76: 148–156.

Jiang W, Zhou H, Bi H, Fromm M, Yang B and Weeks DP (2013) Demonstration of CRISPR/Cas9/sgRNA-mediated targeted gene modification in Arabidopsis, tobacco, sorghum and rice. Nucleic Acids Research 41: e188.

Jiang XL, Chen GF, Li XJ, Li G, Ru ZG and Tian JC (2018) Genome-wide association study for flour color-related traits and polyphenol oxidase activity in common wheat. Cereal Research Communications 46: 1027.

Jin H, Wen W, Liu J, Zhai S, Zhang Y, Yan J, et al. (2016) Genome-wide QTL mapping for wheat processing quality parameters in a Gaocheng 8901/Zhoumai 16 recombinant inbred ilne population. Frontiers in Plant Science 7: 1032–1032.

Jin H, Yan J, Peña R, Xia X, Morgounov A, Han L, et al. (2011) Molecular detection of high- and low-molecular-weight glutenin subunit genes in common wheat cultivars from 20 countries using allele-specific markers. Crop and Pasture Science 62: 746–754.

Juliana P, Singh RP, Singh PK, Crossa J, Huerta-Espino J, Lan C, et al. (2017) Genomic and pedigree-based prediction for leaf, stem, and stripe rust resistance in wheat. Theoretical and Applied Genetics 130: 1415–1430.

King J, Grewal S, Yang CY, Hubbart S, Scholefield D, Ashling S, et al. (2017) A step change in the transfer of interspecific variation into wheat from *Amblyopyrum muticum*. Plant Biotechnology Journal 15: 217–226.

Kristensen PS, Jahoor A, Andersen JR, Cericola F, Orabi J, Janss LL et al. (2018) Genome-wide association studies and comparison of models and cross-validation strategies for genomic prediction of quality traits in advanced winter wheat breeding lines. Frontiers in Plant Science 9: 69.

Lado B, Vázquez D, Quincke M, Silva P, Aguilar I, et al. (2018) Resource allocation optimization with multi-trait genomic prediction for bread wheat (*Triticum aestivum* L.) baking quality. Theoretical and Applied Genetics 131: 2719–2731.

Lei ZS, Gale KR, He ZH, Gianibelli C, Larroque O, Xia XC, et al. (2006) Y-type gene specific markers for enhanced discrimination of high-molecular weight glutenin alleles at the *Glu-B1* locus in hexaploid wheat. Journal of Cereal Science 43: 94–101.

Li G, He Z, Lillemo M, Sun Q and Xia X (2008) Molecular characterization of allelic variations at *Pina* and *Pinb* loci in Shandong wheat landraces, historical and current cultivars. Journal of Cereal Science 47: 510–517.

Li J, Liu J, Wen WE, Zhang P, Wan Y, Xia X, et al. (2018) Genome-wide association mapping of vitamins B1 and B2 in common wheat. Crop Journal 6: 263–270.

Li J, Rasheed A, Guo Q, Dong Y, Liu J, Xia X, Zhang Y and He Z (2017) Genome-wide association mapping of starch granule size distribution in common wheat. Journal of Cereal Science 77: 211–218.

Liang D, Tang J, Pena RJ, Singh R, He X, Shen X, et al. (2010) Characterization of CIMMYT bread wheats for high- and low-molecular weight glutenin subunits and other quality-related genes with SDS-PAGE, RP-HPLC and molecular markers. Euphytica 172: 235–250.

Liu J, Feng B, Xu Z, Fan X, Jiang F, Jin X, et al. (2017a) A genome-wide association study of wheat yield and quality-related traits in southwest China. Molecular Breeding 38: 1–11.

Liu J, He Z, Rasheed A, Wen W, Yan J, Zhang P, et al. (2017b) Genome-wide association mapping of black point reaction in common wheat (*Triticum aestivum* L.). BMC Plant Biology 17: 220–231.

Liu J, He Z, Wu L, Bai B, Wen W, Xie C et al. (2016) Genome-wide linkage mapping of QTL for black point reaction in bread wheat (*Triticum aestivum* L.). Theoretical and Applied Genetics 129: 2179–2190.

Liu L, Ikeda TM, Branlard G, Peña RJ, Rogers WJ, Lerner SE, et al. (2010) Comparison of low molecular weight glutenin subunits identified by SDS-PAGE, 2-DE, MALDI-TOF-MS and PCR in common wheat. BMC Plant Biology 10: 124–124.

Liu S, Chao S and Anderson JA (2008) New DNA markers for high molecular weight glutenin subunits in wheat. Theoretical and Applied Genetics 118: 177–183.

Liu S, Sehgal S, Li J, Lin M, Trick H, Yu J, et al. (2013) Cloning and characterization of a critical regulator for preharvest sprouting in wheat. Genetics 195: 263–273.

Liu Y, He Z, Appels R and Xia X (2012) Functional markers in wheat: Current status and future prospects. Theoretical and Applied Genetics 125: 1–10.

Long YM, Chao WS, Ma GJ, Xu SS and Qi LL (2017) An innovative SNP genotyping method adapting to multiple platforms and throughputs Theoretical and Applied Genetics 130: 597–607.

Lozano-Juste J and Cutler SR (2014) Plant genome engineering in full bloom. Trends in Plant Science 19: 284–287.

Ma W, Zhang W and Gale KR (2003) Multiplex-PCR typing of high molecular weight glutenin alleles in wheat. Euphytica 134: 51–60.

Maphosa L, Langridge P, Taylor H, Chalmers KJ, Bennett D, Kuchel H and Mather DE (2013) Genetic control of processing quality in a bread wheat mapping population grown in water-limited environments. Journal of Cereal Science 57: 304–311.

Maphosa L, Langridge P, Taylor H, Emebiri LC and Mather DE (2015) Genetic control of grain protein, dough rheology traits and loaf traits in a bread wheat population grown in three environments. Journal of Cereal Science 64: 147–152.

Maphosa L, Langridge P, Taylor H, Parent B, Emebiri LC, Kuchel H, et al. (2014) Genetic control of grain yield and grain physical characteristics in a bread wheat population grown under a range of environmental conditions. Theoretical and Applied Genetics 127: 1607–1624.

Marcotuli I, Houston K, Waugh R, Fincher GB, Burton RA, Blanco A et al. (2015) Genome wide association mapping for arabinoxylan content in a collection of tetraploid wheats. PLOS ONE 10: e0132787.

Michel S, Kummer C, Gallee M, Hellinger J, Ametz C, Akgöl B, et al. (2018) Improving the baking quality of bread wheat by genomic selection in early generations. Theoretical and Applied Genetics 131: 477–493.

Nakamura T, Vrinten P, Saito M and Konda M (2002) Rapid classification of partial waxy wheats using PCR-based markers. Genome 45: 1150–1156.

Nigro D, Laddomada B, Mita G, Blanco E, Colasuonno P, Simeone R, et al. (2017) Genome-wide association mapping of phenolic acids in tetraploid wheats. Journal of Cereal Science 75: 25–34.

Parker GD, Chalmers KJ, Rathjen AJ and Langridge P (1998) Mapping loci associated with flour colour in wheat (*Triticum aestivum* L.). Theoretical and Applied Genetics 97: 238–245.

Payne PI (1987) Genetics of wheat storage proteins and the effect of allelic variation of bread-making quality. Annual Review of Plant Physiology 38: 141–153.

Prashant R, Mani E, Rai R, Gupta RK, Tiwari R, Dholakia B, et al. (2015) Genotype× environment interactions and QTL clusters underlying dough rheology traits in *Triticum aestivum* L. Journal of Cereal Science 64: 82–91.

Ragupathy R, Naeem HA, Reimer E, Lukow OM, Sapirstein HD and Cloutier S (2008) Evolutionary origin of the segmental duplication encompassing the wheat *GLU-B1* locus encoding the overexpressed *Bx7* ($Bx7^{OE}$) high molecular weight glutenin subunit Theoretical and Applied Genetics 116: 283–296.

Rasheed A, Hao Y, Xia X, Khan A, Xu Y, Varshney RK et al. (2017) Crop breeding chips and genotyping platforms: Progress, challenges, and Perspectives. Molecular Plant 10: 1047–1064.

Rasheed A, Mujeeb-Kazi A, Ogbonnaya FC, He Z and Rajaram S (2018) Wheat genetic resources in the post-genomics era: Promise and challenges. Annals of Botany 121: 603–616.

Rasheed A, Wen W, Gao F, Zhai S, Jin H, Liu J, et al. (2016) Development and validation of KASP assays for genes underpinning key economic traits in bread wheat. Theoretical and Applied Genetics 129: 1843–1860.

Rimbert H, Darrier B, Navarro J, Kitt J, Choulet F, Leveugle M, et al. (2018) High throughput SNP discovery and genotyping in hexaploid wheat. PLOS ONE 13: e0186329.

Roy N, Islam S, Ma J, Lu M, Torok K, Tomoskozi S, et al. (2018) Expressed Ay HMW glutenin subunit in Australian wheat cultivars indicates a positive effect on wheat quality. Journal of Cereal Science 79: 494–500.

Sánchez-León S, Gil-Humanes J, Ozuna CV, Giménez MJ, Sousa C, Voytas DF et al. (2018) Low-gluten, nontransgenic wheat engineered with CRISPR/Cas9. Plant Biotechnology Journal 16: 902–910.

Schwarz G, Felsenstein FG and Wenzel G (2004) Development and validation of a PCR-based marker assay for negative selection of the HMW glutenin allele *Glu-B1-1d* (*Bx-6*) in wheat. Theoretical and Applied Genetics 109: 1064–1069.

Semagn K, Babu R, Hearne S and Olsen M (2014) Single nucleotide polymorphism genotyping using Kompetitive Allele Specific PCR (KASP): Overview of the technology and its application in crop improvement. Molecular Breeding 33: 1–14.

Sun DJ, He ZH, Xia XC, Zhang LP, Morris CF, Appels R, et al. (2005) A novel STS marker for polyphenol oxidase activity in bread wheat. Molecular Breeding 16: 209–218.

Tadesse W, Ogbonnaya FC, Jighly A, Sanchez-Garcia M, Sohail Q, Rajaram S et al. (2015) Genome-wide association mapping of yield and grain quality traits in winter wheat genotypes. PLOS ONE 10: e0141339.

Torada A, Koike M, Ogawa T, Takenouchi Y, Tadamura K, Wu J, et al. (2016) A causal gene for seed dormancy on wheat chromosome 4A encodes a MAP Kinase Kinase. Current Biology 26: 782–787.

Valluru R, Reynolds MP and Salse J (2014) Genetic and molecular bases of yield-associated traits: a translational biology approach between rice and wheat. Theoretical and Applied Genetics 127: 1463–1489.

Wang J, He X, He Z, Wang H and Xia X (2009) Cloning and phylogenetic analysis of phytoene synthase 1 (*Psy1*) genes in common wheat and related species. Hereditas 146: 208–256.

Wang L, Li G, Peña RJ, Xia X and He Z (2010) Development of STS markers and establishment of multiplex PCR for *Glu-A3* alleles in common wheat (*Triticum aestivum* L.). Journal of Cereal Science 51: 305–312.

Wang S, Wong D, Forrest K, Allen A, Chao S, Huang BE, et al. (2014) Characterization of polyploid wheat genomic diversity using a high-density 90,000 single nucleotide polymorphism array. Plant Biotechnology Journal 12: 787–796.

Wang Y, Wang XL, Meng JY, Zhang YJ, He ZH and Yang Y (2016) Characterization of *Tamyb10* allelic variants and development of STS marker for pre-harvest sprouting resistance in Chinese bread wheat. Molecular Breeding 36: 148.

Wei J, Geng H, Zhang Y, Liu J, Wen W, Zhang Y, et al. (2015) Mapping quantitative trait loci for peroxidase activity and developing gene-specific markers for *TaPod-A1* on wheat chromosome 3AL. Theoretical and Applied Genetics 128: 2067–2076.

Winfield M, Allen A, Burridge A, Barker G, Benbow H, Wilkinson P, et al. (2016) High-density SNP genotyping array for hexaploid wheat and its secondary and tertiary gene pool. Plant Biotechnology Journal 14: 1195–1206.

Xu Q, Xu J, Liu CL, Chang C, Wang CP, You MS, et al. (2008) PCR-based markers for identification of HMW-GS at *Glu-B1x* loci in common wheat. Journal of Cereal Science 47: 394–398.

Yan L, Helguera M, Kato K, Fukuyama S, Sherman J and Dubcovsky J (2004) Allelic variation at the *VRN-1* promoter region in polyploid wheat. Theoretical and Applied Genetics 109: 1677–1686.

Yang L, Zhao D, Yan J, Zhang Y, Xia X, Tian Y, et al. (2016) QTL mapping of grain arabinoxylan contents in common wheat using a recombinant inbred line population. Euphytica 208: 205–214.

Yang Y, Zhao X, Xia L, Chen X, Xia X, Yu Z, et al. (2007) Development and validation of a *Viviparous-1* STS marker for pre-harvest sprouting tolerance in Chinese wheats. Theoretical and Applied Genetics 115: 971–980.

Yi X, Jiang Z, Hu W, Zhao Y, Bie T, Gao D, et al. (2017) Development of a Kompetitive Allele-Specific PCR marker for selection of the mutated *Wx-D1d* allele in wheat breeding. Plant Breeding 136: 460–466.

Zhai S, He Z, Wen W, Jin H, Liu J, Zhang Y, et al. (2016) Genome-wide linkage mapping of flour color-related traits and polyphenol oxidase activity in common wheat. Theoretical and Applied Genetics 129:377–394.

Zhai S, Liu J, Xu D, Wen W, Yan J, Zhang P, et al. (2018) A Genome-wide association study reveals a rich genetic architecture of flour color-related traits in bread wheat. Front Plant Sci 1136–1148.

Zhang C, Dong C, He X, Zhang L, Xia X and He Z (2011) Allelic variants at the *TaZds-D1* locus on wheat chromosome 2DL and their association with yellow pigment content. Crop Science 151: 1580–1590.

Zhang F, Chen F, Wu P, Zhang N and Cui D (2015) Molecular characterization of lipoxygenase genes on chromosome 4BS in Chinese bread wheat (*Triticum aestivum* L.). Theoretical and Applied Genetics 128: 1467–1479.

Zhang Y, Miao X, Xia X and He Z (2014) Cloning of seed dormancy genes (*TaSdr*) associated with tolerance to pre-harvest sprouting in common wheat and development of a functional marker. Theoretical and Applied Genetics 127: 855–866.

Zhang Y, Xia X and He Z (2017) The seed dormancy allele *TaSdr-A1a* associated with pre-harvest sprouting tolerance is mainly present in chinese wheat landraces. Theoretical and Applied Genetics 130: 81–89.

Zhou S, Zhang J, Che Y, Liu W, Lu Y, Yang X, et al. (2018) Construction of *Agropyron Gaertn.* genetic linkage maps using a wheat 660K SNP array reveals a homoeologous relationship with the wheat genome. Plant Biotechnology Journal 16: 818–827.

Durum Wheat Products, Couscous

Rifka Hammami and Mike Sissons

Abstract Durum wheat is one of the most important cereals grown in semiarid zones such as North Africa, South Europe and Middle East. Pasta and couscous are the most common paste products made from durum wheat. Couscous is one of the most ancient staple foods prepared by Berbers, the native inhabitants of North Africa. Couscous has a symbolic meaning and represents a part of their identity. Couscous quality is determined by two main factors, raw material composition and dough proprieties. Visual appearance, test weight, weight of 1000 kernels, vitreuousness, yellow pigment and grain proteins percentage strongly impacts couscous quality. After milling, particle size, high protein semolina, glutenins and gliadins, dough strength have been shown to influence couscous characteristics.

This chapter summarizes the origin of couscous, and traditional processing methods. The key grain and semolina quality characteristics required for good couscous quality will be discussed. Methods used to assess the eating quality of couscous will also be covered.

1 Introduction

This chapter is a general review of durum wheat products, mainly couscous, including the history, homemade production and the influence of raw material.

The domestication of cereals marked a dramatic turn in the development and evolution of human civilization as it enabled the transition from a hunter-gatherer and nomadic pastoral society to more sedentary agrarian one (Fuller 2007). For wheat and barley the center of origin is located in the Fertile Crescent, and more

R. Hammami (✉)
Field Crop Laboratory, National Institute of Agronomic Research of Tunisia – Carthage University, Ariana, Tunisia

M. Sissons
NSW Department of Primary Industries, Tamworth Agricultural Institute, Calala, NSW, Australia

G. Igrejas et al. (eds.), *Wheat Quality For Improving Processing And Human Health*, https://doi.org/10.1007/978-3-030-34163-3_15

particularly in the mountain regions that surround the fertile alluvial plains of the Tigris and Euphrates rivers (Braidwood et al. 1969).Wheat was one of the first crops to be domesticated by man more than 10,000 years ago in the Middle East (Charmet 2011). Cultivation of wild emmer and einkorn wheat started in the early Neolithic period (Stone Age). Later, from the end of the Neolithic period through the Bronze Age (6000–3000 years ago), the naked tetraploid sub-species of *Triticum turgidum* slowly replaced wild emmer wheat, and wheat cultivation spread around the world (Charmet 2011).

Cereals are a primary source of food for humans and animals and use the majority of the cultivated land in the world. Durum wheat (*Triticum turgidum* ssp. *durum* Desf., 2n = 4x = 28, AABB) is an important wheat crop with global cultivation area estimated ~ 16.7 million hectares (International Grain Council) and production ~ 32.1 million tonnes in 2010/11 season increasing to ~ 36.9 million tonnes in 2018 (Ranieri 2015; Taylor and Koo 2015). Durum wheat is mostly grown in West Asia, North, and East Africa, the North American Great Plains, Canada, India, Eastern and Mediterranean Europe, Mexico and Australia (Cantrell 1987; International Wheat Council 1991). With the exception of Europe, North Africa (Algeria, Morocco, Tunisia, and Libya) is the largest import market for durum wheat (Bonjean et al. 2016). Durum wheat is the commodity of choice for production of high-quality pasta in many countries. However, couscous, another important product obtained from this crop, is very popular in the tradition of many Mediterranean regions, together with bread, as pointed out by Guezlane (1994). In this context, Quaglia (1988) mentioned that durum wheat has found traditional use in flat and specialty breads, particularly in Mediterranean countries but now is also experiencing increasing application in the Mediterranean region for breads of all types (Palumbo et al. 2000).

The vast array of homemade foods derived from durum grains is the result of its long history as part of human diets, which dates back to the origin of civilization in the Fertile Crescent (Mac Key 2005). In the next section we present a short description of some durum wheat products.

2 Durum Wheat Products

The wheat preferred for making pasta products is durum (Sissons 2016). Durum wheat semolina and durum flour are used to manufacture pasta and non-pasta food products. Pasta products are manufactured by mixing water with semolina or durum wheat flour to form unleavened dough, which is formed into different shapes either by lamination or more commonly extrusion. The resulting product can be cooked and eaten, so called fresh pasta, or dried under controlled temperature and humidity conditions as dried pasta for later consumption after its cooking. Pasta and couscous are paste products (Pollini et al. 2012). Other products from durum wheat include Bulgur (cracked durum wheat) and Frekeh (parched immature wheat kernel), both on-paste food products (Boggini et al. 2012; Dick and Matsuo 1988).

2.1 Pasta

Pasta frequently associated with Italian cuisine, is made from durum wheat semolina and contains a large proportion of gluten (elastic protein). Pasta is formed into ribbons, cords, tubes and various special shapes, all originally developed for specific characteristics, such as ability to retain heat or hold sauce. Indications are that pasta originated from China, although there is evidence of pasta use in Italy during the Etruscan civilization (Sissons 2016). In 1800, mechanical devices for making pasta appeared in Italy (Banasik 1981). Italians categorize pasta into four main groups: long goods (spaghetti, vermicelli, and linguine); short goods (elbow, macaroni, rigatoni and ziti); egg noodles; and specialty items (lasagna, manicotti, jumbo shells) and stuffed pasta (Dick and Matsuo 1988). Several reviews on pasta processing are available in the literature (Kill and Turnbull 2008; Lucisano et al. 2008; Manthey 2002; Pollini et al. 2012; Turnbull et al. 2001). Typically, semolina and sometimes other ingredients are mixed together and conveyed to a mixer, which is under vacuum. Once in the mixer, warm water is injected as a spray over the mixture to achieve 28% to 32% water content. Paddles in the mixer continuously agitate the wetted semolina mixture while moving the hydrated mixture toward the extrusion auger. The retention time in the mixer is adjusted to allow full hydration of the semolina before it enters the extrusion auger. Full hydration of the semolina particles is very important for the development of the protein (gluten) matrix during pasta extrusion. Development of the protein matrix does not occur during mixing, since the energy supplied by the mixer is insufficient to develop the protein matrix.

Dough develops as it moves along the extrusion auger, which kneads the hydrated semolina and exerts pressure on the dough as it progresses through the extrusion barrel toward the die. The back pressure in the extrusion barrel helps to produce a dense product where starch granules are deeply embedded within the protein matrix. The extrusion process occurs under vacuum. Extruding under vacuum is important in dried pasta, as air trapped in pasta will expand during drying particularly during high and ultra-high temperature drying. These expanded air pockets are points of weakness and detract from the desired uniform, translucent, yellow color. Removing air also reduces pigment loss catalyzed by the enzyme, lipoxygenase. Fresh or frozen pasta manufacturers generally do not use a vacuum system during the extrusion process. The air bubbles in the product do not seem to have any significant impact on the end product appearance or cooking quality in such products.

2.2 Bread

Bread is a food produced using simple ingredients such as wheat flour, salt, yeast and water, and is one of the most consumed cereal products in many countries and a food at the basis of the diet of many people around the world. Durum wheat for bread production occurs mainly in the Near East, Middle East, and Italy (Williams et al. 1984; Williams 1985). In some Middle Eastern countries, 70% to 90% of durum wheat is used for bread. Several types of bread are made from durum wheat: two-layered flat bread (Fig. 1A), called Khobz, is the most popular bread in Syria, Lebanon, and Jordan. In Egypt, two-layered bread is called Baladi and Shami. Single-layer bread (Fig. 1B) also is popular, including Tannur and Saaj (Syria and Lebanon), Mountain bread and Markouk (Lebanon) and Mehrahrah. In Tunisia, the most popular durum breads are Tabouna, Mlaoui (Fig. 2A) and M'Besis (Fig. 2B).

In Turkey, flat bread, Tandir Ekmegi, is made from durum wheat. Thirty percent and 18% of durum wheat in the Near East is used to make two-layered and single-layer breads, respectively (Williams 1985).

Several kinds of bread are made in Italy from durum wheat, depending on the shape of the bread and the region of the country (Abecassis et al. 2012; Quaglia 1988; Sada 1982). The common breads include Fresedde in the province of Bari, Fasella in the province of Foggia, and Fasedda, Frisedda, and Frisa in the province of Salerno. A round, flat bread, Cafone, is produced in Bari. A wheel-shaped durum wheat bread, rote, is produced in the Bari and Foggia provinces. Sckanate is large durum bread typically made in Minervino, Altamura, Bitonto, and Gargano.

Although some countries use durum wheat to produce different kinds of bread, the inferior loaf volume and appearance of durum bread compared to bread made from common wheat has restricted its wider use. Based on the characteristics of certain proteins in the kernel, the differences between bread wheat and durum wheat bread can be attributed largely to their gluten protein properties, with durum wheat normally having weaker and less extensible gluten than bread wheat. However, the development of strong gluten durum cultivars has improved the cooking quality of pasta products and improved the bread making quality. In this regards, Sissons (2008) reported that in bread making, the gluten must have adequate extensibility and elasticity of the dough which expands and retains carbon dioxide that is formed during the fermentation and baking (Buche 2011; Liu et al.1996).

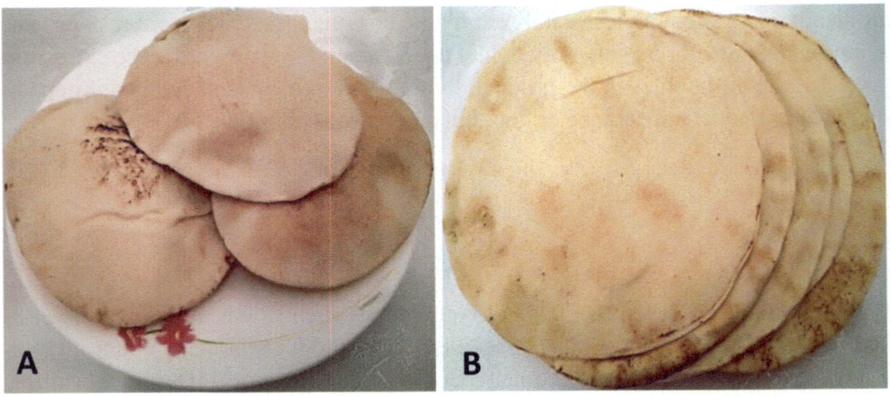

Fig. 1 (**A**): Two layered flat bread; (**B**): Single layered flat bread

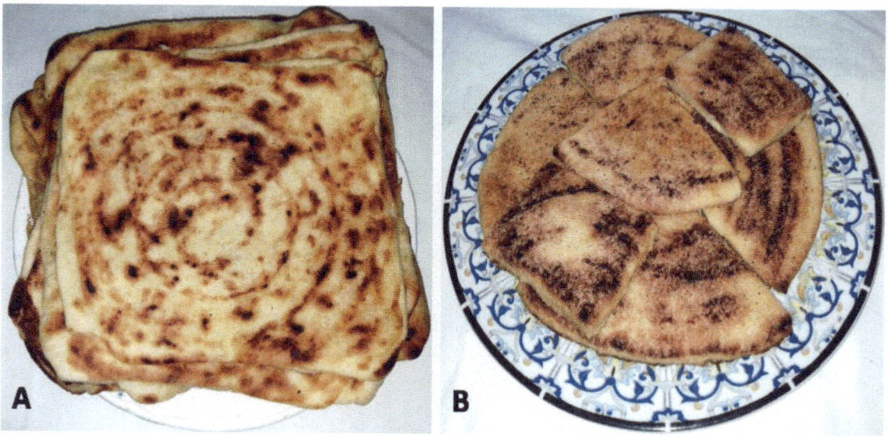

Fig. 2 North African Bread. (**A**): Mlaoui; (**B**): M'Beses

2.3 *Bulgur*

Bulgur (bulghur or burghul), is coarse or fine cracked wheat that has been partially cooked. It is one of the oldest cereal-based foods. Bulgur is used as a main dish or as one of the ingredients in most food consumed in Turkey, Syria, Jordan, Lebanon, and Egypt. Bulgur can be made from bread wheat, durum, barley, and maize. However, durum is preferred because of its hardness and amber color. An estimated 15% of durum wheat in the Near East is used to make bulgur (Williams 1985). Burlgur is used in salads such as tabouleh, soups and baked products such as kibbeh, soups or boiled and consumed like rice. As enzymes are inactivated during the cooking process, bulgur has a long shelf life (Miskelly 2017).

In the Middle East, Turkey, Greece, and Cyprus, bulgur made at home or commercially follows the same steps with one exception: both hard red wheat and durum wheat are used for commercial bulgur; only durum wheat is used for homemade bulgur (Fig. 3). Bulgur has a nutty flavour and can be served as a side dish, similar to rice or couscous. As a wheat product, bulgur is a good source of dietary fiber, protein, iron, and vitamin B6.Traditional artisan preparation utilises sun drying and debranning by hand before crushing with a stone or coarse mill. Modern production consists of cleaning, washing, tempering to 40% moisture, cooking to 95 °C, drying, de-branning, cracking and sifting. Starch is gelatinised during the cooking process (Belibagli et al. 2009). Bulgur is often confused with cracked wheat, which is made from crushed wheat grains which have not been parboiled. Important quality characteristics are moisture content, particle size, shape, colour, texture and flavour. Maximum moisture content of 10–12% is necessary for keeping quality. Bulgur is sold in both coarse and fine particle sizes for different food applications (Miskelly 2017).

To cook bulgur, the product is usually boiled, similar to rice or other grains, but it can also be fried, roasted, baked, or simply soaked. Since bulgur is already partially cooked, it takes less time to prepare than other whole grains and has a longer shelf life.

Fig. 3 Artisanal bulgur made from durum wheat grains

2.4 Frekeh

Frekeh is also known as firik. It is a non-paste durum wheat product, and a staple food in North Africa and the Middle East, especially Syria. Frekeh is green wheat that is used in the same way as rice, bulgur, and couscous (Elias 1995). In the Near East, 2% of durum wheat is used to make frekeh. In contrast to bulgur, frekeh making is a localized village industry. In many villages in northwestern Syria, frekeh is one of the most important sources of income. Although, it is a small industry, an estimated 194,000 to 291,000 tonnes of frekeh are made every year in the Middle East (Williams 1985). The best frekeh is made from the largest and greenest grains. Therefore, durum wheat, especially cultivars with large kernels are the most suitable for making frekeh. The season for making frekeh is only one to two weeks when the durum wheat is in the grain filling stage (Elias 1995). Frekeh is more appetizing than that processed at the full-ripe stage, probably due to higher contents of free simple sugars.

Frekeh is produced by two different procedures roasting or boiling. In the roasting procedure, fire is used to burn the awns, lemma, and palea from immature spikes. Care is taken to avoid excessive parching of the kernels. The fire scorches the grain, giving the frekeh a characteristic flavor. In the boiling procedure, the immature spikes are boiled in water for about 20 min. In either process, the scorched or boiled spikes are dried in the sun. The heads are either hand (small-scale) or mechanically (large-scale) threshed to separate the grain from the chaff. Winnowing in the wind cleans the threshed grain. Finally, the grain is stored in bulk before it is bagged.

Frekeh is prepared for eating by cooking in water (1:2) for 20 minutes and allowed to cool for 5 minutes. A minimum amount of water is used to avoid leaching soluble nutrients. Frekeh is used as a substitute for rice and bulgur in pilav. Frekeh can be either boiled or steamed and is served with lamb or poultry (Özkaya et al. 1999).

3 Couscous

3.1 History of Couscous

Couscous, from the Berber word k'seksu, is the staple food of Northern Africa people and the national dish of Maghreb countries, such as, Tunisia, Algeria and Morocco. Historians have conflicting opinions regarding its origin. However many references mentioned that couscous was invented by the Berbers, the first inhabitants of North Africa. Couscous spread from this area, where it originated, to Libya, Mauritania, Egypt, and sub-Saharan countries. Couscous is also consumed in the Middle East.

Couscous is an iconic food in northern Africa for dietary and cultural reasons. Similar to rice, pasta, or bread, couscous is an inexpensive and highly nutritious product made from durum wheat endosperm with a long shelf life. The famed Arab traveler Leo Africanus (c. 1465–1550), also mentioned couscous with some delight: "Of all things to be eaten once a day it'salcuzcuçu (Arabic for couscous-authors) because it costs little and nourishes a lot". For centuries, black African women were employed as couscous cooks while today in Morocco the dada young black Saharan and sub-Saharan women serve as domestics, especially as cooks are often employed to prepare couscous. The Tuareg, a Muslim Berber tribe of the Sahara, also employed young black servant women to make couscous. Black slaves were also prominent as cooks in medieval Egyptian households and up until the nineteenth century. In Muslim Spain, too, black slaves would prepare meals in aristocratic homes while the wives would prepare the food in poorer homes (http://www.cliffordawright.com/caw/food/entries/display.php/id/34/).

Couscous is an iconic food because it permits the expression of national identities and ways of life, and it has religious and symbolic meanings. According to Habib Bourguiba, ex-leader of Tunisia, the border of Maghreb, the oriental region of North Africa, is marked by an "imaginary line", corresponding not to a geographic but to a cultural boundary: east of that line the staple food is rice, and west of the line the staple-food is couscous (D'Egidio and Pagani 2010).

Women usually prepare the grain known as couscous during a family celebration, and the dish named couscous is eaten during a family feast, thereby associating both the product and the dish with solidarity. Couscous accompanies traditional Arabic weekend (Friday and Saturday), then end of Ramadan celebrations, Muslim year, birth and wedding feasts. The association of couscous with these festivities also attaches it to the concepts of abundance, fertility, fidelity, and Barakah (God's blessing). For example, while preparing couscous, women have to make an invocation and converse about religious facts, prosperity, and positive feelings. According to Namoune et al. (2004), the woman make the couscous alone at home or sometimes she asks her cousins or neighbours for help. In Tunisia, women choose a sunny day and dedicate it to making a large quantity of couscous and call it the 'Oula day'. For years, couscous-preparing knowledge was passed from mother to daughter and played an essential role in North Africa's patriarchal society. Indeed, the know-how was an important "intangible" element of a young woman's dowry.

In the Arab world, the preparation of couscous is one that symbolizes "happiness and abundance". Couscous is prepared in a couscousière, a large covered pot with a lower compartment used to cook the sauce and an upper portion called "keskes" with a pierced bottom in which the couscous steams. The grains must be sprinkled with liquid, stirred to separate the clumps, and steamed two or three times. While the grain is steaming, a stew of lamb, chicken, chickpeas, and vegetables cook in the lower portion of the pot. Alternatively, couscous can be eaten as a sweet dish with dry fruits and milk or as a breakfast with butter. With a basic cooking system, it is possible to prepare an everyday meal or a luxury feast, a main course or a dessert.

According to Galiba et al. (1987), couscous preparation is time consuming but the final product is perfectly suited for the migrant life of Sahelian pastoralists because of its long shelf life. Couscous is known by different names in different countries. It is known as Kuskus in Turkey, Couscous in Morocco, Maftoul in Jordan, Moghrabieh in Lebanon, Seksu in Berber, Kusksi in Libya, Keskesu in Tuareg, Kouskousaki in Greece and Attiéké in West African (Anonymous 2013).

From the early twentieth century on, however, such traditional know-how would start to fade. At the time, couscous production would become increasingly mechanized, thanks in part to innovations made by families like the Italian industrialists, the Ferreros, who had settled in Algeria and set up mills in Blida and Bou Saada. The story of couscous continued nonetheless. First, thanks to the earliest generation of North African immigrants to Europe (mainly to France and Belgium) during World War. Then, with the return of colonial populations to their native countries, following the decolonization of the Maghreb. In less than 50 years, couscous would become one of France's three favorites savory dishes. Today, couscous is produced and eaten around the world, thanks to globalization. EBRO food, based in Spain is the world's leading producer of couscous (Goumeziane 2018). Experts in Algeria are working on a project to include North African couscous on UNESCO's world heritage list. Though in practice the dish is already enjoyed around the world, its origins remain hotly disputed and unsurprisingly so, given the prevalence of couscous throughout much of the Middle East and North Africa, not to mention the various traditions of preparing, cooking and eating the grain (Goumeziane 2018). In North Africa couscous plays the role that semolina pasta has in Italy. Algeria is the leader of couscous production (about 0.90 million tonnes/year), above all consisting of homemade or artisanal production. In Algeria couscous consumption reaches 50 kg per capita/year, while in Tunisia it is about 20 kg per capita/year (D'Egidio and Pagani 2010).

3.2 Couscous Production

Compared with pasta, the industrial production capacity of couscous is low, about 430,000 tonnes/year. This capacity is mainly provided by Maghreb countries whose diet relies on cereals. The couscous estimated production capacity (Cahier du CEPI N°23) by each country (tonnes/year):

- Maghreb countries ~203–204,000, distributed between Tunisia (~77,000), Algeria (~45,000), Morocco (~73,000) and Mauritania (~8,600)
- Europe (~115,000–116,000), distributed between France (~102,000), Italy (~13,700)
- America: ~105,000
- Canada: ~6,000

In Addition, there are small production capacities in Senegal and Israel around 5000 tonnes/year (Jude 2004). In Europe, couscous is still mainly consumed by immigrant populations. However, the undertaken in recent years, make the consumption of couscous gradually seduces the rest of the European population (Jude 2004). In particular, consumer awareness has widened the scope of uses of couscous to other culinary recipes.

3.3 Homemade Production of Couscous in North Africa

The best raw material for couscous preparation is durum wheat semolina although other cereals, such as sorghum, millet, maize or fonio (a cereal cultivated in West Africa without gluten) are used, especially in West and sub-Saharan Africa (Aboubacar and Hamaker 1999; Galiba et al. 1987). However, durum wheat couscous gives the food excellent texture, taste and nutritional qualities. Couscous appears to be unique among cereal grain food products. It is distinguished by the special way it can be and traditionally has been prepared for consumption, namely by a series of simple hydrating and steaming steps (Donnelly et al. 1994). Until recently, the traditional method has been the only known one for making couscous. Kaup and Walker (1986) reported that the steps required to prepare industrial couscous are the same as traditional couscous. Maghreb countries used the same material for producing homemade couscous. However, the names of some of the tools are different. Fig. 4 shows the material required for making couscous. It is composed by the “Guessâa”, a wide bowl (diameter 80 to 120 cm) commonly used for producing couscous or kneading other kinds of durum wheat products such as bread. Four different sieves named “Tallâa”, “Thannaya”, “Manfda” and “Sakkat” are utilized in sieving or sizing operations and correspond, respectively, to 500, 1000, 1200 and 1300 μm mesh opening. The keskes (Couscousière) is the cooking utensil of couscous; it is made from aluminum and consists of an upper part containing identical holes that allow the passage of steam to couscous particles and a lower part larger than the upper part in which is placed the water to boil.

Fig. 4 Utensils and material used for producing homemade couscous

Generally, the manufacture of artisanal couscous is carried out in summer (May–September) at home in a clean and well ventilated room at 25 °C. The main similarity found between the different types of artisanal couscous making is the gradient of semolina in two products of different particle sizes: the fine semolina traditionally called "Dkak" and a coarse semolina called "Fetla". The other common points are essentially steaming the couscous and drying in the open air (Bahchachi 2002; Benatallah et al. 2006; Yousfi 2002). In Tunisia, experienced women are dedicated to manufacturing couscous following the steps described in Fig. 5:

- Hydration (Fig. 5A): this operation makes it possible to prepare coarse semolina agglomeration by adding cold salted water. The hydration establishes bonds between the semolina particles and allows their agglomeration (Hebrard et al. 2003). This step is very delicate; on the one hand, it is necessary to ensure the wetting of the semolina and on the other hand, to avoid the over-agglomeration which leads to the formation of a paste due to the excessive moistening of the semolina. Cold water helps to avoid the formation of large agglomerates.
- Rolling (Fig. 5B): this operation is done in a large bowl named "Guessâa" and provided by the movement of back and forth palm open hands down. This is the beginning of the agglomeration process. In order to obtain the desired semolina agglomeration a small amount of fine semolina is added gradually (Fig. 5C-F) sieving: the desired homogeneity and particle size are ensured by the choice of sieve mesh openings. To meet these quality criteria four sieves are used (Fig. 5).
- Steaming (Fig. 5G, H): Tallâa's refusal (couscous of the desired particle size) is put in an aluminum colander of a couscousière containing boiled water. Couscous is precooked with steam for 15 minutes at a temperature of 95 °C. This time varies according to the granulometry. As the particle size increases, the cooking time decreases because the water vapor circulates more rapidly between coarser grains of couscous (Anga and Belhouchet 2002) and the thickness of the layer of couscous put in the colander. Generally, the precooking time is determined by the fact that water vapor is on the surface of couscous. Couscous seeds break apart between fingers in the form of patôns and have a yellow color.
- Drying (Fig. 5I): the drying of couscous is done in two phases: the couscous is spread out on a clean sheet in the shade at ambient temperature. The drying time

Fig. 5 Steps used for manufacturing dried couscous according to the traditional procedure of Tunisia. A: Semolina hydration with salted water (about10g/l); B: Rolling (circular movement); C, D, E: sizing process by sieving agglomerates; F: humid couscous with desired particle size; G, H: steaming couscous; I: drying couscous

is a function of the ambient temperature and the relative humidity of the air. When the couscous is well dried, the couscous is then dried in the sun. Couscous is stirred occasionally for a good drying process. The drying step is strictly related to the climatic conditions that account for the production of home-made couscous during the sunny summer months (Kaup and Walker 1986). Sun-dried couscous has a long shelf life. In some regions of Tunisia couscous does not undergo steaming and it is dried immediately. However before serving, it has to

Fig.5b (continued)

be steamed two or three times. It is called the moist couscous and it is prepared in the same way as the dry couscous but is not dried in the sun and is usually prepared and consumed the same day (Derouiche 2003).

- Grading: the couscous is separated into fine, medium and coarse. The final product is classified in three different sizes: small couscous, recommended for desert preparation, principally formed by aggregates with diameters lower than 1.5 mm; medium couscous, 1.7–2.0 mm diameter, the most appreciated for traditional dishes; coarse couscous, with particle size up to 2.5 mm used to prepare couscous with vegetables sauce (Kaup and Walker 1986).
- Storage: the couscous is stored until utilization. In Maghreb countries, the processes of making artisanal couscous differ from one region to another or even from one person to another. The industrial process appears as a mechanization of the manual operations however, the general principle of making couscous is the same (Lefkir 2005). Couscous is a versatile food in North Africa that is served in many different ways and with a variety of other foods.

3.4 Couscous Consumption

In Northern Africa regions, couscous is consumed in a sauce with different types of food. The couscous can have several flavours depending on the method of preparation. Generally, the couscous sauce could be associated with different types of vegetables according to the seasons and food preferences. In Tunisia people prefer mutton; other meats are also used such as fish and chicken. Different sources of fat (olive oil, butter or "smen") are added into couscous grain. Salt and some spices (black pepper, Harissa) are also included. Couscous is becoming more and more present in some European countries, particularly in the Mediterranean area: this trend is surely related to the growing interest towards the so-called "ethnic foods" and the increase in the population of Arabic–origin people living in Europe (D'Egidio and Pagani 2010).

3.5 Raw Material and Quality Parameters

Durum wheat is a raw material of choice for the manufacture of durum wheat products in Maghreb countries and the Mediterranean region because of the ideally suited color and cooking quality. However, couscous is made from pearl millet in Senegal and corn in Togo (Kaup and Walker 1986). Couscous from corn, sorghum and millet are traditional foods of several countries in West, Central and Eastern Africa (Galiba et al. 1988). Their fabrication process resembles that of Northern Africa countries (Aluka et al. 1985). About 10% of durum wheat in the Middle East is used to make couscous (Williams 1985). The manufacturing steps of commercial couscous are very similar to the artisanal one (Kaup and Walker 1986). Home-made couscous is generally prepared by coarse semolina: this preference could be due to the higher product yield observed when this type of semolina is used, despite its lower water absorption during mixing (Debbouz et al. 1994). The quality of such foods in terms of texture, colour, flavour and appearance are determined by raw material quality, processing methods and other components (Abecassis 2012). In this context, Lefkir et al. (2017) reported that couscous quality depends on the raw material implementation and the manufacturing conditions: wheat milling, shaping (hydration and rolling), precooking and drying couscous.

In addition to yield potential, grain quality is very important in wheat markets because of the demand for high quality end products such as pasta, couscous and bulgur wheat (Toscano et al. 2014). In the agro-food domain, quality is a term frequently used, but not easily definable, as it is intended to describe a set of features: sanitary, technological, nutritional, and sensorial that meet customer requirements. The sanitary quality is a pre-requisite that must be guaranteed for consumer's health; the technological quality refers to the fitness of raw materials for a specific industrial process whereas the sensory quality relates to consumer acceptability. Lastly, the nutritional quality is linked to the chemical composition and the presence of specific elements and/or bioactive compounds suitable to satisfy the nutritional needs of consumers and contribute to their welfare and health. These compounds have to be present in the raw material and maintained during the technological process (D'Egidio and Pagani 2010).

The quality of durum wheat semolina required for making good couscous is similar to that of other pasta products (Kaup and Walker 1986). However, compared to pasta, very little work has been done on determining quality requirements for couscous and other durum wheat products (Anga and Belhouchet 2002; Bellocq et al. 2018; Debbouz et al. 1994; Debbouz and Donnelly 1996; Derouiche 2003; Dick and Matsuo 1988; Guezlane et al. 1986; Guezlane 1993; Guezlane and Abecassis 1991; Idir 2000; Kaup and Walker 1986; Lefkir 2005; Ounane et al. 2006; Quaglia, 1988; Tigroudja and Bendjoudiouadda 1998; Yettou 1998; Yousfi 2002). The results of the study conducted by Elias (1995) suggested that good quality couscous should have the following characteristics: absorb the sauce well, uniform particle size, individual particles maintain their integrity during steaming or sauce application, and particles are not sticky. All these factors affect the taste of couscous. Abecassis et al. (1994) mentioned that the criteria required for durum wheat and semolina quality can also be analyzed in terms of suitability to be submitted to different heat treatments: drying, steaming and drying, or baking. The main consequence of an unsuitable product during these treatments is the firmness for pasta and the disaggregation for couscous. Abecassis et al. (1994) have noted that a high protein content, strong gluten and low or medium starch damage are key manufacturing traits for high quality durum wheat products. Other authors such as Quaglia (1988) reported that kernel hardness, semolina protein, gluten and particle size are important to the quality of couscous. According to Elias (1995) hardness, protein content, gluten content and granulometry of semolina influences the couscous quality. In this regard, Derouiche 2003 and Chemache et al. (2018) reported that characteristics of good couscous are: amber yellow color, uniform particle size, high sauce absorption capacity, ability to keep its integrity during steaming or sauce application, and particles are non-sticky and good taste. Although the consumption of couscous is increasing, few studies to date have investigated the role of raw materials and quality parameters. Moreover, the definition of quality parameters is still not clear. Uniform size, pleasant colour and no unusual flavour can be used to determine the quality of the couscous grain (Debbouz et al. 1994; Debbouz and Donnelly 1996; Ounane et al. 2006). The cooking behaviour considers rehydration and cooking times, sauce absorption capacity (Debbouz and Donnelly 1996; Ounane et al. 2006) without aggregation of granules (Guezlane and Abecassis 1991) and sensory indices related to texture, as stickiness and mouth feel (firmness and smootheness) (Debbouz et al. 1994; Kaup and Walker 1986). Recently, Bellocq et al. (2018) investigated the effect of different initial water contents (0.32 or 0.48 g/g dry matter) during the steaming stage and different temperatures during the drying stage on durum wheat couscous grain and they suggested an analytical drying model that describes the kinetics of average water contents during the drying stage and allows modeling the changes in compactness and the subsequent changes in diameter for the agglomerates of couscous grains.

3.5.1 Durum Wheat Hard Vitreous Kernels

The grain hardness of durum wheat kernels allows them to be used following a conventional milling process to produce semolina as the raw material for several end products especially pasta and couscous. Kernel hardness is an important factor to produce high quality couscous (Quaglia 1988). Kernel hardness in wheat is primarily conditioned by the hardness locus on 5DS chromosome, which is comprised of Puroindoline a and Puroindoline b (Morris 2002; Morris and Bhave 2008). Puroindolines, which act to soften the endosperm, are completely lacking in durum (Morris and Fuerst 2015). Hard vitreous kernels (HVK), an important grading specification in durum wheat because it is associated with hardness and semolina yield, is a tedious visual procedure (ICC 1999). Shahin and Symons (2008) noted that HVK content is an internationally recognized specification which, along with other factors, is used to determine the value of durum wheat. Vitreousness is the natural hard glossy translucent appearance of wheat kernels which is associated with high protein content. Kernels having an externally visible starch area of any size are considered non-vitreous or starchy. Wheat lots with a high percentage of HVK as contrasted to starchy kernels show better milling performance in terms of a higher yield of granular semolina and lower yield of fine flour (Shahin and Symons 2008). According to Fu et al. (2018) vitreous kernels have consistently shown higher level of wheat protein in comparison to their non-vitreous or mealy counterparts, suggesting the importance of protein content in vitreous kernel development (Dexter et al. 1988, 1989; Samson et al. 2005; Sieber et al. 2015). In this context, Owens (2001) mentioned that a minimum hard vitreous kernel (HVK) content is an important trading criteria.

3.5.2 Protein Content

Protein content in durum wheat is controlled by fertilizer, environment, and genetics. Semolina protein content generally is about one percent less than whole wheat protein (Baum et al. 1995). Cooking quality of semolina products is related to both quantity and quality of the proteins present in the endosperm. A moderately high protein (12% or more) is required to produce an acceptable product such as pasta and couscous. Wheat proteins are responsible for 30–40% of the variability of the culinary quality of pasta and couscous despite relatively low levels in grain (Dexter and Matsuo 1980). According to Boudreau et al. (1992), the couscous value of semolina depends on its protein content (13.5%). The role of protein quantity in determining couscous quality is controversial: some authors referred to a decrease in stickiness as protein content increased (Debbouz et al. 1994) while others showed no significant relationship between couscous characteristics and semolina protein or gluten quantity and quality (Ounane et al. 2006). In contrast, Dexter and Matsuo (1980) reported that wheat proteins would be responsible for 30–40% of the variability of culinary quality of couscous. House wives attribute the yellow color and the purity of the semolina to high protein content (Yousfi 2002). Moreover, Debbouz et al. (1994) observed that wheat varieties with strong gluten expressed better yield of couscous than cultivars with weak gluten. According to Quaglia (1988), semolina proteins, gluten and the particle size are the most important factors in describing the quality of couscous. Several studies have been done by D'Egidio et al. (1990) and Novaro et al. (1993) showing that the protein content and the gluten quality are very important indicators for culinary quality of durum wheat products. In the same way, Kaup and Walker (1986) suggested that the quality of the semolina required for the couscous preparation is similar to that of pasta. However, it is proved that the semolina of high quality used for the manufacture of pasta is not required for the production of couscous and semolina with inferior quality can be used (Quaglia 1988). In addition, Abecassis et al. (2012) mentioned that the insolubilization of gluten proteins could be associated to crosslinking reaction with the formation of covalent bonds. Even if the proteins are not structures as a continuous network inside the couscous grains (in contrast with the structure of the durum wheat pasta), the crosslinking reactions could contribute to decrease the stickiness of the couscous grains after hydration.

3.5.3 Yellow Pigments

The naturally rich yellow color of the durum endosperm gives couscous its golden colour. Colour is of prime importance in semolina, since the consumer generally expects yellow pasta products. The colour of couscous is the most important factor that consumers associate with quality. Consumers believe that couscous with a golden appearance has higher quality than a paler one. High yellow content is desirable to ensure that the end-product has an intense amber colour. The couscous grains are characterized by a light-yellow color (Guezlane, 1993). According to Lepage and Sims (1968) cited by Trono et al. (1999); Hentschel et al. (2002); Guarda et al. (2004), the yellow color of pasta, made from durum wheat semolina, is due to carotene pigments mainly xanthophylls. The color of semolina is usually expressed using the L^* a^* b^* color system. L^* is a measure of brightness and it can go from 21 to 80. The a^* value vary from −7 to +6 the b^* value range from 21 to 37. For durum wheat semolina the higher the $b*$ value the more yellowness. Good quality durum has a b^* of approximately. 27 or more. The artisanal couscous is characterized by higher b^* (30.7) and L^* (71.3) than commercial couscous ($b^* = 27.1$ and $L^* = 68.9$) because the industrial couscous loses more carotene pigments during processing (Debbouz and Donnelly 1996; Guezlaneet al. 1986).

Kobrehel et al. (1972), Feillet et al. (1974) and Kherrif (1996) demonstrated that couscous and pasta color is affected by the acceleration of the browning products during the hydration process. Rolling and semolina hydration are responsible for a notable oxidation of the carotene pigments through the action of lipases, lipoxygenases, peroxidases and polyphenoloxidases, which cause the development of the brown component that reduces couscous clarity (Kim et al. 1986; Taha and Sagi 1987). In addition, yellowness increases during the precooking of couscous

(Bekradouma 1992; Belaid et al. 1994; Boudreau and Matsuo 1992; Guezlane 1993). Yousfi (2002) observed that samples not precooked lose more carotenoid pigments than precooked ones. In fact, non-precooked industrial couscous lose ~39% of the carotenoid content compared to 33% lost by the precooked one. Similarly, the reduction of carotenoid pigments is estimated at 49% in non-precooked samples against only 33% in precooked ones.

3.5.4 Granulometry

The quality of the semolina used for manufacturing couscous is similar to that required to produce pasta. Beside their physicochemical properties, size and shape of semolina particles are of special importance to determine the properties of couscous. However, Abecassis (1991) and Feillet (2000) reported that often semolina intended for couscous production should have a high granulometry. The granulometry of couscous increases with the semolina hydration (Aluka et al. 1985; Guezlane 1993; Saad et al. 2011). In this context, Senator (1983) observed that the semolina hydration during the manufacture of couscous doesn't only depend on the raw material moisture and its granulometry but also the desired particle of couscous. Guezlane (1993) confirmed this result and demonstrated that the high particle size of couscous is a consequence of high hydration. In addition, Tigroudja and Bendjoudiouadda (1998) indicated that the hydration rate is positively correlated with the equivalent diameter of couscous. Abecassis et al. (1994) demonstrated that the production process mainly affects the cooking quality of couscous, being the two important stages agglomeration and pre-cooking. To obtain a high production yield, it is necessary to achieve a homogenous hydration of the particles to agglomerate. Towards this end a sufficiently high hydration rate and long kneading time are recommended. The Codex Alimentarius (Codex Standard 202–1995) indicates that the particle size of couscous should be between 630 and 2000 μm. Industrial couscous is usually sold under three different types depending on the particles size (fine, medium and coarse).

4 Conclusions

Couscous has a long history and a strong symbolic meaning, symbol of the culinary identity of North African countries, where the couscous plays the role that semolina pasta has in Italy or rice in the SE Asia and the South of China. Couscous is an ethnic food because of the simplicity of the raw materials and recipes, inexpensive processing and preparation, long shelf-life and versatility of manufacturing. Homemade couscous is very complex despite the simplicity of the raw materials. From coarse semolina to dry couscous grains there are several steps and each one depends on several factors including semolina quality.

Acknowledgments The authors would like to greatly acknowledge Miss Baya Mzoughui Hammami who greeted us and agreed to take pictures while she handmade couscous.

References

Abecassis J (1991) La mouture du blé dur. Les industries de première transformation des céréales. In: Godon B. et Willim C (ed) Tec et Doc- Apria, p 362–393.

Abecassis J, Abbou R, Morel MH, Vernoux P (1994) Influence of extrusion conditions speed, temperature and pressure in the extruder and on pasta quality. Cereal chemistry 71:247–253.

Aboubacar A, Hamaker BR (1999) Physicochemical properties of flours that relate to sorghum couscous quality. Cereal Chemistry 76: 308–313.

Aluka K, Miche JC, Faure J (1985) Conditions d'une fabrication mécanique du couscous du maïs en Afrique de l'Ouest. Industries Agricoles et Alimentaires: 448–451.

Anonymous (2013) Moroccan Couscous Traditions & Recipes. Available on. http://travelexploration.com/subpage.cfm/Couscous.

Anga O, Belhouchet L (2002) Granulometrie du couscous: relation avec quelques paramétres de fabrication et la qualité culinaire. Mémoire d'Ingénieur. DNATAA. Université de Constantine. 53 p.

Bahchachi N (2002) Incorporation du gluten de maïs dans la fabrication de deux produits céréaliers traditionnels: Trida et Couscous. Thèse de Magister. DNATAA. UniversitéMentouri de Constantine. 134 p

Banasik OJ (1981) Pasta processing. Cereal Foods World 26: 166–169.

Baum M, Impiglia A, Ketata H, Nachit M (1995) Studies on some grain quality traits in durum wheat grown in Mediterranean environments. In: Di Fonzo N, Kaan F, Nachit M, editors. Durum Wheat Quality in the Mediterranean Region. Zaragoza, Spain: CIHEAM, pp. 181–187.

Bekradouma D (1992) Influence des traitements hydrothermiques de précuisson et de séchage sur la valeur nutritionnelle du couscous de blé dur. MémoireDESS. Montpellier.

Belaid MR, Belarbi A, Khendek D (1994) Rôle des monoglycérides dans l'expression de la qualité du couscous du blé dur. MémoireIngénieur. INA, El Harrach Alger.

Belibagli KB, Vardn H, Dalgiç AC (2009) Application of food safety and quality management systems to burlgur processing. Italian Journal of Food Science. 21: 499–515.

Bellocq B, Thierry R, Cuq B (2018) Contribution of cooking and drying to the structure of couscous grains made from durum wheat semolina. Cereal chemistry 95: 646–659.

Benatallah L, Zidoune MN, Oulamara H, Agli A (2006) Formulation et fabrication de couscous à base de riz et de légumes secs pour malades coeliaques, Actes SAR GP3A, Tunis: 160–164.

Boggini G, Namoune H, Abecassis J, Cuq B (2012) Other traditional durum derived products. In: Durum wheat chemistry and technology, pp. 177–199, 2nd edition. M. Sissons, J. Abecassis, B. Marchylo and M. Carcea Eds. AACC International Press.

Bonjean AP, Angus WJ, Van Ginkel M (2016) Theworld wheat book: A history of wheat breeding Vol. 3. Lavoisier, Paris.

Boudreau A, Matsuo R, Laing (1992) L'industrie des pâtes alimentaires. In: Le blé éléments fondamentaux et transformation Boudreau A. et Menard G (ed) Les presses de l'Université Laval, Canada, pp. 193–223.

Braidwood H, Cambel H, Watson PJ (1969) Prehistoric investigations in south-western Turkey Science 164:1275–1276.

Buche F (2011) Influence de la fermentation de pâtes sur leur consommation d'oxygène et de leur production de dioxyde de carbone au cours du pétrissage et de la fermentation: Conséquences biochimiques et rhéologiques. Thèse doctorale en sciences des aliments. AgroParisTech, France.

Cahier du CEPI N°23. Etude de positionnement stratégique de la branche "pâtes alimentaires et couscous".

Cantrell RG (1987) Breeding and genetics of durum wheat. In: Plant Breeding Reviews Janick J (ed) Van NostrandReinhol Co, New York (5), p 11–40.
Charmet G (2011) Wheat domestication: Lessons for the future. Comptesrendusbiologies 334 (3): 212–220.
Chemache L, Kehal F, Namoune H et al (2018) Couscous: Ethnic making and consumption patterns in the Northeast of Algeria. Journal of Ethnic Foods. 5: 211–219.
Codex standard for couscous 202 (1995). Codex Alimentarius 7: 75-78.
D'Egidio MG, Mariani BM, Nardi S et al (1990) Chemical and technological variables and their relationships: A predictive equation for pasta cooking quality. Cereal chemistry 67: 275–281.
D'Egidio MG, Pagani MA (2010) Pasta and couscous: basic foods of Mediterranean tradition. TecnicaMolitoria International 61: 105–114.
Debbouz A, Dick JW, Donnelly BJ (1994) Influence of raw material on couscous quality Cereal Foods World 39: 231–236.
Debbouz A, Donnelly BJ (1996) Process effect on couscous quality. Cereal Chemistry 73: 668–671.
Derouiche M (2003) Couscous: enquête de consommation à Constantine, fabrication deux processus de fabrication. Mémoire d'Ingénieur. INA, El-Harrach, Alger.
Dexter JE, Marchylo BA, Macgegor AW, Tkachuk R (1989) The structure of protein composition of vitreous, piebald and starchy durum wheat kernels. Journal of Cereal Sciences 10: 19–32.
Dexter JE, Matsuo R R (1980) Relationship between durum wheat protein properties and pasta dough rhéology and spaghetti cooking quality. Journal Agriculture Food Chemistry 36–135–140.
Dexter JE, Williams PC, Edwards NM, Martin DG (1988). The relationships between durum wheat vitreousness, Kernel hardness and processing quality. Journal of Cereal Sciences 7: 169–181.
Dick J W, Matsuo R R (1988) Durum wheat and pasta products. in: Wheat Chemistry and Technology, 3rd ed., Vol. 1. Y. Pomeranz, ed. Am. Assoc. Cereal Chem.: St. Paul, MN: pp. 507–547.
Donnelly B J, Debbouz A, Hagen KR (1994) Couscous. U.S. patent 5,334,407.
Elias M (1995) Durum wheat products. In di Fonzo N. di (ed.), Kaan F. (ed.), Nachit M. (ed.) Durum wheat quality in the Mediterranean region Paper presented on Durum Wheat Quality in the Mediterranean Region seminar, 17–19 Nov 1993, Zaragoza (Spain).
Feillet P (2000) Le grain de blé, composition et utilisation. INRA. Paris. 308 p.
Feillet P, Jeanjean MF, Kobrehel K et al (1974) Le brunissement des pâtes alimentaires. Bull. E.N.S.M.I.C.262: 190–194.
Fuller DQ (2007) Contrasting patterns in crop domestication and domestication rates: recent archaeobotanical insights from the Old World. Annal of Botany (Lond) 100: 903–924.
Fu B X, Wang K, Dupuis B, Taylor D, Nam S (2018) Kernel vitreousness and protein content: Relationship, interaction and synergistic effects on durum wheat quality. Journal of Cereal Sciences 79: 210–217.
Galiba M, Ronney L W, Waniska R D et al (1987). The preparation of sorghum and millet couscous in West Africa. Cereal Foods World 30 (12): 878–884.
Galiba M, Waniska RD, Rooney LW, Miller FR (1988) Couscous quality of sorghum with different kernel characteristics. Journal of Cereal Sciences 7: 183–193.
Goumeziane (2018) Le couscous au cœur de l'identité Maghrebienne. Available on. https://www.middleeasteye.net/fr/opinions/le-couscous-au-c-ur-de-l-identit-maghr-bine-1507772255.
Guarda G, Padovan S, Delogu G (2004) Grain yield, nitrogen-use efficiency and baking quality of old and modern Italian bread-wheat cultivars grown at different nitrogen levels. European Journal of Agronomy 21: 181–192.
Guezlane L, Abecassis J (1991) Méthodes d'appréciation de la qualité caulinaire du couscous du blé dur. Industries Alimentaires et Agricoles 108: 966–971.
Guezlane L (1994) Evaluation et amélioration de la qualité du couscous de blé dur. Séminaire Européen "Qualité du blé dur et des pâtes alimentaires", Montpellier, 26–28 Janvier.
Guezlane L (1993) Mise au point de méthodes de caractérisation et étude des modifications physico-chimiques sous l'effet des traitements hydrothermiques en vue d'optimiser la qualité du couscous de blé dur. Thèse de Doctoratd'Etat. INA, El Harrach, Algérie. 89 p.

Guezlane L, Selselat-Attou G, Senator A (1986) Etude comparée du couscous de fabrication industrielle et artisanale. Industrie des Céréales 43: 25–29.

Hebrard A, Oulahna D, Galet L, Cuq B, Abecassis J, Fages J (2003) Hydration properties of durum wheat semolina: influence of particle size and temperature. Powder Technology 130: 211–218.

Hentschel V, Kranl K, Hollmann J, Lindhauer MG, Bohm V, Roland B (2002) Spectrophotometric determination of yellow pigment content and evaluation of carotenoids by high-performance liquid chromatography in durum wheat grain. Journal Agriculture et Food Chemistry 50: 6663–6668.

International Association for Cereal Science and Technology (1999) Standard method 129. Method for determination of vitreousness of Durum Wheat. ICC: Vienne Austria

International Wheat Council (1991) World Grain Statistics. London.

Jude (2004) Projet de diagnostic de la compétitivité de la branche industrielle des pates alimentaires et couscous au Maroc pour l'étude pâtes et couscous AMIPAC.

Kaup SM, Walker CE (1986) Couscous in North-Africa. CFW31: 179–182.

Kherrif A (1996) Effet de la variabilité protéique sur l'expression de la qualité technologique du couscous. MémoireIngénieur. INA, El-Harrach, Alger. 61 p

Kill R, Turnbull K (2008) Pasta and semolina technology Ed. John Wiley and Sons ISBN 978-0-4-470-99936-3.

Kim H, Sieb P, Deyde G, Yang H (1986) Milling hard red winter wheat to farina: Comparison of cooking quality and color of farina and semolina spaghetti. Cereal Foods World 31: 810–819.

Kobrehel K, Laignelet B, Feillet P (1972) Relation entre les activités peroxydasiques et poly phénol oxydasiques des blés durs et le brunissement des pâtes alimentaires. Comptes rendues de l'académie d'agriculture de la France 58: 1099–1106.

Lefkir S (2005) Influence des conditions de l'hydratation sur la qualité technologique du couscous. Mémoire Magister. INA, El-Harrach, Alger. 100 p.

Lefkir S, Yahiaoui K, Yesli A, Ounane G (2017) Hydratation rate influence on the couscous quality. Journal of Food, Agriculture and Environment 15: 5–11.

Lepage M, Sims RPA (1968) Carotenoids of wheat flour: their identification and composition. Cereal chemistry 45: 600–604.

Liu CY, Shephred KW, Rathjen AJ (1996) Improvement of durum wheat pastamaking qualities. Cereal Chemistry 73: 155–166.

Lucisano M, Pagani MA, Mariotti, Locatelli DP (2008) Influence of die material on pasta characteristics. Food Research International 41: 646–652.

Mac Key J (2005) Wheat, its concept, evolution and taxonomy. In: Royo C, Nachit M, Di Fonzo N, Araus JL, Pfeiffer WH, Slafer GA (eds), Durum wheat breeding: current approaches and future strategies, Ed. Vol. I Howorth Press, New York, pp. 3–62.

Manthey FA (2002) Processing and Quality of Wholewheat Pasta. P: 139–154 in: Whole Grains in Health & Disease. L. Marquart and G. Fulcher, eds. Am. Assoc. Cereal Chemistry. St. Paul, MN.

Miskelly D (2017) Optimization of End –product quality for the consumer. p: 653–688 In: Cereal Grains, 2nd edition. Assessing and Managing Quality: ColinWrigley, Ian Batey, Diane Miskelly. p 830.

Morris CF (2002) Puroindolines: themoleculargeneticbasis of wheat grain hardness. Plant Molecular Biology 481: 633–647.

Morris CF, Bhave (2008) Reconciliation of D-genome puroindoline allele designations with current DNA sequence data. Journal of Cereal Sciences 48: 277–287.

Morris CF, Fuerst EP (2015) Quality characteristics of soft kernel durum; A New cereal crop. Advances in Wheat Genetics: from Genome to Field: 275–278.

Namoune H, Kezih R, Feliachi K, et al (2004). The effect of using some fats during the cooking process on the quality of couscous. The Fourth Scientific Conference of Agricultural Sciences in Egypt.

Novaro P, D'Egidio MG, Mariani BM, Nardi S (1993) Combined effect of protein content andhigh temperature drying systems of pasta cooking quality. Cereal chemistry 70: 7 16–719.

Owens G (2001) Cereals processing technology Eds. Woodhead publishing ISBN 9781855736283: pp. 1–249.
Ounane G, Cuq B, Abecassis J, Yesli, A, Ounane SM (2006) Effects of 614 physicochemical characteristics and lipid distribution in Algerian durum wheat 615 semolinas on the technological quality of couscous. Cereal Chemistry 83: 377–384.
Özkaya B, Özkaya H, Eren N, Ünsal A. S, Köksel H (1999) Effects of wheat maturation stage and cooking method on physical and chemical properties of firiks. Food Chemistry 66: 97–102.
Palumbo M, Spina A, Boggini G (2000) Agronomic and breadmaking characteristics of durum wheat genotypes deriving from interspecific hybridization with bread wheat. Paper presented at the Durum Wheat Improvement in the Mediterranean Region: New Challenges". Zaragoza, Spain, 12–14 April 2000.
Pollini CM, Pantò F, Nespoli A, Sissons M, Abecassis J (2012) Manufacture of Pasta Products. In: Durum wheat chemistry and technology, pp. 161–176, 2nd edition. M. Sissons, J. Abecassis, B. Marchylo and M. Carcea Eds. AACC International Press.
Quaglia GB (1988) Other durum wheat products. In: Durum Chemistry and Technology, Fabriani, G. and Lintas, C. (eds). AACC", St. Paul, Minnesota, pp. 263–282.
Ranieri R (2015) Geography of the durum wheat crop. Pastaria International 6: 24–36.
Saad M, Barkouti A, Rondet E, Ruiz T, Cuq B (2011) Study of agglomeration mechanisms of food powders: Application to durum wheat semolina. Powder Technology 208: 399–408.
Sada L (1982) Grano e panedellaPuglia. Panif. Pasticceria 10: 47–45.
Samson MF, Mabille F, Cheret R, Abecassis J, Morel MH (2005) Mechanical and physicochemical characterization of vitreous and mealy durum wheat endosperm. Cereal Chemistry 82: 81–87.
Senator A (1983) Contribution à l'étude de la valeur couscoussière. Comparaison entre deux processus de fabrication. Mémoire Ingénieur. INA, El Harrach, Alger.
Shahin M, Symons SJ (2008). Detection of hard vitreous and starchy kernels in amber durum wheat samples using hyperspectral imaging (GRL Number M306). NIRS News (5): 16–18.
Sieber A, Wurschum T, Longin CFH (2015) Vitreosity, its stability and relationship to protein content in durum wheat. Journal of Cereal Sciences 61: 71–77.
Sissons, M. (2016). Pasta. In: Wrigley, C., Corke, H., and Seetharaman, K., Faubion, J., (eds.) Encyclopedia of Food Grains, 2nd Edition, pp. 79–89 Oxford: Academic.
Taha SA, Sagi F (1987) Relationships between chemical composition of durum wheat semolina and macaroni quality. II Ash, carotenoid pigments and oxidation enzymes, Cereal Research Communicationd 5: 123–129.
Taylor RD, Koo WK (2015) Outlook of the U.S and World Wheat Industries, 2015–2024. Center for agriculture policy and trade studies department of Agribusiness and applied economics No. 738 North Dakota State University.
Tigroudja F, Bendjoudiouadda A (1998) Influence de la granulométrie de la semoule sur la qualité technologique du couscous artisanal de blé dur. Mémoired'Ingénieurd'Etat. INA El Harrach, Alger. 85 p.
Turnbull K, Kuenzli, Willis M, Giles J (2001) Advances in Durum milling. In: R Kill & K Turnbull (eds.). Pasta and Semolina Technology. Oxford, Blackwell Science Ltd. p. 43–85.
Toscano P, Gioli B, Genesio I, Vaccari FP, Miglietta F, Zaldei A et al (2014) Durum wheat prediction in Mediterranean environments: From local to regional scaleEuropean Journal of Agronomy 61: 1–9.
Trono D, Pastore D, Difonzo N (1999) Carotenoid dependent inhibition of durum wheat lipoxygenase. Journal of Cereal Sciences 29: 99–102.
Williams PC (1985) Survey of wheat flours used in the Near East. Rachis 4: 17–20.
Williams PC, Srivastava JP, Nachit MM, El Haramein FJ (1984) Durum wheat quality evaluation at ICARDA. Rachis 3: 30–33.
Yousfi L (2002) Influence des conditions de fabrication et des modes de préparation sur la qualité du couscous industriel et artisanal. Thèse de magister. Université Mentouri Constantine, Algérie. 140 p.

Understanding the Mechanics of Wheat Grain Fractionation and the Impact of Puroindolines on Milling and Product Quality

Valerie Lullien-Pellerin, Réka Haraszi, Robert S. Anderssen, and Craig F. Morris

Wheat grain milling has for aim to isolate the smaller starchy endosperm particles (*i.e.* flours, semolina) from the larger fragments of peripheral tissues. The mechanical properties of the wheat grain tissues strongly influence how effective the steps of grinding and sieving are during this fractionation process. The grain mechanical resistance determines how much energy is required to fracture it, the particle size of the resulting products, and their biochemical composition. Therefore mechanical properties affect both the durability and the quality of the processed products. Genetic loci, and more precisely the key role of the *Hardness* (*Ha*) locus in the D genome of common wheat (*Triticum aestivum*), are well established determinants of the mechanical properties and behavior of grain, which are also influenced by environmental factors. The key role of genes encoding puroindolines has been confirmed by extensive analysis of mutants and through genetic manipulation. Methods of measuring mechanical resistance are being reconsidered because grain hardness needs to be characterized in ways that capture the different contributions of genetic and environmental factors. In particular, methods to acquire data on the mechanical resistance of each of the grain tissues and their components have been developed. Finally, the promise of using numerical modelling to better understand and predict the effect of changes in the wheat starchy endosperm composition will be discussed.

V. Lullien-Pellerin (✉)
IATE, University Montpellier, CIRAD, INRAE, Montpellier SupAgro, Montpellier, France
e-mail: valerie.lullien-pellerin@inrae.fr

R. Haraszi
Campden BRI, Station Road, Chipping Campden, Gloucestershire, UK

R. S. Anderssen
Data 61, CSIRO, Canberra, Australia

C. F. Morris
USDA-ARS, Western Wheat Quality Lab, Washington State University, Pullman, WA, USA

G. Igrejas et al. (eds.), *Wheat Quality For Improving Processing And Human Health*, https://doi.org/10.1007/978-3-030-34163-3_16

1 Introduction

Wheat cultivation began around 10,000 years ago when human populations made the transition from food gathering to settled agriculture in the Neolithic period (Dubcovsky and Dvorak 2007). Wheat is now the second most cultivated cereal in the world with around 700 million tons harvested annually. Wheat flour is used to make many different products valued for their nutritional energy, specific textures and taste properties, like bread, cakes, biscuits, noodles, and pasta. All these food products rely on wheat grain fractionation, the mechanical fracturing and separation of grain components to access the starchy endosperm. The starchy endosperm, containing starch granules and storage proteins, makes up 80–85% of the grain dry mass, and the germ, containing the embryo, constitutes 2–3% of the grain dry mass (Evers et al. 1999). Covering the starchy endosperm and of the same cellular origin is the aleurone layer that contains a functionally distinct set of proteins, mainly enzymes involved in grain germination, as well as minerals and micronutrients (Antoine et al. 2002; Brouns et al. 2012). The entire grain is protected from micro-organisms, insects and water with a number of layers (13–17% of the grain dry mass) going from the hyaline tissue to the outer pericarp.

The first milling step is the grinding of grain on stone or metal rollers, followed by the classification of the obtained particles according to their size, and in some cases also their density (Campbell 2007a; Gruber and Sarkar 2012). This process allows most of the starchy endosperm to be isolated as flours (the finest particles) or semolina, from the other tissues remaining in the wheat shorts and brans (the largest particles). The sizes of the ground particles depend on the intrinsic mechanical properties of each tissue in the original grain that in turn depend on the tissue structure, composition and water content. Size differences in particles from the starchy endosperm and the outer layers were indeed enhanced following a water-tempering step (Posner 2009), i.e. conditioning grain by increasing the water content by around 3–5%. However, it is difficult to separate endosperm particles from those originating only from the outer layers on size alone. Possibly adhesion between tissues is a factor but this aspect has not yet been characterized. Relative proportions of flour and semolina fractions were found to be higher in common wheats (*T. aestivum* L., genotype AABBDD) than in durum wheats (*T. durum* Desf., genotype AABB), the former being less mechanically resistant than the latter. The common wheats can be further distinguished as being soft or hard types depending on the mechanical resistance of grains. These differential mechanical properties mean that wheat genotypes are suited for certain food end-products, even though there is some additional variability in mechanical resistance due to the environment (Peña et al. 2002). Very early on in cereal research, differences in wheat grain hardness were shown to be related to the nature of the interface between starch granules in the endosperm and the protein network rather than to differences in the main grain constituents (Barlow et al. 1973; Simmonds et al. 1973).

2 Puroindolines and the *Hardness* Locus

Symes (1965) is credited as being among the first to conclusively show that a single major genetic factor determines the development of either soft or hard endosperm in common wheat. In a landmark study, grain texture was quantified using a particle size index, where the mechanical resistance of grain was related to the particle size distribution after grinding (Symes 1965), a method which was later normalized (Williams and Sobering 1986). Mattern et al. (1973) showed that when chromosome 5D from the hard wheat cultivar Cheyenne was substituted in the soft variety Chinese Spring, the grain of the progeny behaved like Cheyenne when milled. In 1978, Law et al. localized the locus to the short arm 5DS and designated it *Hardness (Ha)*. It was further shown that an ancestral *Ha*-like gene emerged between 60 and 50 million years ago when the Pooideae (Triticeae and Brachypoidieae) diverged from the Panicoideae (Charles et al. 2009). Further genome rearrangement occurred which notably led to a hard grain phenotype in common wheat (Chantret et al. 2005).

Greenwell and Schofield (1986) described a positive association between the *Ha* locus and a protein fraction around 15 kDa recovered from the surface of water-washed starch granules that was later called "friabilin" due to its supposed "non-stick" role (Morrison et al. 1992). In common wheat, more of this protein fraction was recovered from the starch granule surface of soft genotypes compared with hard genotypes (Darlington et al. 2000). Over the subsequent decade, the friabilin fraction was found to be composed of different proteins (Morris et al. 1994; Morris 2002; Oda and Schofield 1997) with two of them, puroindoline A (PINA) and puroindoline B (PINB), sharing N-terminal tryptophan-rich sequences and a common pattern of cysteine residues (Blochet et al. 1993; Gautier et al. 1994, Rahman et al. 1994). Another protein evolutionarily close to the puroindolines, the so-called Grain softness protein-1 (GSP-1), may be co-purified with the PINs and its gene was mapped to the *Ha* locus next to the puroindoline genes. However, while corresponding genes have been found in all A, B and D genomes, GSP has not been shown to play any notable role in kernel texture variation (Bhave and Morris 2008; Morris et al. 2013).

2.1 *Diversity of* Pin *Alleles*

In *T. aestivum*, the soft phenotypes were found to contain both of the wild-type alleles *Pina-D1a* and *Pinb-D1a* of the genes encoding PINA and PINB respectively, whereas different mutations within these genes all result in harder texture (Bhave and Morris 2008; Giroux and Morris 1997, 1998; Morris and King 2008). A number of different sequences of puroindoline genes have been recorded in Morris and Bhave (2008), as summarized in Table 1. Markedly more sequence variants exist in *Pinb-D1* than in *Pina-D1*, indicating that *Pinb-D1* is likely to be under less stringent regulatory control.

Table 1 Most frequent alleles (x and y types) of genes encoding PINA (*Pina-D1x*) and PINB (*Pinb-D1y*) in wheat from Europe and North America with corresponding mutations (in the amino-acid sequence or in the open reading frame, ORF) and wheat grain hardness phenotypes

Pina-D1x-Pinb-D1y	x	y	Mutation description	Phenotype
	a	a	None	Soft
	b	a	Deletion, no PinA expression	Hard
	a	b	Amino acid substitution G46S	Hard
	a	c	Amino acid substitution L60P	Hard
	a	d	Amino acid substitution W44R	Hard
	a	e	Stop codon at position 202 in the ORF	Hard
	a	f	Stop codon at position 217 in the ORF	Hard
	a	g	Stop codon at position 253 in the ORF	Hard

The single and predominant mutation of *Pina-D1* is a null mutation (*Pina-D1b*) that results in the absence of PINA from the endosperm (Morris and King 2008).

A novel puroindoline variant, *Puroindoline b-2,* was recently identified on the homoeologous group 7 chromosomes in bread wheat but was shown to not significantly influence kernel texture (Chen et al. 2013; Geng et al. 2012).

2.2 Characterization of Puroindoline Biochemistry, Distribution and Properties

PINA and PINB belong to a large set of low molecular weight cysteine-rich proteins that interact *in vitro* with lipids or membranes and that have antimicrobial properties (Lullien-Pellerin and Marion 2002). PINA and PINB are both basic proteins with isoelectric points around 10–11 (Branlard et al. 2003) and the amino acid sequences are 60% similar (Gautier et al. 1994). They each contain ten cysteine residues that are arranged to form five disulfide bonds that stabilize the three-dimensional structure (Le Bihan et al. 1996). They both have a specific tryptophan-rich domain between amino acids 39 and 45 near the N-terminus, somewhat truncated in PINB (Gautier et al. 1994).

These tryptophan-rich sequences appear to be important for the puroindoline function in grain hardness as mutations in these regions lead to a change in grain texture (Feiz et al. 2009). Overexpression of the gene encoding PINB was shown to increase the total amount of puroindolines attached to starch and decrease grain hardness more than the overexpression of PINA (Swan et al. 2006). This observation confirmed that PINB has the predominant role in puroindoline association with starch. However, Capparelli et al. (2003) found that expression of PINA was essential for puroindolines to bind starch.

Puroindolines are synthesized as preproproteins with N- and C-terminal propeptides characteristic of signal peptides targeting proteins to subcellular structures (Gautier et al. 1994). Recently, Lesage et al. (2011) detected puroindolines in the protein bodies of immature wheat grains, confirming that cellular trafficking occurs during grain maturation. Both PINA and PINB are only found in the mature seeds, accumulating in the starchy endosperm as detected by the use of specific antibodies or deduced from the expression of the PIN gene promoters coupled to a reporter gene (Capparelli et al. 2005; Dubreil et al. 1998; Wiley et al. 2007). However, there are some reports that puroindolines, especially PINB, are located in other grain tissues (Digeon et al. 1999; Dubreil et al. 1998; Capparelli et al. 2005).

Darlington et al. (2000) described puroindoline distribution within the starchy endosperm as being more abundant at the starch granule surfaces in soft wheat phenotypes. Wall et al. (2010) identified the tryptophan-rich peptide of PINB by mass spectrometry after *in situ* trypsin digestion of intensively washed starch granules. This suggests that this peptide sequence contributes to starch adhesion. PINA and PINB were also found to interact with each other, both *in vitro* using purified proteins or *in vivo* using the yeast two-hybrid system (Alfred et al. 2014; Geneix et al. 2015).

The mechanism by which puroindolines could act as a non-stick factor between starch granules and the protein matrix is still largely unresolved but they are thought to bind starch either directly or through polar lipids. Puroindolines and polar lipids were found at the starch surface and while it seems that if PINs need lipids to bind with the starch granule, lipids can bind starch independently (Pauly et al. 2014). Nevertheless the content of polar lipids associated with the starch surface was found to be dependent on the presence of puroindoline and the type of alleles (Finnie et al. 2010; Greenblatt et al. 1995).

Mechanical measurements using atomic force microscopy (AFM) showed that starch was four-fold harder than gluten in wheat grains *in situ*, whereas both polymers behave similarly whatever the wheat genetic background (Chichti et al. 2013). Comparing the biological materials to minerals, the mechanical hardness of gluten appears more similar to soft minerals such as talc, whereas starch is as resistant as calcite. Therefore, differences in the mechanical properties of grain must be due either to the way starch granules and the filling protein matrix are organized or to the different forces acting at the interfaces between starch granules and the protein matrix. Further AFM experiments indeed revealed that the interface between starch granules and the protein matrix in a soft wheat background behaved like a sliding surface. This distinctive behavior was observed at particular sites on the starch surface suggesting that a lipid interface was acting as a lubricant which would be consistent with both the non-continuous local distribution of puroindolines at the starch surface (Darlington et al. 2000) and puroindoline interactions with lipids (Chichti et al. 2015). Recently, contact resonance AFM was also used to measure the contact modulus of starch granules *in situ*, pioneering the nano-scale study of the mechanical properties of starch sub-structures (Heinze et al. 2018).

2.3 Demonstration of Puroindoline Involvement in Hardness

The effects of wild-type or mutated puroindoline alleles were clearly demonstrated by comparing the soft or hard phenotypes observed in near-isogenic lines (NIL) differing only in these genes (Rogers et al. 1993; Morris et al. 2001). The role of puroindoline was further demonstrated by complementation of null (Wanjugi et al. 2007) or mutated alleles (Beecher et al. 2002) with the wild-type corresponding allele. Interestingly, the introduction of PIN genes in other cereals lacking equivalent genes, such as durum wheat (Morris et al. 2011), rice (Krishnamurthy and Giroux 2001) or maize (Zhang et al. 2009), was found to reduce grain hardness in the progeny. Gasparis et al. (2011) also showed that RNA interference mediated silencing of one of the PIN genes decreased the expression of the other PIN gene and led to a significant increase in grain hardness.

2.4 PIN Expression during Grain Maturation

Transcripts for puroindoline genes accumulated between 8 and 12 days after flowering and increased markedly to reach a peak before maturation and then declined (Gautier et al. 1994). The corresponding proteins were also detected within the first 10 days after flowering and increased steadily until maturation (Turnbull et al. 2003), with more of both PIN proteins being found in soft compared to hard cultivars. Kim et al. (2012) also described the effect of both PINA and PINB in preventing polar lipid breakdown during seed maturation.

3 Evaluation of Grain Hardness and Prediction of Milling Behavior

With the increasing need to develop predictors of wheat grain processing properties and end-product quality, various protocols to characterize and measure grain hardness have been proposed over the years. Long before the *Ha* locus was characterized and the potential involvement of puroindolines was outlined, Cobb (1896) recognized the value in objectively ascertaining the mechanical resistance of grain and devised a machine with tooth-like "pinchers" to mimic the bite test as a way of classifying wheats according to how much force was necessary to cut the grain in two. Cobb's study (1896) of various wheat varieties illustrates two key points. First, wheat varieties differ in their resistance to mechanical load. Second, this "grain hardness" is related to the grain texture and can only be defined phenotypically.

The phenotype of grain mechanical resistance is in fact commonly measured only indirectly after grinding grain and sieving the fragments through a 75-mm sieve, then establishing the percentage of mass recovered after sieving as the particle size index (PSI) score (Williams and Sobering, 1986). Durum or hard common wheat grains have PSI scores range between 1 and 20, whereas soft common wheat grains have values higher than 20. Near-infrared reflectance spectroscopy (NIRS) methods have been developed as a rapid way to further classify different wheats according to their PSI hardness (Saurer 1978; Norris et al. 1989). To do this, the NIRS values of a set of representative wheats were used to calibrate the grain hardness predictor against which other grain are compared, so is also an indirect method. Like PSI, the basis of the NIRS method exploits the inherent differences in particle size distribution when wheat grain is ground into meal. Therefore wheat grain hardness classification was based on flour production and is an indirect measurement of the physical properties of the grain.

The development of the single kernel characterization system (SKCS) led to a more direct measurement of mechanical resistance to wheat grain rupture (Martin et al. 1993) through a hardness index (HI) based on measuring the incremental change in the force with which a grain resists crushing. Unlike other measurement methods it records data on each individual grain. Classification as soft, hard or in-between types was based on the distribution of data from a population of individual grain measurements on a scale that correlates well with the ranges of PSI and NIRS scores (Osborne and Anderssen 2003). A numerical assessment of the hardness of a set of representative wheats was also done against which other samples could be classified (Morris and Massa 2003). Campbell et al. (2007b) found that SKCS HI (hardness index) is meaningful when considering wheat breakage during roller milling and proposed adding kernel mass to predictions to account for the effect of kernel shape on breakage. The authors also showed that particle size distribution could be predicted accurately from the wheat hardness values obtained (Campbell et al. 2012).

Gaines et al. (1996) used grain crushing data from the SKCS force-deformation profile to calculate "softness equivalent" (SE), another value which unlike SKCS HI reflects the influence of kernel moisture and size. Regression analysis was proposed as a way to predict milling properties from samples of only 300 kernels, and was shown to be particularly useful for soft wheat. The authors were the first to challenge the claimed relationship between milling behaviour and the historic classification of wheat hardness. Using the same parameters from the crush response curve, Osborne et al. (2001) defined similar regression equations to predict values for the maximum stress, the work required to reach the maximum stress, and the modulus of elasticity measured as described by Delwiche (2000) using endosperm samples.

Osborne and Anderssen (2003) summarized the diverse applications of SKCS data for characterizing samples, studying grain variability within and between heads, and predicting milling value. They also developed the idea of using the average overall crush-response profile obtained with an SKCS as a rheological signature of the material relating mechanical information on both the shell and the endosperm properties (Fig. 1).

Osborne et al. (2007) established that an average SKCS crush curve can be used to determine endosperm strength and stiffness so wheat samples can be ranked according to milling performance and flour extraction. Calculation of the rheological

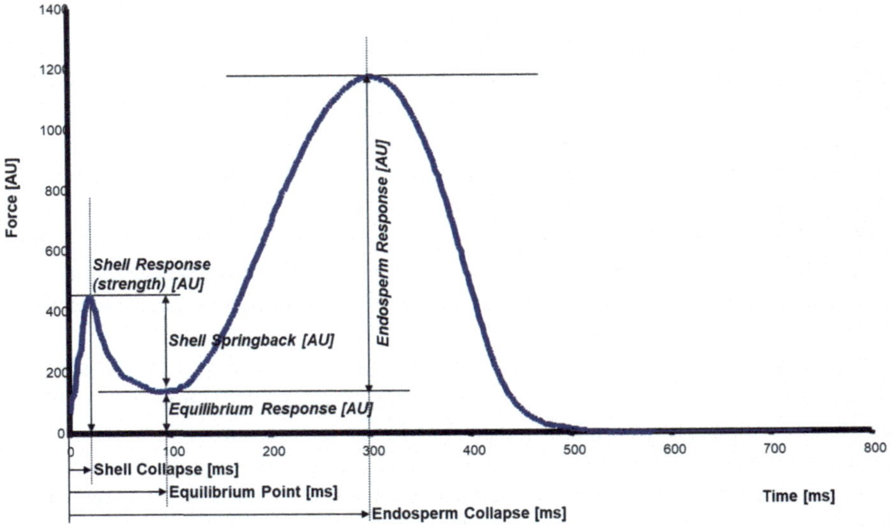

Fig. 1 Typical wheat grain crush curve recovered with a single kernel characterization system and describing the relative force applied depending on time before grain collapse from Anderssen and Haraszi (2009)

parameters of the shell and endosperm from an average SKCS crush curve was further detailed by Anderssen and Haraszi (2009). These parameters were used as potential predictors of the elastic, fragmentation and viscoelastic phases of grain crushing and some were used to establish a prediction model of durum wheat milling behavior (Haraszi et al. 2016). Analysis of crush response profiles led to the conclusion that shell elasticity plays no role in the determination of hardness (Anderssen and Haraszi 2009), but the magnitude of the compressive collapse is directly related to endosperm collapse and repacking, which are quite different for hard and soft wheats.

The geometry of wheat grain structure is complex. Not only is there a number of different tissue layers but there is also a crease, so it is not immediately evident how to interpret grain crush curves in terms of the different grain components. For example, Misailidis and Campbell (2013) interpreted the crush force profile taking account of the influence of the crease, but this approach has been challenged (Anderssen and Haraszi 2013). Anderssen and Haraszi (2009) noted the importance of grain porosity in determining the point of collapse, a factor which was further studied by Oury et al. (2015). A newly defined grain characteristic was proposed, the rheological hardness index (RHI) that is equal to the ratio of the maximum resistance of the endosperm (along the y axis of the crush curve) to the time taken to attain the maximum (Haraszi et al. 2013). RHI was validated as a better way to differentiate between wheat grains according to their brittleness and compactness although there is some overlap between different classes (Haraszi et al. 2013).

Two factors influence the mechanical resistance of grain to rupture. One is the genetic control exerted by the hardness locus *Ha,* which as discussed above, relates to the force of adhesion between starch granules and the protein matrix. The other factor appears to depend on the environmental conditions of plant growth (Oury et al. 2015) and affects the endosperm porosity. Both genetic and environmental factors were shown to influence the mechanical behavior of endosperm during milling (Greffeuille et al. 2006, 2007a; Oury et al. 2017). Indeed, durum and soft or hard common wheat samples can be distinguished by their degree of porosity (Dobraszczyk et al. 2002). The most porous endosperm (mealy) appears white while the most compact (vitreous) appears yellow due to differences in light transmittance through the endosperm which was recently found to obey to the linear relationship described by the Beer-Lambert law (Chichti et al. 2018). In a study using NIL differing only in grain hardness, Oury et al. (2015) demonstrated that grain HI values depended both on genetic factors and environmental ones that affect the porosity. For instance, SKCS HI values between 30 and 50 could correspond to either soft compact grains or hard porous grains, explaining why the different classes overlap when numerical values are used. This can also explain why Osborne et al. (2001) found that the grain hardness of Australian wheat lines evaluated with SKCS was not only associated with the classical *Ha* locus. Oury et al. (2015) showed that the vitreousness of soft grains in which wild-type puroindolines are expressed did not exceed 60% indicating that these proteins lead to an increase in grain porosity. Morris and Beecher (2012) also showed that the *Ha* locus does have a prominent role in determining how vitreous or non-vitreous endosperm may be. Therefore, SKCS HI values for grains from soft NIL were all lower than 45 (Oury et al. 2015). Oury et al. (2017) highlighted the importance of monitoring vitreousness, i.e. the reverse of porosity, either indirectly through SKCS measurement or directly by visual assessment. These authors demonstrated the influence of vitreousness on milling performance together with genetic hardness due to the puroindoline alleles. Values obtained from the SKCS crush curves with durum wheat grains were also found to relate to vitreousness and to be potentially useful in predicting semolina yield (Sissons et al. 2000).

Ways to measure the intrinsic mechanical properties of each tissue more specifically were also developed to better characterize the effect of processing conditions and the influence of biochemical composition and structure. In one approach, millimetre-scale parallelepiped 'bricks' were carved directly from grain endosperm (Haddad et al. 1998). By clearly defining the geometry of the sub-sample, the endosperm mechanical properties could be characterized more accurately using SI units rather than an arbitrary scale. This approach was used to confirm the influence of genetic hardness and vitreousness on the mechanical properties of the starchy endosperm using wheat NIL encoding wild-type or mutated PINB grown in two contrasting environments (Greffeuille et al. 2006). Geometrically defined cylinders of endosperm were also useful for studying how the mechanical properties vary with different water content and for distinguishing between soft and hard genotypes (Delwiche et al. 2012). However, special care has to be taken in calculating and comparing wheat endosperm compressive strength as it may depend on the rigidity of the instrument used and specimen geometry (Delwiche et al. 2012).

By dissecting the wheat grain, the distinct mechanical properties of each grain outer layer were determined (Antoine et al. 2003) and were shown to depend on both the sample water content and the applied temperature along the assay (Hemery et al. 2010; Mabille et al. 2001). Distinct intrinsic mechanical properties of the outer layers were measured between wheat NIL differing in hardness, but it is not clear whether these differences were related to puroindoline expression (Greffeuille et al. 2007b). The mechanical properties of the outer pericarp were found to depend on cell orientation whereas the other outer layers showed an isotropic character (Antoine et al. 2003). Moreover differences in the mechanical properties between layers and genetic origins were correlated to the biochemical composition of their cell walls (Antoine et al. 2003; Greffeuille et al. 2007b).

4 Consequences on Milling Value, Energy Expenditure and Product Properties

NIL differing in the Pinb allele, and thus the hardness of the harvested grain, were grown in a defined environment leading to distinct levels of vitreousness (*i.e.* porosity) so the impact of either the genetic or the environment effect on milling performance could be clearly monitored. These factors were indeed found to affect the energy required to grind a grain population using an instrumented micromill equipped with on-line torque transducers (Greffeuille et al. 2006), in accordance with the SKCS measurement (Oury et al. 2015). Both the presence of wild-type puroindolines and higher endosperm porosity (as evaluated by changes in light transmittance) reduced the amount of energy needed for grinding (Table 2). Additionally, the introduction of wild-type alleles of both puroindoline genes into a durum wheat background confirmed their positive impact on grinding energy reduction (Heinze et al. 2016).

Both genetic and environmental factors also influence flour production. Producing flour from a hard grain genotype with vitreous endosperm is the least efficient, or in other words, to produce 1 kg of flour from this sample requires the most energy (Table 2). The presence of wild-type or mutated puroindoline alleles also affects the

Table 2 Total mechanical energy of grain first break (E1), first flour yield, energy required to produce one kg of first flour (K′) and starch damage measured after milling of mealy and vitreous grain from soft and hard wheat near-isogenic lines differing only in their Pinb-D1a (Soft) or Pinb-D1b (Hard) alleles. Vitreousness of the samples was described in Chichti et al. (2018). Milling was performed on an instrumented micromill equipped with on-line torque transducers to measure the breaking energy (Pujol et al. 2000). Starch damage in first flour was measured with a Megazyme kit (K-SDAM starch damage assay kit, Megazyme Int., Ireland) according to method AACCI N°76–31.01

Grain types	Mealy Soft	Vitreous Soft	Mealy Hard	Vitreous Hard
E1 (kJ/kg)	9.8	10.5	11.6	13.5
First flour (%)	12.3	10.1	13.7	8.1
K′ (kJ/kg flour)	80.0	105.9	85.7	169.3
Starch damage (%)	1.9	2.2	3.4	5.2

granulometric properties of flour (Greffeuille et al. 2006). The particle size (i.e. volume) distribution of flour from a soft wheat NIL was bimodal, with the peak corresponding to smaller particles reflecting the size of starch granules when puroindolines act as a non-cohesive factor between starch granules and the protein matrix. Conversely, the hard NIL displayed a monomodal distribution reflecting how the protein matrix adheres to starch granules. Vitreousness also slightly modulates the particle size distribution (Greffeuille et al. 2006). Due to the differential adhesion within the grain structure and energy required for breaking, the level of starch damage in the flours obtained from different genotypes grown in contrasting environments increases from mealy soft grains to vitreous hard ones (Table 2).

Similar increases in flour production, changes in flour particle size distribution, and lowering of starch damage due to the presence of both puroindolines were demonstrated (Heinze et al. 2016) using a recombinant durum wheat cultivar into which the wild-type *Pina* and *Pinb* alleles were introduced (Morris et al. 2011).

Genetic hardness has an impact on the separation of endosperm from the outer layers. Soft genotypes produce a higher mass of coarse bran after milling compared to hard genotypes (Greffeuille et al. 2005, 2006; Heinze et al. 2016; Oury et al. 2017). Whatever the overall background, the grain fractures according to the puroindoline genotype either in the sub-aleurone part of the endosperm for wild-type alleles, or at or inside the aleurone layer for mutated alleles. This result explains why coarse brans from hard or durum genotypes contain less starch whereas their flours contain more phytic acid, a marker of the aleurone cell content (Greffeuille et al. 2005; Heinze et al. 2016). On the other hand, coarse bran from the genotypes expressing wild-type puroindolines contains non-negligible amounts of starch. Particle size of coarse bran from soft grains was also found to be larger than that from hard grains, a characteristic correlated with the phytic acid content in the respective flours (Greffeuille et al. 2005). Nevertheless, it remains difficult to determine whether the different grain behaviors and the properties of the outer layers are solely the result of their intrinsic mechanical properties (Greffeuille et al. 2007b) or are due to differences in mechanical behavior of the endosperm (Greffeuille et al. 2006; Heinze et al. 2016) or a combination of both.

5 Modelling the Grain Using Micromechanics Data and Simulation

A mathematical model of the wheat endosperm was constructed in order to predict its mechanical behavior. This model, built using the lattice element method (Topin et al. 2007, 2008, 2009), is based on a cohesive granular medium and considers the distinct properties of the starch, protein matrix, and void phases and varying forces acting at the interface between starch and the protein matrix. The model was shown to reflect the different modes of endosperm breakage well and simulate how the level of starch damage depends on the adhesion force at the starch and protein network interface.

The formation of force chains in a cemented granular material during mechanical crushing was used to explain the crushing mechanics of the wheat endosperm only and did not include any effects of the outer layers at this step. This model was further improved using data pertaining to the intrinsic mechanical properties of starch and the protein matrix, measured *in situ* with AFM (Chichti et al. 2013). The model was found to agree with experimental tests, particularly in modelling the elastic response and predicting endosperm failure (Chichti et al. 2016). A new peridynamic modelling approach was recently developed to study the effect of starch granule polydispersity on yield stress and failure properties (Heinze et al. 2017). It helped to further highlight the importance of starch-protein adhesion and porosity, respectively linked to genetic hardness and environmental conditions, far beyond their effects on the starch granule size distribution. Therefore, modelling has helped to enlarge our understanding of natural wheat grain variability which has made it possible to extrapolate and test the effect of different endosperm features on mechanical behavior. Finally, this modelling is a promising tool to guide the selection of plants for cultivation or breeding with particular milling performance or end-use qualities in mind.

6 Conclusions

In recent years, evidence for the role of puroindolines in determining the mechanical resistance of grain has cumulated through the correct identification of the different alleles and the construction of NIL or transgenic plants. Moreover, detailed mechanical studies on individual grains or isolated tissues also led to the direct measurement of differences between wheat grain samples with different puroindoline alleles. Once it was possible to control for the genetic background, the effect of the growing environment could be studied independently to determine its impact on grain fractionation. This revealed why different genetic classes of common wheat grains overlap when evaluated with SKCS measurements. Both the genetic alleles and the level of vitreousness should now be taken into account to better predict the milling behavior. Modelling the endosperm based on a cohesive granular structure was shown to describe the experimental data well. This facilitates the simulation and testing of distinct tissue organizations, even ones not yet to be observed in natural variants. This modelling has already shown that adhesion between the starch granules and the protein matrix, clearly related to the nature of the puroindoline alleles, is the main factor determining grain fractionation behavior.

Acknowledgments The authors thank the doctoral and postdoctoral students who contributed to these studies in Montpellier (France), C. Antoine, E. Chichti, V. Greffeuille, K. Heinze, Y. Hemery, S. Peyron, V. Topin. We also thank our colleagues J. Abecassis, C. Barron, J.Y. Delenne, M. George and F. Mabille in Montpellier and F.-X. Oury in Clermont-Ferrand.

References

Alfred R, Palombo E A, Panozzo J F, Bhave M (2014) The co-operative interaction of puroindolines in wheat texture may involve the hydrophobic domain. J Cereal Sci 60: 323–330.

Anderssen R S, Haraszi R (2009) Characterizing and exploiting the rheology of wheat hardness. European Food Res Technol 229: 159–174.

Anderssen R S, Haraszi R (2013) Interpreting crush response profiles obtained from SKCS 4100. J Cereal Sci 58: 362–363.

Antoine C, Lullien-Pellerin V, Abecassis, J, Rouau X (2002) Nutritional interest of the wheat aleurone layer. Sci des Aliments 22:545–556.

Antoine C, Peyron S, Mabille F, Lapierre C, Bouchet B, Abecassis J, Rouau X (2003). Individual contribution of grain outer layers and their cell wall structure to the mechanical properties of wheat bran. J Agric Food Chem 51: 2026–2033.

Barlow K K, Buttrose M S, Simmonds D H, Vesk M (1973) The nature of the starch-protein interface in wheat endosperm. Cereal Chem 50: 443–454.

Beecher B, Bettge A, Smidansky E, Giroux M J (2002) Expression of wild-type pinB sequence in transgenic wheat complements a hard phenotype. Theor Appl Genet 105: 870–877.

Bhave M, Morris C F (2008) Molecular genetics of puroindolines and related genes: allelic diversity in wheat and other grasses. Plant Mol Biol 66: 205–219.

Blochet J E, Chevalier C, Forest E, Pebay-Peyroula E, Gautier M F, Joudrier P, Pezolet M, Marion D (1993) Complete amino-acid-sequence of puroindoline, a new basic and cysteine rich protein with a unique tryptophan-rich domain, isolated from wheat endosperm by Triton X-114 phase partitioning. FEBS Letters 329: 336–340.

Branlard G, Amiour N, Igrejas G, Gaborit T, Herbette S, Dardevet M, Marion D (2003) Diversity of puroindolines as revealed by two-dimensional electrophoresis. Proteomics 3: 168–174.

Brouns F, Hemery Y, Price R, Anson N M (2012) Wheat aleurone: separation, composition, health aspects, and potential food use. Crit Rev Food Sci Nutr 52: 553–568.

Campbell G M (2007a) Roller milling of wheat. In: Salman A D, Ghadiri M, Hounslow M J (ed) Handbook of Powder Technology, Vol 12, Elsevier, pp. 343–419.

Campbell G M, Fang C, Muhamad I I (2007b) On Predicting Roller Milling Performance VI: Effect of Kernel Hardness and Shape on the Particle Size Distribution from First Break Milling of Wheat. Food Bioprod Process 85: 7–23.

Campbell G M, Sharp C, Wall K, Mateos-Salvador F, Gubatz S, Huttly A, Shewry P (2012) Modelling wheat breakage during roller milling using the Double Normalised Kumaraswamy Breakage Function: Effects of kernel shape and hardness. J Cereal Sci 55: 415–425.

Capparelli R, Amoroso M G, Palumbo D, Iannaccone M, Faleri C, Cresti M (2005) Two plant puroindolines colocalize in wheat seed and in vitro synergistically fight against pathogens. Plant Mol Biol 58: 857–867.

Capparelli R, Borrelio G, Giroux M J, Amoroso M G (2003) Puroindoline A- gene expression is involved in association of puroindolines to starch. Theor Appl Genet 107: 1463–1468.

Chantret N, Salse J, Sabot F, Rahman S, Bellec A, Laubin B, Dubois I, Dossat C, Sourdille P, Joudrier P, Gautier M-F, Cattolico L, Beckert M, Aubourg S, Weissenbach J, Caboche M, Bernard M, Leroy P, Chalhoub B (2005) Molecular Basis of Evolutionary Events That Shaped the *Hardness* Locus in Diploid and Polyploid Wheat Species (Triticum and Aegilops). Plant Cell 17: 1033–1045.

Charles M, Tang H, Belcram H, Paterson A, Gornicki P, Chalhoub B (2009) Sixty Million Years in Evolution of Soft Grain Trait in Grasses: Emergence of the Softness Locus in the Common Ancestor of Pooideae and Ehrhartoideae, after their Divergence from Panicoideae. Mol Biol Evol 26: 1651–1661.

Chen F, Zhang F, Li H, Morris CF, Cao Y, Shang X, Cui D (2013). Allelic variation and distribution independence of Puroindoline b-B2 variants and their association with grain texture in wheat. Mol Breed 32: 399–409.

Chichti E, Carrere M, Delenne J Y, Lullien-Pellerin V (2018) A wheat grain quantitative evaluation of vitreousness by light transmission analysis. J Cereal Sci 83: 58–62.

Chichti E, George M, Delenne J-Y, Lullien-Pellerin V (2015). Changes in the starch-protein interface depending on common wheat grain hardness revealed using atomic force microscopy. Plant Sci 239: 1–8.

Chichti E, George M, Delenne J-Y, Radjai F, Lullien-Pellerin V (2013). Nano-mechanical properties of starch and gluten biopolymers from atomic force microscopy. Europ Polym J 49: 3788–3795.

Chichti E, Lullien-Pellerin V, George M, Radjai F, Affes R, Delenne J-Y (2016). Bottom-up model for understanding the effects of wheat endosperm microstructure on its mechanical strength. J Food Eng 190: 40–47.

Cobb N A (1896) The hardness of the grain in the principal varieties of wheat. Agricultural Gazette of New South Wales 7: 565–578.

Darlington H F, Tecsi L, Harris N, Griggs D L, Cantrell I C, Shewry P R (2000) Starch granule associated proteins in barley and wheat. J Cereal Sci 32: 21–29.

Delwiche S R (2000) Wheat endosperm compressive strength properties as affected by moisture. Trans ASAE 43: 365–373.

Delwiche S R, Morris C F, Mabille F, Abecassis J (2012) Influence of instrument rigidity and specimen geometry compressive strength on calculation of wheat endosperm. Cereal Chem 89: 24–29.

Digeon J-F, Guiderdoni E, Alary R, Michaux-Ferrière N, Joudrier P, Gautier M-F (1999) Cloning of a wheat puroindoline gene promoter by IPCR and analysis of promoter regions required for tissue-specific expression in transgenic rice seeds. Plant Mol Biol 39: 1101–1112.

Dobraszczyk B, Whitworth M, Vincent J, Khan A (2002) Single kernel wheat hardness and fracture properties in relation to density and the modelling of fracture in wheat endosperm. J Cereal Sci 35: 245–263.

Dubcovsky J, Dvorak J (2007) Genome plasticity a key factor in the success of polyploid wheat under domestication. Science 316: 1862–1866.

Dubreil L, Gaborit T, Bouchet B, Gallant D J, Broekaert W F, Marion D, Quillien L (1998) Spatial and temporal distribution of the major isoforms of puroindolines (puroindoline-a and puroindoline-b) and non specific lipid transfer protein (ns-LTP1e-1) of *Triticum aestivum* seeds. relationships with their in vitro antifungal properties. Plant Sci 138: 121–135.

Evers A D, Blakeney A B, O'Brien L O (1999) Cereal structure and composition. Aust J Agric Res 50: 629–650.

Feiz L, Beecher B S, Martin J M, Giroux M J (2009) In planta mutagenesis determines the functional regions of the wheat puroindoline proteins. Genetics 183: 853–860.

Finnie S M, Jeannotte R, Morris C F, Faubion J M (2010) Variation in polar lipid composition among near-isogenic wheat lines possessing different puroindoline haplotypes J Cereal Sci 51: 66–72.

Gaines CS, Finney PF, Fleege LM, Andrews LC (1996) Predicting a hardness measurement using the single-kernel characterization system. Cereal Chem. 73: 278–283.

Gasparis S, Orczyk W, Zalewski W, Nadolska-Orczyk A (2011) The RNA-mediated silencing of one of the Pin genes in allohexaploid wheat simultaneously decreases the expression of the other, and increases grain hardness. J Exp Bot 62: 4025–4036.

Gautier M-F, Aleman M-E, Guirao A, Marion D, Jourdier P (1994) Triticum aestivum, two basic cysteine-rich seed proteins: cDNA sequence analysis and developmental gene expression. Plant Molecular Biology, 25: 43–57.

Geneix N, Dalgalarrondo M, Bakan B, Rollland-Sabaté A, Elmorjani K, Marion D (2015) A single amino acid substitution in puroindoline b impacts its self-assembly and the formation of heteromeric assemblies with puroindoline a. J Cereal Sci 64: 116–125.

Geng H W, Beecher B S, He Z H, Kiszonas A M, Morris C F (2012) Prevalence of *Puroindoline D1* and *Puroindoline b-2* variants in U.S. Pacific Northwest wheat breeding germplasm pools, and their association with kernel texture. Theor Appl Genet 124: 1259–1269.

Giroux M J, Morris C F (1997) A glycine to serine change in puroindoline b is associated with wheat grain hardness and low levels of starch-surface friabilin. Theor Appl Genet, 95: 857–864.

Giroux M J, Morris C F (1998) Wheat grain hardness results from highly conserved mutations in the friabilin components puroindoline a and b. Proc Natl Acad Sci USA 95: 6262–6266.

Greenblatt GA, Bettge AD, Morris CF (1995) Relationship between endosperm texture and the occurrence of friabilin and bound polar lipids on wheat starch. Cereal Chem 72: 172–176.

Greenwell P, Schofield J D (1986) A starch granule protein associated with endosperm softness in wheat. Cereal Chem 63: 379–380.

Greffeuille V, Abecassis J, Bar L'Helgouac'h C, Lullien-Pellerin V (2005) Differences in the aleurone layer fate between hard and soft common wheats at grain milling. Cereal Chem 82: 138–143.

Greffeuille V, Abecassis J, Barouh N, Villeneuve P, Mabille F, Bar L'Helgouac'h C, Lullien-Pellerin V (2007a) Analysis of the milling reduction of bread wheat farina: physical and biochemical characterization. J Cereal Sci 45: 97–105.

Greffeuille V, Abecassis J, Rousset M, Oury F-X, Faye A, Bar L'Helgouac'h C, Lullien-Pellerin V (2006) Grain characterization and milling behavior of near-isogenic lines differing by hardness. Theor Appl Genet 114: 1–12.

Greffeuille V, Mabille F, Rousset M, Oury FX, Abecassis J, Lullien-Pellerin V (2007b) Mechanical properties of outer layers from near-isogenic lines of common wheat differing in hardness. J Cereal Sci 45: 227–235.

Gruber W, Sarkar A (2012) Durum wheat milling. In: Sissons M, Abecassis J, Marchylo B, Carcea M (ed) Durum Wheat: Chemistry and Technology, Chap. 8, Elsevier, pp. 139–159.

Haddad Y, Benet J C, Abecassis J (1998) A rapid general method for appraising the rheological properties of the starchy endosperm of cereal grains. Cereal Chem 75: 673–676.

Haraszi R, Juhasz A, Sissons M, Rakszegi M, Tamas L, Anderssen R S (2013) Rheological hardness index for assessing hardness of hexaploids and durums. Cereal Chem 90: 430–438.

Haraszi R, Sissons M, Juhasz A, Kadkol G, Tamas L, Anderssen R S (2016) Using rheological phenotype phases to predict rheological features of wheat hardness and milling potential of durum wheat. Cereal Chem 93: 369–376.

Heinze K, Arnoud O, Delenne J Y, Lullien-Pellerin V, Ramonda M, George M (2018) On the effect of sample topography during modulus measurements by contact-resonance atomic force microscopy. Ultramicroscopy 194: 78–88.

Heinze K, Frank X, Lullien-Pellerin V, George M, Radjai F, Delenne J-Y (2017) Numerical modeling of the tensile strength of a biological granular aggregate: Effect of the particle size distribution. Conference: Powders and Grains 2017 – 8th International Conference on Micromechanics on Granular Media, in Montpellier. The European Physical Journal Conferences 140. DOI: https://doi.org/10.1051/epjconf/201714008013

Heinze K, Kiszonas AM, Murray JC, Morris CF, Lullien-Pellerin V (2016). Puroindoline genes introduced into durum wheat reduce milling energy and change milling behavior similar to soft common wheats. J Cereal Sci 71: 183–189.

Hemery Y, Mabille F, Martelli M R, Rouau X (2010) Influence of water content and negative temperatures on the mechanical properties of wheat bran and its constitutive layers. J Food Eng 98: 360–369.

Kim K-H, Martin J-M, Giroux M J (2012) Puroindolines are associated with decreased polar lipid breakdown during wheat seed development. J Cereal Sci 56: 142–146.

Krishnamurthy K, Giroux M J (2001) Expression of wheat puroindoline genes in transgenic rice enhances grain softness, Nature Biotech. 19: 162–166.

Law C N, Young C F, Brown J W S, Sanpe J W, Worland A J (1978) The study of grain-protein control in wheat using whole chromosome substitution lines. In: Seed protein improvement by nuclear techniques. Int Atomic Energy Agency, Vienna, pp. 483–502.

Le Bihan T L, Blochet J E, Desormeaux A, Marion D, Pezolet M (1996) Determination of the secondary structure and conformation of puroindolines by infrared and Raman spectroscopy. Biochem 35: 12712–22.

Lesage V S, Bouchet B, Rhazi L, Elmorjani K, Branlard G, Marion D (2011) New insight into puroindoline function inferred from their subcellular localization in developing hard and soft near-isogenic endosperm and their relationship with polymer size of storage proteins. J Cereal Sci 53: 231–238.

Lullien-Pellerin V, Marion D (2002) Plant cysteine-rich antimicrobial proteins that interact with biological membranes. In: "Membrane interacting peptides and proteins" (F. Heitz Ed.), Research Signpost Pub., India, pp. 126–146

Mabille F, Gril J, Abecassis J (2001) Mechanical properties of wheat seed coats. Cereal Chem 78, 231–235.

Martin C R, Rousser R, Brabec D L (1993) Development of single kernel wheat characterization system. Transactions of the ASAE 36: 1399–1404.

Mattern P J, Morris R, Schmidt J W, Johnson V A (1973) Locations of genes for kernel properties in the wheat variety "Cheyenne" using chromosome substitution lines. In: Proceedings of the 4th Int Wheat Genet Symp, Columbia, MO, Eds E.R. Sears and L.M.S. Sears, pp. 703–707.

Misailidis N, Campbell G M (2013) Interpreting crush response profiles from the single kernel characterization system. J Cereal Sci 57: 222–229.

Morris C F (2002) Puroindolines: the molecular genetic basis of wheat grain hardness. Plant Mol Biol 48: 633–647.

Morris CF, Beecher BS (2012) The distal portion of the short arm of wheat (Triticum aestivum L.) chromosome 5D controls endosperm vitreosity and grain hardness. Theor Appl Genet 125: 247–254.

Morris C F, Bhave M (2008) Reconciliation of D-genome puroindoline allele designations with current DNA sequence data. J Cereal Sci 48: 277–287.

Morris C F, Garrison E, King G E, Allan R E, Simeone M C (2001) Identification and characterization of near-isogenic hard and soft hexaploid wheats. Crop Sci. 41: 211–217.

Morris C F, Geng H, Beecher B S, Ma D (2013) A review of the occurrence of *Grain softness protein-1* genes in wheat (*Triticum aestivum* L.). Plant Mol Biol 83: 507–521.

Morris C F, Greenblatt G, Bettge A D, Malkawi H (1994) Isolation and characterization of multiple forms of friabilin. J Cereal Sci 20: 167–174.

Morris C F, King G E (2008) Registration of hard kernel puroindoline allele near-isogenic line hexaploid wheat genetic stocks. J of Plant Registrations 2: 67–68.

Morris CF, Massa A (2003) Puroindoline genotype of the U.S. National Institute of Standards & Technology Reference Material 8441, Wheat Hardness. Cereal Chem 80: 674 678.

Morris C F, Simeone M C, King G E, Lafiandra D (2011) Transfer of soft kernel texture from *Triticum aestivum* to durum wheat, *Triticum turgidum* ssp. *durum*. Crop Sci 51: 114–122.

Morrison W R, Greenwell P, Law C N, Sulaiman B D (1992) Occurrence of friabilin, a low molecular weight protein associated with grain softness, on starch granules isolated from some wheats and related species. J Cereal Sci 15: 143–149.

Norris K H, Hruschka W R, Bean M M, Slaughter D C (1989) A definition of wheat hardness using near infrared reflectance spectroscopy. Cereal Food World 34: 696–705.

Oda S, Schofield J D (1997) Characterisation of friabilin polypeptides. J Cereal Sci 26: 29–36.

Osborne B G, Anderssen R S (2003) Single-kernel characterization principles and applications. Cereal Chem. 80: 613–622.

Osborne B G, Henry R J, Southan M D (2007) Assessment of commercial milling potential of hard wheat by measurement of the rheological properties of whole grain. J Cereal Sci 45, 122–127.

Osborne B G, Jackson R, Delwiche S R (2001) Rapid prediction of wheat endosperm comprehensive strength properties using the single kernel characterisation system Cereal Chem 78: 142–143.

Oury F-X, Lasme P, Michelet C, Dubat A, Gardet O, Heumez E, Rolland B, Rousset M, Abecassis J, Bar L'Helgouac'h C, Lullien-Pellerin V (2017) Bread wheat milling behavior: effects of genetic and environmental factors, and modeling using grain mechanical resistance traits. Theor Appl Genet 130: 929–950.

Oury F-X, Lasme P, Michelet C, Rousset M, Abecassis J, Lullien-Pellerin V (2015) Relationships between wheat grain physical characteristics studied through near-isogenic lines with distinct puroindoline-b allele. Theor Appl Genet 128: 913–929.

Pauly A, Bareyt B, de Brier N, Delcour J A (2014) Incubation of isolated wheat starch with proteolytic or lipolytic enzymes and different extraction media reveals a tight interaction between puroindolines and lipids at its granule surface. Cereal Chem 91: 240–246.

Peña R J, Trethowan R,, Pfeiffer W H, Van Ginkel M (2002) Quality (end-use) improvement in wheat: Compositional, genetic, and environmental factors. J Crop Prod 5: 1–37.

Posner E S (2009) Wheat flour milling. In: Wheat Chemistry and Technology, 4th ed Khan K, Shewry P R. AACC Int., ST Paul, MN pp. 119–152.

Pujol R, Létang C, Lempereur I, Chaurand M, Mabille F, Abecassis J (2000) Description of a micromill with instrumentation for measuring characteristics of wheat grain. Cereal Chem 77: 421–427.

Rahman S, Jolly C J, Skerritt J H, Wallosheck A (1994) Cloning of a wheat 15 kDa grain softness protein (GSP). GSP is a mixture of different puroindoline-like polypeptides. Eur J Biochem 223: 917–925.

Rogers D E, Hoseney R C, Lookhart G L, Curran S P, Lin W D A, Sears R G (1993) Milling and cookie baking quality of near-isogenic lines of wheat differing in kernel hardness. Cereal Chem 70: 183–187.

Saurer W (1978) Use of infra-red reflectance measurement for determination of protein and water content and grain hardness in wheat. Getreide Mehl und Brot 32: 272–276.

Simmonds D H, Barlow K K, Wrigley C W (1973) The biochemical basis of grain hardness in wheat. Cereal Chem 50: 553–562.

Sissons M J, Osborne B G, Hare R A, Sissons S A, Jackson R (2000) Application of the single-kernel characterization system to durum wheat testing and quality prediction. Cereal Chem 77: 4–10.

Swan C G, Meyer F D, Hogg A C, Martin J M, Giroux M J (2006) Puroindoline B limits binding of puroindoline A to starch and grain softness. Crop Sci 46: 1656–1665.

Symes K J (1965) The inheritance of grain hardness in wheat as measured by particle size index. Aust J Agric Research 16: 113–123.

Topin V, Delenne J-Y, Radjai F, Brendel L, Mabille F (2007) Strength and failure of cemented granular matter. Europ Phys J E 23: 413–429.

Topin V, Radjai F, Delenne J-Y, Mabille F. (2009) Mechanical modelling of wheat hardness and fragmentation. Powder Technol 109: 215–220.

Topin V, Radjai F, Delenne J-Y, Sadoudi A, Mabille F (2008) Wheat endosperm as a cohesive granular material. J Cereal Sci 47: 347–356.

Turnbull K M, Marion D, Gaborit T, Appels R, Rahman S (2003) Early expression of grain hardness in the developing wheat endosperm. Planta 216: 699–706.

Wall M L, Wheeler H L, Huebsh M P, Smith J C, Figeys D, Altosaar I (2010) The tryptophan-rich domain of puroindoline is directly associated with the starch granule surface as judged by tryptic shaving and mass spectrometry. J Cereal Sci 52: 115–120.

Wanjugi H W, Hogg A C, Martin J M, Giroux M J (2007) The role of puroindoline A and B individually and in combination on grain hardness and starch association. Crop Sci. 47: 67–76.

Wiley P R, Tosi P, Evrard A, Lovegrove A, Jones H D, Shewry P R (2007) Promoter analysis and immunolocalisation show that puroindoline genes are exclusively expressed in starchy endosperm cells of wheat grain. Plant Mol Biol 64: 125–136.

Williams P C, Sobering D C (1986) Attempts at standardization of hardness testing of wheat. 1 The grinding sieving (particle size index) method. Cereal Food World 31: 359–364.

Zhang J, Martin J M, Beecher B, Morris CF, Hannah L C, Giroux M J (2009) Seed-specific expression of the wheat puroindoline genes improves maize wet milling yields, Plant Biotech J 7: 733–743.

The Impact of Processing on Potentially Beneficial Wheat Grain Components for Human Health

Paola Tosi, Alyssa Hidalgo, and Valerie Lullien-Pellerin

Abstract Wheat based foods, mainly in the form of bread and pasta, are staples of the human diet in many countries around the world. The dry weight of mature wheat grain is composed of 70–75% starch and around 10–14% protein, which has led to the widespread perception of wheat foods as sources of energy and protein. However, whole grains are also important sources of dietary fiber, vitamins and minerals, and contain notable levels of bioactive compounds with potential health benefits like lignans, phenolic acids, alkylresorcinols, phytosterols, folates and tocols.

The prominence of wheat grain in the human diet is largely due to its versatility in being processed into diverse products like flour, semolina, and other bakery products. Processing is a pre-requisite for using cereal grains as food and obtaining end products with various unique properties that are safe and appealing to consume. Processing may also help reduce the amount of hazardous molecules potentially present in harvested wheat, such as pesticides, mycotoxins and heavy metals. Each regulated step in a processing series influences the composition and/or the physicochemical properties of the different grain components, which in turn define the technological quality and the nutritional and health promoting properties of the end product.

The unique textural properties of wheat foods are largely determined by the starch and gluten proteins present in the starchy endosperm, the main constituents of white flour and semolina. The health-promoting effects of wheat-based products are mainly associated with the dietary fiber and bioactive compounds that are enriched in the grain peripheral layers, and mainly the aleurone layer, which is generally a component of the bran fraction after milling. Fractionation by milling and the way the different milling streams are subsequently recombined therefore has a considerable impact on the

P. Tosi (✉)
School of Agriculture Policy and Development, University of Reading, Reading, UK
e-mail: p.tosi@reading.ac.uk

A. Hidalgo
Department of Food, Environmental and Nutritional Sciences (DeFENS), University of Milan, Milan, Italy

V. Lullien-Pellerin (✉)
IATE, Univ. Montpellier, CIRAD, INRAE, Montpellier SupAgro, Montpellier, France
e-mail: valerie.lullien-pellerin@inrae.fr

G. Igrejas et al. (eds.), *Wheat Quality For Improving Processing And Human Health*, https://doi.org/10.1007/978-3-030-34163-3_17

relative abundance of each grain component in the wheat flour/semolina and, consequently, in end products. Further processing steps, such as dough making, microbial fermentation, extrusion, and baking can also affect the relative amounts and bioavailability of grain components. Some examples of the effects of grain processing procedures on the bioavailability of important grain components in wheat foods consumed by humans will be presented in this chapter. Suggestions of how to improve these processes in light of the implications for human health will also be discussed.

1 Introduction

Wheat provides over 20% of the daily protein and food calories for much of the world population and accounts for one-third of worldwide grain production (FAOSTAT). Wheat is also an important source of micronutrients, dietary fibre and other compounds beneficial to human health. All wheat grains consumed undergo some type of processing to increase the palatability, digestibility, safeness and shelf-stability of food products containing wheat ingredients. However, processing can also have positive or negative impacts on the nutritive value and health benefits associated with a specific wheat product by changing the relative content of bioactive components or their physicochemical properties.

Wheat grain is generally not consumed as wholemeal but is transformed by milling to isolate the starchy endosperm from the other tissues (Posner 2009). Wheat macronutrients, the starch and storage proteins (gliadins and glutenins), are the main components of the starchy endosperm, from which white flour and semolina is derived, while micronutrients and bioactive components are mainly located in the embryo and outer layers of the grain (Evers et al. 1999; Hemery et al. 2007). The nutritional and health promoting potential of wheat flour or semolina is inversely related to the degree of refinement determined by the milling specifications and reassembly of different milling fractions. However, milling generally has a positive effect on product safety by removal of potentially toxic compounds like pesticides, mycotoxins and heavy metals.

Secondary processing, where flour or semolina are mixed to prepare a dough that is further processed to make different end products like bread, pasta, biscuits, breakfast cereals, snacks, and so on, also affects the nutritional value of wheat-based foods. The magnitude of the impact depends both on the treatments applied and on whether compounds are added to the dough that may interfere with the stability and release of wheat grain components from the formulated matrix. Preparation of leavened products, for example, requires microbial fermentation of the dough, while pasta preparation steps include extrusion, drying of the extruded product, and cooking by boiling before consumption. Commercial cereals and snacks may be extruded, puffed, or flaked to improve product quality.

In recent years a number of studies have linked the consumption of whole-grain products to better health outcomes in humans (Cooper et al. 2017; Kristensen et al.

2012; Nelson et al. 2016). However, in many of the studies the actual composition of the whole-grain products is poorly described. A definition of whole grain was recently proposed by the American Association of Cereal Chemists (AACC) but is still under discussion (http://www.aaccnet.org/initiatives/definitions/Pages/WholeGrain.aspx).

In this review, we present an inventory of the known effects of the different processing procedures on the amount and properties of wheat grain compounds with potential health benefits in the human diet.

2 Alkylresorcinols (AR)

AR are phenolic lipids or more precisely 1,3-dihydroxy-5-n-alkylbenzenes with an odd number from 17 to 25 of carbon atoms. Different cereals contain mixtures of AR homologues in specific proportions, wheat being enriched in 19 and 21-carbon homologues (Ross et al. 2003). AR are absorbed by the human small intestine and distributed in the blood plasma or incorporated into erythrocyte membranes (reviewed in Landberg et al. 2014). A number of *in vitro* activities of AR have been identified, such as antioxidant activity (Kozubek and Nienartowicz 1995), inhibition of cancer cell growth (Zhu et al. 2012), and inhibition of glycerol-3-phosphate dehydrogenase activity, a key enzyme in the triacylglycerol synthesis in adipocytes (Rejman and Kozubek 2004), reviewed by Kozubek and Tyman (1999) and Landberg et al. (2014). However, the *in vivo* efficacy of these activities is difficult to assess because other molecules with potentially similar activities are consumed with AR in cereal products.

The total AR content of wheat grains varies between 54 and 1489 μg/g (d.m.) depending on the species, cultivar and growing environment, with a mean content around 500–700 μg/g (d.m.) (Andersson et al. 2008, 2010b; Ross et al. 2003; Tluscik et al. 1981). In grains, AR are only present at the boundary between the outer cuticle of the testa and the inner cuticle of the pericarp (Landberg et al. 2014, 2008). Therefore, testing for the presence of AR in milling fractions (Hemery et al. 2009) is a way of monitoring the fate of the tissues between the aleurone layer and the outer pericarp, which represent less than 4% of the grain mass (Barron et al. 2011, 2007).

During milling, the external tissues including the aleurone layer are separated from the starchy endosperm, so AR become concentrated in the bran and shorts fractions (3 to five fold higher AR content than in grains), with only small amounts being recovered in refined flours or semolina, and hence in any ensuing food products (Ross et al. 2003). The AR content of flour can therefore be increased by adding shorts or bran fractions back in, possibly after further grinding, or by adding fractions recovered from grain by pearling, up to the equivalent of between 5 and 10% of the original grain mass (Bordiga et al. 2016). AR content remains stable throughout the transformation chain, as the amount found in final products (bread or pasta) correlates well with the sum of the amounts in different combined fractions if an appropriate solvent is used to extract AR molecules complexed to proteins or starch (Andersson 2010a; Chen et al. 2004, Landberg et al. 2006, Ross et al. 2003). ARs in plasma or AR metabolites in urine are used as markers of whole-grain product intake, even though the same concentration of these compounds might result from diverse composition of the fractions used in the consumed food products and further processing. Depending on the particle size, amount and composition of any fractions containing peripheral tissues, the storage, cooking, texture, color, sensory and safety properties all have to be evaluated to determine if they are acceptable for consumers. For example, Blandino et al. (2013) demonstrated that up to 10% of refined flour can be replaced with a debranning fraction (between 8 and 16%) without modifying too drastically the technological properties and safety (level of the mycotoxin deoxynivalenol) of the resultant bread, while increasing the level of AR more than ten fold.

3 Benzoxazinoids (BX)

Benzoxazolinones, lactams and hydroxamic acids are three structurally different types of BX. The main BX in wheat grains is a hydroxamic acid, the glucoside form of a di-hexose of 2,4-dihydroxy-1,4-benzoxazin-3-one (DIBOA-glc-hexose, around 3 μg/g, Tanwir et al. 2013). Mammals absorb and metabolize BX that may protect health through their anticancer, anti-inflammatory or satiety properties (reviewed by Adhikari et al. 2015). However, potential aneugenic and mutagenic effects of the agluconic hydroxamic acids DIBOA and DIMBOA (2,4-dihydroxy-7-methoxy-1,4-benzoxazin-3) were also reported (Buchmann et al. 2007; Hashimoto and Shudo 1996). Like AR metabolites, BX derivatives were recognized as specific biomarkers in the human urinary metabolome for detecting the consumption of whole grains versus refined flours (Zhu et al. 2016). BX were mainly found concentrated in the milling fractions enriched in peripheral tissues and more specifically containing germ, although other biomarkers specific to this tissue were not used to clearly attribute this fraction to the embryo or scutellum tissues (Tanwir et al. 2013). Only one fifth of the DIBOA was recovered in milled flour. Pedersen et al. (2011) identified changes in the amount and relative proportions of BX due to hydrothermal processing of grains and baking. These changes appeared to reflect the synthesis of hydroxamic acid glucosides, the enzymatic liberation of their aglucone forms, and their subsequent degradation to benzoxazolinones.

4 Betaine and Choline

Wheat grains are a source of glycine betaine (or N,N,N-trimethyl glycine) and its precursor choline, with their respective concentrations ranges of 1000–2940 μg/g and 170–280 μg/g (d.m.) varying due to genotype, environment and genotype-by-environment interactions (Corol et al. 2012). Graham et al. (2009) analyzed fractions from a single wheat cultivar and found higher concentrations of both betaine and choline (respectively around 15 mg/g and 2 mg/g of d.m.) in an enriched aleurone fraction, suggesting that these compounds were specifically located into this tissue. Bran contained around 9 mg/g and 1 mg/g (d.m.) of betaine and choline, respectively. By deduction, refined wheat products obtained after milling contain less of these compounds. For example, Graham et al. (2009) found only 0.2 and 0.3 mg/g (d.m.) of betaine and choline, respectively, in white flour. Betaine and choline are both water-soluble and 40–50% of their content is lost when pasta is cooked (Ross et al. 2014).

Keaveney et al. (2015) and Price et al. (2010, 2012) measured significant increases in the amount of betaine in the plasma of healthy humans who had consumed minimally processed bran, aleurone or aleurone enriched cereal products. The decrease in a marker of the inflammatory reponse, the C-reactive protein, in the plasma was attributed at least partly to the increase in betaine following aleurone consumption.

In the human body, betaine serves as an osmolyte protecting cells from osmotic stress and as a methyl donor lowering the amount of circulating homocysteine by remethylating it to methionine and preventing excess deposition of hepatic fat. Elevated homocysteine levels have now been recognized as a biomarker of elevated vascular disease risk and increased consumption of betaine was proposed to help prevent cardiovascular diseases (Ross et al. 2014). However, recent data indicates that dietary intake and plasma concentrations of choline, but not betaine, are associated with a higher risk of atrial fibrillation (Zuo et al. 2018).

5 Lignans

Lignans were first identified in mammals and their precursors in plants by Axelson et al. (1982). Lignans have a complex diphenolic structure which can be converted by the intestinal microflora in the proximal colon of mammals into phytoestrogens (enterodiol and enterolactone) thus making them agonists or antagonists of endogenous estrogen molecules (Aehle et al. 2011). Plant lignans have been implicated in a number of health effects but the mechanisms of action are not completely understood (Kiyama 2016; Landete 2012). Recently, 7-hydroxymatairesinol (HMR) and its isomer were found to display strong anti-inflammatory properties in human aortic endothelial cells (Spilioti et al. 2014) adding some credence to their potential effect on atherosclerosis.

In grains, lignans are present in either aglycone or glycoside forms (Smeds et al. 2007) so quantifying them accurately is greatly dependent on the method of extraction and the sensitivity of the detection method. Using a less destructive method of extraction, HMR was found to be the major lignan compound in wheat grain, followed by syringaresinol (Smeds et al. 2007). Total lignan content in wheat samples harvested from eight different locations in Finland was in the range of 3–23 μg/g (Smeds et al. 2009). Wheat bran has the highest lignan content with a four to five fold higher content than whole grains (Durazzo et al. 2009; Smeds et al. 2007), so wholemeal flour is richer in lignans than refined flours.

There are only a few studies describing the stability of lignans during processing, but results suggest that lignans such as secoisolariciresinol diglucoside (SDG) isolated from flaxseed would be able to withstand the breadmaking process (Muir and Westcott 2000). A comparison of different lignan-containing cereal products showed that the effet of cooking depends on the particular lignan profile, the chemical structure of each lignan, and on the nature of the food matrix (Durazzo et al. 2013).

6 Tocols

Tocopherols and tocotrienols are two types of tocols and consist of amphipathic and lipo-soluble molecules that have a polar chromanol ring and a hydrophobic 16-carbon side chain, a phytyl or an isoprenoid chain, respectively (Tiwari and Cummins 2009). The side chain of a tocopherol is fully saturated, but the side chain of a tocotrienol has three double bonds. Differing in the number and position of methyl groups in the chromanol ring, four forms of these molecules can be distinguished, α, β, γ, and δ. Tocols act as antioxidants scavenging lipid peroxyl radicals, quenching or reacting with singlet oxygen and other reactive oxygen and nitrogen species. Their *in vivo* vitamin E activity is possibly related to the lipophilicity of the molecule, which is influenced by the number of methyl groups on the phenolic ring and the length and unsaturation of the carbon side chain, also affecting their transport and absorption by the human body (Kamal-Eldin and Appelqvist 1996). For example, the vitamin E activity of β-tocotrienol was found to be 5% of that of α-tocopherol.

The total amount of tocol in wheat grains varies between around 30–88 μg/g d.m., depending both on genotype and environment, with β-tocotrienol and then α-tocopherol being the most abundant (Lampi et al. 2008). However, genotype of durum wheat has little influence on tocol content (Beleggia et al. 2013). Total tocol content was also found to be greatest in germ, followed by bran then flour. Wheat milling fractions enriched in germ were found to mainly contain α-tocopherol whereas bran and flours were enriched in β-tocotrienol (Piironen et al. 1986).

Tocols become unstable when exposed to light and higher temperatures, so losses are incurred in processing (Andersson et al. 2014; Tiwari and Cummins 2009). The level of vitamin E in milled wheat products depends mainly on the extraction rate of the flour (about 50% reduction from whole-grain to white flour). In certain processes, the presence of some germ in flour can also influence the extent of vitamin E loss during storage because it acts as an antioxidant (Nielsen and Hansen 2008). The total loss of vitamin E during storage was 24% for stone-milled wheat flour (which contains a significant amount of germ) but 50% for roller-milled wheat flour (devoid of germ and bran). Oxidation is also an important cause of vitamin E losses during later processing steps. Preparation of gruels and porridges with processes such as extrusion cooking and drum-drying destroyed much of the vitamin E in white flour (Håkansson et al. 1987; Håkansson and Jägerstad 1990). The ratio of tocotrienols to tocopherols was reported to increase after extrusion cooking, indicating that tocotrienols are the main residual isomers of vitamin E (Zielinski et al.

2001), and there is evidence that a higher ratio of tocotrienols to tocopherols in the diet may be important in metabolic regulation.

Substantial losses of vitamin E activity occur during breadmaking, mainly at the stage when dough is made (20–40% loss of vitamin E), as mixing flour with water facilitates lipid oxidation (Galliard 1989), and affects mainly α-tocopherol and α-tocotrienol isomers (Wennermark and Jagerstad 1992).

Losses of vitamin E have also been reported to occur during the kneading step, when oxygen is incorporated into the dough (Leenhardt et al. 2006). Vitamin E losses are a result of both tocol oxidation itself (Slover and Lehmann 1972) and a secondary effect of fatty acid oxidation, i.e. lipoxygenase-catalyzed oxidation of tocols (Galliard 1989; Nicolas and Drapron 1983). The dough-making technique from Chorleywood (UK), which involves particularly high-speed mixing, accentuates these effects. Fermentation of the dough generally only has a minor effect on vitamin E content as any oxygen is rapidly consumed by the baker's yeast (Leenhardt et al. 2006; Wennermark and Jagerstad 1992).

During pasta making, the degradation of tocols is mostly limited to the kneading step (Fratianni et al. 2012) with the contribution of several physicochemical and enzymatic factors. Losses during pasta drying were only significant if the treatment was at high temperature (Beleggia et al. 2011), the average loss of total tocols amounting to about 30% with the highest loss for β–tocopherol (50%) and the lowest for α-tocopherol and α-tocotrienol (20%). Important differences were noted in tocopherol changes occurring during pasta making depending on whether refined semolina or wholemeal were used. Tocopherols in the refined samples progressively decrease during the pasta generation steps, whereas for wholemeal, after a significant decrease during the extrusion step (52% tocopherol loss), the total content of tocopherols increased significantly in dried pasta.

Extrusion causes a significant decrease in both tocopherols and tocotrienols (63–94% depending on the original tocol content of the cultivar used). The tocols the least resistant to hydrothermal processing are α-tocopherol and α-tocotrienol.

7 Short-Chain Carbohydrates

The low molecular mass carbohydrate fraction in wheat flour is composed mainly of fructans (fructooligosaccharides or FOS), short-chain carbohydrates of between 3 to 5 fructose units with different structures, sometimes including a single glucose unit (Ritsema and Smeekens 2003; Roberfroid 2005). Small amounts of galacto-oligosaccharides (GOS) like raffinose and stachyose are also present (Huynh et al. 2008; Lineback and Rasper 1988). Fructan content in wheat grains was found to vary depending on the genotype (7–29 mg/g), so it is a characteristic that could be a target for selection (Huynh et al. 2008). Bran and shorts fractions obtained from milling of wheat grains contain higher concentrations of fructans (34–40 mg/g) compared to white flour and germ (14–25 mg/g, depending on the wheat cultivar and environmental conditions) (Haska et al. 2008; Knudsen 1997). However, at a flour extraction level of 79%, the endosperm appeared to contribute around half of the total fructan amount. FOS and GOS belong to the so-called FODMAPs (fermentable oligo-, di- and mono-saccharides and polyols), a group of molecules which are not digested by humans, who lack the necessary hydrolytic enzymes, and are poorly absorbed in the intestinal lumen. FODMAPs are highly fermentable by beneficial bacteria in the gut (Gibson and Shepherd 2005) and are thus considered to be prebiotics. FODMAPs have also been clearly implicated in non-celiac gluten sensitivity (Biesiekierski et al. 2013; Skodje et al. 2018), a condition with symptoms similar to those of celiac disease and that improve when wheat and other gluten containing cereals are eliminated from the diet.

The fructan content of wheat-based products is impacted by milling and processing of flour. Due to the higher concentration of fructans in the outer layer of the grain, white flour has a lower fructan content than the corresponding whole-grain flour, but the difference is only likely to be significant in low-extraction-rate flours. Breadmaking has been shown to have a major impact on fructan content, but not on their structure (Gelinas et al. 2015). Dough mixing with or without baker's yeast leads to about a 20% reduction in fructans. Dough fermentation leads to even greater reductions, with up to 80% of wheat grain fructans degraded by the action of yeast invertase over a 3-h period, but neither standard baking or overbaking have an impact. The chain length of fructans did not change during breadmaking.

The impact of specific pasta making steps on fructan content have also been investigated. Gelinas et al. (2015) did not detect significant differences in the fructan content of pastas produced from the same batch of semolina but with drying temperatures of either 40 or 80 °C. On the contrary, boiling pasta had a major impact with 40–50% of fructans being lost in the boiling water during cooking, irrespective of the semolina used for pasta making, cooking time or cooking loss.

8 Sterols

Sterols and their saturated forms, stanols, are a class of tetracyclic compounds with a cyclopentane perhydrophenanthrene nucleus, a hydroxyl group at position 3 of the A-ring, and a side chain located at carbon 17. Dietary intake of phytosterols from wheat bran and germ was demonstrated to lower levels of total and low density lipoprotein in human serum (Andersson et al. 2004), but those fractions were also enriched in long chain (C20-C30) aliphatic primary alcohols, policosanols, which have been reported to display similar health effects (Irmak et al. 2006).

Total phytosterol contents in wheat grains representing 26 genotypes, 3 growing seasons and 4 growing locations were found to vary from 700 to 928 μg/g (Nurmi et al. 2010a). Both genetic and environmental factors were found to impact the phytosterol content of grain (Nurmi et al. 2010a), but in durum wheat the genotype had only a slight influence (Beleggia et al. 2013). The main sterols in wheat are β-sitosterol and campesterol and their corresponding saturated forms (Nurmi et al. 2010a, 2010b). Around 10% of wheat grain sterols were found to be esterified to a phenolic acid, mainly ferulic acid, in the form of steryl–ferulate (Nurmi et al. 2010b, Nyström et al. 2007b). Another 10% were glycosidically linked to a mono-, di- or oligosaccharide as steryl glycosides, some of which were esterified with a fatty acid forming an acylated steryl glycoside (Nyström et al. 2007b).

In commercial milling fractions analysed by Nyström et al. (2007b) the highest total sterol concentrations were in a fraction enriched in germ, followed by fine and coarse brans. The same authors also observed a similar enrichment of steryl ferulate forms in the bran fractions but not in the germ. Campestanol was preferentially esterified, while sitosterol was preferentially found in glycoside forms and was most concentrated in germ and flours enriched in fiber and ash, as well as in coarse bran. Nurmi et al. (2012) further characterised the sterol and steryl ferulate content in wheat grain fractions obtained by pearling, ground bran separated with electrostatic processing, and pure aleurone fractions, with the proportion of each grain outer layer estimated using specific biological markers. The conclusion is that different phytosterol forms are differentially distributed among the external grain tissues, but without a marker for germ or a mass balance of the different fractions this remains to be confirmed.

Processing affects the release of sterols from the food matrix and therefore their bioaccessibility. Nyström et al. (2007b) reported that the reduction in particle size of wheat bran from an average of 97 μm to an average of 47 μm using centrifugal milling increased the apparent amount of sterols only slightly (about 5% increase), while thermal treatments (roasting or microwave heating) decreased the apparent sterol content. The addition of water alone (without enzymes) may also dramatically decrease the availability of sterols in cereal products. Soaking wheat bran in water results in an apparent decrease in the sterol content, which was thought to be a consequence of the formation of arabinoxylan hydrates, whose viscous structure traps the hydrophobic sterols inside. Subsequent addition of enzymes, such as xylanases or β-glucanases, resulted in only partial release of bound sterols, so that their apparent content in the treated samples remained lower than in the untreated bran.

9 Phytic Acid and Minerals

Phytic acid is a myo-inositol 1,2,3,4,5,6-hexakis dihydrogen phosphate (IP6) and the main storage form of phosphate in seeds and cereal grains. Its concentration varies between 12 and 18 mg/g in wheat grains (Barron et al. 2011) and it is mainly present in phytin globoids inside the protein storage vacuoles in aleurone cells, where most of the magnesium (Mg) and iron (Fe) of the grain is also found (O'Dell et al. 1972). Phytic acid content in the aleurone layer varies between 95 and 190 mg/g (Barron et al. 2011) and phytic acid concentration can therefore be used as an effective marker to monitor the fate of the aleurone during milling (Greffeuille et al. 2005; Hemery et al. 2009). This is of great practical relevance since the redistribution of aleurone cells into flour or semolina is largely affected by grain hardness (Greffeuille et al. 2005). Being a polyvalent anion, phytic acid is able to chelate minerals present in the grain in the form of a mixed salt, called phytate. Its iron chelating properties interrupt the reactions of the Haber–Weiss cycle, and consequently the formation of hydroxyl radicals ($OH^{\cdot}$), which prevents lipid peroxidation. Furthermore, phytic acid has been shown to inhibit xanthine oxidase mediated O_2^- generation (Muraoka and Miura 2004). Recent studies showed a beneficial impact of phytic acid consumption on human health in the form of anti-carcinogenic effects, prevention of coronary disease, and boosting the immune system (Silva and Bracarense 2016). On the other hand, phytic acid has long been considered an anti-nutritional factor, as when it forms insoluble complexes with dietary cations, particularly Mg, Fe, zinc (Zn) and calcium (Ca), mineral absorption is impaired in humans (Zimmermann and Hurrell 2007).

Mg, Fe and Zn are essential for good preventive nutrition and wheat grain contains considerable amounts of these minerals as well as lower levels of many trace elements, such as selenium and manganese. Mg is among the most abundant intracellular cations in the body and is a cofactor in over three hundred biochemical reactions, including protein synthesis, muscle and nerve function, blood glucose control and blood pressure regulation. There is experimental and clinical evidence showing that the amount of Mg in typical Western diets is often insufficient to meet individual needs and low intakes of Mg have been associated with etiological factors in cardiovascular and nervous diseases, bone deterioration, spasmophilia and stress (reviewed by DiNicolantonio et al. 2018).

Fe also plays an essential role in human physiology. The body requires Fe for the synthesis of its oxygen transport proteins, in particular hemoglobin and myoglobin, and for the formation of heme enzymes and other iron-containing enzymes involved in electron transfer and oxidation-reduction. Iron deficiency can affect resistance to infection and cognitive development (Ryan 1997). However, as it can form free radicals, its concentration in the body must be finely controlled (Abbaspour et al. 2014).

Diet is the only source of Zn and Zn deficiency is widespread in human populations. The main biological function of Zn is in the form of Zn finger motifs, which are among the most abundant structural motifs in small eukaryotic proteins with diverse roles in cellular processes, including DNA recognition, RNA packaging, transcriptional activation, regulation of apoptosis, protein folding and assembly, and lipid binding (Matthews and Sunde 2002). Zn is therefore found in several systems and biological reactions, and is essential for immuno-function, cell division, cell growth, wound healing, blood clotting, thyroid function and the breakdown of carbohydrates (reviewed in Roohani et al. 2013). In addition, Zn is also needed for the senses of smell and taste (Henkin 1984).

Screening of modern bread wheat cultivars showed that concentrations ranges of Mg, Fe, and Zn are respectively 600–1400 ppm, 20–60 ppm and 15–35 ppm, and all negatively correlated with grain yield (Oury et al. 2006). Cereal grain minerals are mainly present in the aleurone and therefore enriched in the bran fraction after milling. Cubadda et al. (2009) reported various degrees of mineral loss upon milling of durum wheat grains. Selenium had the highest retention rate with concentrations in semolina equal to 77–85% of those in grain (d.m.), followed by calcium (54–60%), copper (49–53%), potassium, phosphorous (42–47%), Fe (36–38%), Mg, and Zn (32–36%).

In wheat grain, Fe mainly accumulates in aleurone cells as phytate salts and in this form Fe can be either soluble or insoluble, depending on the nature of the bonding. Insoluble forms are not accessible to iron transporters in the human gut, but there is evidence that soluble salts in the form of monoferric phytate may be a bioavailable source of Fe (Sandberg et al. 1999). Balmer et al. (2006) showed that Fe is also present as ferritin in amyloplasts isolated from developing wheat endosperm, and the presence of Fe in the cytoplasm of endosperm cells was subsequently confirmed by nanoscale secondary ion mass spectrometry imaging of wheat grains (Moore et al. 2012). Eagling et al. (2014a) showed that the presence/content of metal chelators in the grain may also influence Fe bioavailability. The chelator nicotianamine (NA), which is involved in the intra- and intercellular transport of metal cations, was reported to enhance iron uptake in a cell model as, to a lesser extent, was 2′-deoxymugineic acid (DMA), which is involved in the solubilization and acquisition of Fe(III) by the plant from the rhizosphere. Iron speciation studies of white flour and whole grain (Eagling et al. 2014b) showed that the content of Fe complexed with NA/DMA in white flour was 4 to five fold higher than in whole grain. Considering the lower phytic acid content of the endosperm, it is possible that, despite having significantly less iron than bran, it may actually be a better source of bioavailable iron.

Mg and Zn are also largely present in the form of phytate salts in the aleurone, with Mg being co-localized within globoid crystals with phosphorous and potassium. Speciation and localization analysis of Zn in wheat grain showed that this cation can also be found in the endosperm associated with small cysteine-rich proteins (apparent size 10–30 kDa) (Persson et al. 2016).

The chelating action of phytic acid, by hindering the bioaccessibility of Fe, Mg and Zn, drastically reduces their bioavailability in wheat-based foods (Das et al. 2012). Furthermore, phytic acid also interacts with proteins, making them less digestible (Kumar et al. 2010) and it promotes Maillard reactions in certain conditions and thus acrylamide formation (Wang et al., 2013). Although the amount of phytic acid in the grain could potentially be decreased by breeding (Gupta et al. 2015; Magallanes-López et al. 2017), processing is currently the most effective strategy for solubilization of Fe, Mg and Zn from phytate salts in wheat flour.

The sourdough process results in drastic degradation of phytate, due both to microbial and wheat endogenous phytase activated by microbial acidification of the dough. Leenhardt et al. (2005) observed a 70% reduction in phytic acid after 4 h of sourdough fermentation, accompanied by a five fold increase in soluble Mg, while Rodriguez-Ramiro et al. (2017) reported that 36 h of sourdough fermentation reduced the phytic acid content of whole-grain flour to below detectable levels, bringing the IP6 to Fe molar ratio to below 1, and resulting in significantly more bioavailable Fe in the bread.

There is a substantial increase in phytase activity during wheat germination (Bartnik and Szafranska 1987), when there is a marked increase in the availability of Zn in the wheat grain, but not of Fe (Luo et al. 2014). High-temperature short-duration extrusion cooking, the process used for the production of a variety of breakfast cereals and salty and sweet cereal snacks, was also shown to efficiently enhance mineral availability in cereal products. Minerals are considered to be stable during heat treatment, but extrusion may hydrolyze the complex between phytic acid and minerals to release phosphate molecules. A 13–35% reduction in phytate content from a wheat bran-starch-gluten extruded mix has been reported (Andersson et al. 1981).

There is also evidence that absorption of minerals from wheat grain may also be impaired in wholemeal products by the presence of tannins and fibres, which can form insoluble complexes with divalent ions in the gastrointestinal tract. Shear forces and high temperatures during extrusion cooking are very effective at destroying polyphenols (Singh et al. 2007) and may cause modification of fibre components (Wang et al. 1993) and their chelating properties, which could also contribute to improving bioavailability of minerals in extruded foods.

10 Phenolic Acids

Phenolic acids are a complex group of secondary metabolites with a large diversity of structures all containing at least one benzene ring with one or more hydroxyl groups. They are the most abundant phytochemicals in cereals and have been shown to function as free-radical scavengers, reducing agents and quenchers of singlet oxygen formation. Their antioxidant properties (reviewed in Laddomada et al. 2015) are mainly attributed to electron donation and hydrogen atom transfer to free radicals, but it has also been suggested that they modify some cellular signalling processes. As antioxidants, phenolic acids may prevent heart disease and lower the incidence of colon cancer and they have also been shown to exert anti-inflammatory action in the gut, which may be significant for maintaining gastrointestinal health (reviewed in Laddomada et al. 2015).

Ferulic acid (FA), a derivative of hydroxycinnamic acid, has the highest antioxidant activity and accounts for 70–90% of the total phenolic acids in the grain, which also include other hydroxycinnamates, such as caffeic acid, chlorogenic acid, sinapic acid, and p-coumaric, and derivatives of hydroxybenzoic acid (Klepacka and Fornal 2006). Through reactions involving their carboxylic and hydroxyl groups, phenolic acids may form both ester and ether linkages that allow them to cross-link with cell wall macromolecules (Bunzel et al. 2004). In wheat, FA is found as a free compound, as a soluble conjugate bound to low molecular weight compounds such as sugars, and as bound forms associated with the fibre fraction, mainly as dimeric esters bound to arabinoxylan (AX) but also to lignin. The biological properties and physiological effects of dietary polyphenols, and notably their antioxidant properties, depend upon how available they are for absorption and subsequent interaction with target tissue, which depends greatly on their degree of polymerization. Early studies highlighted the inherently low bioavailability of FA in wheat grain and suggested it reflects the fact that it is mostly present (up to 80%) in the bound forms (Anson et al. 2009a). While free forms of FA were efficiently absorbed in the intestine, only a small portion of bound phenolics appeared to be metabolised in the stomach and small intestine. However, the bound phenolic fraction had a significantly higher antioxidant capacity *in vitro* in comparison with free and esterified phenolic acids, which suggests they must be included in any evaluation of antioxidant activity in grains in relation to their phenolic acid content. Subsequent studies showed that intestinal microbes are able to cleave the ester or ether bonds crosslinking wheat phenolic acids to cell wall polymers, thus making the phenolic acids nutritionally available (Vitaglione et al. 2008). Furthermore, the structural complexity of bound phenolics, allowing them to reach the colon mostly undigested, is a functional aspect of their ability to exert unique antioxidant and anti-inflammatory activity locally and therefore contribute to reducing the risk of colorectal cancer (Andreasen et al. 2001; Drankham et al. 2003).

Barron et al. (2007) reported that the content and nature of phenolic acids differ among different grain tissues as do the amount and molecular composition of cell wall polymers present in the different layers, particularly the relative amount of arabinose (Ara) and xylose (Xyl). The outer layers of the wheat grain were reported to contain the highest amount and broadest array of phenolic acids, including FA, dehydrodimers (DHD) and dehydrotrimers of FA, sinapic acid, and p-coumaric acid (p-CA) (Barron et al. 2007; Parker et al. 2005). The embryo tissues, and the scutellum in particular, are relatively rich in FA and DHD, while FA and sinapic acid predominate in the starchy endosperm, although the FA concentrations are several fold lower than in the cell walls of the seed coat and aleurone layer. The outer pericarp contains the highest concentration of a trimeric form of FA that lends itself as a marker to monitor this tissue during fractionation (Hemery et al. 2009). Furthermore, the aleurone tissue has the highest antioxidant capacity among the wheat grain layers, possibly resulting from its high FA content (over 60% of the antioxidant capacity) (Anson et al. 2008). Among the tissues constituting the bran fraction, the outer pericarp and tissues in the crease region contain the highest proportion of strongly bound (ether-linked) phenolic acids, while the aleurone and the hyaline layer contain the highest proportion of weakly cross-linked FA (Barron et al. 2007). The AX polymers of the cell walls of the aleurone and hyaline layer also display less substitution with a lower Ara to Xyl ratio (< 0.5). Using debranning, Beta et al. (2005) showed that total phenolic content and total antioxidant activity are strongly correlated, remaining at similar levels in the very first pearling fractions (up to 10% removal), which correspond almost exclusively to pericarp tissues and part of the aleurone layer (Rios et al. 2009), but then progressively decreasing in fractions with increased endosperm content.

Both industrial and domestic processing can affect the content, composition, and stability of phenolic compounds in wheat-based products. Modern milling, which is based on the separation of different tissues of the grain into milling streams that are then recombined to give flours of different extraction rates, has a major impact on the content and composition of phenolic acids of final flours. The relative content of aleurone and pericarp tissues in the flour increases with its extraction rate, and since these tissues are rich in phenolic acids so does the phenolic acid content. The practice of debranning before roller milling (usually removing about 5% of initial grain weight), would impact specifically on the ratio of soluble to bound phenolic acids in the flours, because the outer pericarp removed by debranning contains a higher proportion of bound phenolics than other grain tissues (Barron et al. 2007).

Bakery and pasta-making processes can also modify the content, composition, and bioaccessibility of phenolic acids with respect to the original flour ingredient. FA bioavailability appears to be determined by the percentage of free FA, and since this has been reported to be extensively absorbed in the intestine (Adam et al. 2002), absorption itself is unlikely to be limiting. On the contrary, bioaccessibility, i.e. release from the food matrix, is likely to be a determinant factor in FA bioavailability in wheat-based products, since most FA in the grain is bound to AX and other indigestible polysaccharides, restricting its release in the small intestine (Anson et al. 2009a; Kern et al. 2003).

Bioprocessing techniques to release bound phenolic compounds from wheat bran have been applied with success to increase the content of free phenolic acids in bran-containing breads (Anson et al. 2009b). Bran bioprocessing involves fermentation with baker's yeast or a combination of fermentation with hydrolytic enzyme treatments (mainly with xylanase, β-glucanase, α-amylase, cellulase and FA esterase) to degrade different wheat polymers, thus improving the solubility of the complex cell wall structure in the bran. Bran fermentation alone was reported to increase the amount of free FA in the resultant bread by approximately three fold, while the combination of fermentation and enzymatic treatment brought this increase up to eight fold, which corresponded to a five-fold increase in FA bioaccessibility, as measured *in vitro*. These same bioprocessing techniques also increased the free forms of p-coumaric acid and sinapic acid.

Increases in free phenolic acid as a result of dough fermentation have also been reported for wholemeal and white breads, the increase being higher when sourdough rather than only yeast fermentation was used (Konopka et al. 2014; Moore et al., 2007). The lowering of pH during sourdough fermentation favours the activity of hydrolases (native flour and/or microbial enzymes) and can contribute to the chemical disintegration of AX and hydrolysis of both esters and glycosides of phenolic acids, leading to structural breakdown of the cell wall matrix and the release of free phenolic acids. However, souring of dough to approximately pH 4 has been reported to have an inhibitory effect on the native cinnamoyl esterase activity in wheat flour (Konopka et al. 2014). Cinnamoyl esterase activity in wheat flour has been ascribed to microflora on the grain surface (Dornez et al. 2006) and normally contributes to the degradation of AX in cell walls.

Baking significantly increased the concentration of free FA, particularly in the crumb of sourdough fermented bread (Konopka et al. 2014). However, reductions in the amount of free phenolic acids during the bread production process have also been reported. One possible explanation is that the free phenolic acids are decomposed by microflora. *Saccharomyces cerevisiae* is able to convert trans-FA into 4-hydroxy-3-methoxystyrene with a 96% yield (Huang et al. 1993). A baking temperature of 230 °C has been suggested to cause either FA released by fermentation to rebind, or to be degraded (Han and Koh 2011), a phenomenon that would explain why bread crust has a lower content of free FA compared to bread crumb (Konopka et al. 2014).

Of great interest is the recent development of more sustainable, non-toxic techniques for the extraction of plant phenolics based on microwave and ultrasound-assisted technologies (Tiwari 2015; Wang and Weller 2006) which make it possible to extract the phenolic compounds from bran so they can be incorporated into functional foods without requiring any preliminary chemical hydrolysis and eliminating the use of organic solvents.

11 Dietary Fibre

Dietary fibre (DF) describes the edible parts of plants or analogous carbohydrates that are resistant to digestion and absorption in the human small intestine, and that with complete or partial fermentation in the large intestine promote physiological effects or benefits to health (according to the AACC definition, Cereal Foods World 46: 112–126, 2001).

Mature wheat grain DF includes: cell wall polysaccharides, which are ubiquitous in the grain; lignin, a phenolic polymer only present in the pericarp/seed coat (Antoine et al. 2003; Stone and Morell 2009); fructans, enriched in bran tissues but also present in endosperm cells and therefore white flour (described above); a small amount of resistant starch (RS), derived solely from endosperm cells.

The non-starch polysaccharides (NSP) present in cell walls account for about 11% of the mature wheat grain dry weight (Andersson et al. 2013) and are the main components of the dietary fibre fraction in wheat. The major cell wall polysaccharides of wheat grain are AX and β-glucan ((1 → 3),(1 → 4)-β-D-glucan), with smaller amounts of cellulose ((1 → 4)-β-D-glucan), glucomannan and pectin. AX is composed of a backbone of xylose residues with some residues being substituted with arabinose residues at either one or two positions. Some of the arabinose residues present as single substitutions on xylose may also be substituted with FA, allowing the formation of diferulate cross-links by oxidation of ferulate present on adjacent AX chains. The extent of diferulate cross-linking is important as it affects the physicochemical properties of AX, such as the solubility and viscosity that govern its behaviour in food processing and its health benefits. AX is therefore often divided into two classes, water-extractable (WE-AX) or water-unextractable (WU-AX). Branching and the presence of ionic groups increase the solubility, as do changes in monosaccharide units and their molecular form (α or β forms). β-glucan comprises glucose residues joined by (1 → 3) linkages usually separated by two or three (1 → 4) linkages, but longer stretches of up to 14 (1 → 4) linked glucan units have been reported for wheat bran β-glucan (Li et al. 2006). Only 10–20% of the total β-glucan in wheat wholemeal flour was found to be soluble (Nemeth et al. 2010).

The content and composition of cell wall NSP vary between grain tissues. Compared to the starchy endosperm, the outer layers of the mature wheat grains comprise more cell wall material (at least 50% more) (Shewry and Hey 2015) in the form of 60% arabinoxylan, 25% cellulose, and ~10% lignin (reviewed by Stone and Morell 2009). Cell walls account for about 35–40% of the aleurone dry weight and are composed of 65% AX and 30% β-glucan (Bacic and Stone 1981), while starchy endosperm cells have thinner walls (about 2–3% of the dry weight), which consist of 70% AX and 20% β-glucan (Mares and Stone 1973; Stone and Morell 2009). There are differences in the proportions of AX and β-glucan between regions of the starchy endosperm, with β-glucan being more abundant in the region close to the germ (Saulnier et al. 2009). Depending on the location in central or prismatic cells, AX structure was also found to differ (Saulnier et al. 2009) and the content of arabinose in AX was found to increase from the outside to the inside of the endosperm (Toole et al. 2010). AX from the starchy endosperm contains low levels of FA, 0.2–0.4% (w/w) in WE-AX and 0.6–0.9% (w/w) in WU-AX (Shewry and Hey 2015), while AX from the aleurone are more esterified and cross-linked (Antoine et al. 2003; Parker et al. 2005). In the aleuronic AX polymer, additional esterification with p-coumaric acid and acetyl groups was described (Rhodes and Stone 2002). A complex highly-branched structure is also characteristic of pericarp AX, which often comprises galactose and glucuronic acid residues in addition to high contents of FA and di- or tri-FA (Hemery et al. 2009; Parker et al. 2005) and acetylation (Mandalari et al. 2005) with significant amounts of FA trimer (Barron et al. 2007).

Lignin is a complex polymer of aromatic alcohols tightly associated to cell wall polysaccharides either directly through covalent links with sugar residues or indirectly via FA esterified to polysaccharides (Davin et al. 2008; Iiyama et al. 1994). Lignin, which is insoluble and largely resistant to bacterial degradation, cements and anchor the cellulose microfibrils and other matrix polysaccharides, stiffening the cell walls and making them difficult for microorganisms to degrade in the human large intestine.

RS is the term given to starch that escapes digestion in the small intestine and therefore becomes available as a fermentation substrate for colonic microorganisms which produce short chain fatty acids (SCFA) that have positive effects on health (Topping and Clifton 2001, Topping et al. 2008). Cereals generally contain about 3% RS, including starch entrapped in the food matrix that is physically inaccessible to digestive enzymes, native (undamaged and/or uncooked) starch granules, and retrograded starch formed after starch granules gelatinize (Eerlingen et al. 1993, 1994; Hallström et al. 2011). The amounts of native and retrograded starch are strongly influenced both by amylose content and processing conditions.

A number of mechanisms probably contribute to the beneficial action of dietary fibre on health (Brownlee 2011; Buttriss and Stokes 2008; Theuwissen and Mensink 2008; Topping 2007), both through its physical properties (faecal bulk, water solubility, water-holding capacity, swelling power and viscosity) and fermentation in the colon. The insoluble fraction of dietary fibre activates intestinal peristalsis and is capable of binding bile acids and water. Soluble fibre reduces the blood cholesterol level, the risk of ischemic heart disease and postprandial glycemia. Fermentation of dietary fibre by gut bacteria produces SCFAs which have physiological effects on the colon and other tissues. There is also increasing evidence that mixed-linkage β-glucans are able to regulate the immune responses that are involved in fighting infection, attacking tumours, and various inflammatory conditions (Brown and Gordon, 2001; Rice et al. 2005). More recently, it has also been proposed that AX have immuno-stimulatory effects (Capek and Matulová 2013; Mendis et al. 2016).

Processing can modify the composition and microstructure of fibre and have an impact on its physicochemical properties and nutritional effects (Zhang et al. 2011). Due to the uneven distribution of dietary fibre components within the grain tissues, milling can have a dramatic effect on the amount and composition of dietary fibre in flour. Wholemeal wheat flour contains on average about 13% total dietary fibre (d.m.), half of which is AX but only a small fraction of it being water soluble (~0.57% WE-AX). By comparison white flour has an average total dietary fibre content of 3.5%, about 75% of which is AX and the same amount of WE-AX as wholemeal flour (Shewry and Hey 2015).

Therefore, milling affects the dietary fibre composition of flour according to the extraction rate and the effectiveness of the separation of bran from endosperm tissues. Furthermore, milling parameters determine the particle sizes produced which influence fibre properties. By studying the effect of ultrafine grinding on the physicochemical properties of wheat bran dietary fibre, Zhu et al. (2010) showed that as particle size decreased, the hydration properties (water holding capacity, water retention capacity and swelling capacity) of wheat bran dietary fibre significantly decreased and fibre components were redistributed from insoluble to soluble fractions. Ultra-fine grinding was also shown to increase the antioxidant capacity of wheat bran, probably due to a greater exposure/accessibility of the phenolic acids linked to the fibres (Rosa et al. 2013). The effect persisted in gastric conditions, showing that ultra-fine grinding can be used to produce wheat bran fractions with higher nutritional value. Physicochemical properties of wheat dietary fibre are also significantly affected by the combination of high temperature, pressure and shear force characteristic of extrusion cooking technology that is increasingly used to produce highly expanded and low-density products such as ready-to-eat breakfast cereals and snacks. Extrusion-cooking of white wheat flour at 161–171 °C, 15–20% water content and a screw speed of 100–200 rpm was found to cause a redistribution of insoluble dietary fibre to soluble forms (Björck et al. 1984). Depending on the process conditions, 50–75% of total fibre becomes soluble in the extruded flour versus 40% in the raw flour. Extruded white flour was also more fermentable, as determined by faecal recovery in mass balance experiments in rats, possibly a consequence of its higher solubility. Extrusion, particularly at the highest screw speed, has been successfully used to increase the solubility of wheat bran dietary fibre (Rashid et al. 2015; Wang et al. 1993), although relative fibre solubilisation is significantly lower than for white flour and it does not seem to affect *in vivo* fermentability (Björck et al. 1984). Rashid et al. (2015) examined the suitability of wheat bran for extrusion cooking and checked the effect of different extrusion parameters on the dietary fibre profile, as well as on the water solubility index.

Several studies describe the impact of breadmaking steps on endogenous flour fibre and results suggest that both mechanical effects and enzymatic reactions are involved. Rouau et al. (1994) monitored the amount of WE-AX during breadmaking, reporting increases that they attributed to solubilisation of some of the WU-AX. More than 10% of WU-AX had become extractable by the end of kneading and solubilisation increased to 25% at the end of fermentation. Cleemput et al. (1997) also observed substantial solubilisation of WU-AX during mixing (7 to 12%) and after baking (14 to 15%) phases but only very low levels (0 to 5%) during fermentation. Furthermore, clear changes in the molecular weight distribution of AX during fermentation were observed with no modification of the Ara to Xyl ratio of the AX fractions.

A more recent study of the impact of breadmaking on endosperm flour dietary fibre (Comino et al. 2016) reported an increase of approximately 18.5% in total solubilised NSP (12.5% was WE-AX and 6% β-glucan). This increase results from a 7% yield decrease in wheat flour unextractable NSP during dough fermentation and a 19% decrease during baking, with much smaller increases ascribable to dough preparation. By contrast, the contribution of the dough mixing step to determining the final amount of WE-AX was substantial in a similar study carried out by Gelinas et al. (2015). It is likely that differences in specific parameters used during dough mixing, dough fermentation and baking are the reason for the reported differences. A 35% decrease in insoluble AX from wholemeal flour to the baked bread product was also reported by Hansen et al. (2002).

The breadmaking process, and in particular the type of fermenting inoculum and fermentation parameters, has a significant impact on the content of other types of dietary fibre, namely fructans and RS. Yeast leavening results in major decreases in flour fructans in dough and bread (Gelinas et al. 2015), while sourdough breadmaking results in more RS (Scazzina et al. 2009), probably as a consequence of the organic acids produced during fermentation, which could facilitate debranching of the amylopectin moiety during baking. In fact, debranched amylopectin may form a high level of RS when heated (Berry 1986).

Pasta and noodle making and cooking pasta also affect dietary fibre amount and composition. Pasta extrusion is known to result in products eliciting lower glycemic responses (Monge et al. 1990; Wolever et al. 1986), thus producing food with the metabolic advantages of a low glycemic index (Jenkins et al. 1987). Available data suggest that pasta (both dried and fresh egg pasta) is a comparatively rich source of RS (Brighenti et al. 1998) relatively to other conventional wheat-based foods. The slow-release features of starch in pasta probably relates to the continuous viscoelastic network formed during pasta making, which surrounds the starch granules restricting starch swelling and leaching during boiling, and likely also reducing its accessibility to enzymatic digestion. However, the pasta surface area does not affect the glycemic response (Wolever et al. 1986) and, similarly, the shape of pasta does not seem important in relation to the RS content.

Processing methods based on wet heat and extrusion cooking are also being assessed for their potential to induce the formation of amylose-lipid complexes, a novel form of RS, in cereal based products (reviewed in Panyoo and Emmambux 2017).

Dietary fibre (AX and β-glucan) amounts were not significantly affected by the alkaline and/or boiling (100 °C) conditions used in the production of yellow alkaline noodles (Comino et al. 2016).

12 Anthocyanins

Anthocyanins are abundant secondary metabolites responsible for most blue to blue-black, and red to purple colours of many plant organs. They have antioxidant, photo-protective and defence roles in the plant, and play an important role in reproductive mechanisms (Escribano-Bailòn et al. 2004). Their molecular structure consists in an anthocyanidin (aglycone) with saccharide residues bound at different hydroxylated positions. The differences between individual anthocyanins are related to the number of hydroxyl groups, to the nature, number and position of sugars attached to the molecule, and to the aliphatic or aromatic acids attached to the sugars (Kong et al. 2003). In coloured wheat grains, six anthocyanidins have been observed: cyanidin, delphinidin, malvidin, pelargonidin, peonidin and petunidin.

Garg et al. (2016) identified 22 different anthocyanins in blue, 23 in purple and 26 in black wheats, and determined that the anthocyanin content was highest in the black, followed by the blue, purple and amber wheat lines. The concentration and composition of total anthocyanins in wheat kernels is influenced by the cropping environment (Abdel-Aal and Hucl. 2003; Varga et al. 2013), but Ficco et al. (2014) reported high heritability for anthocyanins, and only minor genotype-by-year effects.

In food, anthocyanins are a viable alternative to artificial colors. Furthermore, they are valued for their bioactivity against oxidative stress, cancer, inflammation, diabetes, and obesity, and in the reduction of postprandial glycemic levels (Krüger and Morlock 2018). Anthocyanin content in wheat kernels can be assessed either spectrophotometrically or by high performance liquid chromatography and liquid chromatography-mass spectrometry. Using the quicker and cheaper spectrophotometry approach, Abdel-Aal and Hucl (1999) found 155.6, 103.8 and 5.2 mg of cyanidin 3-O-glucoside/kg in the wholemeal of blue, purple and red bread wheat lines respectively. Similarly, higher values of cyanidin 3-O-glucoside were observed in blue wheats than in purple wheats, 21.4–157.6 versus 6.7–30.7 mg/kg by Varga et al. (2013), and 111.5–251.8 versus 50.1–171.7 mg/kg by Jaafar et al. (2013). Meanwhile, Liu et al. (2010) recorded 234.5 mg of cyanidin 3-O-glucoside/kg in purple wheat and 9.6 mg/kg in red wheat. Anthocyanins are also present in wild and ancient wheats. The diploid wheats *Triticum monococcum* ssp. *thaoudar* (wild einkorn), *T. monococcum* ssp. *aegilopoides* (feral einkorn), *Triticum urartu* (red wild einkorn) and *T. monococcum* ssp. *monococcum* (domesticated einkorn) contained 43.0 ± 4.66, 17.4 ± 3.01, 15.3 ± 0.85 and 11.6 ± 1.59 mg of anthocyanins/kg d.m., respectively (Brandolini et al. 2015).

Anthocyanins are mainly located in the outer layers of the caryopsis. Abdel-Aal and Hucl (1999) observed that blue, purple and red wheats display different anthocyanin concentrations in flour (22.5, 5.2 and 1.7 mg/kg) than in bran (458.3, 250.7 and 10.4 mg/kg). Furthermore, in blue cultivars anthocyanins are concentrated in the aleurone layer, while in purple cultivars they are located in the pericarp (Knievel et al. 2009; Krüger and Morlock 2018). In bran fractions from 17 purple pericarp, 10 blue aleurone, and 13 deep purple grained genotypes, the ranges of anthocyanin contents were 47.5–502.5, 117.6–879.2, and 359.9–1289.6 mg/kg d.m. respectively (Böhmdorfer et al. 2018).

The impact of breadmaking on the anthocyanin content of purple wheat bread was investigated by Yu and Beta (2015). The total anthocyanin content decreased by 21% after mixing, then gradually increased to 90% of the original level after fermentation, and finally decreased by 55% during baking. The lowest total anthocyanin content was in the bread crust.

Substantial losses of anthocyanin (55 to 80% depending on the particular product) have also been reported to result from high-temperature short-duration extrusion of cereal flour for snack products (Escalante-Aburto et al. 2013).

13 Carotenoids

Carotenoids are lipid-soluble antioxidants formed by most photosynthetic organisms and give the yellow, orange and red colours typical of many flowers, fruits and bird feathers. There are two classes of carotenoids, the carotenes, which are tetraterpenoid hydrocarbons, and the xanthophylls, which have one or more oxygenated parts in the molecule (Van den Berg et al. 2000). The single and double bonds in the polyenic chain confer the antioxidant properties of the molecules, while the presence of polar groups influences their interaction with cell membranes (Britton 1995). In plants, carotenoids behave as light collectors and protectors against photosensitization in chloroplasts. Animals do not produce carotenoids so must obtain them from food.

The main carotenoids found in wheat species are, in order of decreasing concentration, lutein, zeaxanthin, α- and β-carotenes and cryptoxanthin (Abdel-Aal et al. 2007; Hidalgo et al. 2006; Hidalgo et al. 2010). Lutein accounts for 90–100% of the total carotenoids (Abdel-Aal et al. 2007) and may occur in both esterified and non-esterified forms. Ziegler et al. (2015) identified six lutein monoesters and nine diesters, representing 22.2%, 29.7%, and 7.6% of the total lutein in bread wheat, spelt, and einkorn, respectively. Lutein esters are not found in durum and emmer wheats.

All the carotenoids are valued for their antioxidant activity, which protects cells and tissues from free radicals. The α- and β-carotenes are involved in the biosynthesis of vitamin A, which is essential for reproduction, embryo development, visual functions, etc. (Zile 1998), while lutein and zeaxanthin protect the macula region of the retina, prevent cataracts, enhance the immune response, shield against solar radiation, inhibit some type of cancers and contribute to the prevention of degenerative and cardiovascular diseases (Krinsky 1994; Van den Berg et al. 2000).

The carotenoid content of the seeds is influenced by the wheat species and variety (Brandolini et al. 2008; Hidalgo et al. 2006; Paznocht et al. 2018; Ziegler et al. 2015), environmental conditions and stresses (Hidalgo et al. 2009; Lachman et al. 2013), fertilisation (Hidalgo and Brandolini 2017), and post-harvest storage and milling (Hidalgo and Brandolini 2008a, 2008b; Mellado-Ortega and Hornero-Méndez 2016). The carotenoids are easily degraded by oxygen, with a strong effect of heat, light and exposure to hydroperoxides having been reported. During processing, some enzymes (mainly lipoxygenase) catalyse the hydroperoxidation of polyunsaturated fatty acids, creating conjugate hydroperoxides. The radicals formed during this reaction are responsible for the oxidative degradation of the carotenoids (Gardner 1988; Hidalgo and Brandolini 2012; Leenhardt et al. 2006). Flour particles of different sizes coming from distinct wheat species with different grain hardness and moisture content at milling (Posner et al. 2009), have different carotenoid concentrations and colour (Hidalgo et al. 2014; Symons and Dexter 1991).

Carotenoids are scarce in bread wheat, where they range from 0.1 to 2.5 mg/kg d.m., but are more abundant in durum wheat, ranging from 1.5 to 4.8 mg/kg d.m. (Panfili et al. 2004; Zandomeneghi et al. 2000), and the yellow colour of the semolina is perceived as an important quality trait. Recently, Paznocht et al. (2018) reported average total carotenoid contents of 3.60 and 2.41 mg/kg d.m. in purple- and blue-grained wheats, respectively, but observed maximums of 7.46 and 7.04 mg/kg d.m. in other blue- and yellow-grained accessions, respectively. The highest carotenoid content among cultivated wheats was found in einkorn, with an average of 8.5 mg/kg d.m. and a range of 5.3–13.6 mg/kg d.m. (Abdel-Aal et al. 2007; Brandolini et al. 2008; Hidalgo et al. 2006).

Lutein is particularly concentrated in wheat germ but significant amounts are also found in the endosperm (Hidalgo and Brandolini 2008b, Masisi et al. 2015, Ndolo and Beta 2013). As the endosperm represents between 75–85% of the total kernel weight in different wheat species (Hidalgo and Brandolini 2008b; Pomeranz 1988), most of the overall lutein content is retained in the refined flour.

Carotenoid losses during pasta processing have been reported and vary widely according to the extent of lipoxygenase activity in the durum wheat kernel (Borrelli et al. 2003). Relevant losses (up to 48%) were observed during the kneading-extrusion phase while the drying step does not appear to induce significant changes (Hidalgo et al. 2010). By causing structural changes in the food structure, processing also affects the bioaccessibility of carotenoids with reported values of about 70% in durum wheat pasta versus 57% in pasta containing 10% egg (Werner and Böhm 2011).

14 Vitamin B Complex

Vitamins are essential organic micronutrients that are not synthesized by the human body, but by plants and microorganisms. The water-soluble B vitamins in wheat grains include thiamine (B1), riboflavin (B2), niacin (B3), pantothenic acid (B5), pyridoxine (B6) and folate (B9). They act in the human body as coenzymes or their precursors, and/or as factors involved in genetic regulation and genomic stability. Wheat grain is an important source of B vitamins, and there is a high degree of genetic variation in their content among and within wheat species. Additionally, the presence of B vitamins is not only influenced by the genotype but also by the environment and by genotype-by-environment interactions (Davis et al. 1981; Shewry et al. 2011).

Batifoulier et al. (2006) measured concentrations of 2.59–5.41, 0.53–1.07 and 1.44–3.05 mg/kg d.m. of vitamins B1, B2 and B6 respectively in 46 bread wheat cultivars. Similar values, 3.6–5.2, 1.1–1.4 and 2.6–5.7 mg/kg d.m., were found by Davis et al. (1981) in 378 bread wheat accessions. Shewry et al. (2011) reported higher ranges for B1 and B2 in 26 bread wheat lines (5.53–13.55 and 0.77–1.40 mg/kg d.m. respectively). Lower levels of B1 vitamin were observed by Parveen et al. (2015) in eight wheat cultivars (1.22–1.95 mg/kg) and by Witten and Aulrich (2018) in 151 wheat genotypes (1.58–2.96 mg/kg). B2 and B6 vitamins were in the same range, respectively 0.62–1.19 mg/kg (Witten and Aulrich 2018) and 2.23–2.86 mg/kg (Parveen et al. 2015). In durum wheat, Batifoulier et al. (2006) reported concentrations of 4.73, 0.70 and 1.91 mg/kg d.m. for B1, B2 and B6. These are similar to results found by Tekin et al. (2018) in three lines (4.95–5.66, 0.46–0.92 and 2.46–4.07 mg/kg, respectively) and by Davis et al. (1981) in 28 durum genotypes (3.9–4.8, 1.3–1.4 and 3.7–5.1 mg/kg d.m., respectively). Exceptions were vitamins B2 and B6 which were found in higher amounts.

Davis et al. (1981) reported higher levels of vitamin B3 in durum wheat accessions (65.3–75.9 mg/kg d.m.) than in bread wheat (43.3–67.0 mg/kg d.m.), but very low levels of B3 (0.16–1.74 mg/kg d.m.) were observed by Shewry et al. (2011) when analysing 26 bread wheats. Vitamin B5 content was 0.88–4.04 mg/kg in three durum wheats (Tekin et al. 2018), while average B9 concentrations were 0.323–0.774 mg/kg d.m. in 150 bread wheat genotypes (Piironen et al. 2008).

B vitamins in wheat are mostly concentrated in bran and germ. The aleurone layer is particular rich in vitamin B3 (171–741 mg/kg d.m.; Ndolo et al. 2015; Pomeranz 1988) and B9 (4.0–6.0 mg/kg fresh weight; Fenech et al. 1999). In white flour, the vitamin B content is significantly lower than in wholemeal, i.e. 1.46–2.19 versus 2.24–4.16 mg/kg d.m. for B1, 0.43–0.58 versus 0.75–0.96 for B2 and 0.28–0.52 versus 1.31–2.58 mg/kg d.m. for B6 (Batifoulier et al. 2005, 2006). Lebiedzińska et al. (2018) reported a total B6 vitamin concentration of 1.85 mg/kg in spelt flour and 3.27 mg/kg in spelt wholemeal, while Keagy et al. (1980) observed only 32%, 35–42% and 15% of the whole-wheat content of B1, B2 and B6 vitamins in white flours. Given the considerable vitamin loss occurring during milling, the intake of wholemeal products is highly recommended to fulfil the recommended dietary intake of the B vitamins.

In general, during the conventional breadmaking process the loss of B vitamins is significant. According to Nurit et al. (2016), breadmaking incurs significant losses in vitamins B1, B5, and B6, but with a significant increase in vitamin B2. The B3 vitamers vary inversely, as when nicotinic acid decreased, nicotinamide increased. As a way to limit the loss of some B group vitamins, Batifoulier et al. (2005) proposed a long yeast fermentation, which leads to an increase in B1 and B2 concentrations as the result of yeast metabolism.

The impact of processing on vitamin B9 ie. folate, one of the most important vitamins for normal human metabolic function, has also been studied. Germination of grain seems to have the highest impact on the folate content of flours. Koehler et al. (2007) reported a folate content of 0.58 mg/kg d. m. while values of 0.14 mg/kg where reported by Hefni and Witthoft (2012), but both groups observed increases of 3–6 folds of total folate upon grain germination. The breadmaking process does not seem to impact on the native folate content of flour. Gujska and Majewska (2005) reported that although total native folate content increased from flour to proofed dough due to the action of fermenting yeast, it decreased upon baking so values in bread were similar to those in flour. However, significant losses (between 12 to 21%, depending on the specific processing parameters) were observed for folic acid added to flour for fortification purposes.

15 Conclusions

Wheat is largely consumed in the form of breads, pasta and other processed products and the content and properties of specific wheat grain ingredients in these products compared to the native grain can differ greatly. Studies cited in this review underline how major nutritional losses occur at the milling step with the refinement of the flour, which is carried out mainly to improve its processing properties, storability and safety through the removal of undesirable molecules (mycotoxins, pesticides and some heavy metals). It is therefore essential to develop milling methods that minimize the loss of bioactive components in flour while optimizing its commercial and processing quality. Other processing technologies have more specific effects on the different micronutrients and bioactive components. In particular, sourdough fermentation decreases the content of phytic acid and increases the bioavailability of minerals and the content of soluble cell wall polysaccharides and phenolics. Yeast fermentation decreases the amount of dietary fibre in the form of fructans but helps maintain the content of native folates in flour with varying effects on other B group vitamins. Hot high-pressure extrusion used to make breakfast cereals and snacks has a major impact on cell wall polysaccharides, increasing their extractability in water, but reducing the tocol content. Extrusion used in pasta making increases the content of resistant starch, while pasta cooking drastically reduces the content of most soluble components, including fructans, betaine, choline, lignans and sterols. It is clear that improved and targeted processing technologies could bring substantial gains in the nutritional value and health benefits of both whole-grain and refined wheat-based products.

References

Abbaspour N, Richard Hurrell R, Kelishadi R (2014) Review on iron and its importance for human health. J Res Med Sci 19: 164–174.

Abdel-Aal E-SM, Hucl P (1999) A rapid method for quantifying total anthocyanins in blue and purple pericarp wheats. Cereal Chem 76: 350–354.

Abdel-Aal E-SM, Hucl P (2003) Composition and stability of anthocyanins in blue-grained wheat. J Agric Food Chem 51: 2174–2180.

Abdel-Aal E-SM, Young JC, Rabalski I, Hucl P, Fregeau-Reid J (2007) Identification and quantification of seed carotenoids in selected wheat species. J Agric Food Chem 55: 787–794.

Adam A, Crespy V, Levrat-Verny M-A Leenhardt F, Leuillet M, Demigne C et al (2002) The Bioavailability of Ferulic Acid Is Governed Primarily by the Food Matrix Rather than Its Metabolism in Intestine and Liver in Rats. J Nutrition 137: 1962–68.

Adhikari KB, Tanwir F, Gregersen P, Steffensen SK, Jensen BM, Poulsen LK et al. (2015) Benzoxazinoids: Cereal phytochemicals with putative therapeutic and health-protecting properties. Mol Nutr and Food Res. 59: 1324–1338.

Aehle E, Müller U, Eklund PC, Willfor S M, Sippl W, Dräger B (2011) Lignans as food constituents with estrogen and antiestrogen activity. Phytochem 72: 2396–2405.

Andersson AAM, Aman P, Wandel M, Frølich W (2010a) Alkylresorcinols in wheat and rye flour and bread. J Food Composition and Analysis 23: 794–801.

Andersson AAM, Andersson R, Piironen V Lampi A-M, Nystrom L, Boros D et al. (2013) Contents of dietary fibre components and their relation to associated bioactive components in whole grain wheat samples from the HEALTHGRAIN diversity screen. Food Chem 136: 1243–1248.

Andersson AAM, Dinberg L, Aman P, Landberg R (2014) Recent findings on certain bioactive components in whole grain wheat and rye. J Cereal Sci 59: 294–311.

Andersson Y, Hedlund B, Jonsson L, Svensson S (1981) Extrusion cooking of a high-fiber cereal product with crispbread character. Cereal Chem 58: 370–374.

Andersson AAM, Kamal-Eldin AK-E, Aman P (2010b) Effects of Environment and variety on Alkylresorcinols in Wheat in the HealthGrain Diversity screen. J Agric Food Chem 58: 9299–9305.

Andersson AAM, Kamal-Eldin AK-E, Fra A, Boros D, Aman P (2008). Alkylresorcinols in wheat varieties in the HealthGrain Diversity screen. J Agric Food Chem 56: 9722–9725.

Andersson SW, Skinner J, Ellegard, Welch A A, Bingham S, Mulligan A, Andersson H, Khaw K-T (2004) Intake of dietary plant sterols is inversely related to serum cholesterol concentration in men and women in the EPIC Norfolk population: a cross-sectional study. Eur J Clinical Nutr 58: 1378–1385.

Andreasen MF, Kroon PA, Williamson G, Garcia-Conesa MT (2001) Intestinal release and uptake of phenolic antioxidant diferulic acids. Free Radic Biol Med 31: 304–314.

Anson NM, Selinheimo E, Havenaar R, Aura AM et al (2009b) Bioprocessing of wheat bran improves in vitro bioaccessibility and colonic metabolism of phenolic compounds. J Agric Food Chem 57: 6148–55.

Anson NM, van den Berg R, Havenaar R, Bast A, Haenen G R (2008) Ferulic acid from aleurone determines the antioxidant potency of wheat grain (Triticum aestivum L.). J Agric Food Chem 56: 5589–5594.

Anson NM, van den Berg R, Havenaar R, Bast A, Haenen, G R M M (2009a) Bioavailability of ferulic acid is determined by its bioaccessibility. J Cereal Sci 49: 296–300.

Antoine C, Peyron S, Mabille F, Lapierre C, .Bouchet B, Abecassis J, Rouau X (2003) Individual contribution of grain outer layers and their cell wall structure to the mechanical properties of wheat bran. J Agric Food Chem 51: 2026–2033.

Axelson M, Sjövall J, Gustafsson B E, Setchel, KDR (1982) Origin of lignans in mammals and identification of a precursor from plants. Nature 298: 659–660.

Bacic A, Stone BA (1981) Chemistry and organisation of aleurone cell wall components from wheat and barley. Austr J Plant Phys 8: 475–49.

Balmer Y, Vensel W H, DuPont F-M, Buchanan B B, Hurkman W J (2006) Proteome of amyloplasts isolated from developing wheat endosperm presents evidence of broad metabolic capability. J Exp Botany 57: 1591–1602.

Barron C, Samson M-F, Lullien-Pellerin V, Rouau X (2011) Wheat grain tissue proportions in milling fractions using biochemical marker measurements: Application to different wheat cultivars. J Cereal Sci 53: 306–311.

Barron C, Surget A, Rouau X (2007) Relative amounts of tissues in mature wheat (Triticum aestivum L.) grain and their carbohydrate and phenolic acid composition. J Cereal Sci 45:88–96.

Bartnik M, Szafranska I (1987) Changes in phytate content and phytase activity during the germination of some cereals. J Cereal Sci 5: 23–28.

Batifoulier F, Verny M–A, Chanliaud E, Rémésy C, Demigné C (2005) Effect of different bread-making methods on thiamine, riboflavin and pyridoxine contents of wheat bread. J Cereal Sci 42: 101–108.

Batifoulier F, Verny M–A, Chanliaud E, Rémésy C, Demigné C (2006) Variability of B vitamin concentrations in wheat grain, milling fractions and bread products. Europ J Agronomy 25: 163–169.

Beleggia R, Platani C, Nigro F, De Vita P, Cattivelli L, Papa R (2013) Effect of genotype, environment and genotype-by environment interaction on metabolite profiling in durum wheat (Triticum durum desf.) grain. J Cereal Sci 57: 183–192.

Beleggia R, Platani C, Papa R, Di Chio A, Barros E, Mashaba C et al (2011) Metabolomics and food processing: from semolina to pasta. J Agric Food Chem 59: 9366–9377.

Berry CS (1986) Resistant starch: formation and measurement of starch that survives exhaustive digestion with amylolytic enzymes during the determination of dietary fibre. J Cereal Sci 4: 301–314.

Beta T, Nam S, Dexter J E, Sapirstein HD (2005) Phenolic content and antioxidant activity of pearled wheat and roller-milled fractions. Cereal Chem 82: 390–393.

Biesiekierski JR, Peters S L, Newnham ED, Rosella O, Muir JG, Gibson PR (2013) No Effects of Gluten in Patients With Self-Reported Non-Celiac Gluten Sensitivity After Dietary Reduction of Fermentable, Poorly Absorbed, Short-Chain Carbohydrates. Gastroenterology 145: 320–328.

Björck I, Nyman M, Asp N-G (1984) Extrusion cooking and dietary fibre: Effects on dietary fibre content and on degradation in the rat intestinal tract. Cereal Chem 61: 174–179.

Blandino M, Sovrani V, Marinaccio F, Reyneri A, Rolle L, Giacosa S et al. (2013) Nutritional and technological quality of bread enriched with an intermediate pearled wheat fraction. Food Chem 141: 2549–2557.

Böhmdorfer S, Oberlerchner JT, Fuchs C, Rosenau T, Grausgruber H (2018) Profiling and quantification of grain anthocyanins in purple pericarp × blue aleurone wheat crosses by high-performance thin-layer chromatography and densitometry. Plant Methods 14: 29.

Bordiga M, Locatelli M, Travaglia F, Arlorio M (2016) Alkylresorcinol content in whole grains and pearled fractions of wheat and barley. J Cereal Sci 70: 38–46.

Borrelli GM, De Leonardis AM, Fares, Platani C, DiFonzo N (2003) Effects of modified processing conditions on oxidative properties of semolina dough and pasta. Cereal Chem 80: 225–231.

Brandolini A, Hidalgo A, Gabriele S, Heun M (2015) Chemical composition of wild and feral diploid wheats and their bearing on domesticated wheats. J Cereal Sci 63: 122–127.

Brandolini A, Hidalgo A, Moscaritolo S (2008) Chemical composition and pasting properties of einkorn (Triticum monococcum L., subsp monococcum) whole meal flour. J Cereal Sci 47: 599–609.

Brighenti F, Cristina M, Baggio C (1998) Resistant starch in Italian diet. British J Nutr 80: 333–41.

Britton G (1995) Structure and properties of carotenoids in relation to function. FASEB J 9: 1551–1558.

Brown G, Gordon S (2001) Immune recognition: a new receptor for β-glucans. Nature 413: 36–37.

Brownlee IA (2011) The physiological roles of dietary fibre. Food Hydrocolloids 25: 238–250.

Buchmann CA, Nersesyan A, Kopp B, Schauberger D, Darroudi F, Grummt T et al. (2007) Dihydroxy-7-methoxy-1,4-benzoxazin-3-one (DIMBOA) and 2,4-dihydroxy-1,4-benzoxazin-3-one (DIBOA), two naturally occurring benzoxazinones contained in sprouts of Gramineae are potent aneugens in human-derived liver cells (HepG2). Cancer Letters 246: 290–299.

Bunzel M, Ralph J, Lu F, Hatfield RD, Steinhart, H (2004) Lignins and ferulate-coniferyl alcohol cross-coupling products in cereal grains. J Agric Food Chem 52: 6496–6502.

Buttriss L, Stokes CS (2008) "Dietary Fibre and Health: An Overview," Nutrition Bulletin 33: 186–200.

Capek P, Matulová M (2013) An arabino(glucurono)xylan isolated from immunomodulatory active hemicellulose fraction of Salvia officinalis L. Int J Biol Macromol 59: 396–401.

Chen Y, Ross A, Aman P, Kamal-Eldin, A (2004) Alkylresorcinols as markers of whole grain wheat and rye in cereal products. J of Food Agric Chem 52: 8242–8246.

Cleemput G, Booij C, Hessing M, Gruppen H, Delcour J A (1997) Solubilisation and changes in molecular weight distribution of arabinoxylans and protein in wheat flours during breadmaking, and the effects of endogenous arabinoxylan hydrolysing enzymes. J Cereal Sci 26: 55–66.
Comino P, Collins H, Lahnstein J, Gidley M J (2016) Effects of diverse food processing conditions on the structure and solubility of wheat, barley and rye endosperm dietary fibre. J Food Engineering 169: 228–237.
Cooper DN, Kable ME, Marco ML, De Leon A, Rust B, Baker JE et al. (2017) The Effects of Moderate Whole Grain Consumption on Fasting Glucose and Lipids, Gastrointestinal Symptoms, and Microbiota. Nutrients. 9: 173.
Corol DI, Ravel C, Raksegi M, Charmet G, Beale MH, Shewry PR, Ward JL (2012) Effects of genotype and environment on the contents of betaine, choline and trigonelline in cereal grains. J Agric Food Chem 60: 5471–5481.
Cubadda F, Aureli F, Raggi A, Carcea M (2009) Effect of milling, pasta making and cooking on minerals in durum wheat. J Cereal Sci. 49: 92–7.
Das A, Raychaudhuri U, Chakraborty R (2012) Cereal based functional food of Indian subcontinent: a review. J Food Sci Tech. 49: 665–672.
Davin LB, Jourdes M, Patten A M, Kim KW, Vassão D G, Lewis N et al. (2008) Dissection of lignin macromolecular configuration and assembly: comparison to related biochemical processes in allyl/propenyl phenol and lignan biosynthesis. Natural Product Reports 25: 1015–90.
Davis KR, Cain RF, Peters LJ, Tourneau DL, McGinnis J (1981) Evaluation of the nutrient composition of wheat II. Proximate analysis, thiamine, riboflavin, niacin and pyridoxine. Cereal Chem 58: 116–120.
DiNicolantonio JJH, O'Keefe J, Wilson W (2018) Subclinical magnesium deficiency: a principal driver of cardiovascular disease and a public health crisis Open Heart 5: e000668.
Dornez E, Gebruers K, Wiame S, Delcour JA, Courtin CM (2006) Insight into the distribution of arabinoxylans, endoxylanases, and endoxylanase inhibitors in industrial wheat roller mill streams. J. Agr. Food Chem. 54: 8521–8529.
Drankham K, Carter J, Madl R, Klopfenstein C, Padula F, Lu Y et al. (2003) Antitumor activity of wheats with high orthophenolic content. Nutr Cancer 47: 188–194.
Durazzo A, Azzini E, Raguzzini A, Maiani, G, Finocchiaro F, Ferrari B et al. (2009) Influence of processing on the lignans content of cereal based foods. Tecnica Molitoria International 60: 163–173.
Durazzo A, Azzini E, Turfani V, Polito A, Maiani G, Carcea M (2013) Effect of cooking on lignans content in whole-grain pasta made with different cereals and other seeds. Cereal Chem 90: 169–171.
Eagling T, Neal AL, McGrath SP, Fairweather-Tait S, Shewry PR, Zhao F-J (2014b) Distribution and speciation of iron and zinc in grain of two wheat genotypes. J Agric Food Chem 62: 708–716.
Eagling T, Wawer A, Shewry P R, Zhao F J, Fairweather-Tait S (2014a) Iron Bioavailability in Two Commercial Cultivars of Wheat: Comparison between Wholegrain and White Flour and the Effects of Nicotianamine and 2′-Deoxymugineic Acid on Iron Uptake into Caco-2 Cells. J Agric Food Chem 62: 10320–10325.
Eerlingen RC, Deceuninck M, Delcour JA (1993) Enzyme resistant starch II: Influence of amylose chain length on resistant starch formation. Cereal Chem 70: 345–350.
Eerlingen RC, Jacobs H, Delcour JA (1994) Enzyme- resistant starch V. Effect of retrogradation of waxy maize starch on enzyme susceptibility. Cereal Chem 71: 351–355.
Escalante-Aburto A, Ramírez-Wong B, Torres-Chávez PI Figueroa-Cardenas JD, Lopez-Cervantes J, Barron-Hoyos JM et al (2013) Effect of extrusion processing parameters on anthocyanin content and physicochemical properties of nixtamalized blue corn expanded extrudates. CyTA - J Food 11: 29–37.
Escribano-Bailòn MT, Santos-Buelga C, Rivas-Gonzalo JC (2004) Anthocyanins in cereals. J Chromatrogr A 1054: 129–141.

Evers AD, Blakeney AB, O'Brien LO (1999) Cereal structure and composition. Aust J Agric Res 50, 629–650.

Fenech M, Noakes M, Clifton P, Topping D (1999) Aleurone flour is a rich source of bioavailable folate in humans. J Nutr 129: 1114–1119.

Ficco DBM, De Simone V, Colecchia SA, Pecorella I, Platani C et al. (2014) Genetic variability in anthocyanin composition and nutritional properties of blue, purple, and red bread (*Triticum aestivum* L.) and durum (*Triticum turgidum* L. ssp. *turgidum* convar. *durum*) wheats. J Agric Food Chem 62: 8686–8695.

Fratianni A, Di Criscio T, Mignogna R, Panfili G (2012) Carotenoids, tocols and retinols evolution during egg pasta–making processes Food Chem 131: 590–595.

Galliard T (1989) Rancidity in cereal products. In: Allen JC and Hamilton R J (eds), Rancidity in Foods, 2nd ed. Elsevier, London, p 141–157.

Gardner HW (1988) Lipoxygenase pathways in cereals. In: Pomeranz Y (ed). Advances in cereal science and technology, vol 9. American Association of Cereal Chemists, St. Paul, MN, USA, p 161–215.

Garg M, Chawla M, Chunduri V, Kumar R, Sharma S et al. (2016) Transfer of grain colors to elite wheat cultivars and their characterization J Cereal Sci 71: 138–144.

Gelinas P, McKinnon C, Gagnon F (2015) Fructans, water-soluble fibre and fermentable sugars in bread and pasta made with ancient and modern wheat. Intern J Food Sci and Technol 51: 555–564.

Gibson PR, Shepherd SJ (2005) Personal view: food for thought – western lifestyle and susceptibility to Crohn's disease. The FODMAP hypothesis. Aliment Pharmacol Ther 21: 1399–1409.

Graham SF, Holis JH, Migaud M, Browne RA (2009). Analysis of Betaine and Choline Contents of Aleurone, Bran, and Flour Fractions of Wheat (Triticum aestivum L.) Using 1H Nuclear Magnetic Resonance (NMR) Spectroscopy. J Agric Food Chem 57: 1948–1951.

Greffeuille V, Abecassis J, Bar L'Helgouac'h C, Lullien-Pellerin V (2005) Differences in the aleurone layer fate between hard and soft common wheats at grain milling. Cereal Chem 82: 138–143.

Gujska E, Majewska K (2005) Effect of baking process on added folic acid and endogenous folates stability in wheat and rye breads. Plant Foods Hum Nutr 60: 37–42.

Gupta RK, Gangoliya SS, Singh NK (2015) Screening and Characterization of Wheat Germplasms for Phytic Acid and Iron Content J. Agr. Sci. Tech. 17: 747–756.

Håkansson B, Jägerstad M, öste R, Åkesson B, Jonsson L (1987) The effects of various thermal processes on protein quality, vitamins and selenium content in whole grain wheat and white flour. J. Cereal Sci. 6: 269

Håkansson B, Jägerstad M (1990) The effect of thermal inactivation of lipoxygenase on the stability of vitamin E in wheat. J Cereal Sci. 12: 177–185.

Hallström E, Sestili F, Lafiandra D, Björck I,Östman E (2011) A novel wheat variety with elevated content of amylose increases resistant starch formation and may beneficially influence glycaemia in healthy subjects Food Nutr Res 55: https://doi.org/10.3402/fnr.v55i0.7074

Han H-M, Koh B-K (2011) Antioxidant activity of hard wheat flour, dough and bread prepared using various processes with the addition of different phenolic acids. J Sci Food Agr 91: 604–608.

Hansen HB, Andreasen MF, Nielsen MM, Larsen LM, Knudsen KEB, Meyer AS et al. (2002) Changes in dietary fibre, phenolic acids and activity of endogenous enzymes during rye bread-making. European Food Res and Technol 214: 33–42.

Hashimoto Y, Shudo K (1996) Chemistry of biologically active benzoxazinoids. Phytochem 43, 551–559.

Haska L, Nyman M, Andersson R (2008) Distribution and characterisation of fructan in wheat milling fractions. J Cereal Sci 48: 768–774.

Hefni M, Witthoft CM (2012) Enhancement of the folate content in Egyptian pita bread. Food & Nutrition Research. 56: 5566.

Hemery Y, Lullien-Pellerin V, Rouau X, Abecassis J, Samson M-F, Åman P et al. (2009) Biochemical markers: Efficient tools for the assessment of wheat grain tissue proportions in milling fractions. J. Cereal Sci. 49: 55–64.
Hemery Y, Rouau X, Lullien-Pellerin V, Barron C, Abecassis J (2007) Dry processes to develop wheat fractions and products with enhanced nutritional quality. J Cereal Sci 46: 327–347.
Henkin RI (1984) Zinc in taste function - a critical review. Biol Trace Elem Res 6: 263–280.
Hidalgo A, Brandolini A (2008a) Kinetics of carotenoids degradation during the storage of einkorn (Triticum monococcum L. ssp. monococcum) and breadwheat (Triticum aestivum L. ssp. aestivum) flours. J Agric Food Chem 56: 11300–11305.
Hidalgo A, Brandolini A (2008b) Protein, ash, lutein and tocols distribution in einkorn (Triticum monococcum L. subsp. monococcum) seed fractions. Food Chem 107: 444–448.
Hidalgo A, Brandolini A (2012) Lipoxygenase activity in whole meal flours from Triticum monococcum, Triticum turgidum and Triticum aestivum. Food Chem 131: 1499–1503.
Hidalgo A, Brandolini A (2017) Nitrogen fertilisation effects on technological parameters and carotenoid, tocol and phenolic acid content of einkorn (Triticum monococcum L. subsp. monococcum): A two-year evaluation. J Cereal Sci 73: 18–24.
Hidalgo A, Brandolini A, Pompei C (2010) Carotenoids evolution during pasta, bread and water biscuit preparation from wheat flours. Food Chem 121: 746–751.
Hidalgo A, Brandolini A, Pompei C, Piscozzi R (2006) Carotenoids and tocols of einkorn wheat (Triticum monococcum ssp monococcum L.). J Cereal Sci 44: 182–193.
Hidalgo A, Brandolini A, Ratti S (2009) Influence of genetic and environmental factors on selected nutritional traits of Triticum monococcum. J Agric Food Chem 57: 6342–6348.
Hidalgo A, Fongaro L, Brandolini A (2014) Wheat flour granulometry determines colour perception. Food Res Int 64: 363–370.
Huang Z, Dostal L, Rosazza JPN (1993) Microbial transformations of ferulic acid by Saccharomyces cerevisiae and Pseudomonas fluorescens. Appl. Environ. Microbiol. 59: 2244–2250.
Huynh B-L, Lachlan P, Mather DE, Wallwork H, Graham RD, Welch RM, Stangoulis JCR (2008) Genotypic variation in wheat grain fructan content revealed by a simplified HPLC method. J Cereal Sci 48: 369–378.
Iiyama K, Lam TB-T, Stone BA (1994) Covalent Cross-Links in the Cell Wall. Plant Physiol. 104: 315–320.
Irmak S, Dunford NT, Milligan J (2006) Policosanol contents of beeswax, sugar cane and wheat extracts. Food Chem 95: 312–318.
Jaafar SNS, Baron J, Siebenhandl-Ehn S, Rosenau T, Böhmdorfer S, Grausgruber H (2013) Increased anthocyanin content in purple pericarp × blue aleurone wheat crosses. Plant Breeding 132: 546–552.
Jenkins DJA, Wolever TMS, Collier G R (1987) Metabolic effects of a low-glycemic-index diet. Am J Clin Nutr 46: 968–75.
Kamal-Eldin A, Appelqvist LA (1996) The chemistry and antioxidant properties of tocopherols and tocotrienols. Lipids 31: 671–701.
Keagy PM, Borenstein B, Ranum P, Connor MA, Lorenz K, Hobbs WE et al. (1980) Natural levels of nutrients in commercially milled wheat flours. Cereal Chem 57: 59–65.
Keaveney EM, Price R K, Hamill LL, Wallace JMW, McNulty H, Ward M et al. (2015) Postprandial plasma betaine and other methyl donor-related responses after consumption of minimally processed wheat bran or wheat aleurone, or wheat aleurone incorporated into bread. British J Nutrition 113: 445–453.
Kern SM, Bennett RN, Mellon FA, Kroon PA, Garcia-Conesa MT (2003) Absorption of hydroxycinnamates in humans after high-bran cereal consumption. J Agric Food Chem 51: 6050–55.
Kiyama R (2016) Biological effects induced by estrogenic activity of lignans. Trends Food Sci Technol 54: 186–196.
Klepacka J, Fornal L (2006) Ferulic acid and its position among the phenolic compounds of wheat. Critical Reviews in Food Science and Nutrition 46: 639–647.
Knievel DC, Abdel-Aal E-SM, Rabalski I, Nakamura T, Hucl P (2009) Grain color development and the inheritance of high anthocyanin blue aleurone and purple pericarp in spring wheat (*Triticum aestivum* L.). J Cereal Sci 50: 113–120.

Knudsen KEB (1997) Carbohydrate and lignin contents of plant materials used in animal feeding. Animal Feed Science Technology 67: 319–338.

Koehler P, Hartmann G, Wieser H, Rychlik M (2007) Changes of folates, dietary fiber, and proteins in wheat as affected by germination. J Agric Food Chem 55: 4678–4683.

Kong J-M, Chia L-S, Goh N-K, Chia T-F, Brouillard R (2003) Analysis and biological activities of anthocyanins. Phytochem 64: 923–933.

Konopka I, Tańska M, Faron A, Czaplicki S (2014) Release of Free Ferulic Acid and Changes in Antioxidant Properties during the Wheat and Rye Bread Making Process. Food Sci Biotechnol 23: 831–840.

Kozubek A, Nienartowicz B (1995) Cereal grain resorcinolic lipids inhibit H_2O_2-induced peroxidation of biological membranes. Acta Biochim Pol 42: 309–3015.

Kozubek A, Tyman JHP (1999) Resorcinolic Lipids, the Natural Non-isoprenoid Phenolic Amphiphiles and Their Biological Activity. Chem Rev 99: 1–25.

Krinsky NI (1994) The biological properties of carotenoids. Pure Appl Chem 66: 1003–1010.

Kristensen M, Toubro S, Jensen MG, Ross AB, Riboldi G, Petronio M et al. (2012) Whole grain compared with refined wheat decreases the percentage of body fat following a 12-week, energy-restricted dietary intervention in postmenopausal women. J Nutr. 142: 710–716.

Krüger S, Morlock GE (2018) Fingerprinting and characterization of anthocyanins in 94 colored wheat varieties and blue aleurone and purple pericarp wheat crosses. J Chromatogr A 1538: 75–85.

Kumar V, Sinha A K, Makkar HPS, Becker K (2010) Dietary roles of phytate and phytase in human nutrition: a review. Food Chem 120: 945–959.

Lachman J, Hejtmankova K, Kotikova Z (2013) Tocols and carotenoids of einkorn, emmer and spring wheat varieties: selection for breeding and production. J Cereal Sci 57: 207–214.

Laddomada B, Caretto S, Mita G (2015) Wheat Bran Phenolic Acids: Bioavailability and Stability in Whole Wheat-Based Foods. Molecules 20: 15666–15685.

Lampi AM, Nurmi T, Ollilainen V, Piironen V (2008) Tocopherols and tocotrienols in wheat genotypes in the HEALTHGRAIN Diversity Screening. J Agric Food Chem 56: 9716–21.

Landberg R, Kamal-Eldin A, Andersson R, Aman P (2006) Alkylresorcinol Content and Homologue Composition in Durum Wheat (Triticum durum) Kernels and Pasta Products. J Agric Food Chem 54: 3012–3014.

Landberg R, Kamal-Eldin A, Salmenkallio-Marttila M, Rouau X, Aman P (2008) Localization of alkylresorcinols in wheat, rye and barley kernels. J Cereal Sci 48: 401–406.

Landberg R, Marklund M, Kamal-Eldin A, Aman P (2014) An update on alkylresorcinols – Occurrence, bioavailability, bioactivity and utility as biomarkers. J of Functional Foods 7: 77–89.

Landete JM (2012) Plant and mammalian lignans: A review of source, intake, metabolism, intestinal bacteria and health. Food Research International 46: 410–424.

Lebiedzińska A, Marszałł ML, Grembecka M, Czaja J, Szefer P, Kuta J et al. (2018) Detection of B6 vitamers in grain products: experimental and computational studies. Food Anal Methods 11: 725–732.

Leenhardt F, Levrat-Verny M-A, Chanliaud E, Rémésy C (2005) Moderate Decrease of pH by Sourdough Fermentation Is Sufficient to Reduce Phytate Content of Whole Wheat Flour through Endogenous Phytase Activity. J Agric Food Chem 53: 98–102.

Leenhardt F, Lyan B, Rock E, Boussard A, Potus J, Chanliaud E, Remesy C (2006) Wheat lipoxygenase activity induces greater loss of carotenoids than vitamin E during breadmaking. J Agric Food Chem 54, 1710–1715.

Li W, Cui SW, Kakuda Y (2006) Extraction, fractionation, structural and physical characterization of wheat β-D-glucans. Carbohydrate Polymers 63: 408–416.

Lineback DR, Rasper VF (1988) In: Pomeranz Y (ed). Wheat Chemistry and Technology. American Association of Cereal Chemists Inc., St Paul, p. 277–372.

Liu Q, Qiu Y, Beta T (2010) Comparison of antioxidant activities of different colored wheat grains and analysis of phenolic compounds. J Sci Food Agric 58: 9235–9241.

Luo YW, Xie W-H, Qian JX Wang, He, Y-J (2014) Effects of germination on iron, zinc, calcium, manganese, and copper availability from cereals and legumes. J Food 12: 22–26.

Magallanes-López AM, Hernandez-Espinosa N, Velu G, Posadas-Romano G, Ordoñez-Villegas VM G, Crossa J et al. (2017) Variability in iron, zinc and phytic acid content in a worldwide collection of commercial durum wheat cultivars and the effect of reduced irrigation on these traits. Food Chem 237: 499–505.

Mandalari G, Faulds C, Sancho A, Saija A, Bisignano G, LoCurto R, Waldron K (2005) Fractionation and characterisation of arabinoxylans from brewers' spent grain and wheat bran. J Cereal Sci 42: 205–212.

Mares DJ, Stone BA (1973) Studies on wheat endosperm. I. Chemical composition and ultrastructure of the cell walls. Aust J Biol Sci 26: 793–812.

Masisi K, Diehl-Jones, Gordon J, Chapman D, Moghadasian M, Beta T (2015) Carotenoids of aleurone, germ and endosperm fractions of barley, corn and wheat differentially inhibit oxydative stress. J Agric Food Chem 63, 2715–2724.

Matthews JM, Sunde M (2002) Zinc Fingers - Folds for Many Occasions. Life 54: 351–355.

Mellado-Ortega E, Hornero-Méndez D (2016) Carotenoid evolution during short storage period of durum wheat (Triticum turgidum conv. durum) and tritordeum (× Tritordeum Ascherson et Graebner) whole-grain flours. Food Chem 192: 714–723.

Mendis M, Leclerc E, Simsek S (2016) Arabinoxylans, gut microbiota and immunity. Carbohydrate Polymers 139: 159–166.

Monge L, Cortassa G, Fiocchi F, Mussino G, Carta Q (1990) Glyco-insulinaemic response, digestion and intestinal absorption of the starch contained in two types of spaghetti. Diabetes, Nutrition and Metabolism 3: 239–246.

Moore J, Cheng Z, Hao J, Guo G, Liu J-G, Lin C, Yu L (2007) Effects of solid-state yeast treatment on the antioxidant properties and protein and fiber compositions of common hard wheat bran. J Agric Food Chem 55: 10173–10182.

Moore KL, Zhao F-J, Gritsch CS, Tosi P, Hawkesford MJ, McGrath SP et al. (2012) Localisation of iron in wheat grain using high resolution secondary ion mass spectrometry J Cereal Sci 55: 183–187.

Muir AD, Westcott ND (2000) Quantitation of the lignan secoisolariciresinol diglucoside in baked goods containing flax seed or flax meal. J Agric Food Chem 48: 4048–4052.

Muraoka S, Miura T (2004) Inhibition of xanthine oxidase by phytic acid and its antioxidative action. Life Sci 74: 1691–1700.

Nelson K, Mathai ML, Ashton JF, Donkor ON, Vasiljevic T, Mamilla R, Stojanovska L (2016) Effects of malted and non-malted whole-grain wheat on metabolic and inflammatory biomarkers in overweight/obese adults: a randomised crossover pilot study. Food Chem 194: 495–502.

Ndolo VU, Beta T (2013) Distribution of carotenoids in endosperm, germ and aleurone fractions of cereal grain kernels. Food Chem 139, 663–671.

Ndolo VU, Fulcher RG, Beta T (2015) Application of LC-MS-MS to identify niacin in aleurone layers of yellow corn, barley and wheat kernels. J Cereal Sci 65: 88–95.

Nemeth C, Freeman J, Jones HD, Sparks C, Pellny TK, Wilkinson MD et al. (2010) Downregulation of the CSLF6 gene results in decreased (1,3,1,4)-b-D-glucan in endosperm of wheat. Plant Physiol 152: 1209–1218.

Nicolas J, Drapron R (1983) Lipoxygenase and some related enzymes in breadmaking. In: Barnes P J (ed) Lipids in Cereal Technology. Academic, London, p 213–235.

Nielsen MM, Hansen Å (2008) Stability of Vitamin E in Wheat Flour and Whole Wheat Flour During Storage. Cereal Chemistry 85: 716–720.

Nurit E, Lyan B, Pujos-Guillot E, Branlard G, Piquet A (2016) Change in B and E vitamin and lutein, b-sitosterol contents in industrial milling fractions and during toasted bread production. J Cereal Sci 69: 290–296.

Nurmi T, Lampi A-M, Nyström L, Turunen M., Piironen V (2010a) Effects of environment and genotype on phytosterols in wheat in the HEALTHGRAIN diversity screen. J Agric Food Chem 58: 9314–9323.

Nurmi T, Lampi A-M, Nyström L, Turunen M, Piironen V (2010b) Effects of genotype and environment on steryl ferulates in wheat and rye in the HEALTHGRAIN diversity screen. J Agric Food Chem 58: 9332–9340.

Nurmi T, Lampi A-M, Nyström L, Hemery Y, Rouau X, Piironen V (2012) Distribution and composition of phytosterols and steryl ferulate in wheat grain and bran fractions. Cereal Sci 56: 379–388.

Nyström L, Lampi A-M, Rita H, Aura A-M, Oksman-Caldentey K-M, Piiroen V (2007a) Effects of Processing on Availability of Total Plant Sterols, Steryl Ferulates and Steryl Glycosides from Wheat and Rye Bran. J Agric Food Chem 55: 9059–9065.

Nyström L, Paasonen A, Lampi AM, Piironen V (2007b) Total plant sterols, steryl ferulates and steryl glycosides in milling fractions of wheat and rye. J Cereal Sci 45: 106–115.

O'Dell BL, de Boland AR, Koirtyohann SR (1972) Distribution of phytate and nutritionally important elements among the morphological components of cereal grains. J Agr Food Chem 20: 718–721.

Oury F-X, .Leenhardt F, Rémésy C, Chanliaud E, Duperrier B, Balfourier F, Charmet G (2006) Genetic variability and stability of grain magnesium, zinc and iron concentrations in bread wheat European J of Agronomy 25: 277–185.

Parker ML, Ng A, Waldron KW (2005) The phenolic acid and polysaccharide composition of cell walls of bran layers of mature wheat (Triticum aestivum L. cv. Avalon) grains. J Sci Food Agric 85: 2539–2547.

Panfili G, Fratianni A, Irano M (2004) Improved normal phase high liquid chromatography procedure for determination of carotenoids in cereals. J Agric Food Chem 52: 6373–6377.

Panyoo AE, Emmambux MN (2017) Amylose–lipid complex production and potential health benefits: A mini-review. Starch/Stärke 69: 1600203.

Parveen S, Siyal AN, Memon N, Memon SQ, Khuhawar MY (2015) Extraction and determination of phenolic acids and vitamin B of sieved and unsieved wheat grain by MEKC. J Liquid Chrom Related Technol 38: 143–152.

Paznocht L, Kotíková Z, Šulc M, Lachman J, Orsák M, Eliášová M, Martinek P (2018) Free and esterified carotenoids in pigmented wheat, tritordeum and barley grains. Food Chem 240: 670–678.

Pedersen HA, Laursen B, Mortensen A, Fomsgaard IS (2011) Bread from common cereal cultivars contains an important array of neglected bioactive benzoxazinoids. Food Chem 127: 1814–1820.

Persson DP, de Bang TC, Pedas PR, Kutman UB, Cakmak I, Andersen B et al. (2016) Molecular speciation and tissue compartmentation of zinc in durum wheat grains with contrasting nutritional status. New Phytologist 211: 1255–1265.

Piironen V, Edelmann M, Kariluoto S, Bedo Z (2008) Folate in wheat genotypes in the HEALTHGRAIN diversity screen. J Agric Food Chem 56: 9726–9731.

Piironen V, Syväoja E-L, Varo P, Salminen K, Koivistoinen P (1986) Tocopherols and tocotrienols in cereal products from Finland. Cereal Chem 63: 78–81.

Pomeranz Y (1988) Chemical compounds of kernel structures. In: Pomeranz Y (ed) Wheat chemistry and technology. American Association of Cereal Chemists Inc., St. Paul., Minnesota, USA, pp. 97–158.

Posner ES, Khan K, Shewry PR (2009) Wheat flour milling. In: Kahn K, Shewry PR (eds). Wheat: chemistry and technology, 4th ed. AACC International, Minneapolis, MN, USA, p 119–152.

Price RK, Keaveney EM, Hamill LL et al. (2010) Consumption of wheat aleurone-rich foods increases fasting plasma betaine and modestly fasting homocysteine and LDL-cholesterol in adults. J Nutr 140: 2153–2157.

Price RK, Wallace JMW, Hamill LL, Wallace J M W., Ward M, Ueland P M et al. (2012) Evaluation of the effect of wheat aleurone-rich foods on markers of antioxidant status, inflammation and endothelial function in apparently healthy men and women. British J Nutr 108: 1644–1651.

Rashid S, Rakha A, Anjum FM, Ahmed W, Sohail M (2015) Effects of extrusion cooking on the dietary fibre content and Water Solubility Index of wheat bran extrudates International Journal of Food Science and Technology 50: 1533–1537.

Rios G, Pinson-Gadais L, Abecassis J, Zakhia-Rozis N, Lullien-Pellerin V (2009) Assessment of dehulling efficiency to reduce deoxynivalenol and fusarium level in durum wheat grains. J Cereal Sci 49: 387–392.

Rejman J, Kozubek A (2004) Inhibitory effect of natural phenolic lipids upon NAD-dependent deshydrogenases and on Triglyceride Accumulation in 3T3-L1 cells in culture. J Agric Food Chem 52: 246–250.
Rhodes DI, Stone BA (2002) Proteins in walls of wheat aleurone cells. J Cereal Sci 36: 83–101.
Rice PJ, Adams EL, Ozment-Skelton T, Gonzalez AJ, Goldman MP, Lockhart BE et al. (2005) Oral delivery and gastrointestinal absorption of soluble glucans stimulate increased resistance to infectious challenge. J Pharmacol Exp Ther 314: 1079–1086.
Ritsema T, Smeekens S (2003) Fructans: Beneficial for Plants and Humans. Current Opinion in Plant Biology 6: 223–230.
Roberfroid MB (2005) Introducing inulin-type fructans. British J Nutr 93: 513–525.
Rodriguez-Ramiro I, Brearley CA, Bruggraber S F A, Perfecto A, Shewry P R, Fairweather-Tait S (2017) Assessment of iron bioavailability from different breadmaking processes using an in vitro intestinal cell model. Food Chem 228: 91–98.
Roohani N, Hurrell R, Kelishadi R Shulin R (2013) Zinc and its importance for human health: An integrative review Res Med Sci 18: 144–157.
Rosa NN, Barron C, Gaiani C, Dufour C, Micard V (2013) Ultra-fine grinding increases the antioxidant capacity of wheat bran. J Cereal Sci 57: 84–90.
Ross AB, Shepherd MJ, Schüpphaus M, Sinclair V, Alfaro B, Kamal-Eldin A, Aman P (2003) Alkylresorcinols in cereals and cereal products. J Food Agric Chem 51: 4111–4118.
Ross A B, Zangger A, Guiraud SP (2014) Cereal foods are the major source of betaine in the Western diet – Analysis of betaine and free choline in cereal foods and updated assessments of betaine intake. Food Chem 145: 859–865.
Rouau X, El-Hayek ML, Moreau D (1994). Effect of an enzyme preparation containing pentosanases on the bread-making quality of flours in relation to changes in pentosan properties. J Cereal Sci 19: 259–272.
Ryan AS (1997) Iron deficiency anemia in infant development: Implications for growth, cognitive development, resistance to infection, and iron supplementation Am J Phys Anthropol 104: 25–62.
Sandberg AS, Brune M, Carlsson NG, Hallberg L, Skoglund E, Rossander-Hulthén L (1999) Inositol phosphates with different numbers of phosphate groups influence iron absorption in humans. Am J Clin Nutr 70: 240–246.
Saulnier L, Robert P, Grintchenko M, Jamme F, Bouchet B, Guillon F (2009) Wheat endosperm cell walls: Spatial heterogeneity of polysaccharide structure and composition using micro-scale enzymatic fingerprinting and FT-IR microspectroscopy. J Cereal Sci 50: 312–317.
Scazzina F, Del Rio D, Pellegrini N, Brighenti F (2009) Sourdough bread: Starch digestibility and postprandial glycemic response J Cereal Sci 49: 419–421.
Shewry PR, Hey SJ (2015) The contribution of wheat to human diet and health. Food and Energy Security 4: 178–202.
Shewry PR, VanSchaik F, Ravel C, Charmet G, Rakszegi M, Bedo Z, Ward J L (2011) Genotype and environment effects on the contents of vitamins B1, B2, B3, and B6 in wheat grain. J Agric Food Chem 59: 10564–10571.
Silva EO, Bracarense AP FRL (2016) Phytic acid: From antinutritional to multiple protection factor of organic systems. J Food Sci 81, 1357–1362.
Singh S, Gamlath S, Wakeling L (2007) Nutritional aspects of food extrusion: A review. Int J Food Sci Technol 42: 916–929.
Skodje GI, Sarna VK, Minelle I H, Rolfsen, K L, Muir J G, Gibson P R et al. (2018) Fructan, Rather Than Gluten, Induces Symptoms in Patients With Self-Reported Non-Celiac Gluten Sensitivity. Gastroenterology 154: 529–539.
Slover HT, Lehmann J (1972) Effects of fumigation on wheat in storage. IV. Tocopherols. Cereal Chem. 49: 412–415.
Smeds AI, Eklund PC, Sjoholm RE, Willför SM, Nishibe S, Deyama T, Holmbom BR (2007) Quantification of a Broad Spectrum of Lignans in Cereals Oilseeds, and Nuts. J Agric Food Chem 55: 1337–1346.

Smeds AI, Jauhiainen L, Tuomola E, Peltonen-Sainio P (2009) Characterization of Variation in the Lignan Content and Composition of Winter Rye, Spring Wheat, and Spring Oat. J Agric Food Chem 57: 5837–5842.

Spilioti E, Holmbom B, Papavassiliou AG, Moutsatsou P (2014) Lignans 7-hydroxymatairesinol and 7-hydroxymatairesinol 2 exhibit anti-inflammatory activity in human aortic endothelial cells. Mol Nutr Food Res 58: 749–759.

Stone B, Morell MK (2009) Carbohydrates. In: Khan, K., Shewry, P.R. (Eds.), Wheat: Chemistry and Technology, 4th ed. American Association of Cereal Chemists, St Paul, MN, pp. 299–362.

Symons SJ, Dexter JE (1991) Computer analysis of fluorescence for the measurement of flour refinement as determined by flour ash content, flour grade color, and tristimulus color measurements. Cereal Chem 68: 454–460.

Tanwir F, Fredholm M, Gregersen PL, Fomsgaards IS (2013) Comparison of the levels of bioactive benzoxazinoids in different wheat and rye fractions and the transformation of these compounds in homemade foods. Food Chem. 141: 444–450.

Tekin M, Cengiz MF, Abbasov M, Aksoy A, Canci H, Akar T (2018) Comparison of some mineral nutrients and vitamins in advanced hulled wheat lines. Cereal Chem 95: 436–444.

Theuwissen E, Mensink RP (2008) Water-soluble dietary fibers and cardiovascular disease. Physiology and Behavior 94: 285–92.

Tiwari BK (2015) Ultrasound: A clean, green extraction technology. Trends Anal Chem 71: 100–109.

Tiwari U, Cummins E (2009) Nutritional importance and effect of processing on tocols in cereals. Trends Food Sci Technol 20: 511–520.

Tluscik F, Kozubek A, Mejbaum-Katzenellenbogen W (1981) Alkylresorcinols in rye (Secale-Cereale L) grains. 6. Colorimetric micromethod for the determination of alkylresorcinols with the use of diazonium salt, fast blue-B. Acta Societatis Botanicorum Poloniae 50: 645–651.

Toole GA, Le Gall G, Colquhoun IJ, Nemeth C, Saulnier L, Lovegrove A et al. (2010) Temporal and spatial changes in cell wall composition in developing grains of wheat cv. Hereward. Planta 232: 677–689.

Topping D (2007) Cereal complex carbohydrates and their contribution to human health. J Cereal Sci 46: 220–229.

Topping DL, Bajka BH, Bird AR, Clarke JM, Cobiac L, Conlon MA, Morell MK, Toden S (2008) Resistant starches as a vehicle for delivering health benefits to the human large bowel. Microbial Ecology in Health and Disease 20: 103–108.

Topping DL, Clifton PM (2001) Short-chain fatty acids and human colonic function: roles of resistant starch and nonstarch polysaccharides. Physiol Rev 81: 1031–64.

Van den Berg H, Faulks R, Granado HF, Hirschberg J, Olmedilla B, Sandmann G, Southon S, Stahl W (2000) The potential for the improvement of carotenoid levels in foods and the likely systemic effects. J Sci Food Agric 80: 880–912.

Varga M, Banhidi J, Cseuz L, Matuz J (2013) The anthocyanin content of blue and purple coloured wheat cultivars and their hybrid generations. Cereal Res Commun 41: 284–292.

Vitaglione P, Napolitano A, Fogliano V (2008) Cereal dietary fibre: A natural functional ingredient to deliver phenolic compounds into the gut. Trends Food Sci. Technol 19: 451–463.

Wang W-M, Klopfenstein CF, Ponte JG Jr (1993) Effects of Twin-Screw Extrusion on the Physical Properties of Dietary Fiber and Other Components of Whole Wheat and Wheat Bran and on the Baking Quality of the Wheat Bran. Cereal Chem 70: 707–711.

Wang L, Weller CL (2006) Recent advances in extraction of nutraceuticals from plants. Trends Food Sci Technol 17: 300–312.

Wang H, Zhou Y, Ma J, Zhou Y, Jiang H (2013) The effects of phytic acid on the Maillard reaction and the formation of acrylamide. Food Chem 141: 18–22.

Wennermark BH, Jagerstad M (1992) Breadmaking and Storage of Various Wheat Fractions Affect Vitamin E. Journal of Food Science 57: 1205–1209.

Werner S, Böhm V (2011) Bioaccessibility of carotenoids and Vitamin E from pasta: evaluation of an in vitro digestion model. J Agric Food Chem 59: 1163–1170.

Witten S, Aulrich K (2018) Effect of variety and environment on the amount of thiamine and riboflavin in cereals and grain legumes. Anim Feed Sci Technol 238: 39–46.
Wolever TMS, Jenkins DJA, Kalmusky J, Giordano C, Giudici S, Jenkins AL et al. (1986) Glycemic response to pasta: effect of surface area, degree of cooking, and protein enrichment. Diabetes Care 9: 401–404.
Yu L, Beta T (2015) Phenolic Compounds during Production of Bread from Purple Wheat Grains. Molecules 20: 15525–15549.
Zandomeneghi M, Festa C, Carbonaro L (2000) Front-surface absorbance spectra of wheat flour: determination of carotenoids. J Agric Food Chem 48: 2216–2221.
Zhang M, Bai X, Zhang Z (2011) Extrusion process improves the functionality of soluble dietary fiber in oat bran. J. Cereal Sci. 54: 98–103.
Zhu KX, Huang S, Peng W, Qian H F, Zhou H M (2010) Effect of ultrafine grinding on hydration and antioxidant properties of wheat bran dietary fiber. Food Res Int 43: 943–948.
Zhu Y, Soroka DN, Sang S (2012) Synthesis and Inhibitory Activities against Colon Cancer Cell Growth and Proteasome of Alkylresorcinols. J Agric Food Chem 60: 8624–8631.
Zhu Y, Wang P, Sha W, Sang S (2016) Urinary Biomarkers of WholeGrain Wheat Intake Identifiedby Non-targeted and Targeted Metabolomics Approaches. Sci Reports 6: 36278.
Ziegler JU, Wahl S, Würschum T, Longin CF, Carle R, Schweiggert RM (2015) Lutein and lutein esters in whole grain flours made from 75 genotypes of 5 *Triticum* species grown at multiple sites. J Agric Food Chem 63: 5061–5071.
Zielinski H, Kozlowska H, Lewczuk B (2001) Bioactive compounds in the cereal grains before and after hydrothermal processing. Innovative Food Science& Emerging Technologies 2: 159–169
Zile MH (1998) Vitamin A and embryonic development: an overview. J Nutr 128: 455S–458S.
Zimmermann M B, Hurrell R F (2007) Nutritional iron deficiency. The Lancet 370: 511–520.
Zuo H, Svingen GFT, Tell GS, Ueland PM (2018) Plasma Concentrations and Dietary Intakes of Choline and Betaine in Association With Atrial Fibrillation Risk: Results From 3 Prospective Cohorts With Different Health Profiles. J Am Heart Assoc 7: e008190.

Fusarium Species Infection in Wheat: Impact on Quality and Mycotoxin Accumulation

Sofía Noemí Chulze, Juan Manuel Palazzini, Valerie Lullien-Pellerin, María Laura Ramirez, Martha Cuniberti, and Naresh Magan

Abstract Wheat is the most consumed cereal worldwide and can be processed to different products for human consumption. This crop can be infected by *Fusarium* species, among them those within the *Fusarium graminearum* complex causing Fusarium head blight (FHB. The disease can severely reduce grain yield and quality under conditions of high humidity and warm temperatures during anthesis. Moreover the grains can be contaminated with mycotoxin such as trichothecenes, among them deoxynivalenol and their acetyl derivates 3-ADON, 15-ADON and DON-3-glucoside. Some years, depending on the environmental conditions *Fusarium proliferatum* can also infect the grain and fumonisin contamination can be observed. To understand the way of grain infection by *Fusarium* species will help to undertake strategies to reduce the problem both at pre-harvest and during processing to select adequate procedures to manage mycotoxin production. Different strategies at different stages of the wheat chain have been proposed to reduce the impact of FHB and mycotoxin accumulation.

S. N. Chulze (✉) · J. M. Palazzini · M. L. Ramirez
Research Institute on Mycology and Mycotoxicology (IMICO), The National Scientific and Technical Research Council (CONICET)-National University of Río Cuarto (UNRC), Córdoba, Argentina
e-mail: schulze@exa.unrc.edu.ar

V. Lullien-Pellerin
IATE, CIRAD, INRAE, Montpellier SupAgro, University Montpellier, Montpellier, France

M. Cuniberti
Wheat and Soybean Quality Lab, National Institute of Agricultural Research, INTA, Córdoba, Argentina

N. Magan
Cranfield University, Cranfield, UK

G. Igrejas et al. (eds.), *Wheat Quality For Improving Processing And Human Health*, https://doi.org/10.1007/978-3-030-34163-3_18

1 *Fusarium* Infection of Wheat and Mycotoxin Occurrence

Fusarium graminearum is a complex of species (FGSC) at least 16 species identified by multilocus genotyping assays are included in the complex (Sarver et al. 2011; Ward et al. 2008). Fusarium head blight outbreaks on wheat, barley and other small cereal grains are associated with these species throughout the world (McMullen et al. 2012). Despite outbreaks, *F. graminearum sensu lato* is able to produce trichotecenes, mainly from group B, such as deoxynivalenol (DON), nivalenol (NIV) and their acetylated derivatives (15-ADON, 3-ADON, 4-ANIV). Pathogenicity and aggressiveness of the species within the FGSC are often associated with the mycotoxin profile production (so called chemotype/genotype) (Purahong et al. 2014).

In China, Zhang et al. (2012) collected and identified *Fusarium* isolates from 175 sampling sites (covering 15 provinces), which were grouped in 4 regions. The Multilocus genotyping technique (MLGT) was used to identify the isolates, observing eight species, three of them belonging to the FGSC: *F. asiaticum* (n = 275), *F. graminearum sensu stricto* (n = 169) and *F. meridionale* (n = 2). *F. graminearum* (76%) was prevalent in the northern areas and *F. asiaticum* (97%) prevailed in the South. The analysis of chemotypes analyzed by MLGT showed that all *F. graminearum* isolates were 15-ADON meanwhile the *F. asiaticum* isolates exhibited three profiles: NIV (n = 97), 3-ADON (n = 171) and 15-ADON (n = 7).(Ward et al. 2008).

In Japan, *F. graminearum* 15-ADON chemotypes prevails in North and Central areas meanwhile *F. asiaticum* (NIV and 3-ADON) was observed in Central-South areas.(Suga et al. 2008) In Canada, Ward et al.(2008) analyzed trichotecenes profiles of 492 *F. graminearum* isolated from wheat during 1984–2004. It was observed that 15-ADON was the predominant chemotype but the 3-ADON had rapidly spread from East to West in the last 6 years (1998–2004). Sequent, studies conducted in Ontario province, the central part of Canada, showed that *F. graminearum* strains isolated from wheat and corn during 2010–2012 were mainly 15-ADON producers (Burlakoti et al. 2017).

In the upper Midwest of the United States, the predominant chemotype was 15-ADON (95%), although some 3-ADON chemotypes were also observed (Gale et al. 2007). In the Eastern provinces of USA, a survey of 998 isolates in 2006 identified as *F. graminearum* revealed that 92% were 15-ADON genotypes and only 7% had the 3-ADON genotype (Schmale et al. 2011). In Mexico Cerón-Bustamante et al. (2018) carried out a survey from 77 wheat-growing fields during 2013 and 2014 in the Central-South area of the country, and surprisingly, *F. graminearum ss* was not present in the samples, being *F. tricinctum* species complex the most predominant (56%). Among FGSC, only *F. boothii* and *F. meridionale* were identified.

In South America, several surveys have been done in Brazil, Uruguay and Argentina, and *F. graminearum sensu stricto* 15-ADON producers has been associated as the main pathogen causing FHB (Del Ponte et al. 2014; Yerkovich et al. 2017). In Brazil, *F. graminearum* 15-ADON was the predominant genotype in

Southern regions. Del Ponte et al. (2014) analyzed a larger collection in order to increase knowledge of diversity and spatial distribution of trichothecene genotypes. A predominance of *F. graminearum* 15-ADON was observed (83%) and *F. meridionale* (13%), *F. asiaticum* (0.4%) and *F. cortaderiae* (2.5%) all three with NIV chemotype. In Uruguay, the isolation of *F. asiaticum, F. brasilicum, F. cortaderiae*, and *F. austroamericanum* was reported for the first time; but *F. graminearum* 15-ADON accounted for the 86% of the isolates examined, supporting previous studies with 15-ADON dominating over 3-ADON chemotypes (Umpiérrez-Failache et al. 2013). In Argentina, *F. graminearum* chemotype 15-ADON were more frequent. Ramirez et al. (2006) first described the presence of 3-ADON producers in samples collected from the 2002 FHB outbreak. A study done during FHB epidemic 2012/13 and non-epidemic 2014/15 harvest seasons, showed prevalence of 15-ADON chemotype, but with an increased 3-ADON chemotype observed in the non-epidemic year (Yerkovich et al. 2017).

Over the last decades, several surveys were carried out in European countries, where *F. culmorum* was the main pathogen associated to FHB, but in the recent years appears that it has been replaced by *F. graminearum*. In general, it was observed that *F. graminearum* 3-ADON prevailed in Finland and Northern Russia while 15-ADON did on Germany, France and Italy (Van der Lee et al. 2015).

Table 1 shows an overview about *Fusarium* mycotoxins occurrence in wheat and wheat based products collected by several researchers from various countries since 2015. DON was the most frequently studied mycotoxin, on which there is more data, followed by ZEA and NIV. As we can observe, all data presented on Table 1 were obtained using different methodologies, with distinct sensitivity and accuracy thus quantitative comparison is sometimes difficult, as well as the fact that these results derived from target analysis, after a previous selection of few analytes. It is also important to emphasize that prevalence and contamination levels of mycotoxins vary greatly according to several factors, i.e. geographic location, harvest year and commodity. But in general, in all the reports maximum levels of DON found were higher than those maximum levels (MLs) recommended by de EU (Commission Regulation (EC. No 1881/2006).

2 Toxicity of the Main Mycotoxin Detected in Wheat and by Products

Regarding the toxicity of mycotoxins detected in wheat, zearalenone (ZEA) is classified as a Group 3 carcinogen by the International Agency for Research on Cancer (IARC). Public health concern is associated with its strong estrogenic activity. Fumonisins are also prevalent in wheat and classified in Group 2 B (probably carcinogen to human) by IARC. Due to structural similarity to sphinganine and sphingosine, these toxins exerts its toxic effects by disrupting sphingolipid metabolisms due the competition with the ceramide synthase. Among trichothecenes, deoxyniva-

Table 1 Worldwide occurrence of mycotoxins in wheat and wheat by products

			DON		3-ADON		15-ADON		ZEA		NIV			
Country	Year	Sample	Occurrence[1]	Range[2]	Occurrence[1]	Range[2]	Occurrence[1]	Range[2]	Occurrence[1]	Range[2]	Occurrence[1]	Range[2]	Method	Ref
Argentina	2012	Wheat grains	55/69/80/NM[3]	400–8500									HPLC	Palazzini et al., 2015
Argentina	2013–2014	Durum wheat	84/84/100/1750	50–9480									HPLC-MS/MS	Palacios et al., 2017
Bosnia -Herzegovina	2013–2015	Wheat grains	54/84/64/690	38–2123					49/84/58/127	8–189			ELISA	Pleadin et al., 2017
Brazil	2008–2015	Wheat grains	1981/2714/73/ 855	510–1524					1031/2714/38/180	58–503			LC-MS/MS	Mallmann et al. 2017
Brazil	2009–2010	Wheat grains	641/745/86/ 1046	100–8501					417/745/56/82	10–1057	373/745/50/100	100–1329	HPLC	Calori-Domingues et al., 2016
Brazil	2012	Wheat	149/150/99/706	183–2150					48/150/32/64	LOQ[4]–233			LC-MS/MS	Tralamazza et al., 2016
Brazil	2014	Wheat grains	434/668/65/710	LOQ-11800									LC-MS/MS	Machado et al. 2017
		Wheat flour	195/415/47/693	LOQ-11400										
		Wheat bran	102/139/73/640	LOQ-5190										
Brazil	2015	Wheat grains	51/92/55/795	200–2743					36/92/39/79.78	20–425			UHPLC- MS/MS	Duffeck, et al., 2017
China	2013	Wheat grains	150/181/83/500	33–3030									HPLC	Zhao et al., 2018
China	2013–2015	Wheat grains	423 /450/94 /2050	10–18,709	192/450/60/ 49	10–730	147/450/33/ 18	10–260			10/450/2.2/5	30–204	LC-MS/MS	Dong et al., 2016
Croatia	2015–2016	Wheat grains	30/52/58/254	27.1–1220					30/52/58/37	4.72–94.5			ELISA	Pleadin et al. 2017
Czech republic	2014–2015	Wheat grains	33/152/22/413	12–39,900	4/152/3/44	7–3030			8/152/5/57	1–4660	1/152/0.6/1	193	LC-MS	Sumíková et al. 2017

Hungary	2008–2015	Wheat grains	NM/305 /NM/ 750	NM									ELISA/ LC-MS	Tima et al., 2017
		Wheat flour	NM/17 9/NM/ 245	NM										
		Pasta	ND/226/ NM/239	NM										
Iran	2011–2012	Wheat grains	120/162/ 74/NM	151–1651	58/162/3 6/NM	10.2–320.4	79/162/4 9/NM	10–550					HPLC	Abedi-Tizai and Zafari et al., 2015
Iran	2013	Wheat flour	80/96/8 3/630.53	23–1270									ELISA	Darsanaki et al., 2015
Israel	Nd	Wheat	8/34/2 4/256	LOQ-1747					5/34/15/17	LOQ-64.8			LC/MS/MS	Sadhasivam et al., 2017
Italy	2013	Durum wheat	12/74/ 16/75	48–2267	3/74/4/1	16–33	11/74/15/7	31–79			2/74/3/3	50–197	LC-MS/MS	Juan et al., 2016
Japan	ND	Wheat flour	159/163/ 97.7/50	2–386							42/163/25/2	2–43	LC-MS/MS	Yoshinari et al., 2016
Kosovo	2011	Wheat		LOQ-6310		LOQ-74		13–360		280–350		LOQ-31	HPLC-MS/MS	Shala-Mayrhofer et al., 2015
Lithuania	2013–2014	Wheat grains	NM/103 /97/513	43–6804	NM/10 3/84/263	21–1149	NM/103 /85/179	85–1319					HPLC	Janaviciene et al., 2018
Norway	2004–2009	Wheat grains	164/178 /92/290	50–16,000	71/178/4 0/NM	50–540			69/178/3 9/NM	50–350	ND		LC-MS/MS	Hofgaard et al., 2016
Poland	2011–2014	Durum wheat	45/54/83 /2728	LOQ-10880					48/54/89 /37.8	LOQ-307.3	6/54/11 /37.3	<LOQ-155.4	HPLC	Gorczyca et al. 2017
Poland	2016	Wheat	76/92/8 3/140	10–1265							64/92/70/35	5.1–372.5	LC	Bryła et al., 2018
Portugal	2015	Wheat grains	6/13/46 /NM	<31–297							1/13/7.6 /NM	<31.3	HPLC	Trombete et al. 2016a, b
		Wheat flour	4/4/10 0/NM	79–326							4/4/100/ NM	<31–141		

(continued)

Table 1 (continued)

			DON		3-ADON		15-ADON		ZEA		NIV			
Romania	2015	Wheat grains	8/31/26/748	110–1787	ND[5]		ND		4/31/13/669	327–1135	ND		LC-MS/MS	Stanciu et al., 2017
		Wheat flour	1/35/3/190	190	ND		ND		2/35/6/62	51–73	ND			
Slovak republic	2009–2012	Wheat	161/189/85/ 368	20–2652					175/269/65/NM	<100	51/80/64/35	5–181	HPLC	Lacko-Bartošová et al., 2017
Slovenia	2008–2012	Wheat and wheat products	55/80/69/477	NM-3070					19/80/24/36	NM-113			HPLC	Kirinčič et al. 2015
Switzerland	2007–2014	Wheat grains	549/686/80/607	LOD-10600	20/686/3/ NM	NM	20/686/3/ND	NM	219/686/32/39	LOD-3070	144/686/21/ 15	LOD-470	LC-MS/MS	Vogelgsang et al., 2017

[1]Positive samples/ total samples/ incidence % / mean (μg/kg);
[2]μg/kg;
[3]NM: not mentioned;
[4]LOQ: Limit of quantitation;
[5]ND: not detected

lenol (DON) and its derivatives 15-acetyl-DON (15-ADON), 3 acetyl DON (3-ADON) and DON 3 Glucoside (DON-3-Glc) are found as contaminants in wheat and by products. DON is classified by IARC as Group 3, the main mechanism of trichothecenes toxicity is inhibition of ribosome protein synthesis. Ochratoxins, mainly ochratoxin A (OTA) have been found in a great variety of agricultural commodities including wheat in cold regions. OTA have been classified by IARC as Group 2 B (possible human carcinogen) and it has been associated with the Balkan Endemic Nephropathy. This toxin is mainly nephrotoxic and hepatotoxic, but also cause inmunotoxicity, genotoxicity and teratogenicity both in human and animals. (Ostry et al. 2017).

3 Impact of Climate Change on *Fusarium* Head Blight and Mycotoxins

The impact of climate change (CC) factors needs to be considered in the context of the interactions between the three key abiotic and biotic parameters, which are normally considered to be in a balanced triangle between the plant, the pest/disease and the environment (Medina et al. 2017). This will be pushed/pulled in different directions due to interacting conditions of elevated temperature (+2-4 °C), increased CO_2 (400 vs 800–1200 ppm) and episodes of extreme drought and flooding conditions. These interacting factors + the physiological status of the wheat will influence infection by plant pathogens, especially *Fusarium* species, responsible for FHB symptoms and type B trichothecene contamination. This is particularly critical during the 2–3 week period between anthesis and grain filling. Conducive conditions can allow infection, FHB symptoms and contamination with undesirable mycotoxins. There have been few studies of the impact that interacting CC factors have on both the FHB pathogens and on the interface between the pathogen and the wheat plant. Studies of the effect of interacting CC factors of temperature (25 vs 30-35 °C), CO_2 (350 vs 1000 ppm) and intermediate drought stress (0.995 vs 0.98 water activity) was shown to influence and change the growth pattern of *F. graminearum* (Magan and Medina 2016; Medina et al. 2015; Medina et al. 2017). Indeed, growth was slower but occurred under CC conditions even at 35 °C although normally it is inhibited. While a number of studies have used historical data to predict what impact CC scenarios will have on cereals diseases including FHB (Madgewick et al. 2011; West et al. 2012). More recently (Battilani et al. 2016 and Van der Fels-Klerx et al. 2016) examined predicted changes related to temperature of +2 and + 5 °C and what impacts this would have on both wheat and maize in Europe. They showed that earlier flowering of wheat would result in changes in pest infection and perhaps lead to wider regions of Europe where FHB and perhaps mycotoxin contamination would occur. If elevated CO_2 prediction and drought/flooding episodes were taken into account perhaps the impacts would be more severe.

Váry et al. (2015) is one of the few studies where impacts of exposure to 390 and 780 ppmv CO_2 on both wheat plants and on FHB were examined. In addition they examined whether acclimatisation of the *F.graminearum* by sub-culturing in the elevated CO_2 environment would influence pathogenicity and FHB. The FHB disease development increased under CO_2 exposure. The highest FHB disease levels and associated yield losses occurred for elevated CO_2-acclimated *F.graminearum* in elevated CO_2-acclimated wheat. Thus it is important that we consider the implications of CC scenarios on diseases of wheat, especially FHB, because this could lead to increased contamination with mycotoxins for which legislative limits exists in many countries world-wide.

4 Quality Changes of Wheat and by Products by *Fusarium* Infection

A globalized economy has rapidly increased international trade of a large variety of foods and food products. Consumer satisfaction and health are of most importance. International food trade helps promote economic development, but likewise creates risk. Mycotoxins contamination of cereal grains and products is a worldwide problem. The type and quantity of mycotoxin produced by one fungal species could change from one year to other, depending on the environment conditions, the cultivars and storage condition (Martinez et al. 2014). DON is considered as a contamination marker subjected to the European Commission Regulation (EC 2006) and the maximum level was set at 1250 µg/kg for unprocessed cereals and 750 µg/kg for cereal flours. DON is a mycotoxin very stable which is not degraded during storage, milling, processing or cooking of food (Whitney 2018), even during high temperature treatments and it is present in grains without aspect of *Fusarium* damage.

Besides these facts, several authors also mentioned negative effect of *Fusarium* infection on bread making quality of wheat and the reduction of loaf volumes (Cuniberti 2001 and 2013; Gärtner et al. 2008). On the other hand, there are some contradictory studies where a strong *Fusarium* contamination did not significantly influence the bread making properties (Prange et al. 2005). (Kamimura et al. 1979) reported the effects on baking bread and preparing Chinese and Japanese noodles on levels of DON and five other trichothecenes. They also observed that soaking naturally contaminated ground wheat in water removed about 30% of the DON and nivalenol present. To assess human consumption of DON in finished foods, it is essential to extend available data on the stability of DON during food processing. DON was distributed throughout the milled products and was not destroyed on making bread. The highest concentration of DON was found in dockage (16.7 µ g/g); the cleaned wheat contained 4.6 µg/g and the flour and bread (flour weight basis) contained an average of 4.1 and 4.2 µg/g in two milling and baking experiments. In Canada (Preston et al. 1982) showed that flour color had poorer quality than would be predicted from the flour ash. Farinograph development time was short and mixing tolerance index quite high, as is often the case for sprouted wheat flour. Nevertheless, dough properties were satisfactory for baking, and on a unit protein basis loaf volume approached that normally achieved for high-quality hard red spring wheat flour. Cleaning of wheat brings about a slight reduction of DON levels. DON was not destroyed on making bread from naturally contaminated straight grade flour.

Studies done in Argentina by (Cuniberti 2001, 2013) showed important changes in the quality parameters of hard wheat flour and grain, with changes in the baking quality of spring wheat at different infection levels by *Fusarium* spp. Under situation where the percentage of *Fusarium* infected grains was higher than 10% negative effects were observed on test weight, weight per 1000 kernel and ash content. The amylase activity was not modified. Considering quality of wheat a reduction of flour yield, a change in flour color from yellow to grayish was observed. Also, an increase in the ash content and the flour acidity, with an special smell was observed. The alveograph W, water absorption, consistences, farinograph dough stability and bread volume decreased when grain *Fusarium* percentage was larger than 15%. Sticky dough was also observed.

In Manitoba, Canada (Dexter et al. 1996) studied the *Fusarium* damage (FD) at different levels and concluded that as FD increased, deoxynivalenol (DON) levels increased. Straight-grade flour yield was not related to increasing FD, but flour refinement (ash and color) was adversely affected. Gluten from hand-picked FD kernels contained a lower proportion of glutenins than did sound kernels. No qualitative or quantitative differences in gliadins attributable to FD were apparent. Flour from FD wheat showed relatively normal physical dough properties (mixograph and farinograph), but during remix baking, a long straight-dough procedure, dough became sticky and hard to handle. The effect of FD on loaf volume was cultivar depend. Some cultivars showed a moderate decline whereas other showed a drastic decline.

Recent evaluation realized by Whitney (2018) in United States during wheat processing demonstrated that after removal of bran by milling, there was an approximate reduction of 61.8% in DON and 23,7% in D3G contents. Conjugation of DON with glucose results in deoxynivalenol-3-β-D-glucopyranoside (D3G), the main DON metabolite in wheat. DON levels detected during fermentation (3.03 to 3.93 μg/kg) were significantly higher ($P < 0.05$) than in mixed dough (1.97 μg/kg). There were no significant differences ($P < 0.05$) in the D3G in the dough samples. However, the baked bread had significantly ($P < 0.05$) less D3G detected than the dough. There were significant differences ($P < 0.05$) between the wheat treated with α-amylase, cellulase, protease, and xylanase. DON levels were significantly ($P < 0.05$) higher after treatment with protease (16%) and xylanase (39%) compared to the wheat composite. The result suggest that DON may be bound to the cell wall matrix or protein component of the wheat kernel, due to the rise in detection of DON after these enzyme treatments. Overall, processing or enzymes present in wheat or wheat products may result in release of DON and an increase of DON content in food products.

5 Preharvest Strategies to Reduce the Impact

5.1 *Genetic Crop Resistance*

Planting cultivars less susceptible to *Fusarium* infection is one strategy to reduce FHB. At least 5 types of resistance to FHB have been demonstrated. The types of resistance include: **type I** is the resistance to penetration of the pathogen and the onset of disease; **type II** is the resistance to the propagation of the pathogen in the plant once the disease is established; **type III** is the resistance to infection of the grains; **type IV** is the tolerance of the disease, since the yield of the grains is maintained despite the presence of disease; and **type V** is the ability of the plant to degrade and inhibit toxins produced by the fungus (Mesterházy 1995). *Fhb1* a major locus for resistance to FHB was mapped to chromosome 3B of the resistant wheat cultivar Sumai 3 (Rawat et al. 2016). The ability to metabolize DON to the less toxic DON-3G was demonstrated by the expression of HvUGT13248 in transgenic wheat, that convert DON to D3G 24% more efficiently than non-transformed controls (Li et al. 2015a). FHB resistance in durum wheat is similarly inherited in bread wheat showing a continuous variation, the range from high to low susceptibility is smaller than in bread wheat with a high impact of genotype environment interaction (Prat et al. 2017). The causes for the high susceptibility of durum wheat are not well understood yet, since most of the QTL for FHB resistance in hexaploid wheat have been found on the A and B genomes that are shared by durum wheat (Buerstmayr et al. 2009).

5.2 *Agricultural Practices*

The implementation of good agricultural practices during wheat growing periods is clearly important for FHB management (McMullen et al. 2012). There are different agricultural practices useful for controlling the disease and mycotoxin accumulation. Under no tillage conditions the development of the disease can increase, being important to consider the type of crop before and after wheat production (crop rotation). In some areas still maize and wheat rotation is used increasing the risk of FHB and DON accumulation.

5.3 *Chemical Control of FHB*

Chemical control (fungicide application) is an available strategy to reduce the risk of FHB. Several studies have shown a reduction in the disease and mycotoxin contamination on grain in natural or artificial infection by fungicide application. The most widely used fungicides are included in the demethylation inhibitor (DMI) class. Among them are included prothioconazole, tebuconazole, propiconazole, metconazole. Tebuconazole, prothioconazole and metconazole, solely or in mixture of two triazoles, are the most commonly recommended fungicides for FHB control worldwide (Paul et al. 2008). The spray technology (nozzle angle, droplet size, spike coverage) is a crucial aspect during fungicide application. The cultivar planted also can influence fungicide treatment (Hollingsworth et al. 2008). The resistance to fungicides of *Fusarium graminearum sensu stricto* one of the main pathogens associated to FHB can be a challenge in the management of FHB.

5.4 Biological Control of FHB

Different strategies for reducing the impact of FHB have been proposed (Chulze et al. 2015; McMullen et al. 2012; Mesterházy et al. 2011; Wegulo et al. 2015). An integrated management of the disease is the best proposal and the combination of two or more strategies can reduced the disease impact. Bacteria, filamentous fungi and yeast have been identified and evaluated under in vitro, green house and field conditions to reduce FHB and deoxynivalenol (DON) accumulation. Bacteria that were effective in reducing FHB and DON accumulation included *Bacillus* spp. (Palazzini et al. 2015; Palazzini et al. 2018, Schisler et al. 2006) *Brevibacillus* spp., *Pseudomonas* spp. *Streptomyces* spp. (Palazzini et al. 2007). Among fungal antagonists *Trichoderma* species, *Clonostachys rosea* (Palazzini et al. 2013; Xue et al. 2014), these antagonists were evaluated on crop residues to reduce the inoculum potential of the pathogen or on the spike to reduce infection. *Trichoderma gamsii* 6085 applied on spikes reduced disease severity by 10% under field conditions (Sarrocco and Vannacci 2018). *C. rosea* showed reduction sporulation of *Fusarium graminearum* on wheat stubble and reduction in pathogen DNA by 63% (Palazzini et al. 2013). Also, yeast showed promising biocontrol activity to control *F.* head blight. *Aureobasidium pullulans* was evaluated under greenhouse conditions reduced FHB severity by 20%. Co-inoculation of *C. flavesens* and *C. aureus* reduced FHB by 32% under greenhouse conditions. According to EU Directive 2009/128/EC the European Commission set rules for the sustainable used of pesticides to reduce the risk and impact of pesticides use on people's health and on the environment. This situation will force the need to increase the use of biocontrol in the frame of an integrated pest management.

5.5 Predictive Models

Forecasting systems play a key role in the FHB management since allow the farmers to decide each year whether fungicides are needed to control the development of FHB minimizing the risk of food and feed contamination with DON and then to avoid the use of chemical control. Forecasting systems use principally weather data as input (temperature, rainfall and moisture) and have been developed for application in specific regions of the world, where they were developed. In Argentina (Moschini and Fortugno 1996) developed empirical equations to predict FHB incidence (Predictive Index: PI%) associating mean head blight incidence with temperature (maximum and minimum daily temperature) and moisture variables during a critical period length considered as the period of time beginning 8 days before heading date and ending when 530 degree days were accumulated; this period was regarded as the susceptible period for infection. Since wheat growing season 2005–2006, a system for assessing FHB risk was implemented for the Pampas region (Moschini et al. 2013). In Canada, models that predict both FHB and DON contamination risks were developed. DONcast® is a commercially available forecasting system designed to provide growers a means of predicting DON concentration in wheat at harvest. It can predict DON levels above or below the threshold of 1 ppm 80 to 85% of the time (Giroux et al. 2016). The model was validated in Uruguay and France with good results since DON contamination in samples was correctly predicted in 60 to 80% of the times (Schaafsma and Hooker 2005).

In USA, the *Fusarium* Head Blight Risk Assessment Tool (http://www.wheatscab.psu.edu/) uses models that estimate the risk of a FHB epidemic with more than 10% field severity using weather variables observed 15 days prior to flowering. In Europe, a mechanistic model capable of predicting FHB risk and DON contamination in wheat based on weather data and wheat growth stages during the growing season till to harvest was developed by Rossi et al. (2003). The model produces two indices: one for the risk of FHB on wheat and one for mycotoxin content of kernels. The model was validated over 22 wheat-growing areas of northern Italy and the comparison between the actual content of both mycotoxins and the values estimated showed good concordance (Prandini et al. 2009). Mechanistic models works irrespective where they were developed, as opposed to empiric ones (Camardo Leggieri et al. 2013) and can be included in Decision Support System as core of rationale decision making (Rossi et al. 2012).

6 Postharvest Strategies to Reduce the Impact of Mycotoxins

The safe storage of wheat grain is influenced by the interaction between the moisture content of the grain at harvest, the drying efficiency and the temperature when the grain is placed into storage. If wheat is dried to 14–14.5% moisture content during the drying phase, then short- and medium-term storage is ensured with no loss in nutritional or biochemical quality for downstream processing. However, poor post-harvest management of stored grain, either due to poor silo structural quality, ingress of water, or the failure to exclude insect pests, can result in ochratoxin A contamination caused by *Penicillium verrucosum*, and, under wetter conditions, increased DON and ZEA contamination. In Europe, the management of the wheat grain system is done through a HACCP approach. Much of the current knowledge is based on the identification of the Critical Control Points (CCPs) in the wheat chain, both pre- and post-harvest (Fig. 1). The European Union has published a recommendation to prevent *Fusarium* toxin contamination of cereals and cereal

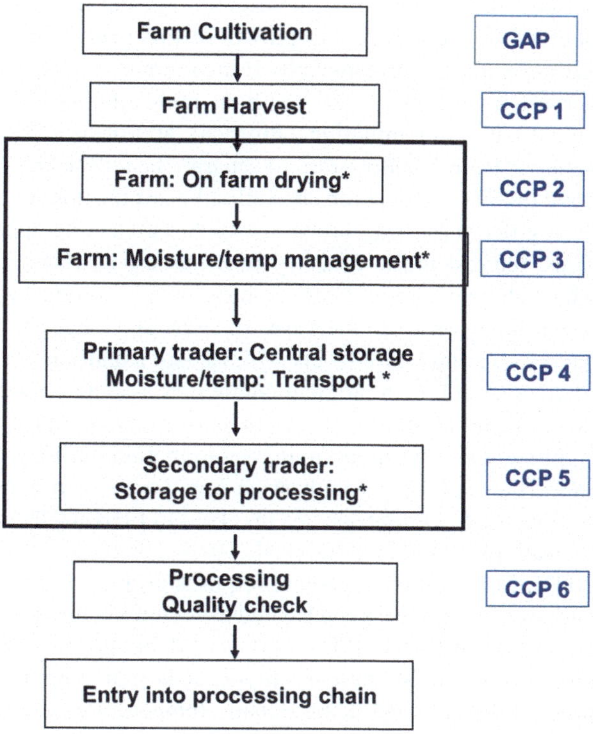

Fig. 1 HACCP scheme developed for mycotoxin control in wheat

products (Commission Recommendation 2006/583/EC). The key focus has to be on risks that occur between harvest and effective short- and medium-term storage in wheat in relation to DON/ZEA and OTA contamination. The most critical factors are the original moisture content and temperature at harvest, which directly influences events that may occur during storage and may result in spoilage, self-heating and increased mycotoxin contamination. Grain is a living material and its respiration is very low when stored at the proper temperature and moisture content (m.c.; <14.5%). Increases in m.c. or temperature also increases respiration and the production of metabolic water, which encourages spoilage to be initiated. Thus, management of the harvesting/drying and subsequent storage phase is critical for conserving wheat grain quality for downstream processing (Magan et al. 2014).

6.1 Risks of Deoxynivalenol and Zearalenone Contamination during Wheat Grain Storage

At harvest, wheat grain enters storage containing a wide range of mycobiota on the grain surface including some potentially spoilage and toxigenic fungi. This community of fungi will depend on conditions just prior to harvest, the harvesting process, and post-harvest management. Sometimes grain is kept for short periods of time, on farm, in buffer storage before drying. This practice can result in conditions conducive to growth of *Fusarium* and possibly increase DON and ZEA contamination especially if the grain is slightly damp. Poor short-term post-harvest management can result in rapid quality loss and increase the risk of mycotoxin contamination. Physical approaches include separating diseased material from healthy grain to separate heavily DON contaminated kernels from good quality ones. Often shrivelled Fusarium-infected grains are lost in the field during the harvesting process, which reduces the apparent contamination level. Thus, harvest is the first key post-harvest Critical Control Point (CCP) in the wheat chain where moisture management becomes the dominant control measure to minimize or avoid increases in DON and ZEA contamination. Wheat often is harvested at moisture levels >14.5% and usually is traded on a wet weight basis. There also are technological challenges associated with bulk drying and storage of grain, in addition to cases of poor practice and negligence. Thus, there is a significant risk for mycotoxin contamination during grain production that may be exacerbated during post-harvest handling and processing. Information on the ecology of *Fusarium* species involved in FHB, DON and for ZEA contamination is available (Garcia-Cela et al. 2018a, b; Hope et al. 2005; Hope and Magan 2003). Moisture management requires prompt, accurate measuring methods and bulk drying as necessary. Heated air-drying is the best method to dry grain to the target m.c. of 14.5%, if for food or feed, but not for seed. When ambient temperature drying is used, there is a higher risk of mycotoxin contamination because the process depends upon the outside temperature and humidity. The drying front also moves slowly upwards through the grain, often over-drying the bottom layers while remoistening the top layers. This moisture distribution is conducive to further DON/ZEA contamination and/or OTA biosynthesis in portions of the stored grain. Hygiene considerations are also important to ensure that insect pests are effectively controlled as they remain viable over a wider range of humidity conditions and can produce metabolic water, which can provide pockets of wet grain which can result in hot spots in silos of wheat grain. In contrast to the *Fusarium* species, *P. verrucosum* is xerotolerant and thus can grow and produce OTA under much direr conditions than the FHB pathogen.

6.2 Post-Harvest Decision Support System Development

With the availability of data on the ecology of the key toxin producing species in wheat it was possible to develop boundary condition models linked to those which just allowed or inhibited growth and those which similar resulted in mycotoxin production or not (Magan et al. 2010). In many present day grain silos in situ sensor cables for temperature monitoring are often used to monitor the quality of the storage. However, recent studies with both wheat and oats have suggested that CO_2 production by both the grain and contaminating spoilage moulds occurs prior to any change in temperature or RH (Garcia-Cela et al. 2018a, b; Mylona et al. 2012). Indeed, CO_2 production can be linked to the actual dry matter losses (DMLs) and the level of mycotoxin contamination and the EU legislative limits. Thus, use of integrated sensors for CO_2 measurement coupled with temperature and R.H. would provide sensitive indicators of potential risks from poorly dried grain or that ventilation is necessary. In addition, they could be linked to biological models to indicate the relative risk of contamination with specific toxins below or above the prevailing legislative limits (Fig. 2). This would provide a potential DSS systems for real time management of stored wheat grain in silos. During any of these stages the grain could become susceptible to fungal spoilage if the storage conditions are not strictly controlled. In most cases the key to adequate storage is drying the freshly harvested material to 14–14.5% m.c. and maintaining the grain in this condition. In general, the cooler and drier the grain, the longer it can be safely stored. Based on the available scientific information on the ecology of *F. graminearum* and related *Fusarium* species and DON/ZEA production, and OTA production by *P. verrucosum* and OTA production, it was generated a simple moisture content/water availability curve combined with the relative safe and unsafe storage conditions and the risk of the different mycotoxins post-harvest (Fig. 3). The so-called "zone of uncertainty" is the area where key and effective post-harvest management is required to ensure that the quality does not deteriorate and that the mycotoxin content remains below the legislative limits.

7 Effect of Processing on Wheat and by Products Quality and Safety

It is important to develop detoxification processes in order to prevent human and animals exposure to mycotoxins, when harvested grains are contaminated above the maximum levels established by the countries. Since the fungal growth could continue during storage and can be linked with toxin production in wheat grains in certain conditions, these methods have to be capable of destroying fungal spores and mycelium in order to avoid mycotoxin formation under favourable conditions. They are also focused on the toxin destruction, inactivation or redistribution depending on its stability, solubility or location. These strategies involve either physical, chemical, or biological methods or a combination of them. They do not have to generate or leave toxic and/or carcinogenic/mutagenic residues in the final products and should not adversely affect desirable physical and sensory properties of the grain or grain products and at last must be technically and economically feasible. Evaluation on the effect of each operating units remains difficult as mycotoxin analysis is dependent on the extraction efficiency, and because reduction of its content may also be due to a decrease in its extractability or its transformation in another molecule.

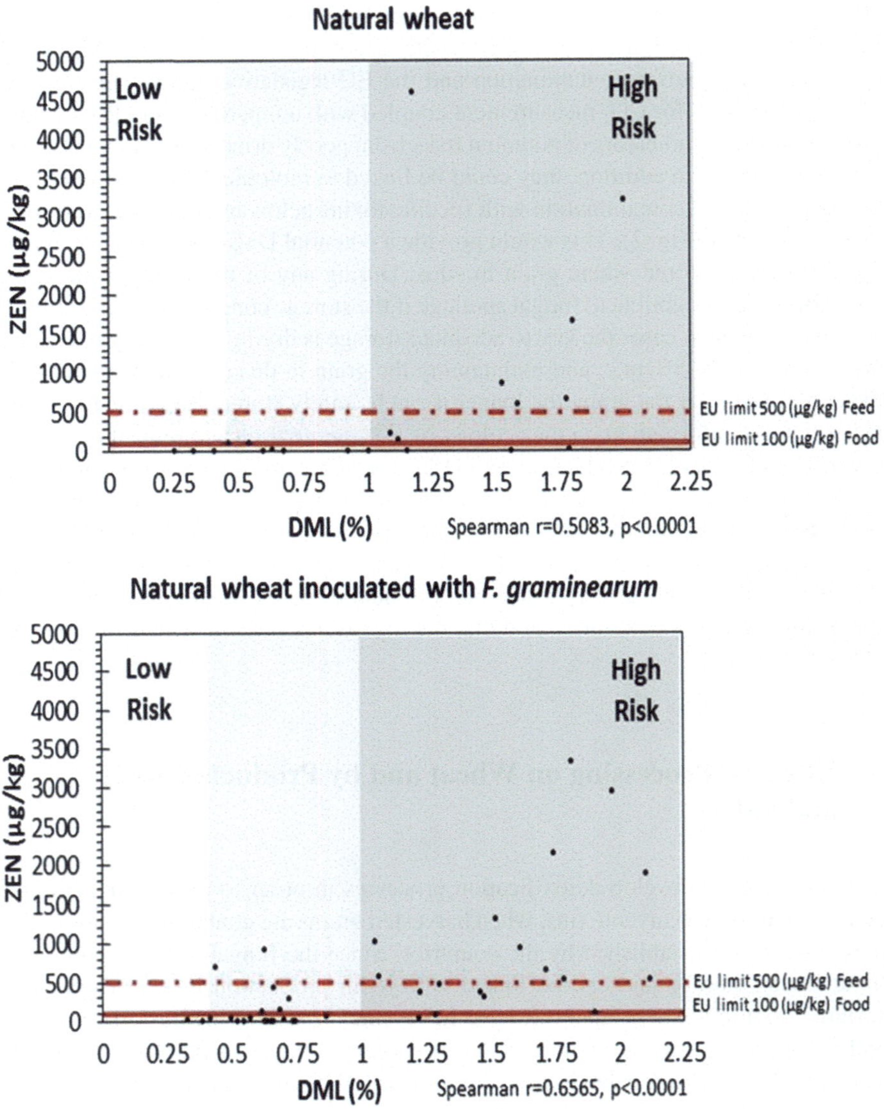

Fig. 2 Scatter-plot of Dry Matter Losses (DMLs) and Zearalenone in stored wheat after 15 days storage under all the environmental conditions examined producing by natural mycobiota (**a**) and both natural mycobiota and F. gramineraum (**b**). The red lines indicate the EU legislative limits for unprocessed wheat for human consumption (solid line) and animal feed (dash line) (EC, 576/2006,1881/2006)

Fig. 3 Ranges for deoxynivalenol (DON) and ochratoxin A (OTA) contamination and the zone of "uncertainty" where conducive conditions exist for increased contamination by ochratoxin A

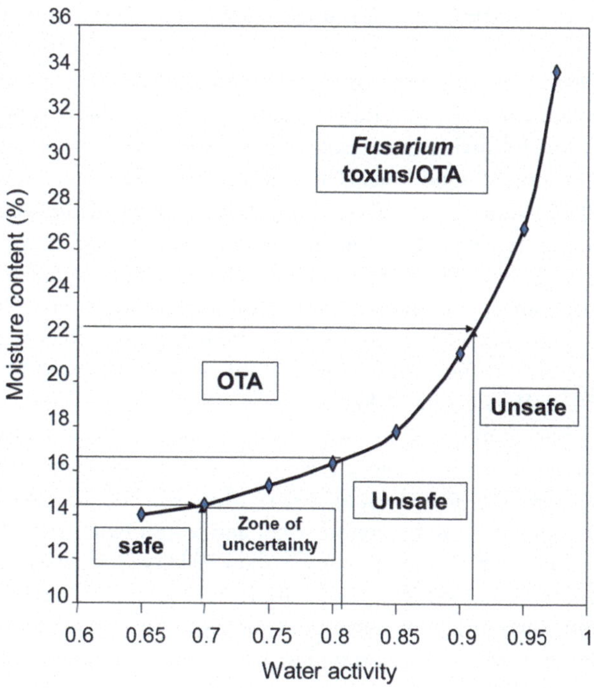

7.1 *Grain Pre-Treatments: Ozone, Irradiation, Microwave, Cold Plasma, Superheated Steam*

A number of chemical treatments were proved to be efficient on mycotoxin reduction but taking into account the human consumption, legislation on chemicals and needs to reduce their uses, we deliberately chose to not discuss the results. Ozone (O_3) properties, either applied as a gas or dissolved in water were largely studied these last 20 years to control the fungi or bacteria growth and insect development, but also to degrade potentially toxic molecules as mycotoxins and pesticides (Lullien-Pellerin 2012; Tiwari et al. 2010). Its main interest is its rapid decomposition into O_2 and thus the absence of any residues (Graham 1997) which classified it as a GRAS (Generally Recognized As Safe) compound recognized by the Food and Drug Administration (FDA 2001). Nevertheless it is a strong oxidant molecule which can induce oxidation of compounds inside the treated material depending on the applied concentration, conditions of treatment, kinetics of diffusion inside the treated material, etc. A decrease in the protein solubility was observed after ozone treatment of wheat grains by a number of authors under different conditions (Desvignes et al., 2008; Goze et al. 2017; Violleau et al. 2012). Depending on the conditions, slight oxidation of starch inside grains can also be shown (Goze et al. 2016). Depending on the applied dose it can moreover affect the grain physiology as demonstrated for germination (Wu et al. 2006) and processing behavior through changes in the grain tissue mechanical properties (Desvignes et al. 2008). Dubois et al. (2006) also described changes in enzyme activity following ozone treatment of wheat grains. Changes in the product properties (mainly rheology and color) following grain exposure to ozone was consequently observed depending on the applied conditions (Goze et al., 2017; Trombete et al. 2016a, b; Violleau et al. 2012; Wang et al. 2016a). A precise control of the ozone treatment conditions is therefore needed to find the better compromise between reduction of the fungi growth and mycotoxin amount and the related induced changes in the raw matter structure and composition. Lastly, more studies have to be undertaken to confirm the absence of toxicity of ozone-treated grains. Recent results appeared encouraging even if these studies were done on animals (rats) far from humans (Gaou et al. 2015; Wang et al. 2017).

Inactivation of the micro-organism growth and insect development was also found efficient using γ-, X-rays or electron-beam irradiation methods and was reviewed in (Calado et al. 2014). More specifically, (Aziz et al. 1997) demonstrated a complete inactivation of fungi and a significant reduction of DON, ZEA and T-2 toxins after γ-irradiation lower than 10 kGy (1 Gy was equal to 1 Joule of irradiation energy/kg of sample, with 1 Joule = 1 $kg.m^2\ s^{-2}$), a level considered too low to significantly affect the nutritional quality of the raw matter (Aziz et al. 2006). However, grain irradiation was found to affect the gluten amount and dough viscosity (Wang and Yu 2009).

Microwave, UV, or pulsed light treatments of wheat grains appeared less efficient in reduction of both fungi contamination and mycotoxin level in comparison with ozone, γ-irradiation or cold plasma (Hojnik et al. 2017; Los et al. 2018; Popovic et al. 2018) but no direct comparison between methods on same contaminated grains presenting a large panel of fungi type and mycotoxin type and level were undertaken. UV treatment was moreover found to change the grain outer layers' mechanical properties affecting the grain milling behavior (Peyron et al. 2002). Recent studies also revealed changes in gliadin polymerization after microwave treatments of wheat grains (Lamacchia et al. 2016). Thus, further investigations to find a compromise between microorganisms inactivation, mycotoxin decrease and nutritional, sensory and technological properties have to be undertaken.

Significant reduction of the DON level after wheat grain treatment for several minutes with superheated steam was also found for temperatures between 160 and 185 °C probably linked to their thermal degradation but again induced changes in the raw matter was not analyzed (Pronyk et al. 2006). Hu et al. (2017a) noted that short treatment of grains (1–3 min) at 110–170 °C led to favorable changes in the corresponding flours (starch damage and α-amylase decrease, brighter color, stronger dough) whereas treatments at 200 °C, 2–3 min decrease the flour quality.

Washing, soaking and boiling in water alone or with the addition of chemicals were also studied by some authors but are not reported here as these operations present too many disadvantages notably in terms of generated costs and raw matter modification for the future use of grains especially for human nutrition.

7.2 *Physical Methods*

Each processing step has for aim not to modify the mycotoxin level but more to retain the less contaminated grains or obtained less contaminated fractions for further use in human nutrition.

7.2.1 Grain Sorting

Grains highly infected with *Fusarium* were generally shrivelled, small size and with a lower mass than healthy grains. Therefore, they can be separated by physical means. For example, if common wheat grains under a size of 2 mm were eliminated, a decrease of the DON concentration by 83% (initial concentration around 5 mg/Kg) was observed but accompanied with a grain mass loss by 55% (Trenholm et al. 1991). Similarly for durum wheat (Rios et al. 2007), selection of grains above 2.4 mm were found to decrease the DON total amount by 22–27% depending on the initial contamination with a moderate loss of the grain mass (2–7%). Selecting grain according to their density on gravity tables can also be used for reduction of DON or HT-2 in a wheat grain batch, respectively by ten fold and 60 fold in comparison with the initial mycotoxin concentration (Tkachuk et al. 1991). Efficiency of the sorting by size or density to reduce mycotoxin concentration however was found variable depending on the wheat batch due to different amount of damaged and infected grains related to impurities and to differences between equipment chosen for sorting (Cheli et al. 2013; Schaarschmidt and Fauhl-Hassek 2018). Moreover, grains infected lately along maturation are generally undistinguished according to mass and size therefore other methods based on differences of spectral properties between infected and uninfected grains were developed. In these last years, equipment of optical sorting enabling separation of grains according to differences color (Delwiche et al. 2005) and also possibly morphologies was developed before processing but will have in the future to be coupled with multispectral imaging (Jaillais et al. 2015; Serranti et al. 2013) at least to track for *Fusarium* infected grains.

7.2.2 Grain Scouring, Pearling-Dehulling-Debranning

These operating steps allow removing part of the grain mass from the outside to the inside using either friction between grains or abrasion against a grid and then classifying obtained particles by sieving, eventually coupled with air classification. Depending on the processing time and type of mechanical solicitations removal of surficial dust (scouring), outer pericarp or more internal tissues (i.e. sequentially inner pericarp-testa, the aleurone, and then part of the starchy endosperm) occurs. However, due to the oblong form of wheat grain, no clear cut between the different tissues removal can be obtained. Different terms pearling, dehulling or debranning were used to describe the removal of these external tissues, but processing was rarely well characterised in terms of percentage of mass removed or identification of the removed tissues. Different authors agreed on the fact that a sharp reduction of both microorganisms and mycotoxins was observed between 4–10% of grain mass removal (Laca et al. 2006; Rios et al. 2009a; Sovrani et al. 2012). The removed tissues correspond to the pericarp and testa and part of the aleurone layer, plus potentially the embryo which was not characterized in these studies but known to also possibly contain mycotoxins (Giménez et al. 2013).

7.2.3 Grain Milling

Aim of this process was to efficiently isolate the starchy endosperm, in the form of flours or semolina depending on the grain hardness, from the outer layers and germ tissues. It combines successive grinding against stone or between rolls and separating steps based only on particle sizes by sieving, for *T. aestivum* (common wheat), or also coupled with particle density on purifiers, for *T. durum* (durum wheat). Therefore due their distribution in grains, fungi and corresponding mycotoxins have to be reduced in the recovered flours or semolina produced by milling in comparison with the initial grains. But, the overall reduction must depend on the extent of the pathogen penetration inside grains and the respective mechanical properties of each grain tissue being themselves influenced by genetic and environmental factors, each operating steps in the process and the water content as recently reviewed in Mayer-Laigle et al. (2018). A number of authors reported a significantly reduced level of mycotoxins into flours in comparison with initial grains after milling of *T. aestivum* grains, and only some recent examples are given in these reports OTA: (Scudamore et al. 2003), (DON or its conjugated form DON-3-Glc: (Kostelanska et al. 2011; Tibola et al. 2016); NIV and zearalenone, ZEA: (Tibola et al. 2016). A similar reduction in the mycotoxin level into semolina in comparison with initial content in *T. durum* grains was observed (T-2 and HT-2: Pascale et al., 2011; DON: Rios et al., 2009b; Visconti et al. 2004). The counterpart is that the other produced fractions can contain up to eight fold more mycotoxins than the initial grains. A compilation between the different authors according to the different toxins was recently reported by Schaarschmidt and Fauhl-Hassek (2018). Due to the way the grain breaks all along the milling process, it can lead to the production of fine particles from the most contaminated part of the grain which can be recovered in final flours or semolina. Indeed comparison between debranning and milling of the same contaminated grains demonstrated a more efficient effect on the former for mycotoxin reduction at a similar extraction rate of semolina (Rios et al. 2009a). Successive operating units able to decrease the grain contamination as the sorting and dehulling before milling was therefore recommended because it will increase the safety of the corresponding products even if the counterpart will be a loss of the initial raw matter.

7.3 *Flour/Semolina Pre-Treatments: Ozone, Irradiation, Cold Plasma, Superheated Steam, Extrusion Cooking*

When flours and semolina were produced, it is also possible to treat them with similar processes than used for grains to reduce their level of mycotoxins, if necessary, before their use for human consumption. However it is very important to ensure that the decontaminated flour was safe for this consumption. In particular, treatment of whole wheat flour by ozone as a gas was found to efficiently reduce the DON level following a first-order kinetic model as in grains but with higher efficiency (Wang et al. 2016b). Thus, DON reduction was improved with increasing ozone concentration and time. Ozone was also efficient in the reduction of total microorganisms but also led to changes in some flour biochemical compounds (Li et al. 2013). For examples, reduction of polyphenol oxidase (PPO) was observed which results in the whitening of the flour (Li et al. 2013). Physical characteristics also appeared different with an observed higher swelling capacity probably linked with higher water binding capacity (Li et al. 2013; Obadi et al. 2018). Opposite effects on pasting properties (viscosity) was observed which may be due to different nature of the flours (whole wheat flour or white flour respectively in (Li et al. 2013 and Obadi et al. 2018). Other authors related different effects of ozone on viscosity as a function of time (Mei et al. 2016). A slight increase in starch damage was observed related with the application time (Mei et al. 2016) whereas grain treatment with ozone was found to have the opposite effect due to a reduction of the grinding energy (Desvignes et al. 2008). But similarly to the grain pre-treatment with ozone, changes in the protein polymerization were observed (Chittrakorn et al. 2014; Sandhu et al. 2011). An increase in the flour whiteness was also observed (Mei et al. 2016; Sandhu et al., 2011).

Similarly to grains, flours treated with γ-radiation at dose level of 6 kGy were shown to be cured of the fungal flora (Aziz et al. 1997). A lower level of DON, ZEA and T-2 mycotoxins was observed after a treatment with 4 kGy whereas a complete elimination was found after a treatment with 8 kGy (Aziz et al. 1997). However depending on the applied dose, changes in the flour properties (color, rheology) were also reported after irradiation (Bashir et al. 2017; Bhat et al. 2016). Flour treatment by low level of cold plasma was found to be inefficient to reduce the micro-organism contamination (Bahrami et al. 2016) but sufficient to already change the flour properties (i.e. lipid and protein oxidation). Superheated steam treatments of flours were found to decrease free sulfhydryl and thus the dough strength, elastic and viscous moduli was increased in relation with the treatment time (Hu et al. 2017b). Therefore again more research was needed to find optimal conditions between reductions of mycotoxin level and the fewer changes in the raw matter composition and properties.

The extrusion cooking is used to produce snacks and breakfast cereals under high temperature, pressure and shear forces. This thermomechanical treatment can be favorable to heat destruction of mycotoxins, but also generates chemical, physico-chemical, and biochemical changes which also could lead to potential changes in the mycotoxin extractability (Schaarschmidt and Fauhl-Hassek 2018; Zhao et al. 2011).

7.4 Baking/Pasta Making and Cooking

Flours produced from first transformation were further transformed after water addition and other additives respectively in fermented products like bread, or non-fermented as biscuits, cakes or Chinese noodles. Similarly, semolina obtained from durum wheat was mixed with water only and transformed into pasta. Bread, biscuits and cakes were further cooked before consumption whereas noodles and pasta were made by extrusion and then eventually dried and cooked by the consumers before eating. Recipes and process conditions (time, temperatures, pH, nature of the additives, enzymes or micro-organisms (yeast, sourdough bacteria), pressure of the screw extruder) varies depending on country uses. Level of contamination and mycotoxin nature and amount were also different in publications. Therefore, comparison between results of different authors on the effect of each operation units is difficult. Moreover, one has to considered that mycotoxin analysis needs first to be extracted from the matrix to be quantified therefore a decrease in the mycotoxin content could also be due to a decrease in extractability of the corresponding compound as for example suggested by Gerenotti et al. (2017) comparing different recipes of biscuits. Additionally, some mycotoxins, as DON, also exist under different modified forms (acetyl, glucoside, de-epoxy, sulphated) displaying potentially contrasted effects (Alizadeh et al. 2016; Del Favero et al. 2018; Freire and Sant'Ana 2018) on human health. Thus, it is also important to precisely characterize these different forms, but appears to be rarely the case.

In baking step few studies were found to isolate the dough fermentation step from the cooking one. Recently Khaneghah et al. (2018) performed a meta-analysis to study the effects of different steps in cereal-based products making, notably bread, on different mycotoxins level and showed discrepancy between authors on the effect of each operating units. Indeed depending on the recipes, operating conditions and the considered mycotoxin, a decrease or an increase amount can be observed. Kostelanska et al. (2011) for example showed that addition of bakery improvers containing glycolytic activities during dough fermentation increase DON-3-Glc amount probably due to its release from polysaccharide-bounded form in the matrix. Indeed, deoxynivalenol glycosides were detected in wheat and barley products (Zachariasova et al. 2012). Kostelanska et al. (2011) also noted the formation of degradation products, as de-epoxidated forms of DON-3-Glc, in the crust during baking due to high temperatures especially at the bread surface. The final reduction between flour and bread was estimated between 10–13% depending on the form of DON. Valle-Algarra et al. (2009) studying different mycotoxins along bread making noted that fermentation only reduced OTA level around 30–35% whereas the other toxins (DON, 3-ADON, NIV) were only reduced by the baking step. Higher levels of mycotoxin reduction were found in this case between 33 and 77% depending on the considered toxin. Important range of DON reduction (from 16.8 to 96.6%) between dough and final products was also noticed in (Neira et al. 1997) depending on the flour initial DON content. Other mycotoxins, called enniatins, were found to be notably reduced first by milling but mostly during bread making probably due to their heat degradation (Vaclavikova et al. 2013). When transformed into noodles, it was found that DON reduction occurs during boiling in water before consumption (Nowicki et al. 1988, Sugita-Konishi et al., 2006).

The effects of ozone treatment of flour on the product (bread, cake, noodles) quality was recently summarized by Zhu (2018). Briefly, the product texture and behaviour can be improved or degraded depending notably on the treatment conditions and the colour was lightened but more work is needed in regards to the sensory properties. Fate of mycotoxins present in semolina, produced from durum wheat milling, along pasta making was also largely studied (Brera et al. 2013; Nowicki et al. 1988; Visconti et al. 2004). During pasta cooking in boiling conditions, around 40–50% of DON is lost in the boiling water. Visconti et al. (2004) as well as Brera et al. (2013) estimated that from uncleaned wheat grains to the cooked pasta, less than 25% of DON was retained in the final consumed product. A model allowing prediction of the final DON concentration in cooked pasta was also established by Vidal et al. (2016). These authors also found that DON-3-Glc was possibly released during cooking from the pasta components and then transferred to water leading to similar amount of this mycotoxin in spaghettis before and after cooking. Moreover the OTA concentrations in spaghettis did not changed after cooking. Contradictory results were obtained however regarding fate of enniatins after pasta cooking. Indeed, De Nijs et al. (2016) found between 0 to 20% mycotoxin losses after pasta cooking whereas Serrano et al. (2016) concluded to loss between 14 to 100% depending on the considered molecule. These last authors also found that a reduction in the water pH to a value of 4 led to enniatin loss equal to 100% whatever the analysed molecule. But these results were obtained with artificially contaminated semolina before pasta making whereas De Nijs et al. (2016) collected real industrial and commercialised pasta samples from the market which appears as a better approach to study enniatin fate.

References

Abedi-Tizaki M, Zafari DM (2015) Natural occurrence of deoxynivalenol and its acetylated derivatives in wheat in north of Iran. Journal of Plant Pathology 97: 431–437.

Alizadeh A, Braber S, Akbari P, Kraneveld A, Garssen J, Fink-Gremmels J (2016) Deoxynivalenol and its modified forms: are there major differences? Toxins 8: 334.

Aziz NH, Attia ES, Farag SA (1997) Effect of gamma-irradiation on the natural occurrence of *Fusarium* mycotoxins in wheat, flour and bread. Nahrung 41: 34–37.

Aziz NH, Souzan RM, Azza AS (2006) Effect of gamma-irradiation on the occurrence of pathogenic microorganisms and nutritive value of four principal cereal grains. Applied Radiation and Isotopes 64: 1555–1562.

Bahrami N, Bayliss D, Chope G, Penson S, Perehinec T, Fisk ID (2016) Cold plasma: A new technology to modify wheat flour functionality. Food Chemistry 202: 247–253.

Bashir K, Swer TL, Prakash KS, Aggarwal M (2017) Physico-chemical and functional properties of gamma irradiated whole wheat flour and starch. LWT - Food Science and Technology 76: 131–139,

Battilani P, Toscano P, Van der Fels-Klerx HJ, Moretti A, Camardo Leggieri M, Brera C et al. (2016) Aflatoxin B1 contamination in maize in Europe increases due to climate change. Scientific Reports 6: 24328.

Bhat NA, Wani IA, Hamdani AM, Gani A, Masoodi FA (2016) Physicochemical properties of whole wheat flour as affected by gamma irradiation. LWT - Food Science Technology 71: 175–183.

Brera C, Peduto A, Debegnach F, Pannunzi E, Prantera E, Gregori E et al. (2013) Study of the influence of the milling process on the distribution of deoxynivalenol content from the caryopsis to cooked pasta. Food Control 32: 309–312.

Bryła M, Ksieniewicz-Woźniak E, Waśkiewicz A, Szymczyk K, Jędrzejczak R (2018) Natural occurrence of nivalenol, deoxynivalenol, and deoxynivalenol-3-glucoside in Polish winter wheat. Toxins, 10: 81.

Buerstmayr H, Ban T, Anderson JA (2009) QTL mapping and marker-assisted selection for *Fusarium* head blight resistance in wheat: a review. Plant Breeding 128: 1–26.

Burlakoti RR, Tamburic-Ilincic L, Limay-Rios V, Burlakoti P (2017) Comparative population structure and trichothecene mycotoxin profiling of *Fusarium graminearum* from corn and wheat in Ontario, central Canada. Plant Pathology 66: 14–27.

Calado T, Venancio A, Abrunhosa L (2014) Irradiation of mold and mycotoxin control: a review. *Comprehensive Reviews* in *Food Science* and Food Safety 13: 1049–1061.

Calori-Domingues MA, Bernardi, CMG, Nardin MS, de Souza GV, dos Santos FGR, Stein M de A et al. (2016). Co-occurrence and distribution of deoxynivalenol, nivalenol and zearalenone in wheat from Brazil. Food Additives and Contaminants Part B, 9: 142–151.

Camardo Leggieri M, Van Der Fels-Klerx HJ, Battilani P (2013) Cross-validation of predictive models for occurrence of deoxynivalenol in wheat at harvest. World Mycotoxin Journal 6: 389–397.

Cerón-Bustamante M, Ward, TJ, Kelly A, Vaughan MM, McCormick SP, Cowger C, Leyva-Mir SG et al. (2018). Regional differences in the composition of Fusarium Head Blight pathogens and mycotoxins associated with wheat in Mexico. International Journal of Food Microbiology 273: 11–19.

Cheli F, Pinotti L, Rossi L, Dell'Orto V (2013) Effect of milling procedures on mycotoxin distribution in wheat fractions: A review. LWT Food Science and Technology 54: 307–314.

Chittrakorn S, Earls D, Mac Ritchie F (2014) Ozonation as an alternative to chlorination for soft wheat flours. Journal of Cereal Science 60: 217–221.

Chulze SN, Palazzini JM, Torres AM, Barros G, Ponsone ML, Geisen R et al. (2015). Biological control as a strategy to reduce the impact of mycotoxins in peanuts, grapes and cereals in Argentina. Food Additives and Contaminants Part A 32: 471–479.

Cuniberti MB (2001). *Fusarium* y su efecto en la calidad de trigo. Información para Extensión INTA-EEA Marcos Juárez, Córdoba, Argentina.

Cuniberti MB (2013) Incidencia del Fusarium en la calidad comercial, molinera e industrial del trigo. Campaña 2012/13. INTA-EEA Marcos Juárez, Córdoba, Argentina.

Darsanaki RK, Issazadeh K, Aliabadi MA, Chakoosari MMD (2015) Occurrence of deoxynivalenol (DON) in wheat flours in Guilan Province, northern Iran. Annals of Agricultural and Environmental Medicine 22: 35–37.

De Nijs M, van den Top H, de Stoppelaar J, Lopez P, Mol H (2016) Fate of enniatins and deoxynivalenol during pasta cooking. Food Chemistry 213: 763–767.

Del Favero G, Woelflingseder L, Braun D, Puntscher H, Kütt M-L, Dellafiora L et al. (2018) Response of intestinal HT-29 cells to the trichothecene mycotoxin deoxynivalenol and its sulfated conjugates. Toxicology Letters 295: 424–437.

Del Ponte, EM, Spolti P, Ward TJ, Gomes LB, Nicolli CP, Kuhnem PR et al. (2014). Regional and field-specific factors affect the composition of Fusarium head blight pathogens in subtropical no-till wheat agroecosystem of Brazil. Phytopathology 105: 246–254.

Delwiche SR, Pearson TC, Brabec DL (2005) High-speed optical sorting of soft wheat for reduction of deoxynivalenol. Plant Disese 89: 1214–1219.

Desvignes C, Chaurand M, Dubois M, Sadoudi A, Abecassis J, Lullien-Pellerin V (2008) Changes in common wheat grain milling behaviour and tissue mechanical properties following ozone treatment. Journal of Cereal Science 47: 245–251.

Dexter JE, Clear RM, Preston KR (1996) Fusarium Head Blight: effect on the milling and baking of some Canadian wheats. Cereal Chemistry 73: 695–701.

Dong F, Qiu J, Xu, J, Yu M, Wang, S, Sun Y, Zhang Z, Shi J (2016) Effect of environmental factors on *Fusarium* population and associated trichothecenes in wheat grain grown in Jiangsu province, China. International Journal of Food Microbiology 230, 58–63.

Dubois M, Coste C, Despres AG, Efstathiou T, Nio C, Dumont E, Parent-Massin D (2006) Safety of Oxygreen, an ozone treatment on wheat grains. Part 2. Is there a substantial equivalence between Oxygreen-treated wheat grains and untreated wheat grains? Food Additives and Contaminants 23: 1–15.

Duffeck MR, Tibola CS, Guarienti EM, Del Ponte EM (2017) Survey of mycotoxins in Southern Brazilian wheat and evaluation of immunoassay methods. Scientia Agricola, 74: 343–348.

European Commission Commission Regulation (EC) (2006) No 1881/2006 of 19 December 2006 setting maximum levels for certain contaminants in foodstuffs. Official. Journal of. European. Communities L364, 5–24.

European Commission EU Directive 2009/128/EC (2009) Sustainable use of pesticides Directive 2009/128/EC of the European Parliament and of the Council of 21 October 2 009 establishing a framework for Community action to achieve the sustainable use of pesticides, OJ L 309, 24.11.2009, p. 71–86.

FDA-United States Food and Drug Administration (2001) Secondary direct food additives permitted in food for human consumption. Federal Register 66: 33829–33830.

Freire L, Sant'Ana A (2018) Modified mycotoxins: An updated review on their formation, detection, occurrence, and toxic effects. Food Chemical Toxicology 111: 189–205.

Gale LR, Ward TJ, Balmas V, Kistler HC (2007) Population subdivision of *Fusarium graminearum sensu stricto* in the upper midwestern United States. Phytopathology 97: 1434–1439.

Gaou I, Dubois M, Pfohl-Leszkowicz A, Coste C, De Jouffrey S, Parent-Massin D (2015) Safety of Oxygreen®, an ozone treatment on wheat grains. Part 1- A four week toxicity study in rats by dietary administration of treated wheat. Food Additives and Contaminants 22: 1113–1119.

Garcia-Cela E, Kiaitsi E, Medina A, Sulyok M, Krska R, Magan N (2018a) Interacting environmental stress factors affects targeted metabolomic profiles in stored natural wheat and that inoculated with *F. graminearum*. Toxins 10: 56.

Garcia-Cela E, Kiaitsi E, Sulyok M, Medina A, Magan N (2018b) *Fusarium graminearum* in stored wheat: use of CO2 production to quantify dry matter losses and relate these to relative risks of Zearalenone contamination under different interacting environmental conditions. Toxins 10, 86.

Gärtner BH, Munich M, Kleijer G, Mascher F (2008) Characterization of kernel resistance against Fusarium infection in spring wheat by baking quality and mycotoxin assessments. European Journal of Plant Pathology, 120: 61–68.

Gerenotti S, Cirlini M, Sarkanj B, Sulyok M, Berthiller F, Dall'Asta C, Suman M (2017) Formulation and processing factors affecting trichothecene mycotoxins within industrial biscuit-making. Food Chemistry 229: 597–603.

Giménez I, Herrera M, Escobar J, Ferruz E, Lorán S, Herrera A, Ariño A (2013) Distribution of deoxynivalenol and zearalenone in milled germ during wheat milling and analysis of toxin levels in wheat germ and wheat germ oil. Food Control 34: 268–273.

Giroux ME, Bourgeois G, Dion Y, Rioux S, Pageau D, Zoghlami S et al. (2016) Evaluation of forecasting models for fusarium head blight of wheat under growing conditions of Quebec, Canada. Plant Disease 100: 1192–1201.

Gorczyca A, Oleksy A, Gala-Czekaj D, Urbaniak M, Laskowska M, Waśkiewicz A, Stępień Ł (2017) *Fusarium* head blight incidence and mycotoxin accumulation in three durum wheat cultivars in relation to sowing date and density. The Science of Nature, 105: 1–2.

Goze P, Rhazi L, Lakhal L, Jacolot P, Pauss A, Aussenac T (2017) Effects of ozone treatment on the molecular properties of wheat grain proteins. Journal of Cereal Science 75: 243–251.

Goze P, Rhazi L, Pauss A, Aussenac T (2016) Starch characterization after ozone treatment of wheat grains. Journal of Cereal Science 70: 207–213.

Graham DM (1997) Use of ozone for food processing. Food Technology 51: 72–75.

Hofgaard IS, Aamot HU, Torp T, Jestoi M, Lattanzio VMT, Klemsdal SS et al. (2016) Associations between *Fusarium* species and mycotoxins in oats and spring wheat from farmers' fields in Norway over a six-year period. World Mycotoxin Journal 9: 365–378.

Hojnik N, Cvelbar U, Tavcar-Kalcher G, Walsh JL, Krizaj I (2017) Mycotoxin Decontamination of Food: Cold Atmospheric Pressure Plasma versus "Classic" Decontamination. Toxins 9: 151.

Hollingsworth CR, Motteberg CD, Wiersma JV, Atkinson LM (2008) Agronomic and economic responses of spring wheat to management of *Fusarium* head blight. Plant Disease 92: 1339–1348.

Hope R and Magan N (2003) Two dimensional environmental profiles of growth, deoxynivalenol and nivalenol production by *Fusarium culmorum* on a wheat-based substrate. Letters Applied Microbiology. 37: 70–74.

Hope R, Aldred D, Magan N (2005) Comparison of the effect of environmental factors on deoxynivalenol production by *F. culmorum* and *F. graminearum* on wheat grain. Letters Applied Microbiology 40: 295–300.

Hu Y, Wang L, Hong Z, Li Z (2017a) Superheated steam treatment improved flour qualities of wheat in suitable conditions. Journal of Food Processing and Preservation 41: e13238.

Hu Y, Wang L, Hong Z, Li Z (2017b) Modification of protein structure and dough rheological properties of wheat flour through superheated steam treatment Journal of Cereal Science 76: 222–228.

Jaillais B, Roumet P, Pinson-Gadais L, Bertrand D (2015) Detection of Fusarium head blight contamination in wheat kernels by multivariate imaging. Food Control 54: 250–258.

Janaviciene S, Mankeviciene A, Suproniene S, Kochiieru Y, Keriene I (2018) The prevalence of deoxynivalenol and its derivatives in the spring wheat grain from different agricultural production systems in Lithuania. Food Additives and Contaminants Part A, 35: 1179–1188.

Juan C, Covarelli L, Beccari G, Colasante V, Mañes J (2016) Simultaneous analysis of twenty-six mycotoxins in durum wheat grain from Italy. Food Control, 62: 322–329.

Kamimura H, Nishijima M, Saito K, Yasuda K, Ibe A, Nagyama T, Ushiyama H, Naoi Y (1979) The decomposition of trichothecene mycotoxins during food processing. Studies on mycotoxins in foods. XII. Journal of the Food Hygienic Society of Japan 20: 352–357.

Khaneghah AM, Fakhri Y, Sant'Ana A (2018) Impact of unit operations during processing of cereal-based products on the levels of deoxynivalenol, total aflatoxin, ochratoxin A and zearalenone: A systematic review and meta-analysis. Food Chemistry 268: 611–624.

Kirinčič S, Škrjanc B, Kos N, Kozolc B, Pirnat N, Tavčar-Kalcher G (2015) Mycotoxins in cereals and cereal products in Slovenia – Official control of foods in the years 2008–2012. Food Control, 50: 157–165.

Kostelanska M, Dzuman Z, Malachova A, Capouchova I, Prokinova E, Skerikova A et al. (2011) Effects of milling and baking technologies on levels of deoxynivalenol and its masked form deoxynivalenol-3-glucoside. Journal of Agricultural Food Chemistry 59: 9303–9312.

Laca A, Mousia Z, Diaz M, Webb C, Pandiella SS (2006) Distribution of microbial contamination within cereal grain Journal of Food Engineering 72: 332–338.

Lacko-Bartošová M, Remža J, Lacko-Bartošová L (2017) *Fusarium* mycotoxin contamination and co-occurrence in Slovak winter wheat grains. Zemdirbyste- Agriculture 104: 173–178.

Lamacchia C, Landriscina L, D'Agnello P (2016) Changes in wheat kernel proteins induced by microwave treatment. Food Chemistry 197: 634–640.

Li M, Peng J, Zhu K-X, Guo X-N, Zhang M, Peng W, Zhou H-M (2013) Delineating the microbial and physical–chemical changes during storage of ozone treated wheat flour. Innovative Food Science & Emerging Technologies 20: 223–229.

Li MM, Guan EQ, Bian K (2015a) Effect of ozone treatment on deoxynivalenol and quality evaluation of ozonised wheat. Food Addititives and Contaminants: Part A 32: 544–553.

Li X, Shin S, Heinen S, Dill-Macky R, Berthiller F, Nersesian N et al. (2015b) Transgenic wheat expressing a barley UDP-glucosyltransferase detoxifies deoxynivalenol and provides high levels of resistance to *Fusarium graminearum*. Molecular Plant-Microbe Interactions 28: 1237–1246.

Los A, Ziuzina D, Bourke P (2018) Current and future technologies for microbiological decontamination of cereal grains. Journal of Food Science 83: 1484–1493.

Lullien-Pellerin V (2012) Ozone in grain processing. In O'Donnell C, Tiwari B K, Cullen P J, Rice R G (ed) Ozone in Food Processing. Wiley-Blackwell, p 81–102.

Madgewick J, West JS, White R, Semenov M, Townsend JA, Turner JA et al. (2011). Future threat: direct impact of climate change on wheat Fusarium ear blight in the UK. European Journal of Plant Pathology 130: 117–131.

Machado LV, Mallmann CA, Mallmann AO, Coelho RD, Copetti, MV (2017). Deoxynivalenol in wheat and wheat products from a harvest affected by *Fusarium* head blight. Food Science and Technology 37: 8–12.

Magan N. and Medina A (2016) Integrating gene expression, ecology and mycotoxin production by *Fusarium* and *Aspergillus* species in relation to interacting environmental factors. World Mycotoxin Journal 9: 863–874.

Magan N, Aldred D, Baxter ES (2014) Mycotoxin Reduction in Grain Chains. Leslie, J. F. and Logrieco, A.F (Eds) John Wiley & Sons, Ltd, Chichester, UK pp. 258–267.

Magan N, Aldred D, Mylona K, Lambert RJW (2010) Limiting mycotoxins in stored wheat. Food Additives and Contaminants Part A 27: 644–650.

Mallmann CA, Dilkin P, Mallmann AO, Oliveira MS, Adaniya ZNC, Tonini C (2017) Prevalence and levels of deoxynivalenol and zearalenone in commercial barley and wheat grain produced in Southern Brazil: an eight-year (2008 to 2015) summary. Tropical Plant Pathology 42: 146–152.

Martinez M, Castañares E, Dinolfo MI, Pacheco W, Moreno MV and Stenglein SA (2014) Presencia de Fusarium graminearum en muestras de trigo destinado al consumo humano. Revista Argentina de Microbiologia 46: 41–44.

Mayer-Laigle C, Barakat A, Barron C, Delenne J-Y, Frank X, Mabille F, Rouau X et al. (2018) DRY biorefineries: Multiscale modeling studies and innovative processing. Innovative Food Science and Emerging Technology 46: 131–139.

McMullen M, Bergstrom G, De Wolf E, Dill-Macky R, Hershman D, Shaner G et al. (2012) a unified effort to fight an enemy of wheat and barley: Fusarium Head Blight. Plant Disease 96: 1712–1728.

Medina A, Akbar A, Baazeem A, Rodriguez A, Magan N (2017) Climate change, food security and mycotoxins: do we know enough? Fungal Biology Reviews 31: 143–154.

Medina A, Rodríguez A, Magan N (2015) Climate change and mycotoxigenic fungi: Impacts on mycotoxin production. Current Opinion in Food Science 5: 99–104.

Mei J, Liu G, Huang X, Ding W (2016) Effects of ozone treatment on medium hard wheat (*Triticum aestivum* L.) flour quality and performance in steamed bread making. CyTA Journal of Food 14: 449–456.

Mesterházy A (1995) Types and components of resistance to *Fusarium* head blight of wheat. Plant Breeding 114: 377–386.

Mesterházy Á, Tóth B, Varga M, Bartók T, Szabó-Hevér Á, Farády L, Lehoczki-Krsjak S (2011). Role of fungicides, application of nozzle types, and the resistance level of wheat varieties in the control of *Fusarium* head blight and deoxynivalenol. Toxins 3: 1453–1483.

Moschini RC, Fortugno C (1996) Predicting wheat head blight incidence using models based on meteorological factors in Pergamino, Argentina. European Journal of Plant Pathology 102: 211–218.

Moschini RC, Martínez MI, Sepulcri MG (2013) Alconada Magliano, TM and Chulze, SN (Eds.). *Fusarium* head blight in Latin America. Springer Netherlands, The Netherlands pp. 205–227.

Mylona K, Sulyok M, Magan N (2012). *Fusarium graminearum* and *Fusarium verticillioides* colonisation of wheat and maize, environmental factors, dry matter losses and mycotoxin production relevant to the EU legislative limits. Food Additives and Contaminants 29: 1118–1128.

Neira MS, Pacin AM, Martinez EJ, Molto G, Resnik SL (1997) The effects of bakery processing on natural deoxynivalenol contamination. International Journal of Food Microbiology 37: 21–25.

Nowicki TW, Gaba DG, Dexter JE, Matsuo RR, Clear RM (1988) Retention of *Fusarium* mycotoxin deoxynivalenol in wheat during processing and cooking of spaghetti and noodles. Journal of Cereal Science 8: 189–202.

Obadi M, Zhu K-X, Peng W, Sulieman AA, Mohammed K, Zhou H-M (2018) Effects of ozone treatment on the physicochemical and functional properties of whole grain flour. Journal of Cereal Science 81: 127–132.

Ostry V, Malir F, Toman J, Grosse Y (2017) Mycotoxins as human carcinogens-the IARC Monographs classification. Mycotoxin Research 33: 65–73.

Palacios SA, Erazo JG, Ciasca B, Lattanzio VMT, Reynoso MM, Farnochi MC et al. (2017). Occurrence of deoxynivalenol and deoxynivalenol-3-glucoside in durum wheat from Argentina. Food Chemistry 230: 728–734.

Palazzini J, Fumero V, Yerkovich N, Barros G, Cuniberti M, Chulze S (2015) Correlation between *Fusarium graminearum* and deoxynivalenol during the 2012/13 wheat Fusarium Head Blight outbreak in Argentina. Cereal Research Communication 43: 627–637.

Palazzini J, Roncallo P, Cantoro R, Chiotta M, Yerkovich N, Palacios S, Echenique V, Torres A, Ramirez M, Karlovsky P, Chulze S (2018) Biocontrol of *Fusarium graminearum* sensu stricto, reduction of deoxynivalenol accumulation and phytohormone induction by two selected antagonists. Toxins 10, 88.

Palazzini JM, Alberione E, Torres A, Donat C, Kohl J, Chulze S (2016). Biological control of *Fusarium gramienarum sensu stricto* causal agent of Fusarium head blight of wheat, using formualted antagonists under field conditions in Argentina. Biological Control 94: 56–61.

Palazzini JM, Groenenboom-de Haas BH, Torres AM, Köhl J, Chulze SN (2013) Biocontrol and population dynamics of *Fusarium* spp. on wheat stubble in Argentina. Plant Pathology 62: 859–866.

Palazzini JM., Ramirez ML, Torres AM, Chulze SN (2007) Potential biocontrol agents for *Fusarium* head blight and deoxynivalenol production in wheat. Crop Protection 26: 1702–1710.

Pascale M, Haidukowski M, Lattanzio VMT, Silvestri M, Ranieri R, Visconti A (2011) Distribution of T-2 and HT-2 toxins in milling fractions of durum wheat. Journal of Food Protection 74: 1700–1707.

Paul PA, Lipps PE, Hershman DE, McMullen MP, Draper MA, Madden LV (2008) Efficacy of triazole-based fungicides for fusarium head blight and deoxynivalenol control in wheat: a multivariate meta-analysis. Phytopathology 98: 999–1011.

Peyron S, Abecassis J, Autran J-C, Rouau X (2002) Influence of UV exposure on phenolic acid content, mechanical properties of bran, and milling behavior of durum wheat (*Triticum durum Desf.*). Cereal Chemistry 79: 726–731.

Pleadin J, Staver MM, Markov K, Frece J, Zadravec M, Jaki V et al. (2017) Mycotoxins in organic and conventional cereals and cereal products grown and marketed in Croatia. Mycotoxin Research 33: 219–227.

Popovic V, Fairbanks N, Pierscianowski J, Biancaniello M, Zhou T, Koutchma T (2018) Feasibility of 3D UV-C treatment to reduce fungal growth and mycotoxin loads on maize and wheat kernels. Mycotoxin Research 34: 211–221.

Prandini A, Sigolo S, Filippi L, Battilani P, Piva G (2009) Review of predictive models for Fusarium head blight and related mycotoxin contamination in wheat. Food and Chemistry Toxicology 47: 927–931.

Prange A, Birzele B, Krämer J, Meier, A, Modrow H, Köhler P (2005) *Fusarium*-inoculated wheat: deoxynivalenol contents and baking Quality in relation to infection time. Food Control, 16: 739–745.

Prat N, Gilbert C, Prah U, Wachter E, Steiner B, Langin T et al. (2017) QTL mapping of Fusarium Head blight resistance in three related durum wheat populations. Theoretical and applied genetics, 130: 13–27.

Preston KR, Kilborn RH, Black HC (1982) The GRL Pilot Mill. II. Physical dough and baking properties of flour streams milled from Canadian red spring wheats. Canadian Institute of Food Science and Technology Journal 15: 29.

Pronyk C, Cenkowski S, Abramson D (2006) Superheated steam reduction of deoxynivalenol in naturally contaminated wheat kernels. Food Control 17: 789–796.

Purahong W, Nipoti P, Pisi A, Lemmens M, Prodi A (2014) Aggressiveness of different *Fusarium graminearum* chemotypes within a population from Northern-Central Italy. Mycoscience 55: 63–69.

Ramirez ML, Reynoso MM, Farnochi, Sofia MC, Chulze S (2006) Vegetative compatibility and mycotoxin chemotypes among *Fusarium graminearum (Gibberella zeae)* isolates from wheat in Argentina European Journal of Plant Pathology 115: 139–148.

Rawat N, Pumphrey MO, Liu S, Zhang X, Tiwari VK, Ando K et al. (2016) Wheat Fhb1 encodes a chimeric lectin with agglutinin domains and a pore-forming toxin-like domain conferring resistance to Fusarium head blight. Nature Genetics 48: 1576–1580.

Rios G, Pinson-Gadais L, Abecassis J, Zakhia-Rozis N, Lullien-Pellerin V (2009a) Assessment of dehulling efficiency to reduce deoxynivalenol and *Fusarium* level in durum wheat grains. Journal of Cereal Science 49: 387–392.

Rios G, Zakhia N, Abecassis J, Chaurand M, Samson M-F, Richard-Forget F, Lullien-Pellerin V (2007) Impact des opérations de transformation sur la répartition du DON dans les produits de fractionnement du blé dur. In Colloque Scientifique « Mycotoxines Fusariennes des Céréales », Arcachon, FRA, (2007-09-11 - 2007-09-13).

Rios G, Zakhia-Rozis N, Chaurand M, Richard-Forget F, Samson MF, Abecassis J, Lullien-Pellerin V (2009b) Impact of durum wheat milling on deoxynivalenol distribution in the outcoming fractions. Food Additives & Contaminants Part A, 26: 487–495

Rossi V, Caffi T, Salinari F (2012) Helping farmers face the increasing complexity of decision-making for crop protection. Phytopathologia Mediterranea 51: 457–479.

Rossi V, Giosuè S, Pattori E, Spanna F, Del Vecchio AA (2003). A model estimating the risk of Fusarium head blight on wheat. EPPO Bulletin 33: 421–425.

Sadhasivam S, Britzi M, Zakin V, Kostyukovsky M, Trostanetsky A, Quinn E, Sionov E (2017) Rapid detection and identification of mycotoxigenic fungi and mycotoxins in stored wheat grain. Toxins 9, 302.

Sandhu HPS, Manthey FA, Simsek S (2011) Quality of bread made from ozonated wheat (*Triticum aestivum* L.) flour. Journal of the Science of Food and Agriculture 91: 1576–1584.

Sarrocco S, Vannacci G (2018) Preharvest application of beneficial fungi as a strategy to prevent postharvest mycotoxin contamination: A review. Crop Protection 110: 160–170.

Sarver BAJ, Ward TJ, Gale LR, Broz K, Kistler HC, Aoki T, Nicholson P et al. (2011). Novel Fusarium head blight pathogens from Nepal and Louisiana revealed by multilocus genealogical concordance. Fungal Genetic Biology 48: 1096–1107.

Schaafsma AW, Hooker DC (2005) Validation of the Doncast prediction tool in wheat across France and Uruguay. In: Canty, S.M., Boring, T., Wardwell, J., Siler, L. and Ward, R.W (Eds.), Proceedings of the National Fusarium Head Blight Forum. December 11-13, 2005. Milwaukee, Wisconsin: Michigan State University, pp. 148.

Schaarschmidt S, Fauhl-Hassek C (2018) The fate of mycotoxins during the processing of wheat for human consumption. Comprehensive Reviews on Food Science and Food Safety 17: 556–593.

Schisler DA, Khan NI, Boehm MJ, Lipps PE, Slininger PJ, Zhang S (2006) Selection and evaluation of the potential of choline-metabolizing microbial strains to reduce *Fusarium* head blight. Biological Control 39: 497–506.

Schmale DG, Wood-Jones AK, Cowger C, Bergstrom GC, Arellano C (2011) Trichothecene genotypes of *Gibberella zeae* from winter wheat fields in the eastern USA. Plant Pathology 60: 909–917.

Scudamore KA, Banks J, MacDonald SJ (2003) Fate of ochratoxin A in the processing of whole wheat grains during milling and bread production. Food Additives & Contaminants 20: 1153–1163.

Serrano AB, Font G, Manes J, Ferrer E (2016) Development a mitigation strategy of enniatins in pasta under home-cooking conditions. LWT Food Science and Technology 65: 1017–1024.

Serranti S, Cesare D, Bonifazi G (2013) The development of a hyperspectral imaging method for the detection of Fusarium-damaged, yellow berry and vitreous Italian durum wheat kernels. Biosystems Engineering 115: 20–30.

Shala-Mayrhofer V, Marjakaj R, Varga E, Berthiller F, Musolli A, Lemmens M (2015) Occurrence of Fusarium head blight and mycotoxins as well as morphological identification o *Fusarium* species in winter wheat in Kosovo. Cereal Research Communication 43: 438–448.

Sovrani V, Blandino M, Scarpino V, Reyneri A, Coïsson JD, Travaglia F et al. (2012) Bioactive compound content, antioxidant activity, deoxynivalenol and heavy metal contamination of pearled wheat fractions. Food Chemistry 135: 39–46.

Stanciu O, Juan C, Miere D, Dumitrescu A, Bodoki E, Loghin F, Mañes J (2017) Climatic conditions influence emerging mycotoxin presence in wheat grown in Romania – A 2-year survey. Crop Protection 100: 124–133.

Suga H, Karugia GW, Ward T, Gale LR, Tomimura K, Nakajima T et al. (2008). Molecular Characterization of the *Fusarium graminearum* Species Complex in Japan. Phytopathology 98, 159–166.

Sugita-Konishi Y, Park BJ, Kobayashi-Hattori K, Tanaka T, Chonan T, Yoshikawas K, Kumagai S, (2006) Effect of cooking process on the deoxynivalenol content and its subsequent cytotoxicity in wheat products. Bioscience, Biotechnology, and Biochemistry 70: 1764–1768.

Sumíková T, Chrpová J, Džuman Z, Salava J, Štěrbová L, Palicová J, Slavíková P et al. (2017) Mycotoxins content and its association with changing patterns of Fusarium pathogens in wheat in the Czech Republic. World Mycotoxin Journal 10: 143–151.

Tibola CS, Fernandes JMC, Guarienti EM (2016) Effect of cleaning, sorting and milling processes in wheat mycotoxin content. Food Control 60: 174–179.

Tima H, Berkics, A, Hannig Z, Ittzés A, Kecskésné Nagy E, Mohácsi-Farkas C, Kiskó G (2017) Deoxynivalenol in wheat, maize, wheat flour and pasta: surveys in Hungary in 2008–2015. Food Additives and Contaminants Part B, 11: 37–42.

Tiwari BK, Brennan CS, Curran T, Gallagher E, Cullen PJ, O'Donnell CP (2010) Application of ozone in grain processing. Journal of Cereal Science. 51: 248–255.

Tkachuk R, Dexter JE, Tipples KH, Nowicki TW (1991) Removal by specific gravity table of tombstone kernels and associated trichothecenes from wheat infected with Fusarium Head Blight. Cereal Chemistry 68: 428–431.

Tralamazza SM, Bemvenuti RH, Zorzete P, de Souza Garcia F, Corrêa B (2016) Fungal diversity and natural occurrence of deoxynivalenol and zearalenone in freshly harvested wheat grains from Brazil. Food Chemistry 196: 445–450.

Trenholm HL, Charmley LL, Prelusky DB, Warner RM (1991) Two physical methods for the decontamination of four cereals contaminated with deoxynivalenol and zearalenone. Journal of Agricultural and Food Chemistry 39: 356–360.

Trombete F, Barros A, Vieira M, Saldanha T, VenÃ¢ncio A, Fraga M, (2016a) Simultaneous determination of deoxynivalenol, deoxynivalenol-3-glucoside and nivalenol in wheat grains by HPLC-PDA with immunoaffinity column cleanup. Food Analytical Methods 9: 2579–2586.

Trombete F, Minguita A, Porto Y, Freitas-Silva O, Freitas-Sa D et al. (2016b) Chemical, Technological, and sensory properties of wheat grains (*Triticum aestivum* L) as affected by gaseous ozonation. International Journal of Food Properties 19: 2739–2749.

Umpiérrez-Failache M, Garmendia G, Pereyra S, Rodríguez-Haralambides A, Ward TJ, Vero S (2013). Regional differences in species composition and toxigenic potential among Fusarium head blight isolates from Uruguay indicate a risk of nivalenol contamination in new wheat production areas. International Journal of Food Microbiology 166: 135–140.

Vaclavikova M, Malachova A, Veprikova Z, Dzuman Z, Zachariasova M, Hajslova J (2013) 'Emerging' mycotoxins in cereals processing chains: Changes of enniatins during beer and bread making. Food Chemistry 136: 750–757.

Valle-Algarra FM, Mateo EM, Medina A, Mateo F, Gimeno-Adelantado JV, Jimenez M (2009) Changes in ochratoxin A and type B trichotecenes contained in wheat flour during dough fermentation and bread-baking. Food Additives and Contaminants Part A 26: 896–906.

Van der Fels-Klerx HJ, Liu C, Battilani P (2016) Modelling climate change impacts on mycotoxin contamination. World Mycotoxin Journal 9: 717–726.

Van der Lee T, Zhang H, van Diepeningen A. Waalwijk C (2015) Biogeography of *Fusarium graminearum* species complex and chemotypes: a review. Food Additives and Contaminants Part A 32: 453–460.

Váry Z, Mullins E, Mcelwain JC, Doohan FM (2015) The severity of wheat diseases increases when plants and pathogens are acclimatized to elevated carbon dioxide. Global Change Biology 21: 2661–2669.

Vidal A, Bendicho J, Sanchis V, Ramos AJ, Marín S (2016) Stability and kinetics of leaching of deoxynivalenol, deoxynivalenol-3-glucoside and ochratoxin A during boiling of wheat spaghettis. Food Research International 85: 182–190

Violleau F, Pernot AG, Surel O (2012) Effect of Oxygreen wheat ozonation process on bread dough quality and protein solubility. Journal of Cereal Science 55: 115–124.

Visconti A, Haidukowski EM, Pascale M, Silvestri M (2004) Reduction of deoxynivalenol during durum wheat processing and spaghetti cooking. Toxicology Letters 153: 181–189.

Vogelgsang S, Musa T, Bänziger I, Kägi A, Bucheli TD, Wettstein FE et al. (2017). Fusarium mycotoxins in Swiss wheat: a survey of growers' samples between 2007 and 2014 shows strong year and minor geographic effects. Toxins 9: 246.

Wang J, Yu Y (2009) Effect of gamma-ray irradiation on the physicochemical properties of flour and starch granule structure for wheat. International Journal of Food Science and Technology 44: 674–680.

Wang L, Luo Y, Luo X, Wang R, Li Y, Li Y et al. (2016b) Effect of deoxynivalenol detoxification by ozone treatment in wheat grains. Food Control 66: 137–144.

Wang L, Shao, H, Luo X, Wang R, Li Y, Li Y, Luo Y, Chen Z (2016a). Effect of ozone treatment on deoxynivalenol and wheat quality. Plos One 11, e0147613.

Wang L, Wang Y, Shao HL, Luo XH, Wang, R, Li YF, Li YN et al. (2017) In vivo toxicity assessment of deoxynivalenol-contaminated wheat after ozone degradation. Food Additives and Contaminants Part A 34: 103–112.

Ward TJ, Clear RM, Rooney A, O'Donnell K, Gaba D, Patrick S et al. (2008) An adaptive evolutionary shift in Fusarium head blight pathogen populations is driving the rapid spread of more toxigenic *Fusarium graminearum* in North America. Fungal Genetic Biology 45: 473–484.

Wegulo SN, Baenziger PS, Hernandez J, Bockus WW, Hallen-Adams H (2015) Management of Fusarium head blight of wheat and barley. Crop Protection 73: 100–107.

West JS, Holdgate S, Townsend JA, Edwards SG, Jennings P, Fitt BDL (2012) Impacts of changing climate and agronomic factors on *Fusarium* ear blight in the UK. Fungal Ecology 5: 53–61.

Whitney K (2018) Fate of deoxynivalenol and deoxynivalenol-3-b-d-glucopyranoside during wheat processing. 4th ICC Latin American Cereals Conference -LACC4- and International Gluten Wheat Seminary -IGW-, pp. 68. 11–14 March, Mexico City, Mexico.

Wu J, Doan H, Cuenca MA (2006) Investigation of gaseous ozone as an anti-fungal fumigant for stored wheat. Journal of Chemical Technology and Biotechnology 81: 1288–1293.

Xue AG, Chen Y, Voldeng HD, Fedak G, Savard ME, Längle T et al. (2014) Concentration and cultivar effects on efficacy of CLO-1 biofungicide in controlling *Fusarium* Head Blight of wheat. Biological Control 73: 2–7.

Yerkovich N, Palazzini JM, Sulyok M, Chulze SN (2017) Trichothecene genotypes, chemotypes and zearalenone production by Fusarium graminearum species complex strains causing Fusarium head blight in Argentina during an epidemic and non-epidemic season. Tropical Plant Pathology 42: 190–196.

Yoshinari T, Suzuki Y, Sugita-Konishi Y, Ohnishi T, Terajima J (2016) Occurrence of beauvericin and enniatins in wheat flour and corn grits on the Japanese market, and their co-contamination with type B trichothecene mycotoxins. Food Additives and Contaminants Part A, 33: 1620–1626.

Zachariasova M, Vaclavikova M, Lacina O, Vaclavik L, Hajslova J (2012) Deoxynivalenol oligoglycosides: new "masked" *Fusarium* toxins occurring in malt, beer, and breadstuff. Journal of Agricultural and Food Chemistry 60: 9280–9291.

Zhang H, Van der Lee T, Waalwijk C, Chen W, Xu J, Xu J et al. (2012). Population analysis of the *Fusarium graminearum* species complex from wheat in China show a shift to more aggressive isolates. PloS one 7: e31722.
Zhao X, Wei Y, Wang Z, Zhang B, Chen F, Zhang P (2011) Mechanochemistry in thermomechanical processing of foods: kinetic aspects. Journal of Food Science 76: R134-R142.
Zhao Y, Guan X, Zong Y, Hua X, Xing F, Wang Y et al. (2018). Deoxynivalenol in wheat from the Northwestern region in China. Food Additives and Contaminants Part B, 1–5.
Zhu F (2018) Effect of ozone treatment on the quality of grain products. Food Chemistry 264: 358–366.

Effects of Environmental Changes on the Allergen Content of Wheat Grain

Angéla Juhász, Réka Haraszi, and Ferenc Békés

Abstract The protein composition of wheat grain is a primary determinant of its end-use quality. Grain proteins are also responsible for food-related disorders in humans, such as celiac disease, wheat dependent exercise-induced anaphylaxis, food allergy and baker's asthma. The effects of environmental conditions on grain protein composition have been extensively studied. Abiotic and biotic stresses can have significant effects on the expression of grain proteins and the overall allergen content of wheat grain. Breeding programs to select wheat varieties with low allergen content represent one step towards eventually improving the quality of life for people affected by gluten intolerance and wheat allergy. This chapter aims to review current understanding of how changes in global environments affect the expression of proteins associated with food-related diseases.

Compared to grain yield, the nature of wheat quality is complex (Basford and Cooper 1998; Fowler and de la Roche 1975; van Lill and Smith 1997). The genetic potential of certain quality traits is manifested through the effects of growing conditions where all of the agronomic treatments, soil and climatic parameters alter the final quality of the grain. The realisation that both qualitative and quantitative aspects of grain composition are important factors determining end-use quality led to research strategies to simultaneously investigate gene (G) and gene products. The level of gene expression is highly dependent on growing conditions, so environmental effects (E) and their interaction with genes (G × E) are essential parts of these

A. Juhász (✉)
School of Science, Edith Cowan University, WA, Joondalup, Australia

Applied Genomics Department, Centre for Agricultural Research, Martonvásár, Hungary
e-mail: a.juhasz@ecu.edu.au

R. Haraszi
Campden BRI, Chipping Campden, UK

F. Békés
FBFD PTY LTD, NSW, North Parramatta, Australia

G. Igrejas et al. (eds.), *Wheat Quality For Improving Processing And Human Health*, https://doi.org/10.1007/978-3-030-34163-3_19

strategies. The effects of G, E and G × E on wheat quality were first comprehensively reviewed by Williams et al. (2008) followed by Georget et al. (2008), Hristov et al. (2010) and Dencic et al. (2011). One of the most important findings is that the effects of E cannot be explained simply by the effect on protein content as the ratios of the different grain protein classes - glutenins, gliadins, and soluble proteins - are altered because of the different independent changes in expression of individual protein-encoding genes.

Quality related wheat research, just as cereal science as a whole, has dramatically changed in recent decades. The introduction of transcriptome and proteome profiling has proved to be valuable for better understanding the biology of the developing grain with the view to improving traits adapted to different environments (Wan et al. 2008). Genomic approaches and tools have become essential in quality related basic research and in applications such as pre-breeding and quality control and assurance in the food industry.

Until recently, the term 'superior quality' mostly covered only wheat quality attributes such as loaf volume and textural parameters directly related to properties of the end-product (Wrigley et al. 2006). Other complex requirements defining good quality are related to the nutritional and health aspects of the end-product, whether real or perceived. Indeed, the general public in most Western countries is now aware of the potential adverse effects of cereals containing 'high calorie', 'toxic' or 'allergic' gluten. Personal accounts and anecdotal evidence appearing in the press or published in books (for example, Ford, 2008; Davis, 2011) promote the health benefits of gluten-free diets but often ignore the importance of an appropriate medical diagnosis for individuals. The key task is nevertheless to identify, quantify and reduce the amounts of the compounds triggering different health related disorders. The detection and particularly the quantification of gluten proteins are critical not only due to their direct effect on end-use quality but also for food safety reasons.

Grain composition varies between cereal genotypes raising an immediate methodological problem in genotype selection for food allergen research, namely how to accurately identify proteins that trigger health problems and determine their genotypic frequency, variability and stability. The high sequence similarity and multi-species origin of prolamins hampers even high-resolution methods (reviewed by Haraszi et al., 2011). For example, while mass spectrometry (MS) can be used, accurate quantitative relationships between prolamin and peptide biomarkers of soluble proteins and the final gluten/prolamin content are needed to relate the detected peptides to their protein sources. These quantitative relationships, however, are difficult to establish due to genotypic and environmental variability.

To assist in peptide biomarker searches, epitope mapping, protein selection and medical studies, a database (ProPepper, https://www.propepper.net) was developed to curate information on members of the prolamin superfamily identified from *Poaceae* species, peptides obtained with single and multi-enzyme *in silico* digestion, and linear epitopes responsible for wheat-related food disorders (Juhász et al. 2015). A similar gluten database tailored for MS studies was also developed to determine the presence of gluten in gluten-free foods (Bromilow et al. 2017). The use of genome sequence and toxic/allergen databases in combination with prediction methodologies, cereal chemistry and industrial processes (e.g. baking) is essential to better understand the level of immunoreactive proteins present in the end-products of wheat flour. Using this approach, the number and distribution of epitopes in a protein fraction have been mapped at genome level and epitopes with the strongest immunoresponse or highest prevalence were identified (Juhász et al. 2018).

Despite over a decade of intensive research, there has been little breakthrough in developing 'celiac-safe' wheat using either conventional or molecular breeding approaches, which may be largely due to the complex multigenic control of gluten protein composition (Shewry and Tatham 2016). For celiac patients a strict gluten-free diet still seems to be the only safe solution. However, cereal products with significantly reduced amounts of harmful components have been developed for the consumption of individuals suffering from wheat related sensitivity (over 15% of the population) although not celiac disease (Dale et al. 2018).

In the process of selecting wheat lines with less immunoreactivity, it is essential to consider the effect of growing conditions that may alter the amounts of the proteins. The aim of this chapter is to provide an overview on how environmental factors influence protein composition of bread wheat, focusing on those proteins that contain known harmful epitopes, whether toxic or allergenic.

1 Wheat Related Food Disorders and the Major Contributing Protein Families

A significant proportion of the human population is either affected by or concerned about food allergy (Lee and Burks 2006; Mills et al. 2007). Food produced from wheat is associated with various immune-mediated responses that can manifest either in the respiratory tract (like occupational or baker's asthma), skin (like atopic dermatitis) or the digestive tract (like celiac disease and food allergy).

Wheat-related food disorders are more prevalent in Western countries affecting 5% of children and ca. 2% of the adult population. The progressive adoption of Western lifestyles has resulted in an increased number of diagnosed patients in Asia and all over the world (Elli et al. 2015). The main components of wheat gluten - glutenins and gliadins - are recognized as major contributors to celiac disease and wheat allergy. In addition, the water-salt soluble fraction of wheat flour containing IgE-binding proteins has also been implicated in wheat allergies (Fig. 1). The genome level distribution of genes encoding these proteins has recently been reviewed by Juhász et al. (2018).

Celiac disease is a chronic inflammatory disorder characterized by wheat gluten induced villous atrophy of the small intestine. Wheat is also one of the causes of baker's asthma which has an overall prevalence of 2% to 3.6% and is among the most common occupational diseases. It affects 10–20% of food allergy sufferers. Baker's asthma is primarily triggered by lipid transfer proteins (LTP) and alpha-

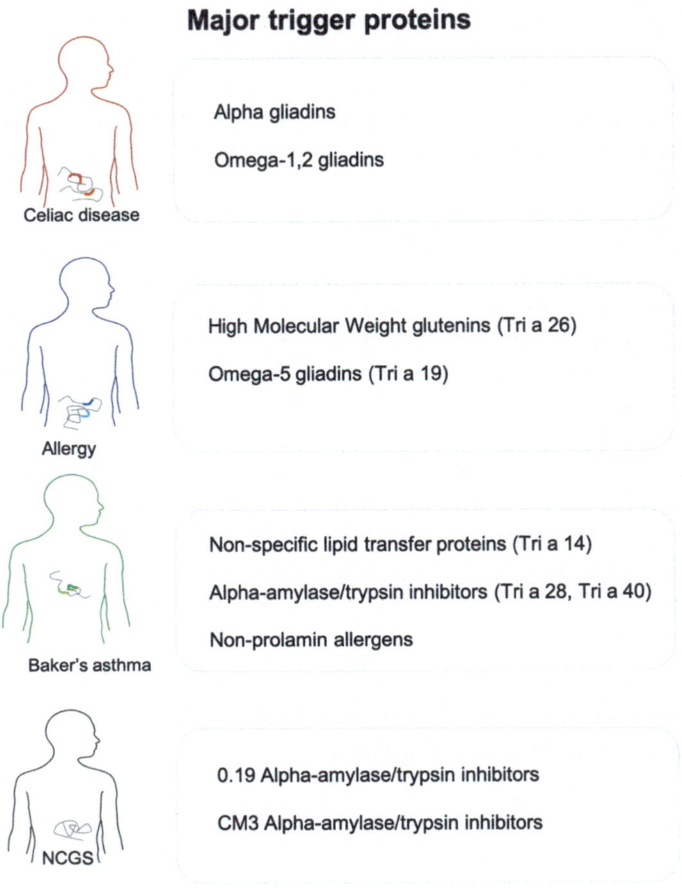

Fig. 1 Wheat related health disorders and their major trigger proteins. Tri a codes represent the allergen nomenclature that is approved by the World Health Organization and International Union of Immunological Societies (WHO/IUIS) Allergen Nomenclature Sub-committee

amylase trypsin inhibitors (ATI), although many non-prolamin proteins encoded in the wheat genome have been linked to baker's asthma. As well as sulphur-rich prolamins, mainly ATIs and LTPs, the spectrum of proteins that trigger allergic reactions includes some non-gluten proteins, such as seed storage globulins, proteins that belong to the pathogenesis-related (PR) families and metabolism related enzymes like peroxidases and chitinases. Wheat-dependent exercise-induced anaphylaxis (WDEIA) is a rare, but potentially severe, food allergy exclusively occurring when wheat ingestion is accompanied by physical activity. There has been a worldwide increase in the number of individuals claiming to be sensitive to dietary gluten without confirmed symptoms of celiac disease or wheat allergy (Hoffmann-Sommergruber 2005). Such patients may complain of functional gastrointestinal symptoms and extra-intestinal symptoms similar to those caused by irritable bowel syndrome. This condition, termed non-celiac wheat sensitivity, is mainly triggered by indigestible oligosaccharides (fructans and galacto-oligosaccharides) and has a prevalence of 0.6–6%. Some ATIs can activate the innate immune system and promote intestinal inflammation via activation of myeloid cells (Zevallos et al. 2017).

2 Climate Change and its Effect on Allergens and Antigens in the Wheat Grain

Global climate change has a significant impact on all biological systems, including cereal crops. Worldwide the unpredictable timing and amount of precipitation and more frequent extreme hot and cold events can have direct effects on grain yield and grain protein content (Lopes et al. 2015). Many studies are now focusing on the impacts of climate change on the crop life cycle and the associated agronomically important traits like grain yield or end-use quality, but not on cereal grain allergens. Only a few studies have been published on other plant food allergens such as peanut, tree nuts or soy (Arbes et al. 2005; Grundy et al. 2002; Hoffmann-Sommergruber 2005; Lee and Burks 2006; Mills et al. 2007; Rancé 2003). It might be expected that allergenic cereals respond to environmental and nutritional changes as peanuts do.

Changes in the global environment strongly affect the yield and composition of the cereal grain. Studies focusing on various abiotic stress effects have described the direct effect of environmental stress factors like heat or drought on the composition of grain proteins and dough quality (Dupont et al. 2006, Hurkman et al. 2009). Food processing and end-use quality cause further changes in protein content and composition whether or not they are involved in adverse immunogenic reactions in the human body.

Comparative genomics and proteomics have become the most powerful tools to obtain information on protein expression in response to abiotic and biotic stresses and can thus also be used to investigate the effect of these environmental changes on grain allergen composition and content. A significant amount of data is already available in published wheat proteomics studies mostly done to monitor changes in end-use quality or to identify stress responsive grain proteins, which can now be analyzed and explored in relation to grain allergen content and composition. Taking advantage of the availability of the high-quality reference genome for bread wheat, Juhász et al. (2018) precisely mapped the known food allergen and antigen protein sequences to generate the first bread wheat allergen reference map. A new resource for breeding and food production, this disease-specific knowledge base can now be used to measure the impact of abiotic stress, nutrient deficiency or pathogen attacks on grain allergen content. Biotic and abiotic stresses have major impacts on protein types that are responsible for wheat allergenicity, for instance, through upregulation of stress response pathways and their corresponding proteins and reorganization of the composition of storage proteins.

3 Effect of Drought on Wheat Allergens and Antigens

Drought stress is one of the most adverse environmental constraints to plant growth and productivity. Comparative proteomics of drought-tolerant and sensitive wheat genotypes is a strategy used to understand the complexity of the molecular mechanisms of the wheat response to drought (Cheng et al. 2016). Water deficit was found to affect technological quality and protein composition differently depending on the timing and duration of the stress. When there was a water deficit throughout the growing season, increases in protein content and in the ratio of high molecular weight (HMW) to low molecular weight (LMW) glutenin subunits (GS) were observed, consistent with a decrease in grain yield. When terminal water stress occurred during grain filling, an improvement in gluten strength was observed, which was consistent with an increase in the amount of glutenin macropolymers associated with a general modulation of HMW-GS and LMW-GS levels (Flagella et al. 2010). Seed storage globulins, alpha gliadins, HMW and LMW glutenins, avenin-like proteins, and both allergy related and celiac disease associated omega gliadins were upregulated in conditions of water deficiency (Begcy and Walia 2015; Jiang et al. 2012) (Table 1).

Wheat allergen expression levels were shown to be genotype dependent and primarily related to the extent of drought stress tolerance (Hajheidari et al. 2007). While alpha amylase inhibitors were mostly down-regulated, the 0.19 ATI group members, which are also implicated in celiac disease, were expressed more abundantly when drought stress was applied during grain development. Expression levels of baker's asthma related proteins like 1-cys-peroxiredoxins, glutathione S-transferases and serpins varied, while thioredoxins and the 0.28 ATI protein levels were downregulated. Stress-related proteins (e.g. glutathione S-transferase) or defence proteins (e.g. thaumatin-like protein) were upregulated in the early stages of drought stress, then gradually downregulated after 24 h of stress (Cheng et al. 2016). Glyceraldehyde-3-phosphate dehydrogenase, a baker's asthma related protein that fulfils a key function in the glycolysis pathway, was also initially upregulated in a drought-sensitive cultivar then downregulated at the same time as energy-producing metabolism was significantly altered. The opposite effect was seen in a drought-tolerant cultivar (Ge et al. 2012). Allergens with defence related function, like ATIs, were more induced in drought-tolerant cultivars, possibly accounting for the higher starch accumulation and contributing to drought tolerance itself (Cheng et al. 2016; Jiang et al. 2012). The expression levels of these defence related proteins were lower in drought-sensitive cultivars; for instance, serpins are not abundant in the drought-sensitive cultivar Janz (Jiang et al. 2012).

Table 1 Summary of environmental effects on expression of known allergens in wheat grain

Protein type	Related disorder	Drought stress	Heat stress	Nitrogen -	Nitrogen +	Sulphur -	Sulphur +	Pathogen effect Fusarium headblight Powdery mildew
0.19 ATIs	BA, CD	Up: [1]	Down: [2]		Down: [3]	Down: [4]		Up: [5]
		Down: [6]						
ATI 0.28	BA	Down: [1]; [7]	Down: [2]		Down: [3]	Down: [4]		Up: [5]
ATI CM17			Down: [2]; [8]			Down: [4]		
ATI CM1			Down: [2]; [8]					
ATI CM3	BA, CD	Down: [1]	Down: [2]; [8]		Down: [3]	Down: [4]		Up: [5]
		Up and down: [7]; [9]						
nsLTP			Up: [2]			Down: [10]		
1-cys-peroxi redoxin	BA	Up and down: [1]	Up: [8]; [11]					Down: [12]
		Down: [6]						
Glutathione S transferase	BA, AD, PA	Up and down: [1]	Up and down: [2]		Down: [3]			Up: [12]
Serpin	BA, CD, AD	Up and down: [1]	Up: [2]; [8]; [11]; [13]		Up: [3]; [33]	Up: [4]		Up: [16]
		Down: [6]; [9]						
Thioredoxin	BA	Down: [1]	Down: [8]					
			Up: [2]					
Beta-amylase	BA	Up and down: [1]	Up: [2]; [20]		Down: [3]			Up: [12]
		Up: [9]						
		Down: [6];[7]						

(continued)

Table 1 (continued)

Protein type	Related disorder	Drought stress	Heat stress	Nitrogen -	Nitrogen +	Sulphur -	Sulphur +	Pathogen effect Fusarium headblight Powdery mildew
Glyceraldehyde 3-phosphate dehydrogenase	BA	Up and down: [1]	Up: [2]; [11]; [17]		Down: [3]	Down: [10]		
		Up: [7]; [18]						
		Down: [6]						
Class II chitinase		Up: [7]			Down: [3]			Down: [12]
Peroxidase		Up: [6]; [9]	Up and down: [17]		Down: [3]	Down: [10]		Down: [12]
1S globulin		Up: [6]; [9]	Up: [2]; [11]; [31]		Down: [3]			Up: [12]; [15]; [17]
Globulin - 3	CD	Down: [7]	Down: [8]; [2]; [31]		Down: [3]	Up: [4]		
			Up: [11]					
19 kDa globulin	CD, OA, AD	Up and down: [1]	Up: [11]		Down: [3]			Up: [12]; [15]; [17]
Alpha gliadin	CD	Up: [6];[19]	Up: [2]; [8]; [21]; [32]	Up: [13]; [22]; [23]; [24]	Up: [13]; [33]	Up: [4]	Up: [25]	Balanced: [26]
			Down: [2]; [8]; [21]					Down: [12]
								Up: [15]; [28]
Gamma gliadin	CD	Up: [6];[19]	Down: [2]; [32]	Down: [13]; [24]	Up and down: [13]	Down: [4]; [27]	Up: [25]	Balanced: [26]
			Up: [8]; [2]	Up: [22]; [24]				Up: [15]; [28]
Omega 1,2	CD	Up: [19]	Down: [8]	Up: [3]; [13]; [22]; [23]	Up: [13]; [33]	Up: [4]; [27]	Down: [25]	Balanced: [26]
			Up: [2]	Down: [24]				Up: [15]; [28]

Omega 5 gliadin	WDEIA	Up: [19]	Down: [8]	Up: [3]; [13]; [22]; [23]; [29]	Up: [13]; [33]	Up: [4]; [27]	Down: [25]	Balanced: [26]
			Up: [2]	Down: [24]				Up: [15]; [28]
HMW glutenin	CD, WA, WDEIA	Up: [6]; [19]	Up: [2]	Down: [10]; [23]; [24]	Up: [13]; [33]	Up: [4]; [27]	Down: [25]	Up: [16]
				Balanced/no change: [23]				Up: [15]; [28]
				Up: [22]; [13]				
LMW glutenin	CD	Up: [19]	Down: [2]	Down: [3]; [13]; [23]	Up and down: [13]; [33]	Down: [4]; [27]	Up: [25]	Up: [15]; [28]
				Up: [10]; [22]				
Avenin-like protein	CD	Up: [6]; [9]	Down: [2]	Up: [10]		Down: [4]		Up: [15]; [17]; [30]

Footnotes: BA - baker's asthma; CD - celiac disease; AD – athopic dermatitis; PA- pollen allergy; OA – oral allergy syndrome; Up, upregulation of protein levels; Down, downregulation of protein levels; –, absence or deficit; +, presence or excess. References: [1] Hajheidari et al. (2007); [2] Zhang et al. (2017); [3] Altenbach et al. (2011); [4] Dai et al. (2015); [5] Perlikowski et al. (2014); [6] Gu et al. (2015); [7] Ge et al. (2012); [8] Yang et al. (2011); [9] Jiang et al. (2012); [10] Yu et al. (2018); [11] Laino et al. (2010); [12] Eggert et al. (2011); [13] Hurkman et al. (2013); [16] Gao et al. (2018); [17] Wang et al. (2012); [18] Cheng et al. (2016); [19] Begcy and Walia (2015); [20] Majoul et al. (2004); [22] Triboï et al. (2000); [24] Cho et al. (2018); [25] Zörb et al. (2009); [26] Dexter et al. (1996); [27] Wieser et al. (2004); [28] Li et al. (2018); [29] Wan et al. (2014); [30] Zhang et al. (2018); [31] Wang et al. (2018); [32] Dupont et al. (2006); [33] Zheng et al. (2018)

4 Effect of High Temperature Stress on Wheat Allergens and Antigens

High temperature stress can have a significant effect on yield and quality of wheat affecting both the major seed storage and metabolic proteins (Laino et al. 2010). The individual gluten protein genes have different basal levels of expression and do not respond identically to stress effects. In general, high temperature regimens increase celiac disease associated protein and gluten protein content, mainly due to the marked decrease in the amount of starch (Hurkman et al. 2013). A shorter grain filling period due to heat stress may result in earlier accumulation of HMW and LMW glutenins. The most notable changes in gluten protein types were in omega-1,2 and omega-5-gliadins, alpha gliadins, and HMW glutenin subunits (Hurkman et al. 2013). These results were confirmed by Juhász et al. (2018) who compared the reference genome cultivar Chinese Spring with two modern Norwegian wheat cultivars. The observed changes demonstrate that gluten protein accumulation is a complex process that is subject to spatial and temporal regulation as well as environmental signalling (Hurkman et al. 2013, Juhász et al. 2018).

It is also notable that the ω-5 gliadins associated with WDEIA increased significantly in response to temperature. Among the gluten proteins, the α-gliadins are considered to be the most immunogenic in celiac disease. Several α-gliadins do not contain the major celiac epitopes (31–32% of the total α-gliadins) so temperature did not influence the overall immunogenic potential of wheat with respect to celiac disease (Hurkman et al. 2013). Analysis of the bread wheat reference genome also indicated that alpha gliadins with the most immunogenic peptides are not strongly expressed and respond only moderately to high temperature stress (Juhász et al. 2018). Because remobilization of nitrogen reserves to the developing grain occurs over a significantly shorter period of time, the temperature treatment may essentially mimic high nitrogen conditions, resulting in changes in the same complement of gluten proteins (Hurkman et al. 2013). ATIs were mostly downregulated (Table 1) except for the ATI CM3 sub-class that were upregulated along with nsLTPs (Yang et al. 2011; Zhang et al. 2017).

Non-prolamin allergens, such as serpins, 1-cys-peroxiredoxins, glyceraldehyde 3-phosphate dehydrogenases, and glutathione-S-transferases found in the soluble fraction of wheat seeds, are upregulated in response to high temperature stress (Laino et al. 2010), while thioredoxins have been reported to be both up and downregulated (Yang et al. 2011; Zhang et al. 2017) due to individual genotype effects.

5 Effect of Nutrient Deficiency or Excess on Wheat Allergens and Antigens

Protein expression is well known to be regulated by nutrient deficiency, so it would be expected to also affect allergen content. In several studies the direction of change in protein composition was investigated as a result of either nutrient deficiency or excess. Up and down regulation of the different protein groups relevant to wheat allergy by N and S fertilisation are summarised in Table 1.

Change in the availability of one nutrient can interfere with the utilisation of another nutrient during grain maturation. For example, high N supply increases the N to S ratio to such an extent that the grain could be considered to be S-deficient. The ratio of N to S from fertiliser has an impact on the relative proportions of the various S-rich and S-poor proteins that accumulate due to altered concentrations of free amino acids (Dai et al. 2015). Consequently, wheat proteins containing Cys and Met are highly susceptible to changes in available nutrients (Altenbach et al. 2011). Moderate N fertilisation is sufficient to trigger the synthesis of S containing amino acids, while high levels of available N may induce S deficiency. Post anthesis fertilization (PAF) using S, analogous to late N fertilisation, is recommended to improve N and S nutritional management of wheat (Zörb et al. 2009).

5.1 Nitrogen Deficiency and Increased Nitrogen Availability

The available nutrients in the plant are distributed among the plant organs according to the plant's needs. The process of N distribution is not significantly affected by post-anthesis N nutrition and at maturity as variations in protein fraction composition are mainly due to differences in the total N content (Triboï et al. 2003).

The total protein content of flour, and particularly the total storage protein content, increased significantly with PAF (Altenbach et al. 2011; Cho et al. 2018; Yu et al. 2018). HMW/LMW and gliadin to glutenin ratio also increased with PAF (Altenbach et al. 2011; Triboï et al. 2000). The overall increase in protein quantity is a consequence of alteration of the expression levels of certain genes involved in protein biosynthesis and storage protein expression (Yu et al. 2018). The genetic determination of the plant protein content has an impact on how N fertilisation or deficiency may regulate the various protein groups. Responses to different N treatments were shown to depend on the wheat variety, especially for omega-gliadins (Triboï et al. 2000; Wan et al. 2014). Similarily, if the wheat has genetically low gliadin content, N fertilisation has a different effect on the plant N distribution compared to wheat with an average gliadin content. Low gliadin transgenic lines showed an increased glutenin content with increasing N (García-Molina and Barro, 2017).

Application of higher concentrations of N fertilizer has a significant effect both on prolamin and non-prolamin allergens. While glutenins and gliadins show a significant variation in the direction of change, ATIs are mostly downregulated. Allergens with metabolic functions such as beta-amylases, glutathione-S transferases or cupin-1 domain containing seed storage globulins are mostly downregulated.

5.2 *Sulphur Deficiency and Application of Excess Sulphur Fertilizer*

Just as N fertilisation is influenced by other nutrients, the use of S fertilisation also has to take into account the level of N in the soil. The effect of S on the protein content of wheat grains can lead to either up or down regulation (Jarvan et al. 2008; Yu et al. 2018). Moderate S deficiency mostly affects protein composition rather than yield, while severe S deficiency affects total protein content (Yu et al. 2018). Besides the nutrient levels and genetic determination of the varieties, the timing of S application may lead to different results. During grain filling, S fertilisation increased the level of S-poor proteins more substiantially than S-rich proteins. The addition of post-anthesis S had no effect on protein composition (DuPont et al. 2006).

Even when the amount of total proteins and total gluten proteins are not influenced by S deficiency, the quantities and proportions of single protein types may be. S fertilisation can strongly affect protein composition depending on how much Cys and Met, the S-containing amino acids, each protein type contains (Wieser et al. 2004). Usually, S deficiency leads to enrichment of S-poor proteins and lower molecular weight proteins at the expense of high molecular weight ones (Zhao et al. 1998). S deficiency results in increased gliadin to glutenin ratio and omega-gliadin content but a decrease in gamma-gliadins. The HMW to LMW ratio is overall not affected but the proportions of individual HMW and LMW glutenins change (Wieser et al. 2004). On the contrary, use of high amounts of S fertilizer results in the downregulation of S-poor prolamin proteins and increased amounts of LMW glutenins and gamma gliadins capable of storing more S.

Being aware of how wheat allergen protein groups are up and down regulated as a consequence of NPK (nitrogen (N), phosphorus (P), potassium (K)) fertilisation provides a guide for breeders when tailoring varieties to have reduced allergenicity. These relationships also map how end-use quality may be affected by such changes caused by the environment.

6 Effect of Plant Pathogens and Fungal Diseases on Wheat Allergens and Celiac Disease Associated Proteins

Plants exposed to biotic stresses (e.g. bacterial and fungal pathogens) produce high amounts of protective molecules such as pathogenesis related proteins and disease associated proteins (Kamal et al. 2010). Such stresses limit productivity and alter wheat quality. Like abiotic stresses, biotic stresses have different effects on the different protein groups and have major effects during grain filling (Chetouhi et al. 2015). Biotic stress regulation of the different protein groups relevant to wheat allergy is summarised in Table 1.

6.1 *Fusarium Head Blight*

Fusarium head blight caused by *Fusarium graminearum* is a common crop disease that has an economic impact on the cereal industry and a potential health impact. Yield and quality of infected crops are reduced and the crops are not suitable for human consumption due to the mycotoxins produced (Eggert et al. 2011).

F. graminearum infection does not alter the grain proteome but has an impact on starch synthesis, signalling, transport and storage proteins during grain filling (Chetouhi et al. 2015). The plant response to *Fusarium* stress is reflected in the significantly increased levels of enzymes and proteins involved in stress responses (serine protease inhibitor), fungal growth reduction (thaumatin-like protein) and starch hydrolysis (beta-amylase) pathways, and in the decreased levels of certain stress-related compounds (peroxidase, peroxiredoxin) and enzymes involved in starch synthesis (glycosyltransferase) and fungal cell wall degradation (chitinase) (Eggert et al. 2011). These changes mostly result in the upregulation of baker's asthma related allergens.

Changes in storage protein composition of wheat as a result of *Fusarium* infection was described as causing no qualitative or quantitative differences in gliadin composition (Dexter et al. 1996). In emmer wheat grains, it caused a decrease in alpha-gliadins and an increase in globulins (Eggert et al. 2011). *Fusarium* resistant wheat lines accumulate alpha-amylase inhibitors (Perlikowski et al. 2014). This might have a direct effect on the accumulation of celiac disease related proteins but might also increase the amounts of ATIs related to non-celiac gluten sensitivity. Overall, *Fusarium* stress changed proportions of the lower molecular weight functional proteins rather than the higher molecular weight storage proteins. Similarly, significant increases in the pathogenesis related prolamin proteins were detected. Avenin-like proteins, related to pathogen defense mechanisms were upregulated, as were ATIs that inhibit plant proteases (Gao et al. 2012; Zhang et al. 2018).

6.2 *Powdery Mildew*

Wheat powdery mildew is caused by *Blumeria graminis* species and affects photosynthesis of stems and leaves reducing the amount of biomass produced (Gao et al. 2018). Powdery mildew increases the level of molecular chaperones, protein synthesis enzymes and some stress response proteins. Increased levels of triticin, serpin, 1-cys peroxiredoxin and storage protein content were reported (Feng et al. 2014; Gao et al. 2014, 2018; Li et al. 2018). As total protein content increased, globulins, avenin-like proteins and enzymes were upregulated, and in particular glutenins, especially HMW-GS, and gliadins (Gao et al. 2014; Li et al. 2018; Wang et al. 2012).

Compared to *Fusarium*, powdery mildew infection has a greater effect on protein composition, affecting more functional and storage proteins of various molecular sizes.

7 Summary and Future Perspectives

After thousands of years of evolution and cultivation, there is a significant pool of variability in wheat available as a resource for breeders. Until recently, breeders had only phenotypic data to rely on to uncover that diversity. The developments in modern omics technologies now enable diversity to be approached from the fundamental genetics point of view. Less than a decade ago when the sequencing of the human genome and the rice genome was almost completed, the possibility of sequencing a genome of the size and complexity of hexaploid wheat was difficult to envisage. However, the advances in generating BAC libraries from flow-sorted chromosomes combined with new high throughput DNA sequencing techniques provided the platforms for the rapid advances in wheat genome analysis that have been achieved recently (IWGSC 2018).

Complementing the structural features of genomes, the advances in the transcriptomics (Liang and Kelemen 2006; Ramírez-González et al. 2018) and proteomics (Chen et al. 2007; Dumur et al. 2004) of wheat provide the basis for functional analyses when combined with the unique genetic resources established over the past 40 years (Lukaszewski et al. 2004). The ability to link phenotypes to specific regions of the genome is a major landmark in defining new approaches to breeding and meeting the challenges of matching the features of the wheat genome to the requirements of both a changing environment and changing demands of consumers (Appels et al. 2011).

The problems related to some health related quality attributes such as immunoreactivity of certain proteins is a good example of important topics where the combined tools of genomics, transcriptomics and proteomics are starting to make progress (Juhász et al. 2018) and hopefully, in the near future, breakthroughs.

Both abiotic and biotic stresses may result in a significant change in the grain proteome with a direct effect on the expression of grain allergens and antigens. Many of these stress effects result in a shorter grain filling period, loss in seed productivity and grain weight, primarily due to a negative impact on starch accumulation. The relative change in the starch to protein ratio results in the overall upregulation of the amount of grain protein per seed. While nutritional changes, such as N or S deficiency have a significant impact on nitrogen metabolism pathways and show an orchestrated accumulation of S-rich and S-poor gliadins and glutenins, stresses related to water availability or temperature boost the expression of proteins primarily related to the accumulation of reactive oxygen species and general stress responses. These changes primarily result in the upregulation of baker's asthma and food allergy related proteins on a genotype dependent manner. These stress related proteins are upregulated in more stress-tolerant cultivars, while cultivars that are less tolerant of stress express lower amounts of wheat allergens. These relationships might be pertinent when lower allergen contents are targeted. Selection of cultivars that are less resistant and poor at adapting to abiotic and biotic stress conditions could be a way to lower the allergen contents, although the potential cost incurred through loss in grain yield and modified seed storage protein content should be considered at the same time.

Changes in the gliadins and glutenins, the primary contributors to celiac disease, WDEIA and food allergy follow some overall tendencies mainly as a result of the accelerated seed maturation and the disturbed N and S accumulation. Changes in individual protein levels are not yet fully understood and need further investigation. The reference genome and systems biology tools now provide a chance to analyse these changes at a single protein or epitope level.

All of these diverse demands make the development of better wheat varieties and manufactured products extremely complex, where the availability of high quality reference genomes along with tools of computational biology will play an important role. There is nevertheless a need to develop decision-making tools to help the breeder, the product developers and process engineers to select the right targets.

References

Altenbach SB, Tanaka CK, Hurkman WJ, Whitehan LC, Vensel WH, Dupont FM (2011) Differential effects of a post-anthesis fertilizer regimen on the wheat flour proteome determined by quantitative 2-DE. Proteome Science 2011, 9: 46.

Appels R, Barsby T, Risacher T, Békés F (2011) Linking the genome to phenotypes in wheat: advances in technologies and concepts. In: The World Wheat Book – a history of wheat breeding. Vol 2. Eds.: Bonjean A, Angus W, and van Ginkel M, pp. 709–748, Lavoisier, London.

Arbes SJ Jr, Gergen PJ, Elliott L, Zeldin DC (2005) Prevalences of positive skin test responses to 10 common allergens in the US population: results from the Third National Health and Nutrition Examination Survey. The Journal of Allergy and Clinical Immunology 116: 377–383.

Basford KE and Cooper M (1998) Genotypexenvironment interactions and some considerations of their implications for wheat breeding in Australia. Australian Journal of Agricultural Research 49: 153–174.

Begcy K, Walia H (2015) Drought stress delays endosperm development and misregulates genes associated with cytoskeleton organization and grain quality proteins in developing wheat seeds. Plant Science 240: 109–119.

Bromilow S, Gethings LA, Buckley M, Bromley M, Shewry PR, Langridge JI, Mills CL (2017) A curated gluten protein sequence database to support development of proteomics methods for determination of gluten in gluten-free foods. Journal of Proteomics 163: 67–75.

Chen J, Lan P, Tarr A, Yan YM, Francki M, Appels R, Ma W (2007) MALDI-TOF based wheat gliadin protein peaks are useful molecular markers for wheat genetic study. Rapid Communications in Mass Spectroscopy 21: 2913–2917.

Cheng L, Wang Y, He Q, Li H, Zhang X, Zhang F (2016) Comparative proteomics illustrates the complexity of drought resistance mechanisms in two wheat (*Triticum aestivum* L.) cultivars under dehydration and rehydration BMC Plant Biology (2016) 16: 188.

Chetouhi C, Bonhomme L, Lecomte P, Cambon F, Merlino M, Biron DG, Langin T (2015) A proteomics survey on wheat susceptibility to Fusarium head blight during grain development. European Journal of Plant Pathology 141: 407–418.

Cho SW, Kang CS, Kang TG, Cho KM, Park CS (2018) Influence of different nitrogen application on flour properties, gluten properties by HPLC and end-use quality of Korean wheat. Journal of Integrative Agiculture 17: 982–993.

Dai Z, Plessis A, Vincent J, Duchateau N, Besson A, Dardevet M, et al. (2015) Transcriptional and metabolic alternations rebalance wheat grain storage protein accumulation under variable nitrogen and sulfur supply. Plant Journal 83: 326–43.

Dale H, Biesiekierski J, Lied G (2018) Non-coeliac gluten sensitivity and the spectrum of gluten-related disorders: An updated overview. Nutrition Research Reviews 1–10.

Davis W (2011) Wheat belly: Loose the wheat and find your path back to health. Rodale Books Inc, New York.

Dencic D, Mladenov N, Kobiljski B (2011) Effects of genotype and environment on breadmaking quality in wheat. International Journal of Plant Production 5: 71–82.

Dexter JE, Clear RM, Preston KR (1996) Fusarium Head Blight: Effect on the Milling and Bakingof Some Canadian Wheats. Cereal Chemistry 73: 695–701.

Dumur J, Jahier J, Bancel E, Laurière M, Bernard M, Branlard G. (2004) Proteomic analysis of aneuploid lines in the homeologous group 1 of the hexaploid wheat cultivar Courtot. Proteomics. 4: 2685–2695.

Dupont FM, Hurkman WJ, Vensel WH, Chan R, Lopez R, Tanaka CK, Altenbach SB (2006) Differential accumulation of sulfur-rich and sulfur-poor wheat flour proteins is affected by temperature and mineral nutrition during grain development. Journal of Cereal Science 44: 101–112.

Eggert K, Zörb C, Muhling KH, Pawelzik E (2011) Proteome analysis of Fusarium infection in emmer grains (*Triticum dicoccum*). Plant Pathology 60: 918–928.

Elli L, Branchi F, Tomba C, Villalta D, Norsa L, Ferretti F, Roncoroni L, Bardella MT (2015) Diagnosis of gluten related disorders: Celiac disease, wheat allergy and non-celiac gluten sensitivity. World Journal of Gastroenterology 21: 7110–7119.

Feng W, Li X, Liu WD (2014) Effects of powdery mildew infection on grain quality traits and yield of winter wheat. Journal of Triticeae Crops 34: 1706–1712.

Flagella Z, Giuliani MM, Giuzio L, Volpi C, Masci S (2010) Influence of water deficit on durum wheat storage protein composition and technological quality. European Journal of Agronomy 33: 197–207.

Fowler DB, de la Roche IA (1975) Wheat quality evaluation. 3. Influence of genotype and environment. Canadian Jounal of Plant Science 55: 263–269.

Ford R (2008) The Gluten Syndrome. Is wheat causing you harm? RRS Global LT, Christchurch, New Zealand.

Gao HY (2012) Influence of powdery mildew on quality and its molecular mechanism in wheat (*Triticum aestivum* L.) [D]. Doctoral dissertation of Henan agricultural university, Zhengzhou city, Henan province, China.

Gao HY, He DX, Niu JS (2014) The effect and molecular mechanism of powdery mildew on wheat grain prolamins. Journal of Agricultural Science 152: 239–253.

Gao HY, Niu J, Li S (2018) Impacts of Wheat Powdery Mildew on Grain Yield and Quality and Its Prevention and Control Methods. American Journal of Agriculture and Forestry 6: 141–147.

García-Molina MD, Barro F (2017) Characterization of Changes in Gluten Proteins in Low-Gliadin Transgenic Wheat Lines in Response to Application of Different Nitrogen Regimes. Frontiers in Plant Science 8: 257.

Ge P, Ma C, Wang S, Gao L, Li X, Guo G, Ma W, Yan Y (2012) Comparative proteomic analysis of grain development in two spring wheat varieties under drought stress. Analitical and Bioanalytical Chemistry 402: 1297–1313.

Georget DM, Underwood-Toscano C, Powers SJ, Shewry PR, Belton PS (2008) Effect of variety and environmental factors on gluten proteins: an analytical, spectroscopic, and rheological study. Journal of Agriculture and Food Chemistry 56: 1172–1179.

Grundy J, Matthews S, Bateman B, Dean T, Arshad SH (2002) Rising prevalence of allergy to peanut in children: data from 2 sequential cohorts. Journal of Allergy and Clinical Immunology 110: 784–789.

Gu A, Hao P, Lv D, Zhen S, Bian Y, Ma C, Xu Y, Zhang W, Yan Y (2015) Integrated Proteome Analysis of the Wheat Embryo and Endosperm Reveals Central Metabolic Changes Involved in the Water Deficit Response during Grain Development. Journal of Agriculture and Food Chemistry 63: 8478–8487.

Hajheidari M, Eivazi A, Buchanan BB, Wong JH, Majidi I, Salekdeh GH (2007) Proteomics uncovers a role for redox in drought tolerance in wheat. Journal of Proteome Research 6: 1451–1460.

Haraszi R, Chassaigne H, Macquet A, Ulberth F (2011) Analytical methods for detection of gluten in food - last method developments in support to the legislations on labelling of foodstuffs. J AOAC 94: 1–20.

Hoffmann-Sommergruber K, SAFE consortium (2005) The SAFE project: 'plant food allergies: field to table strategies for reducing their incidence in Europe' an EC-funded study. Allergy 60: 436–442.

Hristov N, Mladenov N, Djuric V, Kondic-Spika A, Marjanovic-Jeromela A, Simic D (2010) Genotype by environment interactions in wheat quality breeding programs in south-east Europe. Euphytica 174: 315–324.

Hurkman WJ, Vensel WH, Tanaka CK, Whitehand L, Altenbach SB (2009) Effect of high temperature on albumin and globulin accumulation in the endosperm proteome of the developing wheat grain Journal of Cereal Science 4: 12–23.

Hurkman WJ, Tanaka CK, Vensel WH, Thilmony R, Altenbach SB (2013) Comparative proteomic analysis of the effect of temperature and fertilizer on gliadin and glutenin accumulation in the developing endosperm and flour from *Triticum aestivum* L. cv. Butte 86. Proteome Science 11: 8.

International Wheat Genome Sequencing Consortium (IWGSC) (2018) Shifting the limits in wheat research and breeding using a fully annotated reference genome. Science 361: 661.

Jarvan M, Adamson EA, Lukme L, Akk A (2008) The effect of sulphur fertilization on yield, quality of protein and baking properties of winter wheat. Agronomy Research 6: 459–469.

Jiang SS, Liang XN, Li X, Wang SL, Lv DW, Ma CY, Li XH, Ma WJ, Yan YM (2012) Wheat drought-responsive grain proteome analysis by linear and nonlinear 2-DE and MALDI-TOF mass spectrometry. International Journal of Molecular Sciences 13: 16065–16083.

Juhász A, Belova T, Florides CG, Maulis Cs, Fischer I, Gell Gy, Birinyi Zs, Ong J, Keeble-Gagnere G, Maharajan A, Ma W, Gibson P, Jia J, Lang D, Mayer KFX, Spannagl M, International Wheat Genome Sequencing Consortium, Tye-Din JA, Appels R, Olsen OA (2018) Genome mapping of seed-borne allergens and immunoresponsive proteins in wheat. Science Advances 4, eaar8602.

Juhász A, Haraszi R, Maulis Cs (2015) ProPepper: a curated database for identification and analysis of peptide and immune-responsive epitope composition of cereal grain protein families. Database 2015 1–16.

Kamal AHM, Kim KH, Shin KH, Kim DE, Oh MW, Choi JS, Hirano H, Heo HY, Woo SH (2010) Proteomics-based dissection of biotic stress responsive proteins in bread wheat (*Triticum aestivum* L.), African Journal of Biotechnology 9: 7239–7255.

Laino P, Shelton D, Finnie C, De Leonardis AM, Mastrangelo AM, Svensson B, et al. (2010) Comparative proteome analysis of metabolic proteins from seeds of durum wheat (cv. Svevo) subjected to heat stress. Proteomics 10: 2359–2368.

Lee LA, Burks AW (2006) Food allergies: prevalence, molecular characterization, and treatment/prevention strategies. Annual Review of Nutrition 26: 539–565.

Li J, Liu XH, Yang XW (2018) Proteomic analysis of the impacts of powdery mildew on wheat grain. Food Chemistry 261: 30–35.

Liang Y, Kelemen A (2006) Associating phenotypes with molecular events: recent statistical advances and challenges underpinning microarray experiments. Functional and Integratie Genomics 6: 1–13.

Lopes MS, El-Basyoni I, Baenziger PS, Singh S, Royo C, Ozbek K, et al. (2015) Exploiting genetic diversity from landraces in wheat breeding for adaptation to climate change. Journal of Experimental Botany 66: 3477–3486.

Lukaszewski AJ, Rybka K, Korzun K, Malyshev SV, Lapinski B, Whitkus R (2004) Genetic and physical mapping of homoeologous recombination points involving wheat chromosome 2B and rye chromosome 2R. Genome 47: 36–45.

Majoul L, Bancel E, Triboï E, Hamida JB, Branlard G (2004) Proteomic analysis of the effect of heat stress on hexaploid wheat grain: Characterization of heat-responsive proteins from non-prolamins fraction. Proteomics 4: 505–513.

Mills ENC, Mackie AR, Burney P, Beyer K, Frewer L, Madsen C, Botjes E, Crevel RWR, van Ree R (2007) The prevalence, cost and basis of food allergy across Europe. Allergy 62: 717–722.

Perlikowski D, Wisniewska H, Goral T, Kwiatek M, Majka M (2014) Identification of Kernel Proteins Associated with the Resistance to Fusarium Head Blight in Winter Wheat (*Triticum aestivum* L.). PLoS ONE 9: e110822.

Ramírez-González RH, Borrill P, Lang D, Harrington SA, Brinton J, Venturini L, Davey M, Jacobs J, F. van Ex J (2018) The transcriptional landscape of polyploid wheat. Science 361, eaar6089.

Rancé F (2003) Mustard allergy as a new food allergy. Allergy 58: 287–288.

Shewry PR, Tatham AS (2016) Improving wheat to remove celiac epitopes but retain functionality. Journal of Cereal Science 67: 12–21.

Triboï E, Abad A, Michelena A, Lloveras J, Ollier JL, Daniel C (2000) Environmental effects on the quality of two wheat genotypes: 1. quantitative and qualitative variation of storage proteins. European Journal of Agronomy 13: 47–64.

Triboï, E, Martre P, Triboï-Blondel AM (2003) Environmentally-induced changes in protein composition in developing grains of wheat are related to changes in total protein content. Journal of Experimental Botany 54: 1731–1742.

van Lill D, Smith MF (1997) Aquality assurance strategy for wheat (*Triticum aestivum* L.) where growth environment predominates. South African Journal of Plant and Soil 14: 183–191.

Wan Y, Poole RL, Huttly AK, Toscano R, Feeney K, Welham S, Gooding MK, Mills C, Edwards KJ, Shewry PR, Mitchell RAC 2008. Transcriptome analysis of grain development in hexaploid wheat. BMC Genomics 9: 121.

Wan, Y, Gritsch, CS, Hawkesford MJ, Shewry PR (2014) Effects of nitrogen nutrition on the synthesis and deposition of the omega-gliadins of wheat. Annals of Botany 113: 607–615.

Wang B, Xie, C, Xin M, Song N, Ni Z, Sun Q (2012) Comparative Proteomic Analysis of Wheat Response to Powdery Mildew Infection in Wheat Pm30 Near-Isogenic Lines, Journal of Phytopathology 160: 229–236.

Wang X, Hou L, Lu Y, Wu B, Gong X, Liu M, et al. (2018) Metabolic adaptation of wheat grains contributes to a stable filling rate under heat stress. Journal of Experimental Botany 69: 5531–5545.

Wieser H, Gutser R, von Tucher S (2004) Influence of sulphur fertilisation on quantities and proportions of gluten protein types in wheat flour. Journal of Cereal Science 40: 239–244.

Williams RMA, O'Brien L, Eagles HA, Solah VA, Jayasena A (2008) The influences of genotype, environment, and genotype x environment interaction on wheat quality. Australian Journal of Agricultural Research 59: 95–111.

Wrigley CW, Békés F, Bushuk W (2006) Chapter 1. Gluten: A balance of gliadin and glutenin. In: Gliadin and glutenin. The unique balance of wheat quality. Eds.: Wrigley, C.W., Békés, F., and Bushuk, W., pp. 3–33. AACCI Press, St Paul, Min., USA.

Yang F, Jorgensen AD, Li H, Sondergaard I, Finnie C, Svensson B, Jiang D, Wollenweber B, Jacobsen S (2011) Implications of high-temperature events and water deficits on protein profiles in wheat (*Triticum aestivum* L. cv. Vinjett) grain. Proteomics 11: 1684: 1695.

Yu Z, Juhász A, Islam S, Diepeveen D, Zhang J, Wang P, Ma W (2018) Impact of mid-season sulphur defciency on wheat nitrogen metabolism and biosynthesis of grain protein. Scientific Reports 8: 2499.

Zevallos VF, Raker V, Tenzer S, Jimenz-Calvente C, Ashfaq-Khan M, Russel N, et al. (2017) Nutritional Wheat Amylase-Trypsin Inhibitors Promote Intestinal Inflammation via Activation of Myeloid Cells. Gastroenterology 152: 1100–1113.

Zhang Y, Pan J, Huang X, Guo D, Lou H, Hou Z, et al. (2017). Differential effects of a post-anthesis heat stress on wheat (*Triticum aestivum* L.) grain proteome determined by iTRAQ. Scientific Reports 7: 3468.

Zhang Y, Cao X, Juhász A, Islam S, Qi P, She M et al. (2018) Wheat avenin-like protein and its significant Fusarium Head Blight resistant functions. bioRxiv 406694.

Zhao FJ, Hawkesford MJ, McGrath SP (1998) Sulphur Assimilation and Effects on Yield and Quality of Wheat. Journal of Cereal Science 30: 1–17.

Zheng T, Qi PF, Cao YL, Han YN, Ma HL, Guo ZR, et al. (2018) Mechanisms of wheat (*Triticum aestivum*) grain storage proteins in response to nitrogen application and its impacts on processing quality. Scientific Reports 8: 11928l.

Zörb C, Grover C, Steinfurth D, Mühling KH (2009) Quantitative proteome analysis of wheat gluten as influenced by N and S nutrition. Plant and Soil 327: 225–234.

Health Hazards Associated with Wheat and Gluten Consumption in Susceptible Individuals and Status of Research on Dietary Therapies

Sachin Rustgi, Peter Shewry, and Fred Brouns

Wheat accounts for about 20% to over 50% of the total calorie intake of food in regions where it is grown. However, there is a clear perception that disorders related to the consumption wheat are increasing, particularly in Western Europe, North America, and Australia. We consider here the evidence for this perception and discuss strategies and therapies that may be used to reduce the adverse impacts of wheat on the health of susceptible individuals. First, we will introduce the major groups of wheat grain proteins, focusing on those associated with adverse reactions, and discuss in detail the three major adverse reactions triggered by wheat consumption, namely celiac disease, wheat allergy, and non-celiac gluten/wheat sensitivity. Finally, will discuss other issues associated with the consumption of gluten-free foods focusing on gluten contamination of products purported to be gluten-free, gluten threshold or tolerance among celiac patients, and food labeling.

S. Rustgi (✉)
Department of Plant and Environmental Sciences, Clemson University Pee Dee Research and Education Center, Florence, SC, USA
e-mail: srustgi@clemson.edu

P. Shewry
Department of Plant Sciences, Rothamsted Research, Harpenden, Hertfordshire, AL5 2JQ, UK

F. Brouns
NUTRIM School of Nutrition and Translational Research in Metabolism, Department of Human Biology, Maastricht University, Maastricht, Netherlands

G. Igrejas et al. (eds.), *Wheat Quality For Improving Processing And Human Health*, https://doi.org/10.1007/978-3-030-34163-3_20

1 Introduction

Wheat provides one-quarter of the global annual demand for dietary plant proteins and has been consumed, if not cultivated, for at least 40,000 years, with breadmaking dating back at least 14,000 years (Arranz-Otaegui et al. 2018; Henry et al. 2011, 2014). However, wheat may also have adverse health effects on susceptible individuals, some of which are well documented and understood, such as celiac disease, while others have only recently been described and are still poorly understood. There is a perception that the range and prevalence of these conditions have increased in recent decades. We will, therefore, review the types, mechanisms, and prevalence of adverse responses to wheat consumption and discuss strategies to alleviate the effects in susceptible individuals. Most of these conditions are thought to relate to the protein components of the grain, particularly gluten but also other grain proteins. We will, therefore, begin by summarizing the properties of wheat grain proteins.

2 Wheat Grain Proteins

Wheat grain proteins were among the earliest proteins to be studied, so a vast literature has been amassed dating back almost 300 years. However, until recently, the focus was almost exclusively on the gluten proteins because of their unique role in determining the food processing quality of grain. However, this interest has widened in recent years to include other proteins which may affect grain processing or induce adverse reactions to wheat consumption. Although the total number of proteins present in the mature grain may never be determined, using proteomics almost 500 proteins have been separated from flour (Dupont et al. 2011) and over 1100 from mature whole grain (Skylas et al. 2000). Most of these proteins fall into just a few groups. We will, therefore, focus on these major components, particularly those that have effects, whether positive or negative, on human health. It is only possible to provide a broad overview here, and more detailed accounts are available in other review articles (for example, Shewry et al. 2009; Wrigley et al. 2006).

2.1 Gluten Proteins

Gluten can be defined as the cohesive viscoelastic mass that remains after dough made from wheat flour is washed to remove the starch and other particulate and soluble materials. Gluten consists mainly of proteins, with most of the remainder being lipids and residual starch. Isolated gluten contains small amounts of other proteins, either because they associate with the gluten proteins or are entrapped in the gluten network formed in dough. Based on the traditional fractionation method of Osborne (1924), gluten proteins are classified into two groups, the gliadins (classified as prolamins) and glutenins (classified as glutelins). This classification is based on their sequential extraction in a series of solvents, with prolamins being soluble in aqueous ethanol and glutelins in dilute acid or alkali. However, the amounts and proportions of these fractions vary widely with the precise solvents and extraction conditions, and it is now known that they comprise related proteins which differ in being present as monomers (gliadins) or as components of polymeric complexes which may exceed 1×10^6 Da in molecular mass (glutenins). Hence, it is more usual to consider them as a single protein family. The proportions of gliadins and glutenins in grain vary depending on the genotype and the environment, but the ratio of 30% gliadins to 49% glutenins is typical of 17 European cultivars, as reported by Seilmeier et al. (1991). However, high gluten protein content is generally associated with a higher proportion of monomeric gliadins (Godfrey et al. 2010). The gliadins and glutenins form the major storage proteins of wheat and are deposited in the starchy endosperm cells of the developing grain.

2.1.1 Gluten Protein Types

Gluten proteins are conventionally separated by electrophoresis, as shown in Fig. 1. Comparisons of amino acid sequences show that all of these components are related, but they can be classified into three groups and subgroups thereof. The gliadins are most effectively separated by electrophoresis at low pH (Fig. 1) giving four groups of bands which are called, in order of increasing mobility, α-gliadins, β-gliadins (now considered as one group with α-gliadins), γ-gliadins, and ω-gliadins. Reduction of the disulfide bonds which stabilize glutenin polymers allows the subunits to be separated by sodium dodecyl sulfate-polyacrylamide gel electrophoresis (SDS-PAGE) into two groups of bands, called high molecular weight (HMW) and

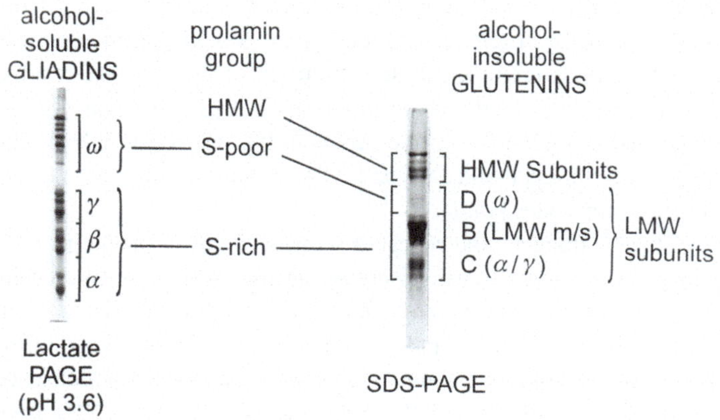

Fig. 1 The classification and nomenclature of wheat gluten proteins separated by SDS-PAGE (right) and electrophoresis at low pH (left). The D-type of LMW subunits are only minor components and are not clearly resolved in the separation shown. Taken from Shewry et al. (1999) with permission

low molecular weight (LMW) subunits. Comparison of sequences shows that the ω-gliadins and HMW glutenin subunits form distinct groups, which have been called sulfur-poor (to reflect their lack of cysteine residues) and HMW prolamins, respectively (Shewry et al. 1986). The α-gliadins, γ-gliadins and the major group of LMW subunits (called B-type) are more closely related to each other than to the ω-gliadins and HMW subunits and have been defined as sulfur-rich prolamins. In addition, two minor groups of LMW subunits, termed C-type and D-type, are most closely related to α- and γ-gliadins and to ω-gliadins, respectively, differing in the presence of unpaired cysteine residues which allow them to be incorporated into gluten in polymers by interchain disulfide bonds.

Although no gluten proteins have been sequenced completely at the protein level, a vast number of sequences derived from the sequences of cDNA and genomic DNA are available in databases. For example, using gluten protein query sequences, Bromilow et al. (2017) retrieved over 24,000 sequences from the UniProt database in May 2015. After removal of redundant, partial, and mis-assigned sequences, a curated database of 630 sequences was assembled. These sequences contain a vast amount of information, and it is only possible to provide a brief summary here, focusing on features which may relate to the role of gluten proteins in health and disease.

Firstly, the comparisons validate the broad classification into the S-rich, S-poor and HMW prolamin groups and into major subgroups of gliadins and LMW glutenin subunits. Secondly, all of the proteins have clearly defined domain structures with a central domain comprising repeated sequences based on one of two short repeated peptide motifs flanked by non-repetitive domains. However, the relative sizes of these domains vary widely between, and to a lesser extent within, the protein types. Notably, whereas the non-repetitive domains are reduced to a few amino acids in the ω-gliadins, the S-rich prolamins contain extensive non-repetitive N-terminal domains.

The sequences of the non-repetitive domains of the three protein groups are clearly related, and it has been suggested that they have been derived from a common ancestral protein by insertion of blocks of repeated peptide sequences. The non-repetitive domains contain most, and often all, of the cysteine residues which form intra-chain disulfide bonds in the monomeric α-gliadins and γ-gliadins, and both interchain and intrachain disulfide bonds in the polymeric HMW and B-type LMW subunits. Most C-type and D-type LMW subunits have single cysteine residues which form inter-chain bonds.

In terms of health impacts, the repeated sequences are of prime interest, as they contain the sequences known to trigger celiac disease (which are often called celiac epitopes). The short peptide motifs range in length from three to about ten amino acid residues and may be repeated in tandem or interspersed with a second motif. The repetitive sequences of some prolamins are highly conserved with clear consensus motifs, for example, PQQPFPQ, PQQPFPQQ, and PQGPQQ+GYYPTSPQQ+GQQ (letters are single letter abbreviations for amino acids: G, glycine; F, phenylalanine; P, proline; Q, glutamine; S, serine; Y, tyrosine; T, threonine). However, in other cases such as the LMW subunits, the sequences are degenerate and consensus motifs are difficult to define. Alignments of the repetitive domains of a range of gluten proteins are presented by Shewry et al. (2009).

The presence of repetitive sequence motifs determines the solubility of the prolamin monomers and skews their amino acid compositions. Unusually glutamine accounts for between 30% and 50% of the total amino acid residues. The numbers of glutamine residues and their contexts within the sequence also define the ability of gluten proteins to elicit a response in celiac disease.

2.2 *Other Storage Proteins*

Prolamins are only found in the grains of cereals and other grasses, where they usually form the major storage protein fraction. Other types of storage protein may also be present, such as globulins related to the 7S and 11S globulins which are the dominant storage proteins in the seeds of most other plants. In wheat, 7S globulins are the major storage protein in the aleurone layer and embryo, and small amounts of 11S globulins (called triticins) are present in the starchy endosperm. A third globulin type, called α-globulin, is low molecular weight member of the prolamin superfamily (discussed below). These proteins are discussed in detail in Shewry et al. (2009). Dupont et al. (2011) reported that total globulins accounted for about 0.4% of total proteins in white flour, although 7S globulins are depleted from this fraction.

2.3 *The Prolamin Superfamily*

Wheat gluten proteins are defined as prolamins based on the characteristic solubility and amino acid composition conferred by their repetitive domains. However, wheat grain also contains several types of proteins whose sequences are related to the non-repetitive domains of prolamins, particularly in the conservation of cysteine residues. They are therefore classified in the prolamin superfamily and together account for most components with molecular masses between about 15,000 and 25,000 Da in SDS-PAGE, although few individual components have been quantified. They include two types of proteins which are considered to be modified forms of prolamins with storage functions. These have been called avenin-like proteins or farinins and LMW gliadins or purinins (Kasarda et al. 2013) and together account for less than 2% of total flour proteins (Dupont et al. 2011). However, three groups of proteins, puroindolines, amylase/trypsin inhibitors (ATIs), and lipid transfer proteins (LTPs) have been shown to have significant impacts on grain functionality and health.

2.4 *Puroindolines*

The wheat grain contains two puroindoline (Pin) proteins, each with a mass of about 15,000 Da and comprising about 120 amino acid residues including six conserved cysteines. However, they differ from the rest of the prolamin superfamily in having a short domain which contains three or five tryptophan residues, which is thought to form a loop region in the folded protein. Pins are encoded by genes at the *Hardness* (*Ha*) locus, and allelic variation in their expression and/or sequences accounts for 60–80% of the variation in endosperm texture (hardness) in bread wheat (Turner et al. 2004). Their biological function is not known.

There is a massive literature on Pins, particularly on the relationship between protein sequences and variation in grain texture, which has been reviewed by Shewry et al. (2009) and Morris (2002). However, Pins have no reported impact on human health and will not be discussed further in this context.

2.5 Lipid Transfer Proteins

Unlike most members of the prolamin superfamily, lipid transfer proteins (LTPs) are not restricted to seeds and have been characterized from a range of plant species and tissues. Two forms occur, LTP1 with a molecular mass of about 9,000 Da and LTP2 with a molecular mass of 7000 Da, both having eight cysteine residues which form four intra-chain disulfide bonds. They were initially defined by their ability to transfer phospholipids from liposomes to mitochondria *in vitro*, but this function lacks specificity, and its relevance to their role *in vivo* is uncertain. LTPs are now more widely considered to be involved in plant defense, as they are concentrated in epidermal tissue and can inhibit fungal growth *in vitro*. LTPs are also a major group of food allergens, as discussed by Marion et al. (2004) and Jenkins et al. (2005). Wheat LTP has also been reviewed by Shewry et al. (2009).

2.6 Amylase/Trypsin Inhibitors (ATIs)

Water-soluble inhibitors of α-amylase and/or trypsin (ATIs) have been known for over 70 years (Kneen and Sandstedt 1946) and are estimated to account for up to 80% of the total albumins and 1% of the total proteins in wheat flour (Cordain 1999). They are readily observed as a group of bands migrating faster than the gluten proteins in SDS-PAGE separations. Early work led to some confusion due to the use of different numbering systems, based either on their electrophoretic mobility at alkaline pH (0.19, 0.28, 0.53) or their selective extraction in chloroform-methanol mixtures (CM1, CM2, etc.). However, with further characterization and the availability of full sequences, eleven subunits which form monomeric, dimeric and tetrameric structures were identified, and their genes to be assigned to chromosomes (as summarized in Table 1). ATIs are described in detail by Carbonero and Garcia-Olmedo (1999), and reviewed by Shewry et al. (2009).

Although all ATIs are inhibitory to human α-amylases, their relative activity varies, being very low for WTAI, higher for WMAI, and highest for WDAI (Salcedo et al. 2004). Similarly, although all three forms are also active against α-amylases

Table 1 Nomenclature and properties of wheat ATIs

Aggregation state	Protein subunit(s)	Synonyms	Gene	Gene location	Inhibitory activity	% total flour protein[a]
Monomeric	WMAI-1	0.28	*Imha-D1*	6DS	human and insect α-amylases	0.5
	WMAI-2		*Imha-B1*	6BS		
Homodimeric	WDAI-1	0.53	*IdhaB1.1*	3BS	human and insect α-amylases	1.0
	WDAI-2	0.19	*IdhaD1.1*	3DS		
	WDAI-3		*IthaB1.2*	3BS		
Tetrameric 1st subunit	WTAI-CM1	CM1	*IthaD1*	7DS	human and insect α-amylases	1.7
	WTAI-CM2	CM2	*IthaB1*	7BS		
2nd subunit	WTAI-CM16	CM16	*IthaB2*	4BS		
	WTAI-CM17	CM16	*IthaD2*	4DS		
3rd subunit (2 copies)	WTAI-CM3B WTAI-CM3D	CM3B/ CM3D	*IthaB3* *IthaD3*	4BS 4D		
Monomeric	CMx[b]	-		4AS 4BS[b] 4DL	trypsin?	0.2[c]

The table is modified from Carbonero and Garcia-Olmedo (1999) to include data from
[a]Dupont et al. (2011) and
[b]Sanchez de la Hoz et al. (1994).
[c]Estimated value based on data from Dupont et al. (2011) and Altenbach et al. (2011)

from insects, WMAI and WTAI inhibit α-amylase from the coleopteran pest *Tenebrio molitor* more strongly than WDAI, which is more active against α-amylase from larvae of the lepidopteran *Ephestia kuehniella* (Salcedo et al. 2004). The CMx proteins are encoded at a single locus on group 4 chromosomes and are proposed to be trypsin inhibitors based on their similarity with the barley trypsin inhibitor BTI-CMe (Sanchez de la Hoz et al. 1994).

Further information on the diversity and relative abundances of ATIs come from two recent proteomic studies (Altenbach et al. 2011; Dupont et al. 2011). Coding sequences for two forms of WMAI, four forms of WDAI, six forms of WTAI and three forms of CMx were identified, and the relative abundances of the inhibitor types (summarized in Table 1) and of the individual subunit forms were estimated (Dupont et al. 2011; Altenbach et al. 2011).

2.7 *Other Bioactive Proteins*

Although ATIs form the major albumin fraction in wheat grain and flour, a number of other proteins are present in very small amounts that could nevertheless contaminate protein preparations used in dietary interventions and other studies. They include proteins related to the well-characterized chymotrypsin inhibitor and amylase/subtilisin inhibitor of barley, serpin-type protease inhibitors, xylanase inhibitors, ribosome-inactivating protein (tritin) related to the well-characterized and highly bioactive ricin of castor bean, pathogenesis-related proteins (proteins produced in response to infection, predation or damage), thionins and, in the embryo only, wheat germ agglutinin (WGA). It is not possible to describe all of these proteins here, but details can be found in review articles (Wrigley and Bietz 1988; Shewry et al. 2009).

3 Adverse Reactions to Wheat or Wheat Components

The contribution of wheat to health and disease in developed countries has been widely debated, including the suggestion that wheat consumption results in obesity and associated diseases. Although these suggestions are widely promoted in the popular press and social media, they are not borne out by detailed scientific studies (Brouns et al. 2013). However, there is clear scientific evidence that three types of the disorder may present in susceptible individuals when exposed to wheat or wheat components, most notably the proteins. These are celiac disease (CD), wheat allergy (WA), and non-celiac wheat sensitivity (NCWS). These disorders are often misclassified and referred to as allergies, which is incorrect. CD, WA, and NCWS differ in the timeframe in which they develop, the mechanisms involved, symptomatology, and options for diagnosis. Many studies have addressed the effects of gluten, and gluten is the assumed cause of a wide range of adverse reactions and symptoms that disappear after the elimination of gluten from the diet. However, it is important to be aware that avoidance of wheat gluten also results in the avoidance of other proteins and components of wheat. Even isolated wheat gluten, which is often used as an ingredient in the food industry, is known to contain many components. A short overview of the most relevant aspects of each of these disorders is given below.

3.1 Celiac Disease

CD is defined by Ludvigsson et al. (2013) as 'a chronic small intestinal immune-mediated enteropathy precipitated by exposure to dietary gluten in genetically predisposed individuals.' CD is triggered by the ingestion of gluten, and over time results in a variable degree of intestinal damage. In most patients with CD, the gut pathology will reverse on the transition to a gluten-free (GF) diet. CD may present in various degrees of 'visibility,' for which different terms are used, described in detail by Ludvigsson et al. (2013).

3.1.1 CD Etiology and Prevalence

CD results from a combination of a specific genetic predisposition and environmental factors. Approximately, 25–40% of the population has a genetic predisposition as shown by the presence of HLA-DQ2 or HLA-DQ8 receptor genes. However, it is estimated that only 4% of these DQ2/8 positive individuals actually develop CD, resulting in a mean CD prevalence of around 1% of the global population but there is a large variation (range 0.3–5.6%) between countries (Stein and Schuppan 2014; Catassi et al. 2015; Lionetti et al. 2015). The disease appears to be more common in women than in men, with a ratio of 2-3 to 1. Factors other than genetic predisposition may trigger the disease such as age (Pinto-Sánchez et al. 2016), the dose of initial gluten exposure (Koning 2012), disease or drug/alcohol-related changes in intestinal permeability, or exposure to viral infections (Lebwohl et al. 2018). Rubio-Tapia et al. (2012) documented an increase in the prevalence of CD over recent decades, which is often suggested to be the result of an increase in the consumption of wheat gluten.

Other factors contributing to the recent increase in the prevalence of gluten-induced disorders are major changes in overall lifestyle, diet, hygiene, and gut microbiota, the use of antibiotics and vaccines, and improvements in diagnosis and medical registrations leading to the recognition of formerly undiagnosed, misdiagnosed and unreported cases (Olivares et al. 2018). A well-known case of misdiagnosis comes from the Punjab province of India, where gluten sensitivity and intolerance were misdiagnosed for a very long time as 'summer diarrhea', which is quite prevalent in the region but is only reported in summer when wheat flour is used for making 'chapatti' flatbreads, instead of maize flour predominantly used in winter (Cataldo and Montalto 2007).

The initial development of CD may be related to the food that a child receives early in life. The influence of breastfeeding and the time of weaning on the disease are not clear (Ludvigsson and Fasano 2012). However, it has been suggested that intake of small quantities of gluten should start gradually before the age of 6 months, preferably simultaneously with breastfeeding (Ivarsson et al. 2013). The reason for this is that the immune-modulatory properties of breastfeeding and the intestinal flora should contribute to the prevention of auto-immune diseases (Agostini et al. 2008). A number of studies indicated a role of gut microbial diversity and related gut-associated immune competence in the etiology of CD (Nadal et al. 2007; Nistal et al. 2012, 2016; Olivares et al. 2018). Whether this link is truly causal or a consequence of altered dietary patterns in individuals that suffer from CD needs further study. It is often suggested that vital wheat gluten, increasingly used as a functional ingredient in food production, also plays a role. However, consumption data over the years are difficult to obtain. Kasarda (2013) calculated that the intake of vital gluten has tripled since 1977, from 136 to 408 g per year, or 0.37–1.12 g per day per capita of the population. However, the impact of this increase is a matter of debate given the many times higher intake of gluten from wheat flour in bread, 5–5.5 kg per year, or 13.7–15.1 g per day. Furthermore, the protein content of wheat has tended to decline as yield has increased (Shewry et al. 2016), and much of the vital gluten is used in breadmaking to compensate for this decrease. Ever since humans began to consume wheat, the ingestion of gluten contained in bread has been many times higher than the amount of gluten that causes CD. Barley and rye contain proteins related to wheat gluten, and hence, consumption of their grains contributes to daily 'gluten' exposure. The high contents of proline and glutamine in gluten proteins may contribute to them resisting digestion in the human gastrointestinal tract. In individuals with altered gut permeability, these undigested gluten fragments (peptides) may enter the lamina propria of the small intestine via transcellular or paracellular routes, leading to a cascade of reactions causing inflammation and immune responses. These mechanisms are described in detail by Fasano (2012) and Lebwohl et al. (2018).

When undigested gluten fragments enter the intestinal wall, the enzyme tissue transglutaminase (tTG or TG2) converts the glutamine in the peptides into glutamic acid, which then binds more strongly with the HLA-DQ2/8 receptor and with T-cells of the immune system, activating an inflammatory response that finally leads to damage to the intestinal villi. These inflammatory reactions allow substances to pass through the intestinal wall and cause problems elsewhere in the body such as dermatitis herpetiformis, which responds to a GF diet (Ludvigsson et al. 2013; Ciacci et al. 2015). In addition, neurological manifestations and ataxia have been described (Ludvigsson et al. 2013). Gluten ataxia is defined as 'idiopathic sporadic ataxia and positive serum antigliadin antibodies even in the absence of duodenal enteropathy'. Wheat gluten consumption has also been suggested to be involved in the etiology of some mental/psychological disorders (Hadjivassiliou et al. 2010).

In the long term, CD-related inflammation and immune reactions will impact on the intestinal wall surface leading to the disappearance of the surface area increasing villi flattening of the gut mucosa (Fig. 2) and resulting in impairment of nutrient uptake, weight loss and, in children, growth retardation (Lebwohl et al. 2018). However, it should be noted that this is not the only manifestation of CD. Some individuals may have the HLA-DQ2 or DQ8 genes and CD-related antibodies in their blood without verifiable damage to the intestinal mucosa but may have one or more general complaints such as chronic fatigue, poor sleep or headaches. Often the symptoms of these people are not recognized as possibly representing CD and, as a result, such individuals are not tested or diagnosed for CD: a phenomenon which is called the subclinical CD, or potential CD. It is estimated that only 1 out of 8 people with CD (10,000–27,000 people) have been diagnosed based on symptomatology, which means that many individuals remain undiagnosed because of unclear or unrecognized symptoms, referred to as 'the celiac iceberg' (Fig. 3).

3.1.2 CD Causing Substances

Wheat gluten proteins all have high contents of proline and glutamine, which are concentrated in their repetitive domains. As a consequence, these domains may only be partially digestible by the enzymes present in the human intestine, producing indigestible gluten fragments which contain short sequences of amino acids known as epitopes. The identification of the fragments which cause immunological reactions have led to rapid progress in CD research (Pastorello et al. 2007; Tatham and Shewry 2008; Mamone et al. 2011; Ludvigsson et al. 2013). It includes the evaluation of epitopes with the greatest ability to stimulate T-cells (immunodominance) (Anderson et al. 2000, 2005; Shan et al. 2002, 2005; Tye-Din et al. 2010) and the

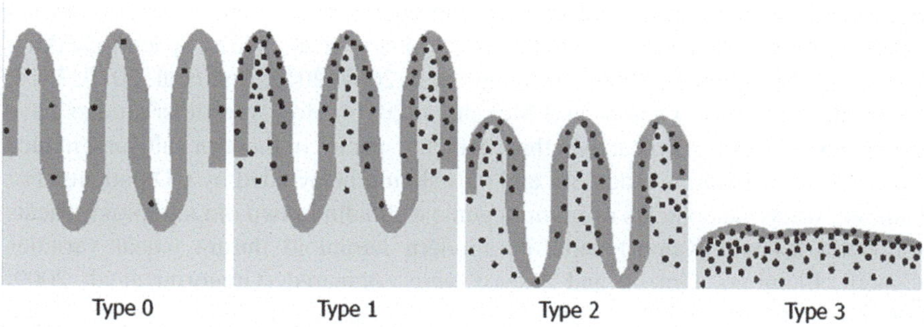

Fig. 2 A schematic illustration of progressive tissue changes in the small intestine leading to flattening of the villi and more intraepithelial lymphocytes (black dots) in the gut mucosa lining, (source Cukrowska et al. 2017)

Fig. 3 The celiac iceberg. (source Fasano 2007)

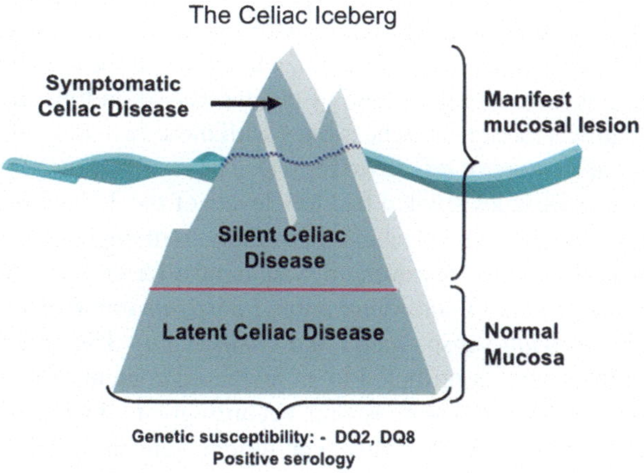

agreement of an internationally accepted list of all epitopes known to play a role in CD (Sollid et al. 2012).

It has also been shown that there are large differences between individuals in the physical responses to single or combinations of gluten fragments. It is also clear that there are broad differences in the celiac activity of different groups of gluten proteins. In particular, the α-gliadins, and to a lesser extent the γ-gliadins, yield large numbers of immunogenic peptides in the form of indigestible protein fragments. People with CD also react to related proteins from rye (secalins) and barley (hordeins), although neither species contains proteins closely related to the α-gliadins. A minority of CD patients react to glutenins.

The number, type, and distribution of toxic epitopes within a wheat genotype are clearly of interest (Shewry and Tatham 2016). Salentijn and coworkers noted the presence of significantly fewer toxic epitopes in several tetraploid wheat species (Salentijn et al. 2009, 2013), while several other reports indicate that 'modern' hexaploid wheat types may induce more immune and inflammatory reactivity than 'ancient' tetraploid and diploid species, and hence result in more gastrointestinal problems in wheat-sensitive individuals (Molberg et al. 2005; Spaenij-Dekking et al. 2005; Pizzuti et al. 2006; Vincentini et al. 2007, 2009; Sofi et al. 2010, 2014; Carnevali et al. 2014; Gelinas and McKinnon 2016). However, other studies have shown that all types of wheat, whether 'ancient' species or modern cultivars, induce some degree of immune reactivity and thus should be avoided by CD patients. For example, when the contents of immunogenic α-gliadin in two old tetraploid wheats (Graziella Ra and Kamut) and four modern tetraploid durum wheat varieties (Capelli, Flaminio, Grazia, and Svevo) were compared (Gregorini et al. 2009; Colomba and Gregorini 2012), the older varieties had significantly more, and so are considered unsuitable for individuals with CD. More recent studies have confirmed these findings (Šuligoj et al. 2013; Escarnot et al. 2018).

By contrast, van den Broeck et al. (2010b) used immunoblotting with antibodies to the Glia-α9 and Glia-α20 epitopes to classify modern bread wheat varieties (1986–1998) and older types (1863–1982) as having low, medium or high reactivity. The Glia-α9 antibody reacts with one of the most harmful epitopes, present in a 33-mer peptide from α-gliadin, while the epitope recognized by Glia-α20 is harmful to a more restricted group of people. Based on this, the authors suggested that the relative reactions of wheat lines with these two antibodies could be used to classify them as more or less toxic for CD. The results showed that only 1 of the 36 modern wheat varieties studied had low levels of the 33-mer peptide, compared with 15 of the 50 older types. The number of lines with high reactivity to Glia-α20 was similar in both groups. However, no systematic differences were found between the modern bread wheat (*T. aestivum* subsp. *aestivum*) and older types of hexaploid wheat (*T. aestivum* subsp. *spelta* and *compactum*). The authors suggested that scientific breeding has contributed to an increased content of highly CD reactive epitopes in modern wheat varieties and accordingly to increased CD prevalence (van den Broeck et al. 2010b). However, more recent studies do not support this suggestion. Kasarda (2013) reviewed the arguments as to whether wheat breeding programs aimed at increasing gluten content may have contributed to the increase in CD in the USA during the latter half of the 20th century, but the available data do not demonstrate an increase in gluten due to breeding.

Ribeiro et al. (2016) compared modern varieties of bread wheat (*Triticum aestivum* subsp. *aestivum*), spelt (*T. aestivum* subsp. *spelta*) and durum wheat (*Triticum turgidum* subsp. *durum*), including modern cultivars and old landraces from different countries. The lines were grown in the same year and location and under identical environmental growing conditions and the amounts of potential CD immune-stimulatory epitopes measured with the R5 monoclonal antibody assay that recognizes several toxic peptides (QQPFP, QQQFP, LQPFP, QLPFP, QLPYP) that occur repeatedly in gluten proteins. Different gliadin types were quantified by reversed-phase high-performance liquid chromatography (RP-HPLC) and acid-polyacrylamide gel electrophoresis (A-PAGE) in order to correlate toxic epitopes with gliadin content. In line with earlier observations (Spaenij-Dekking et al. 2005; van den Broeck et al. 2010a, b), this study confirmed that there is significant heterogeneity in the level of the T-cell–stimulatory epitopes in wheat varieties. For example, the modern *T. aestivum* variety 'Pernel' presented more than 11-fold fewer toxic epitopes than the variety 'Alejo'. In addition, high heterogeneity in the content of toxic epitopes in tetraploid lines was observed with values ranging from 26.40 ± 1.65 g per kg in 'Basto Duro' to 223.14 ± 30.04 g per kg in 'Valenciano'. Spelt varieties proved to have more toxic epitopes compared to other types of hexaploid wheats, with a mean of approximately 311.15 g per kg. The authors concluded that intensive wheat breeding has not resulted in increases in toxic epitopes in modern wheat varieties, so cannot be the reason for increases in CD incidence.

It can, therefore, be concluded that there is no evidence to support the conjecture, particularly in social media, that ancient tetraploid grains and spelt are more tolerated by individuals suffering from CD and WA. In fact, peptides from both spelt and durum result in immune responses and should certainly not be recommended as alternatives to conventional modern wheat for celiac patients (Ribeiro et al. 2016; Vincentini et al. 2007, 2009; Gregorini et al. 2009; Šuligoj et al. 2013; Escarnot et al. 2018). Indeed, a recent comprehensive review of peptides from gluten digestion comparing old and modern wheat varieties concluded that the old varieties may actually contain more immunogenic and toxic sequences than modern varieties (Prandi et al. 2017). The relative proportions of CD-immunogenic and CD-toxic sequences in gluten proteins also vary depending on environmental factors such as climate, soil, fertilization and agricultural practices, and their interactions with the genotype (Ashraf 2014; Hajas et al. 2018).

Gianfrani et al. (2015) provided evidence that extensive *in vitro* gastrointestinal hydrolysis drastically reduced the immune stimulatory properties of *Triticum monococcum* gliadin. The MS-based analysis showed that several *T. monococcum* peptides, including known T-cell epitopes, were degraded during the gastrointestinal treatment, whereas many of *T. aestivum* gliadins survived the gastrointestinal digestion. However, these findings need confirmation *in vivo*, but a crucial observation is that not all peptides were degraded; thus, *T. monococcum* is still not safe for CD patients. More recently, Perez-Gregorio et al. (2018) showed that the composition of gluten hydrolysates depended on the digestion time and structural characteristics of the protein. On the other hand, the main T-cell stimulating epitopes formed during hydrolysis depend on the identity of the precursor protein. Glutenin oligopeptides were degraded faster than gliadins, particularly α-gliadin oligopeptides.

In addition to gluten, it has recently been recognized that non-gluten proteins, notably serpins, purinins, globulins, ATIs, and farinins, may also elicit antibody responses (Huebener et al. 2015). The growth conditions of the plant (shade and plant height) and grain storage conditions appeared to influence the ATI content of three different types of durum wheat grown at three locations in Italy (Prandi et al. 2013), with the effect of the environment being greater than that of genotype.

Recent research has indicated that ATIs may play a role in the development of both CD and non-celiac-related wheat hypersensitivity via the production of inflammatory factors (cytokines) in the intestine (Junker et al. 2012; Zevallos et al. 2017). These studies were carried out in animal models (mice) and *in vitro* assays on human cell lines, using isolated protein fractions enriched in ATIs. These studies indicate a clear need to carry out human trials with oral exposure to ATIs that have also been exposed to heat treatment and simulated gastrointestinal digestion. Zevallos et al. (2017) suggest that both the types and amounts of isoforms affect the degree of bioreactivity of ATIs. ATIs are heat resistant, and active forms may be present in wheat cooked for 5 min at 100°C and may cause an allergic reaction on ingestion (Pastorello et al. 2007). Rye and barley also contain a range of related ATI isoforms (Carbonero and García-Olmedo 1999).

3.1.3 CD Diagnosis and Solutions

CD can be diagnosed based on the pattern of symptoms, blood serology, genetic predisposition for CD and, finally, by the histological screening of small intestinal tissue obtained by biopsy. General guidelines for CD diagnosis are given by Bai and Ciacci (2017) and Lebwohl et al. (2018). A comparative overview of the characteristics of CD, WA, and NCWS is given in Table 2.

A completely GF diet appears to be the only remedy for people with CD. Even traces of wheat should be avoided in foods or food ingredients produced in a 'wheat presence environment' or other 'gluten'-containing grains (barley, rye) present in GF food through cross-contamination that might occur during cultivation, harvest, transport, production (see below for further details). By completely avoiding gluten, the intestinal wall can recover, and symptoms disappear.

3.2 Wheat Allergy (WA)

An allergen is defined as a substance that causes an immediate immune reaction upon exposure by ingestion, inhalation, or skin contact. Depending on the route of allergen exposure and the underlying immunological mechanisms, WA may be classified as (1) an immediate food allergy, (2) wheat-dependent exercise-induced anaphylaxis, (3) a respiratory allergy, or (4) contact urticaria. IgE antibodies play a central role in the pathogenesis of these diseases.

3.2.1 WA Etiology and Prevalence

In WA, the body reacts to the protein as if it was a pathogen that needs to be dealt with. This misrecognition leads to inflammatory reactions and symptoms such as swollen membranes of the mouth and throat, difficulty in swallowing, shortness of breath, diarrhea, vomiting, abdominal pain, asthmatic reactions, and rashes. In some cases, a whole-body reaction may cause a sudden severe drop in blood pressure leading to anaphylactic shock or even death (Hadjivassiliou et al. 2015). The most widely occurring allergy to wheat is bakers' asthma, a respiratory allergy resulting from the inhalation of wheat flour or starch.

Table 2 Different features of celiac disease (CD), wheat allergy (WA), and non-celiac gluten sensitivity (NCGS) according to Catassi (2015)

	CD	NCGS	WA
Time interval between gluten exposure and onset of symptoms	Weeks-years	Hours-days	Minutes-hours
Pathogenesis	Autoimmunity (innate and adaptive immunity)	Immunity? (innate immunity?)	Allergic immune response
HLA	HLA-DQ2/8 restricted (~97% positive cases)	Not HLA-DQ2/8 restricted (50% DQ2/8-positive cases)	Not HLA-DQ2/8 restricted (35-40% positive cases as in the general population)
Autoantibodies	Almost always present	Always absent	Always absent
Enteropathy	Almost always present	Always absent (slight increase in IEL)	Always absent (eosinophils in the lamina propria)
Symptoms	Both intestinal and extraintestinal (not distinguishable from GS and WA with GI symptoms)	Both intestinal and extraintestinal (not distinguishable from CD and WA with GI symptoms)	Both intestinal and extraintestinal (not distinguishable from CD and GS with GI symptoms)
Complications	Co-morbidities, long-term complications	Absence of co-morbidities and long-term complications (long follow-up studies needed to confirm it)	Absence of co-morbidities, short-term complications (including anaphylaxis)

GI = Gastrointestinal; GS = Gluten Sensitivity; IEL = Intraepithelial Lymphocytes

For a schematic representation of allergy mechanisms see: *https://upload.wikimedia.org/wikipedia/commons/e/e7/The_Allergy_Pathway.jpg*. A number of detailed reviews of WA have been published (Tatham and Shewry 2008; Inomata 2009; Sapone et al. 2012; Gilissen et al. 2014; Cianferoni 2016; Pasha et al. 2016).

WA prevalence amongst children varies from <0.1% to 1%, depending on age and country (Hischenhuber et al. 2006; Kotaniemi-Syrjänen et al. 2010; Sapone et al. 2012). A large meta-analysis has shown that the general prevalence is at most approximately 0.2% (Zuidmeer et al. 2008). About one-half of children 'outgrow their food allergy,' and there are reports that more than 80% of children with WA have outgrown it by their 8th year, and 96% before their 16th year (Kotaniemi-Syrjänen et al. 2010). Accordingly, the number of adults with WA is much lower than the number of people with CD. It has also been shown that the prevalence of food allergies in women is generally higher than in men (Afify and Pali-Schöll 2017). Interestingly, the prevalence of NCWS is also higher in women than in men.

3.2.2 WA Causing Substances

Wheat proteins that cause allergic reactions are very diverse and differ between individuals (Mamone et al. 2011; Cianferoni 2016). The two main wheat protein groups causing food allergy are ω5-gliadins and ATIs, and equivalent proteins from other grains. To a lesser extent, reactions to LMW-glutenin subunits, WGA and LTPs have been reported (Gilissen et al. 2014; Baccioglu et al. 2017; Brans et al. 2012; Scherf et al. 2016). Children with wheat allergies react most commonly to α/β- and γ-gliadins (Pastorello et al. 2007; Tatham and Shewry 2008). One-fifth of people with WA are also allergic to barley and rye (Sicherer 2001). It has long been known that ATIs play a role in bakers' asthma (flour dust allergy) and food allergy to wheat (Pastorello et al. 2007; Tatham and Shewry 2008). Recently, Zevallos et al. (2018) and Bellinghausen et al. (2018) showed that ATIs exacerbate allergic reactions in mice. Gélinas and Gagnon (2018) studied the effects of heat treatment on α-amylase inhibition by ATIs and noted that well-heated cereal foods lose the α-amylase activity, compared to those that are only mildly heated. In this respect, it should be noted that enzyme activity may be lost due to heat-induced changes in protein structure but that the protein itself is not degraded. In other words, a change of α-amylase activity does not necessarily mean that there is also a loss of inflammatory and/or immune reactivity in susceptible individuals.

3.2.3 WA Diagnosis and Solutions

WA can be tentatively diagnosed by the combination of a blood test and a skin test. The blood test will demonstrate the presence of specific immunoglobulin E (IgE) antibodies to allergens (in this case wheat proteins) present in the blood, while the skin test will show how the skin reacts via the IgE antibodies after exposure to a very small quantity of subcutaneously injected wheat protein. However, the presence of IgE antibodies against wheat in the blood certainly does not always mean that there is an active (food) allergy nor does the skin test yield a conclusive diagnosis (Sapone et al. 2012). A conclusive diagnosis is only obtained by a food challenge test, executed in a double-blind, placebo-controlled setup. Such tests have shown that WA is a relatively uncommon event, but that many wheat proteins are immunogenic (stimulating the production of IgEs) but have not been found to lead to clinical symptoms (Zuidmeer et al. 2008). It is, therefore, necessary for people with WA to completely avoid products containing wheat (and possibly other grains to which they react). Excellent reviews of WA are available (Battais et al. 2008; Gilissen et al. 2014; Cianferoni 2016).

3.3 *Non-celiac Gluten or Wheat Sensitivity*

In recent years, the third group of people has been identified who experience symptoms after eating wheat products, but who are not diagnosed to be suffering from either WA or CD. These individuals may have gastrointestinal symptoms which are similar to those for irritable bowel syndrome (IBS) and which improve on the transition to a GF diet. This group is referred to as being non-celiac gluten sensitive (NCGS) or more recently non-celiac wheat sensitive (NCWS). Ludvigsson et al. (2013) have defined NCWS as "one or more of a variety of immunological, morphological, or symptomatic manifestations that are precipitated by the ingestion of gluten in individuals in whom CD has been excluded".

Despite the word 'gluten' in the current definition of NCGS/NCWS, it is far from certain that gluten is the main cause of symptoms in this group of people or indeed whether wheat is the direct cause.

3.3.1 NCWS Etiology and Prevalence

NCWS is a condition in which wheat ingestion leads to morphological or symptomatic manifestations despite the absence of CD. NCWS sufferers may show signs of an activated innate immune response, elevations in anti-tTG antibodies, endomysial antibodies, deamidated gliadin peptide antibodies, and increased mucosal permeability, all of which are characteristic of CD but without the associated enteropathy. Although up to 35% of the population may state that they feel better when avoiding gluten and/or wheat, the percentage of individuals suffering from NCWS is probably much lower, but this depends on the region of observation (DiGiacomo et al. 2013; Rubio-Tapia et al. 2012). At present, it is impossible to reliably estimate the number of people suffering from NCWS/NCGS. Although it is expected that this will be higher than the number of people with CD, precise figures are scarce and range from 0.5% to 10% of the population, but the current most likely estimate is 2–5% (Ludvigsson et al. 2013).

3.3.2 NCWS Causing Substances

Gluten and fermentable oligo-, di-, monosaccharides, and polyols (FODMAPs) are most widely discussed as causative agents of NCWS (Barrett and Gibson 2012; Skodje et al. 2018). Several studies with wheat products, not only with gluten, have been reported in which people were also exposed to other wheat components such as LTPs and ATIs (Biesiekierski et al. 2013; Vazquez–Roque et al. 2013). To date, no studies have been carried out to determine how patients diagnosed with NCWS react to individual components or their combinations. Although Kamut, an old type of tetraploid wheat, caused fewer symptoms in IBS patients compared to modern wheat (Sofi et al. 2014), a recent critical review concluded that further studies using well-defined grain samples grown under the same conditions are required to confirm this relationship (Shewry 2018).

FODMAPs are present in all wheats but also in fruits and vegetables, including legumes (Biesiekierski et al. 2011). Considering the large overlap between symptoms associated with IBS and NCWS, wheat-based products are increasingly being listed as foods to avoid since they contain fermentable carbohydrates. Biesiekierski et al. (2011) observed gluten to cause symptoms in one study, but showed that FODMAPs were the cause in a subsequent study (Biesiekierski et al. 2013). Similar results were reported by Skodje et al. (2018). However, the amount

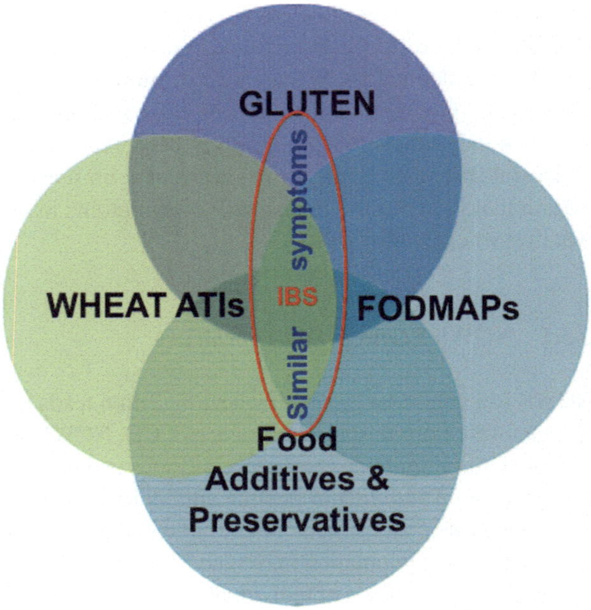

Fig. 4 Significant overlap in symptoms induced by wheat grain components, gluten, amylase/trypsin inhibitors (ATIs), fermentable oligo-, di-, monosaccharides and polyols (FODMAPS), and food additives like vital gluten, wheat starch, etc. This figure is modified from Volta et al. (2013)

of fructans, the major FODMAPs in wheat, in wheat-based foods is low and far below the levels that would cause abdominal distress in normal healthy individuals (Brouns et al. 2017).

3.3.3 NCWS Diagnosis and Solutions

The diagnosis of NCWS/NCGS is difficult because people report symptoms that may indicate CD as well as symptoms that are known to occur with WA. Often these symptoms are self-diagnosed and also overlap with a cluster of symptoms defined as IBS (see Fig. 4). Caio et al. (2014) and Uhde et al. (2016) showed that people who suffer from NCWS symptoms improve significantly on a wheat/gluten free diet. However, well-defined biomarkers and a clinically validated diagnostic test are still lacking for NCWS/NCGS. Excellent reviews comparing aspects of CD and NCWS/NCGS are available (Volta et al. 2013; Schuppan et al. 2015; Scherf et al. 2016; Leonard et al. 2017; Catassi et al. 2017).

4 Hidden Gluten or Gluten Contamination

As discussed above, the only known therapy for gluten-induced disorders is adherence to an abstinent diet. With increased awareness and the growing number of self or medically diagnosed cases of celiac disease, gluten intolerance/sensitivity and wheat allergy, the demand for GF products is also increasing. In fact, the U.S. market for GF products is projected to reach USD 7.59 billion by 2020. Since gluten is widely used in food processing, it is a major challenge to identify truly GF products, as gluten can be found in unexpected products and even in products purported to be GF.

Products containing 'hidden' gluten include sausages, fish fingers, cheese spreads, soups, sauces, mixed seasonings, mincemeats, and even some medications and food supplements such as vitamin preparations. The major causes of contamination are either the use of common machinery during harvest, transportation, and processing or the use of common storage space. Successive cultivation of gluten-containing and GF crops may also lead to contamination of GF grains. Contamination is unavoidable if the same milling equipment is used for GF and gluten-containing grains. Interestingly there is no legislature in place regarding the maximum permitted levels of foreign grains in inherently GF cereals. However, in practice, 2% of other grains would be considered as the upper limit.

The importance of the issue of gluten contamination for celiac patients and the threat to consumers has been acknowledged for some time, and several studies using different detection procedures, such as PCR, quantitative polymerase chain reaction, enzyme-linked immunosorbent assay (ELISA) and mass spectrometry, have determined the levels of contamination in many commodities (Fritschy et al. 1985; Janssen et al. 1991; Van Eckert et al. 1992; Olexová et al. 2006; Hernando et al. 2008; Thompson et al. 2010; Lee et al. 2014; Farage et al. 2017, and references cited in Miranda and Simón 2017 and Do et al. 2018). For example, gluten contamination as high as 3000 mg per kg was found in buckwheat flour claimed to be GF. A literature review revealed that higher than recommended levels of gluten in purportedly GF commodities is a worldwide problem. Although the studies mainly focused on samples collected from industrialized countries where the regulations are much tighter, it is reasonable to extrapolate the results of these studies to the wider world (for detailed reviews, see Miranda and Simón 2017; Do et al. 2018). In a meta-analysis, Bustamante et al. (2017) charted the rise and fall of gluten content in cereal-based GF food products by using ELISA to analyze 3141 samples from 1998 to 2016. Gluten was detected in a total of 11.8% (371) samples, with breakfast cereals and cereal bars being the most highly contaminated commodities. Additionally, the study suggested that the number of contaminated samples containing low levels of gluten had declined whereas those containing over 100 ppm of gluten had increased during the period 2013–2016 (Bustamante et al. 2017). Although the prevalence of foods with detectable gluten has decreased over time in line with the evolving regulations relating to information on food composition and gluten content claims, the problem persists (Miranda and Simón 2017; Do et al. 2018).

5 Gluten Threshold or Tolerance Level

A GF diet does not necessarily mean 'zero gluten' because low levels of gluten may be tolerated by patients. Establishing a patient's threshold for gluten intake is, therefore pertinent. Although a number of studies have determined the effect of low gluten intake in patients with CD, a more detailed study is required to reach a consensus. Initially, Ciclitira et al. (1984) studied the toxicity of gliadin doses and response time in a single patient and concluded that 10 and 100 mg of gliadins induce no or slight changes in the small intestine morphology, whereas 500 mg and 1 g of gliadin caused moderate to extensive damage. Later, a similar daily gluten intake of 10–100 mg was reported to be safe for consumption by CD patients (Hischenhuber et al. 2006). In a separate study with an observation period of 1–6 weeks, 2.4–4.8 mg gluten dose per day was found to cause no damage to jejunal morphometry (Ciclitira et al. 1985). Similarly, Ejderhamn et al. (1988) demonstrated that a daily intake of 4–14 mg of gliadin did not cause morphological changes in the mucosa of the small intestine of CD patients on the abstinent diet. Two groups in Finland also reported similar observations but with slightly higher daily gluten doses of 20–36 mg (Kaukinen et al. 1999; Peräaho et al. 2003). Later, Catassi et al. (1993) demonstrated that 100 mg of gliadins per day caused deterioration of the small intestine architecture, the effects being more pronounced with the higher dose of 500 mg per day. However, despite these studies, a consensus on the critical limit or threshold for gluten intake has not been reached. This uncertainty is to be attributed to the large variability among celiac patients, as demonstrated by an extensive double-blind placebo-controlled multicenter study of the effects of 10–50 mg of gluten per day on 40 CD patients. The patients were administered daily capsules containing 0 mg, 10 mg or 50 mg of gluten for 90 days and studied for clinical, serological, and histological changes in their small intestine. The study reported wide variation among patients in terms of gluten sensitivity. Some patients showed intestinal symptoms after ingesting only 10 mg of gluten daily while other patients showed no histological symptoms even after three months of a daily challenge with 50 mg gluten. In a similar study, it was reported that a daily intake of 50 mg of gluten for three months is sufficient to cause significant damage to the intestinal morphology of CD patients (Catassi et al. 2007a, b).

Because of the variation in the results of such studies, different countries allow different gluten levels in products for consumption by celiac patients. For example, North European countries (Denmark, Estonia, the Faroe Islands, Finland, Iceland, Latvia, Lithuania, Norway, and Sweden) permit up to 200 ppm of gluten in food for celiac patients. By contrast, North American and South European countries (Italy, Spain, Greece, Portugal, Andorra, Vatican City, San Marino, Gibraltar, Corsica, and Malta) adopted a more conservative limit of 20 ppm gluten in GF products (Do et al. 2018). Based on clinical data, an intermediate limit of 100 ppm was adopted by Finland.

Decisions on gluten thresholds depend mainly on two factors, (i) the minimum toxic dose, and (ii) the amount of GF product(s) consumed. The results of a food challenge study indicated that 200 ppm is not a safe threshold as a gluten intake limit of 50 ppm could be reached with the consumption of 250 g of so-called GF product(s) (Catassi and Fasano 2008). A 100 ppm limit that allows 10 mg gluten in 100 g of food is also impractical as the consumption of GF products in European countries could be as high as 500 g per day (Gibert et al. 2006; Catassi and Fasano 2008). However, a threshold of 20 ppm restricts the intake of gluten from GF food well below 50 mg, thus allowing a safety margin for variation in the gluten sensitivities and dietary habits of different patients. This knowledge has led the Codex Alimentarius Commission to define 'any product having <20 ppm gluten could be treated as gluten-free because the gluten levels below this limit, in general, do not invoke a disease response.'

6 Food Labeling

As gluten sensitive individuals rely mostly on product labels to make dietary decisions, it is important to label all food ingredients and specifically those in pre-packed foods. The food ingredients that cause intolerance and/or allergenicity are documented in the 'list of hypersensitivity' assembled by the Codex Commission and include gluten-containing cereals, Crustacea, eggs, fish, peanuts, milk, tree nuts, and product derivatives of the above-listed items, often dubbed the 'big eight'. These labeling regulations were established in various countries to help consumers, specifically people with various dietary allergies, to avoid the consumption of packaged food products that contain the major food allergens. In the USA, the Food Allergen Labeling and Consumer Protection Act (FALCPA), which was implemented in 2004, requires the specific declaration of the 'big eight' when added as ingredients to labeled food products. Similar legislation also exists in other countries, such as Canada, Australia, New Zealand, Japan, and the European Union (Do et al. 2018). More specifically, whereas gluten is not a permitted food additive in the European Union and the United States, other wheat ingredients such as soluble wheat proteins and starches are permitted and therefore can be used without declaration in food products. Therefore, the Codex Commission concluded that gluten ingredients from all wheat, barley and rye species, including spelt, Khorasan, Kamut and durum wheats (all forms of tetraploid *T. turgidum*), einkorn (*T. monococcum*), triticale, tritordeum, and their hybrid varieties are immunogenic, and should be declared. The major concern at present is the misbranding of single/multiple ingredient food products as GF without proper testing, particularly if they are derived from inherently GF grains like millet, flax, buckwheat, and sorghum. Thus, in order to brand them GF, it is important to test to ensure that the gluten level is below the prescribed limit of 20 mg per kg.

7 Potential Therapies

So far, the only reliable therapy available for gluten-related disorders is strict lifelong adherence to a gluten exclusion diet, which is difficult to follow, and is not free of side-effects (see below). The major constraint to the development of new therapies for gluten-induced disorders is the large number of cereal proteins (gluten and non-gluten) that present epitopes (see above), and the fact that different celiac patients differ in the magnitude of reactivity to different gluten epitopes (Koning 2012). Despite extensive efforts to map immunogenic peptides to the prolamins responsible (Juhász et al. 2018; Tye-Din et al. 2010), the repertoire of epitopes is still incomplete (Osorio et al. 2012; Sollid et al. 2012; Comino et al. 2013; Juhász et al. 2015). In general, it would be precautionary to say that gluten peptides more than nine amino acids in length, especially those derived from the repetitive domains or the non-repetitive C-terminal ends of most of the major prolamins, can potentially elicit an immune reaction in susceptible individuals (Osorio et al. 2012; Juhász et al. 2018). Since no extant wheat genotypes either in cultivation or wild are entirely free of gluten, no wheat types can be considered 'celiac-safe' (Mitea et al. 2010; Goryunova et al. 2012; Brouns et al. 2013; Shewry 2018).

7.1 *Reduced-Gluten Cytogenetic Stocks and Mutations in Triticeae Cereals*

An alternative approach to seeking ‘celiac-safe’ types of wheat is to exploit genetic stocks and mutants in which specific groups of gluten proteins are reduced or absent. Early studies compared sets of wheat aneuploid lines in which specific pairs of chromosomes are deleted (and the corresponding chromosomes of other genomes duplicated to improve genetic stability). Thus, some nulli-tetrasomic or deletion lines of Chinese Spring wheat showed low toxicity because certain gliadin loci had been eliminated due to the chromosomal defect (Ciclitira et al. 1980a, b; Frisoni et al. 1995; van den Broeck et al. 2009, 2011). Likewise, wheat mutants lacking α/β-, γ- and/or ω-gliadins and/or showing reduced accumulation or complete elimination of specific gliadins and/or LMW glutenins have been characterized (Moehs et al. 2018; Camerlengo et al. 2017; Waga and Skoczowski 2014; Pogna et al. 1998; Redaelli et al. 1994; Metakovsky et al. 1993). However, these genotypes can be best described as low-toxicity lines, as they still contain gliadins and glutenins. Similar barley mutants with reduced hordein (gluten) content were also identified (Moehs et al. 2018; Cook et al. 2018; Munck 1992; Tallberg 1981a, b, 1982).

Collectively, different celiac patients have different degrees of sensitivity to different prolamin epitopes, so it would be ideal to combine low toxicity alleles of gliadin and glutenin genes in a single genotype (Gil-Humanes et al. 2014). However, given the large number and complexity of the gliadin genes, and their presence and inheritance in tightly linked multigene blocks, the possibility of pyramiding all low toxicity gliadin genes in a single wheat variety is a formidable task through conventional breeding (Koning 2012).

Therefore, it can be concluded that it is not realistic to identify celiac-safe wheat genotypes by screening the extant wheat germplasm or to develop them via conventional breeding. Furthermore, there are logistical issues associated with the release of the reduced-gluten or low-toxicity wheat lines, such as whether they are suitable for individual celiac patients who may differ in their reactions to different gluten proteins, and how derivative food products should be labeled to reflect this.

7.2 *Alternative Cereals and Pseudocereals*

Another way to deal with this problem is to use inherently GF cereals, such as oats, maize, rice, sorghum, fonio, tef, millet, and Job’s tears or pseudocereals such as amaranth, quinoa, and buckwheat. However, none of these cereals and pseudocereals has similar technological and organoleptic properties to wheat. Furthermore, some individuals are sensitive to oat gluten proteins (avenins), and in rare cases even to maize proteins (Comino et al. 2013; Rosella et al. 2014; Ortiz-Sánchez et al. 2013). So far, the two major cereals that have not raised cause for concern are rice and sorghum (Rosella et al. 2014; Pontieri et al. 2013). However, rice grains have low protein and fiber content and are highly enriched in easily digestible carbohydrates that may contribute to less favorable high glycemic responses. Some rice grain storage proteins (other than prolamins and glutelins) are known to trigger a variety of allergenic reactions in susceptible individuals such as asthma, atopic dermatitis, diarrhea, and anaphylaxis (Matsuda et al. 2006; Nambu 2006; Trcka et al. 2012; Gilissen et al. 2014). Similarly, the use of pseudocereals is also controversial, due to the immunotoxicity of some quinoa (Zevallos et al. 2012, 2014) and buckwheat varieties (Panda et al. 2010; Stember 2006).

Other major issues associated with the use of GF commodities are as follows. (i) Unintended contamination of supposedly GF products that are on the market. Accidental contamination of intrinsically GF products could take place at any level from field to fork due to the ubiquitous nature of gluten or gluten-containing grains. Contamination and misbranding of products make it a real challenge to follow a GF diet. (ii) Strict adherence to a diet totally devoid of gluten-containing grains, or based on foods manufactured for celiac patients, may result in poor gut health of the consumer due to a negative effect on the gut microbiota. It has also been shown that this type of diet increases the risk of colon cancer in consumers, owing to the reduced content of dietary fiber and bioactive compounds (De Palma et al. 2009; Gil-Humanes et al. 2014 and references cited therein). (iii) Adaption to a GF diet may initially improve a patient's condition, but long-term adherence results in multiple deficiencies and changes in body mass index (BMI), which increase vulnerability to other disorders (Theethira et al. 2014). As most GF foods are made with starches or refined flours with low fiber content, celiac patients consume more energy in the form of fat than in the form of carbohydrate (Martin et al. 2013). Furthermore, it has been shown in a number of studies that when they are diagnosed, individuals with CD have lower BMI than the regional population, but their BMI increases on the transition to the GF diet, especially in those that adhere closely to it (Kabbani et al. 2012; Sonti and Green 2012).

7.3 Genetically-Engineered Reduced-Gluten or 'Celiac-Safe' Wheat

Recently, several research groups have taken a genetic engineering approach to develop 'celiac-safe' wheat genotypes by either eliminating or detoxifying gluten proteins. In an elimination approach, Becker and co-workers produced a series of transgenic lines where α-gliadin genes were down-regulated using RNA interference (RNAi). In these lines α-gliadins were reduced by over 60% compared to the control cultivar. Compensatory increases in albumins, globulins, other gliadins, and LMW subunits were also reported (Becker et al. 2006, 2012; Becker and Folck 2006; Wieser et al. 2006). Using a similar approach, silencing of the ω5-gliadins was later achieved by Altenbach and co-workers (Altenbach and Allen 2011; Altenbach et al. 2014). More extensive studies were reported by Barro and co-workers in a series of papers which have been reviewed by others (Rosella et al.

2014; Gilissen et al. 2014; Shewry and Tatham 2016; Ribeiro et al. 2018; Jouanin et al. 2018). To summarize, two series of lines were generated downregulating γ-gliadins by between 65% and 97% (Gil-Humanes et al. 2008; Piston et al. 2011) or downregulating all gliadins (α/β, γ and ω) by 60–88% and LMW subunits (Gil-Humanes et al. 2010, 2011). Testing these genotypes with intestinal T-cell clones derived from biopsy samples of CD patients showed almost complete suppression of disease-related T-cell epitopes (Gil-Humanes et al. 2010). More recently, the same researchers used the CRISPR/Cas9 gene editing technology to target conserved regions adjacent to the coding sequence of the 33-mer peptide in the α2-gliadin genes (Sánchez-León et al. 2018). Two single-guide RNAs, dubbed sgAlpha-1 and sgAlpha-2, induced mutations in specific gliadin genes, leading to an 85% reduction in immunoreactivity of mutant lines. This study showed that CRISPR/Cas9 technology can be used to precisely and efficiently reduce the amount of celiac-causing epitopes.

Taking a slightly different approach, the Rustgi and von Wettstein group aimed to silence the wheat *DEMETER* (DME) genes, which are master regulators of gluten protein accumulation (excluding HMW subunits). To suppress *DME* expression, two series of transgenic lines were produced, one with *DME*-specific hairpin RNA and the other with *DME*-specific artificial micro RNA (amiRNA). The former series of transgenic lines expressed a 938-nucleotide hairpin with a 185-nucleotide stem and a 568-nucleotide loop. Using this construct, a total of 118 candidate transformants were obtained, seven of which exhibited 45–76% reductions in the amount of immunogenic gluten proteins (Wen et al. 2012; Rustgi et al. 2014). In the latter series of transgenic lines, three different amiRNAs were expressed to avoid off-target editing. Two amiRNAs were designed to target the active site sequence and one to target the N-terminal sequence of the wheat *DME* gene. The selected amiRNA sequences were subsequently assembled on a rice osaMIR528 template using overlapping primers and cloned under the control of the wheat HMWg *1Dy* promoter. A total of 215 candidate transformants were obtained, 12 of which showed 54–88% reductions in their respective gluten protein contents (Brew-Appiah 2014; Rustgi et al. 2014). More recently, the same group used site-directed insertional mutagenesis to maximize the level of *DME* suppression and gluten elimination. The *DME*-specific transcription activator-like effector (TALE) repressor was introgressed into the wheat *Dre2* (*Derepressed for ribosomal protein S14 expression*) gene using the CRISPR/Cas9 technology. Dre2 is a protein that facilitates deposition of iron-sulfur (Fe-S) clusters in the DME apozyme. Double-stranded breaks introduced into the wheat *Dre2* homoeologues are repaired with the help of a donor construct carrying the *DME*-specific TALE repressor. This approach of simultaneously silencing the *DME* and *Dre2* genes is expected to disrupt *DME* activity at two-time points during transcription and post-translation respectively, to limit the accumulation of immunogenic gluten proteins (Rustgi, unpublished data).

In a gluten detoxification approach, the Rustgi group expressed 'glutenases' in wheat endosperm. This research was inspired by earlier research by the Khosla group, who demonstrated that glutamine-specific proteases from germinating barley grains in combination with bacterial prolyl endopeptidases (PE-Peps) from *Flavobacterium meningosepticum*, *Sphingomonas capsulata* or *Myxococcus xanthus* can completely detoxify the proteolytically resistant proline-glutamine rich peptides (Bethune and Khosla 2012). Later PE-Peps with similar cleavage characteristics (but diverse physicochemical properties) were identified from a number of archeal, fungal and eukaryotic species and their applicability to gluten detoxification was demonstrated (Scherf et al. 2018). Based on these studies, and parameters such as target specificity, substrate length, optimum pH, and site of action, a PE-Pep from *Flavobacterium meningosepticum* and a glutamine specific endoprotease from barley (EP-B2) were selected for expression in wheat endosperm (Osorio et al. 2012, 2019). Wheat transformants expressing these glutenases were obtained, and several exhibited significant reductions in the amounts of indigestible gluten peptides separated by SDS-PAGE gels and RP-HPLC (Fig. 5). The gluten detoxification approach has specific advantages. Firstly, some CD patients are sensitive to the HMW subunit peptides (Dewar et al. 2006). Therefore, wheat transformants which lack specific gliadins and/or LMW subunits are still unsuitable for such patients. Secondly, the combination of enzymes used does not digest the gluten proteins within the grain and therefore, does not affect the end-use quality. The glutamine-specific endoprotease used in this study is encoded as a proenzyme, where the propeptide serves as both an inhibitor and chaperone to facilitate spatio-temporal regulation of the proteolytic activity and proper folding of the proteases, respectively (Bethune et al. 2006; Cappetta et al. 2002; Schilling et al. 2009; Cambra et al. 2012). Both of these properties are of immense importance as they prevent degradation of the prolamins in the protein bodies within the grain or during processing (Osorio et al. 2019). In addition, because the PE-Pep has a strict preference for peptide substrates of ≤33 amino acids in size, it can only degrade peptides generated by the glutamine-specific endoprotease (Gass and Khosla 2007), therefore, permitting both of the enzymes to accumulate within the protein bodies without degrading the gluten proteins. Thirdly, this approach does not require consumers to take dietary supplements, which are under different stages of development, so it is expected to be more acceptable to the public.

7.4 Non-dietary Therapies

In parallel to the efforts to develop dietary therapies for the CD, extensive research has been dedicated to developing non-dietary therapies. These therapies can be largely classified as: (1) luminal therapies based on the detoxification of gluten proteins like enzyme therapy, probiotic therapy, flour/dough pretreatment, and gluten inactivation by polymeric binding; (2) intestinal barrier enhancing therapies, which focus on reducing the permeability of the intestinal epithelial barrier; and (3) immune targeted therapies, which target either CD specific pathways or inflammatory mediators common in gastrointestinal inflammation. These therapies can be

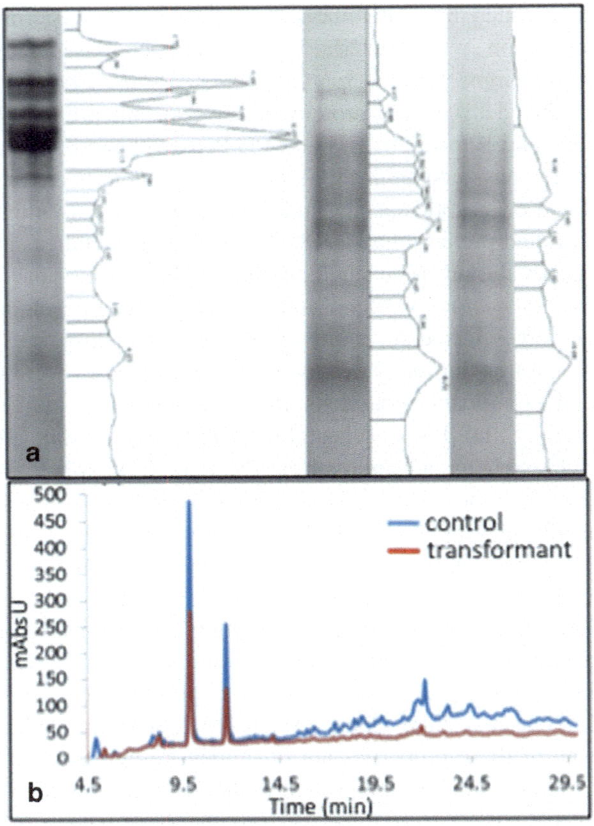

Fig. 5 Differences in the degradation patterns of gluten proteins derived from a transgenic wheat line expressing *Flavobacterium meningosepticum* prolyl endopeptidase (FME108-10) and its untransformed sibling (FME108-12) under simulated gastro-intestinal conditions with endoprotease B isoform 2 (EP-B2). (**a**) Gradient tricine gel and densitometric analyses of the digested gliadin fraction derived from FME108-12 with no EP-B2 (left lane and trace), FME108-12 with EP-B2 (middle lane and trace), and FME108-10 with EP-B2 (right lane and trace). (**b**) Reverse-phase high-performance liquid chromatography of gluten proteins derived from the T_2 grains of FME108-12 with EP-B2 (blue line) and FME108-10 with EP-B2 (red line) on a C18 column (modified after Osorio et al. 2019)

subdivided into tTG2 blockers, human leukocyte antigen (HLA) blockers, T-cell targeted therapies, alteration of inflammatory mediators, and vaccine therapy. An up-to-date list of these non-dietary therapies to treat CD which are in various stages of development, is presented in Table 3 and they have been extensively reviewed by Schuppan et al. (2009), Sollid and Khosla (2011), Osorio et al. (2012), Rashtak and Murray (2012), McCarville et al. (2015) and Ribeiro et al. (2018).

Table 3 A list of nondietary therapies for celiac disease

Therapeutic approach	Mode of action	Compound	Compound class	Company/University	Status	Reference
Anti-inflammatory agent	Glucocorticosteroids with low systemic bioavailability	Budesonide	Small molecule	Generic drug	Approved	Brar et al. (2007); Ciacci et al. (2009)
Intestinal permeability	Rho /Rho kinase inhibition	Fasudil	Small molecule	Generic drug	Approved	Sollid and Khosla (2011)
	RhoA inhibition	BA-210	Recombinant protein	Alseres, USA	Phase II	Sollid and Khosla (2011)
	Zonulin antagonist	AT-1001 (Larazotide)	Peptide	Alba, USA	Phase IIb	Paterson et al. (2007)
	Mitogenic compounds	R-spondin1	Recombinant protein	-	Phase III	Zhao et al. (2007)
Gluten detoxification	Glutamine endopeptidase from barley (EP-B2)	ALV001	Recombinant enzyme	Alvine, USA	-	Bethune et al. (2006)
	Sphingomonas capsulata	ALV002	Recombinant enzyme	Alvine, USA	-	Gass et al. (2007)
	Combination of EP-B2 and PEP from *Sphingomonas capsulata* (SC-PEP)	ALV003	Enzyme	Alvine, USA	Phase II	Siegel et al. (2012); Lähdeaho et al. (2014)
	PEP from *Aspergillus niger* (AN-PEP)	AN-PEP	Enzyme	DSM, Netherlands	Phase I + II	Stepniak et al. (2006); Mitea et al. (2008)
	Aspergillus niger aspergillopepsin (ASP) and *A. oryzae* dipeptidyl peptidase (DPPIV)	STAN1	Enzyme	HeimPal Children's Hospital, Hungary	Phase I + II	Ehren et al. (2009)

(continued)

Table 3 (continued)

Therapeutic approach	Mode of action	Compound	Compound class	Company/University	Status	Reference
	Combination of EP-B2 and PEP from *Flavobacterium meningosepticum* (FM-PEP)	-	Enzyme	Stanford University, USA	Discovery	Siegel et al. (2006)
	Combination of EP-B2 or PEP from *Myxococcus Xanthus* (MX-PEP)	-	Enzyme	Stanford University, USA	Discovery	Siegel et al. (2006)
	Engineered gliadin endopeptidase Kuma030	KumaMax	Enzyme	PVP Biologics, INC and Takeda Pharmaceutical Company Limited	Clinical trial	Wolf et al. (2015)
	Proteolytic components of the pitcher plant (*Nepenthes* spp.)	-	Enzyme	University of Calgary, Canada	Discovery	Lee et al. (2016)
	Peptidase from germinated cereals	-	Enzyme	Deutsche Forschungsanstalt für Lebensmittelchemie, Germany	Discovery	Gessendorfer et al. (2011)
	Probiotic preparation or flour/dough pretreatment	*Lactobacillus alimentarius*, *L. brevis*, *L. hilgardii*, *L. acidophilus*, *L. sanfranciscensis*, *Bifidobacterium longum*, *Streptococcus thermophilus*, VSL#3	Bacteria	-	Preclinical/ phase I clinical trial	Ribeiro et al. (2018); McCarville et al. (2015)

	Transamidation of gliadin	Microbial transglutaminase (mTG); mTG + L-lysine; mTG + L-lysine methyl ester; mTG + L-lysine ethyl ester; mTG + n-butylamine in reducing conditions	Enzyme + small molecule	-	Preclinical	Ribeiro et al. (2018)
	Gluten neutralizing cow's milk antibodies	-	Monoclonal antibody	-	Preclinical	Schuppan et al. (2009)
	Gluten-sequestering polymers	P(HEMA-co-SS) or BL-7010	Polymer resin	University of Montreal, Canada	Discovery	Pinier et al. (2009)
	Sequestering gliadin proteins	Ascorbyl palmitate (or in combination with zinc chloride)	Small molecule	-	Discovery	Engstrom et al. (2017)
Lymphocyte recruitment	CCR9 antagonist	Ccx282-B (Traficet-EN), CCX025	Small molecule	ChemoCentryx, USA	Phase II	Walters et al. (2010); Rashtak and Murray (2012)
	Anti α4β7/MAdCAM-1	Natalizumab	Monoclonal antibody	Tysabri	Preclinical	Berlin et al. (1993); Salmi and Jalkanen (1999); Di Sabatino et al. (2009); Ghosh et al. (2003)
Immune modulation	Parasite Infection	*Necator americanus*	Parasite	Princess Alexandra Hospital, Australia	Phase II	Croese et al. (2006)
	Peptide vaccination	Nexvax2	Peptide	Nexpep, Australia	Phase I	Keech et al. (2009)
	Gluten tolerisation	Genetically modified *Lactococus lactis* (Elafin)	Bacteria	ActoGeniX, Belgium	Discovery	Huibregtse et al. (2009)

(continued)

Table 3 (continued)

Therapeutic approach	Mode of action	Compound	Compound class	Company/University	Status	Reference
Systemic T-cell or inflammatory cytokine blockade	Anti-interleukin-15	AMG714	Monoclonal antibody	Amgen, USA	Phase II in RA	Sollid and Khosla (2011)
	Anti-interleukin-15	Hu-Mik-β-1	Monoclonal antibody	-	Phase I	Yokoyama et al. (2009); Waldmann et al. (2013)
	Anti-interleukin-15	Tofacitinib	Small molecule	-	Approved	Yokoyama et al. (2013)
	Anti TNF –α	Infliximab	Monoclonal antibody	-	Preclinical	Gillett et al. (2002)
	Anti-IFN-γ	Fontolizumab	Monoclonal antibody	PDL and BiogenIdec, USA	Phase II in IBD,	Reinisch et al. (2010)
	Anti-IFN-γ	HuZAF	-	Academic Medical Centre, the Netherlands	Phase II	Schuppan et al. (2009)
	Anti-Jak3	CP-690-550	-	-	Phase II	West (2009)
	Anti-CD52	Alemtuzumab	Monoclonal antibody	-	Phase II	Vivas et al. (2006)
	Anti-CD3	Visilizumab	Monoclonal antibody	Facet, USA	Phase II in UC	Sandborn et al. (2010)
	Anti-CD3	Teplizumab	Monoclonal antibody	MacroGenics, USA	Phase II in T1D	Herold et al. (2009)
	Anti-CD3	Otelixizumab	Monoclonal antibody	Tolerx,USA	Phase III in T1D	Sollid and Khosla (2011)
	Anti-CD20	Rituximab	Monoclonal antibody	BiogenIdec, USA	Approved	Edwards et al. (2004); Hauser et al. (2008); Mei et al. (2010)

	Anti-CD20	Tositumab	Monoclonal antibody	GlaxoSmithKline, USA	Approved	Sollid and Khosla (2011)
	Anti-CD20	Ibritumomab	Monoclonal antibody	Spectrum, USA	Approved	Sollid and Khosla (2011)
	Autologous bone marrow transplantation	-	-	-	Clinical trial	Al-toma et al. (2007); Schuppan et al. (2009)
	Mesenchymal stem cell transplantation (prochymal)	-	-	-	Phase 2	Schuppan et al. (2009)
Antigen presentation suppression	TG2 inhibitor	Dihydroisoxazoles	Small molecule	Stanford University, USA	Discovery	Watts et al. (2006)
	TG2 inhibitor	ZED-101	Small molecule	Zedira, Germany	Discovery	Sollid and Khosla (2011)
	TG2 inhibitor	KCC009	-	-	Preclinical	Rashtak and Murray (2012)
	TG2 inhibitor	L-682777	-	-	Preclinical	Rashtak and Murray (2012)
	TG2 inhibitor	Cinnamoyl triazoles	Small molecule	University of Montreal, Canada	Discovery	Sollid and Khosla (2011)
	HLA-DQ2 blocker	Dimeric analogue of gluten peptide	Peptide	Stanford University, USA & University of Oslo, Norway	Discovery	Xia et al. (2007)
	HLA-DQ2 blocker	Azidoproline analogue of gluten peptide	Peptide	Leiden University, the Netherlands	Discovery	Kapoerchan et al. (2008)

The table is based on Schuppan et al. (2009), Sollid and Khosla (2011), Rashtak and Murray (2012), McCarville et al. (2015) and Ribeiro et al. (2018)

8 Conclusion

The recent availability of the reference wheat genome sequence provides knowledge of the complete gene complement of bread wheat including *cis*-regulatory elements, which will facilitate analysis of the complex transcriptional regulation of the equally complex gene families encoding allergenic and antigenic proteins (International Wheat Genome Sequencing Consortium 2018; Ramírez-González et al. 2018). This knowledge is also expected to facilitate the identification of previously uncharacterized epitopes (Juhász et al. 2018) and the development of novel approaches to produce wheat genotypes safe for all consumers without compromising the organoleptic properties and the end-use quality. The major feat of developing allergen-free and antigen-free celiac-safe wheat could be achieved by a combination of technologies, including genome-editing, genotype and tissue-culture independent genetic transformation procedures, advances in biochemical and immunological detection procedures, and more sensitive and more accurate non-invasive phenotyping methods. All of these approaches are currently being developed and deployed. The challenge will be to bring them together.

Acknowledgements Rothamsted Research receives grant-aided support from the Biotechnology and Biological Sciences Research Council (BBSRC) of the UK and the work forms part of the Designing Future Wheat strategic programme (BBS/E/C/000I0250). Financial support by NIH grants 1R01 GM080749-01A2, Life Sciences Discovery Fund Grant 3143956-01 and Clemson Faculty Succeeds Grant 15-202-EQUIP-5701-430-1502211 to SR is also gratefully acknowledged. Maastricht University NUTRIM receives grants from the Dutch Topsector AgriFood (TKI 1601P01) and from public private partnerships for the 'Well on Wheat?' research consortium addressing the health aspects of wheat consumption and wheat and gluten avoidance (http://www.um-eatwell.nl/wow/).

References

Afify SM, Pali-Schöll I (2017) Adverse reactions to food: the female dominance – A secondary publication and update. World Allergy Organ J 10:43.

Agostini C, Decsi T, Fewtrell M, Goulet O, Kolacek S, Koletzko B, Fleischer Michaelsen K, Moreno L, Puntis J, Rigo J, Shamir R, Szajewska H, Turck D, van Goudoevero J (2008) Medical Position Paper. Complementary Feeding: A Commentary by the ESPGHAN. Journal of Pediatric Gastroenterology and Nutrition 46:99-110.

Al-toma A, Visser OJ, van Roessel HM, von Blomberg BM, Verbeek WH, Scholten PE, Ossenkoppele GJ, Huijgens PC, Mulder CJ (2007) Autologous hematopoietic stem cell transplantation in refractory celiac disease with aberrant T cells. Blood 109:2243–2249.

Altenbach SB, Allen PV (2011) Transformation of the US bread wheat "Butte 86" and silencing of omega-5 gliadin genes. GM Crops 2:66-73.

Altenbach SB, Tanaka CK, Seabourn BW (2014) Silencing of omega-5 gliadins in transgenic wheat eliminates a major source of environmental variability and improves dough mixing properties of flour. BMC Plant Biol 14:1.

Altenbach SB, Vensel WH, Dupont FM (2011) The spectrum of low molecular weight alpha-amylase/protease inhibitor genes expressed in the US bread wheat Butte 86. BMC Research Notes 4:242.

Anderson RP, Degano P, Godkin AJ, Jewell DP, Hill AV (2000) In vivo antigen challenge in celiac disease identifies a single transglutaminase modified peptide as the dominant A-gliadin T-cell epitope. Nat Med 6:337–342.

Anderson RP, van Heel DA, Tye-Din JA, Barnardo M, Salio M, Jewell DP, Hill AVS (2005) T cells in peripheral blood after gluten challenge in coeliac disease. Gut 54:1217–1223.

Arranz-Otaegui A, Carretero LG, Ramsey MN, Fuller DQ, Richter T (2018) Archaebotanical evidence reveals the origins of bread 14.400 years ago in northeastern Jordan. Proc Natl Acad Sci USA 115:7925-7930.

Ashraf M (2014) Stress-induced changes in wheat grain composition and quality. Crit Rev Food Sci Nutr 54:1576–1583.

Baccioglu A, Kalpaklıoglu F, Altan G (2017) Review of wheat dependent exercise induced anaphylaxis with two cases, and a new co-factor – myorelaxant. J Immunol Clin Res 4:1045.

Bai JC, Ciacci C (2017) World gastroenterology organisation global guidelines: celiac disease February 2017. J Clin Gastroenterol 51:755–768.

Barrett JS, Gibson PR (2012) Fermentable oligosaccharides, disaccharides, monosaccharides and polyols (FODMAPs) and nonallergic food intolerance: FODMAPs or food chemicals. Therap Adv Gastroenterol 5:261-268.

Battais F, Richard C, Jacquenet S, Denery-Papini S, Moneret-Vautrin DA (2008) Wheat grain allergies: an update on wheat allergens. Eur Ann Allergy Clin Immunol 40:67-76.

Becker D, Folck A (2006) Inhibierung der α-Gliadingeneexpression in hexaploiden Brotweizen. Getreidetechnologie 30:153-156.

Becker D, Folck A, Knies P, Lörz H, Wieser H (2006) Silencing the a-gliadins in hexaploid bread wheat. In: Lookhart LG, Ng WPK (eds), Gluten Proteins. AACC International, St Paul, MN, pp. 86-89.

Becker D, Wieser H, Koehler P, Folck A, Mühling KH, Zörb C (2012) Protein composition and techno-functional properties of transgenic wheat with reduced α-gliadin content obtained by RNA interference. J Appl Bot Food Qual 85:23-33.

Bellinghausen I, Weigmann B, Zevallos V, Maxeiner J, Reißig S, Waisman A, Schuppan D, Saloga J (2018) Wheat amylase-trypsin inhibitors exacerbate intestinal and airway allergic immune responses in humanized mice. J Allergy Clin Immunol https://doi.org/10.1016/j.jaci.2018.02.041

Berlin C, Berg EL, Briskin MJ, Andrew DP, Kilshaw PJ, Holzmann B, Weissman IL, Hamann A, Butcher EC (1993) Alpha 4 beta 7 integrin mediates lymphocyte binding to the mucosal vascular addressin MAdCAM-1. Cell 74:185–195.

Bethune MT, Khosla C (2012) Oral enzyme therapy for celiac sprue. Methods Enzymol 502:241-271.

Bethune MT, Strop P, Tang Y, Sollid LM, Khosla C (2006) Heterologous expression, purification, refolding, and structural-functional characterization of EP-B2, a self-activating barley cysteine endoprotease. Chem Biol 13:637–647.

Biesiekierski JR, Muir JG, Gibson PR (2013) Is gluten a cause of gastrointestinal symptoms in people without celiac disease? Curr Allergy Asthma Rep 13:631-638.

Biesiekierski JR, Rosella O, Rose R, Liels K, Barrett JS, Shepherd SJ, Gibson PR, Muir JG (2011). Quantification of fructans, galacto-oligosacharides and other short-chain carbohydrates in processed grains and cereals. J Hum Nutr Diet 24:154-176.

Brans R, Sauer I, Czaja K, Pfützner W, Merk HF (2012) Microarray-based detection of specific IgE against recombinant ω-5-gliadin in suspected wheat-dependent exercise-induced anaphylaxis. Eur J Dermatol 22:358-362.

Brar P, Lee S, Lewis S, Egbuna I, Bhagat G, Green PH (2007) Budesonide in the treatment of refractory coeliac disease. Am J Gastroenterol 102:2265–2269.

Brew-Appiah RAT (2014) Epigenetic and post-transcriptional elimination of celiac-causing wheat storage proteins. Ph.D. Dissertation, Washington State University, Pullman, pp. 187.

Bromilow S, Gethings LA, Buckley M, Bromley M, Shewry PR, Langridge JI, Clare Mills EN (2017) A curated gluten protein sequence database to support development of proteomics methods for determination of gluten in gluten-free foods. J Proteomics 163:67-75.

Brouns F, Delzenne N, Gibson G (2017) The dietary fibers – FODMAPs controversy. Cereal Foods World 62:98-103.
Brouns FJPH, van Buul VJ, Shewry PR (2013) Does wheat make us fat and sick? J Cereal Sci 58:209-215.
Bustamante MÁ, Fernández-Gil MP, Churruca I, Miranda J, Lasa A, Navarro V, Simón E (2017) Evolution of Gluten Content in Cereal-Based Gluten-Free Products: An Overview from 1998 to 2016. Nutrients 9:21.
Caio G, Volta U, Tovoli F, De Giorgio R (2014) Effect of gluten free diet on immune response to gliadin in patients with non-celiac gluten sensitivity. BMC Gastroenterol 14:26.
Cambra I, Hernández D, Diaz I, Martinez M (2012) Structural basis for specificity of propeptide-enzyme interaction in barley C1A cysteine peptidases. PLoS One 7:e37234.
Camerlengo F, Sestili F, Silvestri M, Colaprico G, Margiotta B, Ruggeri R, Lupi R, Masci S, Lafiandra D (2017) Production and molecular characterization of bread wheat lines with reduced amount of α-type gliadins. BMC Plant Biol 17:248.
Cappetta M, Roth I, Díaz A, Tort J, Roche L (2002) Role of the prosegment of Fasciola hepatica cathepsin L1 in folding of the catalytic domain. Biol Chem 383:1215-1221.
Carbonero P, Garcia-Olmedo F (1999) A multigene family of trypsin/α-amylase inhibitors from cereals. In: Shewry PR, Casey R (eds) Seed proteins. Kluwer Publishing, Surrey, U.K., pp. 617–634.
Carnevali A, Gianotti A, Benedett S, Tagliamonte MC, Primiterra M, Laghi L, Danesi F, Valli V, Ndaghijimana M, Capozzi F, Canestrari F, Bordoni A (2014) Role of Kamut® brand khorasan wheat in the counteraction of non-celiac wheat sensitivity and oxidative damage. Food Res Int 63:218–226.
Cataldo F, Montalto G (2007) Celiac disease in the developing countries: a new and challenging public health problem. World J Gastroenterol 13:2153-2159.
Catassi C (2015) Gluten Sensitivity. Ann Nutr Metab 67(suppl 2):16–26.
Catassi C, Alaedini A, Bojarski C, Bonaz B, Bouma G, Carroccio A, Castillejo G, De Magistris L, Dieterich W, Di Liberto D, Elli L, Fasano A, Hadjivassiliou M, Kurien M, Lionetti E, Mulder CJ, Rostami K, Sapone A, Scherf K, Schuppan D, Trott N, Volta U, Zevallos V, Zopf Y, Sanders DS (2017) The Overlapping Area of Non-Celiac Gluten Sensitivity (NCGS) and Wheat-Sensitive Irritable Bowel Syndrome (IBS): An Update. Nutrients 9:1268.
Catassi C, Fabiani E, Iacono G, D'Agate C, Francavilla R, Biagi F, Volta U, Accomando S, Picarelli A, De Vitis I, Pianelli G, Gesuita R, Carle F, Mandolesi A, Bearzi I, Fasano A (2007a) A prospective, double-blind, placebo-controlled trial to establish a safe gluten threshold for patients with celiac disease. Am J Clin Nutr 85:160-166.
Catassi C, Fasano A (2008) Celiac disease. In: Arendt EK, Dal Bello F (eds) Gluten-free cereal products and beverages. Elsevier Inc., pp. 1–26.
Catassi C, Gatti S, Lionetti E (2015) World perspective and celiac disease. Epidemiology Dig Dis 33:141–146.
Catassi C, Kryszak D, Louis-Jacques O, Duerksen DR, Hill I, Crowe SE, Brown AR, Procaccini NJ, Wonderly BA, Hartley P, Moreci J, Bennett N, Horvath K, Burk M, Fasano A (2007b) Detection of Celiac disease in primary care: a multicenter case-finding study in North America. Am J Gastroenterol 102:1454-1460.
Catassi C, Rossini M, Rätsch IM, Bearzi I, Santinelli A, Castagnani R, Pisani E, Coppa GV, Giorgi PL (1993) Dose dependent effects of protracted ingestion of small amounts of gliadin in coeliac disease children: a clinical and jejunal morphometric study. Gut 34:1515-1519.
Ciacci C, Ciclitira P, Hadjivassiliou M, Kaukinen K, Ludvigsson JF, McGough N, Sanders DS, Woodward J, Leonard JN, Swift GL (2015) The gluten-free diet and its current application in coeliac disease and dermatitis herpetiformis. United European Gastroenterol J 3:121–135.
Ciacci C, Maiuri L, Russo I, Tortora R, Bucci C, Cappello C, Santonicola A, Luciani A, Passananti V, Iovino P (2009) Efficacy of budesonide therapy in the early phase of treatment of adult coeliac disease patients with malabsorption: an in vivo / in vitro pilot study. Clin Exp Pharmacol Physiol 36:1170–1176.

Cianferoni A (2016) Wheat allergy: diagnosis and management. J Asthma Allergy 9:13-25.
Ciclitira PJ, Cerio R, Ellis HJ, Maxton D, Nelufer JM, Macartney JM (1985) Evaluation of a gliadin-containing gluten-free product in coeliac patients. Hum Nutr Clin Nutr 39:303-308.
Ciclitira PJ, Evans DJ, Fagg NL, Lennox ES, Dowling RH (1984) Clinical testing of gliadin fractions in coeliac patients. Clin Sci (Lond) 66:357-364.
Ciclitira PJ, Hunter JO, Lennox ES (1980a) Clinical testing of bread made from nullisomic-6A wheats in celiac patients. Lancet 2:234-236.
Ciclitira PJ, Hunter JO, Lennox ES (1980b) Clinical testing in celiac patients of bread made from wheats deficient in some α-gliadins. Clinical Sci 59:25.
Colomba MS, Gregorini A (2012) Are ancient durum wheats less toxic to celiac patients? A study of α-gliadin from Graziella Ra and Kamut. ScientificWorldJournal 2012:837416.
Comino I, Moreno M, Real A, Rodríguez-Herrera A, Barro F, Sousa C (2013) The gluten-free diet: Testing alternative cereals tolerated by celiac patients. Nutrients 5:4250-4268.
Cook F, Hughes N, Nibau C, Orman-Ligeza B, Schatlowski N, Uauy C, Trafford K (2018) Barley *lys3* mutants are unique amongst shrunken-endosperm mutants in having abnormally large embryos. J Cereal Sci 82:16-24.
Cordain L (1999) Cereal grains: humanity's double-edged sword. World Rev Nutr Diet 84:19–73.
Croese J, O'Neil J, Masson J, Cooke S, Melrose W, Pritchard D, Speare R (2006) A proof of concept study establishing Necator americanus in Crohn's patients and reservoir donors. Gut 55:136–137.
Cukrowska B, Sowińska A, Bierła JB, Czarnowska E, Rybak A, Grzybowska-Chlebowczyk U (2017) Intestinal epithelium, intraepithelial lymphocytes and the gut microbiota - Key players in the pathogenesis of celiac disease. World J Gastroenterol 23:7505–7518.
De Palma G, Nadal I, Carmen Collado M, Sanz Y, Collado M (2009) Effects of a gluten-free diet on gut microbiota and immune function in healthy adult human subjects. Br J Nutr 102:1154–1160.
Dewar DH, Amato M, Ellis HJ, Pollock EL, Gonzalez-Cinca N, Wieser H, Ciclitira PJ (2006) The toxicity of high molecular weight glutenin subunits of wheat to patients with coeliac disease. Eur J Gastroenterol Hepatol 18:483-491.
Di Sabatino A, Rovedatti L, Rosado MM, Carsetti R, Corazza GR, MacDonald TT (2009) Increased expression of mucosal addressin cell adhesion molecule 1 in the duodenum of patients with active celiac disease is associated with depletion of integrin alpha4beta7-positive T cells in blood. Hum Pathol 40:699–704.
DiGiacomo DV, Tennyson CA, Green PH, Demmer RT (2013) Prevalence of gluten-free diet adherence among individuals without celiac disease in the USA: results from the Continuous National Health and Nutrition Examination Survey 2009–2010. Scand J Gastroenterol 48:921-925.
Do AB, Khuda SE, Sharma GM (2018) Undeclared food allergens and gluten in commercial food products analyzed by ELISA. J AOAC Int 101:23-35.
Dupont FM, Vensel WH, Tanaka CK, Hurkman WJ, Altenbach SB (2011) Deciphering the complexities of the wheat flour proteome using quantitative two-dimensional electrophoresis, three proteases and tandem mass spectrometry Proteome Sci 9:10.
Edwards JC, Szczepanski L, Szechinski J, Filipowicz-Sosnowska A, Emery P, Close DR, Stevens RM, Shaw T (2004) Efficacy of B-celltargeted therapy with rituximab in patients with rheumatoid arthritis. N Engl J Med 350:2572–2581.
Ehren J, Moron B, Martin E, Bethune MT, Gray GM, Khosla C (2009) A food-grade enzyme preparation with modest gluten detoxification properties. PLoS ONE 4:e6313.
Ejderhamn J, Veesess B, Strandvik B (1988) The long term effect of continual ingestion of wheat starch-containing gluten-free products in celiac patients. In: Kumar PJ (ed) Coeliac disease: one hundred years. Leeds, United Kingdom, Leeds University Press, pp. 294–297.
Engstrom N, Saenz-Mendez P, Scheers J, Scheers N (2017) Towards Celiac-safe foods: Decreasing the affinity of transglutaminase 2 for gliadin by addition of ascorbyl palmitate and $ZnCl_2$ as detoxifiers. Sci Rep 7:77.

Escarnot E, Gofflot S, Sinnaeve G, Dubois B, Bertin P, Mingeot D (2018) Reactivity of gluten proteins from spelt and bread wheat accessions towards A1 and G12 antibodies in the framework of celiac disease. Food Chem 268:522–532.

Farage P, de Medeiros Nóbrega YK, Pratesi R, Gandolfi L, Assunção P, Zandonadi RP (2017) Gluten contamination in gluten-free bakery products: a risk for coeliac disease patients. Public Health Nutr 20:413-416.

Fasano A (2007) https://www.slideshare.net/saturni/gliadin-intestinal-permeability-and-celiac-disease-from-innate-immunity-to-autoimmunity (retried August 2018).

Fasano A (2012) Leaky gut and autoimmune diseases. Clinic Rev Allerg Immunol 42:71-78.

Frisoni M, Corazza GR, Lafiandra D, De Ambroggio E, Filipponi C, Bonvicini F, Borasio E, Porcheddu E, Gasbarrini G (1995) Wheat deficient in gliadins: promising tool for coeliac disease. Gut 36:375-378.

Fritschy F, Windemann H, Baumgarten E (1985) Bestimmung von Weizen gliadinen in Lebensmitteln mittels ELISA [Determination of wheat gliadins in foods by ELISA]. Z Lebensm Unters Forsch 181: 379-385.

Gass J, Bethune MT, Siegel M, Spencer A, Khosla C (2007) Combination enzyme therapy for gastric digestion of dietary gluten in patients with celiac sprue. Gastroenterology 133:472-480

Gass J, Khosla C (2007) Prolyl endopeptidases. Cell Mol Life Sci 64:345-355.

Gélinas P, Gagnon F (2018) Inhibitory activity towards human α-amylase in wheat flour and gluten. Food Sci Technol 53:467-474.

Gelinas P, McKinnon C (2016) Gluten weight in ancient and modern wheat and the reactivity of epitopes towards R5 and G12 monoclonal antibodies. Int J Food Sci Technol 51:1801–1810.

Gessendorfer B, Hartmann G, Wieser H, Koehler P (2011) Determination of celiac disease-specific peptidase activity of germinated cereals. Eu Food Res Technol 232:205–209.

Ghosh S, Goldin E, Gordon FH, Malchow HA, Rask-Madsen J, Rutgeerts P, Vyhnálek P, Zádorová Z, Palmer T, Donoghue S; Natalizumab Pan-European Study Group (2003) Natalizumab for active Crohn's disease. N Engl J Med 348:24–32.

Gianfrani C, Camarca A, Mazzarella G, Di Stasio L, Giardullo N, Ferranti P, Picariello G, Rotondi Aufiero V, Picascia S, Troncone R, Pogna N, Auricchio S, Mamone G (2015) Extensive in vitro gastrointestinal digestion markedly reduces the immune-toxicity of *Triticum monococcum* wheat: Implication for celiac disease. Mol Nutr Food Res 59:1844-1854.

Gibert A, Espadaler M, Angel Canela M, Sánchez A, Vaqué C, Rafecas M (2006) Consumption of gluten-free products: should the threshold value for trace amounts of gluten be at 20, 100 or 200 p.p.m.? Eur J Gastroenterol Hepatol 18:1187-1195.

Gil-Humanes J, Piston F, Altamirano-Fortoul R, Real A, Comino I, Sousa A, Rosell CM, Barro F (2014) Reduced-gliadin wheat bread: an alternative to the gluten-free diet for consumers suffering gluten-related pathologies. PLoS One 9:e90898.

Gil-Humanes J, Piston F, Hernando A, Alvarez JB, Shewry PR, Barro F (2008) Silencing of γ-gliadins by RNA interference (RNAi) in bread wheat. J Cereal Sci 48:565-568.

Gil-Humanes J, Pistón F, Shewry PR, Tosi P, Barro F (2011) Suppression of gliadins results in altered protein body morphology in wheat. J Exp Bot 62:4203-4213.

Gil-Humanes J, Piston F, Tollefsen S, Sollid LM, Barro F (2010) Effective shutdown in the expression of celiac disease-related wheat gliadin T-cell epitopes by RNA interference. Proc Natl Acad Sci USA 107:17023-17028.

Gilissen LJWJ, van der Meer IM, Smulders MJM (2014) Reducing the incidence of allergy and intolerance to cereals. J Cereal Sci 59:337-353.

Gillett HR, Arnott ID, McIntyre M, Campbell S, Dahele A, Priest M, Jackson R, Ghosh S (2002) Successful infliximab treatment for steroid-refractory celiac disease: a case report. Gastroenterology 122:800–805.

Godfrey D, Hawkesford MJ, Powers SJ, Millar S, Shewry PR (2010) Effects of crop nutrition on wheat grain composition and end use quality. J Agric Food Chem 58:3012–3021.

Goryunova SV, Salentijn EM, Chikida NN, Kochieva EZ, van der Meer IM, Gilissen LJ, Smulders MJ (2012) Expansion of the gamma-gliadin family in *Aegilops* and *Triticum*. BMC Evol Biol 12:215.

Gregorini A, Colomba M, Ellis HJ, Ciclitira PJ (2009) Immunogenicity characterization of two ancient wheat α-gliadin peptides related to coeliac disease. Nutrients 1:276-290.

Hadjivassiliou M, Sanders D, Grünewald RA, Woodroofe N, Boscolo S, Aeschlimann D (2010) Gluten sensitivity: from gut to brain. Lancet Neurol 9:318–330.

Hadjivassiliou M, Sanders DD, Aeschlimann DP (2015) Gluten-Related Disorders: Gluten Ataxia. Dig Dis 33:264–268.

Hajas L, Scherf KA, Török K, Bugyi Z, Schall E, Poms RE, Koehler P, Tömösközi S (2018) Variation in protein composition among wheat (*Triticum aestivum* L.) cultivars to identify cultivars suitable as reference material for wheat gluten analysis. Food Chem 267:387–394.

Hauser SL, Waubant E, Arnold DL, Vollmer T, Antel J, Fox RJ, Bar-Or A, Panzara M, Sarkar N, Agarwal S, Langer-Gould A, Smith CH; HERMES Trial Group (2008) B-cell depletion with rituximab in relapsing-remitting multiple sclerosis. N Engl J Med 358:676–688.

Henry AG, Brooks AS, Piperno DR (2011) Microfossils in calculus demonstrate consumption of plants and cooked foods in Neanderthal diets (Shanidar III, Iraq; Spy I and II, Belgium). Proc Natl Acad Sci USA 108:486–491.

Henry AG, Brooks AS, Piperno DR (2014) Plant foods and the dietary ecology of Neanderthals and early modern humans. J Human Evolution 69:44-54.

Hernando A, Mujico JR, Mena MC, Lombardía M, Méndez E (2008) Measurement of wheat gluten and barley hordeins in contaminated oats from Europe, the United States and Canada by Sandwich R5 ELISA. Eur J Gastroenterol Hepatol 20:545-554.

Herold KC, Gitelman S, Greenbaum C, Puck J, Hagopian W, Gottlieb P, Sayre P, Bianchine P, Wong E, Seyfert-Margolis V, Bourcier K, Bluestone JA; Immune Tolerance Network ITN007AI Study Group. (2009) Treatment of patients with new onset Type 1 diabetes with a single course of anti-CD3 mAb Teplizumab preserves insulin production for up to 5 years. Clin Immunol 132:166–173.

Hischenhuber C, Crevel R, Jarry B, Mäki M, Moneret-Vautrin DA, Romano A, Troncone R, Ward R (2006) Review article: safe amounts of gluten for patients with wheat allergy or coeliac disease. Aliment Pharmacol Ther 23:559-575.

Huebener S, Tanaka CK, Uhde M, Zone JJ, Vensel WH, Kasarda DD, Beams L, Briani'C, Green PH, Altenbach SB, Alaedini A (2015) Specific nongluten proteins of wheat are novel target antigens in celiac disease humoral response. J Proteome Res14:503–511.

Huibregtse IL, Marietta EV, Rashtak S, Koning F, Rottiers P, David CS, van Deventer SJ, Murray JA (2009) Induction of antigen specific tolerance by oral administration of *Lactococcus lactis* delivered immunodominant DQ8-restricted gliadin peptide in sensitized nonobese diabetic Abo DQ8 transgenic mice. J Immunol 183:2390–2396.

Inomata N (2009) Wheat allergy. Curr Opin Allergy Clin Immunol 9:238–243.

International Wheat Genome Sequencing Consortium (2018) Shifting the limits in wheat research and breeding using a fully annotated reference genome. Science 361:eaar7191.

Ivarsson A1, Myléus A, Norström F, van der Pals M, Rosén A, Högberg L, Danielsson L, Halvarsson B, Hammarroth S, Hernell O, Karlsson E, Stenhammar L, Webb C, Sandström O, Carlsson A (2013) Prevalence of childhood celiac disease and changes in infant feeding. Pediatrics 131:e687-e694.

Janssen FW, Hägele GH, de Baaij JA (1991) Gluten free products, the Dutch experience. In: Mearin ML, Mulder CJJ (eds) Coeliac Disease. Dordrech: Kluwer Academic, pp. 95–100.

Jenkins JA, Griffiths-Jones S, Shewry PR, Breiteneder H, Mills ENC (2005) Structural relatedness of plant food allergens with specific reference to cross-reactive allergens: an in silico analysis. J Allergy Clin Immunol 115:163–170.

Jouanin A, Gilissen LJWJ, Boyd LA, Cockram J, Leigh FJ, Wallington EJ, van den Broeck HC, van der Meer IM, Schaart JG, Visser RGF, Smulders MJM (2018) Food processing and breeding strategies for coeliac-safe and healthy wheat products. Food Res Intl 10:11-21.

Juhász A, Belova T, Florides CG, Maulis C, Fischer I, Gell G, Birinyi Z, Ong J, Keeble-Gagnère G, Maharajan A, Ma W, Gibson P, Jia J, Lang D, Mayer KFX, Spannagl M, International Wheat Genome Sequencing Consortium, Tye-Din JA, Appels R, Olsen O-A (2018) Genome mapping of seed-borne allergens and immunoresponsive proteins in wheat. Sci Adv 4:eaar8602.

Juhász A, Haraszi R, Maulis C (2015) ProPepper: a curated database for identification and analysis of peptide and immune-responsive epitope composition of cereal grain protein families. Database 2015:1–16.

Junker Y, Zeissig S, Kim SJ, Barisani D, Wieser H, Leffler DA, Zevallos V, Libermann TA, Dillon S, Freitag TL, Kelly CP, Schuppan D (2012) Wheat amylase trypsin inhibitors drive intestinal inflammation via activation of toll-like receptor 4. J Exp Med 209:2395-2408.

Kabbani TA, Goldberg A, Kelly CP, Pallav K, Tariq S, Peer A, Hansen J, Dennis M, Leffler DA (2012) Body mass index and the risk of obesity in coeliac disease treated with the gluten-free diet. Aliment Pharmacol Ther 35:723-729.

Kapoerchan VV, Wiesner M, Overhand M, van der Marel GA, Koning F, Overkleeft HS (2008) Design of azidoproline containing gluten peptides to suppress CD4+ T-cell responses associated with coeliac disease. Bioorg Med Chem 16:2053–2062.

Kasarda DD (2013) Can an increase in celiac disease be attributed to an increase in the gluten content of wheat as a consequence of wheat breeding? J Agric Food Chem 61:1155-1159.

Kasarda DD, Adalsteins E, Lew EJ-L, Lazo GR, Altenbach SB (2013) Farinin:characterisation of a novel wheat endosperm protein belonging to the prolamin superfamily J Agric Food Chem 61:2407-2417.

Kaukinen K, Collin P, Holm K, Rantala I, Vuolteenaho N, Reunala T, Mäki M (1999) Wheat starch-containing gluten-free flour products in the treatment of coeliac disease and dermatitis herpetiformis. A long-term follow-up study. Scand J Gastroenterol 34:163-169.

Keech CL, Dromey J, Chen Z, Anderson RP, McCluskey J (2009) Immune tolerance induced by peptide immunotherapy in an HLA-DQ2-dependent mouse model of gluten immunity. Gastroenterology 136:A355.

Kneen E, Sandstedt RM (1946) Distribution and general properties of an amylase inhibitor in cereals. Arch Biochem 9:235–249.

Koning F (2012) Celiac disease: quantity matters. Semin Immunopathol 34:541–549.

Kotaniemi-Syrjänen A, Palosuo K, Jartti T, Kuitunen M, Pelkonen AS, Mäkelä MJ (2010) The prognosis of wheat hypersensitivity in children. Pediatr Allergy Immunol 21(2 Pt 2):e421-e428.

Lähdeaho ML, Kaukinen K, Laurila K, Vuotikka P, Koivurova OP, Kärjä-Lahdensuu T, Marcantonio A, Adelman DC, Mäki M (2014) Glutenase ALV003 attenuates gluten-induced mucosal injury in patients with celiac disease. Gastroenterology 146:1649-1658.

Lebwohl B, Sanders DS, Green PHR (2018) Coeliac disease. Lancet 391:70-81.

Lee HJ, Anderson Z, Ryu D (2014) Gluten Contamination in Foods Labeled as 'Gluten Free' in the United States. J Food Prot 77:1830–1833.

Lee L, Zhang Y, Ozar B, Sensen CW, Schriemer DC (2016) Carnivorous Nutrition in Pitcher Plants (*Nepenthes* spp.) via an Unusual Complement of Endogenous Enzymes. J Proteome Res 15:3108-3117.

Leonard MM, Sapone A, Catassi C, Fasano A (2017) Celiac disease and nonceliac gluten sensitivity: A review. JAMA 318:647-656.

Lionetti E, Gatti S, Pulvirenti A, Catassi C (2015) Celiac disease from a global perspective. Best Pract Res Clin Gastroenterol 29:365-379.

Ludvigsson JF, Fasano A (2012) Timing of introduction of gluten and celiac disease risk. Ann Nutr Metab 60 Suppl 2:22-29.

Ludvigsson JF, Leffler DA, Bai JC, Biagi F, Fasano A, Green PH, Hadjivassiliou M, Kaukinen K, Kelly CP, Leonard JN, Lundin KE, Murray JA, Sanders DS, Walker MM, Zingone F, Ciacci C (2013) The Oslo definitions for coeliac disease and related terms. Gut 62:43-52.

Mamone G, Picariello G, Addeo F, Ferranti P (2011) Proteomic analysis in allergy and intolerance to wheat products. Expert Rev Proteomics 8:95-115.

Marion D, Douliez J-P, Gautier M-F, Elmorjani K (2004) Plant lipid transfer proteins: relationships between allergenicity and structural, biological and technological properties. In: Mills ENC, Shewry PR (eds), Plant Food Allergens, Oxford: Blackwell Publishing, pp. 57-69.

Martin J, Geisel T, Maresch C, Krieger K, Stein J (2013) Inadequate nutrient ixntake in patients with celiac disease: results from a German dietary survey. Digestion 87:240-246.

Matsuda T, Nakase M, Alvarez AM, Izumi H, Kato T, Tada Y (2006) Rice-seed allergenic proteins and hypoallergenic rice. In: Mine Y, Shahidi F (eds) Nutraceutical proteins and peptides in health and disease, CRC Press, pp. 493–511.

McCarville JL, Caminero A, Verdu EF (2015) Celiac treatments, adjuvant therapies and alternatives to the gluten-free diet. In: Arranz E, Fernández-Bañares F, Rosell CM, Rodrigo L, Peña AS (eds) Advances in the understanding of gluten related pathology and the Evolution of Gluten-Free Foods. OmniaScience, Barcelona, Spain, pp. 223-253.

Mei HE, Frölich D, Giesecke C, Loddenkemper C, Reiter K, Schmidt S, Feist E, Daridon C, Tony HP, Radbruch A, Dörner T (2010) Steady state generation of mucosal IgA+ plasmablasts is not abrogated by B cell depletion therapy with rituximab. Blood 116:5181–5190.

Metakovsky EV, Davidov SD, Chernakov VM, Upelniek VP (1993) Gliadin allele identification in common wheat. III. Frequency of occurrence and appearance of spontaneous mutations at the gliadin-coding loci. J Genet Breed 47:221-236.

Miranda J, Simón E (2017) Gluten Content Change Over the Two Last Decades. In: E. Simón et al. (eds) Nutritional and Analytical Approaches of Gluten-Free Diet in Celiac Disease, SpringerBriefs in Food, Health, and Nutrition, Springer, pp. 47-57.

Mitea C, Havenaar R, Drijfhout JW, Edens L, Dekking L, Koning F (2008) Efficient degradation of gluten by a prolyl endoprotease in a gastrointestinal model: Implications for coeliac disease. Gut 57:25–32.

Mitea C, Salentijn EM, van Veelen P, Goryunova SV, van der Meer IM, van den Broeck HC, Mujico JR, Montserrat V, Gilissen LJ, Drijfhout JW, Dekking L, Koning F, Smulders MJ (2010) A universal approach to eliminate antigenic properties of alpha-gliadin peptides in celiac disease. PloS One 5:e15637.

Moehs CP, Austil WJ, Holm A, Large TAG, Loeffler D, Mullenberg J, Schnable PS, Skinner W, van Boxtel J, Wu L, McGuire C (2018) Development of reduced gluten wheat enabled by determination of the genetic basis of the *lys3a* low hordein barley mutant. bioRxiv https://doi.org/10.1101/354548

Molberg Ø, Uhlen AK, Jensen T, Flæte NS, Fleckenstein B, Arentz-Hansen H, Raki M, Lundin KEA, Sollid LM (2005) Mapping of gluten T-cell epitopes in the bread wheat ancestors: implications for celiac disease. Gasteroenterology 128:393-401.

Morris CF (2002) Puroindolines: The molecular genetic basis of wheat grain hardness. Plant Mol Biol 48:633-647.

Munck L (1992) The case of high-lysine barley breeding. In: Shewry PR (ed), Barley: Genetics, Biochemistry, Molecular Biology and Biotechnology. CAB International, Wallingford Oxon, pp. 573-601.

Nadal I, Donat E, Ribes-Koninckx C, Calabuig M, Sanz Y (2007) Imbalance in the composition of the duodenal microbiota of children with coeliac disease. J Med Microbiol 56:1669–1674.

Nambu M (2006) Rice Allergy. Pediatrics 117:2331.

Nistal E, Caminero A, Herrán AR, Pérez-Andres J, Vivas S, Ruiz de Morales JM, Sáenz de Miera LE, Casqueiro J (2016) Study of duodenal bacterial communities by 16S rRNA gene analysis in adults with active celiac disease vs non-celiac disease controls. J Appl Microbiol 120:1691–1700.

Nistal E, Caminero A, Vivas S, Ruiz de Morales JM, Sáenz de Miera LE, Rodríguez-Aparicio LB, Casqueiro J (2012) Differences in faecal bacteria populations and faecal bacteria metabolism in healthy adults and celiac disease patients. Biochimie 94:1724–1729.

Olexová L, Dovičovičová L, Švec M, Siekel P, Kuchta T (2006) Detection of gluten-containing cereals in flours and "gluten-free" bakery products by polymerase chain reaction. Food Control 17:234-237.

Olivares M, Walker AW, Capilla A, Benítez-Páez A, Palau F, Parkhill J, Castillejo G, Sanz Y (2018) Gut microbiota trajectory in early life may predict development of celiac disease. Microbiome 6:36.

Ortiz-Sánchez JP, Cabrera-Chávez F, de la Barca AM (2013) Maize prolamins could induce a gluten-like cellular immune response in some celiac disease patients. Nutrients 5:4174-4183.

Osborne TB (1924) The vegetable proteins, 2nd edition. Longmans Green & Co, London, UK.

Osorio C, Wen N, Gemini R, Zemetra R, von Wettstein D, Rustgi S (2012) Targeted modification of wheat grain protein to reduce the content of celiac causing epitopes. Funct Integr Genomics 12:417-438.

Osorio CE, Wen N, Mejias JH, Liu B, Reinbothe S, von Wettstein D, Rustgi S (2019) Development of wheat genotypes expressing a glutamine-specific endoprotease from barley and a prolyl endopeptidase from *Flavobacterium meningosepticum* or *Pyrococcus furiosus* as a potential remedy to celiac disease. Funct Integr Genomics 19:123–136.

Panda R, Taylor SL, Goodman RE (2010) Development of a Sandwich Enzyme-Linked Immunosorbent Assay (ELISA) for Detection of Buckwheat Residues in Food. J Food Sci 75:T110-T117.

Pasha I, Saeed F, Tauseef Sultan M, Batool R, Aziz M, Ahmed W (2016) Wheat allergy and intolerance; Recent updates and perspectives. Crit Rev Food Sci Nutr 56:13–24.

Pastorello EA, Farioli L, Conti A, Pravettoni V, Bonomi S, Iametti S, Fortunato D, Scibilia J, Bindslev-Jensen C, Ballmer-Weber B, Robino AM, Ortolani C (2007) Wheat IgE-mediated food allergy in European patients: alpha-amylase inhibitors, lipid transfer proteins and low-molecular-weight glutenins. Allergenic molecules recognized by double-blind, placebo-controlled food challenge. International archives of allergy and immunology 144:10–22.

Paterson BM, Lammers KM, Arrieta MC, Fasano A, Meddings JB (2007) The safety, tolerance, pharmacokinetic and pharmacodynamic effects of single doses of AT-1001 in coeliac disease subjects: a proof of concept study. Aliment Pharmacol Ther 26:757–766.

Peräaho M, Kaukinen K, Paasikivi K, Sievänen H, Lohiniemi S, Mäki M, Collin P (2003) Wheat-starch-based gluten-free products in the treatment of newly detected coeliac disease: prospective and randomized study. Aliment Pharmacol Ther 17:587–594.

Perez-Gregorio MR, Días R, Mateus N, de Freitas V (2018) Identification and characterization of proteolytically resistant gluten-derived peptides Food Funct 9:1726–1735.

Pinier M, Verdu EF, Nasser-Eddine M, David CS, Vézina A, Rivard N, Leroux JC (2009) Polymeric binders suppress gliadin-induced toxicity in the intestinal epithelium. Gastroenterology 136:288–298.

Pinto-Sánchez MI, Verdu EF, Liu E, Bercik P, Green PH, Murray JA, Guandalini S, Moayyedi P (2016) Gluten introduction to infant feeding and risk of celiac disease: Systematic review and meta-analysis. J Pediatr 168:132–43.

Piston F, Gil-Humanes J, Rodríguez-Quijano M, Barro F (2011) Down-regulating γ-gliadins in bread wheat leads to non-specific increases in other gluten proteins and has No major effect on dough gluten strength. PLoS One 6:e24754.

Pizzuti D, Buda A, D'Odorico A, D'Incà R, Chiarelli S, Curioni A, Martines D (2006) Lack of intestinal mucosal toxicity of *Triticum monococcum* in coeliac disease patients. Scand J Gastroenterol 41:1305–1311.

Pogna NE, Monari AM, Cacciatori P, Redaelli R, Ng PKW (1998) Development and characterization of bread wheat lines lacking chromosome 1B-, 1D-, 6A- and 6D-encoded prolamins. In: Proc. IXth Intern. Wheat Genetics Symposium, Saskatoon, Saskatchewan, Canada, pp. 265-268.

Pontieri P, Mamone G, De Caro S, Tuinstra MR, Roemer E, Okot J, De Vita P, Ficco DB, Alifano P, Pignone D, Massardo DR, Del Giudice L (2013) Sorghum, a healthy and gluten-free food for celiac patients as demonstrated by genome, biochemical, and immunochemical analyses. J Agric Food Chem 61:2565-2571.

Prandi B, Faccini A, Tedeschi T, Galaverna G, Sforza S (2013) LC/MS analysis of proteolytic peptides in wheat extracts for determining the content of the allergen amylase/trypsin inhibitor CM3: Influence of growing area and variety. Food Chem140:141-146.

Prandi B, Tedeschi T, Folloni S, Galaverna G, Sforza S (2017) Peptides from gluten digestion: A comparison between old and modern wheat varieties. Food Res Int 91:92–102.

Ramírez-González RH, Borrill P, Lang D, Harrington SA, Brinton J, Venturini L, Davey M, Jacobs J, van Ex F, Pasha A, Khedikar Y, Robinson SJ, Cory AT, Florio T, Concia L, Juery C, Schoonbeek H, Steuernagel B, Xiang D, Ridout CJ, Chalhoub B, Mayer KFX, Benhamed M, Latrasse D, Bendahmane A; International Wheat Genome Sequencing Consortium, Wulff

BBH, Appels R, Tiwari V, Datla R, Choulet F, Pozniak CJ, Provart NJ, Sharpe AG, Paux E, Spannagl M, Bräutigam A, Uauy C (2018) The transcriptional landscape of polyploid wheat. Science 361:eaar6089.

Rashtak S, Murray JA (2012) Review article: coeliac disease, new approaches to therapy. Aliment Pharmacol Ther 35:768-781.

Redaelli R, Metakovsky EV, Davydov SD, Pogna NE (1994) Two-dimensional mapping of gliadins using biotypes and null mutants of common wheat cultivar Saratovskaya 29. Hereditas 121:131-137.

Reinisch W, de Villiers W, Bene L, Simon L, Rácz I, Katz S, Altorjay I, Feagan B, Riff D, Bernstein CN, Hommes D, Rutgeerts P, Cortot A, Gaspari M, Cheng M, Pearce T, Sands BE (2010) Fontolizumab inmoderate to severe Crohn's disease: a phase 2, randomized, double-blind, placebo-controlled, multiple-dose study. Inflamm Bowel Dis 16:233–242.

Ribeiro M, Nunes FM, Rodriguez-Quijano M, Carrillo JM, Branlard G, Igrejas G (2018) Next-generation therapies for celiac disease: The gluten-targeted approaches. Trends Food Sci Technol 75:56-71.

Ribeiro M, Rodriguez-Quijano M, Nunes FM, Carrillo JM, Branlard G, Igrejas G (2016) New insights into wheat toxicity: Breeding did not seem to contribute to a prevalence of potential celiac disease's immunostimulatory epitopes. Food Chem 213:8–18.

Rosella CM, Barro F, Sousa C, Mena MC (2014) Cereals for developing gluten-free products and analytical tools for gluten detection. J Cereal Sci 59:354-364.

Rubio-Tapia A, Ludvigsson JF, Brantner TL, Murray JA, Everhart JE (2012) The prevalence of celiac disease in the United States. Am J Gastroenterol 107:1538-1544.

Rustgi S, Wen N, Osorio C, Brew-Appiah RAT, Wen S, Gemini R, Mejias JH, Ankrah N, Moehs CP, von Wettstein D (2014) Natural dietary therapies for the 'gluten syndrome', Scientia Danica, Series B, Biologica 3:1-87.

Salcedo G, Sanchez-Monge R, Garcia-Casado G, Armentia A, Gomez L, Barber D (2004) The cereal α-amylase/trypsin inhibitor family associated with bakers' asthma and food allergy. In Mills ENC, Shewry PR (eds) Plant food allergens. Blackwell Science, Oxford U.K., pp. 70-86.

Salentijn EM, Esselink DG, Goryunova SV, van der Meer IM, Gilissen LJ, Smulders MJ (2013) Quantitative and qualitative differences in celiac disease epitopes among durum wheat varieties identified through deep RNA-amplicon sequencing. BMC Genomics 14:905.

Salentijn EM, Goryunova SV, Bas N, van der Meer IM, van den Broeck HC, Bastien T, Gilissen LJ, Smulders MJ (2009) Tetraploid and hexaploid wheat varieties reveal large differences in expression of alpha-gliadins from homoeologous Gli-2 loci. BMC Genomics 10:48.

Salmi M, Jalkanen S (1999) Molecules controlling lymphocyte migration to the gut. Gut 45:148–153.

Sanchez de la Hoz P, Castagnaro A, Carbonero P (1994) Sharp divergence between wheat and barley at loci encoding novel members of the trypsin/alpha-amylase inhibitors family. Plant Mol Biol 26:1231-1236.

Sánchez-León S, Gil-Humanes J, Ozuna CV, Giménez MJ, Sousa C, Voytas DF, Barro F (2018) Low-gluten, nontransgenic wheat engineered with CRISPR/Cas9. Plant Biotechnol J 16:902-910.

Sandborn WJ, Colombel JF, Frankel M, Hommes D, Lowder JN, Mayer L, Plevy S, Stokkers P, Travis S, Van Assche G, Baumgart DC, Targan SR (2010) Anti-CD3 antibody visilizumab is not effective in patients with intravenous corticosteroid-refractory ulcerative colitis. Gut 59:1485–1492.

Sapone A, Bai JC, Ciacci C, Dolinsek J, Green PH, Hadjivassiliou M, Kaukinen K, Rostami K, Sanders DS, Schumann M, Ullrich R, Villalta D, Volta U, Catassi C, Fasano A (2012) Spectrum of gluten-related disorders: consensus on new nomenclature and classification. BMC Med 10:13.

Scherf KA, Brockow K, Biedermann T, Koehler P, Wieser H (2016) Wheat-dependent exercise-induced anaphylaxis. Clin Exp Allergy 46:10-20.

Scherf KA, Wieser H, Koehler P (2018) Novel approaches for enzymatic gluten degradation to create high-quality gluten-free products. Food Res Int 110:62-72.

Schilling K, Körner A, Sehmisch S, Kreusch A, Kleint R, Benedix Y, Schlabrakowski A, Wiederanders B (2009) Selectivity of propeptide-enzyme interaction in cathepsin L-like cysteine proteases. Biol Chem 390:167-174.

Schuppan D, Junker Y, Barisani D (2009) Celiac disease: from pathogenesis to novel therapies. Gastroenterology 137:1912-1933.

Schuppan D, Pickert G, Ashfaq-Khan M, Zevallos V (2015) Non-celiac wheat sensitivity: Differential diagnosis, triggers and implications. Best Pract Res Clin Gastroenterol 29:469-476.

Seilmeier W, Belitz H-D, Wieser H (1991) Separation and quantitative determination of high-molecular-weight subunits of glutenin from different wheat varieties and genetic variants of the variety Sicco. Z Lebensm Unters Forsch 192:124-129.

Shan L, Molberg Ø, Parrot I, Hausch F, Filiz F, Gray GM, Sollid LM, Khosla C (2002) Structural basis for gluten intolerance in celiac sprue. Science 297:2275–2279.

Shan L, Qiao SW, Arentz-Hansen H, Molberg Ø, Gray GM, Sollid LM, Khosla C (2005) Identification and analysis of multivalent proteolytically resistant peptides from gluten: Implications for celiac sprue. J Proteome Res 4:1732–1741.

Shewry PR (2018) Do ancient types of wheat have health benefits compared with modern bread wheat? J Cereal Sci 79:469-476.

Shewry PR, D'Ovidio R, Lafiandra D, Jenkins JA, Mills ENC, Bekes F (2009) Wheat grain proteins. In: Khan K, Shewry PR (eds)Wheat Chemistry and technology. AACC International Inc., St. Paul, MN, pp. 223-298.

Shewry PR, Pellny TK, Lovegrove A (2016) Is modern wheat bad for health? Nat Plant 2:1-3.

Shewry PR, Tatham AS (2016) Improving wheat to remove coeliac epitopes but retain functionality. J Cereal Sci 67:12-21.

Shewry PR, Tatham AS, Forde J, Kreis M, Miflin BJ (1986) The classification and nomenclature of wheat gluten proteins: A reassessment. J Cereal Sci 4:97-106.

Shewry PR, Tatham AS, Halford NG (1999) The prolamins of the Triticeae. In: Shewry PR, Casey R (eds) Seed Proteins, Kluwer Academic Publishers, Dordrecht, The Netherlands, pp. 37–84.

Sicherer SH (2001) Clinical implications of cross-reactive food allergens. J Allergy Clin Immunol 108:881-890.

Siegel M, Bethune MT, Gass J, Ehren J, Xia J, Johannsen A, Stuge TB, Gray GM, Lee PP, Khosla C (2006) Rational design of combination enzyme therapy for coeliac sprue. Chem Biol 13:649–658.

Siegel M, Garber ME, Spencer AG, Botwick W, Kumar P, Williams RN, Kozuka K, Shreeniwas R, Pratha V, Adelman DC (2012) Safety, tolerability, and activity of ALV003: results from two phase I single, escalating-dose clinical trials. Dig Dis Sci 57:440-450.

Skodje GI, Sarna VK, Minelle IH, Rolfsen KL, Muir JG, Gibson PR, Veierød MB, Henriksen C, Lundin KEA (2018) Fructan, rather than gluten, induces symptoms in patients with self-reported non-celiac gluten sensitivity. Gastroenterology 154:529-539.

Skylas DJ, Mackintosh JA, Cordwell SJ, Basseal DJ, Walsh BJ, Harry J, Blumenthal C, Copeland L, Wrigley CW, Rathmell W (2000) Proteome approach to the characterisation of protein composition in the developing and mature wheat-grain endosperm. J Cereal Sci 32:169–188.

Sofi F, Ghiselli L, Cesari F, Gori AM, Mannini L, Casini A, Vazzana C, Vecchio V, Gensini GF, Abbate R, Benedettelli S (2010) Effects of short-term consumption of bread obtained by an old Italian grain variety on lipid, inflammatory, and haemorheological variables: an intervention study. J Med Food 13: 1–6.

Sofi F, Whittaker A, Gori AM, Cesari F, Surrenti E, Abbate R, Gensini GF, Benedettelli S, Casini A (2014) Effect of *Triticum turgidum* subsp. *turanicum* wheat on irritable bowel syndrome: a double-blinded randomised dietary intervention trial. Br J Nutr 111:1992–1999.

Sollid LM, Khosla C (2011) Novel therapies for coeliac disease. J Int Med 269:604-613.

Sollid LM, Qiao SW, Anderson RP, Gianfrani C, Koning F (2012). Nomenclature and listing of celiac disease relevant gluten T-cell epitopes restricted by HLA-DQ molecules. Immunogenetics 64:455-460.

Sonti R, Green PH (2012) Celiac disease: obesity in celiac disease. Nat Rev Gastroenterol Hepatol 9:247-248.

Spaenij-Dekking L1, Kooy-Winkelaar Y, van Veelen P, Drijfhout JW, Jonker H, van Soest L, Smulders MJ, Bosch D, Gilissen LJ, Koning F (2005) Natural variation in toxicity of wheat: potential for selection of nontoxic varieties for celiac disease patients. Gastroenterology 129:797–806.

Stein J, Schuppan D (2014) Coeliac disease-new pathophysiological findings and their implications for therapy. Viszeralmedizin 30:156-165.

Stember RH (2006) Buckwheat allergy. Allergy Asthma Proc 27:393-395.

Stepniak D, Spaenij-Dekking L, Mitea C, Moester M, de Ru A, Baak-Pablo R, van Veelen P, Edens L, Koning F (2006) Highly efficient gluten degradation with a newly identified prolyl endoprotease: Implications for celiac disease. Am J Physiol Gastrointest Liver Physiol 291:G621–G629.

Šuligoj T, Gregorini A, Colomba M, Ellis HJ, Ciclitira PJ (2013) Evaluation of the safety of ancient strains of wheat in coeliac disease reveals heterogeneous small intestinal T cell responses suggestive of coeliac toxicity. Clin Nutr 32:1043-1049.

Tallberg A (1981a) Protein and lysine content in high-lysine double-recessives of barley. I. Combinations between mutant 1508 and a Hiproly back-cross. Hereditas 94:253-260.

Tallberg A (1981b) Protein and lysine content in high-lysine double-recessives of barley. II. Combinations between mutant 7 and a Hiproly back-cross. Hereditas 94:261-268.

Tallberg A (1982) Characterization of high-lysine barley genotypes. Hereditas 96:229-245

Tatham AS, Shewry PR (2008) Allergens to wheat and related cereals. Clin Exp Allergy 38:1712-1726.

Theethira TG, Dennis M, Leffler DA (2014) Nutritional consequences of celiac disease and the gluten-free diet. Expert Rev Gastroenterol Hepatol 8:123-129.

Thompson T, Lee AR, Grace T (2010) Gluten contamination of grains, seeds, and flours in the United States: a pilot study. J Am Diet Assoc 110:937-940.

Trcka J, Schäd SG, Scheurer S, Conti A, Vieths S, Gross G, Trautmann A (2012) Rice-induced anaphylaxis: IgE-mediated allergy against a 56-kDa glycoprotein. Int Arch Allergy Immunol 158:9-17.

Turner AS, Bradburne RP, Fish L, Snape JW (2004) New quantitative trait loci influencing grain texture and protein content in bread wheat. J Cereal Sci 40:51-60.

Tye-Din JA, Stewart JA, Dromey JA, Beissbarth T, van Heel DA, Tatham A, Henderson K, Mannering SI, Gianfrani C, Jewell DP, Hill AV, McCluskey J, Rossjohn J, Anderson RP (2010) Comprehensive, quantitative mapping of T cell epitopes in gluten in celiac disease. Sci Transl Med 2:41–51.

Uhde M, Ajamian M, Caio G, De Giorgio R, Indart A, Green PH, Verna EC, Volta U, Alaedini A (2016) Intestinal cell damage and systemic immune activation in individuals reporting sensitivity to wheat in the absence of coeliac disease. Gut 65:1930-1937.

van den Broeck H, Hongbing C, Lacaze X, Dusautoir J-C, Gilissen L, Smulders M, van der Meer I (2010a) In search of tetraploid wheat accessions reduced in celiac disease-related gluten epitopes. Mol Biosyst 6:2206-2213.

van den Broeck HC, de Jong HC, Salentijn EMJ, Dekking L, Bosch D, Hamer RJ, Gilissen LJWJ, van der Meer IM, Smulders MJM (2010b) Presence of celiac disease epitopes in modern and old hexaploid wheat varieties: wheat breeding may have contributed to increased prevalence of celiac disease. Theor Appl Genet 121:1527-1539.

van den Broeck HC, Gilissen LJWJ, Smulders MJM., van der Meer IM, Hamer RJ (2011) Dough quality of bread wheat lacking alpha-gliadins with celiac disease epitopes and addition of celiac-safe avenins to improve dough quality. J Cereal Sci 53:206–216.

van den Broeck HC, van Herpen TW, Schuit C, Salentijn EM, Dekking L, Bosch D, Hamer RJ, Smulders MJ, Gilissen LJ, van der Meer IM (2009) Removing celiac disease-related gluten proteins from bread wheat while retaining technological properties: a study with Chinese Spring deletion lines. BMC Plant Biol 9:41.

Van Eckert R, Pfannhauser W, Riedl O (1992) Vienna Food Research Institute, Vienna, Austria. Contribution to quality assessment during production of gluten-free food. *Ernährung/Nutrition* 16:511-512.

Vazquez-Roque MI, Camilleri M, Smyrk T, Murray JA, Marietta E, O'Neill J, Carlson P, Lamsam J, Janzow D, Eckert D, Burton D, Zinsmeister AR (2013) A controlled trial of gluten-free diet in patients with irritable bowel syndrome-diarrhea: effects on bowel frequency and intestinal function. Gastroenterology 144:903-911.

Vincentini O, Borrelli O, Silano M, Gazza L, Pogna N, Luchetti R, De Vincenzi M (2009) T-cell response to different cultivars of farro wheat, *Triticum turgidum* ssp. dicoccum, in coeliac disease patients. Clin Nutr 28:272–277.

Vincentini O, Maialetti F, Gazza L, Silano M, Dessi M, De Vincenzi M, Pogna NE (2007) Environmental factors of celiac disease: Cytotoxicity of hulled wheat species *Triticum monococcum, T. turgidum* ssp. *dicoccum* and *T. aestivum* ssp. *spelta*. J Gastroenterol Hepatol 22:1816–1822.

Vivas S, Ruiz de Morales JM, Ramos F, Suárez-Vilela D (2006) Alemtuzumab for refractory celiac disease in a patient at risk for enteropathy-associated T-cell lymphoma. N Engl J Med 354:2514–2515.

Volta U, Caio G, Tovoli F, De Giorgio R (2013) Non-celiac gluten sensitivity: questions still to be answered despite increasing awareness. Cell Mol Immunol 10:383–392.

Waga J, Skoczowski A (2014) Development and characteristics of ω-gliadin-free wheat genotypes. Euphytica 195:105–116.

Waldmann TA, Conlon KC, Stewart DM, Worthy TA, Janik JE, Fleisher TA, Albert PS, Figg WD, Spencer SD, Raffeld M, Decker JR, Goldman CK, Bryant BR, Petrus MN, Creekmore SP, Morris JC (2013) Phase 1 trial of IL-15 trans presentation blockade using humanized Mikbeta1 mAb in patients with T-cell large granular lymphocytic leukemia. Blood 121:476-484.

Walters MJ, Wang Y, Lai N, Baumgart T, Zhao BN, Dairaghi DJ, Bekker P, Ertl LS, Penfold ME, Jaen JC, Keshav S, Wendt E, Pennell A, Ungashe S, Wei Z, Wright JJ, Schall TJ (2010) Characterization of CCX282-B, an orally bioavailable antagonist of the CCR9 chemokine receptor, for treatment of inflammatory bowel disease. J Pharmacol Exp Ther 335:61–69.

Watts RE, Siegel M, Khosla C (2006) Structure-activity relationship analysis of the selective inhibition of transglutaminase 2 by dihydroisoxazoles. J Med Chem 49:7493–7501.

Wen S, Wen N, Pang J, Langen G, Brew-Appiah RAT, Mejias JH, Osorio C, Yang M, Gemini R, Moehs CP, Zemetra RS, Kogel K-H, Liu B, Wang X, von Wettstein D, Rustgi S (2012) Structural genes of wheat and barley 5-methylcytosine DNA glycosylases and their potential applications for human health. Proc Natl Acad Sci USA 109:20543-20548.

West K (2009) CP-690550, a JAK3 inhibitor as an immunosuppressant for the treatment of rheumatoid arthritis, transplant rejection, psoriasis and other immune-mediated disorders. Curr Opin Investig Drugs 10:491–504.

Wieser H, Koehler P, Folck A, Becker D (2006) Characterization of wheat with strongly reduced α-gliadin content. In: 9th Gluten Workshop, pp. 13-16.

Wolf C, Siegel JB, Tinberg C, Camarca A, Gianfrani C, Paski S, Guan R, Montelione G, Baker D, Pultz IS (2015) Engineering of Kuma030: A Gliadin Peptidase That Rapidly Degrades Immunogenic Gliadin Peptides in Gastric Conditions. J Am Chem Soc 137:13106-13113.

Wrigley CW, Bekes F, Bushuk W (2006) Gluten: A balance of gliadin and glutenin. In Wrigley CW, Bekes F, Bushuk W (eds) Gliadin and glutenin: The unique balance of wheat quality. Am Assoc Cereal Chem, St. Paul, MN, pp. 1-28

Wrigley CW, Bietz JA (1988) Proteins and Amino Acids, In: Pomeranz Y (ed) Wheat Chemistry and Technology, Am Assoc Cereal Chem, St. Paul, MN, pp. 159-275.

Xia J, Bergseng E, Fleckenstein B, Siegel M, Kim CY, Khosla C, Sollid LM (2007) Cyclic and dimeric gluten peptide analogues inhibiting DQ2-mediated antigen presentation in coeliac disease. Bioorg Med Chem 15:6565–6573.

Yokoyama S, Perera PY, Waldmann TA, Hiroi T, Perera LP (2013) Tofacitinib, a janus kinase inhibitor demonstrates efficacy in an IL-15 transgenic mouse model that recapitulates pathologic manifestations of celiac disease. J Clin Immunol 33:586-594.

Yokoyama S, Watanabe N, Sato N, Perera PY, Filkoski L, Tanaka T, Miyasaka M, Waldmann TA, Hiroi T, Perera LP (2009) Antibody-mediated blockade of IL-15 reverses the autoimmune intestinal damage in transgenic mice that overexpress IL-15 in enterocytes. Proc Natl Acad Sci USA 106:15849-15854.

Zevallos VF, Ellis HJ, Suligoj T, Herencia LI, Ciclitira PJ (2012) Variable activation of immune response by quinoa (*Chenopodium quinoa* Willd.) prolamins in celiac disease. Am J Clin Nutr 96:337-344.

Zevallos VF, Herencia LI, Chang F, Donnelly S, Ellis HJ, Ciclitira PJ (2014) Gastrointestinal Effects of Eating Quinoa (*Chenopodium quinoa* Willd.) in Celiac Patients. Am J Gastroenterol 109:270-278.

Zevallos VF, Raker V, Tenzer S, Jimenez-Calvente C, Ashfaq-Khan M, Rüssel N, Pickert G, Schild H, Steinbrink K, Schuppan D (2017) Nutritional wheat amylase-trypsin inhibitors promote intestinal inflammation via activation of myeloid cells. Gastroenterology 152:1100-1113.

Zevallos VF, Raker VK, Maxeiner J, Scholtes P, Steinbrink K, Schuppan D (2018) Dietary wheat amylase trypsin inhibitors exacerbate murine allergic airway inflammation. Eur J Nutr https://doi.org/10.1007/s00394-018-1681-6.

Zhao J, de Vera J, Narushima S, Beck EX, Palencia S, Shinkawa P, Kim KA, Liu Y, Levy MD, Berg DJ, Abo A, Funk WD (2007) R-spondin1, a novel intestinotrophic ameliorates experimental colitis in mice. Gastroentrology 132:1331–1343.

Zuidmeer L, Goldhahn K, Rona RJ, Gislason D, Madsen C, Summers C, Sodergren E, Dahlstrom J, Lindner T, Sigurdardottir ST, McBride D, Keil T (2008) The prevalence of plant food allergies: a systematic review. J Allergy Clin Immunol 121:1210-1218.

FODMAPs in Wheat

Heinrich Grausgruber, Alison Lovegrove, Peter Shewry, and Ferenc Békés

Abstract The dietary intake of fermentable oligo-, di- and monosaccharides, and polyols (FODMAPs) can promote gut health, but also trigger gastrointestinal disorders. Wheat as a staple food is considered a major source of FODMAPs in the daily diet. The most abundant FODMAPs in the wheat grain are fructans, which accumulate during plant development in vegetative tissues and are remobilized during grain filling and synthesized in the developing grain. Abiotic stress can foster the accumulation of fructans. Quantification of fructans and/or other FODMAPs is usually carried out by commercial enzymatic assays or by chromatographic methods. There is evidence for genetic variation in fructan accumulation, remobilization efficiency and concentration in the grain. Heritabilities were shown to be moderate to high. Therefore, breeding for low fructan and/or FODMAPs levels in the grain is feasible and was already successfully demonstrated. A significant reduction in FODMAPs of wheat products, however, can be realised by processing. Therefore, long proofing times, especially sour dough fermentation, are most efficient.

H. Grausgruber (✉)
Department of Crop Sciences, University of Natural Resources and Life Sciences, Vienna, Tulln an der Donau, Austria
e-mail: heinrich.grausgruber@boku.ac.at

A. Lovegrove · P. Shewry
Department of Plant Sciences, Rothamsted Research, Harpenden, Hertfordshire, AL5 2JQ, UK

F. Békés
FBFD Pty Ltd, Beecroft, NSW, Australia

G. Igrejas et al. (eds.), *Wheat Quality For Improving Processing And Human Health*, https://doi.org/10.1007/978-3-030-34163-3_21

1 Introduction

FODMAPs are one of the most recently established topics in wheat research, and currently one of the most active. The term is an acronym for "**F**ermentable **O**ligosaccharides, **D**isaccharides, **M**onosaccharides **A**nd **P**olyols", and was coined by Gibson and Shepherd (2005) in relation to the role of diet in susceptibility to Crohn's disease.

Wheat products are considered to be a major source of FODMAPs (Biesiekierski et al. 2011; Whelan et al. 2011), especially if wheat is a staple food and consumed in large amounts (Moshfegh et al. 1999; van Loo et al. 1995). The only FODMAPs, which are present in significant quantities in the wheat grain are the monosaccharide fructose, the disaccharides sucrose and maltose, the trisaccharide raffinose (galactose-glucose-fructose) and fructo-oligosaccharides (fructans) (Table 1). In addition, breads may contain glycerol and, when sourdough systems are used, mannitol (Costabile et al. 2014). The FODMAP content is highest in the bran fraction, followed by middlings and flour (Bach Knudsen 1997).

2 Structure and Metabolism

The most abundant FODMAPs in wheat grain are polymers of fructose:fructo-oligosaccharides or fructans. The distinction between these two groups is not clear so we will consider them as a single group, which comprise three or more fructose units, with some forms also having a single glucose unit. Wheat fructans are of the graminan (branched)-type, comprising mainly fructose residues but with a single glucose residue and both β-(2,1) linkages and branches through β-(2,6) linkages (Lewis 1993; Ritsema and Smeekens 2003). Variation in the structure may occur between fructans of different developmental stages or from different tissues of the same plant (Carpita et al. 1991).

Table 1 Contents of FODMAPs in whole-wheat grain

Component	Amount (% dm)	References
Glucose	0.10–0.11	Henry (1985); Lineback and Rasper (1988)
Fructose	0.02–0.15	Henry and Saini (1989); Lineback and Rasper (1988); Yasui and Ashida (2011); Ziegler et al. (2016)
Sucrose	0.54–1.69	Bach Knudsen (1997); Henry and Saini (1989); Kuo et al. (1988); Lineback and Rasper (1988); Rakha et al. (2011); Yasui and Ashida (2011); Ziegler et al. (2016)
Maltose	0.05–0.60	Lineback and Rasper (1988); Rakha et al. (2011)
Raffinose	0.12–0.78	Bach Knudsen (1997); Fretzdorff and Welge (2003a); Henry and Saini (1989); Huynh et al. (2008a); Kuo et al. (1988), Lineback and Rasper (1988); Yasui and Ashida (2011); Ziegler et al. (2016)
Stachyose	0.01–0.20	Bach Knudsen (1997); Henry and Saini (1989)
Fructans	0.66–3.20	Detailed information in Table 2

Wheat flour fructans have an average degree of polymerisation (DP) of 4, with 50% of the fraction comprising components of DP 3, 4 or 5. Whole grain fructans have been reported to have a slightly higher DP of 5 (Nilsson et al. 1986; Verspreet et al. 2012). Rakha et al. (2011) reported that 69% of the fructans in a whole grain flour had a DP between 3 and 9 while 31% had a DP of greater than 9.

3 Health Related Issues

FODMAPs are associated with both positive and negative health effects. The consumption of fructans promotes gut health (Roberfroid et al. 2010) and can increase mineral bioavailability (Abrams et al. 2005, 2007). On the other hand, FODMAPs were initially identified as associated with Crohn's disease (Gibson and Shepherd 2005) and, more recently, have been associated with irritable bowel syndrome (IBS) (Gibson and Shepherd 2010; Staudacher et al. 2011).

Preliminary in vitro and animal studies have indicated that FODMAPs, gluten proteins and some non-gluten proteins have a pro-inflammatory effect on intestinal epithelial cells and are therefore theoretically capable of triggering gastrointestinal symptoms in humans (Junker et al. 2012; Valerii et al. 2015; Zevallos et al. 2017). There is an on-going debate as to whether wheat proteins such as α-trypsin/trypsin inhibitors (ATI) and gluten or FODMAPs are the true triggers of wheat-related symptoms in functional gastrointestinal disorders, or whether their combined action, through a possible synergic effect, is required for triggering symptoms.

3.1 Health Benefits

Fructans are recognized as a form of dietary fibre and most of the inulin-type fructans which are consumed survive the passage through the stomach and small intestine (Ellegård et al. 1997; Molis et al. 1996). Based on in vitro and in vivo studies with a rat model, Nilsson et al. (1988) concluded that wheat grain fructans are either not digested, or only digested to a small extent, during their passage through the stomach and small intestine, whereas they are almost completely fermented when reaching the hindgut (Nilsson and Björck 1988).

Inulin-type fructans are also acknowledged as prebiotics and their health promoting potential was demonstrated by clinical studies (for review see e.g. Roberfroid et al. 2010). Inulin-type fructans affect health not only by their impact on the gut microbiota and activity, but may also have immunomodulatory effects and antioxidant properties. However, the latter effects were hitherto only confirmed by in vitro studies (Peshev and van den Ende 2014; van den Ende 2013) and the studies have not been carried out with isolated wheat fructans.

3.2 Gastrointestinal Disorders

FODMAPs such as fructans can trigger gastrointestinal symptoms. These compounds are poorly absorbed in the small intestine, resulting in increased osmotic pressure and hence increased water volume and intestinal motility. In the colon, gas production from bacterial fermentation is increased (Gibson and Shepherd 2010; Mullin et al. 2014; Murray et al. 2014; Shepherd et al. 2013). A number of studies reported an alleviation of IBS symptoms after a low FODMAPs diet (e.g. Gibson and Shepherd 2010; Staudacher et al. 2011). Nevertheless, some researchers remain sceptical of the benefits of a low FODMAPs diet, with Catassi et al. (2017) noting that "the quality of the evidence is lower than optimal" and that drastic reduction of FODMAPs in the diet may have other, poorly understood, consequences.

Numerous individuals have reported adverse effects on health following the consumption of bread wheat (*Triticum aestivum*), but claimed spelt (*T. spelta*) products to be more easily digestible with less adverse effects (Stallknecht et al. 1996), and Biesiekierski et al. (2011) suggested that a lower FODMAP content in spelt products might be the reason for these observations. There have also been anecdotal reports for some time that a proportion of patients suffering from non-celiac wheat-related health disorders can tolerate products made from certain spelt varieties, and Armentia et al. (2012) reported that patients suffering from wheat allergy (bakers' asthma and food allergy) reacted less with spelt than with bread wheat. Vu et al. (2015) also reported that a spelt line with low FODMAP content and unusual albumin/globulin composition showed less immunoreactivity than bread wheat and other spelt lines.

Sourdough fermentation results in activity of proteolytic and fructosidase enzymes in dough and can, thus, decrease the amounts of proteins and fructans in the end products (see "6. Influence of processing"). Sourdough breadmaking could, therefore, theoretically result in products with improved toleration for patients who experience gastrointestinal discomfort on consuming wheat. Laatikainen et al. (2017) reported a pilot study with 26 IBS patients in which sourdough breadmaking with more than 12 h fermentation reduced the quantities of fructan to 0.06% compared to 0.23% in the 2 h yeast fermented bread. Nonetheless, the sourdough bread was not tolerated better than the yeast-fermented bread.

Low FODMAP bread can help to restrict the intake of FODMAPs but at the same time increase the intake of slowly fermentable dietary fibre in IBS patients. Bread which is high in fibre and low in FODMAPs may prevent the depletion of intestinal bifidobacteria that has been observed on other low FODMAP diets (Halmos et al. 2015; Staudacher et al. 2012) and shows promise in reducing symptoms of IBS.

Clinical trials with IBS patients showed that rye bread, which is low in FODMAPs, influences gastrointestinal symptoms and the extent of gas production generated in intestinal fermentation. Laatikainen et al. (2016) showed that rye bread which was low in FODMAPs caused less flatulence, less abdominal pain, fewer cramps, and less stomach rumbling than regular rye bread. Pirkola et al. (2018)

reported that low FODMAP rye bread reduced the generation of hydrogen by colonic fermentation. This study showed significant differences between different types of bread in their postprandial effects. Similar conclusions were reported by Costabile et al. (2014), who used in vitro batch culture experiments to compare the fermentation of breads produced by sourdough, long fermentation with yeast and the "no time" fermentation Chorleywood Baking Process (CBP) by faecal bacteria from healthy subjects and IBS patients. They concluded that breads produced by the traditional long fermentation and sourdough systems were less likely to lead to IBS symptoms.

4 Methods for Quantification

FODMAPs are usually extracted from freeze dried finely ground samples in hot water (boiling or 80 °C), although other solvents may be used depending upon downstream analysis (e.g. acetonitrile/water (Biesiekierski et al. 2011) or 80% ethanol (Haskå et al. 2008)).

Commercial kits are widely used for the measurement of total fructans, such as the Fructan Assay Kit from Megazyme, Bray, Ireland. These kits measure the fructan content following hydrolysis to release D-fructose and D-glucose. Recombinant enzymes of high specificity (exo- and endo-inulinases and endo-levanase) are used to release these monosaccharides, which are then measured spectrophotometrically (AOAC Method 999.03; AACC Method 32.32) (McCleary et al. 2000). Glucose and fructose are determined following reaction with p-hydroxybenzoic acid hydrazide (PAHBAH) reagent or the hexokinase/phosphoglucose isomerase/glucose 6-phosphate dehydrogenase system.

Monosaccharides released following enzyme hydrolysis may also be quantified by HPLC, or by GC (following derivatisation to alditol acetates). For both analyses, authentic sugar standards are required to produce calibration curves for quantitation.

Sugar alcohols (e.g. mannitol) may also be determined by HPLC using a range of columns depending upon the HPLC system available (see below), or by ^{1}H NMR (Costabile et al. 2014).

It should be noted that it has been suggested that the enzymatic kits may not be accurate for samples containing less than 1% fructan (Muir et al. 2007) or that they underestimate some samples that contain high amounts of fructans (Call et al. 2018).

4.1 *Chromatographic Methods*

Samples are extracted with hot water and then centrifuged or filtered. The supernatants can then be analysed by chromatography to separate and quantify mono- and disaccharides as well as sugar alcohols (e.g. mannitol). Depending upon the HPLC system, the columns and solvents vary. The Sugar-Pak columns from Waters separate monosaccharides and polyols based upon strong cation exchange chromatography and use an acetonitrile/water mobile phase and evaporative light scattering (ELSD)/refractive index (RI) detectors. For longer chain fructo-oligosaccharides or galacto-oligosaccharides a second column is required (e.g. Waters High-Performance Carbohydrate Column) (Muir et al. 2007, 2009). Monosaccharides, disaccharides and sugar alcohols can also be separated by high performance anion-exchange chromatography (HPAEC) using, for example, the Thermo MA1 CarboPac columns with sodium hydroxide eluent and pulsed amperometric detector (PAD) (Andersen and Sørensen 2000; Corradini et al. 2004; Cataldi et al. 2000).

Verspreet et al. (2012) extracted fructans with hot water for 60 minutes followed by mild acid hydrolysis with 60 mM HCl. Glucose and fructose were then determined by HPAEC-PAD and an adjustment made for glucose and fructose released from sucrose and raffinose.

Oligosaccharides can also be characterised based on the degree of polymerization, for example, fructans from DP 3 up to DP 19 have been identified in wheat flour (Haskå et al. 2008; Nilsson et al. 1986; Verspreet et al. 2012). They may be separated by HPAEC-PAD and then identified by MALDI-TOF MS and ^{1}H NMR spectroscopy. The raffinose series of oligosaccharides may also be analysed without prior hydrolysis using HPAEC-PAD, and distinguished from fructan oligosaccharides by incubation with α-galactosidase and/or inulinase prior to HPLC analysis. Reference to authentic standards is required or further analysis by mass spectrometry or NMR spectroscopy (see below).

Haskå et al. (2008) used preparative HPAEC to isolate fructans. After extraction by boiling in 80% (v/v) ethanol, the ethanol was removed by rotary evaporation and the isolated fructans dissolved in acetate buffer and separated by HPAEC using sodium hydroxide and sodium acetate eluents. Peaks were collected, desalted and analysed either by ^{1}H NMR spectroscopy in D_2O or dissolved in ethanol, mixed with a suitable matrix and analysed by MALDI-TOF mass spectrometry.

Costabile et al. (2014) reported the determination of sugars (including fructose, maltose, sucrose, raffinose) and mannitol by ^{1}H NMR of polar extracts from flour, dough and bread, without the need for a chromatographic separation, while Verspreet et al. (2015) have used LC-MS/MS to define the fine structure of fructans from cereal grains, permitting rapid identification without the need for purification. Such high throughput analyses should become the methods of choice for fructan and other FODMAPs.

5 Genetic Variation and Breeding

Due to the high daily consumption of wheat in many countries, broad screening to discover high- and low-FODMAPs varieties or species as well as the development of processing strategies to reduce FODMAPs levels are of particular interest to develop foods with low FODMAP contents. Many studies have focused on fructans and the majority have investigated only a limited number of genotypes and/or environments (Table 2).

Table 2 Fructan content in wheat grains (E, number of environments; G, number of studied genotypes; figures in brackets indicate number of samples without information on the variation in genotypes and/or environments; unless indicated by footnotes the values refer to *T. aestivum*)

Fructan (%)	E (n)	G (n)	Reference
1.73–2.46		2	Henry (1985)
1.5		?	Bach Knudsen (1997)
0.9–1.8	5	5	Fretzdorff and Welge (2003a)
1.5–1.7	?	(5)[1]	Fretzdorff and Welge (2003a)
0.9–1.3	?	(5)[4]	Fretzdorff and Welge (2003a)
1.7–3.1	?	(5)[4§]	Fretzdorff and Welge (2003a)
0.87–0.95	8	1	Langenkämper et al. (2006)
1.60–3.20	2	2[1]	Costantini et al. (2008)
1.40–2.70	2	2	Costantini et al. (2008)
2.20–2.30	2	2	Haskå et al. (2008)
0.66–2.84	4	19 + (43)	Huynh et al. (2008a)
0.70–2.60	3	154	Huynh et al. (2008b)
1.61–2.20		4[2]	Brandolini et al. (2011)
1.27–1.29		1	Brandolini et al. (2011)
1.49–1.62	1	2	Ni et al. (2011)
1.8	2	1	Rakha et al. (2011)
1.23–1.88		2	Yasui and Ashida (2011)
1.84–2.65		2[3]	Yasui and Ashida (2011)
2.19–2.80	1	120	Huynh et al. (2012)
0.84–1.85	1	129	Andersson et al. (2013)
0.89–1.82		28[4]	Escarnot et al. (2015)
1.07–2.11		11	Escarnot et al. (2015)
0.80–1.72	1	29 + 2[1]	Morrison et al. (2015)
1.54 + 1.92		1[5] + 1[6]	Gélinas et al. (2016)
1.38–1.97	4	5	Ziegler et al. (2016)
0.97–1.16	4	5[4]	Ziegler et al. (2016)
0.94–1.27	4	5[1]	Ziegler et al. (2016)
0.82–1.08	4	2[5]	Ziegler et al. (2016)
1.67–1.80	4	2[2]	Ziegler et al. (2016)
0.95–6.92	1	19 + 1[4]	Call et al. (2018)

[1]durum wheat (*T. durum*)
[2]einkorn wheat (*T. monococcum*)
[3]waxy (low amylose) wheat (*T. aestivum*)
[4]spelt wheat (*T. spelta*)
[§]'*Grünkern*' samples
[5]emmer wheat (*T. dicoccum*)
[6]Khorasan wheat (*T. turanicum*)

Analysis of a diversity panel of 129 winter wheat samples grown at the same location, as part of the EU HEALTHGRAIN project, revealed a variation in fructan concentration from 0.84% to 1.85% dry weight (Andersson et al. 2013). This is in agreement with an earlier study of five cultivars grown at five locations which showed a range of 0.9–1.8% dry wt (Fretzdorff and Welge 2003a) and with a more recent study from New Zealand of 29 bread wheat genotypes (0.80–1.72% dry wt) (Morrison et al. 2015). However, concentrations up to almost 3% were reported for CIMMYT breeding lines (Huynh et al. 2008a) and double haploid lines of two mapping populations (Huynh et al. 2008b, 2012).

Generally, the fructan content of wheat is lower than that of rye, where values of over twice those in wheat have been reported (Andersson et al. 2009; Henry 1985). Fructans are especially concentrated in the bran (Haskå et al. 2008; Nilsson et al. 1986; Schnyder et al. 1988). However, it must be noted that the concentrations in the pericarp and embryo are especially high during early grain filling and decrease later as assimilate partitioning ceases and the pericarp degenerates (Schnyder et al. 1988).

Fructans and other water-soluble carbohydrates (WSC) accumulate temporarily in vegetative plant tissues such as stems or roots. Fructans stored in vegetative tissues can be remobilized during grain filling and synthesized in the developing grain (Gebbing 2003; McGrath et al. 1997; Schnyder et al. 1993). Fructans also have a direct protective effect under stress conditions with respect to membrane stabilization (Livingston et al. 2009).

Genetic variation has been reported for fructan content in stems and for efficiency of remobilisation (Ehdaie et al. 2006; Ruuska et al. 2006). In the grain, the accumulation of fructan is most rapid between five and nine days after anthesis, reaches a maximum concentration (of between 15 and 30% dry weight) and then decreases (Costantini et al. 2008; Ni et al. 2011; Schnyder et al. 1993; Verspreet et al. 2013a). The decrease in fructan content during grain development coincides with a decrease in the average DP. By contrast, raffinose accumulates late during grain maturation and maltose is present across all grain filling stages (Verspreet et al. 2013a). Fretzdorff and Welge (2003a) confirmed these results and reported fructan and raffinose contents of 1.7% to 3.1% and 0.12% to 0.33%, respectively, in five samples of 'Grünkern' (spelt) harvested at the milk dough stage, compared to 0.9% to 1.3% and 0.22% to 0.47%, respectively, when harvested at grain maturity.

Significant environmental effects have also been reported (Costantini et al. 2008; Ziegler et al. 2016), but with no evidence of strong genotype×environment interactions (Huynh et al. 2008a). No differences were observed between fertilisation levels and management system (organic vs. conventional), but unfertilized grain showed slightly higher fructan concentrations indicating stress due to a lack of nutrients (Langenkämper et al. 2006). Genotypic effects can therefore be considered as most important, with moderate to high heritabilities (h^2 = 0.64 to 0.94) reported for fructan contents in different genetic backgrounds and environments (Huynh et al. 2008a, b; Ziegler et al. 2016). A similarly high heritability (h^2 = 0.69) has also been reported for raffinose content (Ziegler et al. 2016).

Eight quantitative trait loci (QTL) with two pairs of epistatic interactions were found for grain fructan concentration. Two QTL on chromosomes 7A and 6D explained, respectively, 17 and 27% of the total phenotypic variation. Transgressive segregation was observed (Huynh et al. 2008b). Genes encoding the enzymes of fructan biosynthesis (1-SST, 1FFT and 6-SFT) form a functional cluster and have been sequenced and mapped to the major QTL on chromosome 7A (Huynh et al. 2012).

A significant increase in total fructan content of about 60% was observed for waxy isolines compared to wild-type lines by Yasui and Ashida (2011), whereas no effects were observed for raffinose, fructose and sucrose. On the other hand, sweet wheat, a double mutant lacking granule-bound starch synthase I (GBSSI) and starch synthase IIa (SSIIa) genes, accumulates significantly higher amounts of fructans (7.2%), maltose (0.57%), sucrose (4.91%) and glucose (0.16%) compared to waxy or high-amylose mutants, and wild-type wheat (Shimbata et al. 2011).

Escarnot et al. (2015) and Ziegler et al. (2016) reported lower fructan concentrations for spelt, but the ranges of concentrations in spelt and bread wheat were largely overlapping. In Australia, E3 spelt wheat was shown to contain almost half of the usual levels of fructans in bread wheat and over 20% less than any other spelt cultivar grown under the same conditions (Fig. 1). The FODMAP content of bread made from E3 spelt, using an optimised formulation and technology, was considerably lower than the threshold defined for low FODMAP products, showing significantly better responses to IBS patients than the control in a pilot scale single-blind, crossover intervention trial (Muir et al. 2014). Using E3 spelt as control, 105 spelt lines grown in Hungary have been screened for fructan content and seven lines

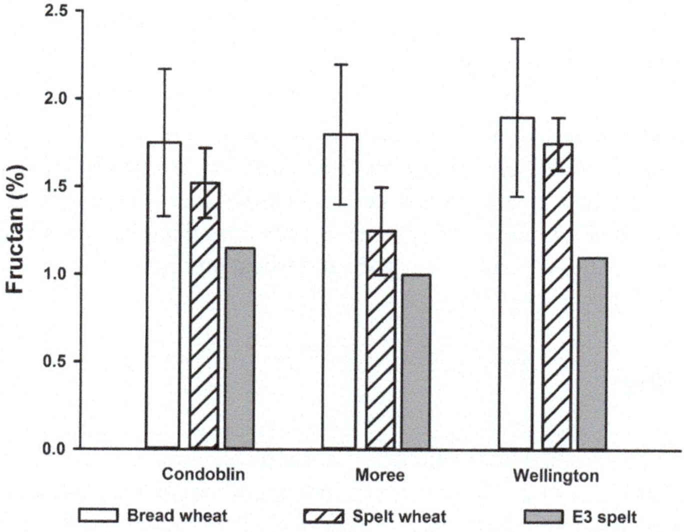

Fig. 1 Total fructan content of 12 bread wheat and 12 spelt wheat grown at three locations in New South Wales, Australia. Data for E3 spelt wheat, which are significantly lower compared to the bread and other spelt wheat genotypes, are shown separately (unpublished data by D. Suter, K. Ács, A. Juhász, G. Brown and F. Békes)

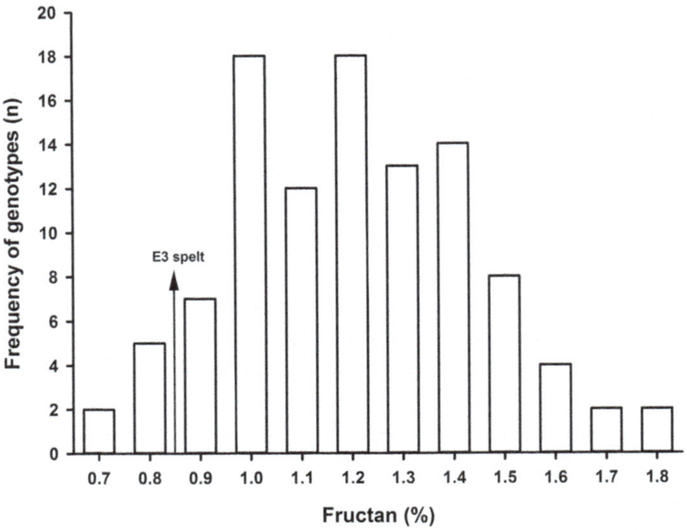

Fig. 2 Variation of fructan content in 105 spelt wheat genotypes grown in Hungary at the same location (Pauk et al. 2019)

showed even lower fructan levels (<0.9%) than the control (Fig. 2) (Békés et al. 2017; Pauk et al. 2019). By contrast, Brandolini et al. (2011) reported higher fructan concentrations in four einkorn (*T. monococcum*) accessions compared to a check bread wheat variety, which was confirmed by Ziegler et al. (2016).

6 Influence of Processing

The FODMAP contents of wheat products can be influenced significantly by different processing technologies. Milling and fractionation can enrich, whereas baking and/or any other heat treatment reduce the FODMAP concentration. Processing technology generally is regarded to be more important than the cereal source with respect to the FODMAP load in the processed wheat product.

6.1 Milling

Fructans can be enriched by fractionation as shorts and bran contain similarly high concentrations of about 2–4%, while refined white flour contains less than half this concentration (Bach Knudsen 1997; Haskå et al. 2008; Kamal-Eldin et al. 2009). Enrichment in the bran was also observed for raffinose and stachyose (Bach Knudsen 1997). The highest fructan content was observed in the bran fraction, followed by shorts, the germ, the middlings flour and finally the refined flour. The total fructan content in the different milling fractions is also significantly correlated with the total dietary fibre and ash content (Haskå et al. 2008).

6.2 Dough Making and Fermentation

Dough mixing reduced the fructan concentration in wheat flours by about 20%, probably due to flour oxidation. Nilsson et al. (1987) first reported that the degradation of fructans during leavened bread production was due to the action of the enzyme invertase in bakers' yeast (*Saccharomyces cerevisiae*). Yeast fermentation of doughs eliminated 40% to 78% of fructans, with higher degradation rates being observed for longer fermentation, whereas fermentation without yeast did not affect the fructan concentration after mixing (Gélinas et al. 2016; Knez et al. 2014; Nilsson et al. 1987; Verspreet et al. 2013b). Individuals sensitive to fructans should, therefore, consume long fermentation breads rather than products arising from processes with short fermentation times.

The reduction or complete prevention of fructan degradation can be achieved by using mutant yeast strains with lower sucrose degradation activity or lacking invertase activity (Verspreet et al. 2013b). Yeast was shown to have a preference for low DP fructans (Nilsson et al. 1987; Rakha et al. 2011) and Praznik et al. (2002) reported that fructan molecules with a higher DP or higher average chain length are more resistant to degradation during the baking process. This may explain why fructan degradation during baking can be higher in wheat than in rye breads (Andersson et al. 2009; Fretzdorff and Welge 2003b; Nilsson et al. 1987).

The importance of proofing (fermentation) time on the FODMAP content of baked products irrespective of the type of wheat was demonstrated by Ziegler et al. (2016). Short proofing of 1 h decreased the raffinose and fructan levels only slightly but significantly increased the 'excess fructose' due to the almost complete hydrolysis of sucrose. Longer proofing times of 2.5 h or 4.5 h also reduced the excess fructose level significantly, resulting in final FODMAP contents of only 29–33% and 10–23% of the initial concentrations present in bread and spelt wheats, respectively. Spelt is often processed by artisan bakers who adhere to traditional recipes and long fermentation times, partly to improve the rather poor technological performance (Frakolaki et al. 2018; Schober et al. 2002). Reports of individuals who tolerate spelt better than bread wheat products may therefore be related to the processing conditions rather than to differences between the compositions of the two wheats. Similar results were obtained in Australia when commercial bread samples made from bread wheat and spelt flours and baked with different baking technologies were analysed (Fig. 3).

It is generally believed that maltose drives yeast-mediated dough fermentation. However, the relative importance of fructose and glucose, released from wheat fructan and sucrose by invertase, is not documented. Investigations by Struyf et al. (2017) revealed that, after 2 h fermentation of wheat flour dough, about 44% of the sugars consumed were generated by invertase-mediated degradation of fructan, raffinose and

Fig. 3 Fructan content of commercial bread samples made from spelt and bread wheat flours with yeasted or sour dough. Figures above bars indicate the fermentation/proofing time in hours and/or minutes (´). Unpublished data by D. Suter, K. Ács, A. Juhász, G. Brown and F. Békés.

sucrose, and the remaining 56% were generated by amylases. In wholemeal dough, 70% of the sugars consumed were released by invertase activity. Invertase-mediated sugar release appeared to be crucial during the first hour of fermentation, while amylase-mediated sugar release was predominant in the later stages of fermentation, which explains why higher amylolytic activity prolonged the productive fermentation time. The recent study of Struyf et al. (2018) evaluated the effects of various factors on the FODMAP levels in wholemeal bread produced with yeast fermentation, including the yeast strain and fermentation parameters such as yeast dosage, fermentation time and the application of enzyme-based technologies. Benítez et al. (2018) demonstrated that dough preparation and baking significantly reduced the fructan content, whereas no effect was observed on the concentrations of fructose, glucose and raffinose, and a significant increase was observed for sucrose and specifically for maltose.

As mentioned above, a significant reduction of FODMAPs in bread can further be achieved by using sourdough. Lactic bacteria are suggested to increase fructan degradation by creating acidic conditions favouring yeast invertase (Kissing Kucek et al. 2015; Nilsson et al. 1987). Loponen and Gänzle (2018) give an overview of the biochemical processes during sourdough fermentation, providing details of the roles of different microorganisms in the culture and the enzymes involved. The main processes in the degradation are: (i) sucrose hydrolysis by yeast invertase or fructosidases of lactic acid bacteria; (ii) oligosaccharide formation by glucansucrases to form isomalto-oligosaccharides or by fructansucrases to form kestose, nystose, and erlose from sucrose; (iii) kestose and nystose degradation by yeast invertase or by intracellular (phospho)-fructosidases of lactic acid bacteria; (iv) raffinose conversion by yeast invertase and levansucrase from lactic acid bacteria; (v) fructose conversion by mannitol-dehydrogenase from heterofermentative lactic acid bacteria; (vi) starch conversion to maltose and glucose by flour amylases and gluco-amylase.

In conclusion, fermentation by yeast or sourdough plays a crucial role in fructan degradation. Selection of the correct strain and the optimal fermentation parameters are of major importance to control fructan degradation. Conventional sourdough baking reduces and converts FODMAPs in rye and wheat flour; however, the extent of FODMAP reduction is dependent on the fermentation organisms, the fermentation process, the grain raw material and the ratio of the sourdough to the final bread dough. The production of low FODMAP bread requires the activity of extracellular fructanase and therefore either sourdough fermentation with lactobacilli expressing fructanases or the use of fructanase-positive yeasts can be used to produce breads with low FODMAP contents.

6.3 Baking and Pasta Making

Although fructan degradation mainly occurs during mixing and fermentation, the final baking step may also have an impact on the fructan content of bread. Boskov Hansen et al. (2002) reported fructan degradation of 26% after dough mixing with a wholemeal rye flour, which increased to 34% after proofing (37 °C, 89% rel. humidity, 37 min) and to 45% after baking (20 s steaming at 250 °C, 35 min at 200 °C). Fretzdorff and Welge (2003b) reported similar results with fructan degradation during baking (60 min at 210 °C) ranging from 10% to 20%. However, when shorter baking times are used, fructan loss may be negligible (Verspreet et al. 2013b). A loss of about 50% was observed in pasta making by drying and cooking (Gélinas et al. 2016).

Acknowledgments H.G. acknowledges funding by the Austrian Research Promotion Agency (FFG) for the project 'ID WHEAT' (No. 858540) and by the OeAD GmbH, Vienna (WTZ project CZ 02/2018). Rothamsted Research (A.L., P.S.) receives grant-aided support from the Biotechnology and Biological Sciences Research Council (BBSRC) of the UK and the work reported here forms part of the Designing Future Wheat Institute Strategic Programme (BB/P016855/1). F.B. acknowledges financial support by the National Research, Development and Innovation Office, Hungary, for parts of the research mentioned in this chapter (project OTKA-K 16-119835).

References

Abrams SA, Griffin IJ, Hawthorne KM, Liang L, Gunn SK, Darlington G, Ellis KJ (2005) A combination of prebiotic short- and long-chain inulin-type fructans enhances calcium absorption and bone mineralization in young adolescents. The American Journal of Clinical Nutrition 82: 471–476.

Abrams SA, Hawthorne KM, Aliu O, Hicks PD, Chen Z, Griffin IJ (2007) An inulin-type fructan enhances calcium absorption primarily via an effect on colonic absorption in humans. The Journal of Nutrition 137: 2208–2212.

Andersen R, Sørensen A (2000) Separation and determination of alditols and sugars by high-pH anion-exchange chromatography with pulsed amperometric detection. Journal of Chromatography A 897: 195–204.

Andersson R, Fransson G, Tietjen M, Åman P (2009) Content and molecular-weight distribution of dietary fiber components in whole-grain rye flour and bread. Journal of Agricultural and Food Chemistry 57: 2004–2008.

Andersson AAM, Andersson R, Piironen V, Lampi A-M, Nyström L, Boros D, Frás A, Gebruers K, Courtin CM, Delcour JA, Rakszegi M, Bedo Z, Ward JL, Shewry PR, Åman P (2013) Contents of dietary fibre components and their relation to associated bioactive components in whole grain wheat samples from the HEALTHGRAIN diversity screen. Food Chemistry 136: 1243–1248.

Armentia A, Martín S, Diaz-Perales A, Palacín A, Tordesillas L, Herrero M, Martín-Armentia M (2012) A possible hypoallergenic cereal in wheat food allergy and baker's asthma. American Journal of Plant Sciences 3: 1779–1781.

Bach Knudsen KE (1997) Carbohydrate and lignin contents of plant materials used in animal feeding. Animal Feed Science and Technology 67: 319–338.

Benítez V, Esteban RM, Moniz E, Casado N, Aguilera Y, Mollá E (2018) Breads fortified with wholegrain cereals and seeds as source of antioxidant dietary fibre and other bioactive compounds. Journal of Cereal Science 82: 113–120.

Békés F, Ács K, Gell G, Lantos C, Kovács AM, Birinyi Z, Pauk J (2017) Towards breeding less allergenic spelt wheat with low FODMAP content. Acta Alimentaria 46: 246–258.

Biesiekierski JR, Rosella O, Rose R, Liels K, Barrett JS, Shepherd SJ, Gibson PR, Muir JG (2011) Quantification of fructans, galacto-oligosacharides and other short-chain carbohydrates in processed grains and cereals. Journal of Human Nutrition and Dietetics 24: 154–176.

Boskov Hansen H, Andreasen MA, Nielsen MM, Larsen LM, Bach Knudsen KE, Meyer AS, Christensen LP, Hansen Å (2002) Changes in dietary fibre, phenolic acids and activity of endogenous enzymes during rye bread-making. European Food Research and Technology 214: 33–42.

Brandolini A, Hidalgo A, Plizzari L, Erba D (2011) Impact of genetic and environmental factors on einkorn wheat (*Triticum monococcum* L. subsp. *monococcum*) polysaccharides. Journal of Cereal Science 53: 65–72.

Call L, Reiter E, Grausgruber H, Schönlechner R, D'Amico S (2018) Fruktane in alten und neuen österreichischen Weizensorten. Getreide, Mehl und Brot 1/2018: 2–6.

Carpita NC, Housley TL, Hendrix JE (1991) New features of plant-fructan structure revealed by methylation analysis and carbon-13 NMR spectroscopy. Carbohydrate Research 217: 127–136.

Cataldi TR, Campa C, De Benedetto GE (2000) Carbohydrate analysis by high-performance anion-exchange chromatography with pulsed amperometric detection: the potential is still growing. Fresenius Journal of Analytical Chemistry 368: 739–758.

Catassi G, Lionetti E, Gatti S, Catassi C (2017) The low FODMAP diet: many question marks for a catchy acronym. Nutrients 9: 292.

Corradini C, Bianchi F, Matteuzzi D, Amoretti A, Rossi M, Zanoni S (2004) High-performance anion-exchange chromatography coupled with pulsed amperometric detection and capillary zone electrophoresis with indirect ultra violet detection as powerful tools to evaluate prebiotic properties of fructooligosaccharides and inulin. Journal of Chromatography A 1054: 165–173.

Costabile A, Santarelli S, Claus S, Sanderson J, Hudspith BN, Brostoff J, Ward JL, Lovegrove A, Shewry PR, Jones HE, Gibson GR (2014) Effect of breadmaking process on in vitro gut microbiota parameters in irritable bowel syndrome. PloS One 9: e111225.

Costantini A, Amoriello T, Cecchini C, D'Egidio MG (2008) Analysis of factors influencing fructans production in winter cereals. Journal of Genetics and Breeding 62: 15–24.

Ehdaie B, Alloush G, Madore M, Waines J (2006) Genotypic variation for stem reserves and mobilization in wheat: II. Postanthesis changes in internode water-soluble carbohydrates. Crop Science 46: 2093–2103.

Ellegård L, Andersson H, Bosaeus I (1997) Inulin and oligofructose do not influence the absorption of cholesterol, or the excretion of cholesterol, Ca, Mg, Zn, Fe, or bile acids but increases energy excretion in ileostomy subjects. European Journal of Clinical Nutrition 51: 1–5.

Escarnot E, Dornez E, Verspreet J, Agneessens R, Courtin CM (2015) Quantification and visualization of dietary fibre components in spelt and wheat kernels. Journal of Cereal Science 62: 124–133.

Frakolaki G, Giannou V, Topakas E, Tzia C (2018) Chemical characterization and breadmaking potential of spelt versus wheat flour. Journal of Cereal Science 79: 50–56.

Fretzdorff B, Welge N (2003a) Fructan- und Raffinosegehalte im Vollkorn einiger Getreidearten und Pseudo-Cerealien. Getreide, Mehl und Brot 57: 3–8.

Fretzdorff B, Welge N (2003b) Abbau von getreideeigenen Fructanen während der Herstellung von Roggenvollkornbrot. Getreide, Mehl und Brot 57: 147–151.

Gebbing T (2003) The enclosed and exposed part of the peduncle of wheat (*Triticum aestivum*) – spatial separation of fructan storage. New Phytologist 159: 245–252.

Gélinas P, McKinnon C, Gagnon F (2016) Fructans, water-soluble fibre and fermentable sugars in bread and pasta made with ancient and modern wheat. International Journal of Food Science and Technology 51: 555–564.

Gibson PR, Shepherd SJ (2005) Personal view: food for thought – western lifestyle and susceptibility to Crohn's disease. The FODMAP hypothesis. Alimentary Pharmacology and Therapeutics 21: 1399–1409.

Gibson P, Shepherd SJ (2010) Evidence-based dietary management of functional gastrointestinal symptoms: The FODMAP approach. Journal of Gastroenterology and Hepatology 25: 252–258.

Halmos EP, Christophersen CT, Bird AR, Shepherd SJ, Gibson PR, Muir JG (2015) Diets that differ in their FODMAP content alter the colonic luminal microenvironment. Gut 64: 93–100.

Haskå L, Nyman M, Andersson R (2008) Distribution and characterisation of fructan in wheat milling fractions. Journal of Cereal Science 48: 768–774.

Henry RJ (1985) A comparison of the non-starch carbohydrates in cereal grains. Journal of the Science of Food and Agriculture 36: 1243–1253.

Henry RJ, Saini HS (1989) Characterization of cereal sugars and oligosaccharides. Cereal Chemistry 66: 362–365.

Huynh B-L, Palmer L, Mather DE, Wallwork H, Graham RD, Welch RM, Stangoulis JCR (2008a) Genotypic variation in wheat grain fructan content revealed by a simplified HPLC method. Journal of Cereal Science 48: 369–378.

Huynh B-L, Wallwork H, Stangoulis JCR, Graham RD, Willsmore KL, Olson S, Mather DE (2008b) Quantitative trait loci for grain fructan concentration in wheat (*Triticum aestivum* L.). Theoretical and Applied Genetics 117: 701–709.

Huynh B-L, Mather DE, Schreiber AW, Toubia J, Baumann U, Shoaei Z, Stein N, Ariyadasa R, Stangoulis JCR, Edwards J, Shirley N, Langridge P, Fleury D (2012) Clusters of genes encoding fructan biosynthesizing enzymes in wheat and barley. Plant Molecular Biology 80: 299–314.

Junker Y, Zeissig S, Kim SJ, Barisani D, Wieser H, Leffler DA, Zevallos V, Libermann TA, Dillon S, Freitag TL, Kelly CP, Schuppan D (2012) Wheat amylase trypsin inhibitors drive intestinal inflammation via activation of toll-like receptor 4. Journal of Experimental Medicine 209: 2395–2408.

Kamal-Eldin A, Nygaard Lærke H, Bach Knudsen K-E, Lampi A-M, Piironen V, Adlercreutz H, Katina K, Poutanen K, Åman P (2009) Physical, microscopic and chemical characterisation of industrial rye and wheat brans from the Nordic countries. Food & Nutrition Research 53: 1912.

Kissing Kucek L, Veenstra LD, Amnuaycheewa P, Sorrells ME (2015) A grounded guide to gluten: how modern genotypes and processing impact wheat sensitivity. Comprehensive Reviews in Food Science and Food Safety 14: 285–302.

Knez M, Abbott C, Stangoulis JCR (2014) Changes in the content of fructans and arabinoxylans during baking processes of leavened and unleavened breads. European Food Research and Technology 239: 803–811.

Kuo TM, Van Middlesworth JF, Wolf WJ (1988) Content of raffinose oligosaccharides and sucrose in various plant seeds. Journal of Agricultural and Food Chemistry 36: 32–36.

Laatikainen R, Koskenpato J, Hongisto SM, Loponen J, Poussa T, Hillilä M, Korpela R (2016) Randomised clinical trial: low-FODMAP rye bread vs. regular rye bread to relieve the symptoms of irritable bowel syndrome. Alimentary Pharmacology & Therapeutics 44: 460–470.

Laatikainen R, Koskenpato J, Hongisto SM, Loponen J, Poussa T, Huang X, Sontag-Strohm T, Salmenkari H, Korpela R (2017) Pilot study: comparison of sourdough wheat bread and yeast-fermented wheat bread in individuals with wheat sensitivity and irritable bowel syndrome. Nutrients 9: 1215.

Langenkämper G, Zörb C, Seifert M, Mäder P, Fretzdorff B, Betsche T (2006) Nutritional quality of organic and conventional wheat. Journal of Applied Botany and Food Quality 80: 150–154.

Lewis DH (1993) Nomenclature and diagrammatic representation of oligomeric fructans - a paper for discussion. New Phytologist 124: 583–594.

Lineback DR, Rasper VF (1988) Wheat carbohydrates. In: Pomeranz Y (ed) Wheat chemistry and technology, 3rd edn, Vol 1. American Association of Cereal Chemists, St. Paul, MN, p 277–372

Livingston DP, Hincha DK, Heyer AG (2009) Fructan and its relationship to abiotic stress tolerance in plants. Cellular and Molecular Life Sciences 66: 2007–2023.

Loponen J, Gänzle MG (2018) Use of sourdough in low FODMAP baking. Foods 7: 96.

McCleary BV, Murphy A, Mugford DC (2000) Measurement of total fructan in foods by enzymatic/spectrophotometric method: Collaborative study. Journal of AOAC International 83: 356–364.

McGrath VB, Blakeney AB, Batten GD (1997) Fructan to nitrogen ratio as an indicator of nutrient stress in wheat crops. New Phytologist 136: 145–152.

Molis C, Flourié B, Ouarne F, Gailing MF, Lartigue S, Guibert A, Bornet F, Galmiche JP (1996) Digestion, excretion, and energy value of fructooligosaccharides in healthy humans. The American Journal of Clinical Nutrition 64: 324–328.

Morrison SC, Delahaye MB, Woods CJ, Larsen NG (2015) Wheat fructans – for better or for worse? 3rd International Conference Food Sturctures, Digestion and Health, Wellington, New Zealand, 28–30 October 2015

Moshfegh AJ, Friday JE, Goldman JP, Ahuja JKC (1999) Presence of inulin and oligofructose in the diets of Americans. The Journal of Nutrition 129: 1407–1411.

Muir JG, Shepherd SJ, Rosella O, Rose R, Gibson PR (2007) Fructan and free fructose content of common Australian vegetables and fruit. Journal of Agricultural and Food Chemistry 55: 6619–6627.

Muir J, Rose R, Rosella O, Liels K, Barrett J, Shepherd S, Gibson PR (2009) Measurement of short chain carbohydrates (FODMAP) in common Australian vegetables and fruit by high performance liquid chromatography. Journal of Agricultural and Food Chemistry 57: 554–565.

Muir JG, Mills J, Suter D, Békés F, Liels K, Yao LK, Gibson PR (2014) FODMAP in gluten-free grains may explain improved gastrointestinal symptoms in IBS on a gluten-free diet. Journal of Nutrition & Intermediary Metabolism 1: 14–15.

Mullin GE, Shepherd SJ, Chander Roland B, Ireton-Jones C, Matarese LE (2014) Irritable bowel syndrome: Contemporary nutrition management strategies. Journal of Parenteral and Enteral Nutrition 38: 781–799.

Murray K, Wilkinson-Smith V, Hoad C, Costigan C, Cox E, Lam C, Marciani L, Gowland P, Spiller RC (2014) Differential effects of FODMAP (fermentable oligo-, di-, mono-saccharides and polyols) on small and large intestinal contents in healthy subjects shown by MRI. The American Journal of Gastroenterology 109: 110–119.

Ni J, Feng B, Xu Z, Wang T (2011) Dynamic changes of wheat quality during grain filling in waxy wheat WX12. Czech Journal of Genetics and Plant Breeding 47: S182-S185.

Nilsson U, Björck I (1988) Availability of cereal fructans and inulin in the rat intestinal tract. The Journal of Nutrition 118: 1482–1486.

Nilsson U, Dahlqvist A, Nilsson B (1986) Cereal fructosans: Part 2 – Characterization and structure of wheat fructosans. Food Chemistry 22: 95–106.

Nilsson U, Öste R, Jägerstad M (1987) Cereal fructans: hydrolysis by yeast invertase, in vitro and during fermentation. Journal of Cereal Science 6: 53–60.

Nilsson U, Öste R, Jägerstad M, Birkhed D (1988) Cereal fructans: in vitro and in vivo studies on availability in rats and humans. The Journal of Nutrition 118: 1325–1330.

Pauk J, Lantos C, Ács K, Gell G, Tömösközi S, Hajdú Búza K, Békés F (2019) Spelt (*Triticum spelta* L.) in vitro androgenesis breeding for special food quality parameters. In: Al-Khayri JM, Jain SM, Johnson DV (eds) Advances in plant breeding strategies: Cereals, Vol 5. Springer, Cham, p 525–557.

Peshev D, van den Ende W (2014) Fructans: prebiotics and immunomodulators. Journal of Functional Foods 8: 348–357.

Pirkola L, Laatikainen R, Loponen J, Hongisto SM, Hillilä M, Nuora A, Yang B, Linderborg KM, Freese R (2018) Low-FODMAP vs regular rye bread in irritable bowel syndrome: randomized SmartPill study. World Journal of Gastroenterology 24: 1259–1268.

Praznik W, Cieślik E, Filipiak-Florkiewicz A (2002) Soluble dietary fibres in Jerusalem artichoke powders: composition and application in bread. Nahrung/Food 46: 151–157.

Rakha A, Åman P, Andersson R (2011) Dietary fiber in triticale grain: Variation in content, composition, and molecular weight distribution of extractable components. Journal of Cereal Science 54: 324–331.

Ritsema T, Smeekens S (2003) Fructans: beneficial for plants and humans. Current Opinion in Plant Biology 6: 223–230.

Roberfroid M, Gibson GR, Hoyles L, McCartney AL, Rastall R, Rowland I, Wolvers D, Watzl B, Szajewska H, Stahl B, Guarner F, Respondek F, Whelan K, Coxam V, Davicco M-J, Léotoing L, Wittrant Y, Delzenne NM, Cani PD, Neyrinck AM, Meheust A (2010) Prebiotic effects: metabolic and health benefits. British Journal of Nutrition 104: S1-S63.

Ruuska SA, Rebetzke GJ, van Herwaarden AF, Richards RA, Fettell NA, Tabe L, Jenkins CL (2006) Genotypic variation in watersoluble carbohydrate accumulation in wheat. Functional Plant Biology 33: 799–809.

Schnyder H, Ehses U, Bestajovsky J, Mehrhoff R, Kühbauch W (1988) Fructan in wheat kernels during growth and compartmentation in the endosperm and pericarp. Journal of Plant Physiology 132: 333–338.

Schnyder H, Gillenberg C, Hinz J (1993) Fructan contents and dry matter deposition in different tissues of the wheat grain during development. Plant, Cell & Environment 16: 179–187.

Schober T, Clarke C, Kuhn M (2002) Characterization of functional properties of gluten proteins in spelt cultivars using rheological and quality factor measurements. Cereal Chemistry 79: 408–417.

Shepherd S, Lomer MCE, Gibson PR (2013) Short-chain carbohydrates and functional gastrointestinal disorders. The American Journal of Gastroenterology 108: 707–717.

Shimbata T, Inokuma T, Sunohara A, Vrinten P, Saito M, Takiya T, Nakamura T (2011) High levels of sugars and fructan in mature seed of sweet wheat lacking GBSSI and SSIIa enzymes. Journal of Agricultural and Food Chemistry 59: 4794–4800.

Stallknecht G, Gilbertson K, Ranney J (1996) Alternative wheat cereals as food grains: einkorn, emmer, spelt, Kamut and triticale. In: Janick J (ed) New crops, new opportunities, new technologies. ASHS Press, Alexandria, VA, p 156–170

Staudacher HM, Whelan K, Irving PM, Lomer MCE (2011) Comparison of symptom response following advice for a diet low in fermentable carbohydrates (FODMAP) versus standard dietary advice in patients with irritable bowel syndrome. Journal of Human Nutrition and Dietetics 24: 487–495.

Staudacher HM, Lomer MC, Anderson JL, Barrett JS, Muir JG, Irving PM, Whelan K (2012) Fermentable carbohydrate restriction reduces luminal bifidobacteria and gastrointestinal symptoms in patients with irritable bowel syndrome. The Journal of Nutrition 142: 1510–1518.

Struyf N, Laurent J, Lefevere B, Verspreet J, Verstrepen KJ, Courtin CM (2017) Establishing the relative importance of damaged starch and fructan as sources of fermentable sugars in wheat flour and whole meal bread dough fermentations. Food Chemistry 218: 89–98.

Struyf N, Verspreet J, Courtin CM (2018) FODMAP reduction in yeast-leavened whole wheat bread. Cereal Foods World 63: 152–154.

Valerii MC, Ricci C, Spisni E, Di Silvestro R, De Fazio L, Cavazza E, Lanzini A, Campieri M, Dalpiaz A, Pavan B, Volta U, Dinelli G (2015) Responses of peripheral blood mononucleated cells from non-celiac gluten sensitive patients to various cereal sources. Food Chemistry 176: 167–174.

van den Ende W (2013) Multifunctional fructans and raffinose family oligosaccharides. Frontiers in Plant Science 4: 247.

van Loo J, Coussement P, De Leenheer L, Hoebregs H, Smits G (1995) On the presence of inulin and oligofructose as natural ingredients in the western diet. Critical Reviews in Food Science and Nutrition 35: 525–552.

Verspreet J, Pollet A, Cuyvers S, Vergauwen R, van den Ende W, Delcour JA, Courtin CM (2012) A simple and accurate method for determining wheat grain fructan content and average degree of polymerization. Journal of Agricultural and Food Chemistry 60: 2102–2107.

Verspreet J, Cimini S, Vergauwen R, Dornez E, Locato V, Le Roy K, De Gara L, Van den Ende W, Delcour JA, Courtin CM (2013a) Fructan metabolism in developing wheat (*Triticum aestivum* L.) kernels. Plant and Cell Physiology 54: 2047–2057.

Verspreet J, Hemdane S, Dornez E, Cuyvers S, Delcour JA, Courtin CM (2013b) Maximizing the concentrations of wheat grain fructans in bread by exploring strategies to prevent their yeast (*Saccharomyces cerevisiae*)-mediated degradation. Journal of Agricultural and Food Chemistry 61: 1397–1404.

Verspreet J, Holmgaard Hansen A, Dornez E, Delcour JA, van den Ende W, Harrison SJ, Courtin CM (2015) LC-MS analysis reveals the presence of graminan-and neo-type fructans in wheat grains. Journal of Cereal Science 61: 133–138.

Vu NT, Chin J, Pasco JA, Kovács A, Wing LW, Békés F, Suter D (2015) The prevalence of wheat and spelt sensitivity in a randomly selected Australian population. Cereal Research Communications 43: 97–107.

Whelan K, Abrahmsohn O, David GJP, Staudacher H, Irving P, Lomer MCE, Ellis PR (2011) Fructan content of commonly consumed wheat, rye and gluten-free breads. International Journal of Food Sciences and Nutrition 62: 498–503.

Yasui T, Ashida K (2011) Waxy endosperm accompanies increased fat and saccharide contents in bread wheat (*Triticum aestivum* L.) grain. Journal of Cereal Science 52: 104–111.

Zevallos VF, Raker V, Tenzer S, Jimenez-Calvente C, Ashfaq-Khan M, Rüssel N, Pickert G, Schild H, Steinbrink K, Schuppan D (2017) Nutritional wheat amylase-trypsin inhibitors promote intestinal inflammation via activation of myeloid cells. Gastroenterology 152: 1100–1113.

Ziegler JU, Steiner D, Longin CFH, Würschum T, Schweiggert RM, Carle R (2016) Wheat and the irritable bowel syndrome – FODMAP levels of modern and ancient species and their retention during bread making. Journal of Functional Foods 25: 257–266.

Epilogue: The Main Activities of the International Collaboration on Wheat Quality and Safety

Tatsuya M. Ikeda

Abstract Improving wheat quality and safety is very important for wheat breeders and related industries. International collaborations are necessary to compare and reproduce results among laboratories. The Expert Working Group (EWG) on improving wheat quality for processing, nutrition and health was established under the Wheat Initiative (WI) and consists of seven subgroups. This expert group focuses on sharing the same research platform among researchers by standardizing methods, selection of master sets covering various quality attributes, reducing the impact of mycotoxin accumulation in grain and processed products. Our EWG helps to coordinate international collaborations within our EWG and with other groups for further collaborations.

Importance of Improving Wheat Quality and Safety

Improving wheat quality and safety is very important for wheat breeders and related industries. For example, glutenin subunit allele combination is a critical determinant of gluten strength as shown in Fig. 1 (Ikeda et al. 2014). Strong gluten is necessary for bread having large loaf volume, but weak gluten is preferable for cake and cookies. Starch property (amylose content) is important factor for springiness of noodle and softness of bread, which are controlled by *Spg-1* and *Wx-1* genes (Fig. 2).

Many research works have been carried out internationally, but collaborations among laboratories were limited and each group often uses different materials, methods and gene nomenclature. Therefore, it has been difficult to compare and reproduce their results with each other (Fig. 3).

T. M. Ikeda
Western Region Agricultural Research Center, NARO, Fukuyama, Japan
e-mail: tmikeda@affrc.go.jp

G. Igrejas et al. (eds.), *Wheat Quality For Improving Processing And Human Health*, https://doi.org/10.1007/978-3-030-34163-3

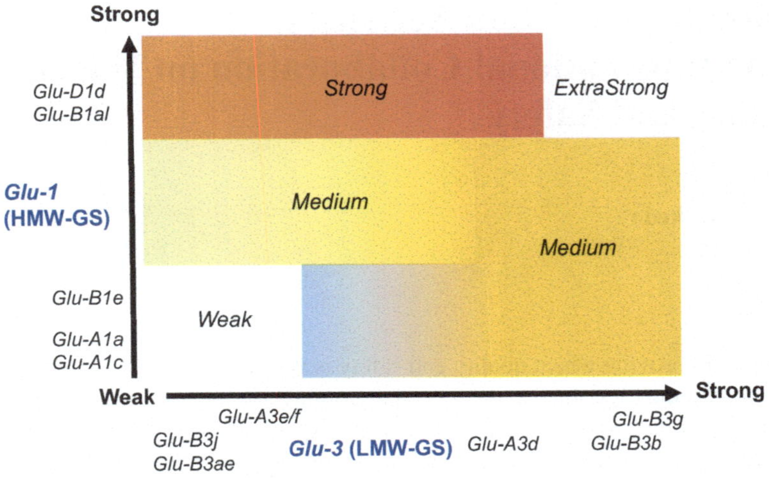

Fig. 1 The effect of *Glu-1* and *Glu-3* alleles on gluten strength

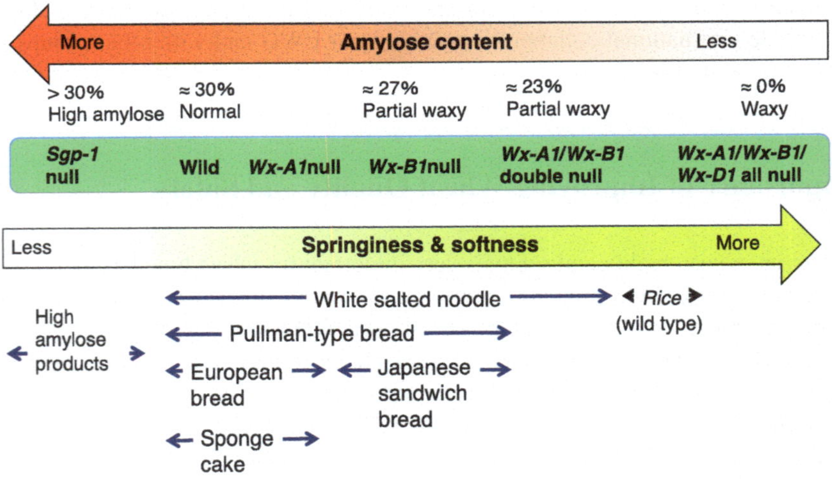

Fig. 2 Genetic control of amylose content of starch and end products

Expert Working Group on Improving Wheat Quality for Processing, Nutrition and Health

The Wheat Initiative (WI) was created to provide a framework to establish strategic research and organization priorities for wheat research at the international level. Under the WI, Expert Working Groups (EWG) for diverse wheat topics are established, bringing together experts to focus on topics of relevance to the WI

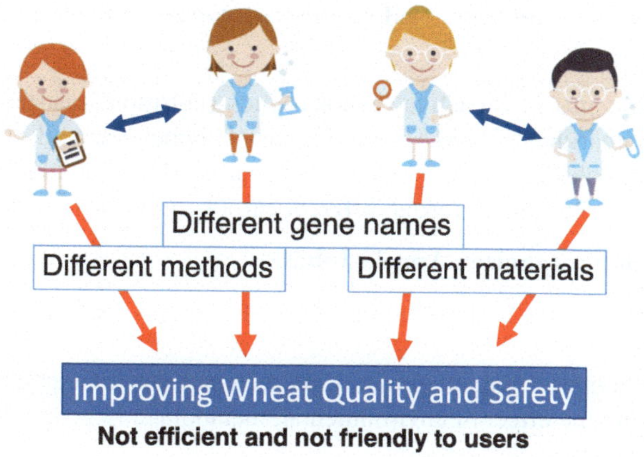

Fig. 3 Limited collaborations among laboratories are not efficient for improving wheat quality and safety

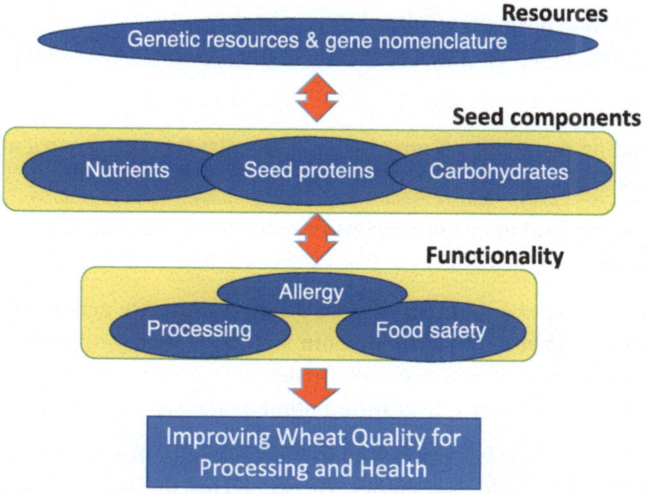

Fig. 4 Subgroups of the EWG for improving wheat quality for processing, nutrition and health

Strategic Research Agenda (2015). The EWG on improving wheat quality for processing, nutrition and health was established in 2015. The EWG is aimed at maintaining and improving wheat quality and safety under varying environmental conditions. Seven subgroups have been formed to cover each of the topics as it is shown in Fig. 4.

Each subgroup has the following objectives:

Seed proteins subgroup (organizer, Tatsuya M. Ikeda)

1. Standardizing methods to examine gluten protein
2. Unifying glutenin alleles between bread wheat and durum wheat

Carbohydrates & Nutrients subgroups (organizer, Regina Ahmed and Peter Shewry)

1. Studying diversity of bioactive carbohydrates (nutrition, resistant starch, cell wall polysaccharides, FODMAPs) and associated phenolic acids
2. Establishing molecular marker systems for major QTLs related to nutrients
3. Selection of master sets covering different carbohydrates and nutrient contents

Allergy subgroup (organizer, Angela Juhász)

1. Improvement of the understanding of seed proteins and their epitopes for allergenicity
2. Selection of a master set covering low allergen and low toxic genotypes
3. Understanding the effect of environmental, metabolic and genetic factors on the expression of wheat allergens
4. Understanding the effect of non-protein factors such as FODMAPs in wheat related food disorders

Food safety subgroup (organizer, Sofia Chulze)

1. Studying toxigenic species isolated from wheat
2. Evaluating potential biocontrol agents to reduce deoxynivalenol in the food chain

Processing subgroup (organizer, Valérie Lullien-Pellerin)

1. Understanding the effects of processing on the distribution of nutrients and toxic substances

Genetic resources and gene nomenclature subgroup (organizer, John W. Rogers and Carlos Guzmán)

2. Updating and Improve Gene Catalogue system for quality genes and protein
3. Development of Master Sets for glutenin and other quality related genes
4. Genome annotation of wheat quality genes in genome sequence (through collaboration with EWG on Wheat Information Systems)

The first meeting of the EWG was held in April 2016 at INRA in Paris, in which participated 31 researchers from 18 countries. The second meeting was held in April 2017 at BOKU in Vienna with 30 researchers from 17 countries. The third meeting was held in March 2018 at CIMMYT in Mexico with 21 researchers from 16 countries. The latest 4th meeting was held in July 2019 in Canada at Saskatoon University with 24 researchers from 17 countries. The EWG currently consists of 77 members from 25 countries.

Selecting Master Sets and Sharing Them

The main output of the EWG so far has been selecting and distributing a set of cultivars representing part of the variability of the low molecular weight glutenin subunit alleles of bread and durum wheat (Liu et al. 2010; Nieto-Taladriz et al. 1997). The idea is that each of these cultivars included in the set works as a standard or

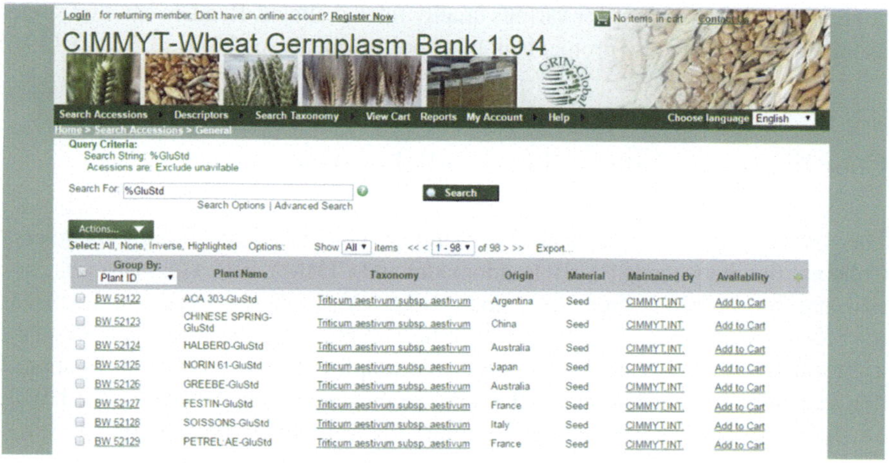

Fig. 5 The web page of the glutenin master set in the CIMMYT Germplasm Bank

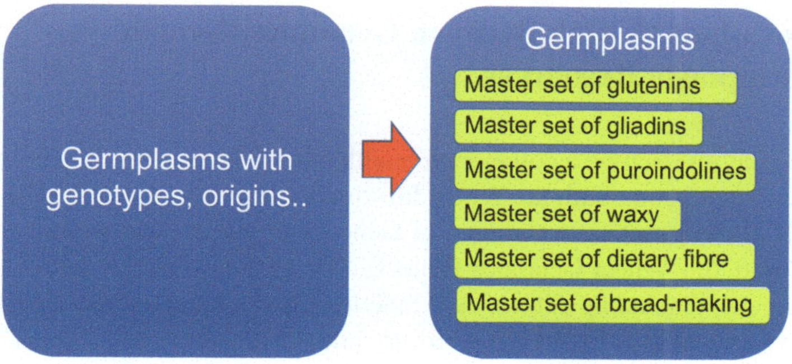

Fig. 6 Reshaping germplasms based on quality traits for more user-friendly gene banks

check for a particular allele. This Master Set is now available at the CIMMYT Germplasm Bank. It is possible to access it through the web site (http://wgb.cimmyt.org/gringlobal/search.aspx) and by searching for "%GluStd" (Fig. 5).

We are also developing master sets for other quality traits including gliadin alleles, dietary fiber and low allergen contents. The plan is to extend these master sets and get sets which represents variability for phenotypic traits, i.e. bread and noodle making. These master sets refine the genetic resources. Sharing these materials should help breeders and researchers working in food industries (Fig. 6).

Standardizing Methods

The EWG is also working on standardizing methods by developing a protocol repository that leads to inter-laboratory trials, including SDS-PAGE protocols to identify *Glu-3* subunits, Acid-PAGE for gliadins, a protocol for arabinoxylan content and DNA markers for various quality related alleles. Readers can find some examples related to these protocols in this book (Appendixes I & II, p132 to p139).

Coordination of International Collaborations

The EWG helps to coordinate international collaborations among members. Collaboration between ISPA (B. Laddomada) and CIMMYT (C. Guzman) was carried out for profiling phenolic acids. INRA (V. Lullien-Pellerin) and USDA (C. Morris) worked together to study the effects of puroindoline genes on flour quality (Heinze et al. 2016). These are only two examples of several bilateral collaboration that have happened thanks to the links established within the EWG. The EWG also plans to apply for international funds for training of students and visiting scientists.

Collaborations to Link Between Genomics, Genetic and Germplasm Resources

The EWG has also started a collaboration with other EWGs of the Wheat Initiative. We had a joint meeting with the Wheat Information System (WheatIS) EWG to fill the gap between genetics (Catalogue of Gene Symbol for Wheats) and genomics (IWGSC and 10+ Wheat Genomes Project). The catalog includes useful information for breeders including alleles, their functions, and related reference papers. The catalog is searchable at KOMUGI web site (https://shigen.nig.ac.jp/wheat/komugi/genes/symbolClassList.jsp). Genome sequence data of various cultivars/lines provide us useful information to identify genes and their alleles related to wheat quality and safety. However, genome databases, such as Ensemble Plants (http://plants.ensembl.org/Triticum_aestivum/Info/Index) cannot retrieve using gene symbols, e.g. a waxy protein gene, *Wx-1*. It means that breeders who use genetic data cannot easily access the genomic sequence data. Therefore, it is very important to connect genome data to the catalog for practical breeding programs to improve wheat quality and other traits (Fig. 7). In our collaboration, the web site of WheatIS was linked to the catalogue (https://urgi.versailles.inra.fr/wheatis/). We have also been working on making links between them by listing Genebank accessions for each allele in the catalog. Standardizing nomenclature and gene symbol annotation among bread and durum wheat and other Triticeae species is also underway. Moreover, the allergy subgroup collaborated with the International Wheat Genome Sequencing Consortium (IWGSC) for genome mapping of seed-borne allergens and immune-responsive proteins in wheat (Juhász et al. 2018). Some of our members will also work with the Durum Wheat Genomics and Breeding EWG to characterize quality of their durum core collection. The safety subgroup collaborated with the Mycokey project to reduce deoxynivalenol accumulation on wheat (Palazzini et al. 2018).

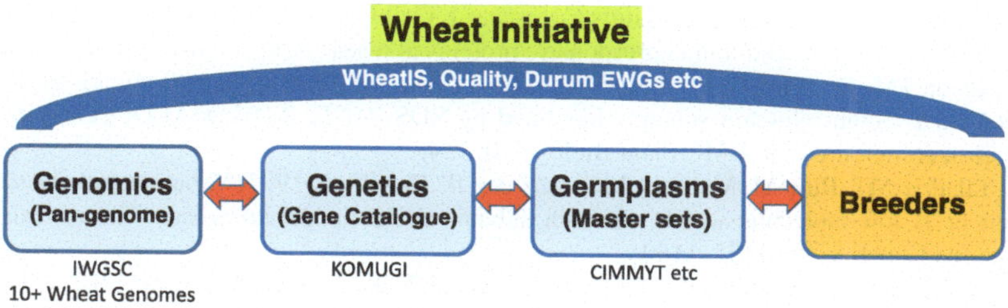

Fig. 7 Importance to link between genomics, genetics, and germplasms in collaboration with other EWGs under the WI

Fig. 8 Further collaborations among laboratories sharing the same platform for improving wheat quality and safety more efficiently

Conclusions

The EWG on improving wheat quality for processing, nutrition and health has been working on various issues since it was created in 2015. The purpose of the EWG is sharing the same platform to conduct further research internationally (Fig. 8). To provide opportunities to do so, our members organized 13th International Gluten Workshop in Mexico (C. Guzman) and the next workshop will be held in an Iberian co-organization with Portugal and Spain (G. Igrejas and P. Giraldo). Active participation to our EWG is very welcome.

References

Ikeda TM, Yanaka M, Takata K (2014) Comparison of Quality-Related Alleles Among Australian and North American Wheat Classes Exported to Japan. Cereal Chemistry 91: 616–622.

Heinze K, Kiszonas AM, Murray JC, Morris CF, Lullien-Pellerin V (2016) Puroindoline genes introduced into durum wheat reduce milling energy and change milling behavior similar to soft common wheats. Journal of Cereal Science 71: 183–189.

Juhász A, Belova T, Florides, CG, Maulis C, Fischer I, Gell G, et al. (2018) Genome mapping of seed-borne allergens and immunoresponsive proteins in wheat. Science Advances 4: eaar8602.

Liu L, Ikeda TM, Branlard G, Pena RJ, Rogers WJ, Lerner SE, et al. (2010) Comparison of low molecular weight glutenin subunits identified by SDS-PAGE, 2-DE. MALDI-TOF-MS and PCR in common wheat. BMC Plant Biology 10: 124.

Nieto-Taladriz MT, Ruiz M, Martínez MC, Vázquez JF, Carrillo J (1997) Variation and classification of B low-molecular-weight glutenin subunit alleles in durum wheat. Theoretical and Applied Genetics. 95: 1155–1160.

Palazzini J, Roncallo P, Cantoro R, Chiotta M, Yerkovich N, Palacios S, et al. (2018) Biocontrol of *Fusarium graminearum sensu stricto*, reduction of deoxynivalenol accumulation and phytohormone induction by two selected antagonists. Toxins 10: 88.

Wheat Initiative (2015) Strategic Research Agenda, Wheat Initiative Ed., Paris, 52.

Index

G. Igrejas et al. (eds.), *Wheat Quality For Improving Processing And Human Health*, https://doi.org/10.1007/978-3-030-34163-3

D

E

Q

The manufacturer's authorised representative in the EU is Springer
Nature Customer Service Centre GmbH, Europaplatz 3, 69115 Heidelberg,
Germany. If you have any concerns regarding our products, please
contact ProductSafety@springernature.com

Printed and bound by CPI Group (UK) Ltd, Croydon, CR0 4YY
30/09/2026
02225647-0016